Electromagnetics

By Arlon T. Adams and Jay Kyoon Lee

Second Edition

cognella
academic publishing

Bassim Hamadeh, CEO and Publisher
Carrie Montoya, Acquisitions Editor
Kaela Martin, Project Editor / Associate Editor
Alia Bales, Associate Production Editor
Jess Estrella, Senior Graphic Designer
Alexa Lucido, Licensing Associate
Natalie Piccotti, Director of Marketing
Kassie Graves, Vice President of Editorial
Jamie Giganti, Director of Academic Publishing

Cover image sources: https://en.wikisource.org/wiki/File:PSM_V17_D008_James_Clerk_Maxwell.jpg; https://commons.wikimedia.org/wiki/File:SS-faraday.jpg

Printed in the United States of America.

ISBN: 978-1-5165-2065-7 (pbk)/ 978-1-5165-2066-4 (br)

cognella
academic publishing

www.cognella.com 800-200-3908

BIOGRAPHIES

Arlon T. Adams (PhD, University of Michigan) is Professor Emeritus of Electrical and Computer Engineering at Syracuse University, where he taught and did research in electromagnetics for many years, with research interests in antennas and microwaves. He served as electronics officer in the U. S. Navy and worked as an engineer for the Sperry Gyroscope Company. He is a Life Fellow of the IEEE from which institution he received eight prize-paper and achievement awards. He has been a Fulbright Scientist in Yugoslavia, a visiting scholar at Berkeley, and was general chairman of the 1988 IEEE AP-S (Antennas and Propagation Society) /URSI International Symposium at Syracuse, New York.

Jay Kyoon Lee (PhD, Massachusetts Institute of Technology) is Professor of Electrical Engineering and Computer Science at Syracuse University, where he teaches Electromagnetics, among other courses. His current research interests are electromagnetic theory, waves in anisotropic media, microwave remote sensing, antennas, and propagation. He was a Research Fellow at Naval Air Development Center, Rome Air Development Center and Naval Research Laboratory, and was an Invited Visiting Professor at Seoul National University in Seoul, Korea. He has received the Eta Kappa Nu Outstanding Undergraduate Teacher Award (1999), the IEEE Third Millennium Medal (2000), the College Educator of the Year Award from the Technology Alliance of Central New York (2002), and the IEEE Region 1 Awards (2002, 2017).

To

Judith Schettler Adams

and

Young Hee Kim Lee

CONTENTS

PREFACE

Electromagnetics is not an easy subject for students. The subject presents a number of challenges, such as new math, new physics, new geometry, new insights, and difficult problems. As a result, every aspect needs to be presented carefully, using thorough mathematics, strong physical insights, and even alternative ways of viewing and formulating the subject. The theoretician James Clerk Maxwell and the experimentalist Michael Faraday, both shown on the cover, had great respect for physical insights.

This book is written primarily as a text for an undergraduate course in electromagnetics, taken by junior and senior engineering and physics students. The book can also serve as a text for beginning graduate courses by including advanced subjects and problems. The book has been thoroughly class-tested for many years for a two-semester Electromagnetics course at Syracuse University for electrical engineering and physics students. It could also be used for a one-semester course, covering up through Chapter 8 and perhaps skipping Chapter 4 and some other parts. For a one-semester course with more emphasis on waves, the instructor could briefly cover basic materials from statics (mainly Chapters 2 and 6) and then go on to Chapters 8 through 12.

The authors have attempted to explain the difficult concepts of electromagnetic theory in a way that students can readily understand and follow, without omitting the important details critical to a solid understanding of the subject. We have included a large number of examples, summary tables, and alternative formulations whenever possible, as well as homework problems. The examples explain the basic approach, leading the students step by step, slowly at first, to the conclusion. Then, special and limiting cases are examined to draw out analogies, physical insights and their interpretation. Finally, a very extensive set of problems enables the instructor to teach the course for several years without repeating assignments. Answers to selected problems at the end allow students to check if their answers are correct.

During our years of teaching electromagnetics, we became interested in its historical aspects and found it useful and instructive to introduce information about the basic discoveries into the classroom. We have included short biographical sketches of some of the leading figures of electromagnetics, including Josiah Willard Gibbs, Charles Augustin Coulomb, Benjamin Franklin, Pierre-Simon de Laplace, Georg Simon Ohm, André-Marie Ampère, Joseph Henry, Michael Faraday, and James Clerk Maxwell.

The text incorporates some unique features that include

- Coordinate transformations in 2D (Figures 1-11, 1-12).
- Summary tables, such as Table 2-1, 4-1, 6-1, 10-1.
- Repeated use of equivalent forms with R (conceptual) and $|\mathbf{r}-\mathbf{r}'|$ (mathematical) for the distance between the source point and the field point as in Eqs. (2-27), (2-46), (6-18), (6-19), (12-21).
- Intuitive derivation of equivalent bound charges from polarization sources, including piecewise approximation to non-uniform polarization (Section 3.3).
- Self-field (Section 3.8).
- Concept of the equivalent problem in the method of images (Section 4.3).
- Intuitive derivation of equivalent bound currents from magnetization sources, including piecewise approximation to non-uniform magnetization (Section 7.3)
- Thorough treatment of Faraday's law and experiments (Sections 8.3, 8.4).
- Uniform plane waves propagating in arbitrary direction (Section 9.4.1).
- Treatment of total internal reflection (Section 10.4).
- Transmission line equations from field theory (Section 11.7.2).
- Presentation of the retarded potential formulation in Chapter 12.
- Interpretation of the Hertzian dipole fields (Section 12.3).

Finally, we would like to acknowledge all those who contributed to the textbook. First of all, we would like to thank all of the undergraduate and graduate students, too numerous to mention, whose comments and suggestions have proven invaluable. In addition, one million thanks go to Ms. Brenda Flowers for typing the entire manuscript and making corrections numerous times. We also wish to express our gratitude to Dr. Eunseok Park, Professor Tae Hoon Yoo, Dr. Gokhan Aydin, and Mr. Walid M. G. Dyab for drawing figures and plotting curves, and to Professor Mahmoud El Sabbagh for reviewing the manuscript. Thanks go to the University of Poitiers, France, and Seoul National University, Korea, where an office and academic facilities were provided to Professor Adams and Professor Lee, respectively, during their sabbatical years. Thanks especially to Syracuse University where we taught for a total of over 50 years. Comments and suggestions from readers would be most welcome.

Arlon T. Adams
Jay Kyoon Lee
leejk@syr.edu
June 2012
Syracuse, New York

Notes on the Second Edition

In the second edition, incorporating the feedback from the students and readers, we revised and added the explanation of concepts and examples to help students understand it better. More examples and problems have been added and the comparison of polarization vs. magnetization is added in Chapter 7.

Jay K. Lee
July 2018

LIST OF TABLES

VECTOR ANALYSIS

CHAPTER 1

1.1 INTRODUCTION

In Chapter 1 we take up the subject of vector analysis. This subject is of fundamental importance to us in the development and the applications of electromagnetics because two major physical quantities that we deal with—the electric field and the magnetic field—are vectors. Vectors and coordinate systems are covered first, including differential elements of length, surface, and volume as well as vector transformations between different coordinate systems. We introduce the basic functions of divergence, gradient, and curl, along with the divergence theorem and Stokes' theorem, and we conclude with Helmholtz's theorem. In this chapter, we may need to refresh our memories on the concepts of partial derivatives and multiple integrals, both of which are fundamental to electromagnetic analysis. We will also strive to gain a deeper understanding of the three-dimensional properties of the coordinate systems.

1.1.1 Josiah Willard Gibbs (1839–1903) and the Development of Vector Analysis

Josiah Willard Gibbs has been called the greatest American scientist of the 19th century. He was a Professor of Mathematical Physics at Yale University from 1871 until his death in 1903. During that period he made major contributions to physical chemistry, statistical mechanics, and vector analysis. Professor Gibbs was the discoverer of the chemical potential, the phase rule, the Gibbs-Dalton Law, the absorption law of chemistry, and the Gibbs' phenomenon of Fourier series.

Willard Gibbs, as he was sometimes called, grew up in New Haven, Connecticut. His father (also named Josiah Willard Gibbs) was a Professor of Theology with an interest in comparative languages. He participated in the famous "Amistad" case of 1839–41 in which it was finally decided that a group of slaves from Africa had the right to mutiny and take over the ship (*The Amistad*) on which they were illegally held. Former President John Quincy Adams was counsel for the defense in this sensational case. After the case was over, Father Gibbs wrote biographies of the slaves involved and compiled a vocabulary of the rare language (Mendi) which they spoke. Father Gibbs died in 1861, leaving Willard and his sisters well provided for.

After completing his graduate studies, Willard Gibbs spent four years in postdoctoral study in Europe, accompanied by his two sisters. He was part of a flood of American students of the time who traveled to Europe for advanced education. He returned in 1869 and was appointed Professor of Mathematical Physics in 1871.

During the 1870s Willard Gibbs accomplished some of his most important work. He published three papers in the Transactions of the Connecticut Academy of Arts and Sciences. The third paper, which was over 300 pages in length, developed the idea of the chemical potential and outlined the phase rule. James Clerk Maxwell immediately recognized the significance of his work, was a strong supporter, and did much to help him gain some recognition. Maxwell corresponded with Gibbs, sent him a plaster cast model* of one of his theories and devoted an entire chapter to Gibbs in his book *Theory of Heat*. Later, scientists in Germany, France, and Holland also recognized Gibbs's work. His third paper was translated and published as a book in Germany in 1892 and in France in 1899. Finally, he received the Copley Medal in 1901.

* Muriel Rukeyser, *Willard Gibbs*, pp. 199–203, Oxbow Press, Woodbridge, CT, 1988, Reprint from Doubleday, Doran, Garden City, NY, 1942.

Gibbs became interested in vector analysis in the early 1880s. He often said that the work gave him more pleasure than any other intellectual activity. He drew up an extensive set of notes on vector analysis for his students during the 1880s but did not publish them. He himself felt that the work was not original since it derived from much earlier work by the German mathematician Grassman. Nonetheless, his work on vector analysis did attract attention. It involved him in extensive and bitter controversy with the champions of quaternions, an extremely complex competing methodology first proposed by Hamilton. Most of this controversy took place in the 1890s. Gibbs finally allowed publication of his lectures in a book *Vector Analysis* authored by E. B. Wilson, one of his students. Why did he not allow his name to appear on the book as author or even coauthor? Certainly part of the reason was that he himself did not regard it as strictly original. Perhaps also he did not want to become involved in further controversy with the fiercely combative quaternionic group.

Much has been written about Gibbs and his relative anonymity. The president of an American university went to Europe to find a physicist and was told that the best man for his purposes was Gibbs. "He can't be a very dynamic person if I haven't heard of him," was the response. It is interesting to note that he was recognized quite early by those within his field. Within 10 years of his initial appointment in 1871, he was recognized and praised extensively by Maxwell, referred to by Lorentz, elected to the National Academy of Sciences, and awarded the Rumford Medal.

Willard Gibbs was a quiet man, by no means an extrovert. He enjoyed an active intellectual life, the affection of a few close friends, and close relationships with his family. He lived out his professional life in the house in which he had been raised, with unmarried sister Anna and the family of sister Julia.

Gibbs's writings were very abstract and unusually concise, making them accessible to only a handful of people. In addition, he published at first only in the relatively obscure *Connecticut Academy Transactions*. Moreover, Gibbs did not fully illustrate his theory with applications. That was largely left to others. Time was required for many of the applications to emerge. So, given his personality and the nature of his work, it should not be surprising that he long remained unknown in the wider sense. He was the last of the classical school, coming, as he did, just after the epochal classical contributions of Maxwell and just before the dramatic appearance of relativity and quantum theory.

1.2. VECTOR ALGEBRA

1.2.1. Basic Operations of Vector Algebra

A **scalar** quantity is characterized by its magnitude. Temperature, charge, and voltage are scalars. Ordinary type is used for scalars, e.g., V for voltage.

A **vector** has both magnitude and direction. Velocity and the electric field are vectors. Boldface type is used for vectors, e.g., $\mathbf{E}$ for electric field. A vector $\mathbf{A}$ may be represented as a *directed line segment* or arrow (Figure 1-1). The length of the arrow represents the magnitude of the vector, denoted by $|\mathbf{A}|$ or A, and the arrow head indicates its direction. Switching the head and tail of arrow changes $\mathbf{A}$ to $-\mathbf{A}$. A vector is changed by rotation, but not by translation.

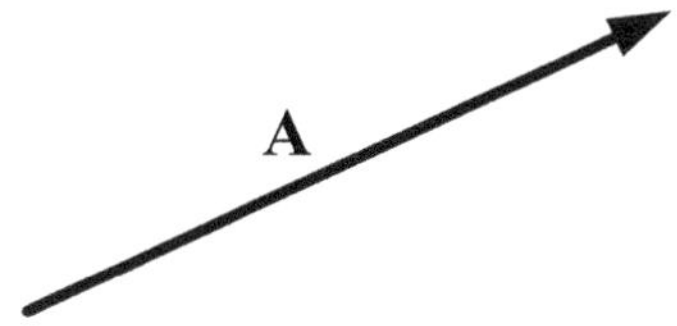

Figure 1-1. The vector $\mathbf{A}$ as a directed line segment.

VECTOR ADDITION

To form the vectorial sum $\mathbf{A} + \mathbf{B}$, place the tail of $\mathbf{B}$ at the head of $\mathbf{A}$ as shown in Figure 1-2(a). The sum $\mathbf{A} + \mathbf{B}$ is the arrow from the tail of $\mathbf{A}$ to the head of $\mathbf{B}$. One may also reverse the order and place the tail of $\mathbf{A}$ at the head of $\mathbf{B}$ (Figure 1-2(b)). Note that both procedures yield the same result and one which is also identical to the parallelogram method (Figure 1-2(c)) with which you may be familiar.

Vector addition is commutative as seen in Figures 1-2(a), (b):

$$\mathbf{A} + \mathbf{B} = \mathbf{B} + \mathbf{A} \tag{1-1}$$

It is also associative

$$(\mathbf{A} + \mathbf{B}) + \mathbf{C} = \mathbf{A} + (\mathbf{B} + \mathbf{C}) \tag{1-2}$$

To subtract $\mathbf{B}$ from $\mathbf{A}$, we add the negative of $\mathbf{B}$ to $\mathbf{A}$ (see Figure 1-2(d)):

$$\mathbf{A} - \mathbf{B} = \mathbf{A} + (-\mathbf{B}) \tag{1-3}$$

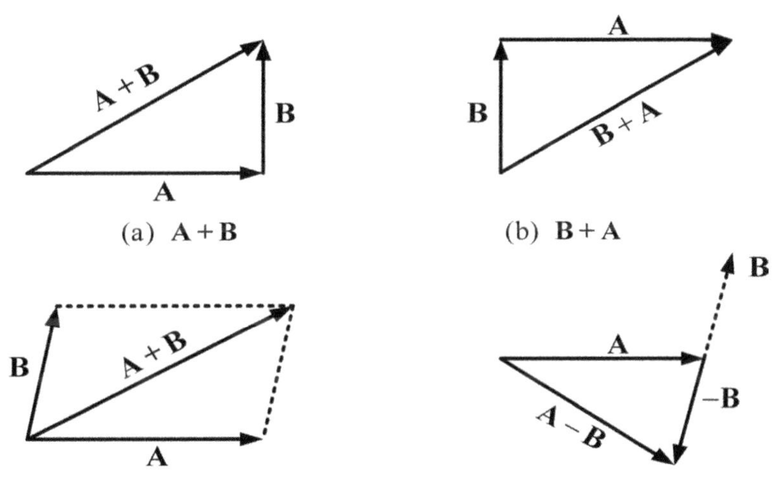

Figure 1-2. Addition of vectors.

MULTIPLICATION OF A VECTOR BY A SCALAR

If a vector is multiplied by a scalar a its magnitude is multiplied by the magnitude of a. The direction of the vector is unchanged if the scalar is positive and real. Scalar multiplication is distributive:

$$a(\mathbf{A}+\mathbf{B}) = a\,\mathbf{A} + a\,\mathbf{B} \tag{1-4}$$

Scalar division by a corresponds to multiplication by the inverse of a:

$$\frac{\mathbf{B}}{a} = \left(\frac{1}{a}\right)\mathbf{B} \tag{1-5}$$

If we divide a vector $\mathbf{A}$ by its magnitude A, we obtain a vector of unit length pointing in the direction of $\mathbf{A}$, i.e., a *unit vector* $\mathbf{a}$.

$$\mathbf{a} = \frac{\mathbf{A}}{\mathrm{A}}, \quad |\mathbf{a}| = 1 \tag{1-6}$$

All unit vectors have the same magnitude; they differ only in *direction*.

DOT OR SCALAR PRODUCT

The **dot product** of two vectors is a scalar and it is defined as

$$\mathbf{A}\cdot\mathbf{B} \equiv \mathrm{A\,B}\,\cos\theta_{\mathrm{AB}} \quad (0 \le \theta_{\mathrm{AB}} \le \pi) \tag{1-7}$$

where $\mathrm{A} = |\mathbf{A}|$, $\mathrm{B} = |\mathbf{B}|$, θ_{AB} is the angle between $\mathbf{A}$ and $\mathbf{B}$ (Figure 1-3).

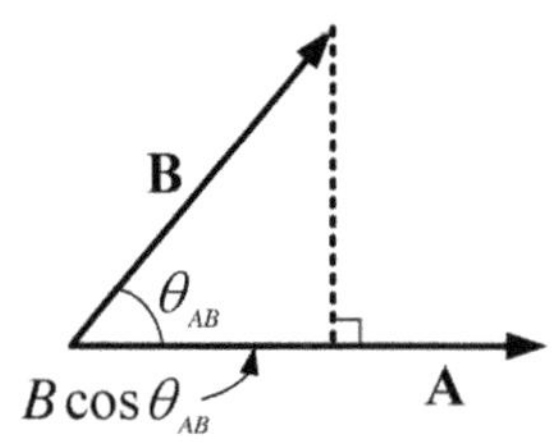

Figure 1-3. The dot product.

The dot product is A times the projection of $\mathbf{B}$ on $\mathbf{A}$ (B $\cos\theta_{\mathrm{AB}}$) or B times the projection of $\mathbf{A}$ on $\mathbf{B}$ (A $\cos\theta_{\mathrm{AB}}$). A projection is negative for $\theta_{\mathrm{AB}} > \pi/2$.

The dot product is commutative:

$$\mathbf{B}\cdot\mathbf{A} = \mathbf{A}\cdot\mathbf{B} \tag{1-8}$$

and distributive:

$$\mathbf{A} \cdot (\mathbf{B} + \mathbf{C}) = \mathbf{A} \cdot \mathbf{B} + \mathbf{A} \cdot \mathbf{C} \tag{1-9}$$

Special cases are:

$$\begin{aligned} \mathbf{A} \cdot \mathbf{B} &= \mathrm{AB} \;\; (\text{for } \theta_{AB} = 0) \\ &= 0 \;\; (\text{for } \theta_{AB} = \pi/2 \text{ when } \mathbf{A} \text{ and } \mathbf{B} \text{ are orthogonal.}) \\ \mathbf{A} \cdot \mathbf{A} &= \mathrm{A}^2 \end{aligned}$$

CROSS OR VECTOR PRODUCT

The **cross product** of two vectors is a vector and it is defined as follows:

$$\mathbf{A} \times \mathbf{B} = \mathbf{a}_n \mathrm{AB} \sin \theta_{AB} \; (0 \le \theta_{AB} \le \pi) \tag{1-10}$$

Note that the magnitude of $\mathbf{A} \times \mathbf{B}$ is $\mathrm{AB} \sin \theta_{AB}$, which is the area (base times height) of the parallelogram formed by $\mathbf{A}$ and $\mathbf{B}$ (see Figure 1-4(a)). Either A or B may be taken as the base. The direction of the cross product is that of $\mathbf{a}_n$ (Figure 1-4(a)), which is a unit vector perpendicular to the plane of the parallelogram (the plane formed by $\mathbf{A}$ and $\mathbf{B}$). The direction of the normal $\mathbf{a}_n$ is determined by the *right-hand rule* (Figure 1-4(b)). Note that $\mathbf{A} \times \mathbf{B}$ is perpendicular to both $\mathbf{A}$ and $\mathbf{B}$. You can also use the *three-finger rule*, by making the first three fingers of your right-hand orthogonal to each other. If you align the thumb with $\mathbf{A}$ and the index finger with $\mathbf{B}$, the middle finger gives you the direction of $\mathbf{A} \times \mathbf{B}$.

The cross product is distributive:

$$\mathbf{A} \times (\mathbf{B} + \mathbf{C}) = \mathbf{A} \times \mathbf{B} + \mathbf{A} \times \mathbf{C} \tag{1-11}$$

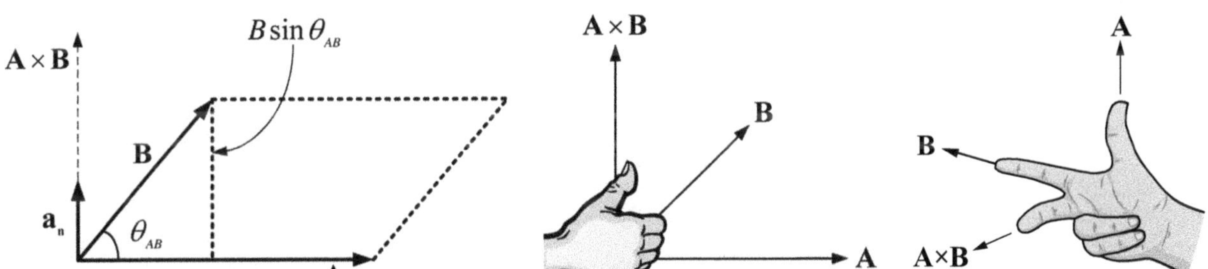

(a) Magnitude. (b) Direction by right-hand rule. (c) Three-finger rule.

Figure 1-4. The cross product.

The cross product is not commutative:

$$\mathbf{A} \times \mathbf{B} = -\,\mathbf{B} \times \mathbf{A} \tag{1-12}$$

Special cases are:

$$\begin{aligned} |\mathbf{A}\times\mathbf{B}| &= \mathrm{AB} \ \ (\text{for } \theta_{AB} = \pi/2) \\ &= 0 \ \ (\text{for } \theta_{AB} = 0) \\ \mathbf{A}\times\mathbf{A} &= 0 \end{aligned}$$

1.2.2. Vector Algebra in Rectangular Coordinates

A vector may be represented in terms of its components in any coordinate system; we choose the rectangular (Cartesian) coordinate system for this section. Figure 1-5(a) shows the Cartesian coordinate system with its unit vectors $\mathbf{a}_x$, $\mathbf{a}_y$, $\mathbf{a}_z$. Figure 1-5(b) shows a vector $\mathbf{A}$ and its projections $\mathbf{a}_x A_x$, $\mathbf{a}_y A_y$, $\mathbf{a}_z A_z$ on the x, y, z axes. By translation, we can place the projections head to tail and show that

$$\mathbf{A} = \mathbf{a}_x A_x + \mathbf{a}_y A_y + \mathbf{a}_z A_z \tag{1-13}$$

A_x, A_y, A_z are the components of $\mathbf{A}$ or the magnitudes of its projections. The magnitudes may either precede or follow the unit vectors.

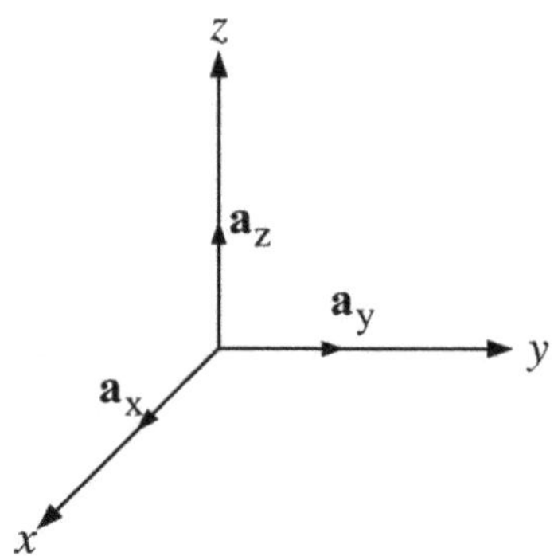

(a) Unit vectors.

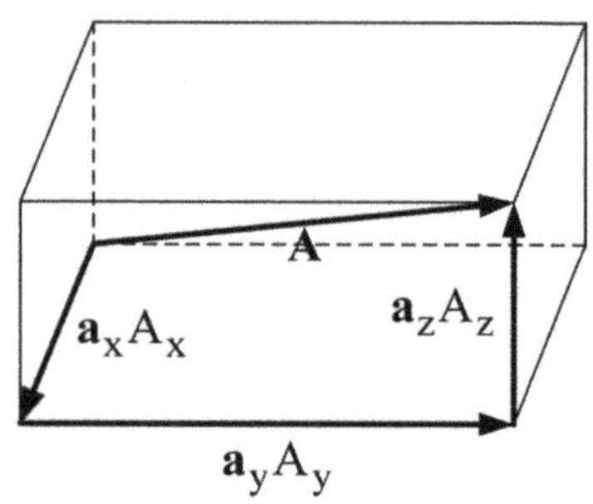

(b) Components of vector **A.**

Figure 1-5. Representation of a vector in rectangular coordinate system.

VECTOR ADDITION

$$\begin{aligned} \mathbf{A}+\mathbf{B} &= (\mathbf{a}_x A_x + \mathbf{a}_y A_y + \mathbf{a}_z A_z) + (\mathbf{a}_x B_x + \mathbf{a}_y B_y + \mathbf{a}_z B_z) \\ &= \mathbf{a}_x (A_x + B_x) + \mathbf{a}_y (A_y + B_y) + \mathbf{a}_z (A_z + B_z) \end{aligned} \tag{1-14}$$

MULTIPLICATION BY A SCALAR

$$\begin{aligned} c\mathbf{A} &= c(\mathbf{a}_x A_x + \mathbf{a}_y A_y + \mathbf{a}_z A_z) \\ &= \mathbf{a}_x (cA_x) + \mathbf{a}_y (cA_y) + \mathbf{a}_z (cA_z) \end{aligned} \tag{1-15}$$

DOT PRODUCT

Unit vector relationships:

$$\mathbf{a}_x \cdot \mathbf{a}_x = \mathbf{a}_y \cdot \mathbf{a}_y = \mathbf{a}_z \cdot \mathbf{a}_z = 1$$
$$\mathbf{a}_x \cdot \mathbf{a}_y = \mathbf{a}_x \cdot \mathbf{a}_z = \mathbf{a}_y \cdot \mathbf{a}_z = 0$$

Thus

$$\begin{aligned}\mathbf{A}\cdot\mathbf{B} &= (\mathbf{a}_x A_x + \mathbf{a}_y A_y + \mathbf{a}_z A_z)\cdot(\mathbf{a}_x B_x + \mathbf{a}_y B_y + \mathbf{a}_z B_z) \\ &= A_x B_x + A_y B_y + A_z B_z \qquad \text{(using the distributive property)}\end{aligned} \tag{1-16}$$

Special case:

$$\mathbf{A}\cdot\mathbf{A} = A_x^2 + A_y^2 + A_z^2 = |\mathbf{A}|^2$$

Thus

$$|\mathbf{A}| = \sqrt{A_x^2 + A_y^2 + A_z^2}$$

CROSS PRODUCT

Unit vector cross product relations:

$$\begin{aligned}&\mathbf{a}_x \times \mathbf{a}_x = \mathbf{a}_y \times \mathbf{a}_y = \mathbf{a}_z \times \mathbf{a}_z = 0 \\ &\mathbf{a}_x \times \mathbf{a}_y = \mathbf{a}_z \qquad \mathbf{a}_y \times \mathbf{a}_x = -\mathbf{a}_z \\ &\mathbf{a}_y \times \mathbf{a}_z = \mathbf{a}_x \qquad \mathbf{a}_z \times \mathbf{a}_y = -\mathbf{a}_x \\ &\mathbf{a}_z \times \mathbf{a}_x = \mathbf{a}_y \qquad \mathbf{a}_x \times \mathbf{a}_z = -\mathbf{a}_y\end{aligned}$$

(Simply use the right-hand rule or consider the cyclic order xyzxyz to determine the cross product)

Thus

$$\begin{aligned}\mathbf{A}\times\mathbf{B} &= (\mathbf{a}_x A_x + \mathbf{a}_y A_y + \mathbf{a}_z A_z)\times(\mathbf{a}_x B_x + \mathbf{a}_y B_y + \mathbf{a}_z B_z) \\ &= \mathbf{a}_x (A_y B_z - A_z B_y) + \mathbf{a}_y (A_z B_x - A_x B_z) + \mathbf{a}_z (A_x B_y - A_y B_x)\end{aligned} \tag{1-17a}$$

The results above are identical to the convenient determinant form of the cross product:

$$\mathbf{A}\times\mathbf{B} = \begin{vmatrix} \mathbf{a}_x & \mathbf{a}_y & \mathbf{a}_z \\ A_x & A_y & A_z \\ B_x & B_y & B_z \end{vmatrix} \tag{1-17b}$$

1.2.3. Triple Products

SCALAR TRIPLE PRODUCT $\mathbf{A} \cdot (\mathbf{B} \times \mathbf{C})$

$\mathbf{A} \cdot (\mathbf{B} \times \mathbf{C})$ is a scalar whose magnitude is equal to the volume of the parallelepiped (box) formed by the three vectors. Figure 1-6(b) shows vectors **A, B, C** and the box which they form. We note that $\mathbf{B} \times \mathbf{C}$ is a vector with direction $\mathbf{a}_n$ and magnitude equal to the area of the base of the parallelepiped: (see Figure 1-6(a))

$$\mathbf{B}\times\mathbf{C} = \mathbf{a}_n(\text{area of Base}_{CB})$$

Then

$$\begin{aligned}\mathbf{A}\cdot(\mathbf{B}\times\mathbf{C}) &= (\mathbf{a}_n\cdot\mathbf{A})\ (\text{area of Base}_{CB}) \\ &= \pm h_a(\text{area of Base}_{CB}) = \pm(\text{volume of the box})\end{aligned} \tag{1-18}$$

$(\mathbf{a}_n \cdot \mathbf{A})$ is positive in the case of Figure 1-6. If **A** were reversed, $(\mathbf{a}_n \cdot \mathbf{A})$ would be negative.

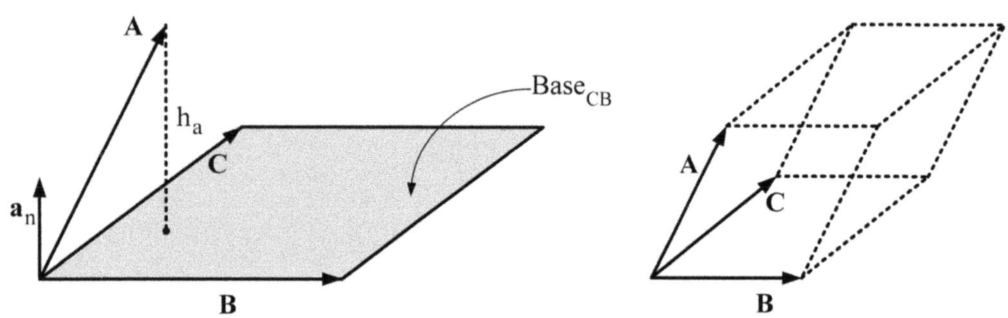

(a) Base_{CB} and height h_a

(b) The parallelepiped formed by vectors **A**, **B**, and **C**.

Figure 1-6. The Triple Product A · (B L C).

Since the volume of the box can be obtained from three different bases and heights:

$$\begin{aligned}\mathbf{A}\cdot(\mathbf{B}\times\mathbf{C}) &= \mathbf{B}\cdot(\mathbf{C}\times\mathbf{A}) = \mathbf{C}\cdot(\mathbf{A}\times\mathbf{B}) \\ &= -\mathbf{A}\cdot(\mathbf{C}\times\mathbf{B}) = -\mathbf{B}\cdot(\mathbf{A}\times\mathbf{C}) = -\mathbf{C}\cdot(\mathbf{B}\times\mathbf{A})\end{aligned} \tag{1-19}$$

Thus we can change the order so long as we retain the cycle ABCABC. We may also interchange dot and cross products in Eq. (1-19) to obtain:

$$\mathbf{A}\cdot(\mathbf{B}\times\mathbf{C}) = (\mathbf{A}\times\mathbf{B})\cdot\mathbf{C} \tag{1-20}$$

Note that all the parentheses of this section may be removed since there is only one possible location.

There is also a determinant form of the scalar triple product:

$$\mathbf{A}\cdot(\mathbf{B}\times\mathbf{C}) = \begin{vmatrix} A_x & A_y & A_z \\ B_x & B_y & B_z \\ C_x & C_y & C_z \end{vmatrix} \qquad (1\text{-}21)$$

This identity may be shown by laboriously expanding both sides in rectangular coordinates.

VECTOR TRIPLE PRODUCT $\mathbf{A} \times (\mathbf{B} \times \mathbf{C})$

$\mathbf{A} \times (\mathbf{B} \times \mathbf{C})$ is a vector. We note that the direction of $\mathbf{B} \times \mathbf{C}$ is that of $\mathbf{a}_n$ (Figure 1-6). Then $\mathbf{A} \times (\mathbf{B} \times \mathbf{C})$ is perpendicular to $(\mathbf{B} \times \mathbf{C})$, i.e., normal to $\mathbf{a}_n$. Thus $\mathbf{A} \times (\mathbf{B} \times \mathbf{C})$ lies in the plane formed by $\mathbf{B}$ and $\mathbf{C}$ and has components in the $\mathbf{B}$ and $\mathbf{C}$ directions. The "BAC-CAB" rule gives these components:

$$\mathbf{A} \times (\mathbf{B} \times \mathbf{C}) = \mathbf{B}\,(\mathbf{A} \cdot \mathbf{C}) - \mathbf{C}\,(\mathbf{A} \cdot \mathbf{B}) \qquad (1\text{-}22)$$

The identity above can be shown by expanding in rectangular components (see Problem 1-5). The parenthesis in $\mathbf{A} \times (\mathbf{B} \times \mathbf{C})$ *is* necessary since it differs from $(\mathbf{A} \times \mathbf{B}) \times \mathbf{C}$.

EXAMPLE 1-1

$$\text{Let}\quad \mathbf{A} = \mathbf{a}_y 2 + \mathbf{a}_z 2$$

$$\mathbf{B} = \mathbf{a}_y 3$$

$$\mathbf{C} = \mathbf{a}_x(-4) + \mathbf{a}_y(4)$$

Find $\mathbf{A} \cdot (\mathbf{B} \times \mathbf{C})$ and $\mathbf{A} \times (\mathbf{B} \times \mathbf{C})$.

SOLUTION

$$\mathbf{A}\cdot(\mathbf{B}\times\mathbf{C}) = \begin{vmatrix} 0 & 2 & 2 \\ 0 & 3 & 0 \\ -4 & 4 & 0 \end{vmatrix} = (-4)\{2\cdot 0 - 2\cdot 3\} = 24$$

$$\begin{aligned}\mathbf{A}\times(\mathbf{B}\times\mathbf{C}) &= \mathbf{B}(\mathbf{A}\cdot\mathbf{C}) - \mathbf{C}(\mathbf{A}\cdot\mathbf{B}) \\ &= \mathbf{a}_y 3\cdot(8) - (-\mathbf{a}_x 4 + \mathbf{a}_y 4)\cdot 6 = \mathbf{a}_X 24\end{aligned}$$

Sketch the vectors and the parallelepiped (box). Show by geometrical considerations that $\mathbf{A} \times (\mathbf{B} \times \mathbf{C})$ has only an $\mathbf{a}_x$ component.

EXAMPLE 1-2

$$\text{Let}\quad \mathbf{A} = \mathbf{a}_x A_x + \mathbf{a}_y A_y + \mathbf{a}_z A_z$$

Find $\mathbf{A} \cdot \mathbf{a}_z$ and $\mathbf{A} \times \mathbf{a}_z$

SOLUTION

$$\mathbf{A} \cdot \mathbf{a}_z = \mathbf{a}_z \cdot (\mathbf{a}_x A_x + \mathbf{a}_y A_y + \mathbf{a}_z A_z) = 0 + 0 + A_z = A_z$$

Taking the dot product of vector $\mathbf{A}$ with any unit vector yields the component of $\mathbf{A}$ in the direction of the unit vector.

$$\mathbf{A} \times \mathbf{a}_z = \begin{vmatrix} \mathbf{a}_x & \mathbf{a}_y & \mathbf{a}_z \\ A_x & A_y & A_z \\ 0 & 0 & 1 \end{vmatrix} = \mathbf{a}_x A_y - \mathbf{a}_y A_x$$

1.3 COORDINATE SYSTEMS

1.3.1. Coordinate System Geometry

Figure 1-7 shows the three basic coordinate systems we will use in this text, namely, the rectangular coordinate system (x, y, z), the cylindrical coordinate system (ρ, ϕ, z), and the spherical coordinate system (r, θ, ϕ). An understanding of the coordinate systems and the geometry involved is most essential in the three-dimensional problems of electromagnetics. Often the geometry is at the heart of the problem.

Note in Figure 1-7 the relationships among the three coordinate systems. The coordinate r is the distance from the origin to a particular point P. The coordinate θ (polar angle) is an angle from the z-axis to the radial line OP. It ranges from 0 to π. The coordinate ρ is the radial distance from P to the z axis. ρ is also the distance from the origin to the projection of P on the xy plane and $\rho = r \sin \theta$. z has the same definition for both rectangular and cylindrical coordinate systems and $z = r \cos \theta$. The coordinate ϕ (azimuth angle) is an angle from the x-axis to the radial line from the origin to the projection of P on the xy plane. It ranges from 0 to 2π. ϕ has the same definition for both cylindrical and spherical coordinate systems. (x, y) and (ρ, ϕ) are related by $x = \rho \cos \phi$ and $y = \rho \sin \phi$.

Table 1-1 is a collection of all the equations of transformation between coordinates. Can you reconstruct some of the relationships, for instance, the first set, from the figures of the coordinate systems?

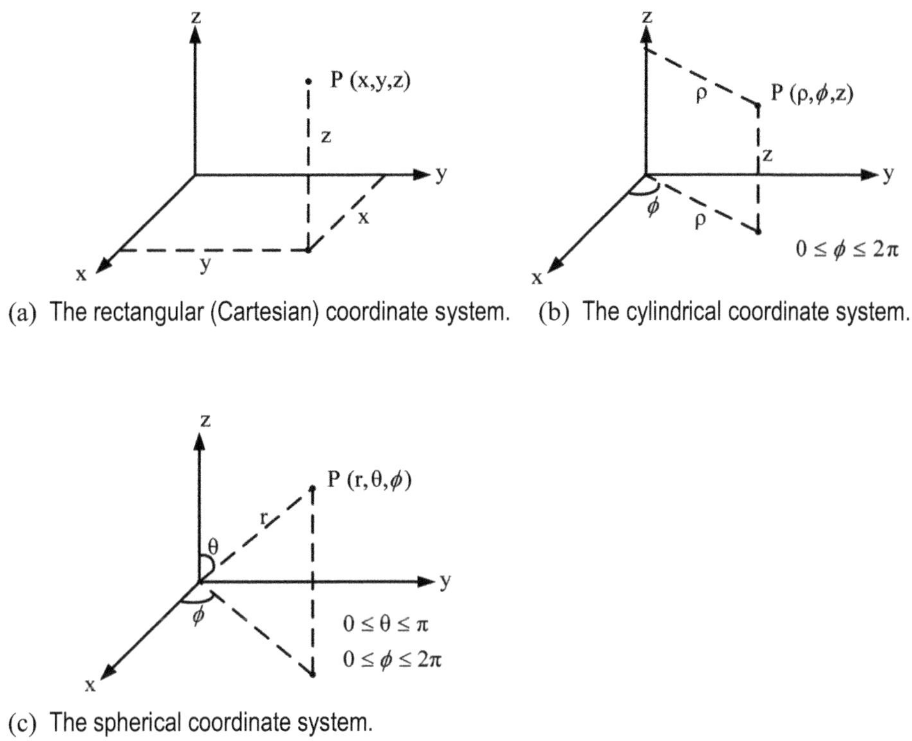

Figure 1-7. The three basic coordinate systems.

Table 1-1. Equations of Transformation between Coordinate Systems

$x = \rho\cos\phi = r\sin\theta\cos\phi$ $y = \rho\sin\phi = r\sin\theta\sin\phi$ $z = r\cos\theta$
$\rho = \sqrt{x^2 + y^2} = r\sin\theta$ $\phi = \tan^{-1}\dfrac{y}{x}$ * $z = r\cos\theta$
$r = \sqrt{x^2 + y^2 + z^2} = \sqrt{\rho^2 + z^2}$ $\theta = \tan^{-1}\dfrac{\sqrt{x^2 + y^2}}{z} = \tan^{-1}\dfrac{\rho}{z}$ $\phi = \tan^{-1}\dfrac{y}{x}$ *
*Note that it is important to consider the signs of x and y to determine the appropriate quadrant of angle ϕ.

Next we consider the orthogonal surfaces in the various coordinate systems. Figure 1-8 shows the orthogonal surfaces and the unit vectors. Note that the surface θ = constant is a cone and the surface ϕ = constant is a half-plane. The surfaces, ρ = constant and r = constant, are cylinders and spheres, respectively. A unit vector is normal to its coordinate surface and points in the direction in which its coordinate is increasing.

Universal Definition of the Unit Vector:

> $\mathbf{a}_\xi$ is a unit vector pointing in the direction of increasing the coordinate ξ and is normal to the surface ξ = constant.

Thus the unit vector $\mathbf{a}_r$ is normal to the surface r = constant (sphere) and points in the direction of increasing r, i.e., radially outward from the origin. The unit vector $\mathbf{a}_\theta$ is normal to the conical surface. It points in the direction of increasing θ and is equal, for example, to $-\mathbf{a}_z$ at $\theta = \pi/2$. Note that only the unit vectors $\mathbf{a}_x$, $\mathbf{a}_y$, $\mathbf{a}_z$ are constant in direction. Thus they can be taken outside the integral when we deal with vector integrals. However, the unit vectors $\mathbf{a}_\rho$, $\mathbf{a}_r$, $\mathbf{a}_\theta$, $\mathbf{a}_\phi$ are *not constant vectors* because their direction depends on the location (see Problem 1-17). Relationships between unit vectors in different coordinates are discussed in Section 1.3.3. Finally, referring to Figure 1-8(c), can you identify the lines that correspond to latitude and longitude lines on the earth?

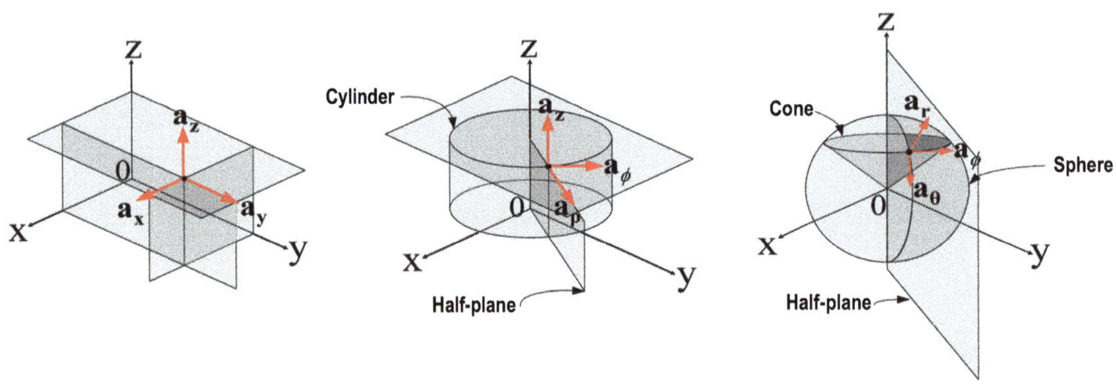

(a) Rectangular coordinates (b) Cylindrical coordinates (c) Spherical coordinates

Figure 1-8. Orthogonal surfaces and unit vectors.

1.3.2 Differential Elements of Length, Surface, and Volume

The integration of scalar and vector functions requires that we know the differential elements of length, surface, and volume. You are already familiar with some of these; others will be new to you. First we consider the vector differential element of length $d\boldsymbol{\ell}$. In moving along the arrow $d\boldsymbol{\ell}$, the changes in x, y, z are dx, dy, dz, respectively. Thus:

$$d\boldsymbol{\ell} = \mathbf{a}_x\,dx + \mathbf{a}_y\,dy + \mathbf{a}_z\,dz \tag{1-23a}$$

For changes in length variables ρ and r, the corresponding differential lengths are $\mathbf{a}_\rho\,d\rho$ and $\mathbf{a}_r\,dr$, respectively. The differential elements of length $d\boldsymbol{\ell}$ in cylindrical coordinates require an additional element associated with angular change ($d\phi$), while in spherical coordinates it requires two additional elements associated with changes in angle variables ($d\theta$ and $d\phi$). These are shown in Figure 1-9. If we consider an element of length in which only θ

changes, the appropriate vector differential length is $\mathbf{a}_\theta\ r\,d\theta$. Note that it is not $\mathbf{a}_\theta\,d\theta$, but multiplied by r, known as the metric coefficient. For changes in ϕ only, the vector differential length is $\mathbf{a}_\phi\ \rho\ d\phi$ or $\mathbf{a}_\phi\ r\sin\theta\ d\phi$ as shown in Figure 1-9. Note again that it is not $\mathbf{a}_\phi\ d\phi$, but multiplied by the metric coefficient $\rho = r\sin\theta$.

Thus, in cylindrical and spherical coordinates, the vector differential elements of length are given as follows:

$$d\ell = \mathbf{a}_\rho\ d\rho + \mathbf{a}_\phi\ \rho\ d\phi + \mathbf{a}_z\ dz \tag{1-23b}$$

$$d\ell = \mathbf{a}_r\ dr + \mathbf{a}_\theta\ r\ d\theta + \mathbf{a}_\phi\ r\ \sin\theta\ d\phi \tag{1-23c}$$

The vector differential elements of length are summarized in Table 1-2(a).

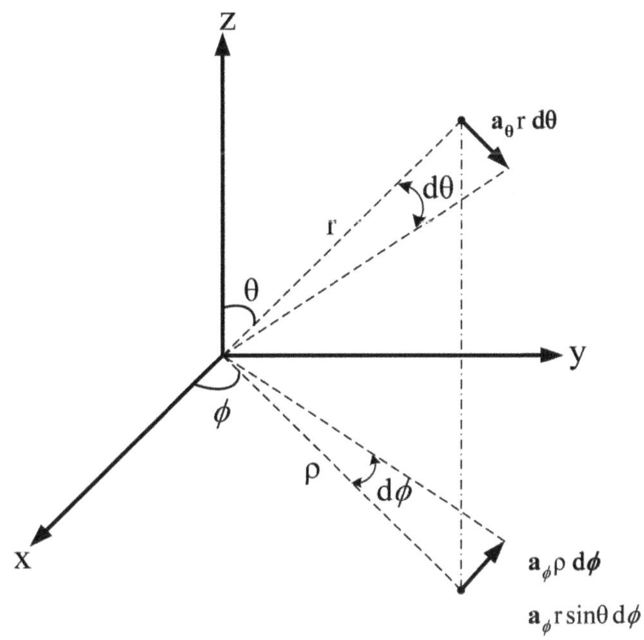

Figure 1-9. Basic elements of differential length in cylindrical and spherical coordinates.

The vector surface elements are readily constructed from the elements of length. They have magnitudes equal to the elementary surface areas and directions corresponding to one of the two normals to the surface. The choice of normal direction depends on the problem. Figure 1-10 shows some typical vector surface elements, namely, a portion of a disk, a curved cylinder, and a sphere.

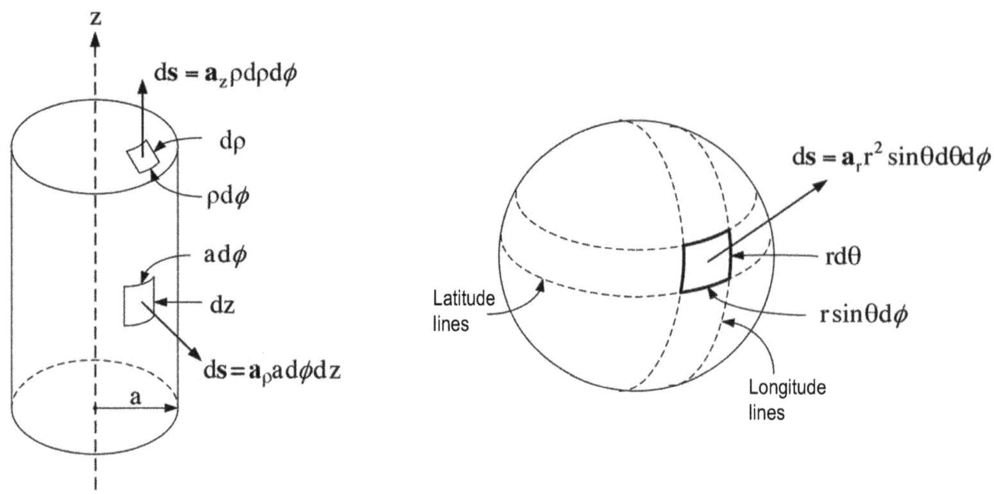

(a) Cylindrical coordinates. (b) Spherical coordinates.

Figure 1-10. Basic surface elements.

The complete set of vector surface elements is given in Table 1-2(b). The elements of volume are given in Table 1-2(c). These may be constructed by multiplying surface elements by the length element associated with the direction normal to the surface. For instance, the disk, curved cylindrical and spherical surfaces of Figure 1-10 are multiplied by dz, dρ, dr, respectively to obtain volume elements.

Table 1-2. Differential Elements of Length, Surface, and Volume

(a) Vector Differential Length Elements

$$
\begin{aligned}
d\ell &= \mathbf{a}_x\,dx + \mathbf{a}_y\,dy + \mathbf{a}_z\,dz \\
&= \mathbf{a}_\rho\,d\rho + \mathbf{a}_\phi\,\rho\;d\phi + \mathbf{a}_z\,dz \\
&= \mathbf{a}_r\,dr + \mathbf{a}_\theta\,r\;d\theta + \mathbf{a}_\phi\,r\,\sin\theta\;d\phi
\end{aligned}
$$

(b) Vector Differential Surface Elements

$$
\begin{aligned}
ds &= \pm\,\mathbf{a}_x\,dy\,dz,\;\mathbf{a}_y\,dz\,dx,\;\mathbf{a}_z\,dx\,dy\;* \\
&= \pm\,\mathbf{a}_\rho\;\rho d\phi\,dz,\;\mathbf{a}_\phi\;d\rho\,dz,\;\mathbf{a}_z\;\rho\,d\rho\,d\phi \\
&= \pm\,\mathbf{a}_r r^2 \sin\theta\;d\theta\;d\phi,\;\mathbf{a}_\theta r\sin\theta\,dr\,d\phi,\;\mathbf{a}_\phi r\,dr\,d\theta
\end{aligned}
$$

(c) Differential Volume Elements

$$\begin{aligned} dv &= dx\,dy\,dz \\ &= \rho\, d\rho\, d\phi\, dz \\ &= r^2 \sin\theta\, dr\, d\theta\, d\phi \end{aligned}$$

* The choice of sign ± depends on the problem.

1.3.3 Coordinate Transformations

There are situations where it is required to change from one coordinate system to another. In order to accomplish this, it is useful to have a systematic method of transforming between different coordinate systems. There are in all six possible vector transformations (rectangular to cylindrical, rectangular to spherical, cylindrical to spherical, and their inverses) for the three basic coordinate systems. Each transformation can be expressed in terms of a matrix relationship.

Scalar transformations are relatively simple since we need only to substitute from the basic equations of transformation in Table 1-1. For example, to transform from cylindrical to rectangular coordinates, a function $F(\rho, \phi, z)$ becomes $F\left(\sqrt{x^2+y^2}, \tan^{-1} y/x, z\right)$.

Vector transformations are also straightforward. All transformations can be obtained from two sets of transformations. Consider first the transformations between rectangular and cylindrical coordinates. Figure 1-11 shows the unit vectors $\mathbf{a}_\rho$, $\mathbf{a}_\phi$ in the xy plane (or $z =$ constant plane) because $\mathbf{a}_z$ is common in both rectangular and cylindrical coordinates. Note that $\mathbf{a}_\rho$ and $\mathbf{a}_\phi$ point in the direction of increasing ρ and ϕ, respectively. We can resolve each into x, y, z components by inspection.

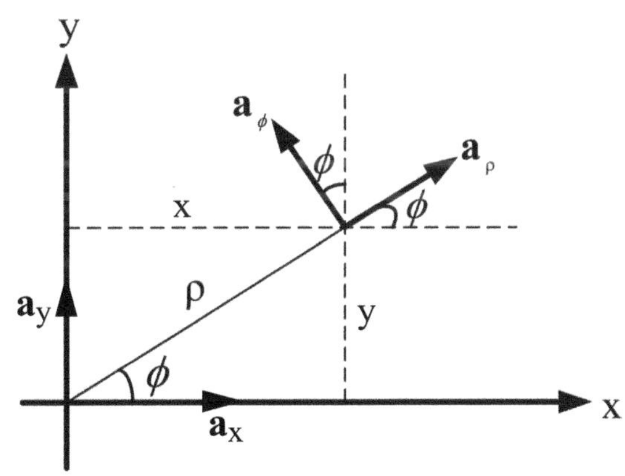

Figure 1-11. The transformation between rectangular and cylindrical coordinates.

$$\mathbf{a}_\rho = \mathbf{a}_x \cos\phi + \mathbf{a}_y \sin\phi + \mathbf{a}_z (0)$$
$$\mathbf{a}_\phi = \mathbf{a}_x (-\sin\phi) + \mathbf{a}_y \cos\phi + \mathbf{a}_z (0)$$
$$\mathbf{a}_z = \mathbf{a}_x (0) + \mathbf{a}_y (0) + \mathbf{a}_z (1)$$

The equations above may be cast in matrix form:

$$\begin{bmatrix} \mathbf{a}_\rho \\ \mathbf{a}_\phi \\ \mathbf{a}_z \end{bmatrix} = \begin{bmatrix} \cos\phi & \sin\phi & 0 \\ -\sin\phi & \cos\phi & 0 \\ 0 & 0 & 1 \end{bmatrix} \begin{bmatrix} \mathbf{a}_x \\ \mathbf{a}_y \\ \mathbf{a}_z \end{bmatrix} \qquad (1\text{-}24a)$$

Similarly, the inverse matrix relationship is:

$$\begin{bmatrix} \mathbf{a}_x \\ \mathbf{a}_y \\ \mathbf{a}_z \end{bmatrix} = \begin{bmatrix} \cos\phi & -\sin\phi & 0 \\ \sin\phi & \cos\phi & 0 \\ 0 & 0 & 1 \end{bmatrix} \begin{bmatrix} \mathbf{a}_\rho \\ \mathbf{a}_\phi \\ \mathbf{a}_z \end{bmatrix} \qquad (1\text{-}24b)$$

The inverse relationship may be determined by inspection of Figure 1-11 or by matrix inversion. The inversion process is particularly simple since only the 2×2 submatrix in the upper left hand corner needs to be considered. The determinant of each matrix is unity.

Consider next the transformations between cylindrical and spherical coordinates.

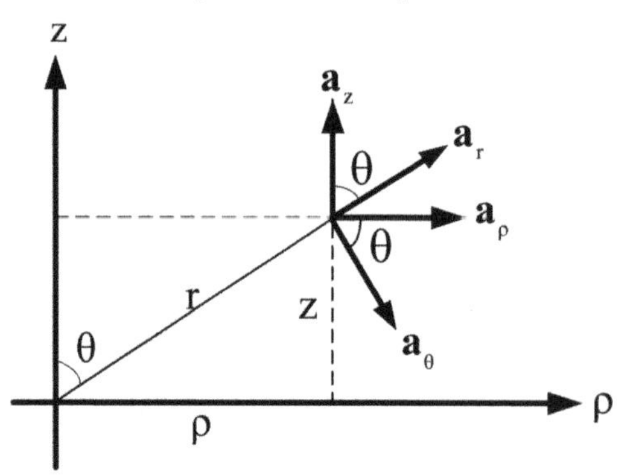

Figure 1-12. The transformation between cylindrical and spherical coordinates.

Figure 1-12 shows the unit vectors $\mathbf{a}_r$, $\mathbf{a}_\theta$ and their relationships to the unit vectors $\mathbf{a}_r$, $\mathbf{a}_z$ in the half plane (ϕ = constant) because $\mathbf{a}_\phi$ is common in both cylindrical and spherical coordinates. Resolving $\mathbf{a}_r$, $\mathbf{a}_\theta$ into cylindrical components yields

$$\mathbf{a}_r = \mathbf{a}_\rho \sin\theta + \mathbf{a}_z \cos\theta + \mathbf{a}_\phi (0)$$
$$\mathbf{a}_\theta = \mathbf{a}_\rho \cos\theta - \mathbf{a}_z \sin\theta + \mathbf{a}_\phi (0)$$
$$\mathbf{a}_\phi = \mathbf{a}_\rho (0) + \mathbf{a}_z (0) + \mathbf{a}_\phi (1)$$

In matrix form:

$$\begin{bmatrix} \mathbf{a}_r \\ \mathbf{a}_\theta \\ \mathbf{a}_\phi \end{bmatrix} = \begin{bmatrix} \sin\theta & 0 & \cos\theta \\ \cos\theta & 0 & -\sin\theta \\ 0 & 1 & 0 \end{bmatrix} \begin{bmatrix} \mathbf{a}_\rho \\ \mathbf{a}_\phi \\ \mathbf{a}_z \end{bmatrix} \tag{1-25a}$$

The inverse relationship is

$$\begin{bmatrix} \mathbf{a}_\rho \\ \mathbf{a}_\phi \\ \mathbf{a}_z \end{bmatrix} = \begin{bmatrix} \sin\theta & \cos\theta & 0 \\ 0 & 0 & 1 \\ \cos\theta & -\sin\theta & 0 \end{bmatrix} \begin{bmatrix} \mathbf{a}_r \\ \mathbf{a}_\theta \\ \mathbf{a}_\phi \end{bmatrix} \tag{1-25b}$$

The transformations given above (rectangular to cylindrical and cylindrical to spherical) are particularly simple because they involve only two coordinates. In each transformation, there is a common coordinate for the transformation between rectangular and cylindrical coordinates, and for the transformation between cylindrical and spherical coordinates. All other transformations can be derived by successive transformation. The six basic transformations are summarized in Table 1-3.

The transformations for unit vectors may also be used for arbitrary vector components. For example, consider a vector $\mathbf{A}$ in rectangular coordinates:

$$\mathbf{A} = \mathbf{a}_x A_x + \mathbf{a}_y A_y + \mathbf{a}_z A_z$$

We introduce the following results from Table 1-3:

$$\mathbf{a}_x = \mathbf{a}_\rho \cos\phi - \mathbf{a}_\phi \sin\phi$$
$$\mathbf{a}_y = \mathbf{a}_\rho \sin\phi + \mathbf{a}_\phi \cos\phi$$

Substitute in $\mathbf{A}$ above to obtain

$$\begin{aligned} A &= \mathbf{a}_\rho (A_x \cos\phi + A_y \sin\phi) + \mathbf{a}_\phi(-A_x \sin\phi + A_y \cos\phi) + \mathbf{a}_z A_z \\ &= \mathbf{a}_\rho A_\rho + \mathbf{a}_\phi A_\phi + \mathbf{a}_z A_z \end{aligned}$$

Thus

$$A_\rho = A_x \cos\phi + A_y \sin\phi$$
$$A_\phi = -A_x \sin\phi + A_y \cos\phi$$

or, in matrix form,

$$\begin{bmatrix} A_\rho \\ A_\phi \\ A_z \end{bmatrix} = \begin{bmatrix} \cos\phi & \sin\phi & 0 \\ -\sin\phi & \cos\phi & 0 \\ 0 & 0 & 1 \end{bmatrix} \begin{bmatrix} A_x \\ A_y \\ A_z \end{bmatrix} \tag{1-26}$$

Note that the matrix for this transformation of components is identical to the corresponding matrix for transformation of unit vectors shown in Eq. (1-24a). Thus, the six matrices of Table 1-3 may be used either for unit vectors or for arbitrary vector components.

Table 1-3. Transformation of Unit Vectors

	RECTANGULAR			CYLINDRICAL			SPHERICAL		
	$\mathbf{a}_x$	$\mathbf{a}_y$	$\mathbf{a}_z$	$\mathbf{a}_\rho$	$\mathbf{a}_\phi$	$\mathbf{a}_z$	$\mathbf{a}_r$	$\mathbf{a}_\theta$	$\mathbf{a}_\phi$
$\mathbf{a}_x$	1	0	0	$\cos\phi$	$-\sin\phi$	0	$\sin\theta\cos\phi$	$\cos\theta\cos\phi$	$-\sin\phi$
$\mathbf{a}_y$	0	1	0	$\sin\phi$	$\cos\phi$	0	$\sin\theta\sin\phi$	$\cos\theta\sin\phi$	$\cos\phi$
$\mathbf{a}_z$	0	0	1	0	0	1	$\cos\theta$	$-\sin\theta$	0
$\mathbf{a}_\rho$	$\cos\phi$	$\sin\phi$	0	1	0	0	$\sin\theta$	$\cos\theta$	0
$\mathbf{a}_\phi$	$-\sin\phi$	$\cos\phi$	0	0	1	0	0	0	1
$\mathbf{a}_z$	0	0	1	0	0	1	$\cos\theta$	$-\sin\theta$	0
$\mathbf{a}_r$	$\sin\theta\cos\phi$	$\sin\theta\sin\phi$	$\cos\theta$	$\sin\theta$	0	$\cos\theta$	1	0	0
$\mathbf{a}_\theta$	$\cos\theta\cos\phi$	$\cos\theta\sin\phi$	$-\sin\theta$	$\cos\theta$	0	$-\sin\theta$	0	1	0
$\mathbf{a}_\phi$	$-\sin\phi$	$\cos\phi$	0	0	1	0	0	0	1

[Note] The unit vectors ($\mathbf{a}_x$, $\mathbf{a}_y$, $\mathbf{a}_z$, $\mathbf{a}_\rho$, $\mathbf{a}_\phi$, $\mathbf{a}_r$, $\mathbf{a}_\theta$) can be replaced by the vector components (A_x, A_y, A_z, A_ρ, A_ϕ, A_r, A_θ). The transformation is still valid.

EXAMPLE 1-3

Apply successively the rectangular to cylindrical and cylindrical to spherical transformations to obtain the rectangular to spherical transformation.

SOLUTION

$$\begin{bmatrix}\mathbf{a}_r\\ \mathbf{a}_\theta\\ \mathbf{a}_\phi\end{bmatrix}=\begin{bmatrix}\sin\theta & 0 & \cos\theta\\ \cos\theta & 0 & -\sin\theta\\ 0 & 1 & 0\end{bmatrix}\begin{bmatrix}\mathbf{a}_\rho\\ \mathbf{a}_\phi\\ \mathbf{a}_z\end{bmatrix}=\begin{bmatrix}\sin\theta & 0 & \cos\theta\\ \cos\theta & 0 & -\sin\theta\\ 0 & 1 & 0\end{bmatrix}\begin{bmatrix}\cos\phi & \sin\phi & 0\\ -\sin\phi & \cos\phi & 0\\ 0 & 0 & 1\end{bmatrix}\begin{bmatrix}\mathbf{a}_x\\ \mathbf{a}_y\\ \mathbf{a}_z\end{bmatrix}$$

$$=\begin{bmatrix}\sin\theta\cos\phi & \sin\theta\sin\phi & \cos\theta\\ \cos\theta\cos\phi & \cos\theta\sin\phi & -\sin\theta\\ -\sin\phi & \cos\phi & 0\end{bmatrix}\begin{bmatrix}\mathbf{a}_x\\ \mathbf{a}_y\\ \mathbf{a}_z\end{bmatrix}$$

which agrees with the relationship in the left bottom box of Table 1-3.

EXAMPLE 1-4

Let $\mathbf{F} = \mathbf{a}_r\, 5\cos\theta - \mathbf{a}_\theta\, 5\sin\theta$. Express $\mathbf{F}$ in rectangular coordinates.

SOLUTION

Use the transformation from spherical to rectangular coordinates in the right top box of Table 1-3 to obtain

$$\begin{aligned}\mathbf{F} &= \mathbf{a}_x[5\cos\theta(\sin\theta\cos\phi) - 5\sin\theta(\cos\theta\cos\phi)]\\ &\quad + \mathbf{a}_y[5\cos\theta(\sin\theta\sin\phi) - 5\sin\theta(\cos\theta\sin\phi)]\\ &\quad + \mathbf{a}_z[5\cos\theta(\cos\theta) - 5\sin\theta(-\sin\theta)]\\ &= \mathbf{a}_z 5(\cos^2\theta + \sin^2\theta) = \mathbf{a}_z 5\end{aligned}$$

where we used, for example,

$$F_x = F_r \sin\theta\cos\phi + F_\theta \cos\theta\cos\phi + F_\phi(-\sin\phi)$$

The following relationship is found to be useful:

$$\boxed{\mathbf{a}_z = \mathbf{a}_r \cos\theta - \mathbf{a}_\theta \sin\theta}$$

1.3.4 Integrals of Vector Functions

There are many types of integrals which occur in electromagnetics. In particular, the *line integral* $\int \mathbf{F}\cdot d\boldsymbol{\ell}$ over a path and the *surface integral* $\iint \mathbf{F}\cdot \mathbf{ds}$ over a surface occur quite frequently. First, we consider the line integral.

LINE INTEGRALS

$\int_C \mathbf{F}\cdot d\boldsymbol{\ell}$ is called the **line integral** of vector $\mathbf{F}$ over path C. It is a scalar quantity. C may be open or closed. Both $\mathbf{F}$ and $d\boldsymbol{\ell}$ may be expressed in rectangular, cylindrical, and spherical coordinates:

$$\begin{aligned}\mathbf{F} &= \mathbf{a}_x F_x + \mathbf{a}_y F_y + \mathbf{a}_z F_z \\ &= \mathbf{a}_\rho F_\rho + \mathbf{a}_\phi F_\phi + \mathbf{a}_z F_z \\ &= \mathbf{a}_r F_r + \mathbf{a}_\theta F_\theta + \mathbf{a}_\phi F_\phi\end{aligned} \qquad \begin{aligned}d\boldsymbol{\ell} &= \mathbf{a}_x dx + \mathbf{a}_y dy + \mathbf{a}_z dz \\ &= \mathbf{a}_\rho d\rho + \mathbf{a}_\phi\, \rho\, d\phi + \mathbf{a}_z dz \\ &= \mathbf{a}_r\, dr + \mathbf{a}_\theta\, r\, d\theta + \mathbf{a}_\phi\, r \sin\theta\, d\phi\end{aligned}$$

Taking the dot product of $\mathbf{F}$ and $d\boldsymbol{\ell}$:

$$\int_C \mathbf{F}\cdot d\boldsymbol{\ell} = \int_C F_x\,dx + F_y\,dy + F_z\,dz \qquad \text{(rectangular coordinates)} \qquad (1\text{-}27a)$$

$$= \int_C F_\rho\,d\rho + F_\phi\,\rho\,d\phi + F_z\,dz \qquad \text{(cylindrical coordinates)} \qquad (1\text{-}27b)$$

$$= \int_C F_r\,dr + F_\theta\,r\,d\theta + F_\phi\,r\sin\theta\,d\phi \qquad \text{(spherical coordinates)} \qquad (1\text{-}27c)$$

In order to carry out the integration it is necessary to introduce the constraints imposed by the path C of integration.

EXAMPLE 1-5

Let $\mathbf{F} = \mathbf{a}_x x^2 + \mathbf{a}_y xy + \mathbf{a}_z 2yz$. Calculate the line integral $\int_C \mathbf{F}\cdot d\boldsymbol{\ell}$ where C is a straight line between the two points (0, 0, 0) and (2, 3, 4) in rectangular coordinates (Figure 1-13). Calculate also $\int_{C_a} \mathbf{F}\cdot d\boldsymbol{\ell}$ where $C_a = C_x + C_y + C_z$ is the integration path shown in Figure 1-13.

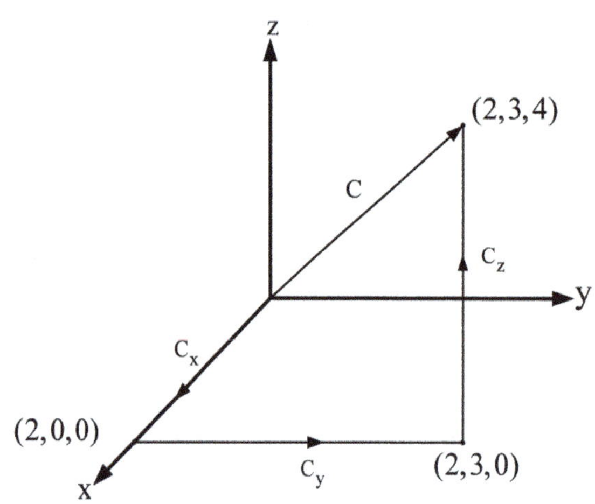

Figure 1-13. Line integrals.

SOLUTION

First we find the equation of the straight line for path C:

$$\frac{x-0}{2-0} = \frac{y-0}{3-0} = \frac{z-0}{4-0} \quad \text{or} \quad \frac{x}{2} = \frac{y}{3} = \frac{z}{4} \qquad \text{(constraints on variables)}$$

Then we differentiate to obtain

$$\frac{dx}{2} = \frac{dy}{3} = \frac{dz}{4} \qquad \text{(constraints on differentials)}$$

Using the above equations of constraint we can obtain an integral in one variable. We choose x as the variable.

$$\int_C \mathbf{F}\cdot d\ell = \int_C F_x dx + F_y dy + F_z dz = \int_C x^2 dx + xydy + 2yzdz$$

$$= \int x^2 dx + x\left(\frac{3x}{2}\right)dy + 2\left(\frac{3x}{2}\right)\left(\frac{4x}{2}\right)dz \qquad \text{(using constraints on variables)}$$

$$= \int x^2 dx + \frac{3}{2}x^2 \cdot \frac{3}{2}dx + 6x^2 \cdot 2dx \qquad \text{(using constraints on differentials)}$$

$$= \int_0^2 \frac{61}{4} x^2\, dx = \frac{122}{3} \text{ or } 40.67$$

Now calculate the integral over $C_a = C_x + C_y + C_z$ which is a different path (Figure 1-13) between the same two points (0, 0, 0) and (2, 3, 4):

$$\int_{C_a} \mathbf{F}\cdot d\ell = \int_{C_x} \mathbf{F}\Big|_{\substack{y=0\\z=0}} \cdot \mathbf{a}_x\, dx + \int_{C_y} \mathbf{F}\Big|_{\substack{x=2\\z=0}} \cdot \mathbf{a}_y dy + \int_{C_z} \mathbf{F}\Big|_{\substack{x=2\\y=3}} \cdot \mathbf{a}_z dz$$

$$= \int_0^2 F_x(y=0,\ z=0)\ dx + \int_0^3 F_y(x=2,\ z=0)\ dy + \int_0^4 F_z(x=2,\ y=3)\ dz$$

$$= \int_0^2 x^2\ dx + \int_0^3 2y\ dy + \int_0^4 6z\ dz = 59\,{}^2\!/_3 \text{ or } 59.67$$

Note that the two results differ. In general, the line integral between two points depends on the path taken. In the integrals above note that the direction of path C is taken into account by the limits of integration. Upper and lower limits correspond to the head and tail, respectively. If the direction of C is reversed, then the upper and lower limits are interchanged.

INDEPENDENCE OF PATH

There is a special class of vector fields $\mathbf{F}$ which are **conservative**, i.e.,

$$\oint_C \mathbf{F}\cdot d\ell = 0 \qquad \text{for all closed paths C.} \qquad (1\text{-}28a)$$

The circle on the integral sign denotes a *closed* path. The line integral on the left-hand side of Eq. (1-28a) is called the **circulation** of the vector $\mathbf{F}$ along the closed path C. For conservative fields, the line integral is independent of path.

To see this, consider the line integral around path C of Figure 1-14(a) where $C = C_1 - C_2$ (Figure 1-14(b)). **F** is a conservative field and therefore

$$\oint_C \mathbf{F} \cdot d\ell = \int_{C_1} \mathbf{F} \cdot d\ell - \int_{C_2} \mathbf{F} \cdot d\ell = 0$$

then

$$\int_{C_1} \mathbf{F} \cdot d\ell = \int_{C_2} \mathbf{F} \cdot d\ell = \int_{C_n} \mathbf{F} \cdot d\ell \tag{1-28b}$$

We may conclude that the line integral between a and b (or between any two points) is *independent of path* for conservative fields. For this special class of vector fields, we may evaluate the line integral by choosing any convenient path. See Problem 1-27 for an example of the conservative vector field.

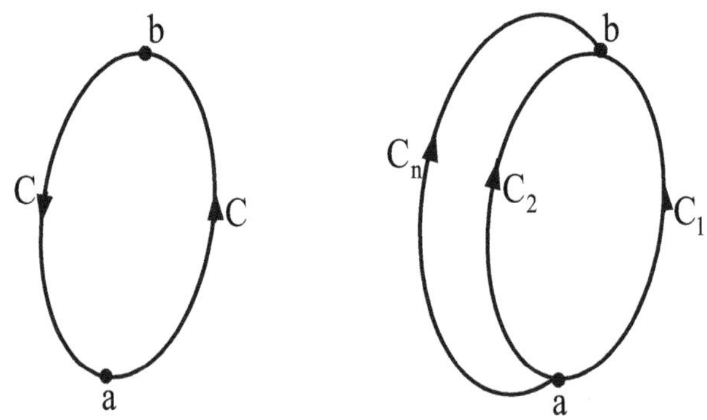

(a) A closed path. (b) Path from a to b

Figure 1-14. Independence of path.

SURFACE INTEGRALS: FLUX

The integral $\iint_S \mathbf{F} \cdot \mathbf{ds}$ occurs often in electromagnetics. It is called the **flux** of the vector **F** through the surface S.

d**s** is defined as $\mathbf{a}_n$ ds, where $\mathbf{a}_n$ is the unit vector normal to the surface and ds is the element of differential surface area. Thus

$$\iint_S \mathbf{F} \cdot \mathbf{ds} = \iint_S \mathbf{F} \cdot \mathbf{a}_n \, ds = \iint_S F_n \, ds \tag{1-29}$$

where F_n is the component of **F** normal to the surface S.

The surface S may be closed or open. For closed surfaces, such as that of Figure 1-15(a), the normal is defined as the *outward* normal. Figure 1-15(a) shows how the vector $\mathbf{a}_n$ varies over the surface. If the surface is open, then the normal direction is not unique and a direction must be chosen as part of the problem. We usually define the normal by a right-hand rule relationship with a path C around the periphery. Figure 1-15(b) shows an open planar surface lying in the xy plane with perimeter C. In this case the normal is constant ($\mathbf{a}_n = \mathbf{a}_z$) with a right-hand rule relationship with the periphery C when C follows the counter clockwise direction. Now we perturb the surface of 1-15(b) by pushing on its central portion. The result is a hat-shaped open surface (Figure 1-15(c)). $\mathbf{a}_n$ varies over the surface and the right-hand rule relationship is maintained.

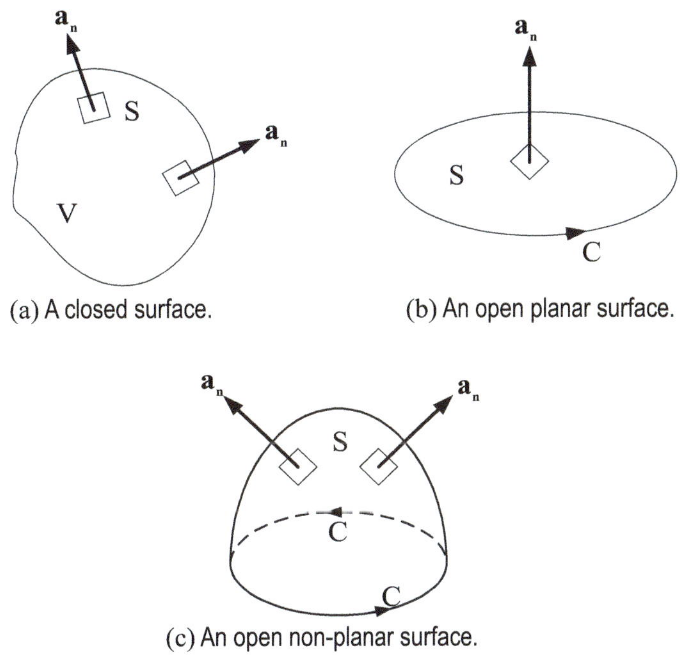

Figure 1-15. Surface integrals.

EXAMPLE 1-6

Let $\mathbf{F} = \mathbf{a}_r \frac{K_1}{r^2}$ (a) Find the flux of the vector $\mathbf{F}$ through the entire surface of the sphere of radius r_0, centered at origin. (b) Find the flux of $\mathbf{F}$ through a surface of area S_0 and arbitrary shape on the surface of the sphere of radius r_0.

SOLUTION

(a) On the surface of the sphere of radius r_0, $r = r_0$. It is a closed surface and $d\mathbf{s} = \mathbf{a}_n\, ds = \mathbf{a}_r\, r^2 \sin\theta\, d\theta\, d\phi$. So

$$\oint\!\!\oint_S \mathbf{F} \cdot d\mathbf{s} = \int_0^{2\pi}\int_0^{\pi} F_r\, r^2 \sin\theta\, d\theta\, d\phi = \int_0^{2\pi}\int_0^{\pi} K_1 \sin\theta\, d\theta\, d\phi = K_1[-\cos\theta]_0^{\pi}\,[\phi]_0^{2\pi} = 4\pi\, K_1$$

Note that the circle on the integral signs denotes a *closed* surface.

(b) The surface S is an open surface.

$$\iint_{S_0} \mathbf{F}\cdot d\mathbf{s} = \iint_{S_0} F_r\, ds = \iint_{S_0} F_r(r=r_0)\, r_0^2 \sin\theta\, d\theta\, d\phi$$

$$= F_r(r=r_0) \iint_{S_0} ds \qquad \text{[because } F_r \text{ is a constant on } S_0\ (r=r_0)\text{, i.e., independent of } \theta \text{ and } \phi]$$

$$= \frac{K_1}{r_0^2} S_0$$

EXAMPLE 1-7

Let $\mathbf{F} = \mathbf{a}_\rho\, \rho \sin^2\phi + \mathbf{a}_\phi\, \rho \cos^2\phi + \mathbf{a}_z\, z$.

Find the flux of the vector $\mathbf{F}$ through the (closed) surface of a cylinder of height 2ℓ meters and radius $\rho = \rho_0$ meters, centered at the origin (see Figure 1-10(a)).

SOLUTION

The closed surface consists of three open surfaces: the side (cylindrical) S_1, the top S_2 and the bottom surface S_3.

$$\text{On } S_1,\ d\mathbf{s} = \mathbf{a}_\rho\, \rho\, d\phi\, dz;\ \rho = \rho_o,\ 0 \le \phi \le 2\pi,\ -\ell \le z \le \ell.$$
$$\text{On } S_2,\ d\mathbf{s} = \mathbf{a}_z\, \rho\, d\rho\, d\phi;\ z = \ell,\ 0 \le \rho \le \rho_o,\ 0 \le \phi \le 2\pi.$$
$$\text{On } S_3,\ d\mathbf{s} = -\mathbf{a}_z\, \rho\, d\rho\, d\phi;\ z = -\ell,\ 0 \le \rho \le \rho_o,\ 0 \le \phi \le 2\pi.$$

Therefore,

$$\oiint_S \mathbf{F}\cdot d\mathbf{s} = \iint_{\text{SIDE}} F_\rho\, \rho_0\, d\phi\, dz + \iint_{\text{TOP}} F_z\, \rho\, d\rho\, d\phi - \iint_{\text{BOTTOM}} F_z\, \rho\, d\rho\, d\phi$$

$$= \int_{-\ell}^{\ell}\int_0^{2\pi} \rho_0 \sin^2\phi\, \rho_0\, d\phi\, dz + \int_0^{2\pi}\int_0^{\rho_0} \ell\, \rho\, d\rho\, d\phi + \int_0^{2\pi}\int_0^{\rho_0} (-\ell)(-\rho\, d\rho\, d\phi)$$

$$= \rho_0^2 \big[z\big]_{-\ell}^{\ell} \left[\frac{1}{2}\left(\phi - \frac{1}{2}\sin 2\phi\right)\right]_0^{2\pi} + 2\ell \int_0^{2\pi} d\phi \int_0^{\rho_0} \rho\, d\rho$$

$$= \rho_0^2\, 2\ell\pi + 2\ell \cdot 2\pi \frac{\rho_0^2}{2} = 4\pi\, \rho_0^2\, \ell$$

1.4 VECTOR CALCULUS

1.4.1 Definitions

In this section we define three new derivative quantities, namely, the gradient of a scalar function and the divergence and curl of a vector function. They describe how the scalar and vector fields vary as a function of position. First we define the vector differential operator ∇ read **del** or **nabla**:

$$\nabla = \mathbf{a}_x \frac{\partial}{\partial x} + \mathbf{a}_y \frac{\partial}{\partial y} + \mathbf{a}_z \frac{\partial}{\partial z} \tag{1-30}$$

∇ operates on a vector or scalar. The partial derivatives operate on the vector or scalar operand in a manner determined by the distributive property.

$$\boxed{\text{grad } V = \nabla V = \mathbf{a}_x \frac{\partial V}{\partial x} + \mathbf{a}_y \frac{\partial V}{\partial y} + \mathbf{a}_z \frac{\partial V}{\partial z}} \tag{1-31}$$

The divergence of a vector function $\mathbf{A}$ (x, y, z) is defined as

$$\boxed{\text{div } \mathbf{A} = \nabla \cdot \mathbf{A} = \frac{\partial A_x}{\partial x} + \frac{\partial A_y}{\partial y} + \frac{\partial A_z}{\partial z}} \tag{1-32}$$

The curl of a vector function $\mathbf{A}$ is defined as

$$\text{curl } \mathbf{A} = \nabla \times \mathbf{A} = \begin{vmatrix} \mathbf{a}_x & \mathbf{a}_y & \mathbf{a}_z \\ \frac{\partial}{\partial x} & \frac{\partial}{\partial y} & \frac{\partial}{\partial z} \\ A_x & A_y & A_z \end{vmatrix}$$

$$\boxed{\nabla \times \mathbf{A} = \mathbf{a}_x \left(\frac{\partial A_z}{\partial y} - \frac{\partial A_y}{\partial z} \right) + \mathbf{a}_y \left(\frac{\partial A_x}{\partial z} - \frac{\partial A_z}{\partial x} \right) + \mathbf{a}_z \left(\frac{\partial A_y}{\partial x} - \frac{\partial A_x}{\partial y} \right)} \tag{1-33}$$

The partial derivatives of the second row of the determinant operate on the quantities in the third row. Note the cycle xyz in the final equation for the curl. Note also that ∇V, $\nabla \times \mathbf{A}$ are vectors and that $\nabla \cdot \mathbf{A}$ is a scalar.

The grad, div, and curl have been succinctly defined in this section in rectangular coordinates. The corresponding forms in cylindrical and spherical coordinates as well are given in Table 1-4. The operations described can readily be performed. In succeeding sections we introduce physical interpretations of grad, div, curl, and their forms in other coordinate systems.

Table 1-4. Vector Derivatives

RECTANGULAR COORDINATES (x, y, z)

$$\nabla V = \mathbf{a}_x \frac{\partial V}{\partial x} + \mathbf{a}_y \frac{\partial V}{\partial y} + \mathbf{a}_z \frac{\partial V}{\partial z}$$

$$\nabla \cdot \mathbf{A} = \frac{\partial A_x}{\partial x} + \frac{\partial A_y}{\partial y} + \frac{\partial A_z}{\partial z}$$

$$\nabla \times \mathbf{A} = \mathbf{a}_x \left(\frac{\partial A_z}{\partial y} - \frac{\partial A_y}{\partial z} \right) + \mathbf{a}_y \left(\frac{\partial A_x}{\partial z} - \frac{\partial A_z}{\partial x} \right) + \mathbf{a}_z \left(\frac{\partial A_y}{\partial x} - \frac{\partial A_x}{\partial y} \right)$$

$$\nabla^2 V = \frac{\partial^2 V}{\partial x^2} + \frac{\partial^2 V}{\partial y^2} + \frac{\partial^2 V}{\partial z^2}$$

CYLINDRICAL COORDINATES (ρ, ϕ, z)

$$\nabla V = \mathbf{a}_\rho \frac{\partial V}{\partial \rho} + \mathbf{a}_\phi \frac{1}{\rho} \frac{\partial V}{\partial \phi} + \mathbf{a}_z \frac{\partial V}{\partial z}$$

$$\nabla \cdot \mathbf{A} = \frac{1}{\rho} \frac{\partial}{\partial \rho} (\rho A_\rho) + \frac{1}{\rho} \frac{\partial A_\phi}{\partial \phi} + \frac{\partial A_z}{\partial z}$$

$$\nabla \times \mathbf{A} = \mathbf{a}_\rho \left(\frac{1}{\rho} \frac{\partial A_z}{\partial \phi} - \frac{\partial A_\phi}{\partial z} \right) + \mathbf{a}_\phi \left(\frac{\partial A_\rho}{\partial z} - \frac{\partial A_z}{\partial \rho} \right) + \mathbf{a}_z \frac{1}{\rho} \left[\frac{\partial}{\partial \rho} (\rho A_\phi) - \frac{\partial A_\rho}{\partial \phi} \right]$$

$$\nabla^2 V = \frac{1}{\rho} \frac{\partial}{\partial \rho} \left(\rho \frac{\partial V}{\partial \rho} \right) + \frac{1}{\rho^2} \frac{\partial^2 V}{\partial \phi^2} + \frac{\partial^2 V}{\partial z^2}$$

SPHERICAL COORDINATES (r, θ, ϕ)

$$\nabla V = \mathbf{a}_r \frac{\partial V}{\partial r} + \mathbf{a}_\theta \frac{1}{r} \frac{\partial V}{\partial \theta} + \mathbf{a}_\phi \frac{1}{r \sin\theta} \frac{\partial V}{\partial \phi}$$

$$\nabla \cdot \mathbf{A} = \frac{1}{r^2} \frac{\partial}{\partial r} (r^2 A_r) + \frac{1}{r \sin\theta} \frac{\partial}{\partial \theta} (A_\theta \sin\theta) + \frac{1}{r \sin\theta} \frac{\partial A_\phi}{\partial \phi}$$

$$\nabla \times \mathbf{A} = \mathbf{a}_r \frac{1}{r \sin\theta} \left[\frac{\partial}{\partial \theta} (A_\phi \sin\theta) - \frac{\partial A_\theta}{\partial \phi} \right] + \mathbf{a}_\theta \frac{1}{r} \left[\frac{1}{\sin\theta} \frac{\partial A_r}{\partial \phi} - \frac{\partial}{\partial r} (r A_\phi) \right] + \mathbf{a}_\phi \frac{1}{r} \left[\frac{\partial}{\partial r} (r A_\theta) - \frac{\partial A_r}{\partial \theta} \right]$$

$$\nabla^2 V = \frac{1}{r^2} \frac{\partial}{\partial r} \left(r^2 \frac{\partial V}{\partial r} \right) + \frac{1}{r^2 \sin\theta} \frac{\partial}{\partial \theta} \left(\sin\theta \frac{\partial V}{\partial \theta} \right) + \frac{1}{r^2 \sin^2\theta} \frac{\partial^2 V}{\partial \phi^2}$$

1.4.2 Gradient

The **gradient** of a scalar function has a very simple physical description. It is a vector whose direction is that of *maximum rate of change* of the scalar with distance. Its magnitude is that of maximum rate of change. Consider the temperature $T(x, y, z)$ in a room.

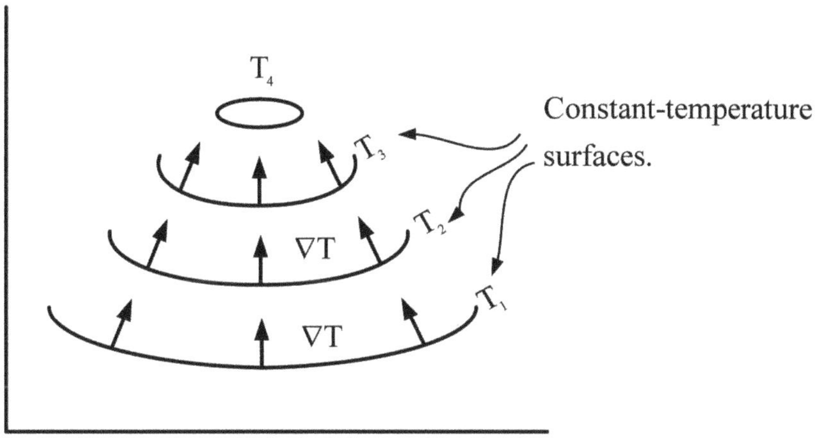

Figure 1-16. Temperature (T) and temperature gradient (∇T) in a room.

Figure 1-16 shows the constant temperature surfaces, called the **level surfaces**. The temperature generally increases with height; there is a "hot spot" near the ceiling caused by a light bulb. ∇T is normal to the constant temperature surfaces as shown. To understand this, consider a point on a constant temperature surface. Construct a tangent plane at that point. The rate of change of T with distance is zero for any direction lying in the tangent plane. It is non-zero only for directions with a normal component. But for any direction with both normal and tangential components, the rate of change is increased by dropping the tangential component; therefore, the maximum rate of change occurs along the normal direction. This means that the gradient (∇T) is perpendicular to the level surfaces where $T = \text{constant}$. Meanwhile, the backpackers among us are rather amused by all this; they knew all along that the gradient, i.e., the steepest path up the mountain, is normal to the contour lines.

It is easy to see that the above description corresponds to our previous definition of ∇T:

$$\nabla T = \mathbf{a}_x \frac{\partial T}{\partial x} + \mathbf{a}_y \frac{\partial T}{\partial y} + \mathbf{a}_z \frac{\partial T}{\partial z} \tag{1-34}$$

Since we are free to choose our coordinate system, choose $\mathbf{a}_z$ normal to the tangent plane to the constant temperature surfaces at (x, y, z). Then $\mathbf{a}_x$, $\mathbf{a}_y$ lie in the tangent plane. $\frac{\partial T}{\partial x}, \frac{\partial T}{\partial y}$ are zero, since $\mathbf{a}_x$, $\mathbf{a}_y$ lie in the tangent plane. ∇T is in the $\mathbf{a}_z$ direction normal to the tangent plane. Its magnitude is $\frac{\partial T}{\partial z}$, the rate of change in a direction normal to the tangent plane.

These concepts can be related to the so-called **directional derivative**, the rate of change in an arbitrary direction. By a chain rule relationship,

$$dT = \frac{\partial T}{\partial x}dx + \frac{\partial T}{\partial y}dy + \frac{\partial T}{\partial z}dz \tag{1-35}$$

$$= \left(\mathbf{a}_x\frac{\partial T}{\partial x} + \mathbf{a}_y\frac{\partial T}{\partial y} + \mathbf{a}_z\frac{\partial T}{\partial z}\right)\cdot(\mathbf{a}_x dx + \mathbf{a}_y dy + \mathbf{a}_z dz)$$

$$= \nabla T \cdot d\boldsymbol{\ell} \tag{1-36}$$

The directional derivative is given by

$$\frac{dT}{d\ell} = \frac{\partial T}{\partial x}\frac{dx}{d\ell} + \frac{\partial T}{\partial y}\frac{dy}{d\ell} + \frac{\partial T}{\partial z}\frac{dz}{d\ell}$$

$$= \nabla T \cdot \left(\frac{d\boldsymbol{\ell}}{d\ell}\right) = \nabla T \cdot (\mathbf{a}_\ell) \tag{1-37}$$

The gradient represents the maximum rate of change. The dot product with any arbitrary direction $\mathbf{a}_\ell$ yields the directional derivative or the rate of change in that particular direction.

The gradient will be used in relating the electric field to the electric potential in Chapter 2.

EXAMPLE 1-8

Calculate the gradient of the following scalar functions:

(a) $V = \frac{K}{\sqrt{x^2+y^2+z^2}} = \frac{K}{r}$.   (b) $V = \frac{K\cos\theta\sin\phi}{r^2}$.

SOLUTION

(a) In rectangular coordinates,

$$\begin{aligned}\nabla V &= \mathbf{a}_x \frac{\partial V}{\partial x} + \mathbf{a}_y \frac{\partial V}{\partial y} + \mathbf{a}_z \frac{\partial V}{\partial z} \\ &= \frac{-K}{\left[x^2+y^2+z^2\right]^{3/2}}(\mathbf{a}_x x + \mathbf{a}_y y + \mathbf{a}_z z) \\ &= \frac{-K}{r^3}\mathbf{r} = \mathbf{a}_r\left(\frac{-K}{r^2}\right)\end{aligned}$$

where we have used $\mathbf{a}_x x + \mathbf{a}_y y + \mathbf{a}_z z = \mathbf{r} = \mathbf{a}_r r$. $\mathbf{r}$ may be called the **position vector**.

In spherical coordinates, $\nabla V = \mathbf{a}_r \frac{\partial V}{\partial r} = \mathbf{a}_r\left(\frac{-K}{r^2}\right)$. The level surfaces on which V = constant or $\frac{K}{r} = C$ are given by $r = \frac{K}{C} = \text{constant}$, which is a family of spherical surfaces. It is clear from the above result that $\nabla V = -\mathbf{a}_r \frac{K}{r^2}$ is perpendicular to the level surfaces.

(b) $$\begin{aligned}\nabla V &= \mathbf{a}_r \frac{\partial V}{\partial r} + \mathbf{a}_\theta \frac{1}{r}\frac{\partial V}{\partial \theta} + \mathbf{a}_\phi \frac{1}{r\sin\theta}\frac{\partial V}{\partial \phi} \\ &= \mathbf{a}_r\left(\frac{-2K\cos\theta\sin\phi}{r^3}\right) + \mathbf{a}_\theta\left(\frac{-K\sin\theta\sin\phi}{r^3}\right) + \mathbf{a}_\phi\left(\frac{K\cot\theta\cos\phi}{r^3}\right)\end{aligned}$$

1.4.3 Divergence

The **divergence** of a vector function $\mathbf{A}$ is a measure of the net outward flux or flow at a given point. It is also a measure of how much a vector spreads out or diverges from a point. The divergence may be defined as the *net outward flux per unit volume.* To find the divergence at some given point, we construct a small volume Δv around the point and calculate the vector flux flowing out of the volume divided by Δv as Δv goes to zero:

$$\nabla \cdot \mathbf{A} = \lim_{\Delta v \to 0} \frac{1}{\Delta v} \oiint_S \mathbf{A} \cdot d\mathbf{s} \qquad (1\text{-}38)$$

where the closed surface S bounds or encloses the small volume Δv (see Figure 1-17(a)). Here we take the limit $\Delta v \to 0$ because the divergence is defined at a point. The definition above is identical to (1-32) of Section 1.4.1. The result is independent of the shape of Δv.

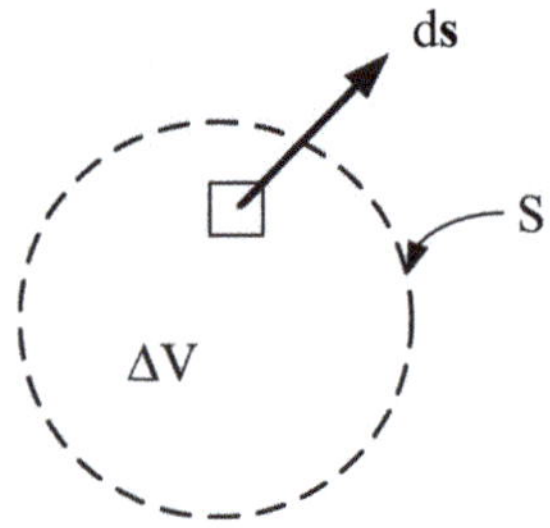

(a) The definition of divergence.

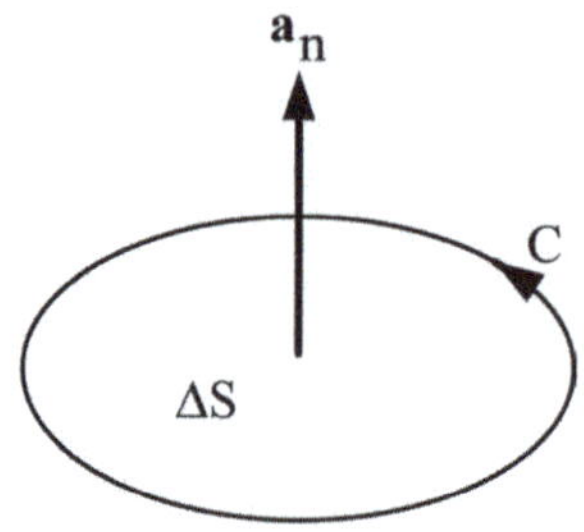

(b) The definition of curl.

Figure 1-17. The definition of divergence and curl.

EXAMPLE 1-9. A VECTOR FIELD THAT HAS DIVERGENCE

Calculate the divergence of the following vector field **F** and explain why **F** has divergence using the definition (1-38).

$$\mathbf{F}(x, y, z) = \mathbf{a}_x \, 2x$$

SOLUTION

$$F_x = 2x, \quad F_y = 0, \quad F_z = 0$$

Using Eq. (1-32),

$$\nabla \cdot F = \frac{\partial F_x}{\partial x} + \frac{\partial F_y}{\partial y} + \frac{\partial F_z}{\partial z} = \frac{\partial}{\partial x}(2x) = 2$$

The vector **F** has only the x component and its magnitude increases as x increases for $x > 0$. The field lines of **F** can be represented by a collection of arrows shown in Figure 1-18(a) (more arrows mean stronger field). When you consider a small spherical volume Δv (shown as dotted surface) around some point P and calculate the net flux flowing out of that volume, you will find that it is positive because the outgoing flux is greater than the incoming flux, i.e., the vector **F** has positive divergence at that point. We can also find the same result by using a small cube rather than a sphere since the divergence is independent of the shape of Δv.

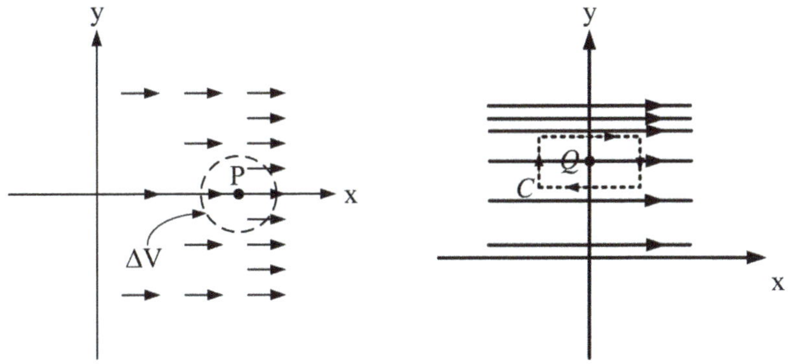

(a) A vector field that has divergence. (b) A vector field that has curl.

Figure 1-18. A vector field that has divergence and curl.

1.4.4 Curl

The **curl** of a vector function is a measure of the circulation about a point. It is also a measure of how much a vector curls around or rotates about a point. The curl may be defined as the *circulation per unit area*. Since the curl is a vector, we need to define both the direction and magnitude. To find the component of curl in a particular direction $\mathbf{a}_n$, we construct a surface Δs, perpendicular to $\mathbf{a}_n$ and bounded by contour C (see Figure 1-17(b)) and find the circulation per unit area as Δs goes to zero:

$$(\nabla \times \mathbf{A})_n = \lim_{\Delta s \to 0} \frac{1}{\Delta s} \oint_C \mathbf{A} \cdot d\boldsymbol{\ell} \tag{1-39}$$

where the closed contour C and the unit vector $\mathbf{a}_n$ are related by a *right-hand rule.* (see Figure 1-17(b)). If $\mathbf{a}_n$ is successively chosen in three orthogonal directions, we obtain the three components of the vector $\nabla \times \mathbf{A}$. The definition above is identical to (1-33) of Section 1.4.1. The result is independent of the shape of Δs.

EXAMPLE 1-10. A VECTOR FIELD THAT HAS CURL

Calculate the curl of the following vector field $\mathbf{G}$ and explain why $\mathbf{G}$ has curl using the definition (1-39).

$$\mathbf{G}(x, y, z) = \mathbf{a}_x\, 2y$$

SOLUTION

$$G_x = 2y,\ G_y = 0,\ G_z = 0$$

Using Eq. (1-33),

$$\nabla \times \mathbf{G} = \mathbf{a}_y \frac{\partial G_x}{\partial z} - \mathbf{a}_z \frac{\partial G_x}{\partial y} = -\mathbf{a}_z \frac{\partial}{\partial y}(2y) = -\mathbf{a}_z 2$$

The vector $\mathbf{G}$ has only the x component and its magnitude increases as y increases for $y > 0$. The field lines are shown in Figure 1-18(b). When you consider a small closed rectangular contour C in the xy plane (shown as a dotted line) around some point Q on the y axis and calculate the circulation clockwise, you will find a nonzero (positive) quantity because the upper portion of $\int \mathbf{G} \cdot d\ell$ is positive and the lower portion is negative, but of smaller magnitude. Using the right-hand rule, we see that $\mathbf{G}$ has a curl in the negative z direction (into the paper). You will find the same result with the circular loop because the curl is independent of the shape of the contour. Note that $\mathbf{G}$ has no divergence and $\mathbf{F}$ in Example 1-9 has no curl.

The basic concepts of divergence and curl are subtle and complex. Convenient analogies from our past experience are lacking. Therefore we should recognize that we have a continuing task. Our understanding of divergence and curl will grow as we proceed in our study of electromagnetics. You will find later that the electric field has a divergence nature and the magnetic field has a curl nature.

1.4.5 The Divergence Theorem and Stokes' Theorem—Solenoidal and Conservative Fields

The two important integral theorems—the divergence theorem and Stokes' theorem—can be obtained by applying the physical or geometrical definitions of divergence and curl, (1-38) and (1-39), to the real (not infinitesimally small) volume and surface, respectively.

The **divergence theorem** is stated as follows:

$$\iiint_V \nabla \cdot \mathbf{A} \, dv = \oiint_S \mathbf{A} \cdot d\mathbf{s} \tag{1-40}$$

where the closed surface S bounds the volume V. The theorem is valid for any vector $\mathbf{A}$, and for any volume V and its boundary surface S. It allows us to replace a volume integral with a surface integral, or vice versa. Note that $d\mathbf{s}$ points in the direction *normal* to the surface S *outward* from the volume V.

If a vector $\mathbf{A}$ is *divergenceless*, i.e., if $\nabla \cdot \mathbf{A}$ is zero everywhere, then $\oiint_S \mathbf{A} \cdot d\mathbf{s} = 0$ for any *closed* surface S.

Such a vector is also called **solenoidal**. For a solenoidal field, $\iint_S \mathbf{A} \cdot d\mathbf{s}$ is independent of the surface, as long as the surface S is bounded by the same closed contour C. The magnetic field is known to be a solenoidal field, as it will be discussed in Chapter 6.

Stokes' theorem is stated as follows:

$$\iint_S \nabla \times \mathbf{A} \cdot d\mathbf{s} = \oint_C \mathbf{A} \cdot d\ell \tag{1-41}$$

where the closed path C bounds the open surface S and $d\ell$ along C and $\mathbf{ds} = \mathbf{a}_n$ ds are related by a right-hand rule as in Figure 1-15(b), (c) and Figure 1-17(b).

If $\mathbf{A}$ is *curl free*, i.e., if $\nabla \times \mathbf{A}$ is zero everywhere, then $\oint_C \mathbf{A} \cdot d\boldsymbol{\ell} = 0$ for any *closed* path C and therefore $\mathbf{A}$ is a *conservative* field.* Furthermore the line integral of $\mathbf{A}$ between two points is independent of path (see Section 1.3.4). The static electric field is known to be a conservative field, as it will be shown in Chapter 2. The formal proof of (1-40) and (1-41) can be found in Cheng (1989).⁺

EXAMPLE 1-11. ILLUSTRATION OF THE DIVERGENCE THEOREM

Consider the vector field

$$\mathbf{F} = \mathbf{a}_z z^2$$

(a) Find the surface integral, $\oint\!\!\oint_S \mathbf{F} \cdot \mathbf{ds}$, i.e., the flux through the closed surfaces of the half-sphere of radius a as shown in Figure 1-19.

(b) Verify the divergence theorem, by calculating $\iiint_V \nabla \cdot \mathbf{F}\, dv$ for the volume enclosed.

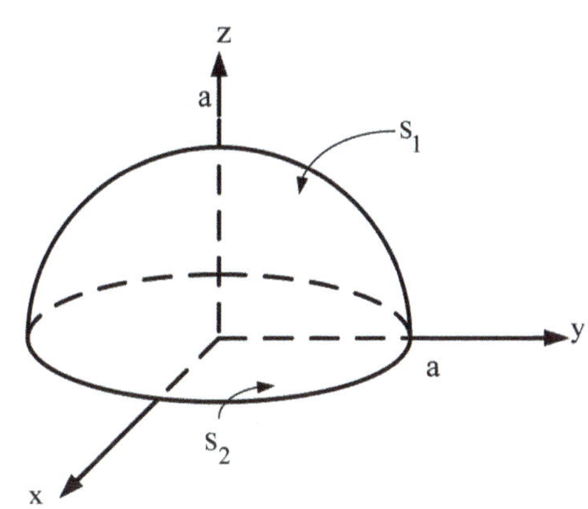

Figure 1-19. A hemispherical volume.

SOLUTION

(a) We first recognize that there are two open surfaces: the upper half-spherical surface S_1 and the lower base surface S_2.

For the surface S_1, r = a, $0 \le \theta \le \pi/2$, $0 \le \phi \le 2\pi$

$$\mathbf{ds} = \mathbf{a}_r r^2 \sin\theta\, d\theta\, d\phi = \mathbf{a}_r a^2 \sin\theta\, d\theta\, d\phi.$$

* Conservative and curlfree are identical characterizations.

⁺ D. K. Cheng, *Field and Wave Electromagnetics*, Addison-Wesley, 1989, 2nd Ed., Chapter 2.

Since $\mathbf{F}$ is given in Cartesian coordinates, we need to transform it into spherical coordinates:

$$\mathbf{F} = \mathbf{a}_z z^2 = (\mathbf{a}_r \cos\theta - \mathbf{a}_\theta \sin\theta)(r\cos\theta)^2$$
$$= \mathbf{a}_r r^2 \cos^3\theta - \mathbf{a}_\theta r^2 \sin\theta \cos^2\theta = \mathbf{a}_r F_r + \mathbf{a}_\theta F_\theta$$

$$\iint_{S_1} \mathbf{F}\cdot d\mathbf{s} = \iint_{S_1} (\mathbf{a}_r F_r + \mathbf{a}_\theta F_\theta)\cdot \mathbf{a}_r ds = \int_0^{2\pi}\int_0^{\frac{\pi}{2}} F_r\big|_{r=a}\, a^2 \sin\theta\, d\theta\, d\phi$$
$$= \int_0^{2\pi}\int_0^{\frac{\pi}{2}} a^2\cos^3\theta\, a^2 \sin\theta\, d\theta\, d\phi = 2\pi\cdot a^4\left[-\frac{1}{4}\cos^4\theta\right]_0^{\frac{\pi}{2}} = \frac{1}{2}\pi a^4$$

For the surface S_2, $z = 0$. Thus $\mathbf{F} = \mathbf{a}_z z^2 = 0$.

$$\iint_{S_2} \mathbf{F}\cdot d\mathbf{s} = 0$$

So,

$$\oint\!\!\!\oint_S \mathbf{F}\cdot d\mathbf{s} = \iint_{S_1} \mathbf{F}\cdot d\mathbf{s} + \iint_{S_2} \mathbf{F}\cdot d\mathbf{s} = \frac{1}{2}\pi a^4$$

(b) $$\nabla\cdot\mathbf{F} = \frac{\partial F_x}{\partial x} + \frac{\partial F_y}{\partial y} + \frac{\partial F_z}{\partial z} = 0 + 0 + \frac{\partial}{\partial z}(z^2) = 2z = 2r\cos\theta$$

For the hemispherical volume, $dv = r^2 \sin\theta\, dr\, d\theta\, d\phi,\ 0 \le r \le a,\ 0 \le \theta \le \frac{\pi}{2},\ 0 \le \phi \le 2\pi$.

$$\iiint_V \nabla\cdot\mathbf{F}\, dv = \iiint_V 2z\, dv = \int_0^{2\pi}\int_0^{\frac{\pi}{2}}\int_0^{a} 2(r\cos\theta)\, r^2 \sin\theta\, dr\, d\theta\, d\phi$$
$$= 2\cdot 2\pi\left[-\frac{1}{2}\cos^2\theta\right]_0^{\frac{\pi}{2}}\left[\frac{1}{4}r^4\right]_0^a = 4\pi\cdot\frac{1}{2}\frac{a^4}{4} = \frac{1}{2}\pi a^4$$

Therefore, $\iiint_V \nabla\cdot\mathbf{F}\, dv = \oint\!\!\!\oint_S \mathbf{F}\cdot d\mathbf{s}$; the divergence theorem is verified.

EXAMPLE 1-12 ILLUSTRATION OF STOKES' THEOREM

Consider the vector field

$$\mathbf{F} = \mathbf{a}_x y - \mathbf{a}_y x$$

(a) Evaluate the circulation $\oint_C \mathbf{F} \cdot d\boldsymbol{\ell}$ along a closed contour of a half-circle of radius a as shown in Figure 1-20.

(b) Verify Stokes' theorem by calculating $\iint_S \nabla \times \mathbf{F} \cdot \mathbf{ds}$ for the surface bounded by the contour.

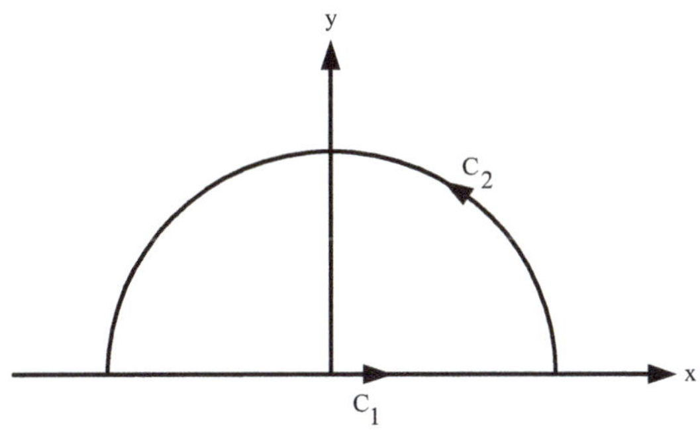

Figure 1-20. A semi-circular contour.

SOLUTION

(a) Define the straight line path C_1 and the half-circular path C_2.
Along C_1, $d\boldsymbol{\ell} = \mathbf{a}_x\, dx$, $y = 0$, $-a \le x \le a$

$$\int_{C_1} \mathbf{F} \cdot d\boldsymbol{\ell} = \int_{-a}^{a} \{\mathbf{a}_x 0 - \mathbf{a}_y x\} \cdot \mathbf{a}_x dx = 0$$

Along C_2, $d\boldsymbol{\ell} = \mathbf{a}_\phi\, a\, d\phi$, $\rho = a$, $0 \le \phi \le \pi$.

We also need to express $\mathbf{F}$ in cylindrical coordinates:

$$\mathbf{F} = \mathbf{a}_x y - \mathbf{a}_y x = \mathbf{a}_x\, \rho \sin\phi - \mathbf{a}_y\, \rho \cos\phi$$

$$= \rho(\mathbf{a}_x \sin\phi - \mathbf{a}_y \cos\phi) = -\,\mathbf{a}_\phi\, \rho$$

$$\int_{C_2} \mathbf{F} \cdot d\boldsymbol{\ell} = \int_0^{\pi} \left\{-\mathbf{a}_\phi \rho\right\}_{\rho=a} \cdot \mathbf{a}_\phi\, a\, d\phi = -a^2 [\phi]_0^{\pi} = -\pi a^2$$

Thus

$$\oint_C \mathbf{F} \cdot d\ell = \int_{C_1} \mathbf{F} \cdot d\ell + \int_{C_2} \mathbf{F} \cdot d\ell = 0 - \pi a^2 = -\pi a^2$$

(b)

$$\nabla \times \mathbf{F} = \begin{vmatrix} \mathbf{a}_x & \mathbf{a}_y & \mathbf{a}_z \\ \frac{\partial}{\partial x} & \frac{\partial}{\partial y} & \frac{\partial}{\partial z} \\ y & -x & 0 \end{vmatrix} = \mathbf{a}_z \left\{ \frac{\partial}{\partial x}(-x) - \frac{\partial}{\partial y}(y) \right\} = -\mathbf{a}_z 2$$

For the surface S bounded by C, $\mathbf{ds} = \mathbf{a}_z ds = \mathbf{a}_z \rho\, d\rho\, d\phi$, $0 \le \rho \le a$, $0 \le \phi \le \pi$.

$$\iint_S \nabla \times \mathbf{F} \cdot \mathbf{ds} = \int_0^{\pi} \int_0^{a} \{-\mathbf{a}_z 2\} \cdot \mathbf{a}_z \rho\, d\rho d\phi = -2\pi \frac{a^2}{2} = -\pi a^2$$

Therefore, $\iint_S \nabla \times \mathbf{F} \cdot \mathbf{ds} = \oint_C \mathbf{F} \cdot d\ell$; Stokes' theorem is verified.

EXAMPLE 1-13. THE ELECTRIC FIELD OF A POINT CHARGE

Consider the field of Figure 1-21

$$\mathbf{E} = \mathbf{a}_r \frac{K}{r^2}$$

where K is a constant.

(a) Find $\nabla \cdot \mathbf{E}$ everywhere. What happens to $\nabla \cdot \mathbf{E}$ at the origin $(r = 0)$?

(b) Find $\nabla \times \mathbf{E}$ everywhere.

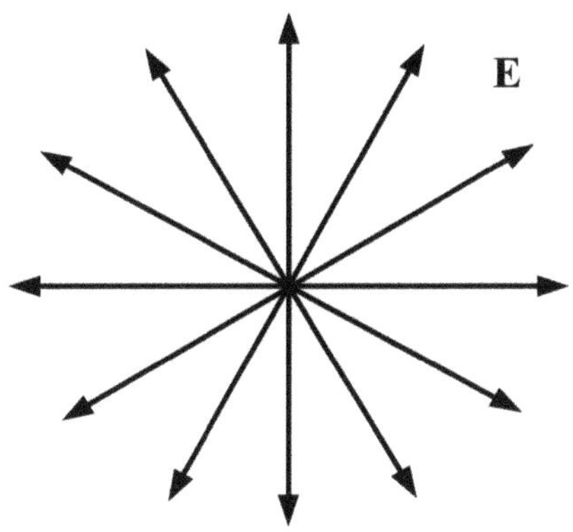

$$\mathbf{E} = \mathbf{a}_r \frac{K}{r^2}$$

$$\nabla \times \mathbf{E} = 0 \text{ everywhere}$$

$$\nabla \cdot \mathbf{E} = 0 \ (r \neq 0)$$

$$= \infty \ (r = 0)$$

Figure 1-21. The electric field of a point charge.

SOLUTION

(a) Consider $\nabla \cdot \mathbf{E}$ in spherical coordinates as given in Table 1-4.

$$\mathbf{E} = \mathbf{a}_r \frac{K}{r^2}, \quad E_\theta = E_\phi = 0$$

$$\nabla \cdot \mathbf{E} = \frac{1}{r^2}\frac{\partial}{\partial r}(r^2 E_r) = \frac{1}{r^2}\frac{\partial}{\partial r}(r^2 \frac{K}{r^2}) = \frac{1}{r^2}\frac{\partial}{\partial r}(K) = 0 \qquad (r \neq 0)$$

However, at $r = 0$, $\nabla \cdot \mathbf{E} = \dfrac{\frac{\partial}{\partial r}(K)}{r^2}$ is not necessarily zero. To examine the behavior of $\nabla \cdot \mathrm{E}$ at the origin, we use the divergence theorem. Construct a small spherical volume V of radius r about the origin and consider the following integral as $r \to 0$:

$$\iiint_V \nabla \cdot \mathbf{E} \ dv = \oiint_{S(r \to 0)} \mathbf{E} \cdot d\mathbf{s} = \int_0^{2\pi}\int_0^{\pi} \frac{K}{r^2} \mathbf{a}_r \cdot \mathbf{a}_r \ r^2 \sin\theta d\theta d\phi = \int_0^{2\pi}\int_0^{\pi} K \sin\theta d\theta d\phi = 4\pi K$$

Since there is no contribution to $\iiint_V \nabla \cdot \mathbf{E} dv$ for $r \neq 0$, the integrand $\nabla \cdot \mathbf{E}$ must be infinite at the origin. Thus $\nabla \cdot \mathbf{E}$ is zero when $r \neq 0$ and infinite at $r = 0$.

(b) To find $\nabla \times \mathbf{E}$, refer to Table 1-4. Since E_r is independent of θ, ϕ,

$$\nabla \times \mathbf{E} = \mathbf{a}_\theta \frac{1}{r\sin\theta}\frac{\partial E_r}{\partial \phi} - \mathbf{a}_\phi \frac{1}{r}\frac{\partial E_r}{\partial \theta} = 0 \qquad (r \neq 0,\ \sin\theta \neq 0)$$

To examine the behavior at $r = 0$, $\sin\theta = 0$, we apply Stokes' theorem. Construct a hemisphere of radius r centered about the origin, $0 \leq \theta \leq \pi/2$. S is the surface of the hemisphere and C is the periphery.

$$\iint_S \nabla\times\mathbf{E} \cdot d\mathbf{s} = \oint_C \mathbf{E} \cdot d\ell = \int_0^{2\pi} \mathbf{E} \cdot \mathbf{a}_\phi r\, d\phi = E_\phi\ 2\pi\ r = 0$$

because $E_\phi = 0$. Thus there is no contribution to $\iint_S \nabla \times \mathbf{E} \cdot d\mathbf{s}$ from the axis, for finite or zero r, and therefore

$\nabla \times \mathbf{E}$ may be considered to be zero everywhere.

The vector field $\mathbf{E}$ above represents the electric field of a point charge as will be shown in Chapter 2. Note that the electric field is *conservative* and has divergence.

EXAMPLE 1-14. THE MAGNETIC FIELD OF A FILAMENT OF CURRENT

Consider the field $\mathbf{H}$ of Figure 1-22.

$$\mathbf{H} = \mathbf{a}_\phi \frac{K}{\rho}$$

where K is a constant.

(a) Find $\nabla \cdot \mathbf{H}$ everywhere.

(b) Find $\nabla \times \mathbf{H}$ everywhere. What happens to $\nabla \times \mathbf{H}$ at points along the z-axis ($\rho = 0$)?

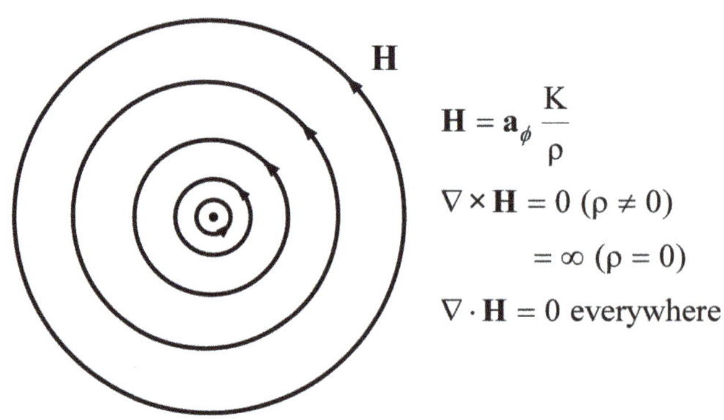

Figure 1-22. The magnetic field of a current filament.

SOLUTION

(a) Consider $\nabla \cdot \mathbf{H}$ in cylindrical coordinates as given in Table 1-4.

$$H_\rho = 0, \quad H_\phi = \frac{K}{\rho}, \quad H_z = 0$$

$$\nabla \cdot \mathbf{H} = \frac{1}{\rho}\frac{\partial H_\phi}{\partial \phi} = \frac{1}{\rho}\frac{\partial}{\partial \phi}\left(\frac{K}{\rho}\right) = 0 \quad (\rho \neq 0)$$

To determine the behavior of $\nabla \cdot \mathbf{H}$ along the z axis ($\rho = 0$), we use the divergence theorem again. Construct a cylinder of radius ρ, length ℓ, whose axis is the z axis:

$$\iiint_V \nabla \cdot \mathbf{H}\, dv = \oiint_{S(\rho \to 0)} \mathbf{H} \cdot d\mathbf{s} = \iint_{SIDE} H_\rho ds + \iint_{TOP,\, BOTTOM} H_z ds = 0$$

since $H_\rho = H_z = 0$. Thus there is no contribution to $\iiint_V \nabla \cdot \mathbf{H}\, dv$ from the axis ($\rho = 0$) and $\nabla \cdot \mathbf{H}$ is zero everywhere.

(b) To find $\nabla \times \mathbf{H}$, refer to Table 1-4.

$$\nabla \times \mathbf{H} = -\mathbf{a}_\rho \frac{\partial H_\phi}{\partial z} + \mathbf{a}_z \frac{1}{\rho}\frac{\partial}{\partial \rho}(\rho H_\phi) = \mathbf{a}_z \frac{1}{\rho}\frac{\partial}{\partial \rho}\left(\rho \frac{K}{\rho}\right) = \mathbf{a}_z \frac{1}{\rho}\frac{\partial}{\partial \rho}(K) = 0 \quad (\rho \neq 0)$$

To determine the behavior of $\nabla \times \mathbf{H}$ along the z axis ($\rho = 0$), we apply Stokes' theorem. Construct a small circular disk (of radius $\rho \to 0$) centered on and perpendicular to the z axis. S is the surface of the disk and C is the circumference. Then

$$\iint_S \nabla \times \mathbf{H} \cdot d\mathbf{s} = \oint_{C(\rho \to 0)} \mathbf{H} \cdot d\boldsymbol{\ell} = \lim_{\rho \to 0} H_\phi 2\pi\rho = \frac{K}{\rho} 2\pi\rho = 2\pi K$$

On the disk, there is no contribution to $\iint_S \nabla \times \mathbf{H} \cdot d\mathbf{s}$ for $\rho \neq 0$. Therefore the integrand $\nabla \times \mathbf{H}$ must be infinite on the z axis ($\rho = 0$). Thus $\nabla \times \mathbf{H}$ is zero when $\rho \neq 0$ and infinite at $\rho = 0$. The vector field $\mathbf{H}$ above represents the magnetic field of a long current-carrying filament along the z axis as will be shown in Chapter 6. Note that the magnetic field is *solenoidal* and has curl.

1.4.6 Vector Identities

A set of vector identities is presented in Table 1-5. These identities can be verified by a process which is straightforward, although laborious in some cases. We merely expand both sides of the identity in rectangular coordinates and show that they are identical. In this section, we will go through the process for certain commonly used vector identities.

First, consider the gradient of a product of two scalar functions:

$$\begin{aligned}\nabla(VW) &= \mathbf{a}_x \frac{\partial}{\partial x}(VW) + \mathbf{a}_y \frac{\partial}{\partial y}(VW) + \mathbf{a}_z \frac{\partial}{\partial z}(VW)\\ &= \mathbf{a}_x V\frac{\partial W}{\partial x} + \mathbf{a}_y V\frac{\partial W}{\partial y} + \mathbf{a}_z V\frac{\partial W}{\partial z} + \mathbf{a}_x W\frac{\partial V}{\partial x} + \mathbf{a}_y W\frac{\partial V}{\partial y} + \mathbf{a}_z W\frac{\partial V}{\partial z}\end{aligned}$$

(using the rule for the derivative of a product)

then

$$\nabla(VW) = V\nabla W + W\nabla V \tag{1-42}$$

Next, consider the divergence of the product of a scalar and a vector:

$$\begin{aligned}\nabla\cdot(V\mathbf{A}) &= \nabla\cdot\{\mathbf{a}_x VA_x + \mathbf{a}_y VA_y + \mathbf{a}_z VA_z\}\\ &= \frac{\partial}{\partial x}(VA_x) + \frac{\partial}{\partial y}(VA_y) + \frac{\partial}{\partial z}(VA_z)\\ &= V\frac{\partial A_x}{\partial x} + V\frac{\partial A_y}{\partial y} + V\frac{\partial A_z}{\partial z} + A_x\frac{\partial V}{\partial x} + A_y\frac{\partial V}{\partial y} + A_z\frac{\partial V}{\partial z}\end{aligned}$$

Then

$$\nabla\cdot(V\mathbf{A}) = V(\nabla\cdot\mathbf{A}) + \mathbf{A}\cdot\nabla V \tag{1-43}$$

Table 1-5. Vector Identities

TRIPLE PRODUCTS	
(1) $\mathbf{A}\cdot(\mathbf{B}\times\mathbf{C})=\mathbf{B}\cdot(\mathbf{C}\times\mathbf{A})=\mathbf{C}\cdot(\mathbf{A}\times\mathbf{B})$	
(2) $\mathbf{A}\times(\mathbf{B}\times\mathbf{C})=\mathbf{B}(\mathbf{A}\cdot\mathbf{C})-\mathbf{C}(\mathbf{A}\cdot\mathbf{B})$	
PRODUCT RULES	
(3) $\nabla(\mathrm{V\,W})=\mathrm{V}(\nabla\,\mathrm{W})+\mathrm{W}(\nabla\,\mathrm{V})$	
(4) $\nabla(\mathbf{A}\cdot\mathbf{B})=\mathbf{A}\times(\nabla\times\mathbf{B})+\mathbf{B}\times(\nabla\times\mathbf{A})+(\mathbf{A}\cdot\nabla)\,\mathbf{B}+(\mathbf{B}\cdot\nabla)\,\mathbf{A}$	
(5) $\nabla\cdot(\mathrm{V}\,\mathbf{A})=\mathrm{V}\,(\nabla\cdot\mathbf{A})+\mathbf{A}\cdot(\nabla\,\mathrm{V})$	
(6) $\nabla\cdot(\mathbf{A}\times\mathbf{B})=\mathbf{B}\cdot(\nabla\times\mathbf{A})-\mathbf{A}\cdot(\nabla\times\mathbf{B})$	
(7) $\nabla\times(\mathrm{V}\,\mathbf{A})=\mathrm{V}(\nabla\times\mathbf{A})-\mathbf{A}\times(\nabla\,\mathrm{V})$	
(8) $\nabla\times(\mathbf{A}\times\mathbf{B})=(\mathbf{B}\cdot\nabla)\,\mathbf{A}-(\mathbf{A}\cdot\nabla)\,\mathbf{B}+\mathbf{A}\,(\nabla\cdot\mathbf{B})-\mathbf{B}\,(\nabla\cdot\mathbf{A})$	
SECOND DERIVATIVES	
(9) $\nabla\cdot(\nabla\times\mathbf{A})=0$	
(10) $\nabla\times(\nabla\,\mathrm{V})=0$	
(11) $\nabla\times(\nabla\times\mathbf{A})=\nabla\,(\nabla\cdot\mathbf{A})-\nabla^2\mathbf{A}$	
INTEGRAL THEOREMS	
(1) Gradient Theorem:	$\int_a^b(\nabla\,\mathrm{V})\cdot d\boldsymbol{\ell}=\mathrm{V(b)}-\mathrm{V(a)}$
(2) Divergence Theorem:	$\iiint_V(\nabla\cdot\mathbf{A})\,dv=\oiint_S\mathbf{A}\cdot d\mathbf{s}$
(3) Stokes' Theorem:	$\iint_S(\nabla\times\mathbf{A})\cdot d\mathbf{s}=\oint_C\mathbf{A}\cdot d\boldsymbol{\ell}$

Finally consider the curl of the product of a scalar and a vector:

$$\nabla\times(\mathrm{V}\mathbf{A})=\begin{vmatrix}\mathbf{a}_x & \mathbf{a}_y & \mathbf{a}_z\\ \dfrac{\partial}{\partial x} & \dfrac{\partial}{\partial y} & \dfrac{\partial}{\partial z}\\ \mathrm{VA}_x & \mathrm{VA}_y & \mathrm{VA}_z\end{vmatrix}$$

Expanding the determinant, using the product rule and collecting terms yields

$$\nabla\times(\mathrm{V}\,\mathbf{A})=\mathrm{V}\,\nabla\times\mathbf{A}+(\nabla\,\mathrm{V})\times\mathbf{A} \tag{1-44}$$

1.4.7 Higher Order Functions of Vector Calculus

The gradient (∇V) is a vector so we can take its divergence to form $\nabla \cdot (\nabla V)$ which is a scalar called the **Laplacian**. There are five such quantities involving two successive del (∇) operations:

$$\nabla \cdot (\nabla V) \quad \nabla(\nabla \cdot \mathbf{A}) \quad \nabla \cdot (\nabla \times \mathbf{A})$$
$$\nabla \times (\nabla V) \qquad\qquad \nabla \times (\nabla \times \mathbf{A})$$

Two of these quantities, namely, $\nabla \times (\nabla V)$ and $\nabla \cdot (\nabla \times \mathbf{A})$, are always zero, for any scalar function V and for any vector function $\mathbf{A}$.

$$\nabla \times \nabla V = \begin{vmatrix} \mathbf{a}_x & \mathbf{a}_y & \mathbf{a}_z \\ \dfrac{\partial}{\partial x} & \dfrac{\partial}{\partial y} & \dfrac{\partial}{\partial z} \\ \dfrac{\partial V}{\partial x} & \dfrac{\partial V}{\partial y} & \dfrac{\partial V}{\partial z} \end{vmatrix}$$

$$= \mathbf{a}_x \left(\frac{\partial^2 V}{\partial y \partial z} - \frac{\partial^2 V}{\partial z \partial y}\right) + \mathbf{a}_y \left(\frac{\partial^2 V}{\partial z \partial x} - \frac{\partial^2 V}{\partial x \partial z}\right) + \mathbf{a}_z \left(\frac{\partial^2 V}{\partial x \partial y} - \frac{\partial^2 V}{\partial y \partial x}\right)$$

If the function V and its derivatives are continuous, the order of differentiation is immaterial, and thus each component of $\nabla \times \nabla V$ is zero. The divergence of the curl, $\nabla \cdot (\nabla \times \mathbf{A})$, is zero for the same reason, as may be shown by expansion in rectangular coordinates.

Thus

$$\boxed{\nabla \times \nabla V = \nabla \cdot (\nabla \times \mathbf{A}) = 0} \qquad (1\text{-}45)$$

$\nabla \cdot (\nabla V)$ is written as $\nabla^2 V$ and is called the *Laplacian* of the scalar V. Expanding in rectangular coordinates:

$$\boxed{\nabla \cdot (\nabla V) = \nabla^2 V = \frac{\partial^2 V}{\partial x^2} + \frac{\partial^2 V}{\partial y^2} + \frac{\partial^2 V}{\partial z^2}} \qquad (1\text{-}46)$$

The Laplacian of a scalar is itself a scalar and is called the *scalar Laplacian.*

The Laplacian of a vector is written as $\nabla^2 \mathbf{A}$ and is defined as follows (see the vector identity (11) of Table 1-5):

$$\nabla^2 \mathbf{A} = \nabla(\nabla \cdot \mathbf{A}) - \nabla \times (\nabla \times \mathbf{A}) \qquad (1\text{-}47)$$

The Laplacian of a vector is itself a vector and is called the *vector Laplacian*. It has a particularly simple form in rectangular coordinates:

$$\nabla^2 \mathbf{A} = \mathbf{a}_x \ \nabla^2 A_x + \mathbf{a}_y \ \nabla^2 A_y + \mathbf{a}_z \nabla^2 A_z \tag{1-48}$$

where $\nabla^2 A_x$ is the scalar Laplacian, etc., from whence derives the concept of the vector Laplacian. In other coordinate systems there is no such simple result and one must, consequently, return to Equation (1-47) to obtain the vector Laplacian.

1.5 HELMHOLTZ'S THEOREM

Helmholtz's theorem states that any vector $\mathbf{F}$ can be split into two parts, one of which is a conservative vector and the other of which is solenoidal, as follows:

$$\boxed{\mathbf{F} = -\nabla V + \nabla \times \mathbf{A}} \tag{1-49}$$

where V and $\mathbf{A}$ are defined by

$$V = \frac{1}{4\pi} \iiint_V \frac{(\nabla \cdot \mathbf{F})' dv'}{R} \tag{1-50a}$$

$$\mathbf{A} = \frac{1}{4\pi} \iiint_V \frac{(\nabla \times \mathbf{F})' dv'}{R} \tag{1-50b}$$

The definitions of V and $\mathbf{A}$ are of interest to us primarily in that they indicate that V derives from $\nabla \cdot \mathbf{F}$ and $\mathbf{A}$ from $\nabla \times \mathbf{F}$. R is the distance between points (x, y, z) and (x', y', z').

$(\nabla \cdot \mathbf{F})'$ merely indicates that we take the divergence of $\mathbf{F}(x, y, z)$ and then change (x, y, z) to (x', y', z'). Now consider Equation (1-49)

$$\mathbf{F} = -\nabla V + \nabla \times \mathbf{A}$$

$-\nabla V$ is the curlfree (conservative) portion of $\mathbf{F}$ since $\nabla \times \nabla V = 0$. $\nabla \times \mathbf{A}$ is the divergenceless (solenoidal) portion since $\nabla \cdot (\nabla \times \mathbf{A}) = 0$. Substituting the definitions into Eq. (1-49), we see that Helmholtz's theorem may be stated as a vector identity involving only the arbitrary vector $\mathbf{F}$:

$$\mathbf{F} = -\nabla \underbrace{\left[\iiint \frac{(\nabla \cdot \mathbf{F})' \, dv'}{4\pi R} \right]}_{V} + \nabla \times \underbrace{\left[\iiint \frac{(\nabla \times \mathbf{F})' \, dv'}{4\pi R} \right]}_{\mathbf{A}} = -\nabla V + \nabla \times \mathbf{A}$$

Now we can see that $\mathbf{F}$ is completely determined by its divergence and curl. The portion $-\nabla V$ is curlfree (conservative) and is determined solely by $\nabla \cdot \mathbf{F}$. The portion $\nabla \times \mathbf{A}$ is divergenceless (solenoidal) and is determined solely by $\nabla \times \mathbf{F}$. If both $\nabla \cdot \mathbf{F}$ and $\nabla \times \mathbf{F}$ are zero, then vector $\mathbf{F}$ is zero (or a constant vector since a constant vector has no divergence or curl).

Furthermore, we note two special cases:

If $\nabla \times \mathbf{F} = 0$, the second integral vanishes and $\mathbf{F} = -\nabla V$, i.e., $\mathbf{F}$ can be expressed as the gradient of a scalar.

If $\nabla \cdot \mathbf{F} = 0$, the first integral vanishes and $\mathbf{F} = \nabla \times \mathbf{A}$, i.e., $\mathbf{F}$ can be expressed as the curl of another vector.

The electrostatic field is an example of the first type (conservative field) and the magnetostatic field is an example of the second type (solenoidal field). A proof of Helmholtz's theorem may be found in Arfken.*

In summary:

(a) A vector field $\mathbf{F}$ is completely determined if its divergence and curl are known everywhere.

(b) Every vector field $\mathbf{F}$ can be broken into two parts, one of which is conservative and the other of which is solenoidal.

(c) The conservative part is determined by $\nabla \cdot \mathbf{F}$.

(d) The solenoidal part is determined by $\nabla \times \mathbf{F}$.

(e) If $\nabla \times \mathbf{F} = 0$, then $\mathbf{F} = -\nabla V$. (1-51a)

(f) If $\nabla \cdot \mathbf{F} = 0$, then $\mathbf{F} = \nabla \times \mathbf{A}$. (1-51b)

EXAMPLE 1-15

Apply Helmholtz's theorem to the vector field of Figure 1-21 (see Example 1-13).

SOLUTION

$$\mathbf{E} = \mathbf{a}_r \frac{K}{r^2}$$

In Example 1-13, we have shown that $\nabla \times \mathbf{E} = 0$ everywhere. Thus, $\mathbf{E} = -\nabla V$ and

$$V = \frac{1}{4\pi} \iiint \frac{(\nabla \cdot \mathbf{E})'\, dv'}{R}$$

$\nabla \cdot \mathbf{E} = 0$ except at the origin (Example 1-13). At the origin ($x' = y' = z' = 0$), R reduces to r which can be taken outside the integral:

$$V = \frac{1}{4\pi r} \iiint (\nabla \cdot \mathbf{E})'\, dv'$$

* G. Arfken, *Mathematical Methods for Physicists*, Academic Press, 1985, 3rd Ed., Section 1.15.

But $\iiint (\nabla \cdot \mathbf{E})' \, dv' = \iiint (\nabla \cdot \mathbf{E}) \, dv = 4\pi K$ (Example 1-13) and thus

$$\mathbf{E} = -\nabla V; \quad V = \frac{K}{r}$$

This result will be determined in another way in Chapter 2 (see Eq. (2-43)).

PROBLEMS

1-1. Given

$$\mathbf{A} = \mathbf{a}_x - \mathbf{a}_y + \mathbf{a}_z 2$$
$$\mathbf{B} = -\mathbf{a}_x + \mathbf{a}_y + \mathbf{a}_z 4$$
$$\mathbf{C} = \mathbf{a}_y 3 - \mathbf{a}_z 2,$$

find

(a) $|\mathbf{A}|$ (magnitude), (b) $\mathbf{a}_A$ (unit vector in the direction of $\mathbf{A}$),
(c) $\mathbf{B} \cdot \mathbf{A}$, (d) the component of $\mathbf{B}$ in the dire ction of $\mathbf{A}$,
(e) θ_{AB}, (f) $\mathbf{A} \times \mathbf{B}$, and (g) $(\mathbf{A} \times \mathbf{B}) \cdot \mathbf{C}$ and $\mathbf{A} \cdot (\mathbf{B} \times \mathbf{C})$.

1-2. Repeat Problem 1-1, given

$$\mathbf{A} = \mathbf{a}_x 2 - \mathbf{a}_y + \mathbf{a}_z 3$$
$$\mathbf{B} = -\mathbf{a}_x + \mathbf{a}_y + \mathbf{a}_z 5$$
$$\mathbf{C} = \mathbf{a}_x 3 - \mathbf{a}_z 2$$

1-3. Repeat Problem 1-1, given

$$\mathbf{A} = \mathbf{a}_x 3 + \mathbf{a}_y 4 - \mathbf{a}_z 5$$
$$\mathbf{B} = -\mathbf{a}_x 4 + \mathbf{a}_y 3$$
$$\mathbf{C} = \mathbf{a}_x - \mathbf{a}_y 2 + \mathbf{a}_z 3$$

1-4. Prove the identity (1-21) by expanding both sides in rectangular coordinates.

1-5. Prove the identity (1-22) (the "BAC-CAB" rule) by expanding both sides in rectangular coordinates.

1-6. For the vectors given in Problems 1-1, 1-2, and 1-3

(a) calculate $\mathbf{A} \times (\mathbf{B} \times \mathbf{C})$.
(b) calculate $(\mathbf{A} \cdot \mathbf{B})$ and $(\mathbf{A} \cdot \mathbf{C})$.
(c) confirm the "BAC-CAB" rule in Eq. (1-22).

1-7. Points $(0, 0, 0)$, $(1, 1, 1)$, $(0, 2, 0)$ form a triangle in space. Find the area of the triangle using the cross product of two vectors.

1-8. Consider the vector equation $\mathbf{C} = \mathbf{A} - \mathbf{B}$. By taking the squared magnitude of both sides of the equation and interpreting the result in a triangular geometry, prove the *law of cosines*:

$$C^2 = A^2 + B^2 - 2AB\cos\theta_{AB}$$

1-9. Prove the *law of sines* for a triangle, using the cross product of two vectors:

$$\frac{A}{\sin\theta_{BC}} = \frac{B}{\sin\theta_{CA}} = \frac{C}{\sin\theta_{AB}}$$

1-10. Let $\mathbf{a}_A$ and $\mathbf{a}_B$ be unit vectors in the directions of vectors $\mathbf{A}$ and $\mathbf{B}$ in the xy-plane that make angles α and β, respectively, with the x-axis.

(a) Obtain a formula for the expansion of the cosine of the difference of two angles, $\cos(\alpha-\beta)$, by taking the dot product $\mathbf{a}_A \cdot \mathbf{a}_B$.
(b) Obtain a formula for $\sin(\alpha-\beta)$, by taking the cross product $\mathbf{a}_B \times \mathbf{a}_A$.

1-11. A tetrahedron is determined by three edge vectors $\mathbf{A}$, $\mathbf{B}$, $\mathbf{C}$ as shown in Figure 1-23. Find the volume of the tetrahedron with $\mathbf{A} = \mathbf{a}_x + \mathbf{a}_y + \mathbf{a}_z$, $\mathbf{B} = \mathbf{a}_x 2 - \mathbf{a}_y$, $\mathbf{C} = \mathbf{a}_y 3 + \mathbf{a}_z 2$. Note that the volume of the tetrahedron is $\frac{1}{6}$ of that of the parallelepiped formed by $\mathbf{A}$, $\mathbf{B}$, $\mathbf{C}$.

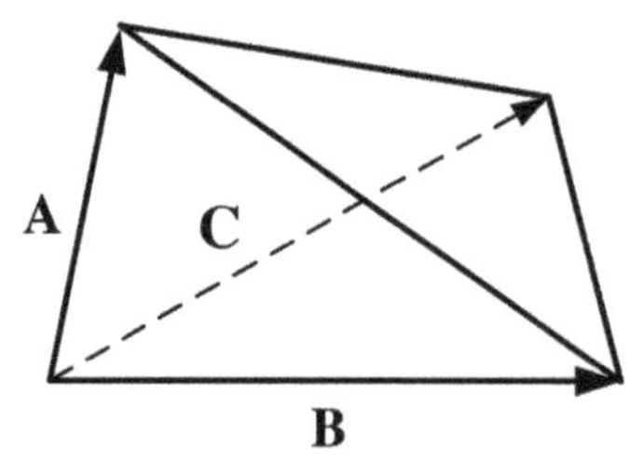

Figure 1-23. A tetrahedron.

1-12. (a) Prove that the vector $\mathbf{A} = \mathbf{a}_x a + \mathbf{a}_y b$ is perpendicular to the straight line: $ax + by = c$.
(b) Find a unit vector in the xy-plane, perpendicular to the straight line $2x - y = 2$.

1-13. (a) Prove that the vector $\mathbf{A} = \mathbf{a}_x a + \mathbf{a}_y b + \mathbf{a}_z c$ is normal to the plane: $ax + by + cz = d$.
(b) Find a unit vector normal to the plane $3x + 4y + 5z = 12$.

1-14. By dot-multiplying both sides of Eq. (1-22) by the vector $\mathbf{D}$, prove the following *identity of Lagrange*:

$$(\mathbf{A} \times \mathbf{D}) \cdot (\mathbf{B} \times \mathbf{C}) = (\mathbf{A} \cdot \mathbf{B})(\mathbf{C} \cdot \mathbf{D}) - (\mathbf{A} \cdot \mathbf{C})(\mathbf{B} \cdot \mathbf{D})$$

1-15. The *position vector* $\mathbf{r}$ of a point $P(x, y, z)$ is a vector pointing from the origin to the point P. Thus

$$\mathbf{r} = \mathbf{a}_x x + \mathbf{a}_y y + \mathbf{a}_z z \text{ in rectangular coordinates}$$

Show that

(a) $\mathbf{r} = \mathbf{a}_\rho \rho + \mathbf{a}_z z$ in cylindrical coordinates (ρ, ϕ, z)

(b) $\mathbf{r} = \mathbf{a}_r r$ in spherical coordinates (r, θ, ϕ)

Note that $\mathbf{r}$ has no components of $\mathbf{a}_\theta$ or $\mathbf{a}_\phi$.

1-16. Prove the relationships in Table 1-1 from Figures 1-7 (a), (b), (c) of the coordinate systems.

1-17. While $\mathbf{a}_x$, $\mathbf{a}_y$, $\mathbf{a}_z$ are constant unit vectors, $\mathbf{a}_\rho$, $\mathbf{a}_r$, $\mathbf{a}_\theta$, $\mathbf{a}_\phi$ depend on the location. Using the definitions of these unit vectors shown in Figure 1-8, answer the following questions.

(a) What is the equation of the half xz-plane $(x > 0)$ in cylindrical or spherical coordinates? What is $\mathbf{a}_\phi$ on this surface? Answer in terms of $\mathbf{a}_x$, $\mathbf{a}_y$, $\mathbf{a}_z$.

(b) What is the equation of the half yz-plane $(y > 0)$ in cylindrical or spherical coordinates? What is $\mathbf{a}_\phi$ on this surface?

(c) What is the equation of the xy-plane in spherical coordinates? What is $\mathbf{a}_\theta$ on this surface?

1-18. Consider the following points given in rectangular coordinates. Find the corresponding cylindrical and spherical coordinates.

(a) $(1, 1, 1)$ (b) $(3, 4, 0)$ (c) $(1, 0, 1)$

(d) $(0, -1, 2)$ (e) $(-1, 2, -3)$

1-19. Consider the following points given in cylindrical coordinates. Find the corresponding rectangular and spherical coordinates.

(a) $(1, 45°, 2)$ (b) $(1, 180°, -2)$ (c) $(2, 270°, 3)$

1-20. Consider the following points given in spherical coordinates. Find the corresponding rectangular and cylindrical coordinates.

(a) $(1, 45°, 45°)$ (b) $(2, 90°, 90°)$ (c) $(3, 120°, 135°)$

When $\theta = 0°$, can you define the ϕ coordinate?

1-21. Derive the transformation relationships (from cylindrical to rectangular) in the middle top box of Table 1-3, using Figure 1-11.

1-22. Derive the transformation relationships (from spherical to cylindrical) in the right middle box of Table 1-3, using Figure 1-12.

1-23. Derive the transformation relationships (from spherical to rectangular) in the right top box of Table 1-3, by successive transformation.

1-24. Consider the following integral:

$$\int_0^{2\pi} \mathbf{a}_\rho \, d\phi$$

(a) Evaluate the integral by expressing $\mathbf{a}_\rho$ in terms of $\mathbf{a}_x$ and $\mathbf{a}_y$ and then integrating.
(b) Can you take the unit vector $\mathbf{a}_\rho$ out of the integral sign? Why or why not?

1-25. Repeat the same questions in Problem 1-24 for the following integral

$$\int_0^{2\pi} \mathbf{a}_\phi \, d\phi$$

1-26. Repeat the same questions in Problem 1-24 for the following integral

$$\int_0^{2\pi}\int_0^{\pi} \mathbf{a}_r \sin\theta \, d\theta \, d\phi$$

1-27. Given a vector field

$$\mathbf{F} = \mathbf{a}_x y + \mathbf{a}_y x,$$

evaluate the line integral $\int_C \mathbf{F} \cdot d\boldsymbol{\ell}$ from $P_1(1, 1, 0)$ to $P_2(4, 2, 0)$

(a) along the straight line joining the two points, and
(b) along the parabola $x = y^2$.

Is this $\mathbf{F}$ a *conservative* field? Explain.

1-28. Given a vector field

$$\mathbf{F} = \mathbf{a}_x y - \mathbf{a}_y x,$$

evaluate the line integral $\int_C \mathbf{F} \cdot d\boldsymbol{\ell}$ from $P_1(0, 1, 0)$ to $P_2(1, 0, 0)$

(a) along the straight line joining the two points, and
(b) along the circle $x^2 + y^2 = 1$.

Is this $\mathbf{F}$ a *conservative* field? Explain.

1-29. For a vector field

$$\mathbf{F} = \mathbf{a}_r\, r^2 \cos\theta - \mathbf{a}_\theta\, r^2 \sin\theta,$$

find the flux of the vector $\mathbf{F}$ through
(a) the upper (spherical) surface of a half-sphere of radius r_0 ($r = r_o$, $0 \le \theta \le \pi/2$), and
(b) the lower (flat) surface of the same half-sphere. Let $d\mathbf{s}$ point upward.

Is this $\mathbf{F}$ a *solenoidal* field? Explain.

1-30. For a vector field

$$\mathbf{F} = \mathbf{a}_\rho \rho \sin\phi + \mathbf{a}_\phi \rho + \mathbf{a}_z\, z \cos^2\phi,$$

find the flux of the vector $\mathbf{F}$ through the closed surface of the right half ($0 \le \phi \le \pi$) of a cylinder of length 2ℓ and radius ρ_o, centered at the origin. Note that the closed surface consists of five open surfaces in cylindrical coordinates. Is this $\mathbf{F}$ a *solenoidal* field? Explain.

1-31. For the following scalar function V, evaluate its gradient ∇V.

(a) $V = e^{-x} \sin 2y \cos 3z$
(b) $V = \rho^2 \sin 2\phi$
(c) $V = \dfrac{1}{r^3} \cos^2\theta$

1-32. Repeat Problem 1-31 for the following scalar functions:

(a) $V = \ln(x^2 + y^2 + z^2)$
(b) $V = \rho \cos\phi + z \sin^2\phi$
(c) $V = r \sin\theta$

1-33. For the following vector function $\mathbf{A}$, evaluate its divergence $\nabla \cdot \mathbf{A}$ and curl $\nabla \times \mathbf{A}$.

(a) $\mathbf{A} = \mathbf{a}_x (2yz + 3y^2) + \mathbf{a}_y (yz^2)$

(b) $\mathbf{A} = \mathbf{a}_\rho \rho \sin\phi + \mathbf{a}_z z \cos\phi$

(c) $\mathbf{A} = \mathbf{a}_r r^2 \sin\theta - \mathbf{a}_\theta r^2 \cos\theta \sin\phi + \mathbf{a}_\phi r^2 \cos\theta$

1-34. Repeat Problem 1-33 for the following vector functions:

(a) $\mathbf{A} = \mathbf{a}_x y^2 + \mathbf{a}_y xy + \mathbf{a}_z z^2$

(b) $\mathbf{A} = \mathbf{a}_\rho \rho^2 + \mathbf{a}_\phi \rho^2 \sin\phi - \mathbf{a}_z \rho z \cos\phi$

(c) $\mathbf{A} = \mathbf{a}_r r^2 \cos\theta + \mathbf{a}_\theta r^2 \sin\theta$

1-35. The height of Onondaga Hill (in meters) is given by

$$h(x, y) = 8\,(2xy - 4x^2 - 3y^2 - 16x + 26y + 12)$$

where x is the distance (in km) east, y the distance north of St. James Church.

(a) Where is the top of the hill located?
(b) How high is the hill?
(c) How steep is the slope (in meter per km) at a point 1 km west and 2 km north of St. James Church? In what direction is the slope steepest at that point?

1-36. Find the directional derivative of a scalar field $f(x, y, z) = x^2 + y^2 + z^2$ at the point (1, 2, 3) in the direction of the vector $\boldsymbol{\ell} = \mathbf{a}_x - \mathbf{a}_y + \mathbf{a}_z 2$.

1-37. In a temperature field, heat flows in the direction of maximum decrease of temperature T. Find the direction of heat flow at the point (1, –2, 3) if $T(x, y, z) = x^2y - 2y^3$. What is the maximum rate of decrease of T at that point?

1-38. In an electrostatic field, the electric force points in the direction of maximum decrease of the electric potential V. (a) When the potential between two concentric cylinders of radii 1 and 3 is $V(\rho, \phi) = 10 + 2\,\ell n\,\rho$, what is the direction of the electric force at the point P $(2, 30°, 0)$? (b) Can you find the equation for the equipotential surface on which $V(\rho, \phi)$ has the same value as the potential of point P? Sketch this surface. (c) Show that $\nabla^2 V = 0$.

1-39. Let

$$\mathbf{F}(x, y, z) = \mathbf{a}_x x^2 y + \mathbf{a}_y y^2 z + \mathbf{a}_z x z^2$$

Consider a cube with side length 1 meter situated in the first octant, with one corner at the origin, whose faces are constant x, y, z surfaces.

(a) Evaluate the net outward flux through the *closed* surface of the cube, i.e., calculate $\oint\!\!\oint_S \mathbf{F} \cdot d\mathbf{s}$.

(b) Evaluate also $\iiint_V \nabla \cdot \mathbf{F}\, dv$ and check if the divergence theorem is satisfied.

1-40. Repeat Problem 1-39 for the following vector field

$$\mathbf{F} = \mathbf{a}_x\, yz^2 + \mathbf{a}_y\, xy^2 + \mathbf{a}_z\, x^2 z$$

Let the side of the cube be b meters.

1-41. Let

$$\mathbf{F}(\rho, \phi, z) = \mathbf{a}_\rho\, \rho^2 \sin^2 \phi + \mathbf{a}_\phi\, \rho^2 \cos^2 \phi + \mathbf{a}_z\, \rho z$$

Consider a cylinder of length 2 meters and radius 2 meters, centered at the origin, whose axis is the z axis.

(a) Evaluate $\oiint_S \mathbf{F} \cdot \mathbf{ds}$, i.e., the flux $\mathbf{F}$ through the closed surface of the cylinder.

(b) Calculate $\iiint_V \nabla \cdot \mathbf{F}\, dv$ and verify the divergence theorem.

1-42. Consider a vector field

$$\mathbf{F}(r, \theta, \phi) = \mathbf{a}_r \frac{\cos^2 \theta}{r^2}$$

for the volume V of a spherical shell, centered at the origin, with inner and outer radii 1 and 2 meters.

(a) Evaluate $\iiint_V \nabla \cdot \mathbf{F}\, dv$.

(b) Calculate also $\oiint_S \mathbf{F} \cdot \mathbf{ds}$ and verify the divergence theorem.

1-43. Consider a vector function

$$\mathbf{F}(r, \theta, \phi) = \mathbf{a}_r\, r\cos\phi + \mathbf{a}_\theta\, r\cos^2\phi - \mathbf{a}_\phi\, 3r\sin\theta\sin\phi$$

(a) Find $\oiint_S \mathbf{F}\cdot d\mathbf{s}$ i.e., the flux $\mathbf{F}$ through the closed surface of the half-sphere of radius a (see Figure 1-19).

(b) Verify the divergence theorem for the volume given.

(c) Can $\mathbf{F}$ be expressed as the curl of a vector? Explain.

1-44. Consider a vector function

$$\mathbf{F}(\rho, \phi, z) = \mathbf{a}_\rho\, \rho\sin\phi - \mathbf{a}_z\, 2z\sin\phi$$

for the volume V, which is a quarter of the cylinder of length ℓ meters and radius ρ_0 meters as shown in Figure 1-24.

(a) Evaluate $\oiint_S \mathbf{F}\cdot d\mathbf{s}$, i.e., the flux $\mathbf{F}$ through the closed surface of the quarter-cylinder. Note that there are five open surfaces.

(b) Calculate $\iiint_V \nabla\cdot\mathbf{F}\, dv$ and verify the divergence theorem.

(c) Can $\mathbf{F}$ be expressed as the curl of a vector? Explain.

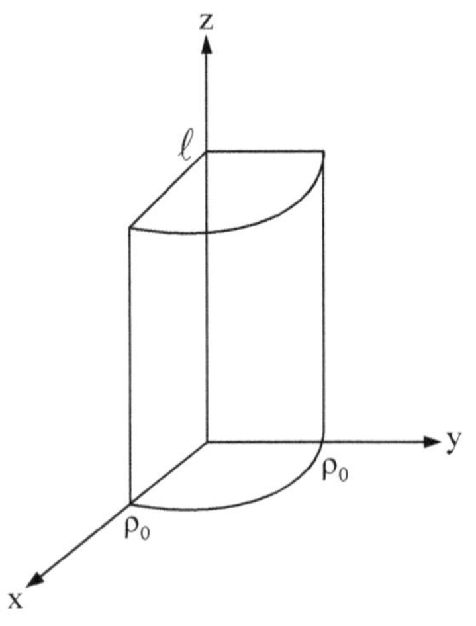

Figure 1-24. A quarter-cylinder.

1-45. Consider a vector field

$$\mathbf{F} = \mathbf{a}_z z$$

(a) Evaluate $\oiint_S \mathbf{F} \cdot d\mathbf{s}$ over the *closed* surface that encloses the volume of one octant of the sphere of radius 1 meter as shown in Figure 1-25. Note that there are four open surfaces.
[HINT] You need to transform $\mathbf{F}$ into spherical coordinates.

(b) Verify the divergence theorem for the volume enclosed.

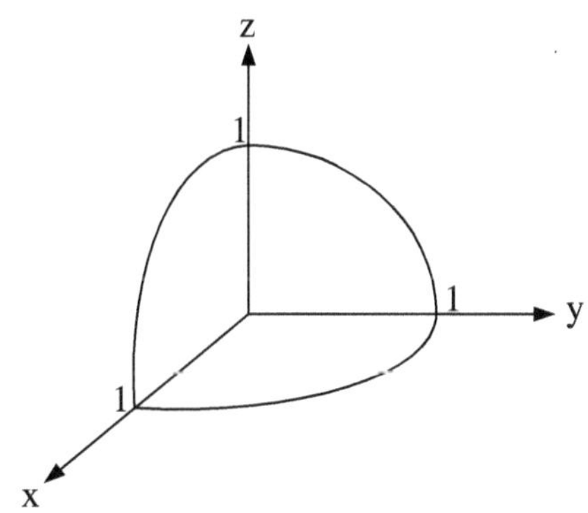

Figure 1-25. An octant of a sphere.

1-46. Repeat Problem 1-45, using the volume of the cone-sphere (ice-cream cone) shown in Figure 1-26.

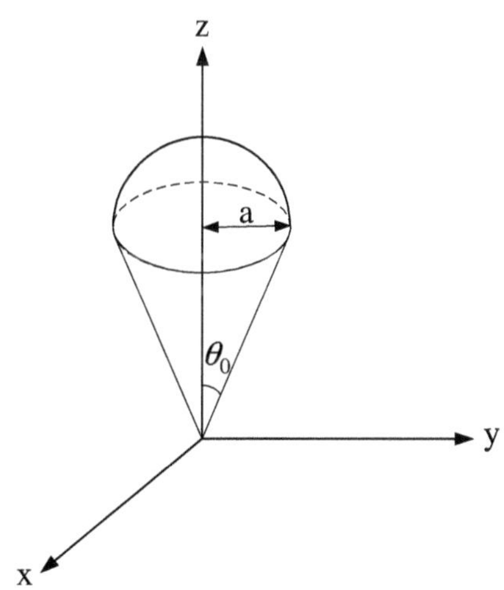

Figure 1-26. A cone-sphere.

1-47. A vector field $\mathbf{F}$ is given as

$$\mathbf{F} = -\mathbf{a}_x y + \mathbf{a}_y x$$

(a) Find $\oint_C \mathbf{F} \cdot d\boldsymbol{\ell}$ around the closed triangular contour C shown in Figure 1-27.

(b) Find $\iint_S (\nabla \times \mathbf{F}) \cdot d\mathbf{s}$ over the triangular surface (bounded by C) and verify Stokes' theorem.

(c) Can $\mathbf{F}$ be expressed as the gradient of a scalar? Explain why.

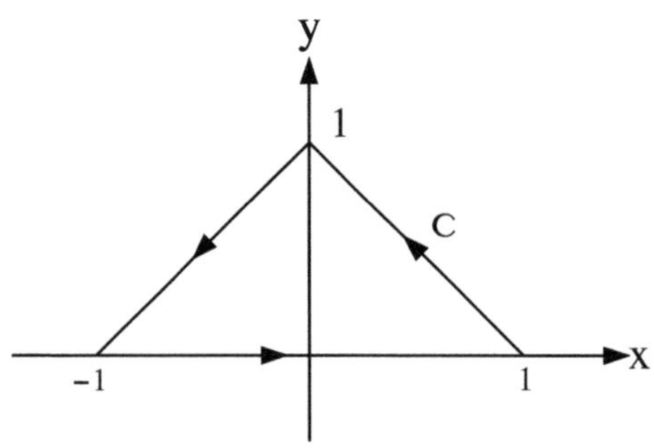

Figure 1-27. A triangular contour.

1-48. For the vector field

$$\mathbf{F} = \mathbf{a}_x y + \mathbf{a}_y x,$$

(a) Evaluate $\oint_C \mathbf{F} \cdot d\boldsymbol{\ell}$ along a *closed* contour of a half-circle of radius a as shown in Figure 1-20.

(b) Verify Stokes' theorem for the surface bounded by the contour.
(c) Can $\mathbf{F}$ be expressed as the gradient of a scalar? Explain why.

1-49. For the vector field $\mathbf{F}(\rho, \phi, z) = \mathbf{a}_\phi\, 3\, \rho$,

(a) Find the circulation $\oint_C \mathbf{F} \cdot d\boldsymbol{\ell}$ along a *closed* contour of a circle of radius a, centered at the origin in the xy plane.
(b) Verify Stokes' theorem for the surface bounded by the contour.

1-50*. For the vector field

$$\mathbf{F}(x, y, z) = \mathbf{a}_x y^2,$$

(a) Find the surface integral $\iint_S \mathbf{F} \cdot d\mathbf{s}$ over the triangular surface bounded by the xy-, yz- and zx- planes as shown in Figure 1-28.

(b) Calculate the line integral $\oint_C \mathbf{F} \cdot d\boldsymbol{\ell}$ along the three sides of the triangle and verify Stokes' theorem.

(c) Find the surface integral $\oiint_S \mathbf{F} \cdot d\mathbf{s}$ over the *closed* surface of the triangular pyramid with its apex at the origin.

(d) Verify the divergence theorem for the volume of the pyramid.

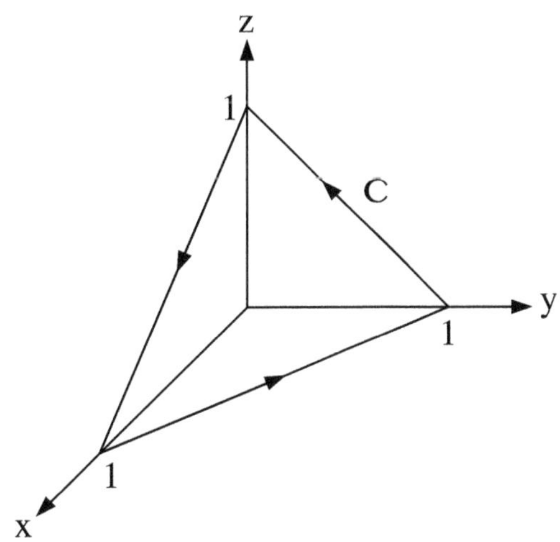

Figure 1-28. A triangular pyramid.

1-51. For the vector field

$$\mathbf{F}(x, y, z) = \mathbf{a}_x (x^2 + y^2),$$

(a) Find $\oint_C \mathbf{F} \cdot d\boldsymbol{\ell}$ for the (closed) quarter-circular path C shown in Figure 1-29.

[HINT] For the circular portion, transform $\mathbf{F}$ into cylindrical coordinates.

(b) Evaluate $\iint_S (\nabla \times \mathbf{F}) \cdot d\mathbf{s}$ over the quarter-circular surface (bounded by C) and verify Stokes' theorem.

*In this book, the starred problems are more challenging problems.

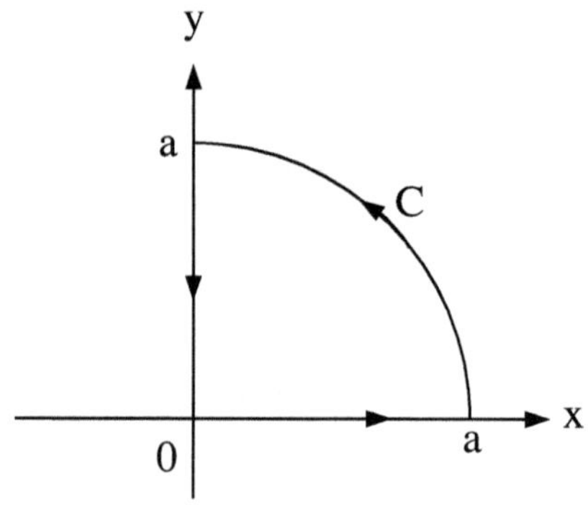

Figure 1-29. A quarter-circle.

1-52. Refer to Figure 1-19 and consider $\mathbf{F} = \mathbf{a}_x y^2$.

(a) Find $\oint_C \mathbf{F} \cdot d\boldsymbol{\ell}$ along the circular contour (counter-clockwise) on the base.

(b) Find $\iint_S (\nabla \times \mathbf{F}) \cdot d\mathbf{s}$ for the planar circular surface (bounded by C).

(c) Repeat (b) for the hemi-spherical surface (also bounded by C).

(d) Show that Stokes' theorem is satisfied for either surface.

ELECTROSTATIC FIELDS AND POTENTIALS

CHAPTER 2

2.1 INTRODUCTION

In Chapter 2, we are concerned with stationary electric charges in free space (vacuum) and the resultant electric fields and potentials. We discuss first the concepts of electric charge and the particular forms (volume, surface, line and point charge distributions) which it takes. We consider next the electric fields, which are related to *forces* exerted by the charges. Then we take up electric potentials which are related to *work* (integrals of the electric field). General methods are described for the determination of fields and potentials due to arbitrary charge distributions. A specialized method, Gauss' law, is used to obtain the electric fields and potentials of certain symmetric charge distributions. Finally, we consider some special arrangements of charges that form dipoles and higher-order multipoles.

In this chapter, we consider only charges and conductors in free space. The charges are stationary and the resultant electric fields and potentials are constant (they do not vary with time). Thus we are concerned with **static electric** or **electrostatic** fields. Chapters 2 through 4 of this text are limited to stationary charges.

In the real world nothing is constant, and so what application can electrostatics have to reality? First, there are situations in which fields are varying slowly with time (low-frequency variations). This is the so-called *quasi-static* case; the electrostatic analysis is very useful here. Second, we have situations in which transients occur. As the transients die out the electric field may approach the electrostatic solution. Finally, there are many situations in which there are only very minor random changes in electric fields with time. Thus there are many cases in which the electrostatic solution is useful. The most important result, however, of our study of electrostatic problems is, undoubtedly, the insight which we gain into electrical phenomena.

2.2 ELECTRIC CHARGE

Perhaps the most striking characteristic of charge is the incredible, nearly perfect balance between positive and negative charges. Most objects have only a tiny net or excess charge. For instance, if two persons each had 0.1% excess charge and were 1 meter apart, the force of repulsion between them would be about 10^{20} tons. Since no such force is observed, the excess charge must be tiny indeed.

Similarly, it takes only a tiny excess charge to exert forces which can be observed. The effects of charged objects were noted very early in history. About 600 B.C., Thales of Miletus observed and recorded the fact that the material amber, when rubbed with silk, attracted particles of straw. The Greek word for amber is *elektron* from whence come our words electron, electricity, etc.

We know now that silk removes electrons from the amber, leaving it positively charged. Similarly, rubbing a glass rod with silk also leaves it positively charged whereas rubbing a plastic comb with wool leaves it negatively charged. The names *positive* and *negative* came from the work of Benjamin Franklin. He recognized that there were two kinds of contributions to the charged state. He arbitrarily called that of the rubbed glass rod positive from which definition the charge of electron becomes negative.

The unit of charge is the coulomb (C). The charge on an electron is $-e$ where

$$e = 1.602 \times 10^{-19} \quad [\mathrm{C}] \tag{2-1}$$

The charge and mass of the electron are known. Little is known concerning its internal structure. From Eq. (2-1) above, we note that –1 coulomb contains $.625 \times 10^{19}$ electrons. This is a very large amount of charge. A lightning stroke carries about 20 coulombs. There are three possible types of charge distributions: volume (ρ_v), surface (ρ_s), and line (ρ_ℓ) charge densities. When charge is distributed over a volume, we define a volume charge density ρ_v. To determine ρ_v at a point, we construct a small volume Δv around the point, measure the charge Δq within Δv, compute the ratio $\frac{\Delta q}{\Delta v}$ and then find the limit as Δv becomes infinitesimal, i.e.

$$\rho_v = \lim_{\Delta v \to 0} \frac{\Delta q}{\Delta v} \quad \left[\frac{C}{m^3}\right] \tag{2-2a}$$

where $[C/m^3]$ represents the unit of the volume charge density. Similarly,

$$\rho_s = \lim_{\Delta s \to 0} \frac{\Delta q}{\Delta s} \quad \left[\frac{C}{m^2}\right] \tag{2-2b}$$

$$\rho_\ell = \lim_{\Delta \ell \to 0} \frac{\Delta q}{\Delta \ell} \quad \left[\frac{C}{m}\right] \tag{2-2c}$$

where Δs is the area of a small surface and $\Delta \ell$ is the length of a small line segment. The total charge q in a volume V, a surface S, and a line L may then be expressed as

$$q = \iiint_V \rho_v \, dv \text{ (for volume charge)} \tag{2-3a}$$

$$= \iint_S \rho_s \, ds \text{ (for surface charge)} \tag{2-3b}$$

$$= \int_L \rho_\ell \, d\ell \text{ (for line charge)} \tag{2-3c}$$

Figure 2-1 illustrates some of the concepts of volume, surface, and line charge densities. Figure 2-1(a) shows a cube of continuous volume charge distribution $\rho_v(x,y,z)$.

Total charge q is defined by Eq. (2-3a). Now if we squeeze all of the charge into a zero thickness layer (Figure 2-1(b)), we have a surface charge density $\rho_s(x,y)$ with total charge q, which is unchanged, determined by Eq. (2-3b). If we compress the surface charge into a line of zero cross section (Figure 2-1(c)), we obtain a line charge distribution $\rho_\ell(x)$ with total charge q determined by Eq. (2-3c). Finally if we compress the line charge down to a point, we have a point charge of q coulombs (Fig. 2-1(d)).

The different types of charge densities are related to different physical situations. For instance, charge in an electron beam may be represented as volume charge density ρ_v. Charge on the surface of a good conductor may be represented by surface charge density ρ_s. Charge on a thin wire is modeled as line charge density ρ_ℓ. Finally, the electron is usually treated as a point charge.

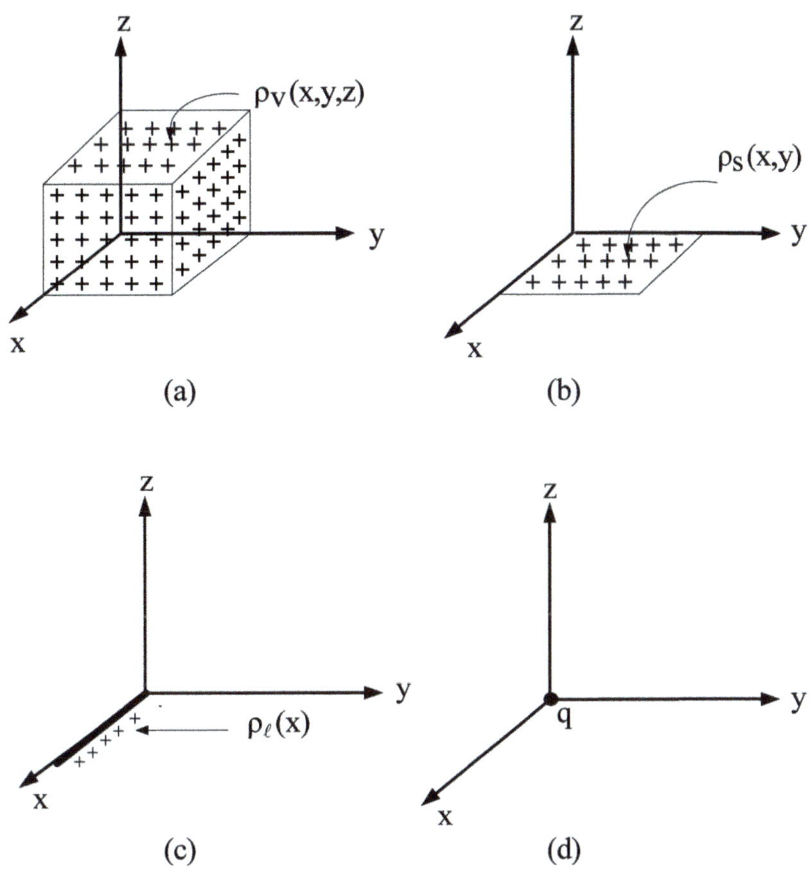

Figure 2-1. Electric charge distribution (a) Volume charge density ρ_v. (b) Surface charge density ρ_s. (c) Line charge density ρ_ℓ. (d) A point charge q.

Surface and line charge densities, and point charge may all be represented as special cases of volume charge density. Those of you familiar with delta functions can probably see how to accomplish this.

In Eqs. (2-2) and (2-3) we are neglecting the microscopic variations of charge density associated with atomic dimensions. We are interested in a smoothed-out or space-averaged distribution. This is the *macroscopic* rather than the *microscopic* point of view.

EXAMPLE 2-1 UNIFORM CHARGE DISTRIBUTION

Consider a spherical cloud of uniform volume charge density:

$$\rho_v = \rho_{vo} \ (0 < r < a) \quad (2\text{-}4a)$$

$$= 0 \text{ elsewhere} \quad (2\text{-}4b)$$

Often we will omit the latter statement (zero elsewhere). Find the total charge contained within the cloud.

SOLUTION

$$
\begin{aligned}
Q &= \iiint \rho_v \, dv \\
&= \int_0^{2\pi}\int_0^{\pi}\int_0^{a} \rho_{vo}\, r^2 \sin\theta \, dr\, d\theta\, d\phi \\
&= \rho_{vo}\; 2\pi\left[-\cos\theta\right]_0^{\pi}\left[\frac{r^3}{3}\right]_0^{a} \\
&= \rho_{vo}\left(\frac{4}{3}\pi a^3\right)
\end{aligned}
\tag{2-5}
$$

Note that we can obtain this result merely by multiplying the volume charge density (ρ_{vo}) by the volume $\left(\frac{4}{3}\pi a^3\right)$ because the charge density is uniformly distributed within the volume. It is useful to recognize the following integral which occurs often:

$$
\int_0^{2\pi}\int_0^{\pi} \sin\theta \, d\theta \, d\phi = 4\pi \tag{2-6}
$$

EXAMPLE 2-2 A NON-UNIFORM CHARGE DISTRIBUTION

Consider a spherical cloud of charge of *non-uniform* volume charge density:

$$
\rho_v = \rho_{vo}\left(\frac{r}{a}\right)^2 \quad (0 < r < a) \tag{2-7}
$$

It is assumed that ρ_v is zero elsewhere. Find the total charge contained within the cloud.

SOLUTION

$$
\begin{aligned}
Q &= \iiint \rho_v \, dv \\
&= \int_0^{2\pi}\int_0^{\pi}\int_0^{a} \rho_{vo}\frac{r^2}{a^2} r^2 \sin\theta \, dr\, d\theta\, d\phi \\
&= \frac{\rho_{vo}}{a^2} 2\pi\left[-\cos\theta\right]_0^{\pi}\left[\frac{r^5}{5}\right]_0^{a} \\
&= \rho_{vo}\left(\frac{4\pi\ a^3}{5}\right)
\end{aligned}
\tag{2-8}
$$

In this case, integration is necessary since the charge density depends on position.

2.3 THE ELECTRIC FIELD IN FREE SPACE AND ITS BASIC LAWS

When electric charges ("source charges") are present, they produce an electric field in the surrounding space. When you bring another charge ("test charge") in this space, the electric field exerts a force on the test charge. The **electric field intensity** $\mathbf{E}$ at a point (x,y,z) is defined in terms of the force $\mathbf{F}$ on a small test charge q located there:

$$\mathbf{E}(x,y,z) = \lim_{q \to 0} \frac{\mathbf{F}}{q} \left[\frac{N}{C}\right] \text{or} \left[\frac{V}{m}\right] \tag{2-9a}$$

Thus the electric field is the force per unit charge. The unit of the electric field is Newtons per coulomb $[N/C]$ but is also given by volts per meter $[V/m]$ since the voltage is related to the electric field by Eq. (2-42). In this book, the SI (Standard International) units or MKSA units are used. A small test charge q is used in order that the determination of $\mathbf{E}$ not be affected by q itself.

The force on a charge q in an electric field $\mathbf{E}$ is then

$$\mathbf{F} = q\mathbf{E} \ [N] \tag{2-9b}$$

Eq. (2-9b) is known as the Lorentz force equation.

Basic Laws of Electrostatics

The basic laws of electrostatics, in point form (or differential form), specify the divergence and curl of the static electric field $\mathbf{E}$. From Helmholtz's theorem, we know that the specification of $\nabla \times \mathbf{E}$ and $\nabla \cdot \mathbf{E}$ completely determines the vector field intensity $\mathbf{E}$.

The first law is

$$\boxed{\nabla \times \mathbf{E} = 0} \qquad \mathbf{E} \text{ is a conservative field} \tag{2-10}$$

Equation (2-10) is the point form of the law, i.e., $\nabla \times \mathbf{E} = 0$ at every point (x,y,z). We can derive an equivalent integral form by integrating both sides of Eq. (2-10) over an arbitrary open surface S and applying Stokes' theorem:

$$\iint_S \nabla \times \mathbf{E} \cdot d\mathbf{s} = \oint_C \mathbf{E} \cdot d\boldsymbol{\ell} = 0$$

for any closed path C which bounds the surface S and where $d\mathbf{s}$ and $d\boldsymbol{\ell}$ are related by a right-hand rule. S is arbitrary and thus C is also arbitrary.

$$\boxed{\oint_C \mathbf{E} \cdot d\boldsymbol{\ell} = 0} \tag{2-11}$$

We have derived Eq. (2-11) from Eq. (2-10). It is also possible to derive Eq. (2-10) from Eq. (2-11) and thus the two equations are identical statements.

The second law is

$$\boxed{\nabla \cdot \mathbf{E} = \frac{\rho_v}{\varepsilon_0}} \tag{2-12}$$

where $\varepsilon_0 = 8.854 \times 10^{-12}$ [C²/N·m²] or [F/m]. ε_0 is a universal constant, called the *permittivity of free space.* ρ_v is the volume charge density in C/m^3. The physical interpretation of this law is that the electric field diverges or spreads out from the (positive) charge. We can derive an integral form by integrating both sides of Eq. (2-12) over a volume V and applying the divergence theorem:

$$\iiint_V \nabla \cdot \mathbf{E}\, dv = \oint\!\!\!\oint_S \mathbf{E} \cdot d\mathbf{s} = \frac{1}{\varepsilon_0} \iiint_V \rho_v\, dv = \frac{Q_{enc}}{\varepsilon_0}$$

S and V are both arbitrary. Q_{enc} is the *total charge enclosed within the closed surface S.*

$$\boxed{\oint\!\!\!\oint_S \mathbf{E} \cdot d\mathbf{s} = \frac{Q_{enc}}{\varepsilon_0}} \quad \text{Gauss' Law} \qquad (2\text{-}13)$$

Eqs. (2-12) and (2-13) are identical statements. The second law is called **Gauss' law**, which will be explained in more detail in Section 2.5. The basic laws above may also be derived directly from Coulomb's law.

2.4 CHARLES AUGUSTIN COULOMB (1736–1806) AND THE DISCOVERY OF COULOMB'S LAW

Charles Augustin Coulomb was one of the leading scientists of the latter half of the 18th century. He made significant contributions to a wide variety of fields, including civil engineering, mechanical engineering, soil mechanics, and electricity and magnetism. He established, in an extremely careful and thorough manner, the inverse square law for electricity, which is known as *Coulomb's law*. In addition to these contributions to engineering and physics, he made several important practical contributions, among which was the invention of the torsion balance, a device capable of measuring very small forces.

His contributions to engineering and physics laid the foundations for later discoveries. He discovered one of the basic laws of friction. His theory of earth pressure and the related *Coulomb's Equation* are still used as the most fundamental principles of soil mechanics. His contributions to electricity form the basis of our theory. All of the basic laws of electrostatics can be derived from Coulomb's law.

Charles Augustin Coulomb was born in 1736 in the town of Angouleme in southern France. His father, Henry Coulomb, grew up in the family home in Montpelier where the Coulombs had lived for several generations. They were a bourgeois or middle class family and many of the Coulombs were lawyers in the Montpelier region. Henry Coulomb started out in the military and then became a minor governmental administrator, an inspector of the king's domains. He later became an administrator of the tax-farm system, a system much-hated by the French people. Many of the tax-farm administrators were guillotined during the French Revolution, some years after Henry's death.

The family moved to Paris during Coulomb's adolescence. He attended mathematical lectures at the Royal College of France, where he became very interested in mathematics and decided to become a mathematician. This decision displeased his mother, who had set her heart on his becoming a doctor. At this point, Coulomb's father Henry lost his entire fortune in speculation and returned to the family home in Montpelier. His mother, Catherine Bajet Coulomb, who had resources of her own, stayed in Paris with her two daughters and disowned her son Charles who was forced to join his father in Montpelier.

Later biographers have sometimes promoted Coulomb to the ranks of the nobility by adding the prefix "de" to his name and calling him "de Coulomb". There is, however, no true basis for this. Neither Coulomb nor any of his relatives ever signed themselves "de Coulomb," and, ironically, if he had been of the nobility, he might not have survived the revolution.

Coulomb presented papers on mathematics and astronomy in 1757, 1758 to the Royal Society of Sciences of Montpelier. He was certainly developing his mathematical and experimental skills, but as a young man of 22, he had no source of income and was forced to seek gainful employment. He decided to join the military corps of engineers and was accepted in its engineering school at Mézières, graduating at the age of 25 in 1761.

The next eleven years Coulomb spent in active service in the engineering corps. He was posted to Martinique for eight years and was given responsibility for the complete construction of a major fortress, a project that involved overseeing several hundred workers. During the eight-year duration of the project, Coulomb faced basic problems of the design of vaults, arches, and retaining walls. He had to understand the strength of materials and to predict bending, compression, and rupture. Rather than relying on rules of thumb already available to accomplish these tasks, Coulomb went back to the most basic questions, developed his own theories, and carried out his own experiments in order to proceed on a firmer basis. All of this was done in his spare time, amid outbreaks of malaria and cases of sunstroke in the equatorial climate of Martinique. This work, in his own words, "was meant at first only for my own use, in the different tasks for which I am responsible in my occupation." Nonetheless, the impact of his work is still felt centuries later. Upon returning to France, Coulomb presented his theory and experiment on these construction problems to the Academy of Sciences in 1773 in a historic paper on arches, strength of materials, and soil mechanics. This paper has been called the most brilliant engineering paper read at the Academy during the last half of the eighteenth century.

In 1777, the Academy offered a special prize for improvements in magnetic compasses. The basic problem is one which all Girl Scouts and Boy Scouts should recognize. The force on the compass needle is so weak that friction in the compass may give us an incorrect reading. Coulomb suspended the needle on a silk or metal thread so that it encountered virtually no friction. By these means he was able to determine magnetic declination to within a few minutes of arc. By the addition of a micrometer at the other end of the thread, one could also determine the torque on the thread and thus the magnitude of the earth's magnetic field. This device is called the torsion balance. Coulomb studied the phenomena of torsion in detail and discovered that, within the linear region of operation, the angle through which a wire is twisted is proportional to the torque exerted. The magnitude of the torque required to twist the wire through a given angle is inversely proportional to the wire length and is proportional to the fourth power of the wire radius. Thus a long, thin wire in a torsion balance is very sensitive. It is capable of measuring very small forces or torques.

Coulomb won the 1777 compass prize. However, he realized that friction theory was needed for a study of conventional compasses. The Academy was interested in friction for naval applications. A prize was proposed for a study of friction. Coulomb undertook an extremely thorough mathematical and experimental study of friction which won him a second prize in 1781. Once again, Coulomb asked himself very basic questions and reformulated the entire theory, earning himself the title of "The Father of Friction." He was elected to the Academy in 1781.

During the period 1785–1791, Coulomb conducted a series of very elegant experiments which established the inverse square force law for electric charges. He verified the law by three completely different methods. For like charges (both positive or both negative), he measured the force of repulsion directly using the torsion balance.

These measurements are extremely difficult, even now, especially because of the influence of the experimenter. For unlike charges, he determined the force of attraction by timing the pendulum swing of a movable charge attracted towards a fixed charge. The third method involved a demonstration that there is no charge inside a conducting body as required by the inverse square law. Maxwell, writing much later, said of his third method, "It is impossible to overestimate the delicacy and ingenuity of his apparatus, the accuracy of his observations, and the sound scientific method of his researches."

Coulomb first reported his results in 1785. They were soon accepted by the French scientific community and the inverse square law was named *Coulomb's Law*. He also established the inverse square force law for isolated magnetic poles. This required a careful choice of magnets and distances. He chose long thin magnets which were closely spaced so that the predominant force was that between nearby poles.

For charged objects, Coulomb noted that the charge leaked off rather rapidly with time. His first experiments had to be carried out in two minutes to avoid error due to leakage. Coulomb found that the charge lost per second was proportional to the charge present, determined the differential equation governing charge decay, and deduced that the charge decay was therefore *exponential*, the exponent depending on factors such as humidity. Thus he was the first person to describe the effect which we will later call *relaxation time* (See Chapter 5).

In some of his later papers, Coulomb attacked the difficult problem of determining charge distributions over the surface of a charged metallic object. He showed that the surface charge density at a point is proportional to the electric field normal to the surface and devised a standard disk which could be used to measure the charge density at different points on the surface of a conductor.

Several scientists preceded Coulomb in the discovery of the inverse square laws of electricity and magnetism. In fact, there were over ten investigators who provided some evidence for the electric and/or magnetic laws over the forty-year period preceding Coulomb's 1785 paper. Some of these contributions were relatively minor, mere statements or assumptions of the inverse square law. In addition there were a number of claims for the inverse first and third powers. The situation was confused and a definitive, comprehensive study was called for. Henry Cavendish had carried out very careful experiments and analysis to show that the charge inside a conductor is zero, but his results were not published until a century later. John Robison had made thorough direct measurements of force between charges but he also did not publish. Thus it fell to Coulomb to carry out the definitive, comprehensive experiments and to make the arguments which convinced the French academic community.

Why was it so difficult to establish the inverse square laws? Why was Coulomb's work accepted over those who went before and why was his name attached to the law?

It is important for us to understand the time in which Coulomb worked. It was the time of the enlightenment, with new ideas being posited in every discipline of study. Science itself was slowly emerging from the imprecise methodology of the middle ages. The concept of action at-a-distance was still difficult for many to accept. It was believed that objects can interact only through contact and so the motion of planets was explained by vortices of invisible matter existing between sun and planets. The clever ideas of Descartes and Nollet involving *vortices* and *effluvia* of hidden invisible matter explained the solar system as well as magnetism and electricity. Thus, one's intellect and imagination were used to create complex systems to explain the known world, but these systems were not tested. In some respects they were untestable. What can one do with material which is invisible, untouchable, unweighable? Such theories were typically accepted on faith. In the Middle Ages, imagination and speculation, however intelligent, formed the basis of science. The discoveries made in one country did not always cross the border. The French did not fully accept Newton and the Germans did not fully accept Coulomb for many years.

Many different forms of the force law echoed through the halls of science in the latter half of the 18th century. $1/r$, $1/r^2$, $1/r^3$ and other forms were seriously pursued, so that the many statements of the inverse square law did not immediately lead to acceptance. Finally, Coulomb comes at the right time. He reports his results and confirms them by three entirely different methods. He was by then a highly respected figure, a leader of 18th-century science within the academy. Because of the thoroughness of his methods and his position in the Academy, his work was immediately accepted in France. The scientific community did not accept the inverse square laws before Coulomb because the situation was confused and competing theories existed. Many scientists were not disposed to accept any theorem involving action at-a-distance. Some of the inverse square law results were merely statements or conjecture. Finally, the time was ripe. Vortices and effluvia were on the wane. New ideas were in the wind. The American revolution was over and the French Revolution was only four years away. Modern scientists were in the ascendancy. Coulomb himself was gaining an unmatched reputation and the confidence of his colleagues. When the new experiments were reported, they were immediately accepted and Coulomb's name was soon attached to the inverse square law.

We learn a very valuable lesson from the history of Coulomb's law. Priority is certainly important. With two equally strong contributions, the prior one is recognized. However, thoroughness and completeness, and depth of understanding may be more important than priority. In addition, the discoverer must be aware of the importance of his discovery and must be able to present his results in a convincing manner.

In the eyes of his colleagues, Coulomb was a great leader in French science. Jean-Baptiste Biot, of whom we will hear later, said, "It is to Borda and to Coulomb that one owes the renaissance of true physics in France, not a verbose and hypothetical physics, but that ingenious and exact physics which observes and compares all with rigor."*

Coulomb's Law

Figure 2-2 shows two point charges q_1, q_2 located at arbitrary positions separated by a distance R. The force on q_2 due to q_1 is

$$\mathbf{F}_2 = \text{force on } q_2 = \mathbf{a}_R \frac{q_1 q_2}{4\pi\varepsilon_0 R^2} = \mathbf{R}\frac{q_1 q_2}{4\pi\varepsilon_0 R^3} \tag{2-14}$$

where $\mathbf{R}$ is the vector pointing from q_1 to q_2, R is the distance from q_1 to q_2, and $\mathbf{a}_R = \dfrac{\mathbf{R}}{R}$ is a unit vector in the direction of $\mathbf{R}$.

The electric field $\mathbf{E}$ at q_2 due to q_1 is defined as $\mathbf{F}_2$ divided by q_2 or the *force per unit charge*:

$$\mathbf{E} = \frac{\mathbf{F}_2}{q_2} = \mathbf{a}_R \frac{q_1}{4\pi\varepsilon_0 R^2} = \mathbf{R}\frac{q_1}{4\pi\varepsilon_0 R^3} \tag{2-15}$$

Think of q_1 and q_2 as source charge and test charge, respectively. Coulomb's law is the first law in Electromagnetics that is discovered by experiment.

* C. Stewart Gilmour, *Coulomb and the Evolution of Physics and Engineering in Eighteenth-Century France*, Princeton University Press, p. 230, 1971.

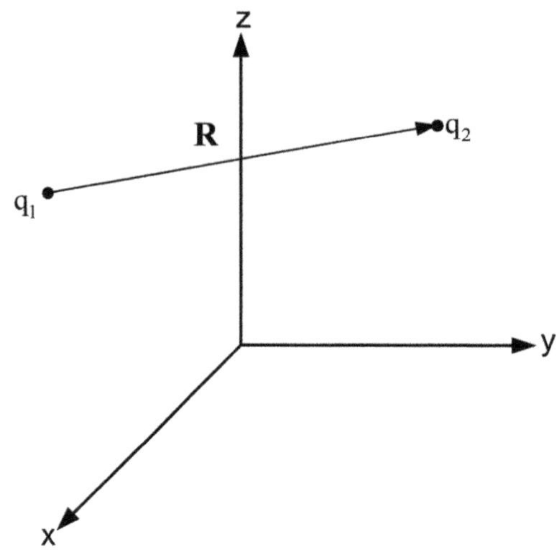

Figure 2-2. Coulomb's law.

2.5 GAUSS' LAW

Gauss' law may be used to obtain the electric fields of certain charge distributions with a high degree of symmetry. It is therefore a specialized method, but it is very useful for that class of problems to which it can be applied. At this stage, Gauss' law will help us to gain insight into the forms of electric fields due to charge distributions. First let us look again at Gauss' law:

$$\oiint_S \mathbf{E} \cdot d\mathbf{s} = \oiint_S E_n \, ds = \frac{Q_{enc}}{\varepsilon_0} \qquad (2\text{-}13)$$

where the surface S is an arbitrary closed surface, chosen for convenience, which will be called the *Gaussian surface.* Q_{enc} is the *charge enclosed within the Gaussian surface S.* E_n is the component of $\mathbf{E}$ normal to S.

We will assume that charges are specified and so the right hand side of Eq. (2-13) is known or can be determined. E_n is unknown. If E_n were constant over the surface S, then we could take it outside the integral and solve for E_n. The key to a Gauss' law problem is to choose S so that it is normal to the electric field lines. Thus we need to have some knowledge of the field lines. The steps then are as follows:

1. Recognize the symmetry.
2. Sketch the electric field lines.
3. Choose S normal to the field lines. Form a closed surface.
4. Solve for E_n.

Electric Field of a Point Charge

Consider first a point charge q located at the origin (Figure 2-3(a)). We can observe spherical symmetry ($\partial/\partial\theta = \partial/\partial\phi = 0$). In other words, the electric field components do not vary with angles θ, ϕ. E_θ, E_ϕ are ruled out because of the first basic law $\oint_C \mathbf{E} \cdot d\boldsymbol{\ell} = 0$ where the contour is taken around the spherical surface. Therefore, the electric field is *purely radial,* i.e., $\mathbf{E} = \mathbf{a}_r E_r$, and is a function only of r. This is because the electric field diverges away from the positive charge q. We construct a Gaussian surface S normal to the electric field, i.e., a sphere of radius r centered at the origin. E_r will be constant over the surface S due to *spherical symmetry*, i.e., E_r will not depend on the position (θ, ϕ) on the sphere.

Then

$$\oiint_S \mathbf{E} \cdot d\mathbf{s} = \oiint_S E_r \mathbf{a}_r \cdot \mathbf{a}_r ds = \oiint_S E_r ds = E_r \int_0^{2\pi} \int_0^{\pi} r^2 \sin\theta \, d\theta \, d\phi = 4\pi r^2 E_r = \frac{Q_{enc}}{\varepsilon_0} = \frac{q}{\varepsilon_0}$$

$$\boxed{E_r = \frac{q}{4\pi\varepsilon_0 r^2}} \tag{2-16}$$

Note that $d\mathbf{s} = \mathbf{a}_r\, ds$ and distinguish between the vector surface $d\mathbf{s}$ and the scalar surface ds. It can be shown that the electric field $\mathbf{E}$ in Eq. (2-16) satisfies both basic laws in differential form, Eqs. (2-10) and (2-12). See also Example 1-13.

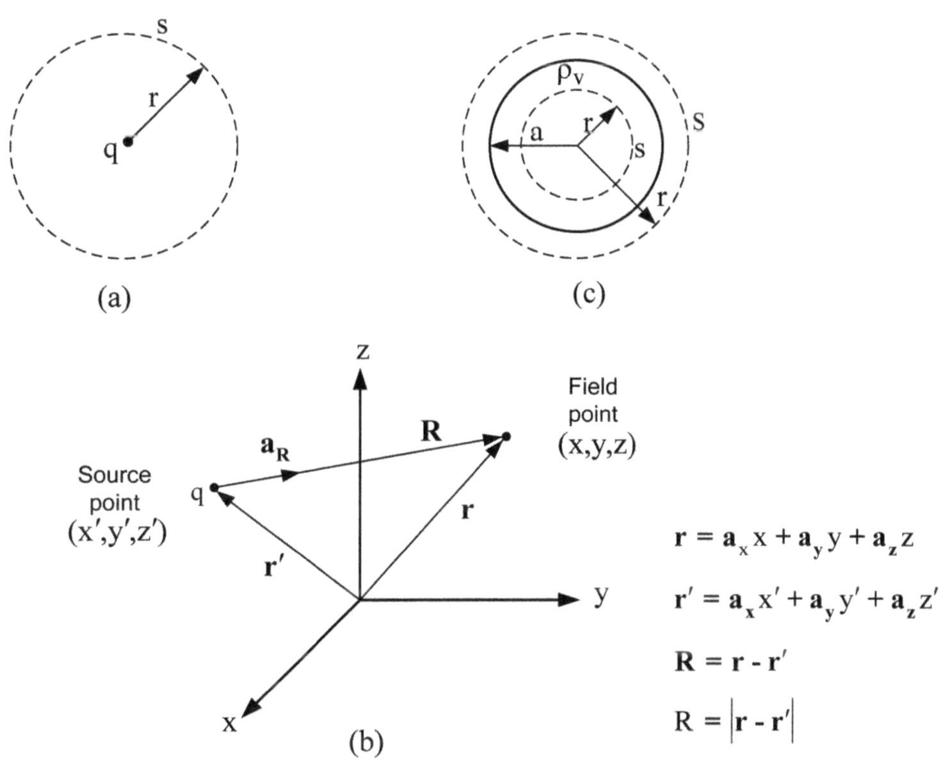

Figure 2-3. Spherical charge distributions. (a) A point charge at the origin. (b) A point charge at source point (x', y', z'). (c) A uniform volume charge density ρ_V.

For a point charge located at an arbitrary *source point* (x', y', z') as in Figure 2-3(b), the electric field at the *field point* or the *observation point* (x, y, z) is given as follows:

$$\mathbf{E} = \mathbf{a}_R \frac{q}{4\pi\varepsilon_0 R^2} = \frac{\mathbf{R}q}{4\pi\varepsilon_0 R^3} = \frac{(\mathbf{r}-\mathbf{r}')q}{4\pi\varepsilon_0 |\mathbf{r}-\mathbf{r}'|^3} \tag{2-17}$$

where $\mathbf{R}$ is a vector from the source point to the field point and $\mathbf{a}_R$ is a unit vector in the direction of $\mathbf{R}$. Note that $\mathbf{R}$ is identical to $\mathbf{r} - \mathbf{r}'$ and that R is identical to $|\,\mathbf{r} - \mathbf{r}'\,|$ as indicated in Figure 2-3(b). $\mathbf{r}$ and $\mathbf{r}'$ are the position vectors of the field point and source point, respectively. Note also that Eq. (2-17) is identical to Eq. (2-15), Coulomb's law.

EXAMPLE 2-3 VOLUME CHARGE IN A SPHERICAL SHAPE

Consider a uniform volume charge distributed within a sphere of radius a (Fig. 2-3c):

$$\rho_v = \rho_0 \left[\frac{C}{m^3}\right] \quad (r \le a) \qquad (2\text{-}18)$$

Find the electric field everywhere.

SOLUTION

The total charge Q within the sphere is $\frac{4}{3}\pi a^3 \rho_o$. Rotation of the spherical cloud of charge produces no change in charge distribution and therefore $\partial/\partial\theta = \partial/\partial\phi = 0$. The charge distribution has *spherical symmetry*. E_θ, E_ϕ are ruled out because of the first law. Since the electric field diverges away from the charge distribution, we expect that $\mathbf{E} = \mathbf{a}_r E_r$. We can construct two Gaussian surfaces (a sphere of radius r), one inside ($r \le a$) and one outside ($r \ge a$) as in Figure 2-3(c). The charge enclosed within the Gaussian surface is $\frac{4}{3}\pi r^3 \rho_o$ when $r < a$ (inside the charged sphere), and is Q or $\frac{4}{3}\pi a^3 \rho_o$ when $r > a$ (outside the charged sphere).

Then

$$\oint\!\!\oint_S \mathbf{E}\cdot d\mathbf{s} = 4\pi r^2 E_r = \frac{Q_{enc}}{\varepsilon_0} = \frac{4\pi r^3 \rho_0}{3\varepsilon_0} \ (r \le a)$$
$$= \frac{4\pi a^3 \rho_0}{3\varepsilon_0} \ (r \ge a) \qquad (2\text{-}19a)$$

$$E_r = \frac{r\rho_0}{3\,\varepsilon_0} = \frac{r}{a}\frac{Q}{4\pi\varepsilon_0 a^2} \ (r \le a) \quad \text{inside the sphere}$$
$$E_r = \frac{\rho_0 a^3}{3\varepsilon_0 r^2} = \frac{Q}{4\pi\varepsilon_0 r^2} \ (r \ge a) \quad \text{outside the sphere} \qquad (2\text{-}19b)$$

The electric field outside is the same as if the total charge Q were concentrated at the origin. Note that the electric field E_r increases linearly with distance r inside ($r < a$) and decreases as the inverse square of r outside ($r > a$).

EXAMPLE 2-4 SURFACE CHARGE ON A SPHERE

Consider a uniform surface charge density ρ_{so} distributed on a sphere of radius a:

$$\rho_s = \rho_{so} \left[\frac{C}{m^2}\right] \quad (r = a) \tag{2-20}$$

Find the electric field everywhere.

SOLUTION

The total charge Q on the sphere is $4\pi a^2 \rho_{so}$. Note the spherical symmetry. Rotating the sphere in θ, ϕ does not change the charge distribution and therefore $\partial/\partial\theta = \partial/\partial\phi = 0$. E_θ, E_ϕ are again ruled out because of the first law and $\mathbf{E} = \mathbf{a}_r E_r$. We construct two Gaussian surfaces, one inside ($r < a$) and one outside ($r > a$). The charge enclosed is zero for $r < a$ and is $4\pi a^2 \rho_{so}$ for $r > a$.

Then

$$\oint\!\!\oint_S \mathbf{E} \cdot d\mathbf{s} = 4\pi r^2 E_r = \frac{Q_{enc}}{\varepsilon_0} = 0 \quad (r < a)$$
$$= \frac{4\pi a^2 \rho_{so}}{\varepsilon_0} \quad (r > a) \tag{2-21a}$$

$$E_r = 0 \quad (r < a) \qquad \text{inside the sphere}$$
$$E_r = \frac{Q}{4\pi\varepsilon_0 r^2} = \frac{\rho_{so}\, a^2}{\varepsilon_0 r^2} \quad (r > a) \quad \text{outside the sphere} \tag{2-21b}$$

Note that the electric field inside the hollow sphere with uniform surface charge is zero.

EXAMPLE 2-5 ELECTRIC FIELD OF A LINE CHARGE OF INFINITE LENGTH

Consider an infinitely long line charge with a uniform charge density ρ_ℓ [C/m], located on the z axis (Figure 2-4a). Find the electric field everywhere.

SOLUTION

Rotation in ϕ and translation in z do not change the charge distribution and therefore $\partial/\partial z = \partial/\partial\phi = 0$. We say that the charge distribution has *cylindrical symmetry*. E_ϕ is ruled out by the first law. E_z is also ruled out. We expect that the electric field is *purely radial* from the z-axis (E_ρ only), i.e., $\mathbf{E} = \mathbf{a}_\rho E_\rho$. Construct a cylindrical Gaussian surface S, of radius ρ and length ℓ, whose axis is the z axis. Note that the line charge is of infinite length whereas the Gaussian surface is of finite length. The electric field E_ρ is normal to the curved surface and parallel to the end surfaces of the Gaussian cylinder.

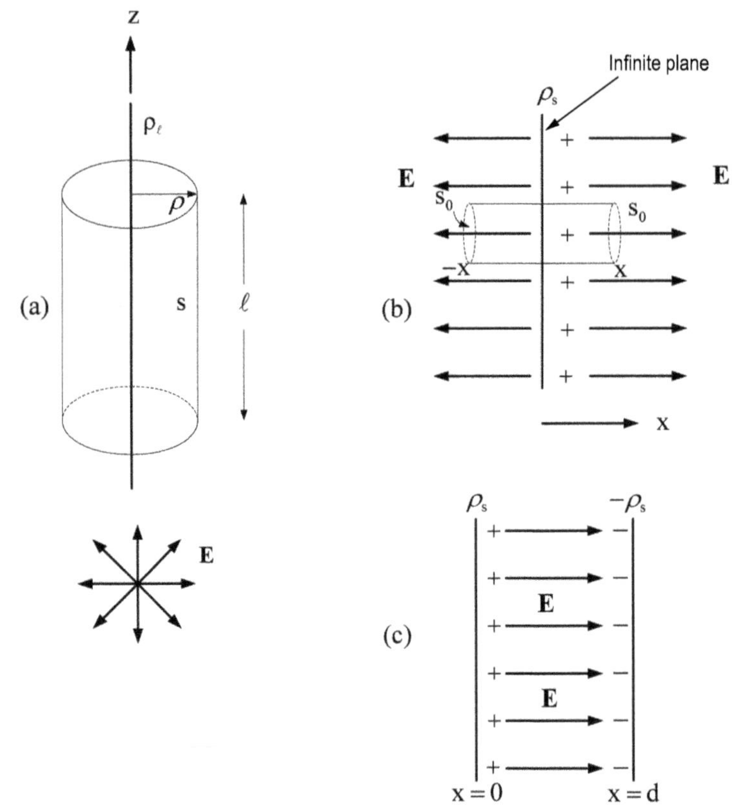

Figure 2-4. Cylindrical and planar charge distributions. (a) A uniform line charge density ρ_ℓ. (b) A uniform planar surface charge density ρ_s. (c) Parallel plates

Then

$$\oiint_S \mathbf{E}\cdot d\mathbf{s} = \iint_{SIDE} \mathbf{E}\cdot\mathbf{a}_\rho\, ds + \iint_{TOP} \mathbf{E}\cdot\mathbf{a}_z\, ds + \iint_{BOTTOM} \mathbf{E}\cdot(-a_z)ds$$

$$= E_\rho \iint_{SIDE} ds + 0 + 0 = E_\rho\, 2\pi\rho\,\ell = \frac{Q_{enc}}{\varepsilon_0} = \frac{\rho_\ell\,\ell}{\varepsilon_0}$$

$$\boxed{E_\rho = \frac{\rho_\ell}{2\pi\varepsilon_0}\left(\frac{1}{\rho}\right)} \tag{2-22}$$

Note that ρ, with no subscripts, is a cylindrical coordinate (a distance from the z axis), whereas ρ_ℓ is the line charge density.

Note also that the length ℓ above is arbitrary and the field doesn't depend on ℓ. The electric field decreases as the *inverse* of ρ for a very long line charge while the field of a point charge decreases as the *inverse square* of r.

EXAMPLE 2-6 PLANAR SURFACE CHARGE DISTRIBUTIONS

Consider a uniform surface charge density ρ_s [C/m^2] over the infinite plane $x = 0$ as shown in Figure 2-4(b). Find the electric field everywhere.

SOLUTION

Translation in y, z directions produces no change in charge distribution and thus $\partial/\partial y = \partial/\partial z = 0$. We say that the charge distribution has *planar symmetry*. We expect that E_y and E_z are zero and that E_x is odd $(E_x(x) = -E_x(-x))$ since $\mathbf{E}$ will point *away* from the charge source as shown in Figure 2-4(b).

$$\begin{aligned}\mathbf{E} &= \mathbf{a}_x E_x \quad (x>0)\\ &= -\mathbf{a}_x E_x \quad (x<0)\end{aligned}$$

Construct a cylindrical Gaussian surface consisting of length 2x and end surfaces S_o as shown in Figure 2-4(b). The cylinder is of arbitrary cross section with area S_o. The electric field E_x is normal to the end caps and parallel to the side surface. Thus there is no contribution to the electric flux from the side surface. The charge enclosed in the cylinder is $\rho_s S_o$. Then

$$\begin{aligned}\oiint_S \mathbf{E}\cdot d\mathbf{s} = 2\iint_{END} \mathbf{E}\cdot d\mathbf{s} = E_x 2S_0 = \frac{Q_{enc}}{\varepsilon_0} = \frac{\rho_s S_0}{\varepsilon_0}\\ E_x = \frac{\rho_s}{2\varepsilon_0}\end{aligned} \tag{2-23a}$$

$$\begin{aligned}\mathbf{E} &= \mathbf{a}_x \frac{\rho_s}{2\varepsilon_o} \quad (x > 0)\\ &= -\mathbf{a}_x \frac{\rho_s}{2\varepsilon_o} \quad (x < 0)\end{aligned} \tag{2-23b}$$

Next consider two parallel plates, separated by distance d, with uniform surface charge distributions $\pm\rho_s$ as shown in Figure 2-4(c). Now we add the electric fields due to each charged plate by superposition. The fields add inside and cancel outside, yielding

$$\mathbf{E} = \mathbf{a}_x \frac{\rho_s}{\varepsilon_0} \, (0 < x < d) \tag{2-23c}$$

We see from Examples 2-3 through 2-6 that the solution for the electric field by Gauss' law is very straightforward for the class of symmetrical problems to which it can be applied.

The problems treated are of three types:

1. Spherical symmetry: Spherical charge distributions which are independent of θ and ϕ such as

$$\rho_v = f(r) \quad \left(\frac{\partial}{\partial\theta} = \frac{\partial}{\partial\phi} = 0\right) \tag{2-24a}$$

2. Cylindrical symmetry: Infinitely long cylindrical charge distributions which are independent of z and ϕ such as

$$\rho_v = f(\rho) \quad \left(\frac{\partial}{\partial z} = \frac{\partial}{\partial\phi} = 0\right) \tag{2-24b}$$

3. Planar symmetry: Infinite planar or slab charge distributions which are independent of y and z such as

$$\rho_v = f(x) \quad \left(\frac{\partial}{\partial y} = \frac{\partial}{\partial z} = 0\right) \tag{2-24c}$$

Let's look at a problem which cannot be treated by Gauss's law. Consider a line charge of finite length with a uniform charge distribution:

$$\rho_\ell = \rho_{\ell o} \quad \left(-\frac{\ell}{2} \le z \le \frac{\ell}{2},\ x = y = 0\right) \tag{2-25a}$$

$$= 0 \quad \text{elsewhere} \tag{2-25b}$$

Figure 2-5 shows the charge distribution and a sketch of the electric field lines. In this case $\partial/\partial\phi = 0$ but $\partial/\partial z \neq 0$. If we try a cylindrical Gaussian surface, we find that two problems arise: (a) the cylindrical surface is not normal to the electric field $\mathbf{E}$ and (b) E_n is not constant over the surface. We will need more general methods to treat this problem (see Example 2-8).

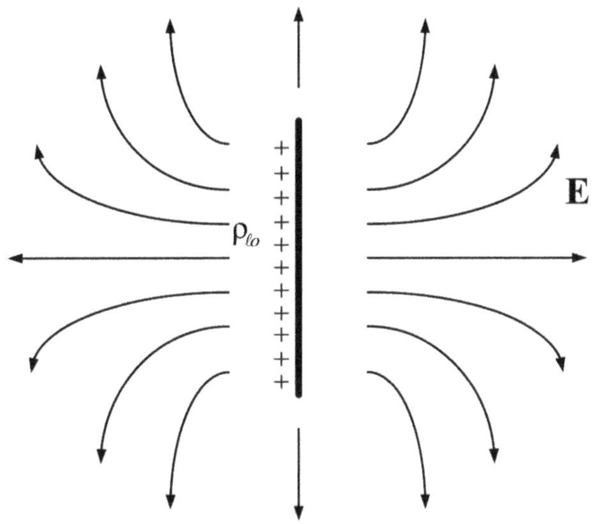

Figure 2-5. A uniform line charge density $\rho_{\ell o}$ of finite length.

2.6. THE ELECTRIC FIELDS OF ARBITRARY CHARGE DISTRIBUTIONS

In this section we consider the electric fields due to arbitrary charge distributions. The methods developed are completely general but more difficult to apply than are those of Gauss' law. They can handle the problems of charge distributions that do not have symmetry.

The concepts of this section are particularly difficult for many students. We introduce two independent sets of variables (x, y, z) and (x', y', z') for the *field point* and the *source point*, respectively. We integrate over the primed variables (x', y', z') of the source leaving our electric field in terms of the unprimed variables (x, y, z) of the field point. These concepts are new and require a careful introduction.

To begin, consider an arbitrary volume charge distribution $\rho_v(x', y', z')$ as shown in Figure 2-6. The volume charge is contained within a volume V. We want to find the electric field $\mathbf{E}(x, y, z)$ due to the entire charge distribution at a field point located at (x, y, z).

First we consider the elementary contribution due to a typical volume element dv' with volume charge density $\rho_v(x', y', z')$ located at typical source point (x', y', z'). Treating the volume charge element as a point charge (dQ), we can use the results already obtained for the field at (x, y, z) due to a point charge at (x', y', z') shown in (2-17):

$$d\mathbf{E} = \mathbf{a}_R \frac{dQ}{4\pi\varepsilon_0 R^2} = \mathbf{a}_R \frac{\rho_V\, dv'}{4\pi\varepsilon_0 R^2} = \frac{\mathbf{R}\rho_V\, dv'}{4\pi\varepsilon_0 R^3} = \frac{(\mathbf{r}-\mathbf{r}')\,\rho_V\, dv'}{4\pi\varepsilon_0 |\mathbf{r}-\mathbf{r}'|^3} \tag{2-26}$$

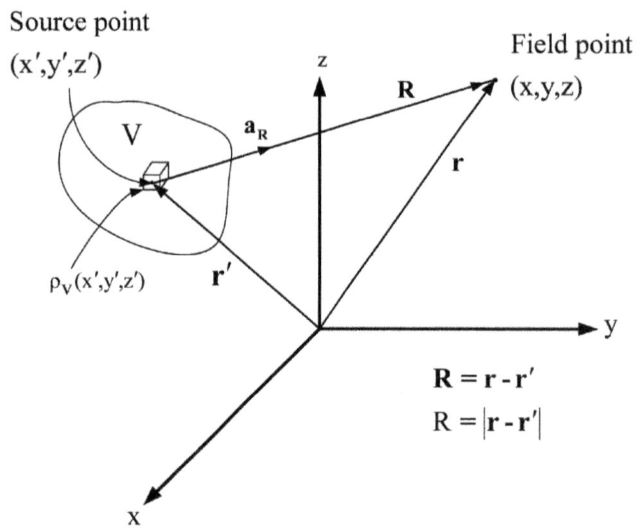

Figure 2-6. An arbitrary volume charge distribution ρv(x′, y′, z′) in the basic source point-field point representation.

Note that $dQ = \rho_v\, dv'$. Then, summing all such typical contributions over volume V, we obtain

$$\begin{aligned}\mathbf{E}(x,y,z) &= \frac{1}{4\pi\varepsilon_0}\iiint_V \frac{\mathbf{a}_R\, \rho_v(x',y',z')\, dv'}{R^2} \\ &= \frac{1}{4\pi\varepsilon_0}\iiint_V \frac{\mathbf{R}\rho_v(x',y',z')\, dv'}{R^3} \\ &= \frac{1}{4\pi\varepsilon_0}\iiint_V \frac{(\mathbf{r}-\mathbf{r}')\,\rho_v(x',y',z')\, dv'}{|\mathbf{r}-\mathbf{r}'|^3} \\ &\text{(for volume charge density } \rho_v\text{)}\end{aligned} \tag{2-27a}$$

where

$$\mathbf{R} = \mathbf{r} - \mathbf{r}' = \mathbf{a}_x(x-x') + \mathbf{a}_y(y-y') + \mathbf{a}_z(z-z')$$
$$R = |\mathbf{r}-\mathbf{r}'| = \sqrt{(x-x')^2 + (y-y')^2 + (z-z')^2}$$

For the special cases of surface or line charge densities, dQ is replaced with $\rho_s\, ds'$, $\rho_\ell\, d\ell'$, respectively, and integration is carried out over the surface or line to obtain:

$$\mathbf{E} = \frac{1}{4\pi\varepsilon_0} \iint_S \frac{\mathbf{R}\rho_s\, ds'}{R^3} \quad \text{(for surface charge density } \rho_s\text{)} \tag{2-27b}$$

$$= \frac{1}{4\pi\varepsilon_0} \int_L \frac{\mathbf{R}\rho_\ell\, d\ell'}{R^3} \quad \text{(for line charge density } \rho_\ell\text{)} \tag{2-27c}$$

Eq. (2-27) is called the **generalized Coulomb's law**. For a collection of point charges q_i at source points (x_i, y_i, z_i) we merely sum the contributions to obtain

$$\mathbf{E}(x, y, z) = \frac{1}{4\pi\varepsilon_0} \sum_i \frac{\mathbf{R}_i q_i}{R_i^3} \tag{2-28}$$

where

$$\mathbf{R}_i = \mathbf{a}_x(x - x_i) + \mathbf{a}_y(y - y_i) + \mathbf{a}_z(z - z_i)$$

$$R_i = \sqrt{(x - x_i)^2 + (y - y_i)^2 + (z - z_i)^2}$$

Note that the inverse square relationship may be represented in terms of $\frac{\mathbf{a}_R}{R^2}$ or $\frac{\mathbf{R}}{R^3}$.

Note that $\mathbf{R}$ of Eqs. (2-27a,b,c) has a very simple physical interpretation. $\mathbf{R}$ is a vector from the variable source point to the fixed field point. R is the distance from the source point to the field point.

Note also that in Eqs. (2-27a,b,c), there are six independent variables (x, y, z, x', y', z'). The integrands are functions of all six variables. The densities $\rho_v(x', y', z')$, $\rho_s(x', y', z')$, $\rho_\ell(x', y', z')$ are expressed in primed coordinates. Integration is carried out over the primed variables. The other independent variables, namely, the unprimed variables (x, y, z), may be considered as constants for purposes of the integration. Integration removes the primed variables leaving an electric field $\mathbf{E}(x, y, z)$ which is a function of unprimed variables only. To reiterate, the field point (x, y, z) is fixed during the integration while the source point (x', y', z') varies over volume V. Depending on the geometry of the problem, these coordinates may also be cylindrical or spherical.

EXAMPLE 2-7 ON-AXIS ELECTRIC FIELD OF A CHARGED DISK

Consider a disk of uniform surface charge density, with radius a, centered at the origin and lying in the xy plane (Figure 2-7).

$$\rho_s = \rho_{so} \quad (\rho \le a, z = 0) \tag{2-29}$$

Find the electric field at points along the z-axis.

SOLUTION

The electric field is given by

$$\mathbf{E} = \frac{1}{4\pi\varepsilon_o} \iint \frac{\rho_{so} \mathbf{R}\, ds'}{R^3} \tag{2-27b}$$

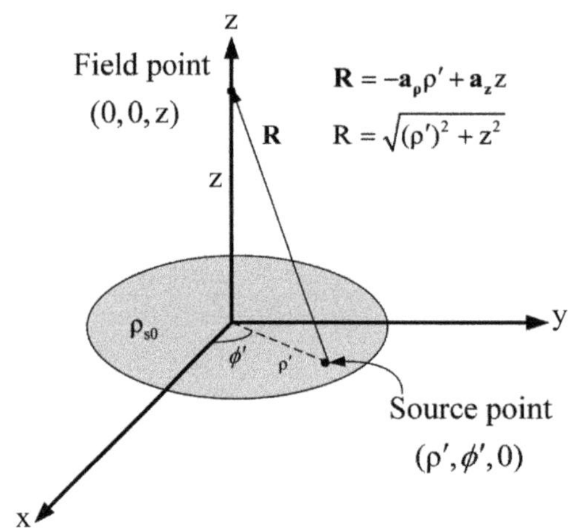

Figure 2-7. A uniformly charged disk.

The electric field along the z axis is to be found. Since the disk is circular, it is better to use cylindrical coordinates. We identify the field point (0, 0, z) and the source point (ρ', ϕ', 0) in cylindrical coordinates.

Then

$$\mathbf{r} = \mathbf{a}_z z,$$

$$\mathbf{r}' = \mathbf{a}_\rho \rho' \text{ (Note that } \mathbf{r}' \text{ has no } \mathbf{a}_\phi \text{ component.)}$$

$$\begin{aligned}\mathbf{R} &= \mathbf{a}_z z - \mathbf{a}_\rho \rho' \text{ (Note that } \mathbf{a}_\rho \text{ is not a constant vector, i.e., it varies with } \phi'.) \\ &= \mathbf{a}_z z - (\mathbf{a}_x \cos\phi' + \mathbf{a}_y \sin\phi')\, \rho' \\ &= -\mathbf{a}_x \rho' \cos\phi' - \mathbf{a}_y \rho' \sin\phi' + \mathbf{a}_z z\end{aligned}$$

$$ds' = \rho' d\rho' d\phi', \quad R = \sqrt{(\rho')^2 + z^2}$$

First we consider the z component of $\mathbf{E}$:

$$E_z = \frac{1}{4\pi\varepsilon_o}\int_0^{2\pi}\int_0^{a}\frac{\rho_{so}\, z\rho'\, d\rho'\, d\phi'}{[(\rho')^2+z^2]^{3/2}} = \frac{\rho_{so}\, z}{2\varepsilon_o}\left[\frac{-1}{\sqrt{(\rho')^2+z^2}}\right]_0^a$$

$$= \frac{\rho_{so}}{2\varepsilon_o}\left[\frac{z}{\sqrt{z^2}} - \frac{z}{\sqrt{a^2+z^2}}\right]$$

Note that $\sqrt{z^2} = |z| = \pm z \begin{Bmatrix} +(z>0) \\ -(z<0) \end{Bmatrix}$ and thus E_z may also be represented as follows:

$$E_z = \frac{\rho_{so}}{2\varepsilon_o}\left[\pm 1 - \frac{z}{\sqrt{z^2+a^2}}\right], \tag{2-30}$$

Note that E_z is odd ($E_z(-z) = -E_z(z)$) as it should be.

As $z \to 0$ and the field point approaches the center of the disk along the positive axis, the disk appears large and $E_z \to \frac{\rho_{so}}{2\varepsilon_o}$ which agrees, as it should, with the electric field of a charged infinite planar surface (Eq. (2-23a)).

As $z \to \infty$ along the positive z axis, the disk appears small and the electric field E_z should approach that of a point charge at the origin:

$$E_z = \frac{\rho_{so}}{2\varepsilon_o}\left(1 - \frac{1}{\sqrt{1+a^2/z^2}}\right) \approx \frac{\rho_{so}}{2\varepsilon_o}\left(1-(1-\frac{a^2}{2z^2})\right)\left(\text{as } \frac{a}{z} << 1\right)$$

$$= \frac{\rho_{so}(\pi a^2)}{4\pi\varepsilon_o z^2} = \frac{Q}{4\pi\varepsilon_o r^2}$$

Thus the electric field is reduced to the appropriate form of Coulomb's law, with the entire charge concentrated at the origin. Here we have used the approximation

$$\frac{1}{\sqrt{1+x}} \approx 1 - \frac{x}{2} \quad (|x| << 1) \tag{2-31}$$

which represents the first two terms of the binomial series $(n = -\frac{1}{2})$ or the Taylor series for small x:

$$(1+x)^n = 1 + nx + \frac{n(n-1)}{2!}x^2 + \frac{n(n-1)(n-2)}{3!}x^3 + \ldots \tag{2-32a}$$

$$\approx 1 + nx \quad (|x| << 1) \tag{2-32b}$$

The other components, E_x and E_y, are zero along the z axis. This can be determined analytically (see Eq. (2-34) below) or by observing the symmetry of the problem.

Partial Disk

As an extension of Example 2-7, we can also calculate the field along the axis of a partial disk:

$$\rho_s = \rho_{so} \ (\rho_1 \le \rho \le \rho_2,\ \phi_1 \le \phi \le \phi_2) \tag{2-33}$$

$$\mathbf{E} = \frac{1}{4\pi\varepsilon_o} \int_{\phi_1}^{\phi_2} \int_{\rho_1}^{\rho_2} \frac{\rho_{so}\,[-\mathbf{a}_x \rho' \cos\phi' - \mathbf{a}_y \rho' \sin\phi' + \mathbf{a}_z z]\,\rho'\,d\rho' d\phi'}{[(\rho')^2 + z^2]^{3/2}} \tag{2-34}$$

First we note that integration of $\cos\phi'$, $\sin\phi'$ over the full range ($0 \le \phi' \le 2\pi$) yields zero for E_x and E_y. The z component of $\mathbf{E}$ is given by

$$E_z = \frac{\rho_{so} z}{4\pi\varepsilon_o} \left[\frac{1}{\sqrt{z^2 + \rho_1^2}} - \frac{1}{\sqrt{z^2 + \rho_2^2}} \right] (\phi_2 - \phi_1) \tag{2-35}$$

You can solve also for E_x, E_y and find the electric fields of an annular disk or a pie-shaped partial disk.

EXAMPLE 2-8 ELECTRIC FIELD OF A LINE CHARGE OF FINITE LENGTH

Consider a uniform line charge density along the z axis (Figure 2-8a):

$$\rho_\ell = \rho_{\ell o} \ (z_1 \le z \le z_2;\ x = y = 0) \tag{2-36}$$

Find the electric field at an arbitrary point (ρ, ϕ, z).

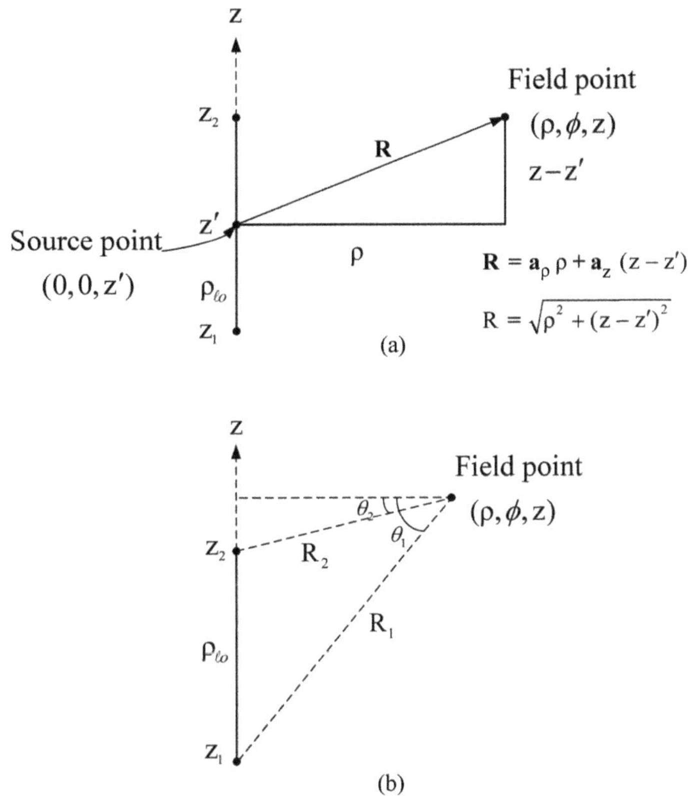

Figure 2-8. A uniform line charge density of finite length. (a) Source point-field point representation. (b) Alternative geometry.

SOLUTION

The electric field is given by

$$\mathbf{E} = \frac{1}{4\pi\varepsilon_o} \int \frac{\mathbf{R}\,\rho_\ell\, d\ell'}{R^3} \qquad (2\text{-}27c)$$

where

$$\mathbf{r} = \mathbf{a}_\rho \rho + \mathbf{a}_z z, \quad \mathbf{r}' = \mathbf{a}_z z' \text{ (Note that } \mathbf{r}' \text{ has no } \mathbf{a}_\phi \text{ component)}$$

$$\mathbf{R} = \mathbf{a}_\rho \rho + \mathbf{a}_z (z - z') \text{ (}\mathbf{a}_\rho \text{ is fixed for this problem, i.e., independent of } z'\text{)}$$

$$R = \sqrt{\rho^2 + (z - z')^2}; \; d\ell' = dz'$$

Here it is very important to distinguish between z (field point) and z′ (source point) for this problem. The integration is done over z′.

Thus,

$$\mathbf{E} = \mathbf{a}_\rho E_\rho + \mathbf{a}_z E_z \tag{2-37}$$

where

$$E_z = \frac{1}{4\pi\varepsilon_o} \int_{z_1}^{z_2} \frac{\rho_{\ell o}(z-z')\,dz'}{[\rho^2+(z-z')^2]^{3/2}}$$

Let $u = z - z'$, $du = -\,dz'$ (z constant)

$$\begin{aligned} E_z &= -\frac{1}{4\pi\varepsilon_o} \int_{z-z_1}^{z-z_2} \frac{\rho_{\ell o}\, u\, du}{[\rho^2+u^2]^{3/2}} = \frac{\rho_{\ell o}}{4\pi\varepsilon_o}\left[\frac{1}{\sqrt{\rho^2+u^2}}\right]_{z-z_1}^{z-z_2} \\ &= \frac{\rho_{\ell o}}{4\pi\varepsilon_o}\left[\frac{1}{\sqrt{\rho^2+(z-z_2)^2}} - \frac{1}{\sqrt{\rho^2+(z-z_1)^2}}\right] \\ &= \frac{\rho_{\ell o}}{4\pi\varepsilon_o}\left(\frac{1}{R_2} - \frac{1}{R_1}\right) \qquad \text{(see Figure 2-8b)} \end{aligned} \tag{2-38}$$

$$\begin{aligned} E_\rho &= \frac{1}{4\pi\varepsilon_o} \int_{z_1}^{z_2} \frac{\rho_{\ell o}\rho\, dz'}{[\rho^2+(z-z')^2]^{3/2}}; \qquad \text{Let } u = z - z' \\ &= \frac{\rho_{\ell o}}{4\pi\varepsilon_o}\rho\left[\frac{-(z-z')}{\rho^2\sqrt{\rho^2+(z-z')^2}}\right]_{z'=z_1}^{z'=z_2} \\ &= \frac{\rho_{\ell o}}{4\pi\varepsilon_o}\left(\frac{1}{\rho}\right)\left[\frac{z-z_1}{\sqrt{\rho^2+(z-z_1)^2}} - \frac{z-z_2}{\sqrt{\rho^2+(z-z_2)^2}}\right] \end{aligned} \tag{2-39}$$

which may also be represented as

$$E_\rho = \frac{\rho_{\ell o}}{4\pi\varepsilon_o}\left(\frac{1}{\rho}\right)(\sin\theta_1 - \sin\theta_2) \qquad \text{(see Fig. 2-8(b))} \tag{2-40}$$

Note that ρ, with no subscript, is cylindrical distance, whereas ρ_v, ρ_s, ρ_ℓ are all charge densities.

2.7 THE SCALAR ELECTRIC POTENTIAL V

We noted in Section 2.3 that the electric field is conservative ($\nabla \times \mathbf{E} = 0$). Therefore it can be expressed, as noted in Section 1.5, as the gradient of a scalar potential V:

$$\boxed{\mathbf{E} = -\nabla V} \qquad \mathbf{E} \text{ is the gradient of a scalar potential} \qquad (2\text{-}41)$$

As discussed in Section 1.4.2, the gradient represents the maximum rate of change; thus the electric field, which is the negative gradient of V, points in the direction of maximum rate of *decrease* of the scalar potential V. The electric field lines are perpendicular to the level surfaces where $V =$ constant, called the **equipotential surfaces**, and they flow from high potential to low potential.

Eq. (2-41) is a point relationship. We can obtain the corresponding integral relationship by taking the line integral of both sides between two points. The lower limit is an arbitrary *reference point* (x_o, y_o, z_o) at which the potential is zero. The upper limit is the field point (x,y,z). Since $\mathbf{E}$ is conservative, the line integral is independent of path

$$-\int_{(x_o,y_o,z_o)}^{(x,y,z)} \mathbf{E}\cdot d\boldsymbol{\ell} = \int_{\text{ref. point}}^{\text{field point}} \nabla V\cdot d\boldsymbol{\ell}$$

$$= \int_{\text{Ref.}}^{(x,y,z)} \left(\mathbf{a}_x \frac{\partial V}{\partial x} + \mathbf{a}_y \frac{\partial V}{\partial y} + \mathbf{a}_z \frac{\partial V}{\partial z}\right)\cdot(\mathbf{a}_x\, dx + \mathbf{a}_y\, dy + \mathbf{a}_z\, dz)$$

$$= \int_{\text{Ref.}}^{(x,y,z)} \frac{\partial V}{\partial x}dx + \frac{\partial V}{\partial y}dy + \frac{\partial V}{\partial z}dz = \int_{\text{Ref.}}^{(x,y,z)} dV = V(x,y,z) - V(x_o,y_o,z_o) = V(x,y,z)$$

$$\boxed{V(x,y,z) = -\int_{\text{Ref.}}^{(x,y,z)} \mathbf{E}\cdot d\boldsymbol{\ell}\ [V]} \qquad (2\text{-}42)$$

(We have proved here the gradient theorem of Table 1-5). The unit of the potential is volts [V], thus giving the unit of $\mathbf{E}$ volts per meter [V/m].

$V(x,y,z)$ is thus the voltage or potential difference between two points: the field point and the reference point. The potential is seen as the work per unit charge just as the field $\mathbf{E}$ is the force per unit charge since $W = -\int \mathbf{F}\cdot d\boldsymbol{\ell}$ (see Section 3.9.1). Whether mentioned or not, there is always an implied reference point. Note the effect of raising all potentials by a constant. The electric field is unchanged (Eq. (2-41)) and the potential difference between any two points is unchanged. Therefore, one may set the reference point potential at zero as we have done. The location of the reference point (x_o,y_o,z_o) is arbitrary. It may be chosen for convenience. It should not be changed during the course of a problem. The zero potential reference point may be chosen at the origin, at infinity, or at a finite location. It may be chosen on a ground plane. It can be identified in some cases with a location which we call *ground*. We have made two arbitrary choices, i.e., the level and the location of the reference.

Consider an automobile which is electrostatically charged so that its chassis is 20 volts above the earth potential. The negative terminal of the 12 volt battery chassis is connected to the chassis. Thus the positive end of the battery is 32 volts above earth. What do we call zero potential, the chassis or the earth? The choice is up to you.

Consider a battery of voltage V connected to parallel conducting plates as shown in Figure 2-9.

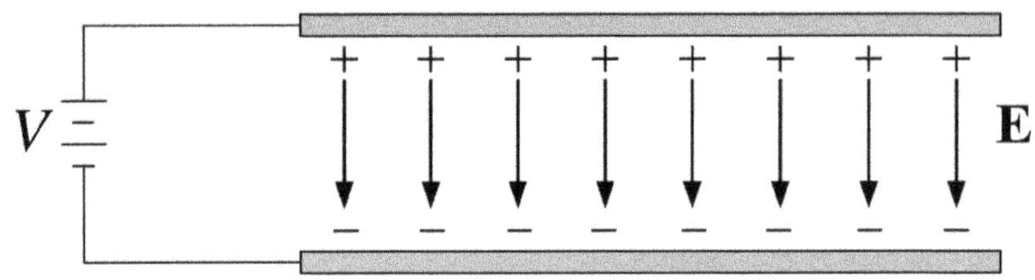

Figure 2-9. Parallel plates with applied voltage.

The electric field $\mathbf{E}$ points downward as shown in agreement with the force law $\mathbf{F} = q\mathbf{E}$. The $\mathbf{E}$ field points in the direction of decreasing potential (from high to low potential) in agreement with the minus signs in Eqs. (2-41) and (2-42).

Potential of a Point Charge

Now let us apply these concepts to determine the potential of a point charge q at the origin [Figure 2-10(a)] and at an arbitrary point (x', y', z') [Figure 2-10(b)].

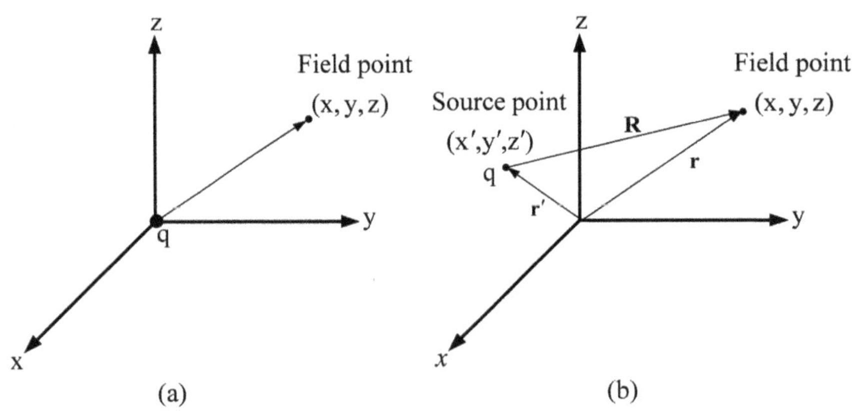

Figure 2-10. Electric potential of a point charge q (a) at the origin, and (b) at the source point (x', y', z').

For a point charge at the origin:

$$E_r = \frac{q}{4\pi\varepsilon_o r^2}$$

$$V(r) = -\int_{r_o}^{r} \mathbf{E}\cdot d\ell = -\int_{r_o}^{r} E_r\,dr = -\int_{r_o}^{r} \frac{q\,dr}{4\pi\varepsilon_o r^2} = \left[\frac{q}{4\pi\varepsilon_o r}\right]_{r_o}^{r} = \frac{q}{4\pi\varepsilon_o}\left\{\frac{1}{r} - \frac{1}{r_o}\right\}$$

$$\Rightarrow \frac{q}{4\pi\varepsilon_o r} \text{ as } r_o \to \infty \quad \text{(reference point at infinity)}$$

$$V(r) = \frac{q}{4\pi\varepsilon_0 r} \tag{2-43}$$

For a point charge at (x',y',z'):

The potential at the field point (x,y,z) depends only on the distance between the field point and the source point (x',y',z'), hence

$$\boxed{V(x, y, z) = \frac{q}{4\pi\varepsilon_0 R}} \tag{2-44}$$

where $R = |\mathbf{r} - \mathbf{r}'|$.

2.8. POTENTIAL OF AN ARBITRARY CHARGE DISTRIBUTION

In this section we consider the potential due to arbitrary charge distributions. The method developed here represents an alternative to that of Section 2.6. Figure 2-6 shows an arbitrary volume charge distribution $\rho_v(x',y',z')$ contained within a volume V. We wish to find the potential $V(x,y,z)$ due to the entire charge distribution. First consider the elementary contribution due to a typical volume dv′ located at (x',y',z'). We use the results obtained previously for the potential due to a point charge:

$$dV = \frac{dQ}{4\pi\varepsilon_0 R} = \frac{\rho_v dv'}{4\pi\varepsilon_0 R} = \frac{\rho_v dv'}{4\pi\varepsilon_0 |\mathbf{r}-\mathbf{r}'|} \tag{2-45}$$

Summing all such typical contributions dv′:

$$\boxed{\begin{aligned} V(x,y,z) &= \frac{1}{4\pi\varepsilon_0} \iiint_V \frac{\rho_v(x',y',z')dv'}{R} \\ &= \frac{1}{4\pi\varepsilon_0} \iiint_V \frac{\rho_v(x',y',z')dv'}{|\mathbf{r}-\mathbf{r}'|} \\ &\text{(for volume charge density } \rho_v) \end{aligned}} \tag{2-46a}$$

For the cases of surface or line charge densities, dQ is replaced by $\rho_s\, ds'$, $\rho_\ell\, d\ell'$ and integration is carried out over the surface or line to obtain

$$V = \frac{1}{4\pi\varepsilon_0} \iint_S \frac{\rho_s ds'}{R} \text{(for surface charge density } \rho_s) \tag{2-46b}$$

$$= \frac{1}{4\pi\varepsilon_0} \int_L \frac{\rho_\ell d\ell'}{R} \text{(for line charge density } \rho_\ell) \tag{2-46c}$$

For a collection of point charges q_i located at points (x_i, y_i, z_i) we sum contributions to obtain

$$V(x,y,z) = \frac{1}{4\pi\varepsilon_0}\sum_i \frac{q_i}{R_i} \tag{2-47}$$

where $R_i = \sqrt{(x-x_i)^2+(y-y_i)^2+(z-z_i)^2}$. The potential integral formulation of Eqs. (2-46a,b,c) represents an alternative to that of Eqs. (2-27a,b,c). Given the charge distribution ρ_v we have two choices for the determination of the electric field:

1. Find $\mathbf{E}$ directly from Eqs. (2-27a, b, c)
2. Find V directly from Eqs. (2-46a, b, c), then $\mathbf{E} = -\nabla V$.

Each method has advantages and may be simpler for a particular problem. Method 1 involves vector integration but has only one step. Method 2 involves scalar integration but has two steps. We should be prepared to use either method.

To find the potential from the charge we have a choice of two methods:

1. Find V directly from Eqs. (2-46a, b, c)
2. Find $\mathbf{E}$ directly from Eqs. (2-27a, b, c) or Gauss' law. Then $V = -\int \mathbf{E}\cdot d\boldsymbol{\ell}$.

EXAMPLE 2-9. ON-AXIS POTENTIAL OF A CHARGED DISK

Consider the disk of surface charge of Example 2-7 and Figure 2-7:

$$\rho_s = \rho_{so} \ (\rho \le a, z = 0) \tag{2-48}$$

Find the potential at points along the z-axis.

SOLUTION

$$V(z) = \frac{1}{4\pi\varepsilon_0}\iint \frac{\rho_{so}\, ds'}{R}$$

$$= \frac{1}{4\pi\varepsilon_0}\int_0^{2\pi}\int_0^a \rho_{so}\frac{\rho' d\rho' d\phi'}{\sqrt{(\rho')^2+z^2}} = \frac{\rho_{so}}{4\pi\varepsilon_0}2\pi\left[\sqrt{(\rho')^2+z^2}\,\right]_0^a$$

$$= \frac{\rho_{so}}{2\varepsilon_0}\left[\sqrt{a^2+z^2} - |z|\right] \tag{2-49}$$

The electric field E_z along the positive z axis may be obtained by using the relationship $\mathbf{E} = -\nabla V$:

$$E_z = -\frac{\partial V(z)}{\partial z} = \frac{\rho_{so}}{2\varepsilon_o}\left[-\frac{z}{\sqrt{a^2+z^2}}+1\right], \ z > 0$$

which agrees with Eq. (2-30).

As $z \to \infty$ the potential approaches that of a point charge at the origin:

$$V(z) = \frac{\rho_{so}}{2\varepsilon_0}\left[z\sqrt{1+\frac{a^2}{z^2}} - z\right] \to \frac{\rho_{so}}{2\varepsilon_0}\left[z\left(1+\frac{a^2}{2z^2}\right) - z\right]\left(\text{as } \frac{a}{z} \ll 1\right)$$
$$= \frac{\rho_{so}(\pi a^2)}{4\pi\varepsilon_o z} = \frac{Q}{4\pi\varepsilon_o R} \qquad (2\text{-}50)$$

where we have used the approximation

$$\sqrt{1+x} \approx 1 + \frac{x}{2} \; (|x| \ll 1) \qquad (2\text{-}51)$$

which represents the first two terms of the binomial series $\left(n = \frac{1}{2}\right)$:

$$(1+x)^n = 1 + nx + \frac{n(n-1)}{2!}x^2 + \frac{n(n-1)(n-2)}{3!}x^3 + \ldots \qquad (2\text{-}32a)$$

$$\approx 1 + nx \; (|x| \ll 1) \qquad (2\text{-}32b)$$

EXAMPLE 2-10. POTENTIAL OF A LINE CHARGE OF FINITE LENGTH

Consider the line charge of Example 2-8 and Figure 2-8:

$$\rho_\ell = \rho_{\ell 0} \; (z_1 \le z \le z_2; \; x = y = 0) \qquad (2\text{-}52)$$

Find the electrostatic potential.

SOLUTION

$$V = \frac{1}{4\pi\varepsilon_0}\int \frac{\rho_\ell \, d\ell'}{R}$$

where $\quad R = \sqrt{\rho^2 + (z - z')^2}, \; d\ell' = dz'$.

$$V(\rho, z) = \frac{1}{4\pi\varepsilon_0}\int_{z_1}^{z_2} \frac{\rho_{\ell o}\, dz'}{\sqrt{\rho^2 + (z-z')^2}} \quad \text{Let } u = z - z', du = -dz'$$

$$= \frac{-1}{4\pi\varepsilon_0}\int_{z-z_1}^{z-z_2} \frac{\rho_{\ell o}\, du}{\sqrt{\rho^2 + u^2}} = \frac{-\rho_{\ell o}}{4\pi\varepsilon_0}\left[\ln\left(u + \sqrt{u^2+\rho^2}\right)\right]_{z-z_1}^{z-z_2}$$

$$= \frac{\rho_{\ell o}}{4\pi\varepsilon_0}\ln\left[\frac{z - z_1 + \sqrt{\rho^2 + (z-z_1)^2}}{z - z_2 + \sqrt{\rho^2 + (z-z_2)^2}}\right] \qquad (2\text{-}53)$$

You can show that $\mathbf{E} = -\nabla V$ will lead to the results in Eqs. (2-38) and (2-39) (see Problem 2-30). The following two examples show cases in which the potential can be easily found from the line integral of the electric field.

EXAMPLE 2-11. POTENTIAL OF SPHERICAL VOLUME CHARGE

Consider the charge distribution of Example 2-3:

$$\rho_v = \rho_o \ (r \le a) \tag{2-54}$$

Find the electrostatic potential.

SOLUTION

The electric field is obtained by Gauss' law in Example 2-3 and is given by Eq. (2-19):

$$\begin{aligned} E_r &= \left(\frac{r}{a}\right)\frac{Q}{4\pi\varepsilon_0 a^2} \ (r \le a) \\ &= \frac{Q}{4\pi\varepsilon_0 r^2} \ (r \ge a) \end{aligned} \tag{2-19}$$

where $Q = \frac{4}{3}\pi a^3 \rho_o$. To find the potential $V(r)$, we integrate the electric field. Since $\mathbf{E}$ is in the radial ($\mathbf{a}_r$) direction, we choose the integration path to be also in the radial direction: $d\ell = \mathbf{a}_r dr$.

When $r \ge a$,

$$\begin{aligned} V &= -\int_{r_o}^{r} \mathbf{E} \cdot d\ell = -\int_{\infty}^{r} E_r\big|_{r\ge a} dr \\ &= -\int_{\infty}^{r} \frac{Q\,dr}{4\pi\varepsilon_0 r^2} = \frac{Q}{4\pi\varepsilon_0 r} \ (r \ge a) \end{aligned} \tag{2-55a}$$

When $r \le a$,

$$V = -\int_{\infty}^{r} E_r\,dr = -\int_{\infty}^{a} E_r\big|_{r\ge a} dr - \int_{a}^{r} E_r\big|_{r\le a} dr$$

$$= -\int_{\infty}^{a} \frac{Q\,dr}{4\pi\varepsilon_0 r^2} - \int_{a}^{r}\left(\frac{r}{a}\right)\frac{Q\,dr}{4\pi\varepsilon_0 a^2} = \frac{Q}{8\pi\varepsilon_0 a^3}(3a^2 - r^2) \ (r \le a) \tag{2-55b}$$

EXAMPLE 2-12. POTENTIAL OF A LINE CHARGE OF INFINITE LENGTH

Consider the infinite-length line charge density of Example 2-5 and Figure 2-4(a). The electric field is

$$E_\rho = \frac{\rho_\ell}{2\pi\varepsilon_0}\left(\frac{1}{\rho}\right) \tag{2-22}$$

Then the potential is given by

$$V(\rho) = -\int_{\rho_o}^{\rho} \mathbf{E}\cdot d\boldsymbol{\ell} = -\int_{\rho_o}^{\rho} E_\rho d\rho = \frac{\rho_\ell}{2\pi\varepsilon_o}\ln\left(\frac{\rho_o}{\rho}\right) \tag{2-56}$$

where ρ_o is the reference point. Note that, for this charge distribution of infinite extent, the reference point must remain finite. Infinity cannot be chosen as a reference point because the potential will not be finite at infinity.

Table 2-1 summarizes the electric fields and potentials of certain simple charge distributions. Electric fields are obtained by Gauss' law. Potentials are obtained by the line integral $-\int \mathbf{E}\cdot d\boldsymbol{\ell}$. Note the simple dependence on distance for the particular cases shown here:

$$\text{For spherical symmetry}\left(\frac{\partial}{\partial\theta} = \frac{\partial}{\partial\phi} = 0\right): |\mathbf{E}| \sim \frac{1}{r^2};\ V \sim \frac{1}{r} \tag{2-57}$$

$$\text{For cylindrical symmetry}\left(\frac{\partial}{\partial\phi} = \frac{\partial}{\partial z} = 0\right): |\mathbf{E}| \sim \frac{1}{\rho};\ V \sim \ln\rho \tag{2-58}$$

$$\text{For planar symmetry}\left(\frac{\partial}{\partial y} = \frac{\partial}{\partial z} = 0\right): \mathbf{E}\text{ is constant; } V \sim x \tag{2-59}$$

Table 2-1 Basic Electrostatic Fields and Potentials

Charge Distribution	Electric Fields	Electrostatic Potential
Point Charge at origin	$E_r = \frac{q}{4\pi\varepsilon_0 r^2}$	$V(r) = \frac{q}{4\pi\varepsilon_0 r}$
Spherical Shell Surface Charge $\rho_s = \rho_{so}$ (r=a) $q = 4\pi a^2 \rho_{so}$	$E_r = \frac{q}{4\pi\varepsilon_0 r^2}$ (r > a) $= 0$ (r < a)	$V(r) = \frac{q}{4\pi\varepsilon_0 r}$ $(r \geq a)$ $= \frac{q}{4\pi\varepsilon_0 a}$ $(r \leq a)$

(Continue)

Table 2-1 Basic Electrostatic Fields and Potentials (Continued)

Charge Distribution	Electric Fields	Electrostatic Potential
Long Line Charge on z axis	$E_\rho = \frac{\rho_\ell}{2\pi\varepsilon_0}\left(\frac{1}{\rho}\right)$	$V(\rho) = -\frac{\rho_\ell}{2\pi\varepsilon_0}\ln\left(\frac{\rho}{\rho_o}\right)$ Reference Point: $\rho = \rho_o$
Long Cylindrical Shell Surface Charge $\rho_s = \rho_{so}\ (\rho = a)$ $\rho_\ell = 2\pi a\rho_{so}$	$E_\rho = \frac{\rho_\ell}{2\pi\varepsilon_0}\left(\frac{1}{\rho}\right)(\rho > a)$ $= 0\ (\rho < a)$	$V(\rho) = -\frac{\rho_\ell}{2\pi\varepsilon_0}\ln\left(\frac{\rho}{a}\right)(\rho \geq a)$ $= 0 \quad (\rho \leq a)$ Reference Point: $\rho = a$
Large Plane of Surface Charge Surface Charge Density ρ_s at $x = 0$	$E_x = \pm\frac{\rho_s}{2\varepsilon_0},\ \begin{matrix}+(x>0)\\-(x<0)\end{matrix}$	$V(x) = \mp\frac{\rho_s}{2\varepsilon_0}x$ Reference Point: $x = 0$
Parallel Plates ρ_s at $x = 0$ $-\rho_s$ at $x = d$	$E_x = \frac{\rho_s}{\varepsilon_0}\ (0 < x < d)$ $= 0\ (x < 0, x > d)$	$V(x) = -\frac{\rho_s}{\varepsilon_0}x\ (0 \leq x \leq d)$ Reference Point: $x = 0$

2.9. CONDUCTORS

Conductors are materials which possess a large number of electrons which are free to move under the influence of an electric field. The conductor has the property that, under static conditions, there is no charge or electric field within the conductor:

$$\mathbf{E} = 0 \qquad \text{inside a conductor} \tag{2-60}$$

$$\rho_v = 0 \tag{2-61}$$

Let's see how this comes about. Consider a solid conducting sphere. Suppose that some excess charges are placed at the center of the sphere. An electric field is produced and the field exerts a force on the charges, making them move away from one another. Eventually charges will reach the surface in a symmetrical surface charge distribution (Figure 2-11(a)) so that both the charge and the field vanish inside. Gauss' law reveals that the electric field is zero inside (see Example 2-4).

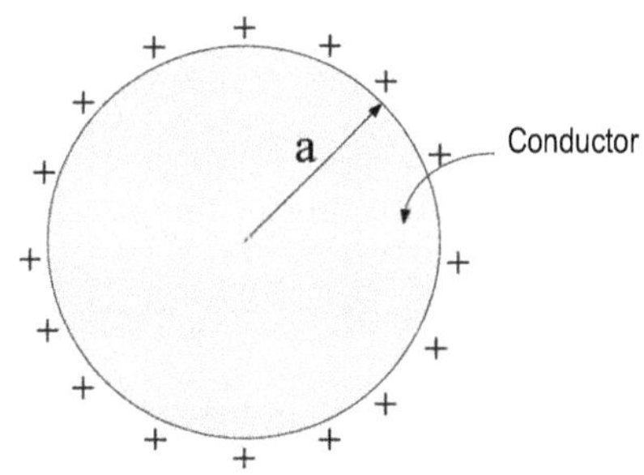

Figure 2-11(a). A symmetric surface charge distribution for a spherical conductor.

For an arbitrarily-shaped conductor (Figure 2-11(b)) charges placed inside lead to a non-uniform surface charge distribution which produces zero electric field inside. If the electric field were not zero inside, charges would continue to move until equilibrium ($\mathbf{E} = 0$) was reached.

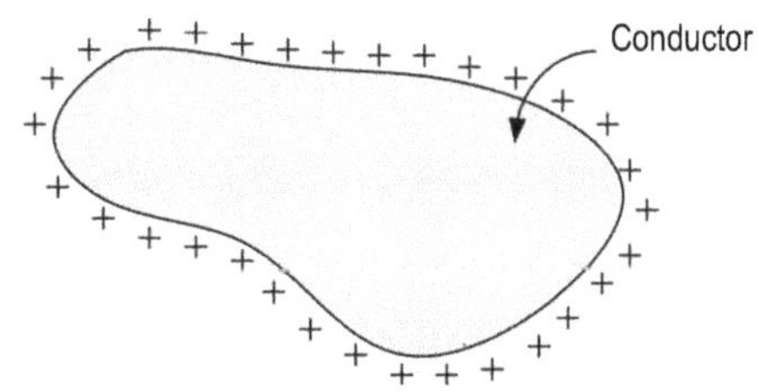

Figure 2-11(b). A surface charge distribution for an arbitrarily-shaped conductor.

Finally, consider a conductor in the presence of an applied electric field $\mathbf{E}_1$ (Figure 2-11(c)). Charge movement due to $\mathbf{E}_1$ sets up an opposing field. The process continues until the opposing field exactly cancels the applied field $\mathbf{E}_1$. The total electric field is then zero inside the conductor.

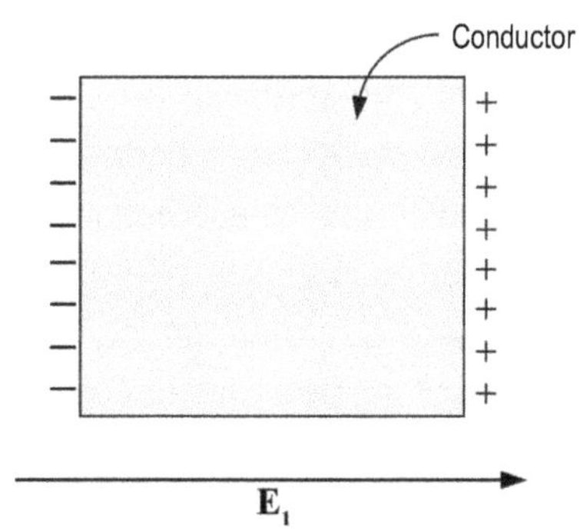

Figure 2-11(c). A surface charge for a conductor in the presence of an applied field.

Thus, if charges are placed within a conductor or if external electric fields are applied, charges move quickly to set both $\mathbf{E}$ and ρ_v equal to zero within the conductor. The process is nearly instantaneous as we shall see later in Chapter 5.

For an ideal conductor, we may also draw the following conclusion.

$$\boxed{\text{V = constant throughout the conductor}}$$

The potential is constant throughout a conductor including the surface because the potential difference, i.e., the line integral $-\int_{P_1}^{P_2} \mathbf{E} \cdot d\ell$, between any two points P_1 and P_2 in the conductor, is zero since $\mathbf{E} = 0$ in the conductor. Thus, the conductor surface is an *equipotential surface.*

Let's consider some of the characteristics of conductors and their charges. First we will consider a solid conductor with a net positive charge as shown in Figure 2-12(a).

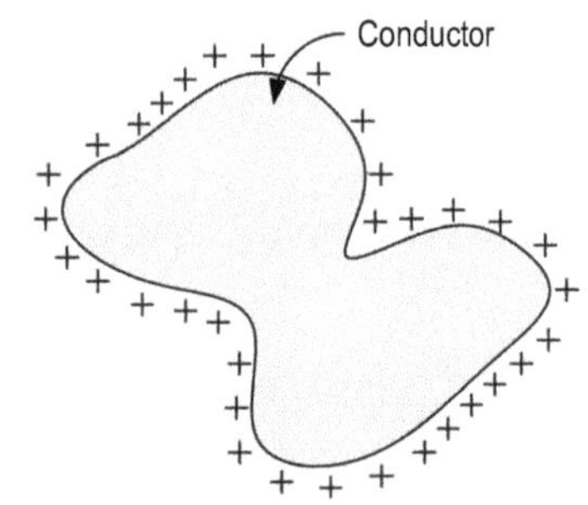

Figure 2-12(a). A charged conductor.

The charge will distribute itself over the surface of the conductor in such a way as to produce zero field inside. In the absence of external sources or applied fields the surface charge density will be positive everywhere. A negative charge density is ruled out because it implies electric field lines beginning and ending on the same conductor which contradicts the assumption of a constant potential. The surface charge density may vary considerably over the conductor. We may encounter very large densities near sharp edges or corners, but the charge density cannot be negative anywhere. We can of course reverse the words positive and negative above.

If, however, there are external sources near the conductor, or external applied fields such as that of Figure 2-11(c), then the charge density may be positive in some locations and negative in others. For example, consider a point charge q located near an uncharged conductor as in Figure 2-12(b). If q is positive, negative charge is drawn to the near vicinity of q leaving a positive charge elsewhere on the conductor as shown in Figure 2-12(b). The conductor provides a brief interruption of some of the field lines of $\mathbf{E}$ all of which go to infinity eventually. Can you explain what will happen if the conductor of Figure 2-12(b) carries a small net charge?

Now consider a charged hollow conductor, i.e., a conductor with a cavity V_0 bounded by surface S_0 (Figure 2-13(a)).

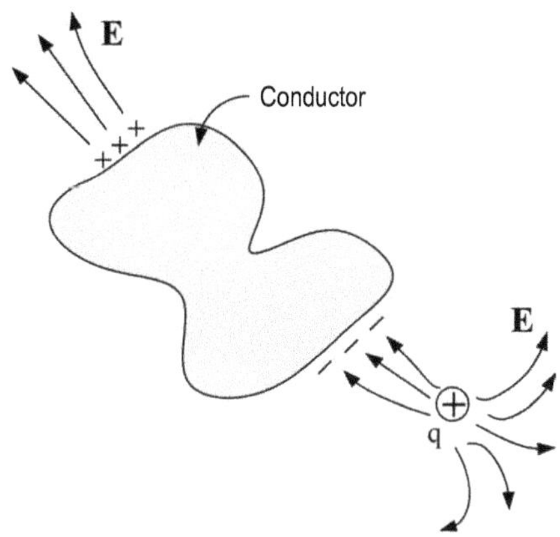

Figure 2-12(b). An uncharged conductor with charge source nearby.

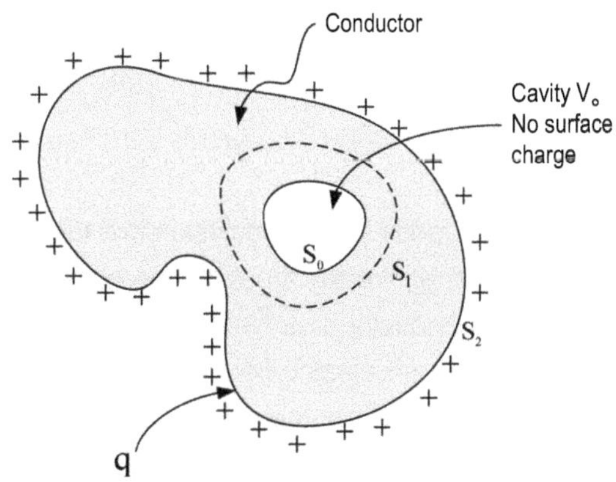

Figure 2-13(a). A charged hollow conductor.

Let there be a net charge q on the conductor; there is no charge in the empty cavity V_0. Now where does the charge reside? Is there any on the inner surface S_0? First we apply Gauss' law over the surface S_1 to conclude that the total net charge enclosed by S_1 is zero because $\oiint \mathbf{E} \cdot d\mathbf{s} = 0$ ($\mathbf{E} = 0$ inside a conductor). Could there be both positive and negative surface charge densities on surface S_0? This would imply field lines beginning and ending on S_0 which is impossible since it is an equipotential surface. Therefore there is no charge on S_0. Now what about the electric field in V_0? Is it zero also? If there were electric field lines in V_0 they must begin and end on surface S_0, leading again to a contradiction. Therefore the electric field is zero in V_0. Finally the net charge q must reside on the outer surface S_2.

Thus the electric field is zero inside a hollow charged conductor. This curious fact was first noted in 1755 by Benjamin Franklin who recognized that the fact was significant and communicated it to Joseph Priestley. Priestley concluded that the electric field satisfied an inverse square law similar to that of gravity; this conclusion was reached long before Coulomb's discovery.

Now consider the same problem (the hollow conductor has a net charge q) but with a charge q_o placed inside the cavity (Figure 2-13(b)). The charge q_o will produce an electric field in the cavity, which will draw the negative charge on the inner surface S_o.

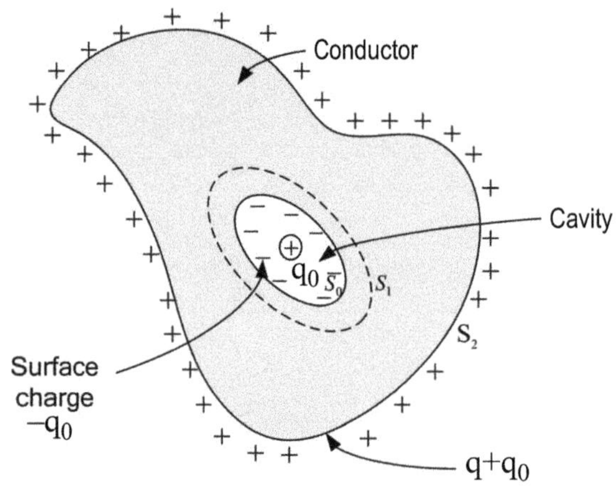

Figure 2-13(b). A charged hollow conductor with source inside.

Now an application of Gauss' law over S_1 again leads to the conclusion that the net charge enclosed by S_1 is zero. Therefore a charge $-q_0$ must reside on surface S_0. Finally a net charge $q + q_0$ must reside on the outer surface S_2 so that the net total charge carried by the conductor is q. Note that in the problems shown in Figure 2-13 the electric field is zero in the conductor. Therefore, no electric field lines can pass through the conductor and the interior problem is therefore isolated to some degree from the exterior problem.

Figure 2-14 shows an air-conductor interface. We can determine the behavior of the electric field at the surface. The tangential component is zero and the normal component is ρ_s/ε_0.

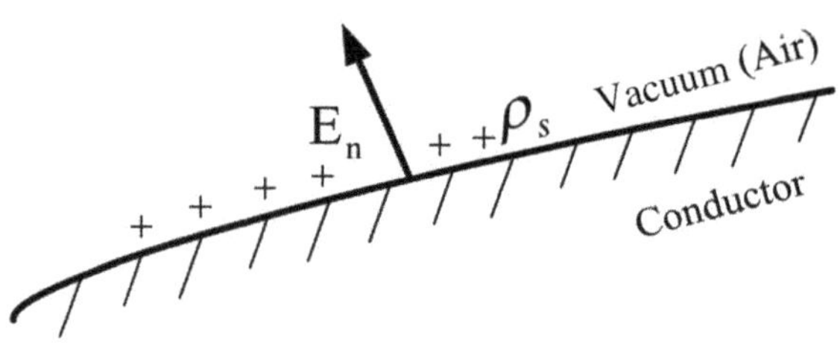

Figure 2-14. An air-conductor interface.

$$E_t = 0 \qquad \text{at an air-conductor interface} \tag{2-62}$$

$$E_n = \frac{\rho_s}{\varepsilon_o} \tag{2-63}$$

Since $\mathbf{E} = -\nabla V$ and the conductor surface is an equipotential surface, the electric field lines are normal to the surface of the conductor. Therefore the tangential component of the electric field is zero and only the normal component of $\mathbf{E}$ exists at the conductor surface. A rigorous derivation of these results will be presented in Chapter 3.

This section has been devoted to the characteristics of ideal or perfect conductors. Metals such as copper, gold, silver, brass, aluminum are almost perfect conductors. In Chapter 5, *conductivity* is defined. The **perfect conductor** corresponds to an *infinite conductivity*. Imperfect conductors or resistors are treated in Chapter 5. In our next chapter we discuss **perfect dielectrics** or insulators which represent the other extreme of conductivity, namely, *zero conductivity*.

EXAMPLE 2-13. POINT CHARGE WITHIN A CHARGED CONDUCTING SPHERICAL SHELL

Consider a point charge Q_1 placed at the center of a conducting spherical shell of inner and outer radii a, b. The shell carries a net charge Q_2. Find the surface charge and the electric field everywhere.

SOLUTION

First we find the surface charge which is distributed as shown in Figure 2-15.

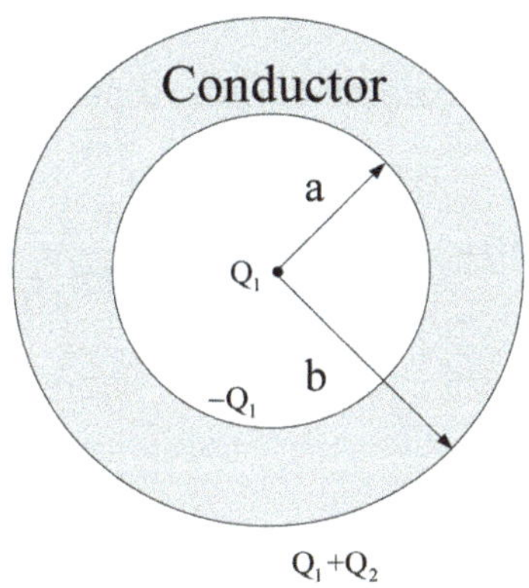

Figure 2-15. A point charge within a conducting shell.

A surface charge $-Q_1$ is drawn to the surface $r = a$ to terminate the field lines emanating from the point charge Q_1. The charge $Q_1 + Q_2$ must reside on the surface $r = b$ to make the net total charge in the conductor Q_2. The surface charge is symmetrically distributed because of a spherical shape and

$$\rho_s = -\frac{Q_1}{4\pi a^2} \quad \text{on the inner surface } (r = a) \tag{2-64a}$$

$$= \frac{Q_1 + Q_2}{4\pi b^2} \quad \text{on the outer surface } (r = b) \tag{2-64b}$$

Applying Gauss' law for a sphere of radius r, we obtain

$$E_r = \frac{Q_1}{4\pi\varepsilon_o r^2} \quad (r < a) \tag{2-65a}$$

$$= 0 \quad (a < r < b) \text{ inside the conductor} \tag{2-65b}$$

$$= \frac{Q_1 + Q_2}{4\pi\varepsilon_o r^2} \quad (r > b) \tag{2-65c}$$

The electrostatic potential can also be obtained from $\mathbf{E}$ (see Problem 2-51).

2.10 THE ELECTRIC DIPOLE

Figure 2-16 shows an **electric dipole** consisting of charges $\pm\mathbf{q}$ separated by a small distance d. The *dipole moment* $\mathbf{p}$ is defined as follows:

$$\mathbf{p} = \mathbf{a}_z qd = \mathbf{a}_z p \tag{2-66}$$

The direction of $\mathbf{p}$ is defined to point from $-q$ to $+q$. We are interested in the electric potential V and the electric field $\mathbf{E}$ at far field point (r, θ, ϕ). We assume that $r >> d$, i.e., $\frac{d}{r} << 1$ since we are interested in the potential and fields at distances r large compared to the separation d. Using superposition of potentials of two point charges:

$$V(r, \theta) = \frac{q}{4\pi\varepsilon_o}\left(\frac{1}{r_1} - \frac{1}{r_2}\right) = \frac{q\,(r_2 - r_1)}{4\pi\varepsilon_o\, r_1 r_2}$$

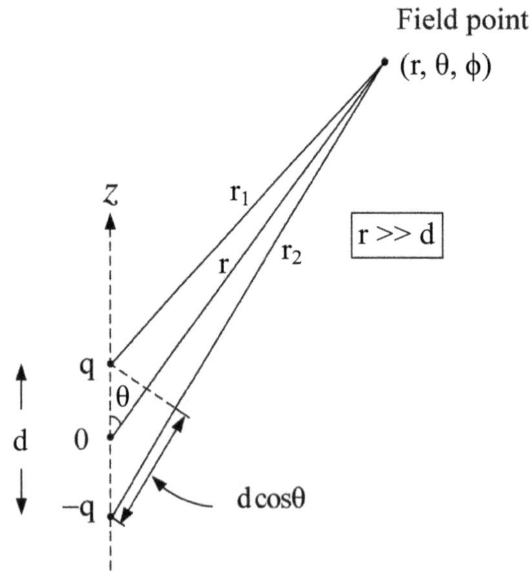

Figure 2-16. A dipole.

We can see in Figure 2-16 that $r_2 - r_1 \approx d\cos\theta$. However, let's show this by an alternative method. Using the approximation $\sqrt{1 \pm x} \approx 1 \pm \frac{x}{2}\ (x << 1)$,

$$r_1 = \sqrt{r^2 + \left(\frac{d}{2}\right)^2 - 2r\frac{d}{2}\cos\theta} \quad \text{(using the law of cosines)}$$

$$= r\sqrt{1 - \frac{d}{r}\cos\theta + \frac{d^2}{4r^2}}$$

$$\approx r\left(1 - \frac{d}{2r}\cos\theta\right) = r - \frac{d}{2}\cos\theta \quad \left(\text{we used } \sqrt{1+x} \approx 1 + \frac{x}{2},\ |x| \ll 1\right)$$

$$r_2 = \sqrt{r^2 + \left(\frac{d}{2}\right)^2 - 2r\frac{d}{2}\cos(\pi - \theta)} \approx r - \frac{d}{2}\cos(\pi - \theta) = r + \frac{d}{2}\cos\theta$$

Note that r_1 and r_2 are arbitrarily close to r (percentage wise) and that the difference $r_2 - r_1$ is equal to $d\cos\theta$ for $\frac{d}{r} << 1$.

$$V(r,\theta) = \frac{qd\cos\theta}{4\pi\varepsilon_o r^2} = \frac{p\cos\theta}{4\pi\varepsilon_o r^2} = \frac{\mathbf{p}\cdot\mathbf{r}}{4\pi\varepsilon_o r^3}$$

where we have used $\mathbf{p} = \mathbf{a}_z qd$ and $\mathbf{r} = \mathbf{a}_r r$.

$$V(r,\theta) = \frac{p\cos\theta}{4\pi\varepsilon_o r^2} \quad \text{Potential of an electric dipole} \qquad (2\text{-}67)$$

The electric field $\mathbf{E}$ can be found by taking the gradient of the potential in spherical coordinates:

$$\mathbf{E} = -\nabla V = -\left[\mathbf{a}_r\frac{\partial V}{\partial r} + \mathbf{a}_\theta\frac{1}{r}\frac{\partial V}{\partial\theta} + \mathbf{a}_\phi\frac{1}{r\sin\theta}\frac{\partial V}{\partial\phi}\right]$$

$$\mathbf{E} = \frac{qd}{4\pi\varepsilon_o r^3}(\mathbf{a}_r 2\cos\theta + \mathbf{a}_\theta\sin\theta) \quad \mathbf{E}\ \text{field of a dipole} \qquad (2\text{-}68)$$

We note that the potential and the electric field of a dipole varies with distance r as $\frac{1}{r^2}$ and $\frac{1}{r^3}$, respectively. Thus both the potential and electric field fall off more rapidly with distance (by a factor of $\frac{1}{r}$) than do those of a point charge. The reason for this is that the dipole consists of positive and negative charges whose potentials nearly cancel, the cancellation becoming more nearly perfect as r increases. Secondly, we note the variation with angle θ. The potential is maximum, minimum at $\theta = 0, \pi$, respectively. It is zero in the plane $\theta = \frac{\pi}{2}$. The electric field also depends strongly on the dipole orientation. $E_\theta = 0$ along the polar axis ($\theta = 0$) and $E_r = 0$ in the equatorial plane ($\theta = \frac{\pi}{2}$).

Higher order multipoles may be formed merely by combining lower order multipoles, one reversed with respect to the other. A *quadrupole* is shown in Figure 2-17(a). It is formed by placing one dipole on the left and bringing up a reversed dipole from the right. Figure 2-17(b) shows a linear quadrupole. Can you describe how it is formed? Figure 2-18 shows the atomic structure of crystal salt which forms an *octopole*. It is constructed by forming a quadrupole on the front and then bringing up a reversed quadrupole from the back. The quadrupole and octopole have potentials which fall off as $\frac{1}{r^3}$ and $\frac{1}{r^4}$, respectively, because of higher order cancellation of potentials. If we consider the sequence of elements: point charge, dipole, quadrupole, octopole, then each element has potentials and fields which fall off faster by the factor $\left(\frac{1}{r}\right)$ than those of the preceding element.

For dielectrics, which are treated in the next chapter, the dipole is the principal contributor to the electric potential and electric fields.

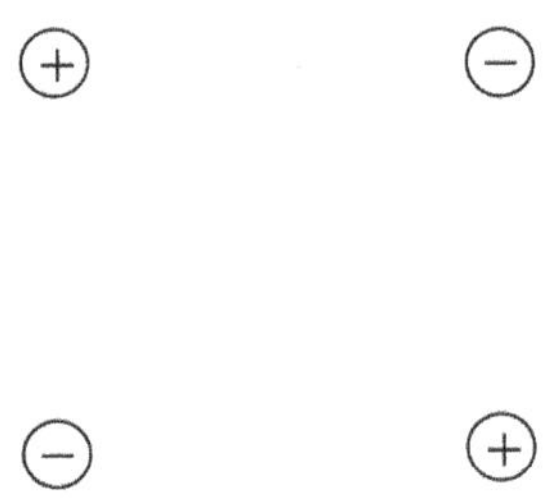

Figure 2-17 (a). A quadrupole.

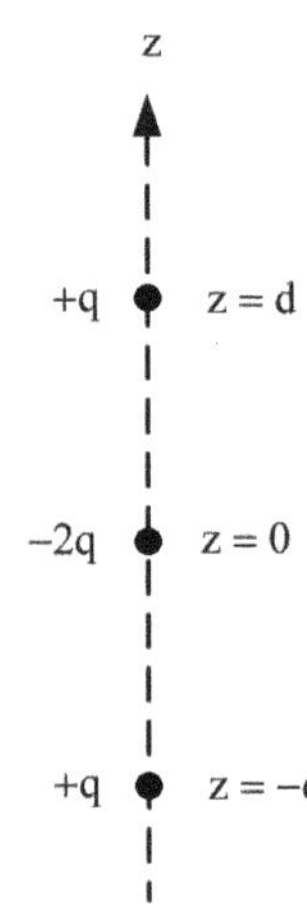

Figure 2-17(b). A linear quadrupole.

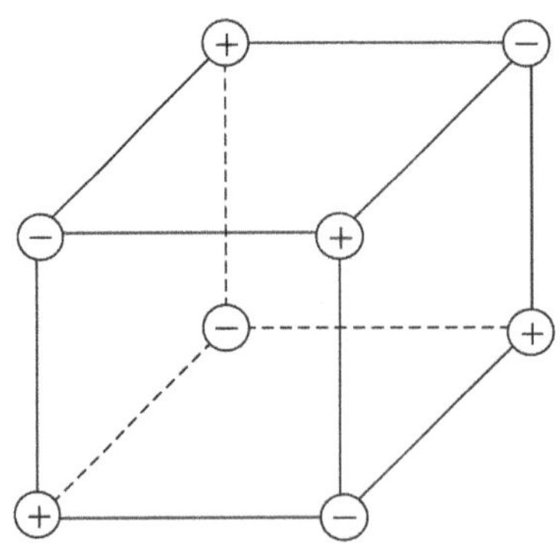

Figure 2-18. An octopole.

PROBLEMS

2-1. Consider the following volume, surface, and line charge distributions. Find the total charge in each case:

(a) A spherical shell:

$$\rho_v = \rho_{vo} \quad (a < r < b)$$

(b) A charged disk:

$$\rho_s = \rho_{so} \quad (0 < \rho < a,\ z = 0)$$

(c) A charged loop at the equator:

$$\rho_\ell = \rho_{\ell o} \cos^2\phi \quad (\rho = a,\ z = 0)$$

2-2. Find the total charge in each case:

(a) A cylindrical cloud of infinite length:

$$\rho_v = \rho_{vo} \frac{\rho^2}{a^2} \quad (0 < \rho < a)$$

Find the charge per unit length.

(b) A rectangular surface:

$$\rho_s = \rho_{so}\, x^2 y^2 \quad (a < x < b,\ c < y < d)$$

(c) A charged straight wire:

$$\rho_\ell = \rho_{\ell o} \left(x = \frac{y}{2} = \frac{z}{3},\ 0 < x < 1 \right)$$

2-3. Find the total charge associated with each of the following charge distributions:

(a) A spherical cloud of volume charge:

$$\rho_v = \rho_{vo} \left(\frac{r}{a} \right) \cos^2\theta \sin^2\phi \quad (0 < r < a)$$

(b) A charged spherical surface:

$$\rho_s = \rho_{so} \cos\frac{\theta}{2} \sin\frac{\theta}{2} \cos^2\phi \quad (r = a)$$

(c) A charged loop at latitude $\left(\frac{\pi}{2} - \theta_0 \right)$:

$$\rho_\ell = \rho_{\ell o} \quad (\theta = \theta_o,\ r = a)$$

2-4. Find the total charge in each case:

(a) A cylindrical cloud of charge:

$$\rho_v = \rho_{vo} \quad (0 < \rho < a, -\frac{\ell}{2} < z < \frac{\ell}{2})$$

(b) A charged cylindrical surface:

$$\rho_s = \rho_{so} \; (1 - \sin\phi) \, (\rho = a \, , -\frac{\ell}{2} < z < \frac{\ell}{2})$$

(c) A charged helix (see Figure 2-19):

$$\rho_\ell = \rho_{\ell o} \left(\rho = a, -\frac{\ell}{2} < z < \frac{\ell}{2}, z = \alpha a \phi \right)$$

[HINT] $\left| d\,\ell \right| = \sqrt{(\rho \, d\phi)^2 + (dz)^2}$

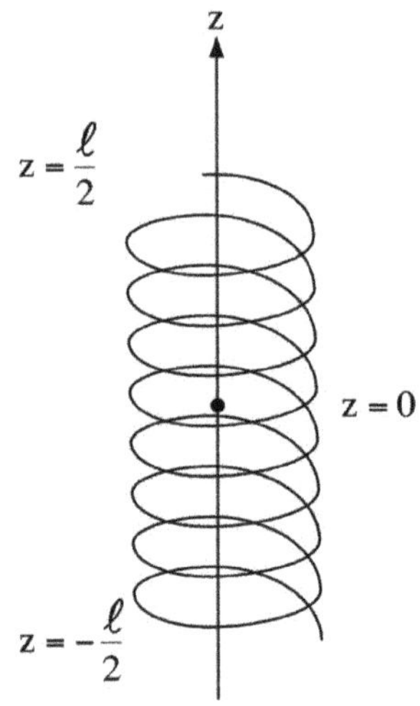

Figure 2-19. A uniformly charged helical coil.

2-5. Find the total charge associated with each of the following charge distributions:

(a) A charged conical surface (see Figure 2-20):

$$\begin{aligned} \rho_s &= \rho_{s1} \, (\theta = \theta_o, 0 < r < a) \\ &= \rho_{s2} \, (r = a, 0 < \theta < \theta_o) \end{aligned}$$

(b) A charged spherical shell:

$$\rho_v = \rho_{vo} \left(\frac{r}{a} \right) \quad (a < r < b)$$

(c) A charged Archimedean spiral:

$$\rho_\ell = \rho_{\ell o} \quad (\rho = a\,\phi,\ \phi_1 < \phi < \phi_2, z = 0)$$

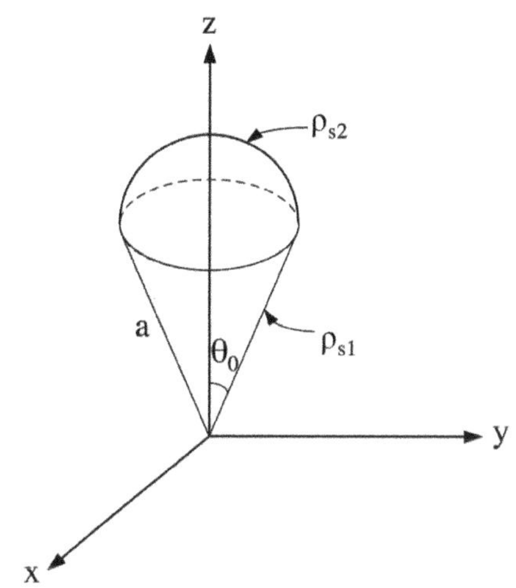

Figure 2-20. A conical surface terminated in a sphere.

2-6. A charge q is located at each corner of a square of side a, as shown in Figure 2-21.

(a) Find the electric field $\mathbf{E}$ at $(x = a/2,\ y = a/2,\ z = 0)$ due to the other charges.
(b) Find the direction and magnitude of the force on each charge. Sketch the directions of forces.

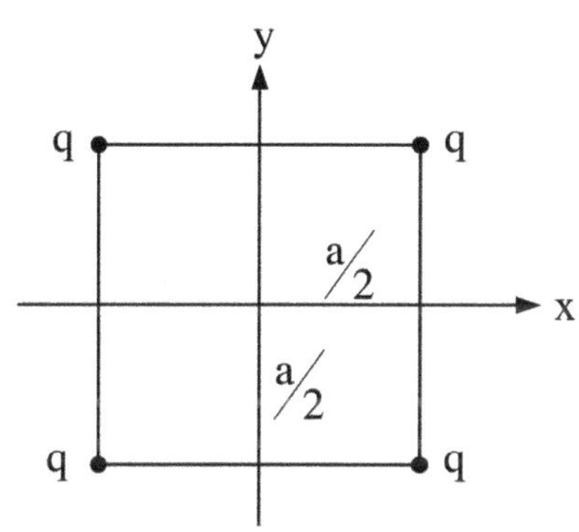

Figure 2-21. Four identical charges in the plane $z = 0$.

2-7. A charge q is located at each corner of an equilateral triangle of side a. Find the magnitude of the force on each charge. Sketch the directions of forces.

2-8. A charge q is located at each corner of a rectangle of sides a, b. Find the force on each charge.

2-9. A *nonuniform* volume charge ρ_v is distributed within a sphere of radius a as follows:

$$\rho_v = \begin{cases} \rho_{vo}\left[1-\dfrac{r}{a}\right], & 0 \le r \le a \\ 0 \quad , & r \ge a \end{cases}$$

where ρ_{vo} is a constant.

(a) Find the electric field everywhere (both for $r \le a$ and $r \ge a$).
(b) Confirm that the two basic laws are satisfied, that is, $\nabla \times \mathbf{E} = 0$ and $\nabla \cdot \mathbf{E} = \rho_v/\varepsilon_o$ in both regions.

2-10. Consider a cloud of volume charge density in the shape of a spherical shell:

$$\rho_v = \rho_{vo}\frac{r}{a} \ (a < r < b)$$

where ρ_{vo} is a constant.

(a) Find the electric field everywhere: (i) $r \le a$, (ii) $a \le r \le b$, (iii) $r \ge b$.
(b) Show that the two basic laws are satisfied in each of the three regions.

2-11. Consider a spherical cloud of volume charge density as given below:

$$\rho_v = \rho_{vo}e^{-kr} \quad (0 < r < a)$$

where ρ_{vo} and k are constants. Find the electric field everywhere.
[HINT] Use integration by parts.

2-12. Spherical layers of surface charge density exist at $r = a, b \ (b > a)$:

$$\begin{aligned} \rho_s &= \rho_{s1} \qquad (r = a) \\ &= \rho_{s2} \qquad (r = b) \end{aligned}$$

(a) Find the electric field everywhere.
(b) Under what condition will the electric field be zero for $r > b$?

2-13. Spherical layers of surface charge density surround a spherical cloud of volume charge density (Figure 2-22):

$$\rho_v = \rho_{vo} \, (a < r < b)$$
$$\rho_s = +\rho_{s1} \, (r = a)$$
$$= -\rho_{s1} \, (r = b)$$

Find the electric field everywhere.

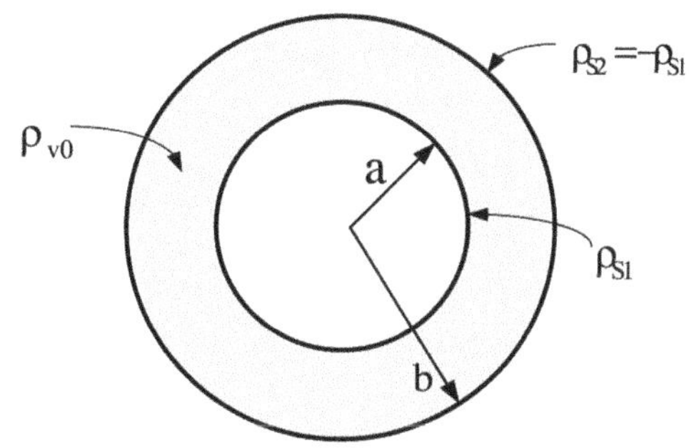

Figure 2-22. Spherical volume and surface charge.

2-14. Find the electric field everywhere due to an infinitely long, cylindrical cloud of electrons with uniform volume charge density:

$$\rho_v = -\rho_{vo} \, (0 < \rho < a)$$

Show that the two basic laws are satisfied.

2-15. Consider separately the following cylindrical volume charge distributions:

(a) $\rho_v = \rho_{vo} \quad (a < \rho < b)$

(b) $\rho_v = \dfrac{k_1}{\rho} \quad (a < \rho < b)$

(c) $\rho_v = \dfrac{k_2}{\rho^2} \quad (a < \rho < b)$

Find the electric field everywhere for each of the three cases listed above. ρ_{vo}, k_1 and k_2 are constants.

2-16. Consider the following infinitely long cylindrical volume charge distribution:

$$\rho_v = \rho_{vo} \, e^{-k\rho^2} \quad (0 < \rho < a)$$

where ρ_{vo} and k are constants. Find the electric field everywhere.

2-17. Two infinitely long cylindrical layers of surface charge density exist at $\rho = a, b$ $(b > a)$:

$$\rho_s = \rho_{s1} \quad (\rho = a)$$
$$= \rho_{s2} \quad (\rho = b)$$

(a) Find the electric field everywhere: (i) $\rho < a$, (ii) $a < \rho < b$, (iii) $\rho > b$.
(b) Under what condition is the electric field zero for $\rho > b$?

2-18. Two cylindrical layers of surface charge density surround a cylindrical cloud of volume charge density:

$$\rho_v = \rho_{vo} \quad (a < \rho < b)$$
$$\rho_s = +\rho_{s1} \quad (\rho = a)$$
$$= -\rho_{s2} \quad (\rho = b)$$

Find the electric field everywhere.

2-19. A perfectly conducting cylinder of radius a and infinite length carries a uniform surface charge density ρ_{so}. The cylinder is surrounded by a plasma layer of uniform volume charge density ρ_{vo}:

$$\rho_s = \rho_{so} \quad (\rho = a)$$
$$\rho_v = \rho_{vo} \quad (a < \rho < b)$$

Find the electric field everywhere.

2-20. Consider a slab of uniform volume charge density. The slab is of thickness d in the x-direction and is of infinite extent in the y, z directions:

$$\rho_v = \rho_{vo} \quad (-\frac{d}{2} < x < \frac{d}{2})$$

(a) Find the electric field everywhere: (i) $x > \frac{d}{2}$, (ii) $-\frac{d}{2} < x < \frac{d}{2}$, (iii) $x < -\frac{d}{2}$.

It is expected that the electric field is antisymmetric about $x = 0$, i.e., $\mathbf{E}(x) = -\mathbf{E}(-x)$.

(b) Show that the two basic laws are satisfied everywhere.

2-21. Consider an infinite slab of non-uniform symmetric volume charge density:

$$\rho_v = \rho_{vo}\frac{|x|}{d} \quad (-\frac{d}{2} < x < \frac{d}{2})$$

Find the electric field everywhere. When the charge distribution is even (symmetric) about $x = 0$, it is expected that the electric field is antisymmetric, i.e., $\bar{\mathbf{E}}(-x) = -\bar{\mathbf{E}}(x)$.

2-22. Consider an infinite slab of non-uniform anti-symmetric volume charge density:

$$\rho_v = \rho_{vo}\frac{x}{d}\left(-\frac{d}{2} < x < \frac{d}{2}\right)$$

When the charge distribution is odd about $x = 0$, the electric field will be symmetric, i.e., $\mathbf{E}(x) = \mathbf{E}(-x)$. Find the electric field everywhere.

2-23. Calculate the electric potential distribution for each of the following problems, using the electric field previously obtained:

(a) 2-9 (b) 2-12 (c) 2-18 (d) 2-20

2-24. Calculate the electric potential distribution for each of the following problems:

(a) 2-10 (b) 2-14 (c) 2-17 (d) 2-22

2-25. Consider a uniformly charged circular loop of radius a, centered at origin, and lying in the xy plane as shown in Figure 2-23.

$$\rho_\ell = \rho_{\ell o}\ (\rho = a,\ z = 0)$$

(a) Find the electric field along the z axis.
(b) Find the electrostatic potential along the z axis.
(c) Show that electric field approaches the appropriate form of Coulomb's law as $z \to \infty$.
(d) Show that the potential approaches the appropriate form as $z \to \infty$.
(e) Calculate $\mathbf{E}$ using the result for V from (b) and show that it agrees with the result of (a).

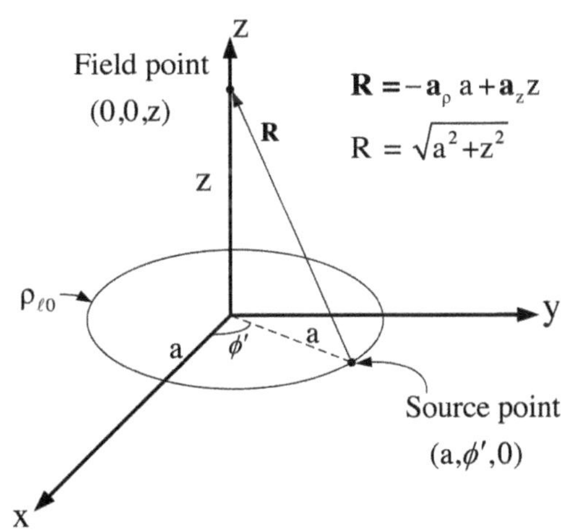

Figure 2-23. A uniformly charged loop.

2-26. Consider a partial loop of radius a, centered at origin, and lying in the xy plane:

$$\rho_\ell = \rho_{\ell o}\ (\rho = a,\ \phi_1 < \phi < \phi_2,\ z = 0)$$

(a) Find the electric field components E_x, E_y, E_z along the z axis.
(b) Find the electrostatic potential along the z axis.

2-27. Consider the uniformly charged loop (Figure 2-23). Find the maximum of the on-axis electric field. Explain the existence of this maximum.

2-28. Consider the charged disk (Example 2-7). Find the value of the electric field at the center. Compare this result with the electric field of a planar surface charge distribution (Example 2-6). Can you explain the agreement between these two results?

2-29. Consider the line charge of finite length (Example 2-8).

(a) Find the fields in the perpendicular bisector plane (Let $z_1 = -\ell/2$, $z_2 = +\ell/2$, $z = 0$).
(b) Find the fields of semi-infinite line (Let $z_1 = -\infty$, $z_2 = 0$).
(c) Find the fields of a semi-infinite line in the perpendicular end-plane (Let $z_1 = -\infty$, $z_2 = 0$, $z = 0$). Sketch the fields in this plane. Note that E_ρ is exactly half that of the infinite line. Can you explain this?
(d) Obtain the fields of an infinite line ($z_1 = -\infty$, $z_2 = +\infty$). Show that the result agrees with that of Example 2-5.

2-30. Show that the electric field and the potential of a line charge of finite length (Eqs. (2-38, 39) and (2-53)) are consistent, i.e., show that $\mathbf{E} = -\nabla V$.

2-31. Consider a thin, uniformly charged ring (annulus) of inner, outer radii a, b, respectively, as shown in Figure 2-24:

$$\rho_s = \rho_{so}\ (z = 0,\ a < \rho < b)$$

(a) Find the electric field at a distance z above the center of the charged ring.
(b) Find an approximate expression for $\mathbf{E}$ when $z >> b$ (or $z \to \infty$) and show that it is identical to the field of a point charge.

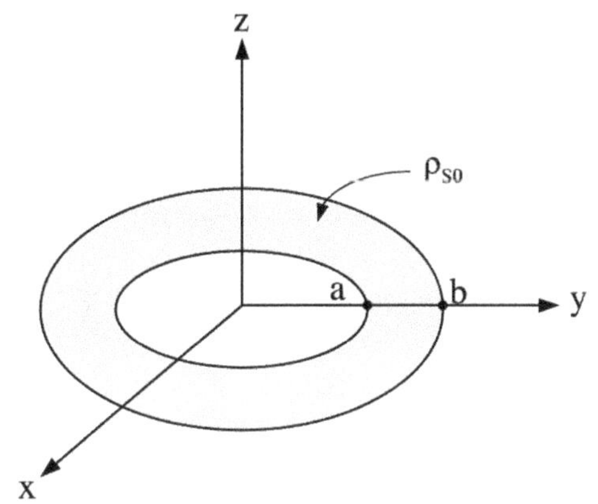

Figure 2-24. A thin, uniformly charged annulus.

2-32. Consider the charge distribution in Problem 2-31.

(a) Find the electric potential $V(z)$ above the center of the charged annular ring.
(b) Calculate the electric field from the potential using $\mathbf{E} = -\nabla V$ and show that it agrees with the result of Problem 2-31.
(c) Find the potential $V(z)$ at distant points ($z >> a$) and show that it has the proper form.

2-33. Consider two charged disks:

$$\rho_s = \rho_{so} \left(z = -\frac{d}{2},\ 0 < \rho < a \right)$$
$$= -\rho_{so} \left(z = \frac{d}{2},\ 0 < \rho < a \right)$$

(a) Find the electric field along the z axis.
(b) Consider the special case $a >> d$ and show that the on-axis field reduces to that of parallel plates in Eq. (2-23c).

2-34. A disk carries a non-uniform surface charge distribution:

$$\rho_s = \rho_{so} \frac{\rho}{a} \quad (0 < \rho < a, z = 0)$$

Find the electric field on the axis of the disk.

2-35.* A thin wire of length ℓ carries a non-uniform line charge distribution:

$$\rho_\ell = \rho_{\ell o} \left(1 + \frac{|z|}{\ell} \right) \left(-\frac{\ell}{2} < z < \frac{\ell}{2} \right)$$

Find the electric field everywhere.

2-36. Consider a uniformly charged square loop of side a lying in the plane $z = 0$ and centered at the origin (Figure 2-25). The line charge density is $\rho_{\ell o}$.

(a) Find the electric field E_z along the z axis. Show by symmetry that $E_x = E_y = 0$ along the axis and that each side contributes equally to E_z along the axis.
(b) Find the point of maximum field intensity along the z axis. Explain why there is a maximum.

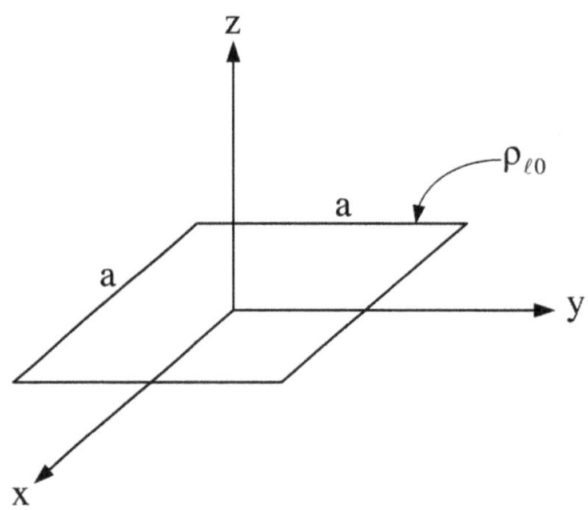

Figure 2-25. A uniformly charged square loop.

2-37. Consider the square loop of Fig. 2-25.

(a) Find the electric potential of the uniformly charged square loop along the z axis.
(b) Calculate the on-axis electric field using the potential and compare with the result of Problem 2-36.

2-38.* A regular polygonal loop of n sides, each of length a, lies in the xy plane and is centered at the origin. The polygon carries a uniform line charge density $\rho_{\ell o}$.

(a) Find the electric field **E** along the z axis.
(b) Find the potential V along the z axis.

2-39*. A square disk of sides $a \times a$, lying in the xy plane and centered at the origin, carries a uniform surface charge density ρ_{so}. Find the electric potential and the electric field along the z axis.

2-40. A thin-wire helix is wound on a cylinder of radius a and is determined by the equation $z = \alpha a \phi$. The starting and ending points of the helix are z_1 and z_2, respectively and the helix is uniformly charged (see Figure 2-19):

$$\rho_\ell = \rho_{lo} \quad (z = \alpha a \phi,\ z_1 < z < z_2)$$

Find the potential $V(z)$ along the z axis and the z-component of the electric field ($E_z(z)$) along the z axis.

2-41. Figure 2-26 shows a thin cylindrical shell which carries a uniform surface charge distribution. The cylinder is of length ℓ and radius a:

$$\rho_s = \rho_{so}\left(\rho = a,\ -\frac{\ell}{2} < z < \frac{\ell}{2}\right)$$

(a) Find the electric field along the z axis.

(b) Show that the electric field reduces to the proper form of Coulomb's law at distant points as $z \to \infty$ or $z >> a, \ell$.

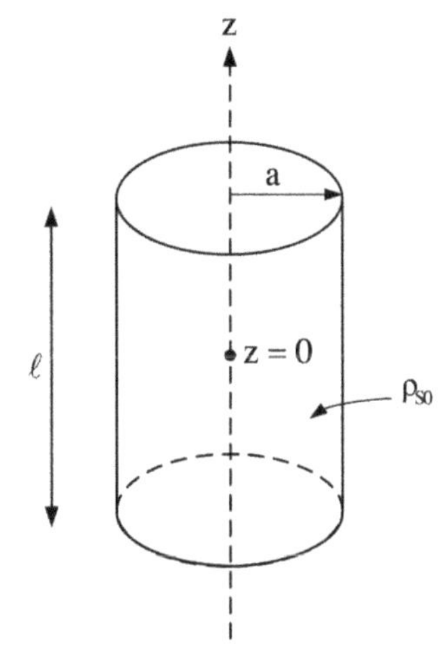

Figure 2-26. A thin, uniformly charged cylindrical shell.

2-42.* A thin cylindrical shell carries a non-uniform surface charge distribution:

$$\rho_s = \rho_{so}\left(1 + \frac{|z|}{\ell}\right)\ (\rho = a, -\frac{\ell}{2} < z < \frac{\ell}{2})$$

Find the electric field along the z axis.

2-43. Two parallel thin wires of finite length carry uniform line charge densities:

$$\rho_\ell = -\rho_{\ell o}\left(x = -\frac{d}{2};\ -\frac{\ell}{2} < z < \frac{\ell}{2}\right)$$
$$= +\rho_{\ell o}\left(x = \frac{d}{2};\ -\frac{\ell}{2} < z < \frac{\ell}{2}\right)$$

(a) Find the (attractive) force between the wires. [HINT] Consider the electric field, due to one wire, at the location of the other.

(b) Show that the force between wires has the appropriate form of Coulomb's law if the wires are far apart ($d >> \ell$).

2-44.* Two parallel semi-infinite wires are terminated in a semi-circular loop of radius a (Figure 2-27). The wires lie in the plane $z = 0$ and are uniformly charged:

$$\begin{aligned} \rho_\ell &= \rho_{\ell o} \ (z=0, y=a, 0<x<\infty) \\ &= \rho_{\ell o} \ (z=0, y=-a, 0<x<\infty) \\ &= \rho_{\ell o} \left(z=0, \rho=a, \frac{\pi}{2}<\phi<\frac{3\pi}{2} \right) \end{aligned}$$

Find the electric field at points along the z axis.

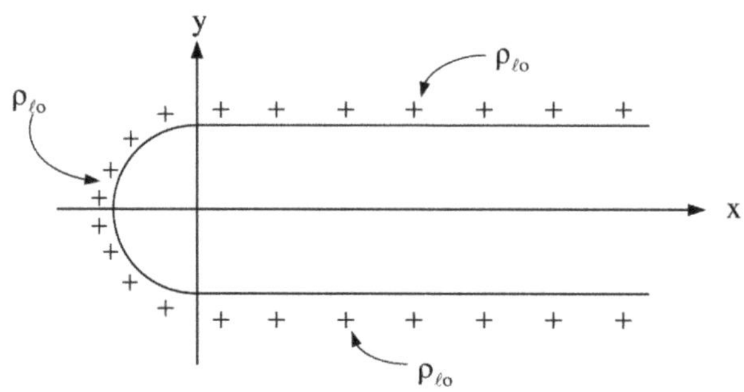

Figure 2-27. Uniformly charged semi-infinite wires terminated in a semi-circular loop.

2-45. Consider a uniformly charged strip of width w and infinite length in y direction, lying in the xy plane. The strip carries a uniform surface charge density ρ_{so}:

$$\rho_s = \rho_{so} \left(z=0, \ -\frac{w}{2}<x<\frac{w}{2} \right)$$

Find the electric field everywhere.

[HINT] Consider the strip as a collection of thin wires of infinite length each of width dx carrying a line charge density

$$\rho_\ell = \rho_{so} dx$$

2-46. Consider two parallel, uniformly charged infinite strips at $z = \pm\frac{d}{2}$:

$$\begin{aligned} \rho_s &= \rho_{so} \left(z=-\frac{d}{2}, \ -\frac{w}{2}<x<\frac{w}{2} \right) \\ &= -\rho_{so} \left(z=\frac{d}{2}, \ -\frac{w}{2}<x<\frac{w}{2} \right) \end{aligned}$$

(a) Find the electric fields everywhere, using the results of Problem 2-45.
(b) Plot the electric field along the z axis.
(c) Find the force per unit length between the two strips.

2-47. Figure 2-28 shows two charges q located on the z-axis at $z = -\frac{d}{2}, \frac{d}{2}$. The charges are released and allowed to fly apart due to mutual repulsion. The mass of each charge is m. Find the magnitude and directions of final velocities.

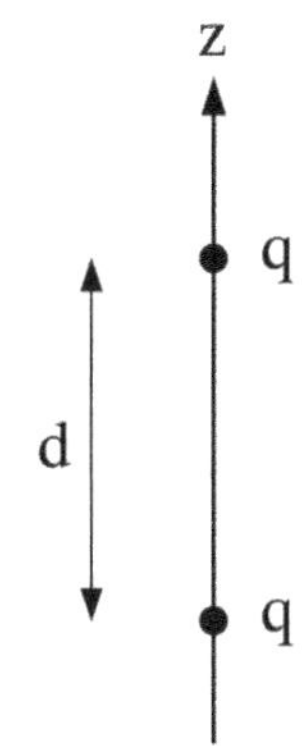

Figure 2-28. Two charges.

2-48. Consider a cylindrical cloud of uniform volume charge density. The cylinder is of radius a and length ℓ:

$$\rho_v = \rho_{vo} \left(0 < \rho < a,\ -\frac{\ell}{2} < z < \frac{\ell}{2}\right)$$

Find the electric field $E_z(z)$ along the z axis.

2-49. A charge q is located at each corner of a regular polygon of n sides. Each side is of length a. The polygon is centered at the origin and lies in the xy plane. Find the electrostatic potential V(z) along the z axis.

2-50. A uniform line charge of infinite length is located along the z axis within a perfectly conducting cylindrical shell which extends from $\rho = a$ to $\rho = b$:

$$\rho_\ell = \rho_{\ell o}\ (\rho = 0)$$

The conducting shell carries no net charge. Find the electric field everywhere and the surface charges at $\rho = a, b$.

2-51. Consider the problem of a point charge within a charged conducting spherical shell in Example 2-13. Find the electrostatic potential everywhere.

2-52. A spherical cloud of volume charge of radius a is centered at the origin within a perfectly conducting spherical shell which extends from $r = b$ to $r = c$:

$$\rho_v = \rho_{vo}\ (r \le a)$$

where $c > b > a$.

(a) Find the surface charge densities at $r = b, c$.
(b) Find the electric fields and the electrostatic potential everywhere.

2-53. Figure 2-29 shows a charge q located at the origin. The charge is surrounded by two thick, perfectly conducting shells. The first shell occupies the space $a \le r \le b$ and carries a net charge q_1. The second shell occupies the space $c \le r \le d$ and carries a net charge q_2.

(a) Find the surface charge densities at $r = a, b, c, d$.
(b) Find the electric field everywhere (in each of the five regions).
(c) Find the conditions on q_1, q_2 such that $\mathbf{E} = 0$ for $r > d$.

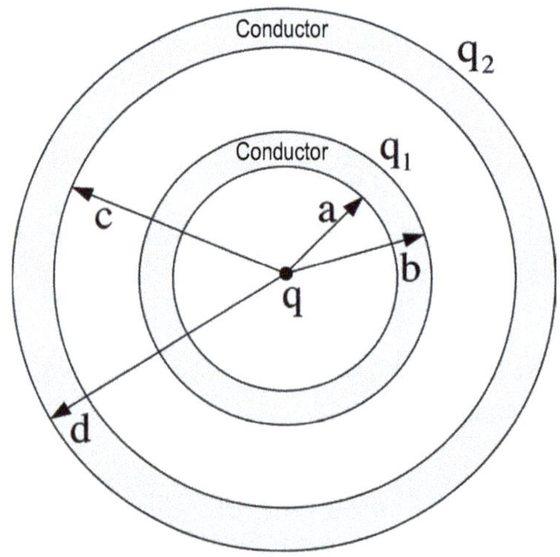

Figure 2-29. A charge q surrounded by two thick, charged, spherical conducting shells.

2-54. A uniform line charge distribution of infinite length exists along the z axis:

$$\rho_\ell = \rho_{\ell 1} \quad (\rho = 0)$$

The line charge is surrounded by a thick, perfectly conducting cylindrical shell. The shell occupies the space $a \le \rho \le b$ and carries a net charge $\rho_{\ell 2}$ per unit length.

(a) Find the surface charge densities at $\rho = a, b$.
(b) Find the electric field everywhere.
(c) Find the conditions on $\rho_{\ell 1}$, $\rho_{\ell 2}$ such that no electric fields exist beyond $\rho = b$.

2-55. Consider Problem 2-33.

(a) Find the electric field along the z axis at distant point as $z \to \infty$ or $z >> a, d$.
(b) Find the electric potential along the z axis at distant point.
(c) Show that the results above agree with those of the electric dipole in Eqs. (2-67, 68).

2-56. Sketch the equipotential surfaces (on which $V = \text{constant}$) and the electric field lines of the electric dipole, using the computer program with the expression given in Eqs. (2-67) and (2-68).

2-57.* Figure 2-17(b) shows a linear quadrupole. Show that the electric potential at distant point is

$$V(r,\theta) = \frac{qd^2}{4\pi\varepsilon_o}\left(\frac{3\cos^2\theta - 1}{r^3}\right) \quad (r \gg d)$$

[HINT] You need to use a better approximation: $\sqrt{1+x} \approx 1 + \frac{x}{2} - \frac{x^2}{8}$ $(x \ll 1)$. The quadrupole of Figure 2-17(b) may be considered as four separate point charges or as two dipoles.

2-58. Figure 2-23 shows a circular loop of radius a with uniform line charge density $\rho_{\ell o}$.

(a) Formulate an integral expression for the electric potential $V(\rho, \phi, z)$ at an arbitrary point.

(b) Find the electric potential at a distant point $\left(\sqrt{\rho^2 + z^2} \gg a\right)$.

(c)* Calculate numerically the potential $V(\rho)$ in the xy plane $(z = 0)$.

2-59. Consider a nonuniformly charged circular loop of radius a in the yz plane as shown in Figure 2-30.

(a) Calculate the electric field at the centre of the loop.
(b) Find the electric potential and the electric field at points (r, θ) in the yz plane far away from the loop, i.e., assume $r >> a$. You can treat it as an electric dipole.

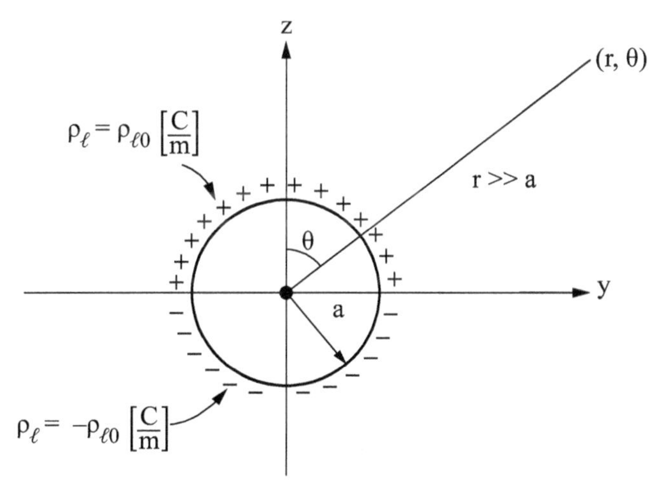

Figure 2-30. A nonuniformly charged loop.

ELECTRIC FIELDS IN DIELECTRICS

CHAPTER 3

3.1 INTRODUCTION

We turn our attention now to ideal (perfect) dielectrics. Wood, glass, chalk, plastics, rubber, paper, quartz, and distilled water are all considered close to perfect dielectrics. Ideal dielectrics do not contain *free* charges which can move from molecule to molecule. Instead, they possess *bound* charges which are tightly bound to the atomic or molecular structure.

Now what happens when an electric field is applied to a dielectric as in Figure 3-1? The bound charges are not free to move from molecule to molecule but they can move over very small distances. Positive and negative charges tend to move in opposite directions. A typical result is shown in Figure 3-1 with dipoles existing in the body of the dielectric and surface charges on two surfaces. Dipoles are lined up with their dipole moments parallel to $\mathbf{E}$. We say that the dielectric is *polarized.* We see that in Figure 3-1 bound surface charge density is formed on the right and left hand surfaces of the dielectric. Bound volume charge density may or may not exist inside the dielectric, depending on the type of polarization.

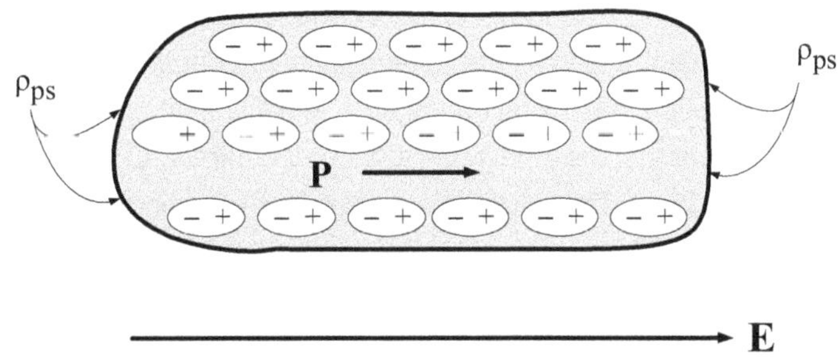

Figure 3-1. A polarized dielectric.

We distinguish bound charges from free charges by adding a subscript p. Thus bound or polarization charges may exist as bound volume charge density ρ_{pv} or as bound surface charge density ρ_{ps}. The net bound charge in a volume is represented as Q_p. The definitions for ρ_{pv}, ρ_{ps} are identical to Eqs. (2-2) except that ΔQ is replaced with ΔQ_p. Note that ρ_v, ρ_s, Q represent *free* charges; we add the subscript p for *bound* or *polarization* charge densities.

3.2. POLARIZATION

We have seen an example of a polarized dielectric with dipoles existing within the body of a dielectric. Since dipoles give rise to electric fields and potentials, we need to characterize the strength of the dipoles within the dielectric. How many dipoles are there per unit volume and what are the dipole moments? We define a **polarization vector** $\mathbf{P}(x,y,z)$ to characterize the polarized state of the dielectric material. To calculate $\mathbf{P}$ at a particular point, we construct a small volume Δv around the point in question, and add vectorially all the dipole moments $\mathbf{p}_i$ within the volume Δv. Then

$$\mathbf{P} = \lim_{\Delta V \to 0} \frac{1}{\Delta v} \sum_i \mathbf{p}_i \tag{3-1}$$

$\mathbf{P}$ is called the polarization or the *dipole moment per unit volume*. By definition it is automatically zero in vacuum (free space).

In the absence of an applied electric field, most dielectrics are unpolarized ($\mathbf{P} = 0$) either because of an orderly arrangement of dipoles whose moments cancel as in Figures 2-17(a),(b) or because of a random orientation of dipoles. The individual molecules may be polarized or not. *Nonpolar* molecules have no net dipole moment. *Polar* molecules each have a net dipole moment but a random orientation often tends to produce an unpolarized dielectric macroscopically in the absence of an electric field.

There are some dielectrics which are *permanently* polarized, i.e., they may remain polarized in the absence of an electric field. An example is barium titanate, which is one of the ferroelectrics. Such materials are called **electrets**, the electrical analogue of magnets.

If an electric field is applied, dielectrics which were previously unpolarized become polarized ($\mathbf{P} \neq 0$), as in Figure 3-1, with a polarization $\mathbf{P}$ which tends to be in the direction of the electric field $\mathbf{E}$. There are several contributors to the polarization. First, consider an unpolarized dielectric with dipoles whose moments cancel in the absence of an electric field. The cancellation may be orderly or disorderly. Figure 3-2(a) shows an orderly cancellation. The application of an electric field leads to forces on the positive and negative bound charges which make up the dipoles. We assume that the dipoles can rotate slightly about their centers. The dipoles will tend to rotate to the new positions shown dotted, each rotation resulting in a dipole moment with an increased component in the direction of $\mathbf{E}$. We also assume that the dipole can stretch or contract. Consider a dipole pair with moment parallel, antiparallel to $\mathbf{E}$ (Figure 3-2(b)). The applied field stretches one dipole and contracts another, in each case increasing the dipole moment in the direction of $\mathbf{E}$. Figures 3-2(c), (d) show the effect of an applied field on an individual atom. The nucleus is displaced with respect to the center of the electron cloud. This produces a net dipole moment in the direction of $\mathbf{E}$. Thus we see that given an unpolarized dielectric ($\mathbf{P} = 0$), the addition of an applied electric field tends to polarize the dielectric with a $\mathbf{P}$ in the direction of $\mathbf{E}$.

We should emphasize that the movements of the bound charges and the resultant deformation of the structure, i.e., the resultant rotations and stretching of dipoles, are very small indeed. The basic reason for this is that the applied fields produce forces on the bound charges which are very small compared with the strong internal forces

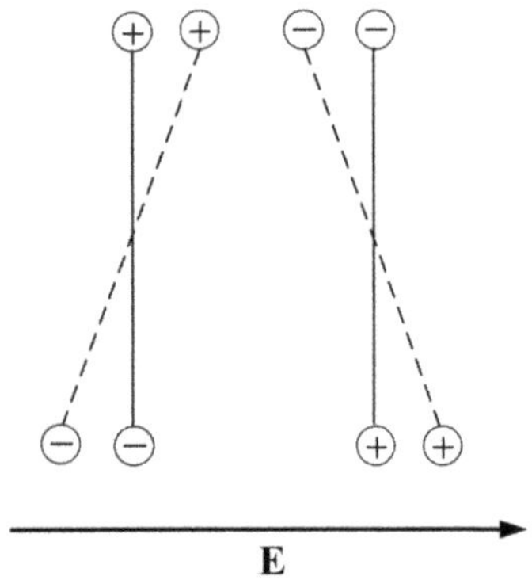

(a) Dipole rotation.

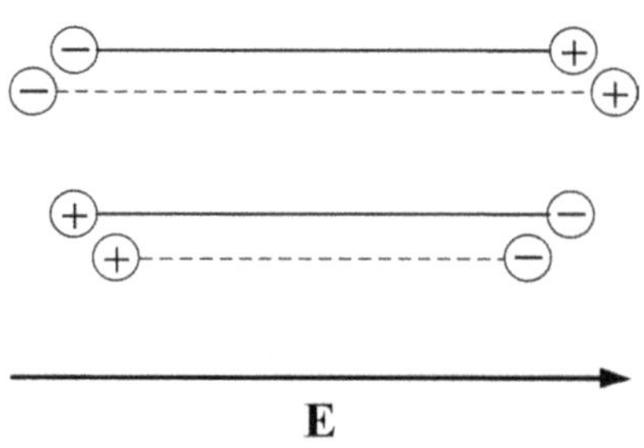

(b) Stretching or contraction of dipoles.

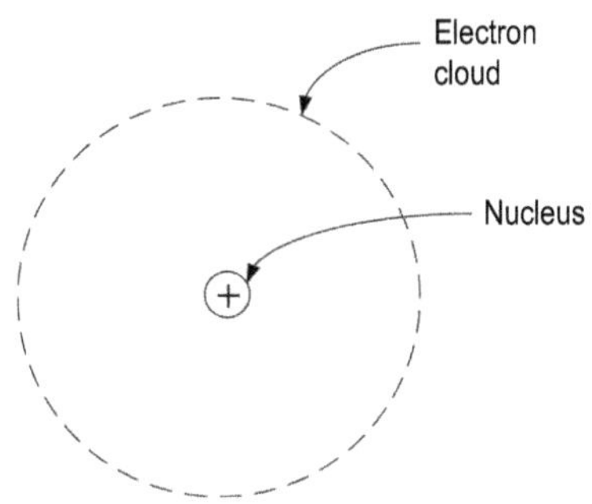

(c) An atom.

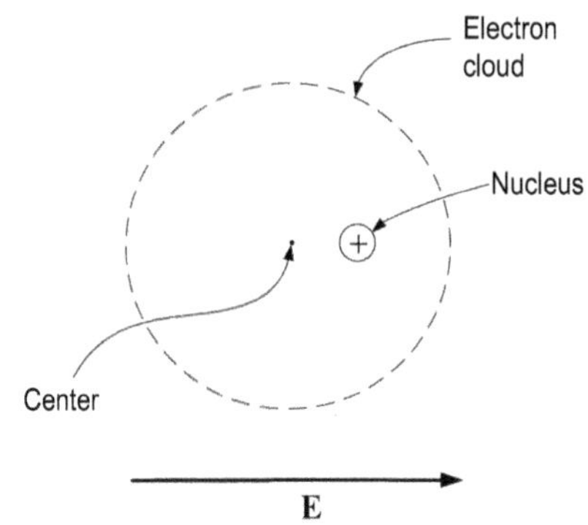

(d) An atom with applied field.

Figure 3-2. Dipoles in an electric field.

holding the structure together. The movement shown in our figures is much greater than that usually obtained. The perturbation or deformation of the structure is usually minimal and for this reason the process is often linear ($\mathbf{P} \sim \mathbf{E}$). On the other hand, if extremely large electric fields are applied, the bound charges may be torn from their molecules and *breakdown* will occur.

3.3. THE ELECTRIC FIELD OF A POLARIZED DIELECTRIC

As we have seen, a dielectric becomes polarized in the presence of an electric field. An orderly arrangement of dipoles is created with a concomitant bound surface charge and perhaps, as we will soon see, a bound volume charge as well. What are the bound electric charge densities associated with the state of polarization and what is the electric field due to those charges?

First, let us consider the charge distribution of a uniformly polarized dielectric. Figure 3-3(a) shows a block of dielectric with *uniform* polarization $\mathbf{P} = \mathbf{a}_z P_o$.

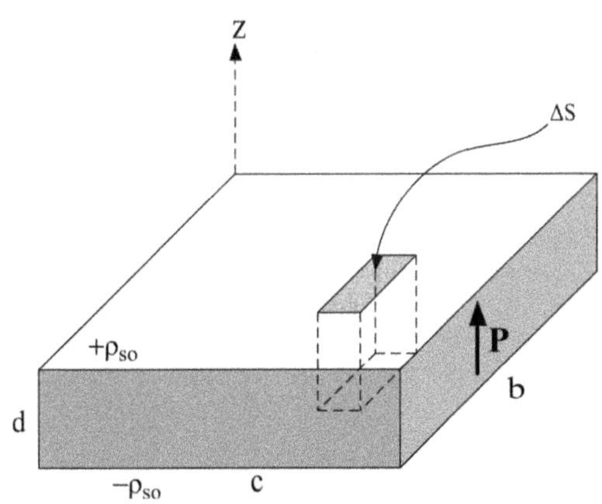

(a) A uniformly polarized block of dielectric.

Figure 3-3(a). A uniformly polarized block of dielectric.

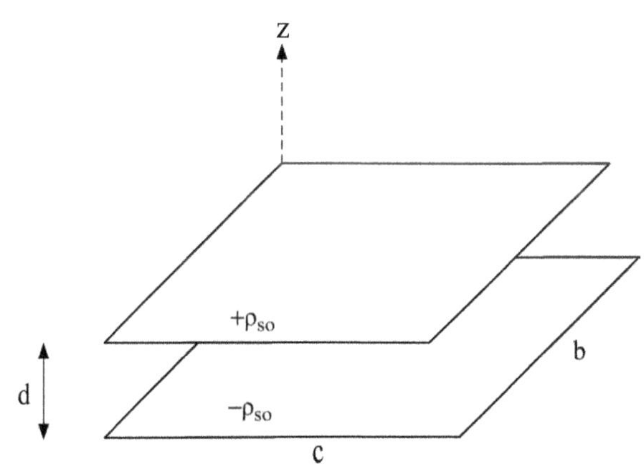

(b) An equivalent charge distribution.

Figure 3-3(b). An equivalent charge distribution.

We note that an unknown *uniform* bound surface charge density $\pm \rho_{so}$ is present on top and bottom surfaces, respectively. There is no bound volume charge density within the block since every small volume contains equal amounts of positive and negative charges, i.e., $\rho_{pv} = 0$. Since there is no volume charge density, we can represent the charges of the system as in Figure 3-3(b), which shows uniform surface charge densities $\pm \rho_{so}$ separated by a distance d. Now we will solve for ρ_{so} by requiring that the dipole moments of Figures 3-3(a), (b) be equal. We consider a small patch of surface Δs above and below and the intervening volume $d(\Delta s)$. The dipole moment of the intervening volume in Figure 3-3(a) is $\mathbf{a}_z (P_o\, d\, \Delta s)$ because P_o is the dipole moment per unit volume. The dipole moment of the same volume in Figure 3-3(b) is $\mathbf{a}_z (q\, d) = \mathbf{a}_z (\rho_{so}\, \Delta s\, d)$. The moments must be equal and therefore

$$\rho_{so} = P_o \tag{3-2a}$$

We can also compare the total dipole moments of Figures 3-3(a), (b), i.e., $\mathbf{a}_z\, P_0\, (bcd) = \mathbf{a}_z\, \rho_{so}\, (bcd)$, with the same result. The bound surface charge density ρ_{ps} of Figure 3-3(a) may be characterized as follows:

$$\begin{aligned} \rho_{ps} &= \pm P_o = \mathbf{P} \cdot (\pm \mathbf{a}_z) \text{ on top and bottom} \\ &= 0 \text{ on vertical sides} \end{aligned} \tag{3-2b}$$

or, in general,

$$\boxed{\rho_{ps} = \mathbf{P} \cdot \mathbf{a}_n} \tag{3-2c}$$

where $\mathbf{a}_n$ is the outward unit normal to the dielectric surface.

It is perhaps surprising that we can throw away all the dipoles in the body of the dielectric since the individual dipoles do produce electric fields. This is essentially the same process as replacing two dipoles laid end-to-end with the two end charges. The dipole moment doubles and two charges in the middle cancel.

Now let us consider what happens if $\mathbf{P}$ is *not uniform* but varies in one dimension. Assume that $\mathbf{P}$ is z-directed, $\mathbf{P} = \mathbf{a}_z P_z$, and let P_z be a continuous function of z as shown in Figure 3-4.

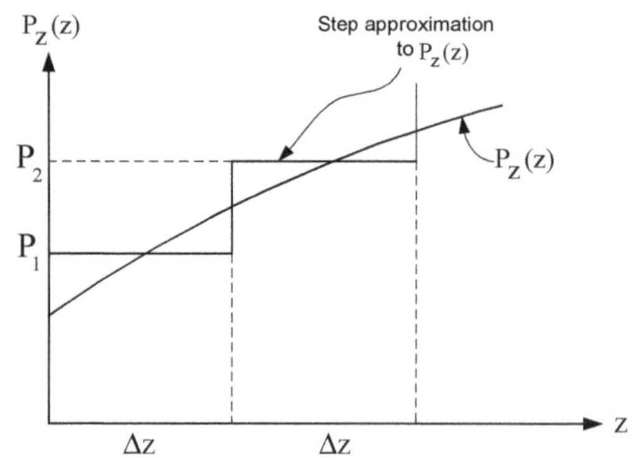

Figure 3-4. A non-uniform polarization $P_z(z)$.

We approximate the continuous function with a series of steps, each of length Δz. We then represent each step by a model such as that of Figure 3-3 with a uniform $\mathbf{P}$ in each step and bound surface charges at the beginning and end of each step. The resulting bound charges are shown in Figure 3-5.

The first two regions are represented by two slabs of uniformly polarized material with polarization P_1 and P_2, respectively. Since the polarization of the two slabs differ, there is a net surface charge density $P_1 - P_2$ which is equal in magnitude to the step discontinuity at $z = \Delta z$. Assume a surface of cross section area of Δs at $z = \Delta z$. Then the total bound surface charge at $z = \Delta z$ is given as follows:

$$\Delta Q_p = \rho_{ps}\ \Delta s = (P_1 - P_2)\,\Delta s = \frac{(-\Delta P)}{\Delta z}\Delta z\,\Delta s = -\frac{\partial P_z}{\partial z}\Delta z\,\Delta s$$

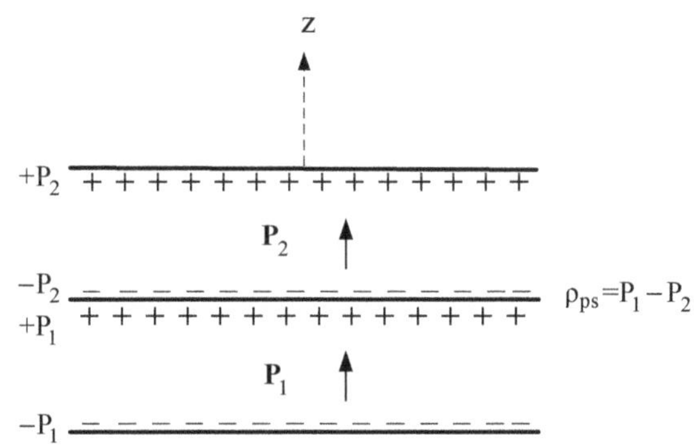

Figure 3-5. The charge distribution for the step approximation to the non-uniform polarization $P_z(z)$.

Now let's spread the excess bound surface charge ΔQ_p over the region (one half subsection (or $\frac{\Delta z}{2}$) to the right and one half subsection to the left) to obtain a representation of the bound volume charge density ρ_{pv}. This is equivalent to taking smaller and smaller steps in the approximation of the continuous function $P_z(z)$. Then $\Delta Q_p = \rho_{pv}\,\Delta s\ \Delta z = -\frac{\partial P_z}{\partial z}\ \Delta s\ \Delta z$ and

$$\rho_{pv} = -\frac{\partial P_z}{\partial z}$$

If we consider additional components of $\mathbf{P}$, then

$$\boxed{\rho_{pv} = -\frac{\partial P_x}{\partial x} - \frac{\partial P_y}{\partial y} - \frac{\partial P_z}{\partial z} = -\nabla \cdot \mathbf{P}} \qquad (3\text{-}3)$$

The result above corresponds to the limit $\Delta z \to 0$. As $\Delta z \to 0$, we obtain an infinite number of steps in the approximation and the surface charge becomes volume charge density.

Table 3-1 gives the equivalent bound charge densities for polarized material. These are *real* bound charges that exist in polarized material. They are *equivalent* in the sense that we can obtain the electric field $\mathbf{E}$ by assuming those charges in free space (so that the formulations of Chapter 2 apply). The bound charges are also called the *polarization charges.*

Table 3-1 Equivalent Bound Charge Densities

Surface Charge [C/m^2]	Volume Charge [C/m^3]
$\rho_{ps} = \mathbf{P} \cdot \mathbf{a}_n$	$\rho_{pv} = -\nabla \cdot \mathbf{P}$

Equations (3-2) and (3-3) above specify the bound charge densities which are associated with the polarization vector $\mathbf{P}$. Note that $\mathbf{P}$ in Eq. (3-2) should be evaluated *at the surface* and $\mathbf{a}_n$ is a unit vector normal *outward* from the dielectric surface. The negative sign in Eq. (3-3) is due to the previous choice that dipole moment $\mathbf{p}$ is directed from negative to positive charges. Eq. (3-2) physically means that the strength of polarization $\mathbf{P}$ determines the bound surface charge density and Eq. (3-3) indicates that the spatial rate of change of $\mathbf{P}$ (i.e., $-\nabla \cdot \mathbf{P}$) determines the volume bound charge density. We note that for uniform polarization $\mathbf{P}$ there is no volume charge:

$$\rho_{pv} = 0 \ \ (\text{if } \mathbf{P} \text{ is constant}) \tag{3-4}$$

Later we will see that uniform $\mathbf{P}$ is sufficient but not *necessary,* in order that $\rho_{pv} = 0$. We will see that $\rho_{pv} = 0$ if the dielectric is linear and homogeneous.

Given polarization $\mathbf{P}$, the bound volume and surface charge densities are known. We can use this information to obtain the electric potential V and the electric field $\mathbf{E}$ merely by substituting ρ_{pv}, ρ_{ps} for ρ_v, ρ_s, respectively, in Eqs. (2-46) and (2-27), to obtain

$$V = \frac{1}{4\pi\varepsilon_0} \iiint \frac{\rho_{pv}\, dv'}{|\mathbf{r}-\mathbf{r}'|} \ \text{or} \ \frac{1}{4\pi\varepsilon_0} \iiint \frac{\rho_{pv}\, dv'}{R} \tag{3-5a}$$

for volume charge

$$V = \frac{1}{4\pi\varepsilon_0} \iint \frac{\rho_{ps}\, ds'}{|\mathbf{r}-\mathbf{r}'|} \ \text{or} \ \frac{1}{4\pi\varepsilon_0} \iint \frac{\rho_{ps}\, ds'}{R} \tag{3-5b}$$

for surface charge

and

$$\mathbf{E} = \frac{1}{4\pi\varepsilon_0} \iiint \frac{\rho_{pv}\,(\mathbf{r}-\mathbf{r}')\, dv'}{|\mathbf{r}-\mathbf{r}'|^3} \ \text{or} \ \frac{1}{4\pi\varepsilon_0} \iiint \frac{\rho_{pv}\, \mathbf{R}\, dv'}{R^3} \tag{3-6a}$$

for volume charge

$$\mathbf{E} = \frac{1}{4\pi\varepsilon_0} \iint \frac{\rho_{ps}(\mathbf{r}-\mathbf{r}')ds'}{|\mathbf{r}-\mathbf{r}'|^3}$$

$$\text{or } \frac{1}{4\pi\varepsilon_0} \iint \frac{\rho_{ps}\mathbf{R}\, ds'}{R^3} \qquad (3\text{-}6b)$$

for surface charge

Note that Eqs. (2-46) and (2-27) can now be interpreted in a more general light. A charge is a charge; *bound and free charges contribute identically to V and $\boldsymbol{E}$. Thus we can interpret ρ_v, ρ_s in Eqs. (2-46) and (2-27) as free charge densities, bound charge densities, or total (bound plus free) charge densities.*

EXAMPLE 3-1. ON-AXIS POTENTIAL OF A PERMANENTLY POLARIZED CYLINDER (ELECTRET)

Consider a uniformly polarized cylinder (Figure 3-6):

$$\mathbf{P} = \mathbf{a}_z P_o \begin{Bmatrix} 0 < \rho < a \\ z_1 \le z \le z_2 \end{Bmatrix}$$

Find the electrostatic potential along the z axis.

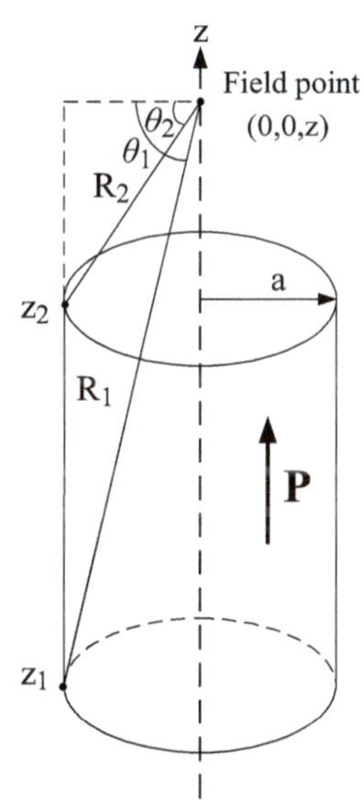

Figure 3-6. A permanently polarized dielectric cylinder (electret).

SOLUTION

First we find the bound charge densities.

$$\rho_{ps} = \mathbf{P} \cdot \mathbf{a}_n = P_o \text{ (upper disk, } z = z_2)$$

$$= -P_o \text{ (lower disk, } z = z_1)$$

We can use directly the results from Example 2-9. Substituting $(z—z_2)$ for z everywhere in Eq. (2-49), we obtain the following contribution to the potential from the upper disk ($\rho_{ps} = P_o$):

$$\frac{P_o}{2\varepsilon_o}\left[\sqrt{(z-z_2)^2 + a^2} - |z-z_2|\right]$$

Adding a similar contribution from the lower disk ($\rho_{ps} = -P_o$), we obtain

$$V(z) = \frac{P_o}{2\varepsilon_o}\left[\sqrt{(z-z_2)^2 + a^2} - \sqrt{(z-z_1)^2 + a^2} - |z-z_2| + |z-z_1|\right] \tag{3-7}$$

which can also be expressed as

$$V(z) = \frac{P_o}{2\varepsilon_o}\left[R_2 - R_1 - R_2|\sin\theta_2| + R_1|\sin\theta_1|\right] \tag{3-8}$$

Can you establish the following identity?

$$-|z-z_2| + |z-z_1| = z_2 - z_1 \quad (z \geq z_2)$$
$$= z_1 - z_2 \quad (z \leq z_1)$$
$$= 2z - z_2 - z_1 \quad (z_1 \leq z \leq z_2)$$

Equation (3-7) can also be obtained by using Eq. (3-5b) for both upper and lower surfaces ($z = z_1, z_2$). The potential $V(z)$ of Eq. (3-7) is continuous along the z axis. The electric field E_z jumps (is discontinuous) at $z = z_1$, z_2 because of the bound surface charges on the upper, lower surfaces of the cylinder.

The on-axis electric field E_z may be obtained in several ways: (a) by using $\mathbf{E} = -\nabla V$, (b) by direct application of Eq. (3-6b), (c) by using the results of Example 2-7 for both upper and lower discs at $z = z_1, z_2$.

3.4. THE DISPLACEMENT VECTOR D

As we have noted before, free and bound charges are identical in their effect on V, **E**. Therefore, in the presence of dielectrics, Gauss' law, the second basic law, should be modified. We replace ρ_v with $\rho_v + \rho_{pv}$ in Eq. (2-12) (Gauss' law) to obtain

$$\nabla \cdot \mathbf{E} = \frac{\rho_v + \rho_{pv}}{\varepsilon_0} \tag{3-9a}$$

or

$$\nabla \cdot \varepsilon_0 \mathbf{E} = \rho_v + \rho_{pv} \tag{3-9b}$$

and note that

$$\nabla \cdot \mathbf{P} = -\rho_{pv} \tag{3-3}$$

Equation (3-9) indicates that **E** lines begin and end on free or bound charges, going from positive to negative charges. Equation (3-3) indicates that **P** lines begin and end on bound charges, going from negative to positive charges. Adding Eqs. (3-9b) and (3-3),

$$\nabla \cdot (\varepsilon_0 \mathbf{E} + \mathbf{P}) = \rho_v \tag{3-10a}$$

The quantity $\varepsilon_o \mathbf{E} + \mathbf{P}$ is defined as the **electric displacement** or **displacement vector D**:

$$\boxed{\mathbf{D} = \varepsilon_0 \mathbf{E} + \mathbf{P}} \tag{3-11}$$

Then Eq. (3-10a) becomes

$$\boxed{\nabla \cdot \mathbf{D} = \rho_v} \tag{3-10b}$$

D is thus a vector whose lines begin and end on free charges. We obtain the integral form of Eq. (3-10b) by integrating both sides over volume V and applying the divergence theorem:

$$\boxed{\oint\!\!\oint_S \mathbf{D} \cdot d\mathbf{s} = Q_f} \tag{3-12}$$

where Q_f is the *free* charge enclosed within the surface S. Equation (3-12) should be recognized as a new form of Gauss' law which is valid even in the presence of the dielectric. Equation (3-12) is more useful since we often know the free charges but not the bound charges. Without knowing bound charges, we can determine **D** first. Table 3-2 summarizes the two forms of Gauss' law for **D** and **E**. Note that

$$Q_f = \iiint_V \rho_v \, dv \quad \text{(total free charge enclosed)}$$

$$Q_p = \iiint_V \rho_{pv} \, dv \quad \text{(total bound charge enclosed)}$$

Table 3-2 Gauss' Law for Electric Fields in Matter

Point Form	Integral Form
$\nabla \cdot \mathbf{D} = \rho_v$	$\oint\!\!\oint_S \mathbf{D} \cdot d\mathbf{s} = Q_f$
$\nabla \cdot \mathbf{E} = \frac{1}{\varepsilon_0}(\rho_v + \rho_{pv})$	$\oint\!\!\oint_S \mathbf{E} \cdot d\mathbf{s} = \frac{Q_f + Q_p}{\varepsilon_0}$

Linear Dielectrics

Many dielectrics are highly *linear*, thus we can assume that **P** is proportional to **E**.

$$\mathbf{P} = \varepsilon_0 \chi_e \mathbf{E} \tag{3-13}$$

Then

$$\mathbf{D} = \varepsilon_0 \mathbf{E} + \mathbf{P} = \varepsilon_0 (1 + \chi_e) \mathbf{E} = \varepsilon_0 \varepsilon_r \mathbf{E} = \varepsilon \mathbf{E} \tag{3-14}$$

χ_e, ε_r, ε are called **electric susceptibility**, relative permittivity (or **dielectric constant**), **permittivity**, respectively. They represent three different ways of specifying the same linear relationship. We will usually specify the permittivity ε. The three parameters (χ_e, ε_r, ε) are not functions of **E** for linear dielectrics but may be functions of position (x,y,z). For *homogeneous* media they are constants. Thus $\varepsilon(x,y,z)$ represents a linear, *inhomogeneous* medium, and ε = constant represents a linear, homogeneous medium.

For a linear dielectric medium,

$$\boxed{\begin{aligned} \mathbf{D} &= \varepsilon \mathbf{E} \\ \mathbf{P} &= (\varepsilon - \varepsilon_0)\mathbf{E} \end{aligned}} \quad \text{Linear dielectric medium} \tag{3-15}$$

and both **D** and **P** are known once **E** is known. Equation (3-15) is known as a **constitutive relation**. Table 3-3 shows some typical values of dielectric constant ε_r. Most of common materials has small dielectric constant except water which has a very high dielectric constant of about 80.

Table 3-3 Dielectric Constant

Material	Dielectric Constant
Air	1.0006
Balsa Wood	1.4
Paper	2 – 4
Teflon	2.1
Polyethylene	2.25
Polystyrene	2.55
Dry Soil	2.6
Glass	4 – 10
Quartz	4.3
Mica	6
Silicon	11.7
Alcohol	28
Water	80
Rutile (TiO_2)	100
$BaTiO_3$	500 – 4500

Linear, Homogeneous Dielectrics

If a medium is both linear and homogeneous, then $\mathbf{D} = \varepsilon\,\mathbf{E}$ and ε is a constant. Thus

$$\nabla \cdot \mathbf{D} = \nabla \cdot (\varepsilon \mathbf{E}) = \varepsilon \nabla \cdot \mathbf{E} + \underbrace{\nabla \varepsilon}_{0} \cdot \mathbf{E} = \rho_v$$

(see Eq. (1-43)). When ε is a constant, $\nabla\varepsilon = 0$.

Then

$$\nabla \cdot \mathbf{E} = \frac{\rho_v}{\varepsilon} = \frac{\rho_v + \rho_{pv}}{\varepsilon_0}$$

and

$$\boxed{\rho_{pv} = -\left(\frac{\varepsilon - \varepsilon_0}{\varepsilon}\right)\rho_v} \qquad \text{Linear homogeneous dielectric} \qquad (3\text{-}16)$$

Also, since an ideal (perfect) dielectric has no free charge ($\rho_v = 0$),

$$\boxed{\rho_{pv} = 0} \qquad \text{Linear, homogeneous, perfect dielectric} \qquad (3\text{-}17)$$

EXAMPLE 3-2. A POINT CHARGE AND A DIELECTRIC SHELL

Figure 3-7 shows a point charge q surrounded by a dielectric shell of radii a, b and permittivity ε_1. Find the electric field in each region and the bound surface charges at $r = a$, b.

SOLUTION

Because of spherical symmetry, we expect that **E** and **D** will be radial.

$$\mathbf{E} = \mathbf{a}_r E_r , \mathbf{D} = \mathbf{a}_r D_r$$

Since we don't know the bound charges, we apply Gauss' law for **D** in Eq. (3-12) for the spherical surface of radius r.

Regardless of where r is chosen,

$$\oiint \mathbf{D} \cdot d\mathbf{s} = 4\pi r^2 D_r = Q_f = q$$

$$D_r = \frac{q}{4\pi r^2} \quad \text{everywhere} \qquad (3\text{-}18)$$

$$(r < a,\ a < r < b, \text{ and } r > b)$$

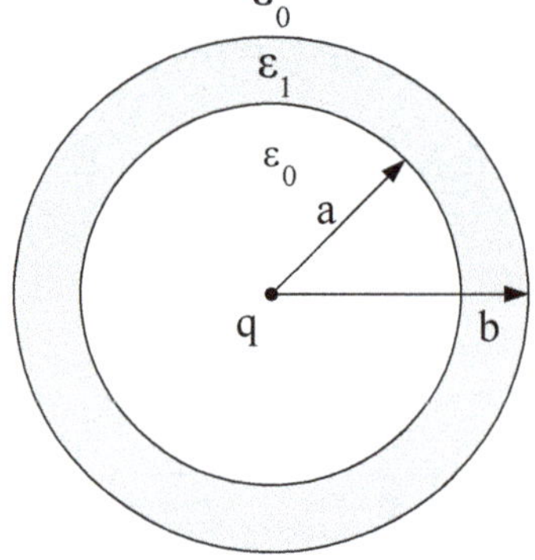

Figure 3-7. A point charge and a spherical dielectric shell.

Since $\mathbf{D} = \varepsilon \mathbf{E}$ in linear dielectric medium,

$$E_r = \frac{D_r}{\varepsilon_1} = \frac{q}{4\pi\varepsilon_1 r^2} \quad (a < r < b) \text{ in the dielectric shell}$$

$$E_r = \frac{D_r}{\varepsilon_0} = \frac{q}{4\pi\varepsilon_0 r^2} \quad (r < a,\ r > b) \text{ in air}$$

The polarization (vector) **P** in the dielectric shell is given by

$$\mathbf{P} = (\varepsilon_1 - \varepsilon_0)\mathbf{E} = \mathbf{a}_r \frac{(\varepsilon_1 - \varepsilon_0)q}{4\pi\varepsilon_1 r^2} \quad (a < r < b) \qquad (3\text{-}19a)$$

Bound charges are:

$$\rho_{pv} = -\nabla \cdot \mathbf{P} = -\frac{1}{r^2}\frac{\partial}{\partial r}\left(r^2 P_r\right) = -\frac{1}{r^2}\frac{\partial}{\partial r}\left\{\frac{(\varepsilon_1 - \varepsilon_0)q}{4\pi\varepsilon_1}\right\} = 0$$

$$\rho_{ps} = \mathbf{P} \cdot \mathbf{a}_n = -\left[\frac{\varepsilon_1 - \varepsilon_0}{\varepsilon_1}\right]\frac{q}{4\pi a^2} \text{ at } r = a \ [\mathbf{a}_n = -\mathbf{a}_r] \qquad (3\text{-}19b)$$

$$= \left[\frac{\varepsilon_1 - \varepsilon_0}{\varepsilon_1}\right]\frac{q}{4\pi b^2} \text{ at } r = b \ [\mathbf{a}_n = \mathbf{a}_r]$$

Note that the electric charge q has drawn up a negative bound charge at $r = a$ and left a positive bound charge at $r = b$. This bound charge is related to free charge (that would be present if there were a conducting shell) by the ratio indicated in Eq. (3-16). It also weakens the electric field within the dielectric. Note also that although $\mathbf{P}$ is not uniform, $\rho_{pv} = 0$, which confirms Eq. (3-17).

EXAMPLE 3-3. A DIELECTRIC-LOADED COAXIAL TRANSMISSION LINE

Figure 3-8 shows a coaxial line of inner radius a, outer radius b, which is loaded with a dielectric of permittivity ε_1. A voltage V_o is applied between inner and outer conductors. The inner conductor is the positive reference. Find the electrostatic fields and the bound and free charge densities.

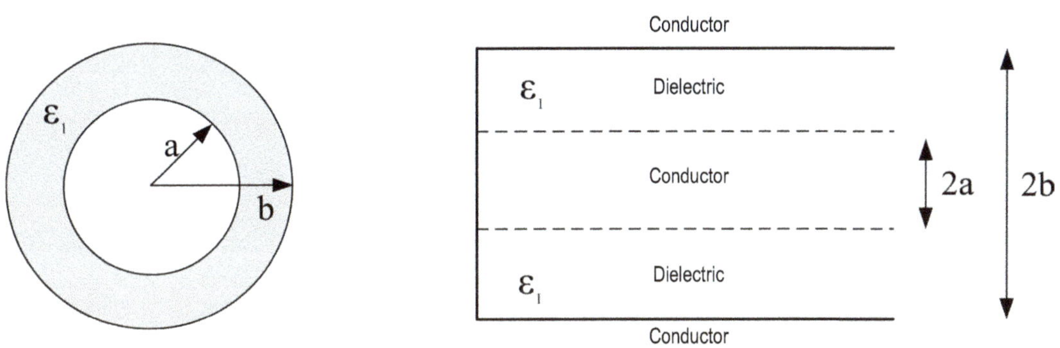

Figure 3-8. A dielectric-loaded coaxial transmission line.

SOLUTIONS:

First we assume an unknown free charge ρ_ℓ per unit length at $\rho = a$, which is drawn up by the application of voltage V_o. Then $-\rho_\ell$ per unit length appears at $\rho = b$. Assuming that the coaxial line is infinitely long, the fields will be radial (ρ-directed) because of cylindrical symmetry. $\mathbf{E} = \mathbf{a}_\rho E_\rho$, $\mathbf{D} = \mathbf{a}_\rho D_\rho$.

Applying Gauss' law for the cylinder of radius ρ and length ℓ,

$$\oiint \mathbf{D} \cdot d\mathbf{s} = 2\pi\rho\, \ell\, D_\rho = Q_f = \rho_\ell \ell$$

$$D_\rho = \frac{\rho_\ell}{2\pi}\left(\frac{1}{\rho}\right), \quad E_\rho = \frac{D_\rho}{\varepsilon_1} = \frac{\rho_\ell}{2\pi\varepsilon_1}\left(\frac{1}{\rho}\right)$$

To determine the (linear) relationship between applied voltage V_o and the resultant free charge ρ_ℓ per unit length:

$$V_o = -\int \mathbf{E}\cdot d\ell = -\int_b^a E_\rho \, d\rho = -\frac{\rho_\ell}{2\pi\varepsilon_1}\int_b^a \frac{1}{\rho} d\rho = \frac{\rho_\ell}{2\pi\varepsilon_1}\ln\left(\frac{b}{a}\right)$$

and

$$\rho_\ell = \frac{2\pi\varepsilon_1 V_o}{\ln\left(\frac{b}{a}\right)}$$

The free surface charge density is

$$\rho_s = \frac{\rho_\ell}{2\pi a} = \frac{\varepsilon_1 V_o}{a \ln\left(\frac{b}{a}\right)} \quad \text{at } \rho = a \tag{3-20a}$$

$$= \frac{-\rho_\ell}{2\pi b} = -\frac{\varepsilon_1 V_o}{b \ln\left(\frac{b}{a}\right)} \quad \text{at } \rho = b \tag{3-20b}$$

The electric field and the polarization are

$$\mathbf{E} = \mathbf{a}_\rho \frac{\rho_\ell}{2\pi\varepsilon_1}\left(\frac{1}{\rho}\right) = \mathbf{a}_\rho \frac{V_o}{\ln\left(\frac{b}{a}\right)}\left(\frac{1}{\rho}\right) \tag{3-21a}$$

$$\mathbf{P} = (\varepsilon_1 - \varepsilon_0)\,\mathbf{E} = \mathbf{a}_\rho \frac{(\varepsilon_1 - \varepsilon_0)\, V_o}{\ln\left(\frac{b}{a}\right)}\left(\frac{1}{\rho}\right) \tag{3-21b}$$

The bound surface charge density is

$$\rho_{ps} = \mathbf{P}\cdot\mathbf{a}_n = -\frac{(\varepsilon_1 - \varepsilon_0)\, V_o}{a \ \mathrm{In}\left(\frac{b}{a}\right)} \quad (\rho = a) \tag{3-22a}$$

$$= \frac{(\varepsilon_1 - \varepsilon_0)\, V_o}{b \ \ln\left(\frac{b}{a}\right)} \quad (\rho = b) \tag{3-22b}$$

Note that the bound and free surface charge densities at both $r = a$ and $r = b$ are related by

$$\rho_{ps} = -\left(\frac{\varepsilon_1 - \varepsilon_0}{\varepsilon_1}\right)\rho_s \tag{3-23}$$

as in Eq. (3-16).

3.5. BOUNDARY CONDITIONS

We now turn to problems involving two or more media. To treat these problems, we need to understand how the vectors $\mathbf{E}$, $\mathbf{D}$ behave at an interface. First, we analyze the tangential component of the electric field $\mathbf{E}$. Consider a particular point P lying on the interface between two arbitrary media 1, 2 (Figure 3-9).

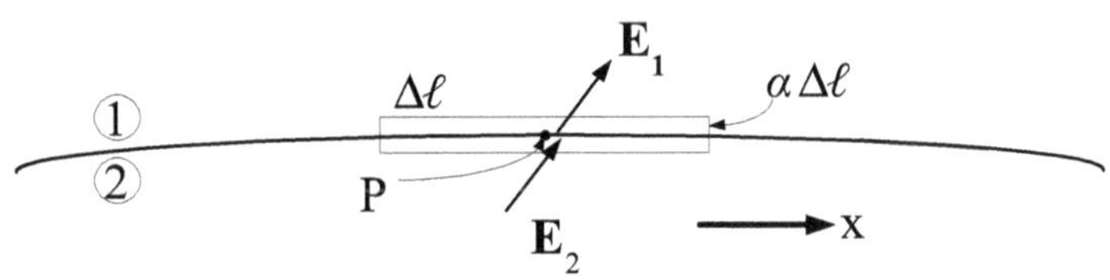

Figure 3-9. Tangential $\mathbf{E}$ at a boundary.

Let x, y be rectangular coordinates of a point P located in the tangent plane to the interface between media 1 and 2. We construct a closed contour C around P as shown in Figure 3-9 and consider the line integral

$$\oint_C \mathbf{E}\cdot d\boldsymbol{\ell}$$

The long side $\Delta\ell$ is parallel to the x direction; $\Delta\ell$ is made small so that $\mathbf{E}$ does not vary over the length. α is made arbitrarily small compared to unity so that the short side does not contribute to the line integral. $\mathbf{E}_1$, $\mathbf{E}_2$ are the electric fields at the interface in regions 1, 2, respectively. Then from the first law, of electrostatics, Eq. (2-11),

$$\oint_C \mathbf{E}\cdot d\boldsymbol{\ell} = (E_{1x} - E_{2x})\,\Delta\ell = 0$$

and $E_{1x} = E_{2x}$; similarly, $E_{1y} = E_{2y}$, therefore

$$\boxed{E_{1t} = E_{2t}} \tag{3-24}$$

i.e., the tangential component of the electric field is continuous across the boundary.

Next we analyze the normal component of the displacement $\mathbf{D}$. Consider Figure 3-10 which shows a small closed surface S about the point P on the interface, with top and bottom (horizontal) surfaces of area Δs parallel to the tangent plane and vertical side of area $\alpha\Delta s$ normal to the tangent plane. Then we apply the second law of electrostatics for matter, Eq. (3-12).

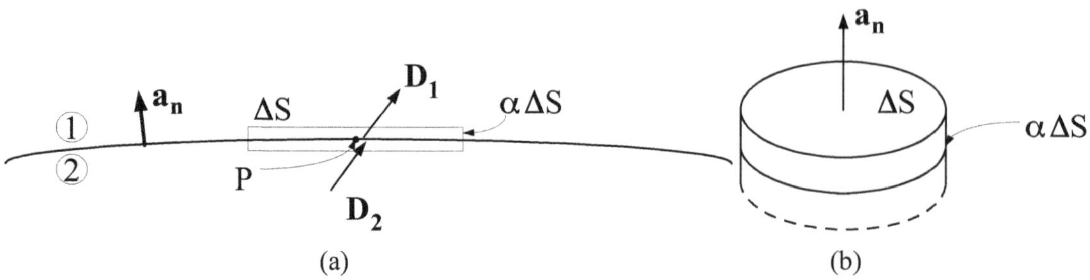

Figure 3-10. (a) Normal D at a boundary (side view).
(b) Full view of surface S.

We consider the integral $\oiint_S \mathbf{D} \cdot d\mathbf{s}$. Δs is made small so that $\mathbf{D}$ does not vary over S and α is made small compared to unity so that the flux through the vertical side does not contribute to $\oiint_S \mathbf{D} \cdot d\mathbf{s}$. A direction $\mathbf{a}_n$ normal to the interface is arbitrarily chosen pointing into region 1. Then if there exists a free surface charge ρ_s at the boundary, Gauss' law, Eq. (3-12), leads to

$$\oiint_S \mathbf{D} \cdot d\mathbf{s} = (D_{1n} - D_{2n})\ \Delta s = Q_f = \rho_s\ \Delta s$$

$$\boxed{D_{1n} - D_{2n} = \rho_s} \qquad (3\text{-}25)$$

Normal $\mathbf{D}$ is discontinuous by an amount equal to the free surface charge density. The polarities can be checked by noting that a normal vector $\mathbf{D}$, upon encountering a positive charge, increases. Note that the normal direction ($\mathbf{a}_n$) was chosen pointing into region 1; Eq. (3-25) will change if the normal is chosen pointing into region 2.

Because we have used the two basic laws $\oint_C \mathbf{E} \cdot d\boldsymbol{\ell} = 0$ and $\oiint_S \mathbf{D} \cdot d\mathbf{s} = Q_f$, which are valid for any electrostatic problem, the boundary conditions are also valid for any electrostatic problem. They hold for any media whatsoever, i.e., dielectrics (perfect and imperfect) and vacuum and any combinations thereof. There are thus many special cases for regions 1, 2. Table 3-4 outlines a few of the most important cases.

Note that for arbitrary media, the jump (discontinuity) in normal **D, E, P** is proportional to ρ_s, $\rho_s + \rho_{ps}$, $-\rho_{ps}$ respectively. At the interface between two ideal dielectrics, normal $\mathbf{D}$ is continuous.

$$D_{1n} = D_{2n} \quad \text{(Dielectric-dielectric interface)} \qquad (3\text{-}25a)$$

because there is no free charge at the dielectric interface ($\rho_s = 0$). At the interface between the conductor (region #2) and the dielectric (region #1), the boundary conditions are

$$E_{1t} = 0, \quad D_{1n} = \rho_s \quad \text{(Dielectric-conductor interface)} \qquad (3\text{-}25b)$$

The boundary conditions at an air-conductor interface, Eqs. (2-62) and (2-63), can be obtained by setting $\varepsilon_1 = \varepsilon_0$ in the last case of Table 3-4.

Table 3-4 Boundary Conditions

Case	Boundary conditions
Arbitrary Media	$E_{1t} = E_{2t}, \quad D_{1n} - D_{2n} = \rho_s$ $\varepsilon_0 (E_{1n} - E_{2n}) = \rho_s + \rho_{ps}$ $P_{1n} - P_{2n} = -\rho_{ps}$
Linear, Ideal Dielectrics ($\rho_s = 0$)	$E_{1t} = E_{2t} \qquad D_{1n} = D_{2n}$ $D_{1t} = D_{2t}\left(\frac{\varepsilon_1}{\varepsilon_2}\right) \qquad E_{1n} = E_{2n}\left(\frac{\varepsilon_2}{\varepsilon_1}\right)$ $P_{1t} = P_{2t}\left(\frac{\varepsilon_1 - \varepsilon_0}{\varepsilon_2 - \varepsilon_0}\right) \qquad P_{1n} = P_{2n}\left(\frac{\varepsilon_1 - \varepsilon_0}{\varepsilon_2 - \varepsilon_0}\right)\frac{\varepsilon_2}{\varepsilon_1}$
Conductor, Dielectric (Region #2 is a conductor)	$\mathbf{E}_2 = \mathbf{D}_2 = \mathbf{P}_2 = 0; \quad \varepsilon_0 E_{1n} = (\rho_s + \rho_{ps}), \; E_{1t} = 0$ $D_{1n} = \rho_s$ $P_{1n} = -\rho_{ps}$
Conductor, Linear Dielectric (Region #2 is a conductor and Region #1 has permittivity ε_1)	$\mathbf{E}_2 = \mathbf{D}_2 = \mathbf{P}_2 = 0; \quad E_{1n} = \frac{\rho_s}{\varepsilon_1}, \; E_{1t} = 0$ $D_{1n} = \rho_s$ $P_{1n} = \rho_s\left(\frac{\varepsilon_1 - \varepsilon_0}{\varepsilon_1}\right)$

EXAMPLE 3-4. FIELDS IN A PARALLEL-PLATE REGION

Consider the parallel-plate region with a homogeneous dielectric and air gaps, as shown in Figure 3-11. Assume that the upper and lower conducting plates carry the *free* surface charges ρ_s and $-\rho_s$. The linear dielectric has permittivity ε_1. Find the electric field $\mathbf{E}$, the polarization vector $\mathbf{P}$, and the displacement vector $\mathbf{D}$ in each region (air gaps and dielectric), using the boundary conditions. Assume that the plates are very large such that the fields between the plates are approximately *uniform*.

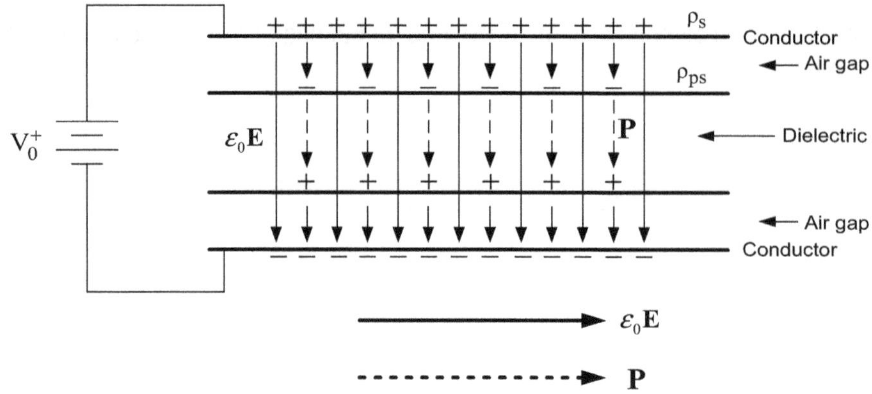

Figure 3-11. Field lines in a dielectric-loaded parallel-plate region with air gaps.

SOLUTION:

We first recognize that inside the dielectric medium there exists the polarization $\mathbf{P}$ and it leads to the bound surface charges (ρ_{ps}) with opposite polarity. Figure 3-11 shows the free and bound surface charges at each boundary and the field lines for $\varepsilon_0\mathbf{E}$ and $\mathbf{P}$ in a parallel plate region with a homogeneous dielectric and air gaps. The fringing fields near the edges are neglected. The electric field $\mathbf{E}$ is stronger in the air gap, due to the fact that some $\mathbf{E}$ lines terminate on bound charge. Note that the sum of $\varepsilon_0\mathbf{E} + \mathbf{P}$ equals $\mathbf{D}$ and that the $\mathbf{D}$ lines begin and end only on free charge. Now we obtain the fields using the boundary conditions.

(i) Upper Air Region:

Let the fields in the air region be $\mathbf{E}_1$, $\mathbf{P}_1$, $\mathbf{D}_1$ and the fields in the upper conductor be $\mathbf{E}_2$, $\mathbf{P}_2$, $\mathbf{D}_2$. The boundary conditions at the air-conductor boundary are, from Table 3-4 or Eq. (3-25b),

$$\mathbf{E}_2 = \mathbf{D}_2 = \mathbf{P}_2 = 0; \quad E_{1n} = \frac{\rho_s}{\varepsilon_0}, \ E_{1t} = 0$$

Therefore, the fields in the upper air region are

$$\mathbf{E} = \mathbf{a}_n \frac{\rho_s}{\varepsilon_0}, \ \mathbf{P} = 0 \text{ (air)}, \ \mathbf{D} = \varepsilon_0\mathbf{E} + \mathbf{P} = \mathbf{a}_n \rho_s$$

where $\mathbf{a}_n$ is a unit normal vector pointing downward.

(ii) Dielectric Region:

Let the fields in the air region be $\mathbf{E}_1$, $\mathbf{P}_1$, $\mathbf{D}_1$ and the fields in the dielectric be $\mathbf{E}_2$, $\mathbf{P}_2$, $\mathbf{D}_2$. The boundary conditions at the air-dielectric boundary are, from Table 3-4 or Eq. (3-25a),

$$E_{1t} = E_{2t}, \quad D_{1n} = D_{2n} \quad (\text{because } \rho_s = 0)$$

Since $E_{1t} = 0$, E_{2t} must be zero. Using $D_{1n} = \varepsilon_0 E_{1n}$ and $D_{2n} = \varepsilon_1 E_{2n}$, we obtain the following fields in the dielectric:

$$\mathbf{E} = \mathbf{a}_n \frac{\rho_s}{\varepsilon_1}, \quad \mathbf{D} = \mathbf{a}_n \rho_s, \quad \mathbf{P} = \mathbf{D} - \varepsilon_0 \mathbf{E} = \mathbf{a}_n \rho_s \left(1 - \frac{\varepsilon_0}{\varepsilon_1}\right)$$

(iii) Lower Air Region:

The fields in the lower air region are the same as those in the upper air region because of the same boundary conditions used in (ii). We note that since $\varepsilon_1 > \varepsilon_0$, the field in the dielectric is weaker than that in the air gaps. The bound surface charges can be obtained by using $\rho_{ps} = \mathbf{P} \cdot \mathbf{a}_n$ ($\mathbf{a}_n$ is a unit vector normal outward from the dielectric). For a linear dielectric with permittivity ε_1, we find $\rho_{ps} = -\left(\frac{\varepsilon_1 - \varepsilon_0}{\varepsilon_1}\right)\rho_s$ similar to Eq. (3-16). Thus, there exist the *negative* bound surface charge on the upper surface and the *positive* bound charge on the lower surface of the dielectric.

3.6. CAPACITANCE

In this section we study the capacitance of a capacitor. A capacitor is a circuit element that stores electric energy and it consists of two conductors filled with dielectric in between. Consider two conductors (1), (2) with charges Q, –Q, and electric field $\mathbf{E}$ and voltage V as shown in Figure 3-12(a).

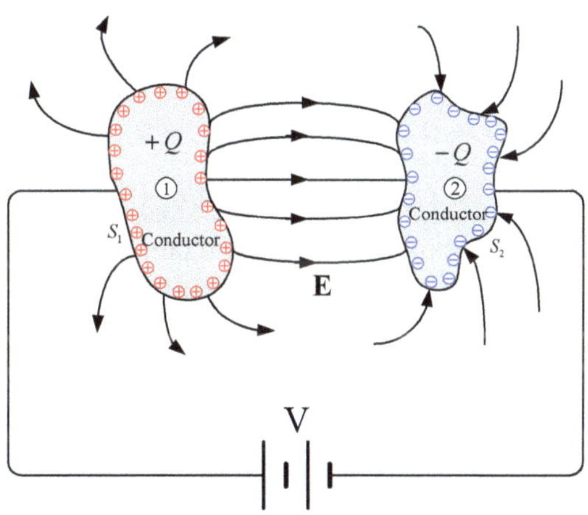

Figure 3-12(a). Capacitance.

A linear dielectric of permittivity ε exists in the region between the conductors. The charge resides on the surfaces S_1, S_2 with a surface charge distribution ρ_s. Because of the linear relationships between ρ_s, $\mathbf{E}$, V, there is a linear relationship between charge and voltage. Doubling one doubles the other. The ratio of charge to voltage is called the **capacitance:**

$$\boxed{C = \frac{Q}{V}} \tag{3-26}$$

Note that the voltage V here is the potential difference between the two conductors. The unit of the capacitance is the farad [F] (named after Michael Faraday). The capacitance may also be expressed in terms of electric field **E** and displacement **D**:

$$C = \frac{Q}{V} = \frac{\iint_{S_1} \rho_s \, ds}{-\int_2^1 \mathbf{E} \cdot d\ell} = \frac{\iint_{S_1} \mathbf{D} \cdot d\mathbf{s}}{-\int_2^1 \mathbf{E} \cdot d\ell} \tag{3-27a}$$

which reduces to the following form for a linear, homogeneous dielectric

$$C = \frac{\varepsilon \iint_{S_1} \mathbf{E} \cdot d\mathbf{s}}{-\int_2^1 \mathbf{E} \cdot d\ell} \tag{3-27b}$$

(for a linear, homonogeneous dielectric)

The capacitance of a capacitor depends on the geometry of the capacitor and the dielectric property of the medium between the conductors. The capacitance C can be calculated by the following steps:

1. Assume charges $+Q$, $-Q$ on the two conductors.
2. Find the electric field **E**, using one of the methods we learned so far.
3. Calculate the potential difference, $V = -\int_{(-)}^{(+)} \mathbf{E} \cdot d\ell$.
4. Find C by taking the ratio $\frac{Q}{V}$.

EXAMPLE 3-5. PARALLEL-PLATE CAPACITOR

Consider a parallel-plate capacitor with dielectric material of uniform permittivity ε shown in Figure 3-12(b). The area of each plate is A and plate separation is d. Neglect the fringing fields. Determine the capacitance.

SOLUTION

First we assume charges Q, $-Q$ on the plates, resulting in a surface charge ρ_s as shown in Figure 3-12(b), where

$$\rho_s = \frac{Q}{A}$$

We expect that the electric field is uniform as in Example 2-6. Applying the boundary condition at $x = 0$, Eq. (3-25b),

$$D_n = D_x = \varepsilon E_x = \rho_s = \frac{Q}{A}; \quad E_x = \frac{Q}{\varepsilon A}$$

and the potential difference V across the capacitor between plates can now be determined.

$$V = -\int_d^0 \mathbf{E} \cdot d\ell = \int_0^d E_x \, dx = \frac{D_x d}{\varepsilon} = \frac{Qd}{\varepsilon A}$$

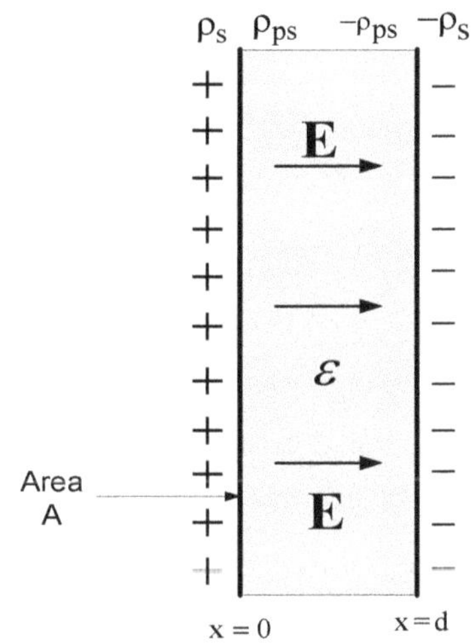

Figure 3-12(b). A parallel-plate capacitor.

The capacitance is

$$\boxed{C = \frac{Q}{V} = \frac{\varepsilon A}{d}} \tag{3-28}$$

The bound charge ρ_{ps} at $x = 0$ is determined from boundary conditions:

$$\rho_{ps} = \mathbf{P} \cdot \mathbf{a}_n = \mathbf{P} \cdot (-\mathbf{a}_x) = -\frac{(\varepsilon - \varepsilon_0)}{\varepsilon} D_x = -\left(\frac{\varepsilon - \varepsilon_0}{\varepsilon}\right) \rho_s$$

as in Eq. (3-16). Note then $\mathbf{a}_n$ is normal *outward* from the dielectric.

Consider what happens when we insert the dielectric between parallel plates. Under constant charge conditions, bound charge is drawn up to weaken $\mathbf{E}$ and V, thereby increasing C. Under constant voltage conditions, the electric field must be maintained. Additional free charge must be drawn up to maintain $\mathbf{E}$ despite the weakening effect of bound charge. Q and C are therefore increased. Insertion of a dielectric increases the capacitance. C is proportional to the area of the plate and inversely proportional to the plate separation.

EXAMPLE 3-6. SPHERICAL CAPACITORS (SEE FIGURE 3-13)

(a) *The Isolated Sphere*

Consider an isolated conducting sphere of radius a in free space. To find the capacitance, we assume a charge Q uniformly distributed over the surface of the sphere. Here we assume that the negative terminal is at infinity. Applying Gauss' law (see Example 2-4):

$$E_r = \frac{Q}{4\pi\varepsilon_0 r^2}(r > a); \quad V(r) = \frac{Q}{4\pi\varepsilon_0 r}; \quad V(a) = \frac{Q}{4\pi\varepsilon_0 a}, \quad V(\infty) = 0$$

Then

$$C = \frac{Q}{V(a)} = 4\pi\varepsilon_0 a \tag{3-29}$$

The capacitance of the earth is therefore less than a millifarad.

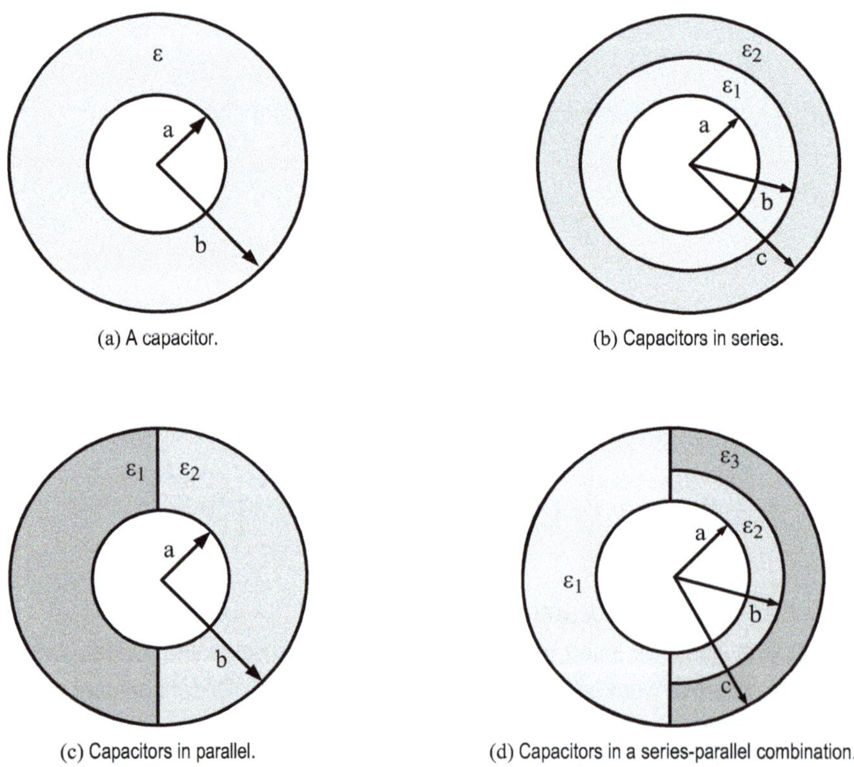

Figure 3-13. Concentric spherical and cylindrical geometries. Inner and outer surfaces are perfect conductors.

(b) *Concentric Spheres* (see Figure 3-13(a))

Consider concentric spheres (perfect conductors at $r = a, b$). The space between spheres is filled with a dielectric of permittivity ε. Assume a charge Q at $r = a$ and $-Q$ at $r = b$. Then

$$E_r = \frac{Q}{4\pi\varepsilon r^2} \ (a < r < b); \quad V_{ab} = -\int_b^a \mathbf{E}\cdot d\ell = \frac{Q}{4\pi\varepsilon}\left(\frac{1}{a} - \frac{1}{b}\right) = \frac{Q}{4\pi\varepsilon}\left(\frac{b-a}{ab}\right)$$

Then

$$\boxed{C = \frac{Q}{V_{ab}} = \frac{4\pi\varepsilon ab}{b-a}} \tag{3-30}$$

$$\rightarrow 4\pi\varepsilon a \quad \text{as } b \rightarrow \infty$$

which agrees with Eq. (3-29).

(c) *Capacitors in Series* (Figure 3-13(b))

Figure 3-13(b) shows a spherical capacitor with two dielectrics. Two conductors are at $r = a, c$. We assume a charge Q at $r = a$ and $-Q$ at $r = c$. Applying Gauss' law for $\mathbf{D}$:

$$D_r = \frac{Q}{4\pi r^2} \quad (a < r < c) \tag{3-31}$$

Note that normal $\mathbf{D}$ is continuous across the boundary of two dielectrics at $r = b$ as required by the boundary conditions. The electric fields are given by

$$\mathbf{E} = \frac{\mathbf{D}}{\varepsilon_1} = \mathbf{a}_r \frac{Q}{4\pi\varepsilon_1 r^2} \quad (a < r < b)$$

$$= \frac{\mathbf{D}}{\varepsilon_2} = \mathbf{a}_r \frac{Q}{4\pi\varepsilon_2 r^2} \quad (b < r < c)$$

V_{ac} is then evaluated as follows

$$V_{ac} = -\int_c^a \mathbf{E}\cdot d\ell = -\int_c^b \frac{Q}{4\pi\varepsilon_2 r^2}\, dr - \int_b^a \frac{Q}{4\pi\varepsilon_1 r^2}\, dr = \frac{Q}{4\pi}\left(\frac{1}{\varepsilon_1}\frac{(b-a)}{ab} + \frac{1}{\varepsilon_2}\frac{(c-b)}{cb}\right)$$

then

$$C = \frac{Q}{V_{ac}} = \left(\frac{1}{C_1} + \frac{1}{C_2}\right)^{-1} = \frac{C_1 C_2}{C_1 + C_2} \tag{3-32}$$

where

$$C_1 = \frac{4\pi\varepsilon_1 ab}{b-a}, \quad C_2 = \frac{4\pi\varepsilon_2 bc}{c-b}$$

Note that the result corresponds to capacitances in series.

(d) *Capacitors in Parallel* (Figure 3-13(c))
Figure 3-13(c) shows a spherical capacitor with two dielectrics in parallel. We try a solution of the following form:

$$E_r = \frac{K}{r^2} \quad (a < r < b) \tag{3-33}$$

in both dielectrics where K is an unknown constant. *Note that tangential* ***E*** *is continuous at the interface of two dielectrics as required by the boundary conditions.* The free charge densities ρ_{s1}, ρ_{s2} (at r = a) differ. Using the boundary condition (3-25b), $D_n = \rho_s$, the total charge at r = a is given by

$$Q = 2\pi a^2 (\rho_{s1} + \rho_{s2}) = 2\pi a^2 \left(\frac{\varepsilon_1 K}{a^2} + \frac{\varepsilon_2 K}{a^2} \right) = 2\pi K (\varepsilon_1 + \varepsilon_2)$$

$$V_{ab} = -\int_b^a \mathbf{E} \cdot d\boldsymbol{\ell} = -\int_b^a \frac{K}{r^2}\, dr = K\left(\frac{1}{a} - \frac{1}{b}\right) = \frac{K(b-a)}{ab}$$

then

$$C = \frac{Q}{V_{ab}} = \frac{2\pi(\varepsilon_1 + \varepsilon_2)}{\left(\dfrac{b-a}{ab}\right)} = C_1 + C_2 \tag{3-34}$$

where

$$C_1 = \frac{2\pi\varepsilon_1 ab}{b-a}, \qquad C_2 = \frac{2\pi\varepsilon_2 ab}{b-a}$$

The result corresponds to capacitors in parallel.

(e) *A Series-Parallel Combination* (Figure 3-13(d))
Figure 3-13(d) shows a capacitor with three dielectrics. We are tempted to call this a series-parallel combination. *There is no simple form for* ***E*** *and normal* ***D*** *which will satisfy continuity of both tangential* ***E*** *and normal* ***D***. In fact, this is a difficult problem involving fringing fields and requiring numerical computation for precise results. However, for thin spherical shells, the fringing is less significant and the capacitance may be approximated by the result given below:

$$C \approx C_1 + \frac{C_2 C_3}{C_2 + C_3} \quad (\text{if } a \gg c - a) \tag{3-35}$$

where

$$C_1 = \frac{2\pi\varepsilon_1\, ac}{c-a}$$

$$C_2 = \frac{2\pi\varepsilon_2\, ab}{b-a}$$

$$C_3 = \frac{2\pi\varepsilon_3\, bc}{c-b}$$

The result in Eq. (3-35) corresponds to a series-parallel combination.

EXAMPLE 3-7. A COAXIAL CYLINDRICAL CAPACITOR

Note that Figure 3-13 applies to both *spherical* and *cylindrical* geometries.

The cross section of a coaxial cylindrical capacitor of infinite length is shown in Figure 3-13(a). Surfaces $\rho = a, b$ are perfect conductors. Find the capacitance per unit length appropriate to this cylindrical geometry.

SOLUTION

We assume a charge ρ_ℓ per unit length uniformly distributed over the surface $\rho = a$ and $-\rho_\ell$ at $\rho = b$. Then, applying Gauss' law (see Example 3-3)

$$E_\rho = \frac{\rho_\ell}{2\pi\varepsilon}\left(\frac{1}{\rho}\right) \quad (a < \rho < b)$$

$$V_{ab} = -\int_b^a \mathbf{E} \cdot d\boldsymbol{\ell} = -\int_b^a E_\rho\, d\rho = \frac{\rho_\ell}{2\pi\varepsilon} \ln\left(\frac{b}{a}\right)$$

The capacitance per unit length is given as follows:

$$\frac{C}{\ell} = \frac{Q/\ell}{V} = \frac{\rho_\ell}{V_{ab}} = \frac{2\pi\varepsilon}{\ln\left(\frac{b}{a}\right)} \quad \left[\frac{F}{m}\right] \tag{3-36}$$

and

$$\boxed{C = \frac{2\pi\varepsilon\ell}{\ln\left(\frac{b}{a}\right)}} \tag{3-37}$$

For a coaxial cylindrical capacitor of finite length ℓ, fringing at the ends invalidates the results given in Eq. (3-37), which will, however, be reasonably accurate if $\ell >> a, b$. Cylindrical capacitors of other geometrics [Figures 3-13(b), (c), (d)] are treated in Problems 3-21, 3-22, 3-23.

3.7 Benjamin Franklin (1706–1790) and the Beginnings of Electrical Science

Benjamin Franklin can accurately be described as a Renaissance man: he did so many things and he did them to perfection. The breadth and depth of his accomplishments are largely unparalleled. History remembers him as the inventor of the lightning rod, bifocals, and the Franklin stove. We recall his role in the American Revolution as Ambassador to France. He appears in numerous famous paintings: performing his kite experiment, presenting the Declaration of Independence to John Hancock, and in his fur cap and spectacles in Paris. As memorable as these well-known images are, they commemorate only a very small number of his accomplishments. To gain some idea of the scope of his activities, we would have to read some of the biographies of Franklin; the 1938 biography by Carl Van Doren is still one of the best*.

Benjamin Franklin was born in 1706, the youngest boy of a very large family. He attended school for about two years and, at age 10, began to work with his father, a soap and candle maker. At age 12 he was apprenticed to his brother James, a printer. He ran away to Philadelphia at age 17, worked as a printer, and eventually set up his own printing shop at age 22. Within four years, he had started writing and printing "Poor Richard's Almanac", a yearly publication which would become immensely popular, eventually selling 10,000 copies a year and continuing for 27 years. By his late twenties, Franklin was well-established as a local printer and businessman, and was well on his way to wider recognition.

Benjamin Franklin wrote two outstanding books during his lifetime. The first was a description of his discoveries in electricity, *Experiments and Observations on Electricity*, published in 1751. This first publication was crucial in establishing Franklin's reputation as a scientist. In it, Franklin put forward the idea that lightning was electricity, proposed a specific experiment (the sentry box experiment) for verification of the theory, and recommended the use of grounded rods to protect buildings from lightning strikes. A year later, several French scientists carried out his sentry box experiment, while Franklin performed his famous kite experiment, resulting in a thorough confirmation of his basic theory. The book and these lightning experiments catapulted Franklin into the limelight and established him as one of the world's foremost scientists and a leading figure in the new science of electricity. Accolades and awards from around the world followed soon after. Two years later, in 1753, Benjamin Franklin received the Copley medal, an award that carried prestige equivalent to today's Nobel Prize.

Benjamin Franklin wrote in a very clear and readable style. His judgment in reaching for a conclusion or drawing an hypothesis is very sure. His style is engaging and frank, and he readily acknowledges his ignorance or uncertainty: "You require the reason. I do not know it. Perhaps you may discover it and then you will be so good as to communicate it to me." His treatment of his own hypotheses was very refreshing. He readily acknowledged them as only tentative models and he was ready to abandon them if they proved inadequate: "I am still at a loss about the manner in which they (clouds) become charged with electricity; no hypothesis I have yet formed perfectly satisfying me."

*Carl Van Doren, *Benjamin Franklin*, Bramhall House, 1987, Reprint of Viking Press, NY, 1938.

Benjamin Franklin's second book is his *Autobiography*, which remains extremely popular to this day. Particularly interesting is the description of his self-education and self-discipline. It is the story of a brilliant and unusual man struggling to emerge from ignorance and poverty. He relates many of his mistakes and failures and does not bask in his successes and honors.

Throughout his life, Benjamin Franklin was also a prolific writer of newspaper articles. These were usually written to advance some social or political purpose. They were entertaining as well as educational, and their impact was always considerable. The undergraduate of today can read with edification and delight any of these famous articles, including "Miss Polly Baker," "Rattlesnakes for Felons," "The King of Prussia," and "Rules by Which a Great Empire May be Reduced to a Small One." In the age of Enlightenment, Franklin's articles led to personal recognition and to profound respect for the power of his arguments.

To return to Franklin's role as a scientist, what were the key scientific accomplishments which established his reputation? To begin with, one of his initial contributions was his analysis of the "Leyden jar," essentially a glass jar filled with water and held in one's hand. We may think of it as a jar which is coated separately with a conductor on the inside and outside. The Leyden jar is thus a "warped" parallel plate capacitor. Because of the large surface area and the thinness of the glass, the capacity is large and considerable electric energy can be stored in the glass. The Leyden jar was a convenient source of energy and was therefore used in a wide variety of experiments after it was discovered by the Dutch physicist Pieter Van Mussenbroek in 1746.

Franklin demonstrated that the conductors inside and outside were oppositely charged, that the total charge was zero and was therefore unchanged as the capacitor was discharged. He wrote: "These two states of electricity, the plus and the minus, are combined and balanced in this miraculous bottle in a manner that I can by no means comprehend." The Leyden jar analysis led to a general theory of positive and negative charges, the conservation of total charge, and the single fluid theory. Franklin explained his experiments with the motion of a single type of charge, namely, free electrons. He also concluded that the energy was contained in the glass itself. He "unwrapped" the Leyden jar and formed the parallel plate capacitor. This was merely a flat glass plate coated with conductors on both sides. He then combined a set of plates in either series or parallel combinations. The series combination was called a "battery." The Leyden jar experiments were crucial because they led directly to Franklin's general theory of electricity. His theory swept aside the vague previous theories of "effluvia" and "affluvia" espoused by the Abbé Nollet, one of the leading figures in the early study of electricity.

Franklin's earliest writings on electricity in 1747 described some of his observations on the effects of pointed objects. His colleague Thomas Hopkinson had carried out some of the early experiments. By 1750 Franklin had proposed the lightning rod as a protection for houses and for ships and had added the very important grounding wire for both. He had also proposed his "sentry box" experiment for determining whether clouds were electrified, and whether lightning was, in fact, electricity. In 1752 the experiment was performed successfully in France with the sentry box and in America with the kite.

The drama of these experiments cannot be overrated. Lightning was proven to be electrical in origin and identical to the small-scale sparks produced in the laboratory. The clouds were shown to be electrically charged and the cloud-earth combination analogous to a giant Leyden jar. Foremost, a solution to the important practical problem of lightning hazard was proposed, namely the grounded lightning rod. The potential usefulness of lightning rods was enormous. It is estimated that in one 33-year period in the 18th century 386 church towers were struck and 120 bell ringers were killed in Germany alone.

Ultimately, Benjamin Franklin's most important contributions to electrical science were

1. The conservation of electric charge.
2. The single-fluid theory.
3. The concept of positive and negative charges.
4. The analysis and extension of the Leyden Jar.
5. The effects of pointed objects.
6. The grounded lightning rod.
7. The electrification of clouds.

Now we begin to see why Franklin, who first organized in a coherent theory the basic ideas of electricity, was regarded in his time, and for some time after, as one of the leading scientists of electricity. His experiments and the book which described them gave order and structure to what had previously been a disorganized collection of theories. The importance of his work was immediately recognized by leading European scientists. He received the highest scientific awards of the day, including the prestigious Copley medal. He was also elected Fellow of the Royal Society and a member of the French Academy of Sciences. Long before the American Revolution, Franklin had established himself as the leading scientist of electrical phenomena.

So Franklin, in his time and for years afterwards, was widely recognized for his pioneering work in electricity. In recent times, however, we seem to overlook Franklin's scientific contributions. It is certainly possible that you do not think of Benjamin Franklin as a scientist, and there are several reasons for this oversight. Franklin left no equations which we recognize as fundamental parts of the theory. We have absorbed many of his contributions and use them unconsciously. As well, he was a prolific inventor, and it is easy to think of his inventions as representing his primary technical accomplishments, whereas they were actually the fruits of his scientific curiosity and scientific theories.

When Benjamin Franklin went to France as Ambassador, his scientific reputation preceded him. He was not an unknown Philadelphian, but an eminent scientist. His greatest experiment, the lightning experiment, had first been carried out by Frenchmen. His book had been translated into French and was in wide circulation. His greatest antagonist, the French Abbé Nollet, wrote two books attacking the Franklinian theory; he was particularly averse to lightning rods. Franklin chose not to reply directly to any of the Abbé's many attacks, and said, "I have an extreme aversion to public altercation on philosophic points, and have never yet disputed with anyone who thought fit to attack my opinions." He was directly defended by several Frenchmen, and, if anything, his reputation was enhanced by this controversy.

How much of his life did Benjamin Franklin devote to his key electrical experiments? We are surprised to note that it was a relatively short period of time, namely, the six years from 1747 to 1752. Having achieved financial independence, Benjamin Franklin retired from the printing business in 1748 in order to devote himself more completely to his electrical research. His book on experiments was first published in 1751 and the lightning rod experiments were performed in 1752.

Franklin was the quintessential public relationist. He was very conscious of his image and strove at an early age to project a positive one. This is not to say that the image necessarily conflicted with reality but that Franklin fully realized its importance. Thus, "To show that I was not above my business, I sometimes brought home the paper I purchased at the stores through the streets on a wheelbarrow." In his scientific writings he projected the image of an open, receptive person. As a result, other scientists flocked to share their results with him and to perform experiments for him. When Franklin first went to France as Ambassador during the American Revolution, he was already known as an eminent scientist. What image then was appropriate for the diplomat Franklin? Could he possibly outdo the other diplomats in their courtly finery? He decided to do just the opposite of what was expected and appeared in simple dress with a fur cap, thereby projecting a unique and lasting image.

Franklin was a very highly skilled diplomat. His efforts on behalf of the fledgling republic led to diplomatic recognition by France, to massive military aid, and to huge financial loans. There are many examples of Franklin's diplomatic skill but one of the most striking occurred at the conclusion of the Peace Treaty negotiations. The Americans had decided to break their agreement with France and their instructions from Congress and carry out separate negotiations with England. Any other course of action would have made the negotiations almost impossible. An extremely favorable treaty had resulted, including extensions of United States Territory to the Mississippi and to Canada. The negotiations could not have gone better. The Americans had come to the table with their most talented statesmen (Franklin, Adams, Jefferson, Jay, Laurens) and had overwhelmed the English negotiators. Now it was necessary to tell the French about the negotiations. Franklin reported to his French counterpart, Charles Gravier, the comte de Vergennes, and sent him a copy of the treaty which had already been signed. Vergennes sent him back a note of reproach. Franklin now sent Vergennes a famous letter in which he put forth the American position in the most favorable light, made a partial apology, and said "the English, I just now learn, flatter themselves they have already divided us." Somehow this did the trick! Not only were fences mended but Vergennes arranged yet another much-needed loan to the United States. The years of careful diplomacy to establish cordial relationships with Vergennes combined with just the right touch in the letter had preserved America's unique relationship with France.

Additionally, Franklin had recognized early the injustice of the institution of slavery and had written many articles on the subject, including those that highlighted unrecognized talents of African Americans. Two months before his death he submitted to Congress a petition for the abolition of slavery. It was turned down.

3.8. THE FORCE ON A CONDUCTOR IN AN ELECTRIC FIELD

We turn now to the calculation of the force on a conductor in an electric field. This would seem to be a straightforward matter, but it's not. There are subtleties involved which are concerned with the *self* field $\mathbf{E}_s$. In the course of the discussion we will learn the very useful fact that in calculating forces we can always subtract the self force. For the force on a conductor, this means that we can use $(\mathbf{E} - \mathbf{E}_s)$ in place of $\mathbf{E}$.

To find the force on a conductor, we need to consider only the forces on the electric charges which exist on the surface of the conductor. Figure 3-14(a) shows a conductor with a typical patch of the surface with area ds, surface charge density ρ_s, and electric field $\mathbf{E}$. The force $d\mathbf{F}$ on the patch is equal to the charge $\rho_s ds$ times the electric field vector. There is just one problem and that is, "which electric field?"

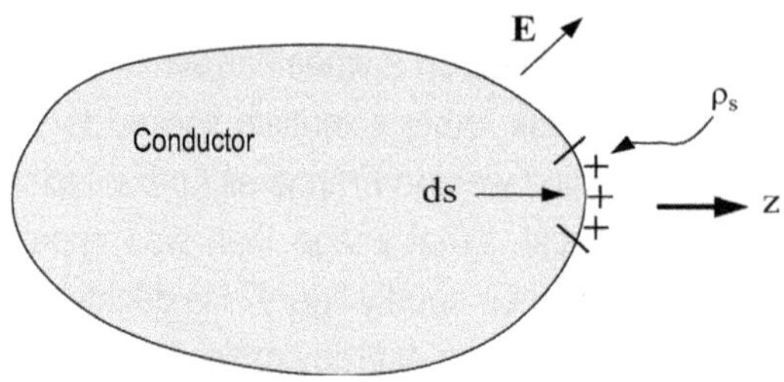

Figure 3-14(a). A conductor in an electric field.

Figure 3-14(b) shows a plot of the electric field near the surface of the conductor. We choose z as the direction normal to the patch. E_z is a step function which jumps from zero to $\frac{\rho_s}{\varepsilon_o}$ at the surface of the conductor. For $\mathbf{E}$ should we use the value (zero) just inside the conductor, the value $\frac{\rho_s}{\varepsilon_o}$ just outside, the average $\frac{\rho_s}{2\varepsilon_o}$, or something else?

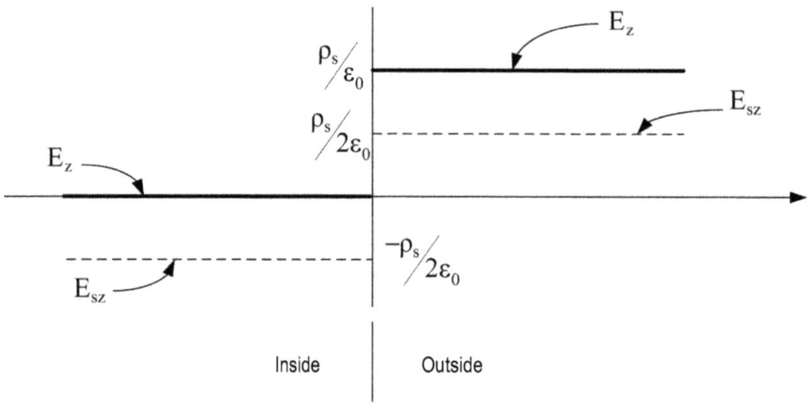

Figure 3-14(b). The normal electric field E_z and a self field E_{sz} at a conductor surface.

Before going any further, let's remind ourselves of the behavior of the electric field $\mathbf{E}$ and also the behavior of the self field $\mathbf{E}_s$ which is the contribution of the patch itself. We are interested in the behavior of these fields just inside and just outside the conductor surface.

$$\mathbf{E} = \mathbf{a}_n \frac{\rho_s}{\varepsilon_0} = \mathbf{a}_z \frac{\rho_s}{\varepsilon_0} \quad \text{(just outside)} \tag{3-38a}$$

$$= 0 \text{ (just inside)} \tag{3-38b}$$

$$\mathbf{E}_{av} = \frac{1}{2}\{\mathbf{E}(\text{outside}) + \mathbf{E}(\text{inside})\} = \mathbf{a}_z \frac{\rho_s}{2\varepsilon_0} \tag{3-39}$$

The self field $\mathbf{E}_s$ due to the small patch itself is the same as that of an infinite plane of charge:

$$\mathbf{E}_s = \pm \mathbf{a}_n \frac{\rho_s}{2\varepsilon_0} = \mathbf{a}_z \frac{\rho_s}{2\varepsilon_0} \quad \text{(just outside)} \tag{3-40a}$$

$$= -\mathbf{a}_z \frac{\rho_s}{2\varepsilon_0} \quad \text{(just inside)} \tag{3-40b}$$

Figure 3-14(b) shows a plot of $\mathbf{E}$ (solid) and $\mathbf{E}_s$ (dotted). We note that $\mathbf{E}$ and $\mathbf{E}_s$ have the same jump (discontinuity) at the surface and therefore we consider using their difference ($\mathbf{E} - \mathbf{E}_s$) in calculating forces. This is legitimate since the conductor exerts no net force on itself. The difference $\mathbf{E} - \mathbf{E}_s$ is continuous:

$$\mathbf{E} - \mathbf{E}_s = \mathbf{a}_z \frac{\rho_s}{2\varepsilon_0} = \mathbf{E}_{av} \tag{3-41}$$

Now we see the culprit. It is the self field $\mathbf{E}_s$ which is discontinuous at the surface. Its presence in $\mathbf{E}$ leads to an ambiguity; its removal resolves the ambiguity. Its removal is also equivalent to taking the average of fields inside and out (since $\mathbf{E}_s$ is odd). $\mathbf{E}_s$ is the field at the patch due to the patch itself; $\mathbf{E} - \mathbf{E}_s$ is the field at the patch due to the rest of the conductor, i.e., everything but the patch.

Thus the force $\mathbf{F}$ on a conductor in an electric field is given as

$$\mathbf{F} = \iint_S (\mathbf{E} - \mathbf{E}_s)\rho_s \, ds = \iint \mathbf{E}_{av} \, \rho_s \, ds = \iint_S \mathbf{a}_n \frac{\rho_s^2}{2\varepsilon_0} ds$$

$$\boxed{\mathbf{F} = \iint_S \mathbf{a}_n \frac{\rho_s^2}{2\varepsilon_0} ds} \tag{3-42}$$

3.9. ENERGY AND FORCES IN THE ELECTROSTATIC FIELD

In this section, we calculate the energy of, or the work required to assemble, electrostatic systems. In addition, we are interested in the forces exerted on portions of the system. First we consider the work required to assemble a collection of point charges.

3.9.1. Energy of a Collection of Point Charges

Figure 3-15 shows a set of N point charges at arbitrary locations. The distance between q_1 and q_2 is r_{12} and the distance between q_i and q_j is r_{ij}. Let us determine the work required to bring up the charges from infinity. We start with all the charges an infinite distance from the origin and from each other. The point charges have been given to us; we do not, for the moment, inquire into the work required to construct them.

No work is required to bring up the first charge (q_1) since there are no forces exerted on it. Charge q_2 is brought up working against the potential of q_1. The potential through which q_2 is moved is $V_1 = \frac{q_1}{4\pi\varepsilon_0 r_{12}}$; the work required to bring q_2 in the presence of q_1 is given by

$$W_2 = -\int \mathbf{F}_1 \cdot d\ell = -\int_{\infty}^{r_{12}} q_2 \mathbf{E}_1 \cdot d\ell = q_2 V_1 = \frac{q_1 q_2}{4\pi\varepsilon_0 r_{12}}$$

Similarly, charge q_3 works against the forces of both q_1 and q_2. The work required to bring up q_3, working against the forces of both q_1 and q_2, is

$$W_3 = q_3 \left(\frac{q_1}{4\pi\varepsilon_0 r_{13}} + \frac{q_2}{4\pi\varepsilon_0 r_{23}} \right)$$

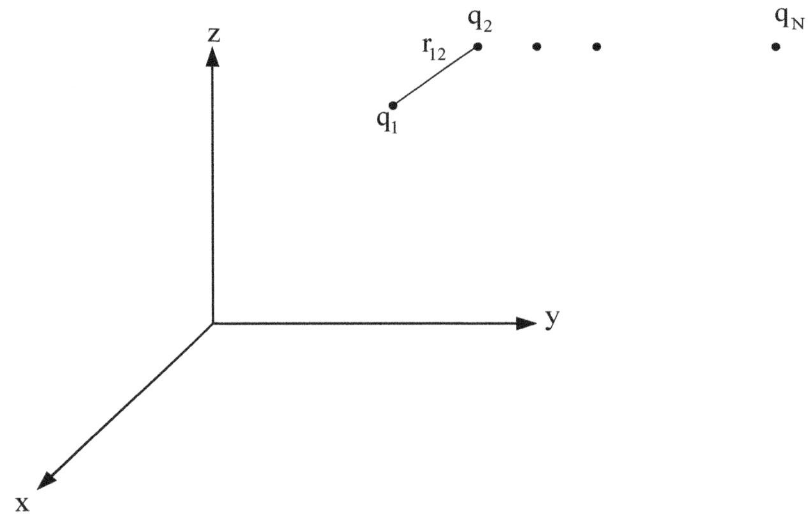

Figure 3-15. A collection of point charges.

Finally, the work required to bring up q_N is

$$W_N = q_N \left(\frac{q_1}{4\pi\varepsilon_0 r_{1N}} + \frac{q_2}{4\pi\varepsilon_0 r_{2N}} + \ldots + \frac{q_{N-1}}{4\pi\varepsilon_0 r_{(N-1)N}} \right)$$

Consider an $N \times N$ matrix of terms $\frac{q_i q_j}{4\pi\varepsilon_0 r_{ij}}$. We have taken into account all the terms above or below the main diagonal and may thus express the total electric energy W_e as follows:

$$W_e = W_2 + W_3 + \ldots + W_N = \sum_{i=1}^{N} \sum_{\substack{j=1 \\ j>i}}^{N} \frac{q_i q_j}{4\pi\varepsilon_0 r_{ij}} \quad \text{(3-43a)}$$

where $j > i$ includes terms above the main diagonal. Because of symmetry about the main diagonal,

$$W_e = \frac{1}{2}\sum_{i=1}^{N}\sum_{\substack{j=1\\ j\neq i}}^{N}\frac{q_i q_j}{4\pi\varepsilon_0 r_{ij}} \tag{3-43b}$$

$$W_e = \frac{1}{2}\sum_{i=1}^{N} q_i\left(\sum_{\substack{j=1\\ j\neq i}}^{N}\frac{q_j}{4\pi\varepsilon_0 r_{ij}}\right) = \frac{1}{2}\sum_{i=1}^{N} q_i V_i \tag{3-44}$$

where V_i is the potential at the location of q_i due to all charges except q_i itself. The factor of $\frac{1}{2}$ in Eq. (3-44) is related to the linear nature of the problem. The first charge is brought up through zero potential, the last charge is brought up through the total potential V, and the average charge is brought up through $\frac{1}{2}$ V.

3.9.2. Energy of Continuous Charge Distribution

For continuous charge distribution $\rho_v(x,y,z)$, we divide it into infinitesimally small volumes and consider the charge of an infinitesimally small volume element to be $dQ = \rho_v dv$. We then use the summation of Eq. (3-44) which becomes an integral. Integrating over all volume elements yields

$$\boxed{W_e = \frac{1}{2}\iiint \rho_v\,(x,y,z)\;V(x,y,z)\,dv} \tag{3-45}$$

Eq. (3-45) is the energy expression for volume charge distribution. For surface charge distribution it reduces to $\frac{1}{2}\iint \rho_s V\,ds$; if the surface is a conductor it reduces to $\frac{1}{2}QV$ where Q is the total charge on the conductor and V is the potential of the conductor, since the potential V is constant on the conductor. The energy stored in the capacitor, which consists of two conductors carrying charges $+Q$, $-Q$ (see Figure 3-12(a)), can be expressed as follows:

$$W_e = \frac{1}{2}QV_1 + \frac{1}{2}(-Q)\,V_2 = \frac{1}{2}Q(V_1 - V_2) = \frac{1}{2}QV = \frac{1}{2}CV^2$$

where V_1 and V_2 are the potentials of the two conductors and V is the potential difference or voltage.

3.9.3 Energy in Terms of Fields

Since the charges produce the electric fields in the surrounding space, the energy can also be expressed in terms of the electric fields. A second general form may be obtained by extending the volume of Eq. (3-45) to infinity. This is permitted since $\rho_v = 0$ outside the charge distribution. Then

$$W_e = \frac{1}{2} \iiint_{\text{all space}} \rho_v V \, dv = \frac{1}{2} \iiint_{\text{all space}} (\nabla \cdot \mathbf{D}) V \, dv$$

$$= \frac{1}{2} \iiint_{\text{all space}} \left\{ \nabla \cdot (V\mathbf{D}) - \mathbf{D} \cdot \nabla V \right\} dv$$

where we have used the vector identity

$$\nabla \cdot (V\,\mathbf{D}) = V\,(\nabla \cdot \mathbf{D}) + (\nabla V) \cdot \mathbf{D}$$

We now use $\mathbf{E} = -\nabla V$ and apply the divergence theorem to the first term of the integral above to obtain

$$W_e = \frac{1}{2}\left\{ \iiint_{\text{all space}} \mathbf{E} \cdot \mathbf{D} \, dv + \oiint_{\text{sphere at}\infty} V\mathbf{D} \cdot d\mathbf{s} \right\}$$

For a charge distribution of finite extent, $V \sim \frac{1}{r}$, $|\mathbf{D}| \sim \frac{1}{r^2}$, $ds \sim r^2$, as $r \to \infty$, and the surface integral above vanishes. Thus

$$\boxed{W_e = \frac{1}{2} \iiint_{\text{all space}} \mathbf{E} \cdot \mathbf{D} \, dv} \qquad (3\text{-}46)$$

In Eq. (3-45), the energy is viewed as being stored with the charge source, whereas Eq. (3-46) shows the energy being stored with the fields. Both equations yield the same result for an electrostatic system, but for time-varying sources and fields, Eq. (3-46) makes more sense because the fields move in space, carrying the energy, as will be shown in Chapter 8.

EXAMPLE 3-8. A SPHERICAL CLOUD OF CHARGE

Consider a spherical cloud of uniform volume charge density ρ_v:

$$\rho_v = \rho_o \qquad (r < a)$$

Find the work required to assemble the cloud of charge or equivalently the electric energy stored in the system.

SOLUTION

Applying Gauss' Law:

$$\oiint_S \mathbf{D} \cdot d\mathbf{s} = 4\pi r^2 D_r = \rho_o \frac{4}{3}\pi r^3 \ (r \le a)$$

$$= \rho_o \frac{4}{3}\pi a^3 \ (r \ge a)$$

$$D_r = \varepsilon_o E_r = \frac{\rho_o r}{3} \quad (r \le a)$$

$$= \frac{\rho_o a^3}{3r^2} \quad (r \ge a)$$

The electric potential is given by

$$V = -\int \mathbf{E} \cdot d\ell = -\int_\infty^r E_r\Big|_{r \ge a} dr = -\int_\infty^r \frac{\rho_o a^3}{3\varepsilon_0 r^2} dr = \frac{\rho_o a^3}{3\varepsilon_0 r} \quad (r \ge a)$$

$$= -\int_\infty^a E_r\Big|_{r \ge a} dr - \int_a^r E_r\Big|_{r \le a} dr = \frac{\rho_o a^2}{2\varepsilon_0} - \frac{\rho_o r^2}{6\varepsilon_0} \quad (r \le a)$$

Using Eq. (3-45), we obtain

$$W_e = \frac{1}{2} \iiint \rho_v \ V \, dv$$

$$= \frac{1}{2} \int_0^{2\pi} \int_0^{\pi} \int_0^{a} \rho_o^2 \left[\frac{a^2}{2\varepsilon_0} - \frac{r^2}{6\varepsilon_0} \right] r^2 \sin\theta \, dr \, d\theta \, d\phi$$

$$= \frac{4\pi \rho_o^2 a^5}{15\varepsilon_0} = \frac{3Q^2}{20\pi\varepsilon_0 a} \qquad (3\text{-}47)$$

(where $Q = \frac{4}{3}\pi a^3 \rho_o$). We can also derive the same result, using the energy in terms of fields, Eq. (3-46) [see Problem 3-31]. Note that

$$W_e \to 0 \text{ as } a \to 0 \ (\rho_o \text{ constant})$$

$$W_e \to \infty \text{ as } a \to 0 \ (Q \text{ constant})$$

This implies that (a) a small volume of uniform finite ρ_o has negligible energy and (b) a point charge Q has infinite energy.

3.9.4. Forces and Torques in Terms of Energy

Let us consider an electrostatic system consisting of conductors and dielectrics. Allow one of the parts of the system to move under the influence of the force $\mathbf{F}$ acting upon it. We allow a small (virtual) displacement Δx of the movable part. Consider two situations: (a) constant charge for which conductors are isolated from each other and the total charge on each conductor remains constant during the displacement, (b) constant potentials for which potentials between conductors are constant during the displacement. Figure 3-16 shows parallel plates with a dielectric block partially inserted between plates. In the constant charge case the plates are unconnected and charge Q remains constant during a small displacement Δx. x is defined as the distance between a point on the fixed plates and the movable block of dielectric. In the constant voltage case an ideal voltage source is connected between plates. In this case, additional charge ΔQ is deposited on the upper plate during the displacement Δx.

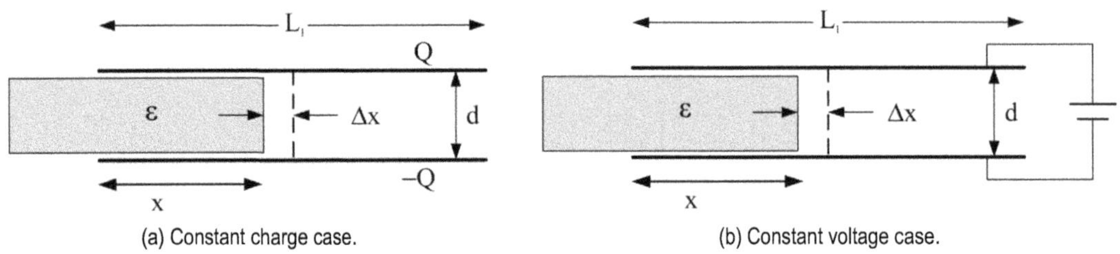

Figure 3-16. A dielectric slab partially inserted between parallel plates.

Constant Charge

In the constant charge case, charge is fixed but potentials and fields may vary. Therefore the electrostatic energy W_e may vary with displacement Δx. As the dielectric block moves through displacement Δx, under the influence of force F_x, mechanical work $F_x \Delta x$ is done. The mechanical energy is thereby changed. It could be stored (as in a spring) or it could show up as kinetic energy $(\frac{1}{2}mv^2)$. In any case, both the mechanical and electrostatic energy may change. Conservation of energy may be expressed as follows:

$$F_x \Delta x + \Delta W_e = 0$$

$$F_x = -\frac{\Delta W_e}{\Delta x} \rightarrow -\frac{\partial W_e(Q,x)}{\partial x} \quad (\text{as } \Delta x \rightarrow 0) \tag{3-48a}$$

Constant Voltage

In the constant voltage case, the voltage is fixed but the charge may vary. If the charge is increased by ΔQ then the electrostatic energy, which is equal to $\frac{1}{2}$ VQ, increases by $\frac{1}{2}$ $V\Delta Q$ to $\frac{1}{2}V(Q+\Delta Q)$. The mechanical work done is $F_x \Delta x$ as before. The battery is a third contributor to energy storage. The battery does work (supplies energy) $V(\Delta Q)$ in moving charge ΔQ through voltage V.

Conservation of energy may be expressed as follows:

$$\Delta W_e + F_x \Delta x = V \Delta Q = 2 \Delta W_e \text{ (since } \Delta W_e = \frac{1}{2} V \Delta Q).$$

$$F_x = \frac{\Delta W_e}{\Delta x} \rightarrow \frac{\partial W_e(V,x)}{\partial x} \quad \text{as} \quad \Delta x \rightarrow 0 \qquad (3\text{-}48b)$$

If the dielectric block is hinged at one end and free to rotate within the parallel plates, the mechanical work done is $T_\theta \Delta\theta$ in each case, where θ is an angle between the fixed plates and the rotatable block of dielectric. T_θ is the torque. The corresponding results are

$$T_\theta = -\frac{\partial W_e(Q,\theta)}{\partial \theta} \quad \text{(constant charge)} \qquad (3\text{-}49a)$$

$$= \frac{\partial W_e(V,\theta)}{\partial \theta} \quad \text{(constant voltage)} \qquad (3\text{-}49b)$$

Forces and displacement in directions other than x may also be considered. The results may be generalized as follows:

$$\mathbf{F}_Q = -\nabla W_e \qquad (3\text{-}50a)$$

$$\mathbf{F}_V = +\nabla W_e \qquad (3\text{-}50b)$$

where the subscripts Q and V represent constant charge and voltage, respectively.

EXAMPLE 3-9. FORCE ON A DIELECTRIC BLOCK BETWEEN PARALLEL PLATES

Consider the configuration of Figure 3-16. The plates are rectangular with plate area equal to L_1L_2. Assuming that $L_1, L_2 >> d$, we neglect fringing fields. In other words, we assume that the electric field is uniform in the region between the plates and zero outside. The electrostatic energy of the system and the forces are given by

$$W_e = \frac{1}{2} QV = \frac{1}{2} CV^2 = \frac{1}{2} \frac{Q^2}{C} \qquad (3\text{-}51)$$

$$F_x = \frac{\partial W_e(V,x)}{\partial x} = \frac{1}{2} V^2 \frac{dC(x)}{dx} \qquad (3\text{-}52a)$$

$$= -\frac{\partial W_e(Q,x)}{\partial x} = -\frac{1}{2} Q^2 \frac{d}{dx}\left[\frac{1}{C(x)}\right] = \frac{1}{2}\frac{Q^2}{C^2}\frac{dC(x)}{dx} \qquad (3\text{-}52b)$$

Note that the constant charge and constant voltage cases yield identical expressions, as they must, for F_x. Considering the configuration as a parallel connection of two parallel-plate capacitors and using the result of Eq. (3-28),

$$C(x) = \frac{\varepsilon L_2 x}{d} + \frac{\varepsilon_0 L_2 (L_1 - x)}{d}$$

$$\frac{dC(x)}{dx} = \frac{L_2(\varepsilon - \varepsilon_0)}{d}$$

Then

$$F_x = \frac{V^2 L_2 (\varepsilon - \varepsilon_0)}{2d} \tag{3-53}$$

Since $\varepsilon > \varepsilon_o$, F_x is always positive and the dielectric slab is pulled in between the plates. Note that we have assumed a uniform electric field in finding the force. However, it is not this uniform field but a *nonuniform fringing field* that pulls the dielectric into the capacitor. Without knowing the fringing field, which is difficult to calculate, we were able to find the force on the dielectric from the change in the electrostatic energy of the system.

PROBLEMS

3-1. In each of the cases below, find the bound surface and volume charge densities ρ_{ps}, ρ_{pv}. In addition, show that the dielectric is electrically neutral, i.e., that the total bound charge is zero.

(a) A uniformly polarized cylinder of length ℓ and radius a:

$$\mathbf{P} = \mathbf{a}_z P_o \quad \left(0 < \rho < a, -\frac{\ell}{2} < z < \frac{\ell}{2}\right)$$

(b) A uniformly polarized sphere of radius a:

$$\mathbf{P} = \mathbf{a}_z P_o \quad (r < a)$$

(c) A radially polarized cube of side ℓ, centered at the origin:

$$\mathbf{P} = (\mathbf{a}_x x + \mathbf{a}_y y + \mathbf{a}_z z)\, P_o = \mathbf{a}_r r P_o$$

3-2. In each of the cases below, find the bound charge densities ρ_{ps}, ρ_{pv} and show that the total bound charge is zero.

(a) A radially polarized sphere of radius a:

$$\mathbf{P} = \mathbf{a}_r P_o \left(\frac{r}{a}\right) \quad (r < a)$$

(b) A radially polarized cylinder of infinite length and radius a:

$$\mathbf{P} = \mathbf{a}_\rho \, P_o \left(\frac{\rho}{a}\right) \quad (\rho < a)$$

(c) A polarized cube of side ℓ, centered at the origin:

$$\mathbf{P} = \mathbf{a}_x \, P_o \left(\frac{x^2}{\ell^2}\right)$$

3-3. In each of the cases below, find the bound charge densities ρ_{ps}, ρ_{pv} and show that the total bound charge is zero.

(a) A cylindrical shell of length ℓ and radii a, b:

$$\mathbf{P} = \mathbf{a}_\rho P_o \left(\frac{a}{\rho}\right) \left(a < \rho < b, \quad -\frac{\ell}{2} < z < \frac{\ell}{2}\right)$$

(b) A spherical shell of radii a, b:

$$\mathbf{P} = \mathbf{a}_r \, P_o \left(\frac{a^2}{r^2}\right) \quad (a < r < b)$$

(c) A polarization rectangular box of sides a, b, c:

$$\mathbf{P} = \left(\mathbf{a}_x \frac{x}{a} + \mathbf{a}_y \frac{y}{b}\right) P_o \; (0 < x < a, \; 0 < y < b, \; 0 < z < c)$$

3-4. Consider a radially polarized cylinder of infinite length and radius a:

$$\mathbf{P} = \mathbf{a}_\rho \, P_o \left(\frac{a}{\rho}\right)$$

Find the bound charge densities ρ_{ps}, ρ_{pv}. Find the bound line charge density $\rho_{p\ell}$ along the axis ($\rho = 0$) by requiring that the total bound charge is zero.

3-5. Consider a radially polarized sphere of radius a:

$$\mathbf{P} = \mathbf{a}_r \, P_o \left(\frac{a^2}{r^2}\right)$$

Find the bound charge densities ρ_{ps}, ρ_{pv}. What type of bound charge exists at the origin? Find it by requiring that the total bound charge is zero.

3-6. Determine the on-axis electric field for the uniformly polarized cylinder of Problem 3-1(a). Compare with the result of Example 3-1.

3-7. Determine the on-axis electric field for the radially polarized cylinder of Problem 3-3(a).

3-8. Determine the electric field at the center of the uniformly polarized sphere of Problem 3-1(b).

3-9. If the bound charges of a polarized dielectric are symmetrically disposed, then a special form of Gauss' law may be applicable.

The integral form of Eq. (3-9a), obtained by integrating both sides of Eq. (3-9a) over volume V bounded by surface S, is

$$\oiint_S \mathbf{E}\cdot d\mathbf{s} = \frac{(Q_f + Q_p)}{\varepsilon_0}$$

For an ideal dielectric with no free charge, the above form is reduced to

$$\oiint_S \mathbf{E}\cdot d\mathbf{s} = \frac{Q_p}{\varepsilon_0}$$

where Q_p represents the bound charge enclosed within surface S.

Using the above results, find the electric field $\mathbf{E}$ and the displacement vector $\mathbf{D}$ for the radially polarized sphere of Problem 3-2(a).

3-10. Find the electric field $\mathbf{E}$ and the displacement vector $\mathbf{D}$ for the radially polarized cylinder of Problem 3-2(b). (See Problem 3-9)

3-11. Find the electric field $\mathbf{E}$ and the displacement vector $\mathbf{D}$ for the radially polarized spherical shell of Problem 3-3(b). (See Problem 3-9)

3-12. Find the electric field $\mathbf{E}$ and the potential V for the polarized cylinder of Problem 3-4. (See Problem 3-9)

3-13. Find the electric field $\mathbf{E}$ and the potential V for the polarized sphere of Problem 3-5. (See Problem 3-9)

3-14. Consider a dipole of arbitrary location (x', y', z') and arbitrary orientation as specified by dipole moment $\mathbf{p}$. Show that the potential may be represented as follows:

$$V(r, \theta) = \frac{\mathbf{p}\cdot\mathbf{R}}{4\pi\varepsilon_0 R^3} = \frac{1}{4\pi\varepsilon_0}\frac{\mathbf{p}\cdot(\mathbf{r}-\mathbf{r}')}{|\mathbf{r}-\mathbf{r}'|^3}$$

where $\mathbf{r}'$ is a vector from the origin to (x', y', z'). Compare the result with Eq. (2-67).

3-15. A simple classical model of an atom consists of a nucleus of positive charge Ne surrounded by a spherical electron cloud of the same total negative charge where N is the atomic number and $e = 1.6 \times 10^{-19}$ [C] is the electron charge. An external electric field E_0 will cause the nucleus to be displaced a distance d from the center of the electron cloud, thus polarizing the atom (see Figure 3-2(d)). Assuming a *uniform* charge distribution within the electron cloud of radius a, find

(a) the electric field $\mathbf{E}_i$ produced by the electron cloud ($r < a$)
(b) the displacement d (the nucleus is to be located at $r = d$)
(c) the net dipole moment p of the atom
(d) the electric susceptibility χ_e of the bulk medium, assuming that the number of atoms per unit volume is *n*. Relate the polarization $\mathbf{P}$ to the field E_0.

[HINT] The external field E_0 balances the internal field produced by the electron cloud at equilibrium. You can use the result of Example 2-3.

3-16. A slab of dielectric material, infinite in the x-, y- directions and extending from $z = 0$ to $z = d$, is permanently polarized with a uniform polarization

$$\mathbf{P} = \mathbf{a}_z\, P_o \quad (0 < z < d)$$

where P_o is a constant.

(a) Find the bound (polarization) charge densities: ρ_{pv} inside the slab and ρ_{ps} at the two surfaces, $z = 0$ and $z = d$.
(b) Find the electric field, caused by the polarization of the slab, everywhere: (i) $z > d$, (ii) $0 \le z \le d$, (iii) $z < 0$.

3-17. Repeat Problem 3-16, when the slab has the following non-uniform polarization:

$$\mathbf{P} = \mathbf{a}_z\, P_o \frac{z}{d} \quad (0 < z < d)$$

3-18. A uniformly charged sphere of radius a and volume charge density ρ_{vo} [C/m^3] is surrounded by a dielectric shell of inner and outer radii, b and c ($c > b > a$). and permittivity ε. i.e., the shell is a *linear* dielectric.

(a) Find $\mathbf{E}$, $\mathbf{D}$ and $\mathbf{P}$ everywhere as functions of r: (i) $r < a$, (ii) $a < r < b$, (iii) $b < r < c$, (iv) $r > c$.
(b) Calculate all the bound charge densities, ρ_{pv} and ρ_{ps}.

[HINT] Find $\mathbf{D}$ first. You may use Gauss' law for $\mathbf{D}$, since the charge distribution has spherical symmetry.

3-19. A very long straight wire, carrying uniform line charge density $\rho_{\ell o}$ [C/m], is placed in a very long dielectric cylinder of radius a. The dielectric constant of the cylinder is ε_r, i.e., it is a *linear* dielectric.

(a) Find $\mathbf{E}$, $\mathbf{D}$ and $\mathbf{P}$ everywhere as functions of ρ (the distance from the axis of the wire): (i) $\rho < a$, (ii) $\rho > a$.
(b) Calculate all the bound charge densities, ρ_{pv} and ρ_{ps}.

[HINT] Find $\mathbf{D}$ first. You may use Gauss' law for $\mathbf{D}$, since the charge distribution has cylindrical symmetry.

3-20. Repeat Problem 3-19, when the long line charge $\rho_{\ell o}$ is surrounded by a very long dielectric cylindrical shell of inner and outer radii, a and b (b > a). Assume that the dielectric shell has permittivity ε.

3-21. Consider the cylindrical geometry of Figure 3-13(b). The surfaces $\rho = a, c$ are perfect conductors.

(a) Find the capacitance per unit length. Show that it satisfies Eq. (3-32).
(b) Find the approximate capacitance for cylinders of length ℓ. Specify limitations on dimensions so that fringing fields may be neglected.

3-22. Consider the cylindrical geometry of Figure 3-13(c). The surfaces $\rho = a, b$ are perfect conductors.

(a) Find the capacitance per unit length. Show that it satisfies Eq. (3-34).
(b) Find the approximate capacitance for cylinders of length ℓ. Specify limitations on dimensions.

3-23. Consider the cylindrical geometry of Figure 3-13(d). The surfaces $\rho = a, c$ are perfect conductors.

(a) Find the approximate capacitance per unit length. Specify limitations on dimensions.
(b) Find the approximate capacitance for cylinders of length ℓ. Specify all limitations on dimensions.

3-24. Figure 3-17 shows parallel plate capacitors in series, parallel and series-parallel combinations. Top and bottom surfaces are conductors.

(a) Find the total capacitance for the width w into the page in each case.
(b) Specify limitations on dimensions so that fringing may be neglected.

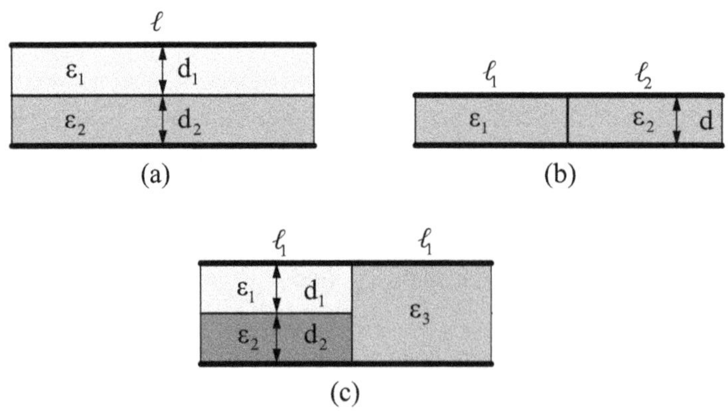

Figure 3-17. Parallel-plate capacitors of width w into the paper in (a) series (b) parallel and (c) series-parallel combinations.

3-25. Figure 3-18 shows two coaxial conducting cylinders of radii a, b, respectively. The inner cylinder is supported by a dielectric wedge of permittivity ε_1 which subtends an angle of 60° from the center.

(a) Determine the capacitance per unit length.
(b) A voltage V is applied between cylinders. Determine the electric field **E** everywhere and the free and bound surface charge densities at $\rho = a$.

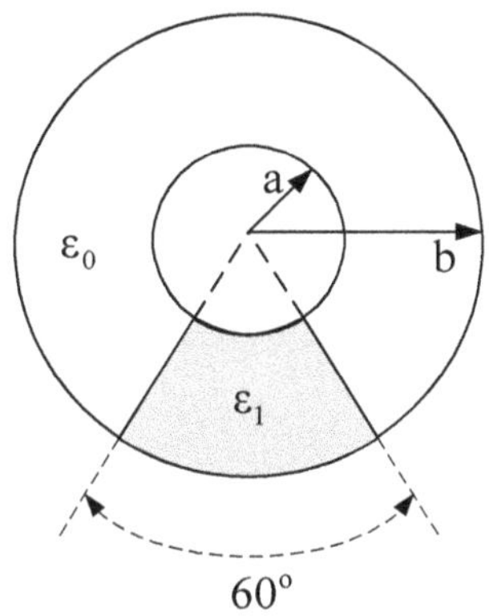

Figure 3-18. A dielectric wedge/cone supporting an inner cylinder/sphere.

3-26. Reconsider Figure 3-18, which is now interpreted to represent concentric conducting spheres of radii a, b. The inner sphere is supported by a dielectric cone of permittivity ε, and cone angle 60°.

(a) Determine the capacitance.
(b) A voltage V is applied between spheres. Determine the electric field E everywhere and the free and bound surface charge densities at $r = a$.

3-27. Consider a spherical capacitor with three dielectrics; ε_1 for $a < r < b$, ε_2 for $b < r < c$, ε_3 for $c < r < d$. Conductors cover the surfaces $r = a, d$. Determine the capacitance.

3-28. Consider a spherical capacitor consisting of N dielectrics in series, with permittivity ε_k for $r_k < r < r_{k+1}$ and $k = 1, 2, 3 \ldots, N$. Conductors cover the surfaces $r = r_1$ and $r = r_{N+1}$. Determine the capacitance.

3-29*. Consider a coaxial line with conducting cylinders at $\rho = a, b$. The coaxial line is filled with an inhomogeneous dielectric of permittivity ε where $\varepsilon = \varepsilon_0 (\rho/a)$. A voltage V is applied between cylinders.

(a) Find the capacitance per unit length.
(b) Find the bound volume and surface charge densities ρ_{pv}, ρ_{ps} everywhere.
[HINT] Assume a free charge density ρ_ℓ on the inner cylinder.

3-30*. A spherical capacitor has conducting spherical surfaces at $r = a, b$. It is filled with an inhomogeneous material of permittivity ε where $\varepsilon = \varepsilon_0 (r/a)$. A voltage V is applied between conductors at $r = a, b$.

(a) Find the capacitance.
(b) Find the bound volume and surface charge densities.
(c) Find the electric field **E** everywhere.

[HINT] Compare with Problem 3-28.

3-31. Calculate the stored energy of a spherical cloud of uniform volume charge density, given in Example 3-8, using the energy in terms of fields, Eq. (3-46). Show that the result agrees with Eq. (3-47).

3-32. Consider a cylindrical double cloud of charge

$$\begin{aligned} \rho_v &= \rho_o \quad (\rho \le a) \\ &= -\rho_o \quad (a \le \rho \le a\sqrt{2}) \end{aligned}$$

Calculate the stored energy per unit length.

3-33. Consider a spherical cloud of nonuniform volume charge density:

$$\rho_v = \rho_o \frac{r}{a}, \quad (0 \le r \le a)$$

where ρ_o is a constant. Calculate the electrostatic energy of this system (a) using Eq. (3-46) (in terms of fields), and (b) using Eq. (3-45) (in terms of charges).

[HINT] Calculate the fields (using Gauss' law) and potentials first.

3-34. Two spherical clouds of charge of radius *a* and volume charge densities ρ_{v1}, ρ_{v2}, respectively, are initially separated by a large distance. Find the work required to bring the clouds into coincidence.

3-35. Two cylindrical clouds of charge of radius *a* and infinite length and volume charge densities ρ_{v1}, ρ_{v2}, respectively, are initially separated by a large distance. Find the work required per unit length to bring the clouds together.

3-36. A dielectric block of material with permittivity ε and thickness d is inserted between parallel plates of separation 2d, leaving an air gap of thickness d (see Figure 3-19). A voltage V_o is applied between plates. Find the force per unit area on the upper plate.

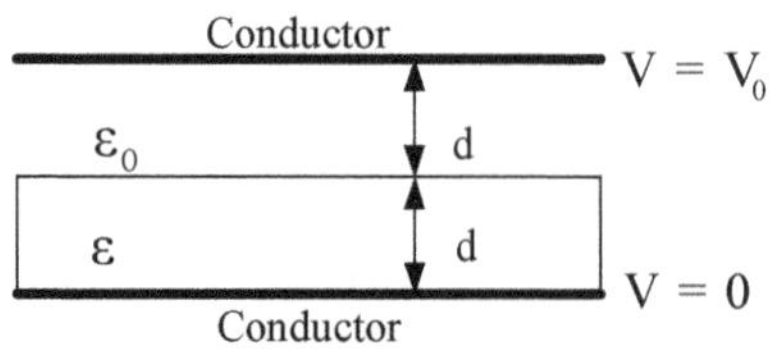

Figure 3-19. Parallel plates partially filled with a dielectric.

3-37. Consider a spherical capacitor with radii a, b. A voltage V is applied between conductors. Find the force of attraction per unit area on the conductors at $r = a, b$.

3-38. A coaxial line of radii a, b is inserted normally into a liquid dielectric of permittivity ε and mass density m (see Figure 3-20). A voltage V is applied between conductors.

(a) Find the electric field $\mathbf{E}$.
(b) Find the force tending to pull the dielectric liquid into the coaxial line.
(c) To what height H does the dielectric liquid rise?

Assume that $b–a << a$, so that fringing fields may be neglected.

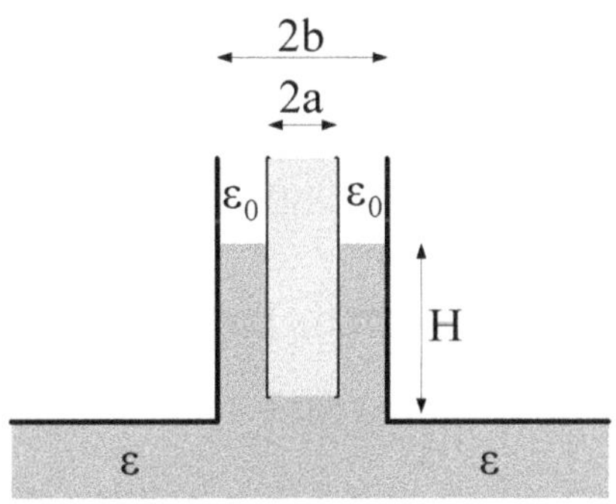

Figure 3-20. A coaxial line partially filled with a dielectric liquid.

3-39. A spherical capacitor has conducting spherical surfaces at $r = a, b$. A voltage V is applied between conductors. A fluid dielectric of permittivity ε is pumped into the capacitor, completely filling it. Find the mechanical work done by the pump and the electrical work done by the constant voltage source.

3-40*. An inhomogeneous block of dielectric is shown in Figure 3-21. The permittivity ε varies linearly from ε_0 to $2\varepsilon_0$. The block is partially inserted between parallel plates of separation d. A voltage V is applied between plates. Find the force on the block per unit width into the page as a function of insertion distance ℓ.

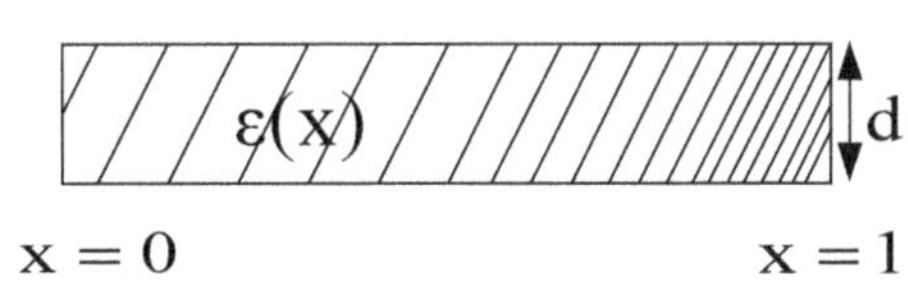

Figure 3-21. An *inhomogeneous* block of dielectric $\varepsilon(x) = \varepsilon_0(1+x)$.

SPECIAL SOLUTION TECHNIQUES

CHAPTER 4

4.1. INTRODUCTION

Thus far we have learned how one could find the electric field, given the stationary charge distribution or the polarization. The latter gives bound charge distribution through Eqs. (3-2c) and (3-3). If the known charge distributions (either free or bound) have certain symmetries, Gauss' law, Eqs. (2-13) and (3-12), is useful in finding the electric field. Otherwise, for arbitrary charge distributions, we use Eqs. (2-27) or (3-6) to calculate $\mathbf{E}$ directly, or use Eqs. (2-46) or (3-5) to calculate the electric potential V from which $\mathbf{E}$ can be easily obtained by $-\nabla V$.

However, in many practical electrostatic problems involving conductors, the exact charge distribution is not known in advance, although we can control the total charge or the potential of each conductor. In such cases, the above methods are not useful in finding the fields; thus we need to seek other methods. In this chapter, we discuss two special techniques that have proven to be very useful in finding the electric potential. They are the method of images and the method of separation of variables. With both methods, one basically solves Laplace's equation, which is the second-order partial differential equation that the potential V must satisfy. These methods can also be used in solving certain magnetostatic problems.

4.2. LAPLACE'S AND POISSON'S EQUATIONS

In this section we derive Poisson's equation and Laplace's equation. To derive these equations, we go back and visit the two basic (fundamental) laws for the electric field.

$$\nabla \times \mathbf{E} = 0 \tag{2-10}$$

$$\nabla \cdot \mathbf{D} = \rho_v \tag{3-10b}$$

In a linear homogeneous dielectric medium of permittivity ε, Eq. (3-10b) reduces to

$$\nabla \cdot \mathbf{E} = \frac{\rho_v}{\varepsilon} \tag{4-1}$$

As discussed in Section 2.7, from (2-10) we have

$$\mathbf{E} = -\nabla V \qquad \text{(2-41) or (4-2)}$$

Substituting Eq. (4-2) into Eq. (4-1), we obtain

$$\nabla \cdot (-\nabla V) = -\nabla \cdot (\nabla V) = -\nabla^2 V = \frac{\rho_v}{\varepsilon}$$

thus we have

$$\nabla^2 V = -\frac{\rho_v}{\varepsilon} \tag{4-3}$$

where $\nabla^2 V$ is the Laplacian of V (see Section 1.4.7). Equation (4-3) is known as **Poisson's equation**. It is a second-order partial differential equation that the electric potential should satisfy when the charge distribution ρ_v is given in a linear dielectric. One may note that it takes only *one* differential equation (Poisson's equation) to determine V, because V is a scalar. On the other hand, for **E** we need *two*, involving the divergence and the curl.

Very often, we are interested in finding V in a region where there is no charge, $\rho_v = 0$, a so-called *source-free* region. In such a region, Eq. (4-3) reduces to

$$\nabla^2 V = 0 \text{ in a source-free region} \tag{4-4}$$

which is called **Laplace's equation.** You may ask how one could have V or **E** if there is no charge, $\rho_v = 0$. Of course, if $\rho_v = 0$ *everywhere*, then $V = 0$. We explain that charges may exist *outside* the source-free region. Laplace's equation, (4-4), is the theme of this chapter. It is fundamental to the subject of electrostatics and also appears in many different branches of physics, such as the gravitational field, magnetostatics, and heat conduction. In Cartesian coordinates, Laplace's equation becomes

$$\frac{\partial^2 V}{\partial x^2} + \frac{\partial^2 V}{\partial y^2} + \frac{\partial^2 V}{\partial z^2} = 0 \tag{4-5}$$

For other coordinates, see Table 1-4.

4.2.1. Laplace's Equation in One Dimension

In this subsection, we consider the solutions of Laplace's equation in one dimension through examples.

EXAMPLE 4-1. A PARALLEL-PLATE CAPACITOR

A voltage V_0 is applied between two parallel conducting plates separated by a distance d as shown in Figure 4-1.

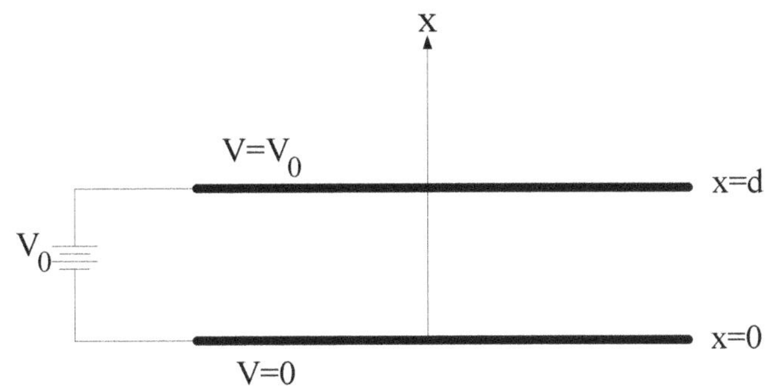

Figure 4-1. A parallel-plate capacitor for Example 4-1.

Find the potential distribution and the electric field in the region ($0 \le x \le d$) between the two plates. Assume both plates are infinitely large.

SOLUTION

Since the region $0 \le x \le d$ is free of charge, V will satisfy Laplace's equation. Because both plates are infinitely large planar surfaces and the potential is uniform at fixed x, we expect that V will not depend on y and z, i.e.,

$\frac{\partial}{\partial y} = \frac{\partial}{\partial z} = 0$. In this case, Laplace's equation becomes one-dimensional:

$$\nabla^2 V = \frac{\partial^2 V}{\partial x^2} = 0 \tag{4-6}$$

Integrating Eq. (4-6) with respect to x twice, we obtain

$$V(x) = Ax + B \tag{4-7}$$

To determine the unknown coefficients A and B, we use the boundary conditions:

$$\begin{aligned} V(x = d) &= V_0 \\ V(x = 0) &= 0 \end{aligned} \tag{4-8}$$

which result in

$$A = \frac{V_0}{d}, \quad B = 0 \tag{4-9}$$

Substituting Eq. (4-9) into Eq. (4-7), we obtain

$$V(x) = V_0 \frac{x}{d}, \quad 0 \le x \le d \tag{4-10}$$

The electric field is given by

$$\mathbf{E} = -\nabla V = -\mathbf{a}_x \frac{\partial V}{\partial x} = -\mathbf{a}_x \frac{V_0}{d} \tag{4-11}$$

The field is uniform as was shown in Example 2-6. The surface charge density at the upper plate can be obtained by Eq. (2-63):

$$\rho_s = \varepsilon_0 E_n = -\varepsilon_0 E_x \Big|_{x=d} = \varepsilon_0 \frac{V_0}{d} \left[\frac{C}{m^2}\right] \tag{4-12}$$

EXAMPLE 4-2. A COAXIAL CABLE

A coaxial cable consists of two coaxial conducting cylinders with radii a and b ($a < b$). It is filled with a linear dielectric of permittivity ε. A voltage V_0 is applied between inner and outer conductors, positive at the inner conductor (Figure 4-2). Find the potential distribution and the electric field in the region $a \le \rho \le b$. Assume that the cylinders are infinitely long.

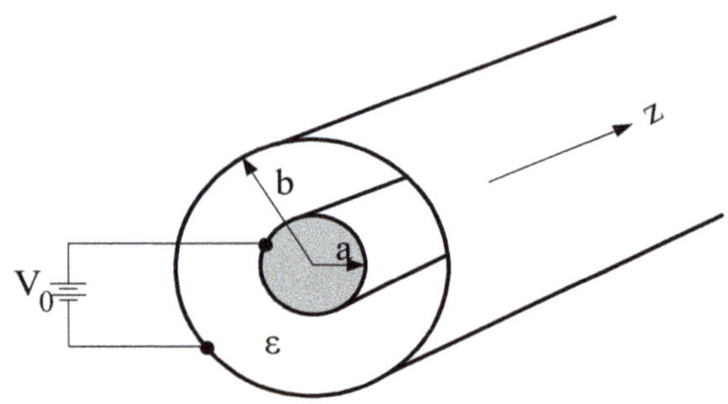

Figure 4-2. A coaxial cable for Example 4-2.

SOLUTION

Since the region $a \le \rho \le b$ is free of charge and the boundary surfaces coincide with the cylindrical coordinate surfaces, V satisfies Laplace's equation in cylindrical coordinates:

$$\nabla^2 V = \frac{1}{\rho}\frac{\partial}{\partial \rho}\left(\rho \frac{\partial V}{\partial \rho}\right) + \frac{1}{\rho^2}\frac{\partial^2 V}{\partial \phi^2} + \frac{\partial^2 V}{\partial z^2} = 0 \tag{4-13}$$

Because the cylinders are infinitely long in the z direction and the potential is uniform (or constant) on the cylindrical surfaces ($\rho = a$ and $\rho = b$), we expect that V will not depend on z and ϕ, i.e., $\frac{\partial}{\partial \phi} = \frac{\partial}{\partial z} = 0$. In this case, Eq. (4-13) becomes

$$\frac{\partial}{\partial \rho}\left(\rho \frac{\partial V}{\partial \rho}\right) = 0 \tag{4-14}$$

Integrating Eq. (4-14) with respect to ρ, we obtain

$$\rho \frac{\partial V}{\partial \rho} = C_1 \quad \text{or} \quad \frac{\partial V}{\partial \rho} = \frac{C_1}{\rho} \tag{4-15}$$

Integrating Eq. (4-15) once more with respect to ρ, we have

$$V(\rho) = C_1 \ln \rho + C_2 \tag{4-16}$$

We apply the two boundary conditions:

$$\begin{aligned} V(\rho = a) &= V_0 \\ V(\rho = b) &= 0 \end{aligned} \tag{4-17}$$

which gives

$$C_1 \ln a + C_2 = V_0 \tag{4-18a}$$

$$C_1 \ln b + C_2 = 0 \tag{4-18b}$$

Solving Eqs. (4-18), we obtain

$$C_1 = \frac{V_0}{\ln \frac{a}{b}}, \quad C_2 = -\frac{V_0 \ln b}{\ln \frac{a}{b}} \tag{4-19}$$

Substituting Eq. (4-19) into Eq. (4-16), we obtain

$$\begin{aligned} V(\rho) &= \frac{V_0}{\ln \frac{a}{b}} \{\ln \rho - \ln b\} \\ &= V_0 \frac{\ln \frac{\rho}{b}}{\ln \frac{a}{b}}, \quad a \le \rho \le b \end{aligned} \tag{4-20}$$

The electric field is given by

$$\begin{aligned} \mathbf{E} &= -\nabla V = -\mathbf{a}_\rho \frac{\partial V}{\partial \rho} \\ &= -\mathbf{a}_\rho \frac{V_0}{\ln \frac{a}{b}} \frac{1}{\rho} = \mathbf{a}_\rho \frac{V_0}{\ln \frac{b}{a}} \frac{1}{\rho} \end{aligned} \tag{4-21}$$

The radial dependence of the electric field agrees with the result of an infinitely long line charge with uniform charge density in Example 2-5. The surface charge density on the surface of the inner conductor can be obtained by

$$\rho_s = \varepsilon E_n = \varepsilon E_\rho \Big|_{\rho=a} = \varepsilon \frac{V_0}{\ln \frac{b}{a}} \frac{1}{a} \qquad (4\text{-}22)$$

The capacitance of a cylindrical capacitor of length ℓ ($\ell >> b$) can also be calculated from Eq. (4-22) as follows:

$$C = \frac{Q}{V_0} = \frac{\rho_s \cdot (2\pi a \ell)}{V_0} = \frac{2\pi\varepsilon \ell}{\ln \frac{b}{a}} \qquad (4\text{-}23)$$

which agrees with Eq. (3-37) in Example 3-7.

4.2.2. Uniqueness Theorem

Laplace's equation itself alone does not determine V uniquely. In addition, a suitable set of boundary conditions must be supplied as shown in Examples 4-1 and 4-2. The general statement about the uniqueness of the solutions of Laplace's equation (or Poisson's equation) is as follows:

> The solution to Laplace's equation in some region is uniquely determined if the value of V or $\frac{\partial V}{\partial n}$ (normal derivative) is specified on all boundaries of the region (Figure 4-3).

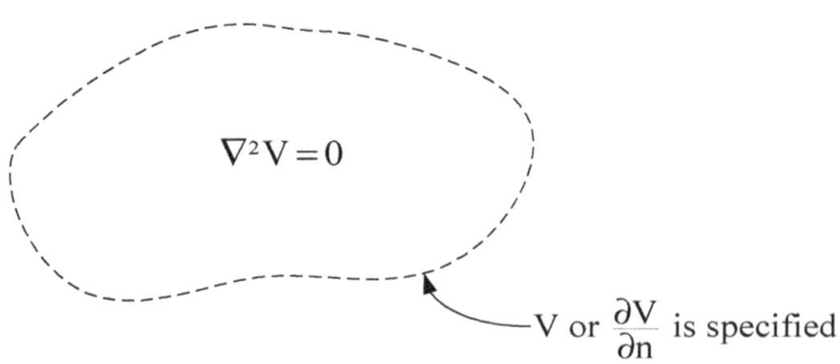

Figure 4-3. Uniqueness theorem.

Due to this uniqueness theorem, a solution of an electrostatic problem satisfying the set of boundary conditions is the only possible solution. It does not matter how you find a solution; as long as it satisfies (i) Laplace's equation and (ii) the correct boundary conditions, it is the only correct solution. The uniqueness theorem is so powerful that even intelligent guessing may give a correct solution easily. The formal proof of this theorem can be found in, for example, Cheng (1989).*

We note from the statement of the uniqueness theorem that there is a choice of specifying boundary values: V or $\frac{\partial V}{\partial n}$. The problem for which the values of V are specified everywhere on the boundaries is called the *Dirichlet* boundary-value problem (BVP). The BVP for which the values of $\frac{\partial V}{\partial n}$ are specified everywhere on the boundaries is called the *Neumann* problem. There is also a mixed BVP for which V is specified on some boundaries and $\frac{\partial V}{\partial n}$ is specified on other remaining boundaries. In the following two sections, we will consider only the Dirichlet BVP's.

4.2.3. Marquis Pierre-Simon de Laplace (1749–1827)

Laplace was a French astronomer and mathematician. He spent most of his working life focusing on astronomy, particularly on the solar system, and the mathematical problems that arose from these studies. He was born Pierre-Simon Laplace into a middle class family in Normandy. The "Marquis" and the "de" were added by King Louis XVIII, who came to power in 1814 when the monarchy was restored after the French Revolution and the Napoleonic Empire.

Laplace's father expected him to establish a career in the Church. He studied theology at the University of Caen for two years, and then, interrupting his plans, went off to Paris. His mathematical talents were quickly recognized in Parisian academic circles, and he obtained a professorship in mathematics at the Military School of Paris at the young age of 20.

Pierre was a young man in a hurry. He proceeded to flood the Academy of Science with four or five papers a year on planetary orbits, satellites, probability, difference equations, differential equations, etc. In this way, he quickly established himself as one of the brightest astronomers of his day and was elected to the Academy at the age of 24. Some of his early work was concerned with the stability of the solar system. Newton had felt that the solar system was unstable over time and that it would not be maintained without some (divine?) intervention. Laplace showed that, within some substantial restrictions, the solar system is, in fact, stable. He also refined the nebular hypothesis, which attributed the formation of the planets to gaseous nebulae from the equatorial rim of the sun that had collapsed and coalesced into these bodies. This was in stark contrast to the Comte de Buffon's hypothesis, which assumed that a comet striking the sun was the cause of planetary formation.

Laplace discovered his eponymous equation, $\nabla^2 V = 0$, while he was studying the gravitational potential and gravitational forces. This equation is also satisfied by the electrostatic potential, the magnetostatic potential, the temperature, the velocity potential of hydrodynamics, etc.

*D. K. Cheng, *Field and Wave Electromagnetics*, Addison-Wesley, 1989, 2nd Ed., Chapter 4.

Laplace is known as the French Newton because of his many contributions to our understanding of the solar system. *Celestial Mechanics* is his best-known astronomical work. It contains 16 books, the last published just two years before his death. In addition to his work on astronomy, he wrote a treatise on probability and is regarded as one of the founders of the discipline. Of course, he also laid the groundwork for the Laplace transform.

Laplace was a very compelling writer. Eventually he was elected to the French (literary) Academy. But he did have certain quirks and foibles. We are all familiar with the expression "It is easy to see … " which then requires several hours of work to fill in the missing proof. Certainly Laplace did not invent this trick, but he was an inveterate user, to the dismay of his many readers. He also failed often to give proper credit to other scientific contributors, thus clouding his own contributions. But, in contrast, he was very generous to neophytes. The story is told by Biot of how he once gave a talk at the Academy. Laplace had actually performed the same work many years before, but rather than embarrassing the young man in public, he showed his notebook to Biot in private and urged him to publish.

In his latter years, Laplace was elected to the French (literary) Academy as noted and became its president in 1817. He died in his seventy-eighth year in 1827. His last words were "we know so little; what we do not know is immense."

4.3. METHOD OF IMAGES

The method of images is very useful in handling the problems of charges in the vicinity of conductors. The idea is to replace the conductors by the so-called image charges to take into account the effect of conductors. For example, a power transmission line hanging above the conducting earth can be treated by the method of images.

4.3.1. A Point Charge Near a Grounded Conducting Plane

Let's consider the problem of a point charge q located a distance d above an infinite grounded conducting plane as shown in Figure 4-4(a). You are asked to find the electric potential everywhere above the plane. Before we solve this problem, think first what happens with respect to the conductor. Since the conductor has many free charges inside, the point charge q, (if positive), will attract negative free charges inside the conductor; thus negative charges will appear on the surface of the conducting plane. Then the total electric potential or the field will be due to the original point charge q and the induced surface charges $\rho_s(x,y)$, which will be a function of position (x,y) on the plane. So we can write the potential as follows:

$$V(x, y, z) = \frac{q}{4\pi\varepsilon_0\sqrt{x^2+y^2+(z-d)^2}} + \iint \frac{\rho_s(x',y')\,dx'dy'}{4\pi\varepsilon_0\sqrt{(x-x')^2+(y-y')^2+z^2}}$$

The trouble is that (i) ρ_s must be obtained first and (ii) the surface integral is very difficult to evaluate. In fact, ρ_s can be found only after the potential or the field is obtained. Therefore, we will seek other methods.

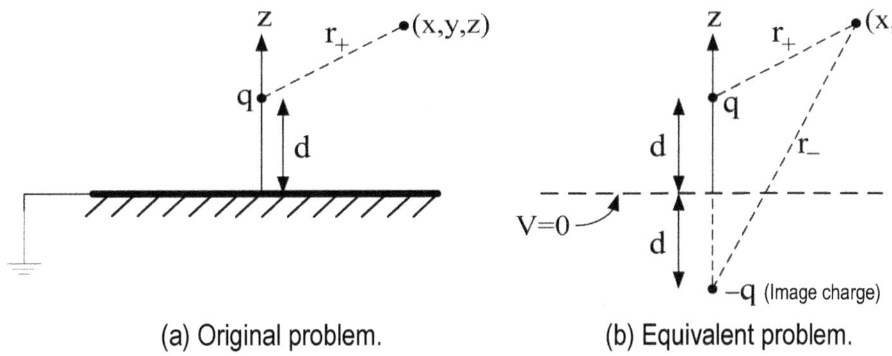

Figure 4-4. A point charge near a grounded conducting plane.

The problem of finding the electric potential $V(x, y, z)$ for $z > 0$ (note that $V = 0$ everywhere for $z < 0$, i.e., for the region inside the conductor) can be formulated as follows:

V satisfies $\nabla^2 V = 0$ everywhere for $z > 0$ except at the point charge, subject to the boundary conditions:

(a) $V = 0$ when $z = 0$ (since the conducting plane is grounded)

(b) $V \to 0$ as $r = \sqrt{x^2 + y^2 + z^2} \to \infty$ (far from the charge)

(c) $V \to \dfrac{q}{4\pi\varepsilon_0 r_+}$ as $r_+ = \sqrt{x^2 + y^2 + (z-d)^2} \to 0$ (very close to the charge)

It appears very difficult to construct a solution for V that satisfies all these conditions. The method of images greatly simplifies the difficulty. Suppose we remove the conductor and replace it by another point charge $-q$ at $z = -d$ as shown in Figure 4-4(b). This is a completely different problem, but it will give the correct solution that we seek. The solution of this new problem (or the equivalent problem) is very easy: the potential due to this charge configuration, i.e., a pair of point charges, is given by

$$\begin{aligned} V(x, y, z) &= \frac{q}{4\pi\varepsilon_0 r_+} + \frac{(-q)}{4\pi\varepsilon_0 r_-} \\ &= \frac{1}{4\pi\varepsilon_0}\left\{\frac{q}{\sqrt{x^2 + y^2 + (z-d)^2}} - \frac{q}{\sqrt{x^2 + y^2 + (z+d)^2}}\right\} \end{aligned} \tag{4-24}$$

Now one can show that V given by Eq. (4-24) satisfies $\nabla^2 V = 0$ everywhere for $z > 0$ except at the point $(0,0,d)$ and satisfies all the boundary conditions (a), (b) and (c) described above. By the uniqueness theorem, this is the only solution to the original problem for $z > 0$. The complete solution for the original problem is

$$V(x, y, z) = \begin{cases} \dfrac{1}{4\pi\varepsilon_0}\left(\dfrac{q}{r_+} - \dfrac{q}{r_-}\right), & z \geq 0 \\ 0, & z \leq 0 \end{cases} \tag{4-25}$$

We call the charge $-q$ at $(0,0,-d)$ the **image charge**. In summary, we find the solution of the original problem by solving the completely different equivalent problem for which

(i) the conductor $(z < 0)$ is replaced by air (the same medium as above)
(ii) an image charge $(-q)$ is placed at $(0, 0, -d)$.

The difference in solutions of the two problems—the original and equivalent problems—lies in the region $z < 0$, as shown in Eqs. (4-25) and (4-24). The latter is a complete solution for the equivalent problem everywhere including $z < 0$. We call the region including the point charge $(z > 0)$ the *region of interest* and the conducting region $(z < 0)$ the *region of no interest*. We note that the image charge must be placed outside the region of interest (i.e., inside the conductor) so as not to disturb the requirement of $\nabla^2 V = 0$ in that region.

Since the potential is obtained, we can now calculate the actual induced surface charge density at $z = 0$ by using Eq. (2-63).

$$\rho_s(x, y) = \varepsilon_0 E_n = -\varepsilon_0 \left.\frac{\partial V}{\partial z}\right|_{z=0} \tag{4-26}$$

From Eq. (4-24)

$$\frac{\partial V}{\partial z} = \frac{1}{4\pi\varepsilon_0}\left\{\frac{-q(z-d)}{[x^2+y^2+(z-d)^2]^{3/2}} + \frac{q(z+d)}{[x^2+y^2+(z+d)^2]^{3/2}}\right\}$$

So

$$\rho_s(x, y) = \frac{-qd}{2\pi(x^2+y^2+d^2)^{3/2}} \tag{4-27}$$

The induced charge is negative (assuming q is positive) as expected. The total induced charge

$$Q = \int_{-\infty}^{\infty}\int_{-\infty}^{\infty} \rho_s(x, y)\, dx\, dy$$

is equal to the magnitude of the image charge, $-q$ (see Problem 4–8).

EXAMPLE 4-3 A POINT CHARGE NEAR TWO CONDUCTING PLATES

A point charge q is in the vicinity of a corner where two grounded conducting plates meet at right angle (Figure 4-5(a)). Find the potential in the region $x > 0$, $y > 0$.

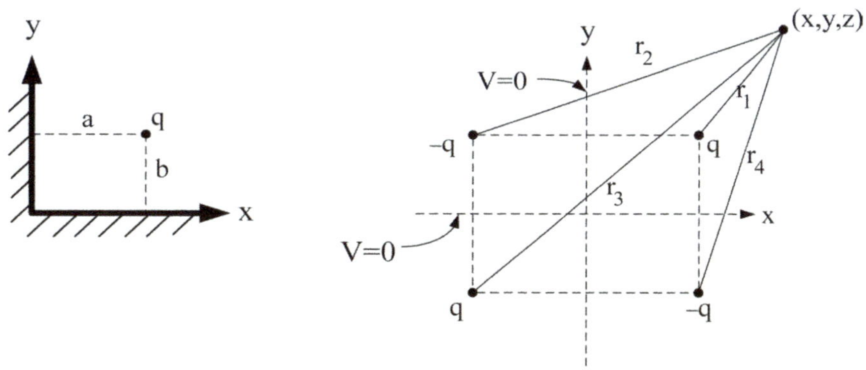

(a) Original problem. (b) Equivalent problem.

Figure 4-5. A point charge in a right-angle conducting corner.

SOLUTION

This problem can be solved by the method of images. The boundary conditions to be satisfied are:

$$V = 0 \text{ at } x = 0 \text{ (vertical plane)} \tag{4-28a}$$

$$V = 0 \text{ at } y = 0 \text{ (horizontal plane)} \tag{4-28b}$$

In order to satisfy Eq. (4-28a), we need an image charge $-q$ at the point $(-a, b)$. In addition, the boundary condition, Eq. (4-28b), requires two more image charges, $-q$ at $(a, -b)$ and $+q$ at $(-a, -b)$ so that each pair of $+q$ and $-q$ guarantee the zero potential on the horizontal plane. Therefore, the equivalent problem is to find the potential due to a group of four point charges as shown in Figure 4-5(b). Then the potential is given by

$$V(x, y) = \frac{1}{4\pi\varepsilon_o}\left\{\frac{q}{r_1} + \frac{(-q)}{r_2} + \frac{q}{r_3} + \frac{(-q)}{r_4}\right\} \tag{4-29}$$

4.3.2. A Line Charge and a Parallel Conducting Cylinder

EXAMPLE 4-4

Consider an infinitely long line charge ρ_ℓ [C/m] located a distance d away from the axis of an infinitely long, parallel, conducting circular cylinder of radius a as shown in Figure 4-6(a). Find the electric potential everywhere.

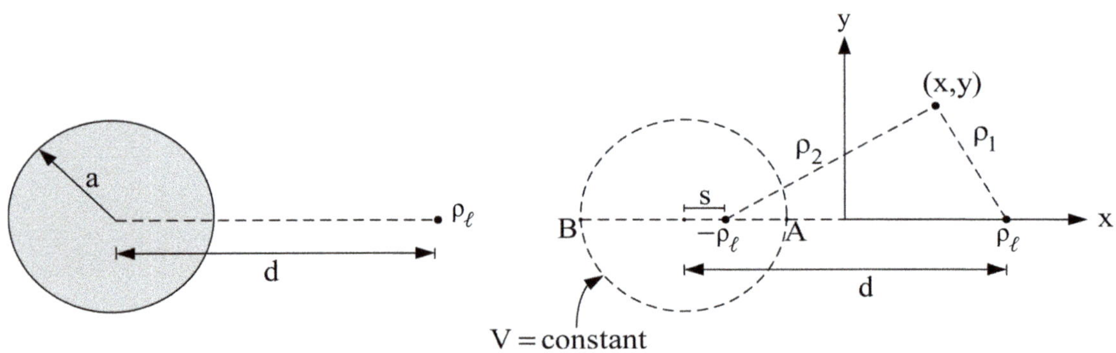

(a) Original problem. (b) Equivalent problem.

Figure 4-6. Cross section of a line charge and a parallel conducting cylinder.

SOLUTION

We again solve this problem by the method of images. The boundary condition is V = constant on the surface of the conducting cylinder. The main task is to replace the conducting cylinder by an image charge ρ_ℓ'.

Let's first consider the potential of a pair of infinitely long, opposite line charges, ρ_ℓ and $-\rho_\ell$ (Figure 4-6(b)). Using Eq. (2-56) for a single line charge, we can write

$$\begin{aligned} V &= \frac{\rho_\ell}{2\pi\varepsilon_0}\ln\left(\frac{\rho_o}{\rho_1}\right) + \frac{(-\rho_\ell)}{2\pi\varepsilon_0}\ln\left(\frac{\rho_o}{\rho_2}\right) \\ &= \frac{\rho_\ell}{2\pi\varepsilon_0}\left\{\ln\left(\frac{\rho_o}{\rho_1}\right) - \ln\left(\frac{\rho_o}{\rho_2}\right)\right\} = \frac{\rho_\ell}{2\pi\varepsilon_0}\ln\left(\frac{\rho_2}{\rho_1}\right) \end{aligned} \tag{4-30}$$

where ρ_o is the reference point. Letting the (x, y) coordinates of ρ_ℓ and $-\rho_\ell$ be $(x_0,0)$ and $(-x_0,0)$, the equipotential surfaces—lines in the cross section—are given by

$$\frac{\rho_\ell}{2\pi\varepsilon_0}\ln\left(\frac{\rho_2}{\rho_1}\right) = \text{constant}, \tag{4-31a}$$

or

$$\frac{\rho_2}{\rho_1} = \frac{\sqrt{(x+x_o)^2+y^2}}{\sqrt{(x-x_o)^2+y^2}} = \text{constant} \equiv \sqrt{C} \tag{4-31b}$$

where $C \geq 0$. Squaring both sides of Eq. (4-31b), we obtain

$$(x+x_o)^2 + y^2 = C(x-x_o)^2 + Cy^2 \tag{4-32a}$$

$$(1-C)x^2 + (1-C)y^2 + 2(1+C)x_0 x + (1-C)x_0^2 = 0 \tag{4-32b}$$

When $C \neq 1$, after algebraic manipulation, we have

$$\left\{x + \frac{1+C}{1-C}x_o\right\}^2 + y^2 = \left(\frac{2\sqrt{C}\,x_o}{1-C}\right)^2 \tag{4-33a}$$

which is the equation of a circle of radius $\frac{2\sqrt{C}\,x_o}{1-C}$, centered at $\left(-\frac{1+C}{1-C}x_o,\ 0\right)$. When $C = 1$, Eq. (4-32b) reduces to

$$4x_0 x = 0 \quad \text{or} \quad x = 0 \tag{4-33b}$$

which is the equation of the plane that bisects the line connecting the two line charges. We conclude from Eqs. (4-33) that the equipotential lines (surfaces) are a family of circles (cylinders) including a bisecting plane.

Consider the original problem now. If the surface of the conducting cylinder coincides with one of these equipotential surfaces, then it will satisfy the boundary condition that $V = \text{constant}$. Therefore, we place the image line charge

$$\boxed{\rho_\ell' = -\rho_\ell} \tag{4-34}$$

at a distance s away from the axis of the cylinder, on the line connecting ρ_ℓ and ρ_ℓ' as shown in Figure 4-6(b). Note that the image charge is on the right side of the axis of the cylinder because $\left(\frac{1+C}{1-C}\right)x_o > x_o$ or a family of circles on the left half plane $(0 < C < 1)$. To determine the value of s, we consider the potential of two points A and B on the surface of the conductor for which it is simple to write the potential or the ratio of ρ_2 and ρ_1 as follows:

$$\left.\frac{\rho_2}{\rho_1}\right|_A = \frac{a-s}{d-a} = \left.\frac{\rho_2}{\rho_1}\right|_B = \frac{a+s}{d+a} = \text{constant} \tag{4-35a}$$

$$\begin{gathered}(a-s)(d+a)=(a+s)(d-a)\\ ad+a^2-sd-sa=ad-a^2+sd-sa\\ 2sd=2a^2\end{gathered} \tag{4-35b}$$

Thus

$$\boxed{s=\frac{a^2}{d}} \tag{4-36}$$

Knowing ρ'_ℓ and s from Eqs. (4-34) and (4-36) and expressing ρ_1 and ρ_2 in terms of the coordinates (x, y), Eq. (4-30) gives the electric potential everywhere in the region outside the conducting cylinder (i.e., in the region of interest).

4.3.3. A Point Charge Near a Grounded Conducting Sphere

EXAMPLE 4-5

Consider a point charge q located a distance d away from the center of a *grounded* conducting sphere of radius a as shown in Figure 4-7 (a). Find the electric potential everywhere.

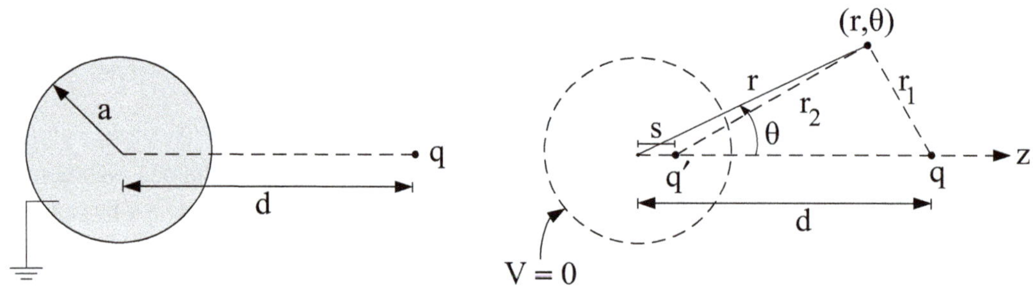

(a) Original problem. (b) Equivalent problem.

Figure 4-7. A point charge near a grounded conducting sphere.

SOLUTION

We can solve this problem by the method of images. We first notice that the boundary condition is $V = 0$ on the surface of the conductor because it is grounded. As mentioned earlier, we replace the conducting sphere by an image charge q' at an appropriate point inside the conductor region (region of no interest) such that the potential at the surface of the conductor is zero, i.e., $V = 0$ at $r = a$ (Figure 4-7(b)). Because of symmetry, we place the image charge at a point on the line connecting the center of the sphere and the location of the original charge q. Let the line be the z axis and the distance of q' from the center of the sphere be s. Now the question becomes, "Can we find the values of q' and s that will satisfy the boundary condition, $V = 0$ at $r = a$?" The potential at (r, θ) (spherical coordinates) due to q and q' is given by

$$V(r, \theta) = \frac{1}{4\pi\varepsilon_0}\left(\frac{q}{r_1} + \frac{q'}{r_2}\right) = \frac{1}{4\pi\varepsilon_0}\left\{\frac{q}{\sqrt{r^2 + d^2 - 2rd\cos\theta}} + \frac{q'}{\sqrt{r^2 + s^2 - 2rs\cos\theta}}\right\} \tag{4-37}$$

where the law of cosines is used.

It can be shown that if

$$\boxed{q' = -\frac{a}{d}q, \quad s = \frac{a^2}{d}} \tag{4-38}$$

then $V(r,\theta)$ vanishes when $r = a$ for any arbitrary value of θ, which corresponds to all points on the surface of the conducting sphere (see Problem 4-18). Thus the boundary conditions for the original problem are satisfied. Then the potential for $\rho > a$ (in the region of interest) is given by Eq. (4-37) with q' and s as given in Eq. (4-38).

4.4. METHOD OF SEPARATION OF VARIABLES

The separation of variables is a typical technique for solving partial differential equations (PDEs). Laplace's equation is a PDE involving three variables. The key idea is to look for solutions that are products of functions each of which depends on only one variable. In the process of separation of variables, a PDE involving n variables is transformed into a set of n ordinary differential equations, each involving one variable; thus, the PDE can be solved more readily. The method is applicable to boundary value problems for which the potential (V) or its normal derivative $\frac{\partial V}{\partial n}$ is specified on the boundaries of some region and the potential in the interior is to be found. We will deal with only two-dimensional problems for which the potential depends on only two coordinates: (x,y), (ρ,ϕ), and (r,θ). The method of separation of variables will also be used in solving the wave equation in Chapter 11.

4.4.1. Boundary Value Problems in Cartesian Coordinates

For an electric potential $V(x, y)$ that depends only on x and y and is independent of z, i.e., $\left(\frac{\partial}{\partial z}\right) = 0$, Laplace's equation (4-5) becomes

$$\frac{\partial^2 V}{\partial x^2} + \frac{\partial^2 V}{\partial y^2} = 0 \tag{4-39}$$

To achieve separation of variables, we assume the solution of Eq. (4-39) for $V(x, y)$ in the following form:

$$V(x,y) = X(x)\,Y(y) \tag{4-40}$$

where $X(x)$ and $Y(y)$ are functions of only x and y, respectively. Substituting Eq. (4-40) in Eq. (4-39), we have

$$Y(y)\frac{\partial^2 X(x)}{\partial x^2} + X(x)\frac{\partial^2 Y(y)}{\partial y^2} = 0 \tag{4-41}$$

Dividing Eq. (4-41) by $X(x)\,Y(y)$, we obtain

$$\frac{1}{X(x)}\frac{\partial^2 X(x)}{\partial x^2} + \frac{1}{Y(y)}\frac{\partial^2 Y(y)}{\partial y^2} = 0 \tag{4-42a}$$

or

$$\frac{1}{X(x)}\frac{d^2X(x)}{dx^2} = -\frac{1}{Y(y)}\frac{d^2Y(y)}{dy^2} \tag{4-42b}$$

Note that the partial derivatives are replaced by the ordinary derivatives because $X(x)$ and $Y(y)$ are functions of one variable. In order for Eq. (4-42) to be satisfied for all values of x and y, the functions on both sides of Eq. (4-42b) must be a constant which is independent of x and y. Setting the constant equal to k^2, we have

$$\frac{1}{X(x)}\frac{d^2X(x)}{dx^2} = -\frac{1}{Y(y)}\frac{d^2Y(y)}{dy^2} = k^2 \tag{4-43}$$

which leads to two ordinary differential equations:

$$\frac{d^2X(x)}{dx^2} - k^2 X(x) = 0 \tag{4-44a}$$

$$\frac{d^2Y(y)}{dy^2} + k^2 Y(y) = 0 \tag{4-44b}$$

k is called the separation constant. We can choose k^2 or $-k^2$ in Eq. (4-43). Note that we have converted a PDE into two ordinary differential equations. Solutions to Eqs. (4-44) are well known as they are second-order ordinary differential equations with constant coefficients.

When $k \neq 0$, the general solutions of Eqs. (4-44a) and (4-44b) are, respectively,

$$X(x) = A_k \sinh kx + B_k \cosh kx \tag{4-45a}$$

or

$$X(x) = A'_k\, e^{kx} + B'_k\, e^{-kx} \tag{4-45b}$$

$$Y(y) = C_k \sin ky + D_k \cos ky \tag{4-46}$$

The solutions for $X(x)$ can be written as either Eq. (4-45a) or Eq. (4-45b) because the hyperbolic functions and the exponential functions are related as follows:

$$\sinh u = \frac{1}{2}(e^u - e^{-u}) \tag{4-47a}$$

$$\cosh u = \frac{1}{2}(e^u + e^{-u}) \tag{4-47b}$$

These functions are plotted in Figure 4-8. Finally, the general solutions for $V(x,y)$ are given by

$$V(x,y) = (A_k \sinh kx + B_k \cosh kx)(C_k \sin ky + D_k \cos ky) \tag{4-48}$$

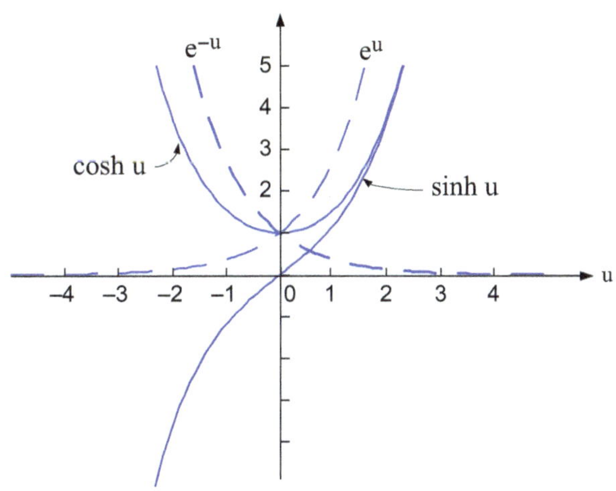

Figure 4-8. Hyperbolic and exponential functions.

There are another set of possible solutions when the separation constant $-k^2$ is chosen in Eq. (4-43). Then k^2 is replaced by $-k^2$ in Eqs. (4-44). It is clear that the new solutions are

$$V(x, y) = (A_k \sin kx + B_k \cos kx)(C_k \sinh ky + D_k \cosh ky) \tag{4-49}$$

The choice of appropriate separable solutions and the determination of coefficients and values of k depend on the specified boundary conditions of the problems. As seen in Eq. (4-48) and Eq. (4-49), one of the characteristics of the solutions to Laplace's equation is that they are oscillatory or harmonic (sine and cosine) in one direction and are non-oscillatory in the other direction.

When $k = 0$, solutions in Eqs. (4-48) and (4-49) are not valid. Letting $k = 0$ and solving Eqs. (4-44), we find

$$X(x) = A_o x + B_o \tag{4-50a}$$

$$Y(y) = C_o y + D_o \tag{4-50b}$$

Then

$$V(x, y) = (A_o x + B_o)(C_o y + D_o) \tag{4-51}$$

Solutions in the form of Eq. (4-50) were shown in Example 4-1. In summary, Eqs. (4-48), (4-49), and (4-51) constitute a complete set of separable solutions to two-dimensional Laplace's equation and they are listed in Table 4-1.

EXAMPLE 4-6

Consider an infinitely long rectangular pipe as shown in Figure 4-9. Three metal strips at $x = 0$, $x = a$ and $y = 0$ are grounded and the metal strip at $y = b$ is insulated from other strips and maintained at a constant potential V_0. Find the potential distribution inside the pipe.

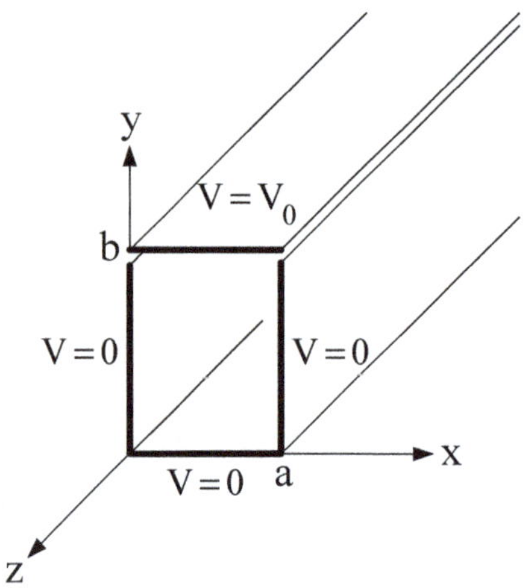

Figure 4-9. Boundary value problem for Example 4-6.

Table 4-1 Solutions of Laplace's Equation in Two-Dimensions

I. Cartesian Coordinates, $\dfrac{\partial}{\partial z} = 0$ (independent of z):

(1) $k \neq 0$,

$V(x, y) = (A_k \sinh kx + B_k \cosh kx)\,(C_k \sin ky + D_k \cos ky)$

or Replace $\sinh kx$ and $\cosh kx$ by e^{kx} and e^{-kx}

or Switch the x-dependence and the y-dependence ($x \leftrightarrow y$).

(2) $k = 0$,

$V(x, y) = (A_o x + B_o)\,(C_o y + D_o)$

II. Cylindrical Coordinates, $\dfrac{\partial}{\partial z} = 0$ (independent of z):

(1) $n \neq 0$,

$V(\rho, \phi) = (A_n \rho^n + B_n \dfrac{1}{\rho^n})(C_n \sin n\phi + D_n \cos n\phi)$

[Note] $\rho \cos\phi$ or $\rho \sin\phi$ ($n = 1$) gives a uniform field.
$\dfrac{1}{\rho}\cos\phi$ or $\dfrac{1}{\rho}\sin\phi$ ($n = 1$) gives a line-dipole field.

(2) $n = 0$,

$V(\rho, \phi) = (A_o \ln\rho + B_o)(C_o \phi + D_o)$

[Note] $\ln \dfrac{\rho_o}{\rho}$ is a potential of an infinite line charge.

III. Spherical Coordinates, $\dfrac{\partial}{\partial \phi} = 0$ (independent of ϕ):

$V(r, \theta) = A_n\, r^n\, P_n(\cos\theta) + B_n \dfrac{1}{r^{n+1}} P_n(\cos\theta)$

where

$P_o(\cos\theta) = 1$, $P_1(\cos\theta) = \cos\theta$, $P_2(\cos\theta) = \dfrac{1}{2}(3\cos^2\theta - 1)$, etc., are the Legendre polynomials.

[Note] (1) $\dfrac{1}{r}$ ($n = 0$) is a potential of a point charge.

(2) $\dfrac{1}{r^2}\cos\theta$ ($n = 1$) gives a dipole field.

(3) $r\cos\theta$ ($n = 1$) gives a uniform field.

SOLUTION

First of all, because the pipe is infinitely long in the z-direction with uniform rectangular cross section, we expect that the potential will not depend on z, i.e. $\frac{\partial}{\partial z}=0$. Since there is no charge inside the pipe, V(x,y) satisfies Laplace's equation subject to the following boundary conditions:

$$\text{(i) } V(x=0, y)=0, \qquad 0 \le y \le b \tag{4-52}$$

$$\text{(ii) } V(x=a, y)=0, \qquad 0 \le y \le b \tag{4-53}$$

$$\text{(iii) } V(x, y=0)=0, \qquad 0 \le x \le a \tag{4-54}$$

$$\text{(iv) } V(x, y=b)=V_o, \qquad 0 < x < a \tag{4-55}$$

We will solve this problem step by step.

Step 1. The first task is to choose between Eq. (4-48) and Eq. (4-49). To answer this question, we look at the boundary conditions and see in which direction V vanishes (or takes the same value) at two points (or more than once). In this problem, V vanishes at $x = 0$ and $x = a$. So the potential is oscillatory in the x direction. Therefore, V(x,y) for this problem can be written as Eq. (4-49).

Step 2. Considering the combination of two solutions for X(x), the boundary condition (i) indicates that sine function is appropriate because it vanishes at the origin $(x = 0)$:

$$X(x)=A\sin kx \tag{4-56}$$

Step 3. Considering the combination of two hyperbolic solutions for Y(y), the boundary condition (iii) indicates the sinh function is appropriate because sinh vanishes at the origin $(y = 0)$:

$$Y(y)=C\sinh ky \tag{4-57}$$

Then we can write

$$V(x,y)=AC\sin kx\sinh ky \tag{4-58}$$

Step 4. Application of the boundary condition (ii) determines the acceptable values of k, the separation constant, as follows:

$$\sin ka \sinh ky = 0 \text{ for } 0 \le y \le b \tag{4-59}$$

i.e.,

$$\sin ka = 0$$

which is satisfied only if

$$ka = n\pi \text{ or } k=\frac{n\pi}{a}, \quad n=0,1,2,3,\ldots \tag{4-60}$$

For this particular set of values of k, Eq. (4-58) is a solution to $\nabla^2 V = 0$ which satisfies the first three homogeneous boundary conditions. Then the general solution that satisfies (i), (ii) and (iii) can be written as a linear combination of solutions with all possible values of k:

$$V(x,y) = \sum_{n=1}^{\infty} A_n \sin\frac{n\pi}{a}x \sinh\frac{n\pi}{a}y \tag{4-61}$$

where the $n = 0$ term is not included because $X(x) = 0$ for $n = 0$.

Step 5. The final task is to determine the unknown arbitrary coefficients A_n, $n = 1, 2, 3, \ldots$ by using the last *inhomogeneous* boundary condition (iv). Application of (iv) in Eq. (4-61) yields

$$V(x, y=b) = \sum_{n=1}^{\infty} A_n \sin\frac{n\pi}{a}x \sinh\frac{n\pi}{a}b = V_o, \quad 0 < x < a \tag{4-62}$$

Rearranging coefficients,

$$\sum_{n=1}^{\infty} C_n \sin\frac{n\pi}{a}x = V_o \equiv f(x),\ 0 < x < a \tag{4-63}$$

where

$$C_n = A_n \sinh\frac{n\pi}{a}b, \quad n = 1,\ 2,\ 3, \ldots \tag{4-64}$$

We have an infinite number of unknowns (A_n or C_n) and one equation: How can we determine all these unknowns? The answer is the powerful technique known as *Fourier's method*. What we see in Eq. (4-63) is basically an expansion of the function $f(x) = V_o$ in terms of a Fourier sine series in the interval $0 < x < a$. The formal procedure is:

Multiply both sides of Eq. (4-63) by another harmonic function $\sin\frac{m\pi}{a}x$ and integrate over x from 0 to a.

We then obtain

$$\int_0^a \sum_{n=1}^{\infty} C_n \sin\frac{n\pi}{a}x \sin\frac{m\pi}{a}x\,dx = \int_0^a V_o \sin\frac{m\pi}{a}x\,dx \tag{4-65}$$

The integral on the right hand side (RHS) is easily evaluated:

$$RHS = \left[V_o \frac{(-a)}{m\pi}\cos\frac{m\pi}{a}x\right]_0^a = \begin{cases} \dfrac{2V_o a}{m\pi}, & \text{if } m \text{ is odd} \\ 0, & \text{if } m \text{ is even} \end{cases} \tag{4-66}$$

To evaluate the sum of integrals on the left hand side (LHS) of Eq. (4-65), we first evaluate the following integral:

$$\int_0^a \sin\frac{n\pi}{a}x\ \sin\frac{m\pi}{a}x\ dx = \frac{1}{2}\int_0^a \left\{\cos\frac{(n-m)\pi}{a}x - \cos\frac{(n+m)\pi}{a}x\right\}dx$$

$$= \begin{cases} \dfrac{1}{2}\left[\dfrac{a}{(n-m)\pi}\sin\dfrac{(n-m)\pi}{a}x - \dfrac{a}{(n+m)\pi}\sin\dfrac{(n+m)\pi}{a}x\right]_0^a = 0 \ \text{ if } \ n \neq m \\ \dfrac{1}{2}\left[x - \dfrac{a}{2m\pi}\sin\dfrac{2m\pi}{a}x\right]_0^a = \dfrac{a}{2} \quad \text{if } \ n = m \end{cases} \qquad (4\text{-}67)$$

This particular integral is zero for all values of n except when n is equal to m. Such a property is known as the *orthogonality* of the Fourier sine and cosine functions.
Using this result,

$$\begin{aligned} \text{LHS} &= \sum_{n=1}^{\infty} C_n \int_0^a \sin\frac{n\pi}{a}x\ \sin\frac{m\pi}{a}x\ dx \\ &= C_1\cdot 0 + C_2\cdot 0 + \cdots + C_m\cdot\frac{a}{2} + C_{m+1}\cdot 0 + \cdots \qquad (4\text{-}68) \\ &= C_m\frac{a}{2} \end{aligned}$$

From Eqs. (4-66) and (4-68), we obtain

$$C_m\frac{a}{2} = \begin{cases} \dfrac{2V_o a}{m\pi}, \text{ if m is odd } (m = 1, 3, 5, \ldots) \\ 0, \ \text{if m is even } (m = 2, 4, 6, \ldots) \end{cases} \qquad (4\text{-}69)$$

Thus

$$C_n = A_n \sinh\frac{n\pi}{a}b = \begin{cases} \dfrac{4V_o}{n\pi}, \text{ if n is odd} \\ 0, \quad \text{if n is even} \end{cases}$$

i.e.,

$$A_n = \begin{cases} \dfrac{4V_o}{n\pi\sinh\dfrac{n\pi}{a}b}, \ \text{ if } n = 1, 3, 5\ldots \\ 0, \quad \text{if } n = 2, 4, 6\ldots \end{cases} \qquad (4\text{-}70)$$

We replaced m by n since it is a dummy integer now. Substituting Eq. (4-70) into Eq. (4-61), we obtain the final answer as follows:

$$V(x, y) = \sum_{n=1,3,\ldots}^{\infty} \frac{4V_o}{n\pi \sinh \frac{n\pi}{a} b} \sin\frac{n\pi}{a} x \sinh\frac{n\pi}{a} y \tag{4-71}$$

Although the final solution is given in the form of an infinite series, we can find a good approximate potential by retaining the first few terms because the coefficient decreases rapidly as n increases. The equipotential lines and the **E** field lines are sketched in Figure 4-10.

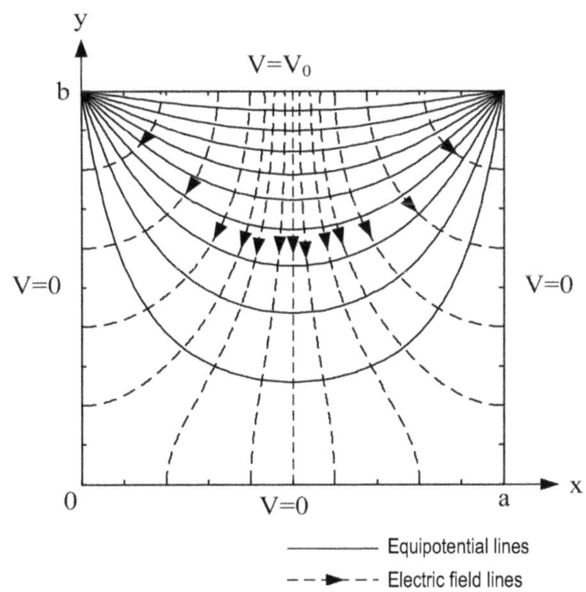

Figure 4-10. The equipotential lines and the electric field lines for the problem of Figure 4-9.

Before we close this example, it is interesting to visit Eq. (4-63) with the substitution of C_n:

$$f(x) = V_o = \sum_{\substack{n=1 \\ n=odd}}^{\infty} \frac{4V_o}{n\pi} \sin\frac{n\pi}{a} x, \quad 0 < x < a \tag{4-72}$$

and see how the sum of the first few terms in this Fourier expansion constructs the constant function. Figure 4-11 shows the plot of the summation of n terms for different values of n. As n increases, the sum forms a better approximation to the function f(x).

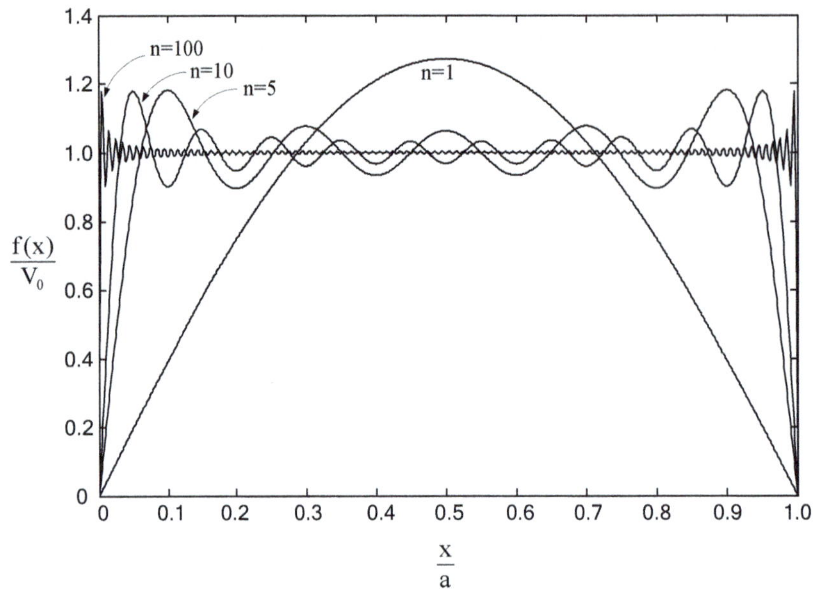

Figure 4-11. Plot of sum of n terms in the Fourier series of Eq. (4-72), (a) $n = 1$, (b) $n = 5$, (c) $n = 10$, (d) $n = 100$.

EXAMPLE 4-7

Consider two half-infinite parallel conducting plates separated by a distance b as shown in Figure 4-12. A very long conducting strip is placed between the plates at the left end and is insulated from the plates. The two plates are connected together and grounded, and a constant potential V_o is applied between the strip and the plates. Find the potential distribution in the region between the plates to the right of the strip.

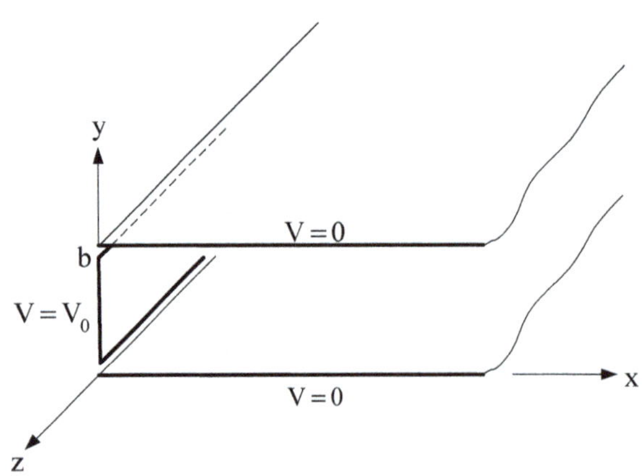

Figure 4-12. Boundary value problem for Example 4-7.

SOLUTION

Setting the rectangular coordinate system as in Figure 4-12, the potential will be independent of z. Since there is no charge inside the region ($x > 0$, $0 \le y \le b$), $V(x, y)$ satisfies Laplace's equation subject to the following boundary conditions:

$$\text{(i)}\ V(x, y = 0) = 0, \quad x > 0 \tag{4-73a}$$

$$\text{(ii)}\ V(x, y = b) = 0, \quad x > 0 \tag{4-73b}$$

$$\text{(iii)}\ V(x = 0, y) = V_o, \quad 0 < y < b \tag{4-73c}$$

$$\text{(iv)}\ V(x,y) \to 0, \quad \text{as } x \to \infty \tag{4-73d}$$

Note that the last condition is included to guarantee physically acceptable solutions. First, since $V = 0$ at $y = 0$ and $y = b$, the potential will be oscillatory in the y direction. Thus $V(x,y)$ for this problem can be written as Eq. (4-48). Secondly, the boundary condition (i) implies

$$Y(y) = C \sin ky \tag{4-74}$$

Thirdly, in order to satisfy the boundary condition (iv), the choice of exponential function is more appropriate than that of the hyperbolic functions. From (4-45b) and (4-73d), we have

$$X(x) = B' e^{-kx} \tag{4-75}$$

Then we can write

$$V(x, y) = B' C e^{-kx} \sin ky \tag{4-76}$$

Apply the boundary condition (ii),

$$e^{-kx} \sin kb = 0 \text{ or } \sin kb = 0 \tag{4-77}$$

which is satisfied only if

$$kb = n\pi \text{ or } k = \frac{n\pi}{b}, \quad n=0, 1, 2, 3, \ldots \tag{4-78}$$

Now the general solution that satisfies (i), (ii), and (iv) can be written as

$$V(x,y) = \sum_{n=1}^{\infty} A_n e^{-\frac{n\pi}{b}x} \sin \frac{n\pi}{b} y \tag{4-79}$$

Applying the remaining inhomogeneous boundary condition (iii),

$$V(x=0,y) = \sum_{n=1}^{\infty} An \sin\frac{n\pi}{b} y = V_o, \quad 0<y<b \tag{4-80}$$

which is the same Fourier series as Eq. (4-63). Following the same procedures used in Example 4-6, we obtain, from Eq. (4-69),

$$A_n = \begin{cases} \dfrac{4V_o}{n\pi}, & \text{if n is odd} \\ 0, & \text{if n is even} \end{cases} \tag{4-81}$$

Therefore,

$$V(x,y) = \sum_{n=1,3,...}^{\infty} \frac{4V_o}{n\pi} e^{-\frac{n\pi}{b}x} \sin\frac{n\pi}{b} y \tag{4-82}$$

4.4.2 Boundary Value Problems in Cylindrical Coordinates

Laplace's equation in cylindrical coordinates is

$$\frac{1}{\rho}\frac{\partial}{\partial \rho}\left(\rho\frac{\partial V}{\partial \rho}\right)+\frac{1}{\rho^2}\frac{\partial^2 V}{\partial \phi^2}+\frac{\partial^2 V}{\partial z^2}=0 \tag{4-13}$$

For the boundary value problems where the geometry is infinitely long in the z direction, V is independent of z, i.e., $\frac{\partial}{\partial z}=0$. In such cases, Laplace's equation becomes

$$\frac{1}{\rho}\frac{\partial}{\partial \rho}\left(\rho\frac{\partial V}{\partial \rho}\right)+\frac{1}{\rho^2}\frac{\partial^2 V}{\partial \phi^2}=0 \tag{4-83}$$

To find the general solutions, we again use the method of separation of variables. We let

$$V(\rho,\phi) = R(\rho)\Phi(\phi) \tag{4-84}$$

and substitute into Eq. (4-83):

$$\Phi(\phi)\frac{1}{\rho}\frac{\partial}{\partial \rho}\left\{\rho\frac{\partial R(\rho)}{\partial \rho}\right\}+\frac{R(\rho)}{\rho^2}\frac{\partial^2 \Phi(\phi)}{\partial \phi^2}=0 \tag{4-85}$$

We divide this equation by $\frac{R(\rho)\,\Phi(\phi)}{\rho^2}$ to obtain

$$\frac{\rho}{R(\rho)}\frac{\partial}{\partial\rho}\left\{\rho\frac{\partial R(\rho)}{\partial\rho}\right\}+\frac{1}{\Phi(\phi)}\frac{\partial^2\Phi(\phi)}{\partial\phi^2}=0 \tag{4-86a}$$

or

$$\frac{1}{R(\rho)}\rho\frac{d}{d\rho}\left\{\rho\frac{dR(\rho)}{d\rho}\right\}=-\frac{1}{\Phi(\phi)}\frac{d^2\Phi(\phi)}{d\phi^2} \tag{4-86b}$$

In order for Eq. (4-86) to be satisfied for all values of ρ and ϕ, the functions on both sides must be a constant. Letting the constant be n^2, we have

$$\frac{1}{R(\rho)}\rho\frac{d}{d\rho}\left\{\rho\frac{dR(\rho)}{d\rho}\right\}=-\frac{1}{\Phi(\phi)}\frac{d^2\Phi(\phi)}{d\phi^2}=n^2 \tag{4-87}$$

which leads to two ordinary differential equations:

$$\rho\frac{d}{d\rho}\left\{\rho\frac{dR(\rho)}{d\rho}\right\}-n^2R(\rho)=0 \tag{4-88a}$$

or

$$\rho^2\frac{d^2R(\rho)}{d\rho^2}+\rho\frac{dR(\rho)}{d\rho}-n^2R(\rho)=0 \tag{4-88b}$$

and

$$\frac{d^2\Phi(\phi)}{d\phi^2}+n^2\Phi(\phi)=0 \tag{4-89}$$

The solutions for $\Phi(\phi)$ are simple: when $n \neq 0$,

$$\Phi(\phi)=C_n\sin n\phi+D_n\cos n\phi\ (n\neq 0) \tag{4-90}$$

where n is an integer because $\Phi(\phi)$ must be a single-valued function for $0 \le \phi \le 2\pi$, i.e., it should satisfy

$$\Phi(\phi+2\pi)=\Phi(\phi) \tag{4-91}$$

When ϕ goes through the full range from 0 to 2π, the function must join smoothly at $\phi = 2\pi$. When $n = 0$, the solutions are

$$\Phi(\phi) = C_o\phi + D_o \quad (n=0) \tag{4-92}$$

where C_o and D_o are arbitrary constants.

Eq. (4-88) is known as the Euler-Cauchy equation.* When $n \neq 0$, the two independent solutions are

$$R(\rho) = A_n \rho^n + B_n \frac{1}{\rho^n} \quad (n \neq 0) \tag{4-93}$$

When $n = 0$, the solutions are

$$R(\rho) = A_o \ln\rho + B_o \quad (n=0) \tag{4-94}$$

Note that ln $\frac{\rho_o}{\rho}$ is a potential of an infinite line charge as shown in Example 2-12.

Combining Eqs. (4-90), (4-92), (4-93) and (4-94), the most general solutions for $V(\rho,\phi)$ are given by

$$\begin{aligned} V(\rho,\phi) &= (A_0 \ln\rho + B_0)(C_0\phi + D_0) \\ &+ \sum_{n=1}^{\infty}\left\{A_n \rho^n + \frac{B_n}{\rho^n}\right\}\{C_n \sin n\phi + D_n \cos n\phi\} \end{aligned} \tag{4-95}$$

A complete set of separable solutions to two-dimensional Laplace's equation in (ρ, ϕ) are summarized in Table 4-1.

EXAMPLE 4-8 A CONDUCTING CYLINDER IN A UNIFORM ELECTRIC FIELD

An infinitely long uncharged cylindrical conductor of radius a is placed in an initially uniform electric field $\mathbf{E}_o = \mathbf{a}_x E_o$. The axis of the cylinder is the z axis (Figure 4-13(a)). Find the electric field at points (ρ, ϕ) exterior to the cylinder.

*See E. Kreyszig, *Advanced Engineering Mathematics*, John Wiley & Sons, 1993, 7th Ed., Section 2.6.

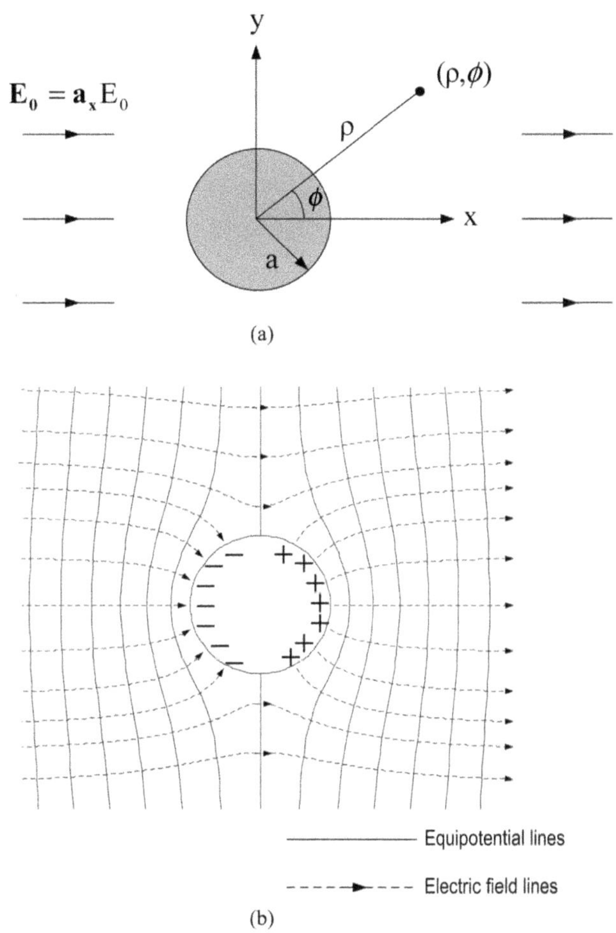

Figure 4-13. Cross section of a conducting cylinder in a uniform electric field.

SOLUTION

Since the conductor is a circular cylinder, it is natural to use cylindrical coordinates. Because the cylinder is infinitely long in the z direction, V will not depend on z, i.e., $\frac{\partial}{\partial z} = 0$. Thus Eq. (4-95) represents a solution of this problem. However, since the conductor is uncharged, i.e., the net charge is zero; we don't expect to have the term $\ln \rho$. Since V has to be single-valued for $0 \le \phi \le 2\pi$, $C_o \phi + D_o$ is not an appropriate solution. Then the general solution for $V(\rho, \phi)$ can be written as

$$V(\rho,\phi) = \sum_{n=1}^{\infty} \left\{ A_n \rho^n + \frac{B_n}{\rho^n} \right\} \{ C_n \sin n\phi + D_n \cos n\phi \} \tag{4-96}$$

Now the boundary conditions (B.C.s) are:

$$\text{(i)}\quad V(\rho = a, \phi) = \text{constant} = 0,\ \ 0 \le \phi \le 2\pi \tag{4-97}$$

$$\text{(ii)}\quad \mathbf{E} = -\nabla V \rightarrow \mathbf{E}_o = \mathbf{a}_x E_o = \nabla\{E_o x\},\quad \text{as } \rho \rightarrow \infty \tag{4-98a}$$

$$V \rightarrow -E_o x = -E_o \rho\ \cos\phi\ \text{ as }\ \rho \rightarrow \infty \tag{4-98b}$$

One could set the potential of the conductor to be any other constant; only the reference potential will change. So we assume it to be zero. The boundary condition (ii) indicates that the solution in the form of $\sin n\phi$ is not appropriate (it can not match the boundary condition). Thus we can rewrite the potential as follows:

$$V(\rho,\phi) = \sum_{n=1}^{\infty}\left\{A_n\, \rho^n \cos n\phi + \frac{B_n}{\rho^n}\cos n\phi\right\} \tag{4-99}$$

where A_n and B_n are redefined. Applying the B.C. (ii),

$$V(\rho = \infty, \phi) = \sum_{n=1}^{\infty} A_n\, \rho^n \cos n\phi = -E_o\, \rho \cos\phi\ ,\ 0 \le \phi \le 2\pi \tag{4-100}$$

In order for this equation to be satisfied for all values of ρ and ϕ, it is necessary that

$$A_1 = -E_o, A_2 = A_3 = \ldots = 0 \tag{4-101}$$

Then

$$V(\rho, \phi) = -E_o\, \rho \cos\phi + \sum_{n=1}^{\infty}\frac{B_n}{\rho^n}\ \cos n\phi \tag{4-102}$$

Application of the B.C. (i) gives

$$V(\rho = a, \phi) = -E_o\ a \cos\phi + \sum_{n=1}^{\infty}\frac{B_n}{a^n}\cos n\phi = 0 \tag{4-103a}$$

or

$$\sum_{n=1}^{\infty} C_n \cos n\phi = E_o\ a \cos\phi \equiv g(\phi), 0 \le \phi \le 2\pi \tag{4-103b}$$

where

$$C_n = \frac{B_n}{a^n},\ n = 1,2,3,\ldots \tag{4-104}$$

Eq. (4-103b) is again the expansion of the function $g(\phi) = E_o$ a cos ϕ in terms of the Fourier cosine series. Multiplying both sides of Eq. (4-103b) by another harmonic function $\cos m\phi$ and integrating over ϕ from 0 to 2π, we can show that

$$C_1 = E_o a, C_2 = C_3 = \ldots = 0 \tag{4-105a}$$

or

$$B_1 = C_1 a = E_o a^2,\ B_2 = B_3 = \ldots = 0 \tag{4-105b}$$

Substituting Eq. (4-105b) into Eq. (4-102), we finally obtain

$$V(\rho,\phi) = -E_o \rho \cos\phi + E_o \frac{a^2}{\rho}\cos\phi \tag{4-106}$$

The electric field is given by

$$\begin{aligned}\mathbf{E} &= -\nabla V = -\mathbf{a}_\rho \frac{\partial V}{\partial \rho} - \mathbf{a}_\phi \frac{1}{\rho}\frac{\partial V}{\partial \phi} \\ &= E_o(\mathbf{a}_\rho \cos\phi - \mathbf{a}_\phi \sin\phi) + E_o \frac{a^2}{\rho^2}(\mathbf{a}_\rho \cos\phi + \mathbf{a}_\phi \sin\phi) \\ &= \mathbf{a}_x E_o + E_o\left(\frac{a}{\rho}\right)^2 (\mathbf{a}_\rho \cos\phi + \mathbf{a}_\phi \sin\phi)\end{aligned} \tag{4-107}$$

We notice that the first term is the initially applied uniform field and the second term is the additional field due to the *induced charge* on the surface of the conductor. The $\frac{1}{\rho^2}$ dependence of the latter indicates that it comes from the dipole configuration of the line charges. This can be easily confirmed by calculating the surface charge density at $\rho = a$:

$$\rho_s = \varepsilon_o E_n = \varepsilon_o E_\rho|_{\rho=a} = 2\varepsilon_o E_o \cos\phi,\ 0 \le \phi \le 2\pi \tag{4-108}$$

Note that $\rho_s > 0$ for the right-half portion of the cylinder surface and $\rho_s < 0$ for the left-half portion. It is a *line dipole*. The field lines and the equipotential lines are plotted in Figure 4-13(b).

4.4.3 Boundary Value Problems in Spherical Coordinates

Laplace's equation in spherical coordinates is

$$\frac{1}{r^2}\frac{\partial}{\partial r}\left(r^2 \frac{\partial V}{\partial r}\right) + \frac{1}{r^2 \sin\theta}\frac{\partial}{\partial \theta}\left(\sin\theta \frac{\partial V}{\partial \theta}\right) + \frac{1}{r^2 \sin^2\theta}\frac{\partial^2 V}{\partial \phi^2} = 0 \tag{4-109}$$

For the boundary value problems where the potential is independent of ϕ (*azimuthally symmetric*), i.e., $\frac{\partial}{\partial \phi} = 0$, Laplace's equation becomes

$$\frac{1}{r^2}\frac{\partial}{\partial r}\left(r^2\frac{\partial V}{\partial r}\right)+\frac{1}{r^2\sin\theta}\frac{\partial}{\partial \theta}\left(\sin\theta\frac{\partial V}{\partial \theta}\right)=0 \tag{4-110}$$

Applying the method of separation of variables, we assume

$$V(r,\theta) = R(r)\Theta(\theta) \tag{4-111}$$

Following the procedures similar to those in the previous two sections, we obtain

$$\frac{1}{R(r)}\frac{d}{dr}\left\{r^2\frac{dR(r)}{dr}\right\} = -\frac{1}{\Theta(\theta)\sin\theta}\frac{d}{d\theta}\left\{\sin\theta\frac{d\Theta(\theta)}{d\theta}\right\} \tag{4-112}$$

If Eq. (4-112) is to be satisfied for all values of r and θ, the functions on both sides must be a constant. Letting the constant be k, we obtain

$$\frac{d}{dr}\left\{r^2\frac{dR(r)}{dr}\right\} - k\ R(r) = 0 \tag{4-113}$$

$$\frac{1}{\sin\theta}\frac{d}{d\theta}\left\{\sin\theta\frac{d\Theta(\theta)}{d\theta}\right\}+k\ \Theta(\theta)=0 \tag{4-114}$$

There exist solutions for $\Theta(\theta)$ that are physically acceptable for $0 \le \theta \le \pi$, when

$$k = n(n+1),\ n = 0, 1, 2, \dots . \tag{4-115}$$

The solutions are called the **Legendre polynomials***

$$\Theta(\theta) = P_n(\cos\theta) \tag{4-116}$$

where $P_n(x)$ is defined by

$$P_n(x) = \frac{1}{2^n n!}\left(\frac{d}{dx}\right)^n (x^2-1)^n \tag{4-117a}$$

The first few Legendre polynomials are listed in Table 4-2. Similar to Eq. (4-67), the Legendre polynomials satisfy the following orthogonality:

*R. V. Churchill and J. W. Brown, *Fourier Series and Boundary Value Problems*, McGraw-Hill, 1987, 4th Ed., Chapter 9.

$$\int_{-1}^{1} P_m(x)\ P_n(x)\ dx = \int_{0}^{\pi} \{P_m(\cos\theta) P_n(\cos\theta)\} \sin\theta\ d\theta$$

$$= \begin{cases} 0 & \text{if } n \neq m \\ \dfrac{2}{2n+1} & \text{if } n = m \end{cases} \qquad (4\text{-}117b)$$

Table 4-2 Legendre Polynomials

$P_o(\cos\theta) = 1$
$P_1(\cos\theta) = \cos\theta$
$P_2(\cos\theta) = \frac{1}{2}(3\cos^2\theta - 1)$
$P_3(\cos\theta) = \frac{1}{2}(5\cos^3\theta - 3\ \cos\theta)$

Substituting Eq. (4-115) into Eq. (4-113), we find the following two independent solutions for $R(r)$:

$$R(r) = A_n r^n + \frac{B_n}{r^{n+1}} \qquad (4\text{-}118)$$

Combining Eqs. (4-116) and (4-118), the general solutions for $V(r, \theta)$ can be written as

$$V(r,\theta) = \sum_{n=0}^{\infty} \left\{ A_n\ r^n + \frac{B_n}{r^{n+1}} \right\} P_n(\cos\theta) \qquad (4\text{-}119)$$

Note that we have seen two of the second solutions with $n = 0$ and $n = 1$ before:

$$\frac{B_o}{r}\ (n = 0),\quad \frac{B_1}{r^2}\cos\theta\ \ (n = 1)$$

They are the potentials of a single point charge and an electric dipole, respectively (see Eqs. (2-43) and (2-68)). The set of separable solutions to two-dimensional Laplace's equation in (r,θ) are summarized in Table 4-1. The following example illustrates the application of these solutions.

EXAMPLE 4-9 A DIELECTRIC SPHERE IN A UNIFORM ELECTRIC FIELD

An uncharged dielectric sphere of radius a and permittivity ε is placed in an initially uniform electric field $\mathbf{E}_o = \mathbf{a}_z E_o$ as shown in Fig. 4-14(a). Find the electric field everywhere.

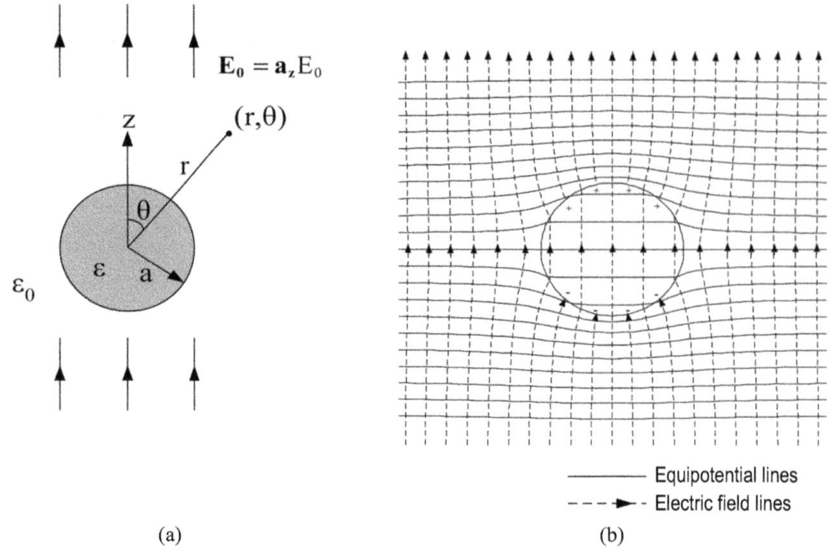

Figure 4-14. A dielectric sphere in a uniform electric field.

SOLUTION

This problem is very similar to that of Example 4-8 except that the object is a dielectric and a sphere. Thus the electric field is not zero inside the sphere and the potential is not a constant on the surface of the sphere. Since the applied electric field is uniform in the z direction, the problem has azimuthal symmetry; the potential will be independent of ϕ. Thus Eq. (4-119) represents a general solution of this problem. We write the potential inside and outside the sphere as follows:

$$V_1(r,\theta) = V_{app} + V_{ind,1}, \quad r \le a \tag{4-120a}$$

$$V_2(r,\theta) = V_{app} + V_{ind,2}, \quad r \ge a \tag{4-120b}$$

where V_{app} is the potential due to an initially applied uniform field and V_{ind} is the *induced* potential due to the presence of dielectric. V_{app} can be easily obtained from

$$\mathbf{E}_o = \mathbf{a}_z E_o = -\nabla V_{app} = -\nabla \{-E_o z\} \tag{4-121}$$

Thus

$$V_{app} = -E_o z = -E_o r \cos\theta = -E_o r P_1(\cos\theta) \tag{4-122}$$

Since V_{ind} must be solutions of $\nabla^2 V = 0$ and it should be finite at the origin $(r = 0)$ and at infinity $(r \to \infty)$, V_{ind} can be formulated as

$$V_{ind,1} = \sum_{n=0}^{\infty} A_n r^n P_n(\cos\theta), \quad r \le a \tag{4-123a}$$

$$V_{ind,2} = \sum_{n=0}^{\infty} \frac{B_n}{r^{n+1}} P_n(\cos\theta), \quad r \ge a \tag{4-123b}$$

Note that the solutions $\frac{1}{r^{n+1}} P_n(\cos\theta)$ are not appropriate in the region $r \le a$ because they blow up at the origin $(r = 0)$. The solutions $r^n P_n(\cos\theta)$ are not appropriate in the region $r \ge a$ because they blow up at infinity $(r \to \infty)$. The unknown coefficients A_n, B_n can be determined by applying the boundary conditions at the dielectric interface $(r = a)$ (see Table 3-4):

$$E_{1t} = E_{2t} \text{ or } V_1 = V_2 \tag{4-124a}$$

$$D_{1n} = D_{2n} \text{ or } \varepsilon \frac{\partial V_1}{\partial r} = \varepsilon_0 \frac{\partial V_2}{\partial r} \tag{4-124b}$$

for all values of θ. Note that the continuity of $\mathbf{E}_t$ is equivalent to the continuity of the potential. The θ-dependence of V_{app} in Eq. (4-122) and the boundary conditions suggest that we need only the term with $n = 1$. Therefore, we rewrite $V(r, \theta)$ as follows:

$$V_1(r, \theta) = -E_o r \cos\theta + A_1 r \cos\theta \equiv A r \cos\theta, \quad r \le a \tag{4-125a}$$

$$V_2(r, \theta) = -E_o r \cos\theta + \frac{B_1}{r^2} \cos\theta, \quad r \ge a \tag{4-125b}$$

Applying the two boundary conditions, Eqs. (4-124), at $r = a$,

$$A \, a \cos\theta = -E_o \, a \cos\theta + \frac{B_1}{a^2} \cos\theta \tag{4-126a}$$

$$\varepsilon A \cos\theta = \varepsilon_0 \left\{ -E_o - \frac{2B_1}{a^3} \right\} \cos\theta \tag{4-126b}$$

Solving these equations for A and B_1, we obtain

$$A = -\frac{3\varepsilon_0}{\varepsilon + 2\varepsilon_0} E_o, \quad B_1 = \frac{\varepsilon - \varepsilon_0}{\varepsilon + 2\varepsilon_0} E_o a^3 \tag{4-127}$$

Substituting Eq. (4-127) into Eq. (4-125),

$$V_1(r,\theta) = -\frac{3\varepsilon_0}{\varepsilon+2\varepsilon_0} E_o r\cos\theta = -\frac{3\varepsilon_0}{\varepsilon+2\varepsilon_0} E_o z, \quad r \le a \tag{4-128}$$

$$V_2(r,\theta) = -E_o r\cos\theta + \frac{\varepsilon-\varepsilon_0}{\varepsilon+2\varepsilon_0} E_o \frac{a^3}{r^2}\cos\theta, \quad r \ge a \tag{4-129}$$

The electric fields are given by

$$\begin{aligned}\mathbf{E}_1 &= -\nabla V_1 = -\mathbf{a}_z \frac{\partial V_1}{\partial z} = \mathbf{a}_z \frac{3\varepsilon_0}{\varepsilon+2\varepsilon_0} E_o, \quad r \le a \\ &= \frac{3\varepsilon_0}{\varepsilon+2\varepsilon_0} E_o(\mathbf{a}_r\cos\theta - \mathbf{a}_\theta \sin\theta)\end{aligned} \tag{4-130}$$

$$\mathbf{E}_2 = -\nabla V_2 = -\mathbf{a}_r \frac{\partial V_2}{\partial r} - \mathbf{a}_\theta \frac{1}{r}\frac{\partial V_2}{\partial \theta}$$

$$= \mathbf{a}_z E_o + \frac{\varepsilon-\varepsilon_0}{\varepsilon+2\varepsilon_0} E_o \left(\frac{a}{r}\right)^3 (\mathbf{a}_r 2\cos\theta + \mathbf{a}_\theta \sin\theta) \tag{4-131a}$$

$$\begin{aligned}&= \mathbf{a}_r E_o \left\{1 + \frac{2(\varepsilon-\varepsilon_0)}{\varepsilon+2\varepsilon_0}\left(\frac{a}{r}\right)^3\right\}\cos\theta \\ &\quad - \mathbf{a}_\theta E_o \left\{1 - \frac{\varepsilon-\varepsilon_0}{\varepsilon+2\varepsilon_0}\left(\frac{a}{r}\right)^3\right\}\sin\theta, \quad r \ge a\end{aligned} \tag{4-131b}$$

It is interesting to note that the field inside the dielectric sphere is uniform at all points as seen in Eq. (4-130). The field outside the dielectric sphere shown in Eq. (4-131a) consists of two parts: the first term is the initially applied uniform field and the second term is the additional field due to the induced bound charge on the surface of the dielectric sphere. The latter is a dipole field. In fact, the induced surface bound charge ρ_{ps} at $r = a$ has a dipole nature because from Eq. (3-2c),

$$\begin{aligned}\rho_{ps} &= \mathbf{P}\cdot\mathbf{a}_n = (\mathbf{D}_1 - \varepsilon_0\mathbf{E}_1)\cdot\mathbf{a}_r = (\varepsilon-\varepsilon_0)\ E_{1r}\big|_{r=a} \\ &= \frac{3\varepsilon_0(\varepsilon-\varepsilon_0)}{\varepsilon+2\varepsilon_0} E_o\cos\theta, \quad 0 \le \theta \le \pi\end{aligned} \tag{4-132}$$

Note that $\rho_{ps} > 0$ for $0 < \theta < \pi/2$ and $\rho_{ps} < 0$ for $\pi/2 < \theta < \pi$. The field lines and the equipotential lines are plotted in Figure 4-14(b).

PROBLEMS

4-1. Show that the potential of a point charge, Eq. (2-43), satisfies Laplace's equation (4-4) at all points except at the origin $(r = 0)$.

4-2. Show that the potential of an infinitely long line charge, Eq. (2-56), satisfies Laplace's equation at all points except on the z-axis $(\rho = 0)$.

4-3. Show that the potential of spherical volume charge in Example 2-11 satisfies Poisson's equation (4-3) everywhere (for $r \leq a$ and $r \geq a$).

4-4. Show that the potential of an infinitely long, cylindrical cloud of electrons in Problem 2-24(b) satisfies Poisson's equation everywhere (for $\rho \leq a$ and $\rho \geq a$).

4-5. Two infinite conducting plates form a wedge with an angle ϕ_0 (Figure 4-15). A voltage V_0 is applied between the two plates. Find the potential distribution and the electric field in the region between the plates $(0 \leq \phi \leq \phi_0)$. [HINT] V satisfies a one-dimensional Laplace's equation.

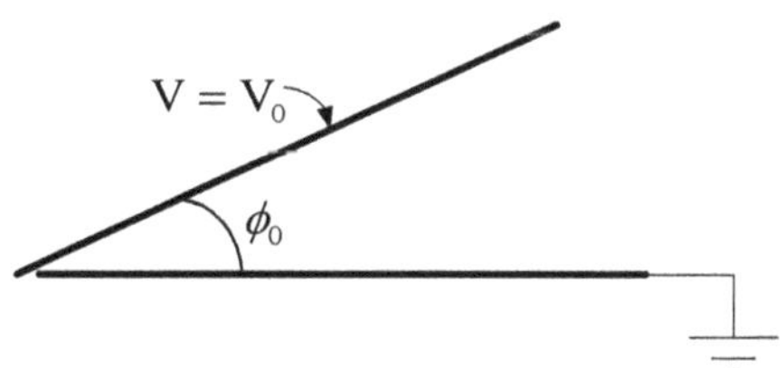

Figure 4-15. Cross section of two infinite conducting plates forming a wedge

4-6. Consider a spherical capacitor in Example 3-6 and Figure 3-13(a). A voltage V_0 is applied between inner and outer conducting spheres, positive at the inner conductor. Find the potential distribution and the electric field in the region $a \leq r \leq b$ by solving Laplace's equation. Also find the surface charge densities on the two spherical conducting surfaces, $r = a, b$.

4-7. A potential V_0 is applied between an infinite conducting cone of polar angle θ_0 and an infinite grounded conducting plane (Figure 4-16). Find the potential distribution and the electric field in the region between the two conductors $(\theta_0 \leq \theta \leq \pi/2)$.
[HINT] V satisfies a one-dimensional Laplace's equation.

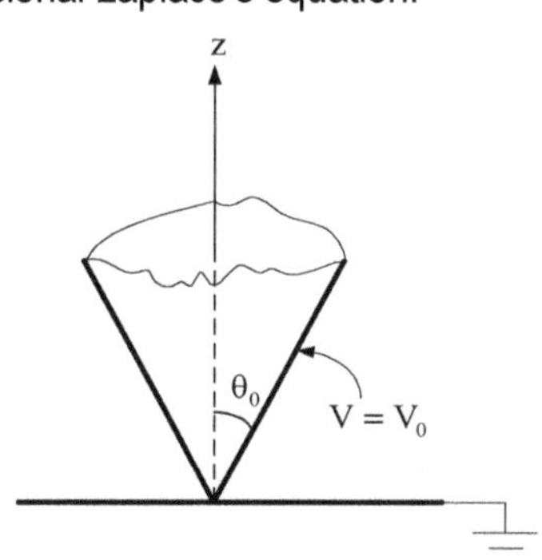

Figure 4-16. An infinite conducting cone and a conducting plane.

4-8. Show that for $\rho_s(x,y)$ in Eq. (4-27), the total induced charge

$$\int_{-\infty}^{\infty}\int_{-\infty}^{\infty} \rho_s(x,y)\, dx\, dy$$

is equal to $-q$. [HINT] Change the integration variables to (ρ, ϕ).

4-9. Consider the problem of a point charge q located a distance d above an infinite grounded conducting plane [Figure 4-4(a)]. Find the approximate electric potential at far points ($r_+ >> d$). Express V in terms of spherical coordinates (r, θ). What kind of potential do you expect?

4-10. A point charge q is placed midway between two large, intersecting conducting plates forming a 45-degree angle, as shown in Figure 4-17. Find the electric potential in the region between the plates by determining the system of image charges that will make the potential zero at the two conducting boundaries.

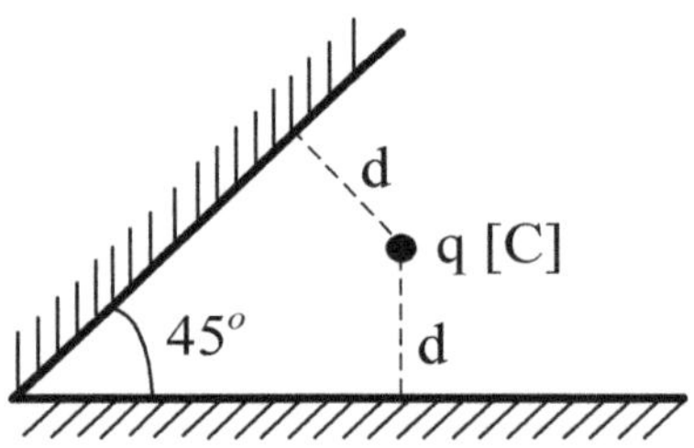

Figure 4-17. A point charge between two intersecting conducting plates.

4-11. A point charge q is located midway between two infinite, grounded, parallel conducting planes as shown in Figure 4-18. Find the potential in the region between the two conducting planes.

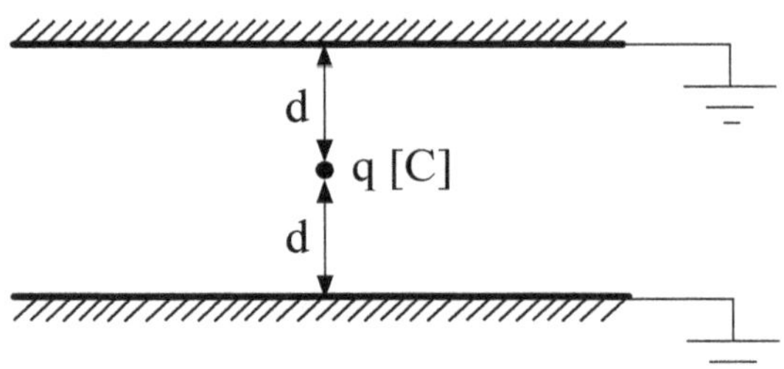

Figure 4-18. A point charge between two infinite, grounded, parallel conducting planes.

4-12. An infinitely long line charge ρ_ℓ [C/m] is located a distance d above an infinite grounded conducting plane (parallel to the plane).

(a) Find the potential in the region above the plane.
(b) Find the surface charge density ρ_s induced on the conducting plane.

4-13. Consider the problem of a line charge and a parallel conducting cylinder in Example 4-4. Find the surface charge density ρ_s on the surface of the conducting cylinder.

[HINT] Introduce a new cylindrical coordinate system (ρ,ϕ,z) with the axis of the conducting cylinder being the z-axis.

4-14. Using the result of Example 4-4, show that the capacitance per unit length between two long, parallel, circular conducting wires of radius a, separated by a distance d (between the axes of the two wires), is given by

$$C = \frac{\pi\varepsilon_0}{\ln\left[\dfrac{d}{2a} + \sqrt{\left(\dfrac{d}{2a}\right)^2 - 1}\right]} = \frac{\pi\varepsilon_0}{\cosh^{-1}\left(\dfrac{d}{2a}\right)} \left[\frac{F}{m}\right]$$

4-15. A straight conducting wire of radius a is placed parallel at a distance d above the surface of the earth. Assuming that the earth is a perfect conductor, calculate the capacitance per unit length between the wire and the earth. What is the force per unit length between the wire and the earth?

4-16. Consider the problem of a point charge in a right-angle conducting corner in Example 4-3 (see Figure 4-5). Let $a = b$. Find the approximate potential at far points ($r_1 >> a$) in the xy plane. Express V in terms of cylindrical coordinates (ρ, ϕ). What kind of potential do you expect?

4-17. Repeat Problem 4-16 when an infinitely long line charge ρ_ℓ is placed in the right-angle conducting corner.

4-18. Consider the problem of a point charge near a grounded conducting sphere in Example 4-5.

(a) Show that V in Eq. (4-37) satisfies the boundary condition at all points on the surface of the conducting sphere with the values of q' and s given by Eq. (4-38).
(b) Derive the results of Eq. (4-38) by applying the boundary condition at two points, e.g., $V(a, 0)$ and $V(a, \pi) = 0$, and solving the resulting two equations for two unknowns q' and s.

4-19. Consider the problem of Example 4-5.

(a) Find the surface charge ρ_s induced on the surface of the conducting sphere, as a function of θ.
(b) Calculate the total induced charge by integrating ρ_s over the spherical surface. What do you expect?

4-20. In Example 4-5, we assumed that the conducting sphere was grounded $(V = 0)$. Suppose the conducting sphere is not grounded. It is electrically neutral, i.e., net charge q is zero.

(a) Find the system of charges—the original charge and the image charge(s)—that will make the potential satisfy the boundary condition and make the conductor neutral.
[HINT] You need one more image charge at an appropriate place.
(b) At what potential is the conducting sphere maintained?

4-21.* A point charge q is placed in the air at a distance d above the planar surface of a half-infinite dielectric medium of permittivity ε as shown in Figure 4-19. Find the electric potential everywhere (in the air and in the dielectric). [HINT] Use the method of images with the appropriate boundary conditions at the air-dielectric interface. You need to consider two different equivalent problems.

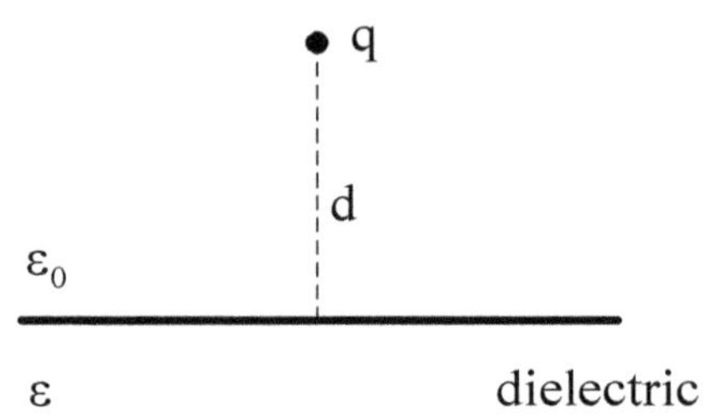

Figure 4-19. A point charge near a dielectric interface.

4-22. Consider the infinitely long rectangular pipe in Figure 4-9. Three metal strips at $x = 0$, $y = 0$ and $y = b$ are grounded and the metal strip at $x = a$ is insulated from other strips and maintained at a constant potential V_o. Find the potential distribution inside the pipe ($0 < x < a$, $0 < y < b$).

4-23. Two infinitely long grounded metal plates at $y = 0$ and $y = b$ are connected at $x = \pm a$ by metal strips maintained at potential V_o as shown in Figure 4-20. Find the potential inside the resulting rectangular pipe. Sketch the electric field lines and the equipotential lines.

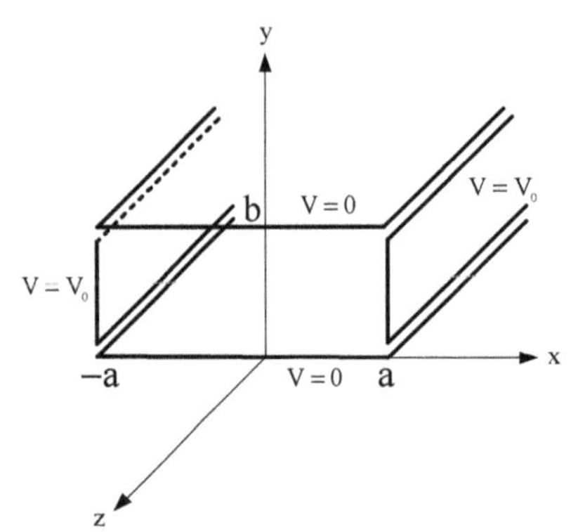

Figure 4-20. Boundary value problem for Problem 4-23.

4-24. A cylindrical region with rectangular cross section extending to $z = \pm\infty$ has the following potential distributions on its four walls:

$$V = 0 \text{ at } x = 0,\ x = a \text{ and } y = 0$$

$$V = V_o \sin\frac{\pi}{a}x \quad \text{at} \quad y = b$$

The interior region is free of charge. Find the potential $V(x, y)$ inside the cylindrical space.

4-25. A cylindrical region with square cross section has the following potential distributions on its four walls:

$$V = 0 \text{ at } x = 0 \text{ and } y = 0$$

$$V = V_o \sin\frac{\pi}{a}x \quad \text{at} \quad y = a$$

$$V = V_o \sin\frac{2\pi}{a}y \quad \text{at } x = a$$

The interior region is free of charge. Find the potential inside the cylindrical space ($0 < x < a$, $0 < y < a$).
[HINT] Use the superposition of two boundary value problems.

4-26. Repeat Problem 4-25 when the walls have the following potential distributions:

$$V = 0 \text{ at } x = 0 \text{ and } y = 0$$

$$V = V_o \frac{x}{a} \quad \text{at} \quad y = a$$

$$V = V_o \frac{y}{a} \quad \text{at} \quad x = a$$

[HINT] Use the solutions of $\nabla^2 V(x,y) = 0$ with $k = 0$.

4-27. Consider the infinitely long rectangular pipe in Figure 4-9. Two metal strips at $x = 0$ and $x = a$ are grounded. The two strips at $y = 0$ and $y = b$ are insulated and maintained at constant potentials V_1 and V_2, respectively. Find the potential distribution inside the pipe.
[HINT] Use the superposition of two boundary value problems.

4-28.* A rectangular box consists of five metal plates which are welded together and grounded (Figure 4-21). The top metal sheet is insulated from the rest and maintained at a constant potential V_o. Find the potential distribution $V(x, y, z)$ inside the box. The sides are a,b and c.
[HINT] You need a three-dimensional solution of $\nabla^2 V = 0$ with the double Fourier series.

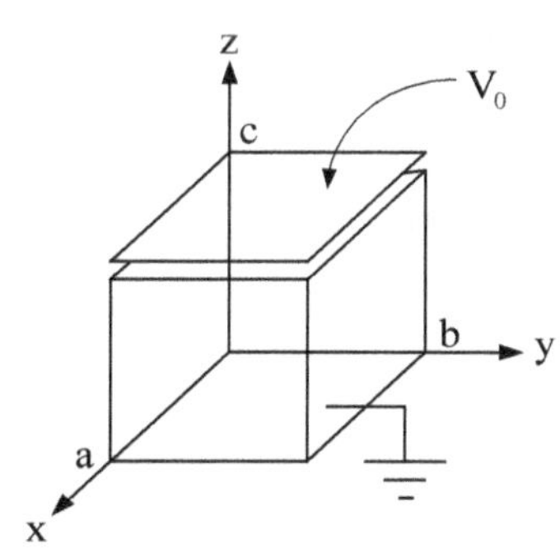

Figure 4-21. A rectangular metal box.

4-29. An infinitely long, thin conducting circular tube of radius a is split into two halves as shown in Figure 4-22. The left half is kept at a potential $V = V_o$ and the right half is kept at a potential $V = -V_o$. Find the potential distribution $V(\rho, \phi)$ both inside and outside the tube ($\rho < a$ and $\rho > a$). Find also the surface charge density on the conducting tube.
[HINT] Inside the tube ($\rho < a$), the potential must be finite at the central axis ($\rho = 0$). Outside the tube ($\rho > a$), the potential must be finite at infinity.

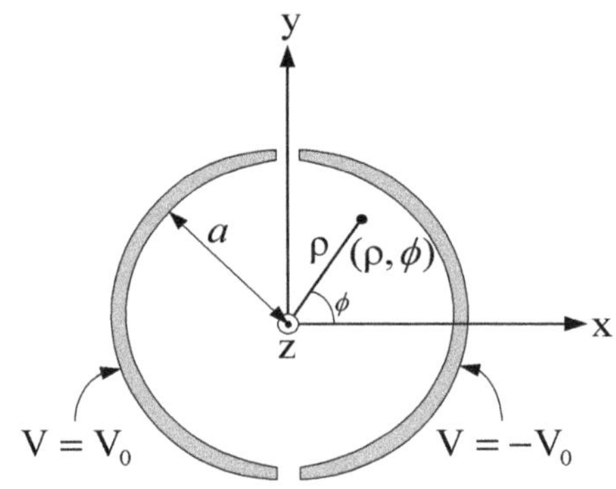

Figure 4-22. Cross section of split circular conducting cylinders.

4-30. An infinitely long dielectric cylinder of radius a and permittivity ε is placed in an initially uniform electric field $\mathbf{E}_o = \mathbf{a}_x E_o$. The axis of the cylinder is the z-axis [Figure 4-13(a)]. Find the potential $V(\rho, \phi)$ and electric field everywhere (both inside and outside the dielectric cylinder).

4-31. An infinitely long uncharged cylindrical conductor of radius a is placed in an initially uniform electric field $\mathbf{E}_o = \mathbf{a}_y E_o$. The axis of the cylinder is the z axis. See Figure 4-13(a) and let the field point in the y direction. Find the electric field at points (ρ, ϕ) exterior to the cylinder.

4-32. The surface of an infinitely long cylinder of radius a has the following prescribed charge distribution:

$$\rho_s(\phi) = \rho_{so} \sin 2\phi$$

where ρ_{so} is a constant. The axis of the cylinder is the z-axis. Find the potential $V(\rho, \phi)$ inside and outside the cylinder.
[HINT] In this problem, the charge density is given, instead of the potential values at the surface. Formulate the potentials in both regions simultaneously and make use of two boundary conditions.

4-33. A conducting sphere of radius a is placed in an initially uniform electric field $\mathbf{E} = \mathbf{a}_z E_o$ [Figure 4-14(a)]. Find the electric field at points (r, θ, ϕ) exterior to the sphere ($r > a$).

4-34. A small spherical cavity of radius a is carved out of a dielectric material with permittivity ε which is sandwiched between two conducting plates. The voltage V_o is applied between the two plates. Assuming the thickness of the dielectric slab is $d >> a$, find the electric field inside the cavity.

4-35. The potential at the spherical surface $r = a$ is given by

$$V(\theta) = V_o \cos 3\theta$$

where V_o is a constant. Assuming that the medium inside and outside the sphere is air, free of charge, find the potential everywhere ($r < a$ and $r > a$) and the surface charge density on the sphere ($r = a$).

4-36.* A thin spherical shell of radius a carries a uniform surface charge ρ_{so} on the upper hemisphere ($0 < \theta < \pi/2$) and $-\rho_{so}$ on the lower hemisphere ($\pi/2 < \theta < \pi$). Find the potential inside and outside the sphere. Calculate the first five coefficients explicitly.
[HINT] Formulate the potentials in both regions simultaneously using Eq. (4-119).

4-37. A dielectric sphere of radius a has a uniform permanent polarization:

$$\mathbf{P} = \mathbf{a}_z P_o (r < a)$$

where P_o is a constant. The polarized sphere gives rise to an electric field.

(a) First find the bound charge densities: ρ_{pv}, ρ_{ps}.
(b) Calculate the electric field at the center of the sphere by using the integral (2-27).
(c) Find the potentials $V(r, \theta)$ inside and outside the sphere using the solutions of Laplace's equation. Sketch the electric field lines and equipotential lines everywhere.
(d) Calculate the electric field inside the sphere and confirm that the field at $r = 0$ is consistent with the result of (b).

4-38. A dielectric sphere of radius a has the following nonuniform permanent polarization:

$$\mathbf{P} = \mathbf{a}_r 2P_o \frac{r}{a} \cos\theta - \mathbf{a}_\theta 3P_o \frac{r}{a} \sin\theta \ (r < a)$$

where P_o is a constant.

(a) Find the bound charge densities: ρ_{pv}, ρ_{ps}.
(b) Find the potentials $V(r, \theta)$ inside and outside the sphere using the solutions of Laplace's equation.

4-39. An infinitely long, dielectric cylinder of radius a has the following permanent polarization:

$$\mathbf{P} = \mathbf{a}_\rho P_o \frac{\rho}{a} \cos\phi - \mathbf{a}_\phi 2P_o \frac{\rho}{a} \sin\phi \ (\rho < a)$$

where P_o is a constant.

(a) Find the bound charges: ρ_{pv}, ρ_{ps}.
(b) Find the potentials $V(\rho, \phi)$ inside and outside the cylinder using the solutions of Laplace's equation.

4-40. Repeat Problem 4-38 when

$$\mathbf{P} = \mathbf{a}_\rho P_o \frac{\rho}{a} \sin\phi + \mathbf{a}_\phi 2P_o \frac{\rho}{a} \cos\phi \ (\rho < a)$$

STEADY CURRENTS AND CONDUCTIVE MATERIALS

CHAPTER 5

5.1. INTRODUCTION

In previous chapters we considered electrostatic problems in which charges were at rest. Now we turn to problems which involve moving charges and resultant current flow. We are particularly interested in steady currents such as those which arise when steady voltages are applied across conductive or resistive materials. In Chapter 3, perfect dielectrics with zero conductivity were studied. Now we consider dielectrics with finite conductivity.

First we consider definitions of current. Then we establish some general relationships, including the equation of continuity, Ohm's law, Joule's law, and relaxation time. These relationships may be applied to the general time-varying case, as well as to the special case, of steady currents. Finally, we consider a new boundary condition which is required for steady currents, namely, the continuity of normal volume current density.

5.2. ELECTRIC CURRENT

Let's consider charges which are moving within a conductor. Consider any surface S within the conductor. Let $Q(t)$ be the charge which has passed through the surface. Then the current $I(t)$ through the surface is defined as

$$\boxed{I(t) = \frac{dQ(t)}{dt}} \tag{5-1}$$

The unit of current is the ampere $[A]$, or coulomb/second, named after A. M. Ampère. The current is thus a measure of the rate at which charges are crossing the surface S.

VOLUME CURRENT DENSITY

Consider charges which are moving through a volume. To define the volume current density $\mathbf{J}$ at a point, construct a small area ds perpendicular to current flow. dI is the current passing through the surface. Then the magnitude of vector $\mathbf{J}$ may be represented as follows:

$$J = \frac{dI}{ds} \quad [A/m^2] \tag{5-2}$$

Vector $\mathbf{J}$ has magnitude J and a direction identical to that of positive charge movement or current flow. If the charges are negative (for example, electrons), the direction of current is assumed to be opposite to the direction of negative charge movement. The current through a differential surface $\mathbf{ds}$ not necessarily perpendicular to current flow (Figure 5-1) is given by

$$dI = \mathbf{J} \cdot \mathbf{ds}$$

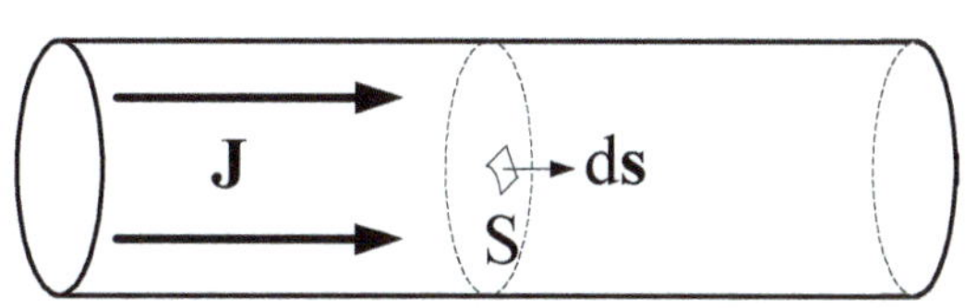

Figure 5-1. Volume current flow through an arbitrary surface.

and the total current I passing through surface S is

$$I = \iint_s \mathbf{J} \cdot d\mathbf{s} \qquad (5\text{-}3)$$

Next consider charges which are moving at a steady drift velocity $\mathbf{v}$ (Figure 5-2). Construct a rectangular solid of volume (ds v dt) as shown with surface ds perpendicular to current flow. In time dt all charges in the box will pass through the surface ds. Let there be N charges per unit volume, each of charge q. The current passing through the surface $d\mathbf{s}$ is then

$$dI = \frac{dQ(t)}{dt} = \frac{Nq(ds\, v\, dt)}{dt} = Nq\, ds\, v$$

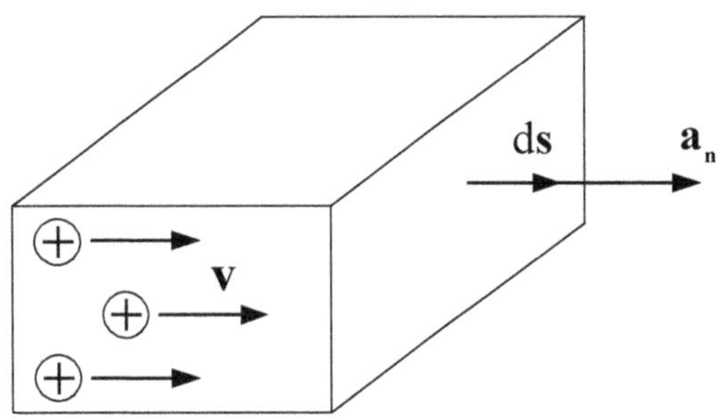

Figure 5-2. Drift velocity.

and

$$J = \frac{dI}{ds} = N q v \quad \text{where} \quad v = |\mathbf{v}|$$

$$\mathbf{J} = N\, q\, \mathbf{v} = \rho_v\, \mathbf{v}$$

$$\mathbf{J} = \rho_v\, \mathbf{v} \qquad (5\text{-}4)$$

This is how the volume current density is related to the volume charge density and its drift velocity. In the most general case where charges are of different types,

$$\mathbf{J} = \sum_i N_i\, q_i\, \mathbf{v}_i = \sum_i \rho_{vi}\, \mathbf{v}_i$$

where N_i, q_i, $\mathbf{v}_i$, ρ_{vi} are number of charges per unit volume, charge per carrier, velocity and volume charge density, respectively, for each type of charge carrier.

SURFACE CURRENT DENSITY

Surface current flows in a zero-thickness layer on a surface (Figure 5-3). To define the surface current density $\mathbf{J}_s$ at a point we construct a line $d\ell_\perp$. The subscript $\perp$ denotes that the line is perpendicular to the direction of current flow. Then

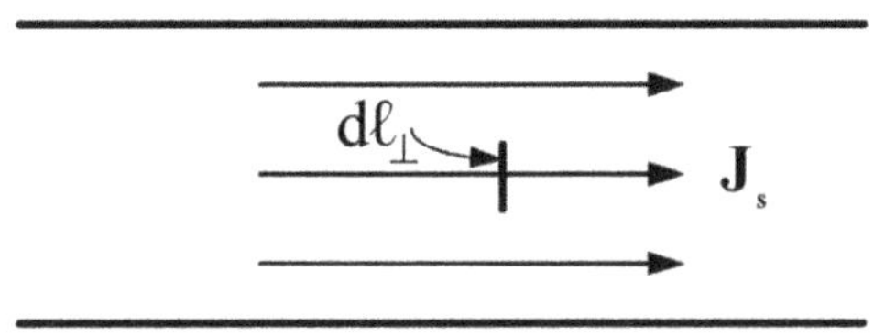

Figure 5-3. Surface current density J_s.

$$J_s = \frac{dI}{d\ell_\perp} \quad [\text{A/m}] \tag{5-5}$$

and the vector $\mathbf{J}_s$ has magnitude J_s and a direction identical to that of current flow. Note that surface current density has the unit of amps/m whereas volume current density has the unit of amps/m^2. The total current passing through the strip (Figure 5-3) is given by

$$\boxed{I = \int J_s d\ell_\perp} \tag{5-6}$$

LINE CURRENT

For line current, that is, current which flows in a filament of zero cross section, we merely count the charges passing through a point to determine the current I there. Then

$$I = \frac{dQ}{dt}$$

$\mathbf{I}$ is the current vector, which has magnitude I and whose direction is that of current flow.

5.3. THE EQUATION OF CONTINUITY

Current is produced by the motion of charges. Current and charge quantities are thus related. The fundamental relationship may be obtained by considering an arbitrary volume V bounded by surface S (Figure 5-4). Current I(t) is flowing inward through the surface and charge Q(t) is accumulating within.

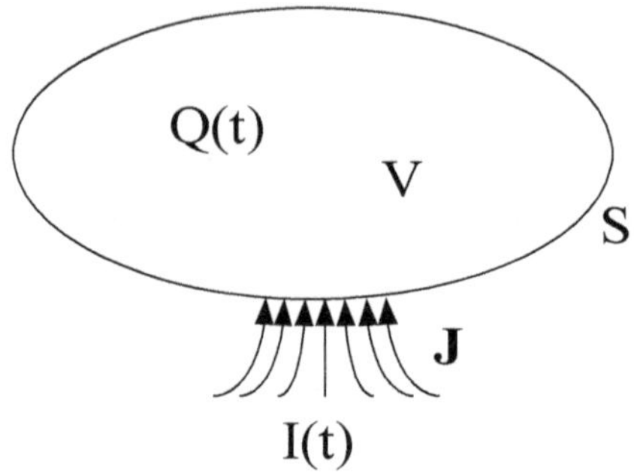

Figure 5-4. The equation of continuity.

The total current flowing into the volume V is

$$I(t) = -\oint\!\!\oint_S \mathbf{J} \cdot d\mathbf{s} = -\iiint_V \nabla \cdot \mathbf{J}\, dv$$

(by the divergence theorem)

$$= \frac{dQ(t)}{dt} = \frac{d}{dt} \iiint_V \rho_v\, dv = \iiint_V \frac{\partial \rho_v}{\partial t}\, dv$$

(by Leibniz's rule for the derivative of an integral)

The above relationship is equivalent to *conservation of charge* within volume V.

$$\iiint_V \nabla \cdot \mathbf{J}\, dv = \iiint_V -\frac{\partial \rho_v}{\partial t}\, dv$$

The volume V of integration is arbitrary and thus the integrands must be equal:

$$\boxed{\nabla \cdot \mathbf{J} = -\frac{\partial \rho_v}{\partial t}} \qquad (5\text{-}7)$$

Equation (5-7) is called the **equation of continuity**. It expresses the fundamental relationship between current and charge densities. It plays an important role for time-varying fields in Chapter 8.

In the steady-state (charge and current do not vary in time), Eq. (5-7) reduces to

$$\boxed{\nabla \cdot \mathbf{J} = 0} \qquad \text{for steady currents} \qquad (5\text{-}8)$$

5.4. OHM'S LAW AND CONDUCTIVE MATERIALS

The electromagnetic form of **Ohm's law** gives the relationship between volume current density $\mathbf{J}$ and the electric field $\mathbf{E}$ in a conducting material:

$$\boxed{\mathbf{J} = \sigma \mathbf{E}} \tag{5-9}$$

where σ is known as the **conductivity** of the material. It has the unit of siemens/m [S/m] or mho/m [℧/m]. The conductivity of various materials is given in Table 5-1. The table includes both good conductors and good insulators. Silver, copper, and gold are the three best conductors. Silver is best but it is expensive and tarnishes readily, which reduces the conductivity. Often materials are silver-plated to improve their conductivity. Copper is much less expensive than silver or gold and so it is the conductor of choice where large quantities are needed. Gold is often used in microelectronics. It is the most corrosion-free of the three. Finally, we note that rubber and fused quartz are two of the best insulators which can be treated as perfect dielectrics ($\sigma = 0$).

Table 5-1 Conductivities

Material	Conductivity (S/m)
Silver	6.2×10^7
Copper	5.8×10^7
Gold	4.1×10^7
Aluminum	3.5×10^7
Brass	1.6×10^7
Iron	10^7
Sea Water	4
Silicon	4.4×10^{-4}
Distilled Water	10^{-4}
Dry Earth	10^{-5}
Glass	10^{-12}
Rubber	10^{-15}
Fused Quartz	10^{-17}

To see that Eq. (5-9) is equivalent to the usual form of Ohm's law in circuit theory, consider a rectangular conductor with uniform $\mathbf{J}$, $\mathbf{E}$, σ throughout (Figure 5-5). Multiply both sides of Eq. (5-9) by S, the cross sectional area, and equate components to obtain

$$\mathrm{JS} = \sigma\,\mathrm{ES} = \frac{\sigma \mathrm{S}}{\ell}(\mathrm{E}\,\ell)$$

or

$$I = \frac{\sigma S}{\ell} V \ (\text{since } I = JS,\ V = E\ell).$$

or

$$V = \frac{\ell}{\sigma S} I$$

where V is the potential difference or the voltage between the two ends.

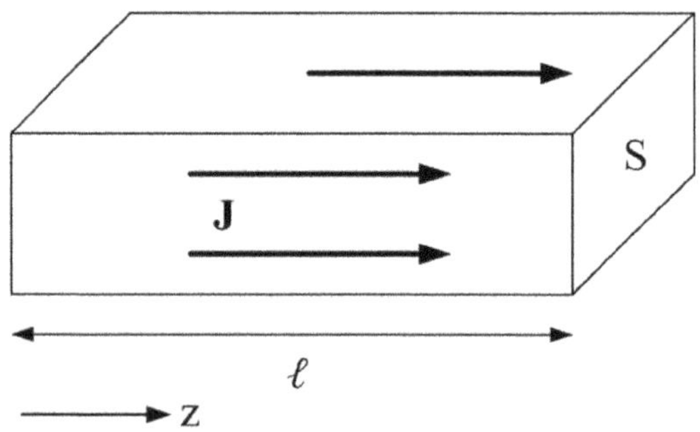

Figure 5-5. Uniform current flow in a resistor.

We have, then, a linear relationship between V and I. The term $\frac{\ell}{\sigma S}$ is called the **resistance** R whose unit is ohm [Ω]. The familiar circuit relation

$$\boxed{V = IR} \tag{5-10}$$

is thus obtained. The resistance,

$$R = \frac{V}{I} = \frac{\ell}{\sigma S} \tag{5-11}$$

applies to any homogeneous resistor of length ℓ and uniform cross section S. $\left(\frac{1}{\sigma}\right)$, the inverse of conductivity, is called *resistivity*.

Why should **J** be proportional to **E**? Why should **J** not increase indefinitely as electrons are accelerated by the electric field **E**? To answer these questions, we need to understand the processes of current flow in a conductor, which can be explained as follows. In the absence of an applied electric field, at room temperature, the conduction electrons are moving at about 10^6 m/s through the "lattice" structure of the material. The movement is in random directions so that the average vector velocity is zero and there is no net current ($\mathbf{J} = 0$). We then apply a voltage across the conductor and an electric field **E** is established within the conductor. The conduction electrons are accelerated by the force $q\mathbf{E} = -e\mathbf{E}$ and produce a current **J** in the direction of **E**. However, the electrons are accelerated for only a very short time about (about 10^{-14} s). They travel only a very short distance (about 10^{-8}m) before colliding with the lattice structure and being scattered in arbitrary directions. Before collision, the electron motion contributes to current $\mathbf{J}$ in the direction of $\mathbf{E}$. Immediately after collision, the direction is arbitrary and so the electron motion does not, on average, contribute to $\mathbf{J}$ in the direction of $\mathbf{E}$. The contribution to current is "lost," so to speak. Then the process starts all over again. The electron is accelerated, contributes to $\mathbf{J}$ between collisions and loses its contribution upon collision. There is however an average "*drift*" velocity contributing to $\mathbf{J}$. This is merely the average velocity acquired during the brief acceleration period between collisions. The drift velocity is very slow, on the order of 10^{-4} m/s.

Thus we have electrons which are moving very rapidly (10^6 m/s) in random directions. Superimposed on this very rapid motion is a very slow drift velocity (10^{-4} m/s) which contributes to $\mathbf{J}$ in the direction of $\mathbf{E}$. If we double the voltage, the time between collisions (the acceleration period) is unchanged since it is determined primarily by the much larger random velocity. The acceleration time is constant, the acceleration force $q\mathbf{E}$ is doubled, and so the average "drift" velocity is doubled. Thus we have a linear system and $\mathbf{J}$ is proportional to $\mathbf{E}$ as in Eq. (5-9). The constant σ is called conductivity.

It should be pointed out that a conductor is usually electrically neutral. The moving negative charges travel through a stationary lattice structure which is oppositely charged. For linear homogeneous media characterized by a constant conductivity σ, the conductor or resistor is uncharged ($\rho_v = 0$) within the volume. This is analogous to the case for the linear homogeneous dielectric ($\rho_{pv} = 0$). Surface charge will in general accumulate at interfaces. Volume charge density may exist ($\rho_v \neq 0$) for inhomogeneous conductors.

5.5. GEORG SIMON OHM (1789–1854) AND THE DISCOVERY OF OHM'S LAW

1827 should have been a very good year for Georg Simon Ohm. He had just published his comprehensive book on resistive circuits. He had investigated the subject very thoroughly and had confirmed his theory with very carefully performed experiments. He was now almost 40 years old. Surely he would receive the recognition which he deserved and with it, perhaps, the university appointment which he had long sought.

In his book, *The Galvanic Circuit Mathematically Investigated*, Ohm had explored the subject as completely as was possible at the time. He had shown that in a simple loop circuit with a battery of voltage V and uniform conductor size, the current in the wire is uniform throughout the length. Thus the circuit can be characterized with a single current, I. He also demonstrated that the potential falls uniformly with distance along the wire. Both of these essential facts, which are taken for granted by modern students, were unknown at the time.

Then, he had established the linear relationship between voltage and current

$$V = IR \tag{5-10}$$

where the resistance R of the circuit is proportional to wire length ℓ and inversely proportional to wire cross section S:

$$R = \frac{k\,\ell}{S} = \frac{\ell}{\sigma S} \tag{5-11}$$

where k, the resistivity and the inverse of conductivity, is a constant which depends on the material of the wire. Ohm also determined the relative conductivity of various metals, the best conductors being silver, copper, gold, in that order, as shown in Table 5-1. Ohm's experiments were very sophisticated and reliable. He used the Seebeck thermoelectric battery, invented in 1821, because of its predictable internal resistance. To measure the current and its magnetic field, he used the torsion balance compass needle galvanometer invented by Coulomb in his measurements of the inverse square law for magnetic poles.

There is a direct analogy between resistive circuits and heat flow. The quantity of heat flowing in a wire is proportional to the temperature difference between the ends of the wire and inversely proportional to thermal resistance, which, in turn, is proportional to the length of the wire and inversely proportional to its cross-sectional area. Ohm had studied Fourier's book, *The Analytical Study of Heat,* and had, with great insight, seen the analogy to electrical current flow. Voltage, current, and electrical resistance are analogous to temperature difference, heat flow, and thermal resistance, respectively.

Note that our familiar rules for combinations of resistances in series and parallel follow directly from Ohm's law. For instance, placing two identical sections of wire in series doubles the resistance since total length ℓ is doubled; placing them in parallel halves the resistance since cross-section area S is doubled.

Georg Ohm was born in Erlangen, Germany, in 1789, the son of a locksmith. He and his younger brother Martin were well-educated and Georg received a Ph.D. from the University of Erlangen in 1811. He then held several teaching positions in secondary schools, finally settling in Cologne, where he taught for about nine years. It was in Cologne that he carried out most of his basic research. It was a long and difficult process. He had to educate himself in the mathematics and science of the times and become an expert in the current experimental techniques of electrical science, all the while carrying the heavy teaching load of a secondary school. We are told that he was a very serious, effective, and innovative teacher.

The early 19th century was a time of great advances in electrical science. In 1800 Alessandro Volta invented the voltaic cell or battery, which supplied a steady source of current. This was much preferable to the Leyden jar or capacitor as a source. The voltaic cell served as a great stimulus to research. Oersted's dramatic discovery in 1820 of the magnetic field of an electric current is described in Chapter 6. Within a year, this discovery led to a method of measuring current by determining its magnetic field. This was done by noting the deflection of a compass needle. In this way, Volta and Oersted provided a steady current and the means to measure it, making it possible for Ohm to carry out his experiments.

Unfortunately, Ohm's book was not well received in Germany. In fact, his theory was widely rejected, condemned as too complicated by some and too simple by others. In general, reaction to his work was very negative and the backward educational establishment criticized him in response. Ohm felt that he must resign his post in Cologne and went to teach at a military school in Berlin.

However, scientists in other countries slowly began to recognize his work and to give it the respect which it deserved. He received the Copley medal in 1841 and finally, in 1849, 22 years after the publication of his book, he was appointed Professor of Physics at the University of Munich, where he continued to teach, honored and revered, until his death at 65 in 1854.

Henry Cavendish and other researchers had preceded Ohm in his discovery of Ohm's law but the thoroughness and completeness of Ohm's work guaranteed that he would receive primary credit for the discovery.

5.6. POWER—JOULE'S LAW

When the current flows in a conducting medium such as a resistor, the power is delivered or dissipated in the medium. The power dissipated in a conductive or resistive material may be calculated from the electric field and the current density as follows:

$$\boxed{P = \iiint_V \mathbf{E} \cdot \mathbf{J}\, dv \quad [W]} \tag{5-12}$$

where V is the volume of the resistor. $\mathbf{E} \cdot \mathbf{J}$ is thus a volume density of power dissipated with units of watts/m^3. Equation (5-12) is known as **Joule's law**.

To see the correspondence with circuit theory, consider the resistor of Figure 5-5 with uniform $\mathbf{E}, \mathbf{J}, \sigma$. Making use of Eq. (5-9), we have

$$\begin{aligned} P &= \iiint \sigma E^2\, dv = \iiint \frac{1}{\sigma} J^2 dv = \sigma E^2 (S\ell) = \frac{J^2}{\sigma}(S\ell) \\ &= (E\ell)^2 \frac{S\sigma}{\ell} = \frac{V^2}{R} \\ &= (JS)^2 \frac{\ell}{S\sigma} = I^2 R \\ &= \int E d\ell \iint J\, ds = (E\ell)\,(JS) = VI \end{aligned}$$

5.7. RELAXATION TIME

In Chapter 2, we noted that charges placed within a conductor move almost immediately by mutual repulsion toward the surface of the conductor. In other words, volume charge density disappears very rapidly giving rise to surface charge density. Now we are in a position to examine the phenomenon quantitatively and ask the question "*how fast*?". Consider a linear homogeneous medium of conductivity σ, permittivity ε, volume V and surface S (Figure 5-6).

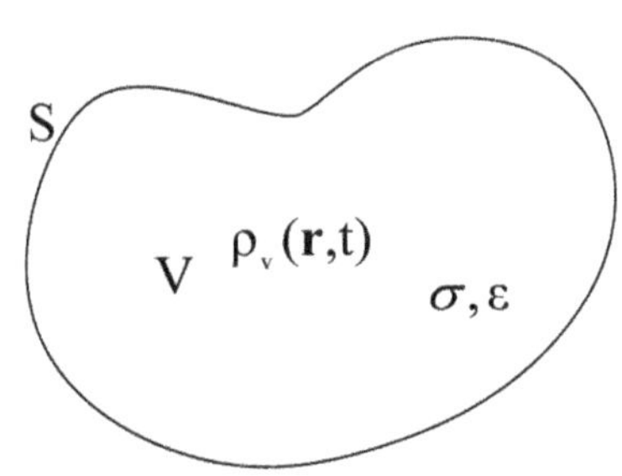

Figure 5-6. Relaxation time.

We start with the equation of continuity:

$$\nabla \cdot \mathbf{J} = -\frac{\partial \rho_v}{\partial t} = \nabla \cdot (\sigma \mathbf{E}) = \nabla \cdot \left(\frac{\sigma}{\varepsilon} \mathbf{D} \right) \quad \text{[Ohm's law and } \mathbf{D} = \varepsilon \mathbf{E}\text{]}$$

$$= \frac{\sigma}{\varepsilon} \nabla \cdot \mathbf{D} = \frac{\sigma}{\varepsilon} \rho_v \quad \text{[Eq.(3-10b) is used.]}$$

$$\frac{\partial \rho_v}{\partial t} = \left(-\frac{\sigma}{\varepsilon} \right) \rho_v$$

The solution to the differential equation above is

$$\boxed{\rho_v(x,y,z,t) = \rho_{vo}(x,y,z)\; e^{-\frac{\sigma}{\varepsilon}t}} \qquad (5\text{-}13)$$

or

$$\rho_v(x,y,z,t) = \rho_{vo}(x,y,z) e^{-t/\tau}$$

where $\tau = \frac{\varepsilon}{\sigma}$ is called the *relaxation time*. We may interpret Eq. (5-13) in the following way. Suppose we start with an initial volume charge density ρ_{vo} (x,y,z) within volume V at time t = 0. Then the volume charge density decays exponentially at every point inside the volume. The net charge will eventually reside on the surface S since charge is conserved. So the surface charge density is increasing as the volume charge density decreases. τ is the time required for the volume charge density to decay to 1/e or about 37% of its value at any point in V. Table 5-2 shows relaxation times for some conductors and insulators. Relaxation time is so short, on the order of 10^{-19} seconds, for good conductors, that we may consider the process to be practically instantaneous for good conductors.

Table 5-2. Relaxation Times

Material	**Relaxation Time**
Copper	1.5×10^{-19} (s)
Sea Water	2×10^{-10} (s)
Distilled Water	10^{-6} (s)
Fused Quartz	10^{5} (s) or about 11 days

5.8. BOUNDARY CONDITIONS FOR STEADY CURRENTS

In the presence of steady currents it is necessary to add one boundary condition, namely, the continuity of normal **J**, to the basic boundary conditions already derived in Chapter 3. Figure 5-7 shows a boundary between two arbitrary media (1), (2) with characteristics (σ_1, ε_1) and (σ_2, ε_2), respectively.

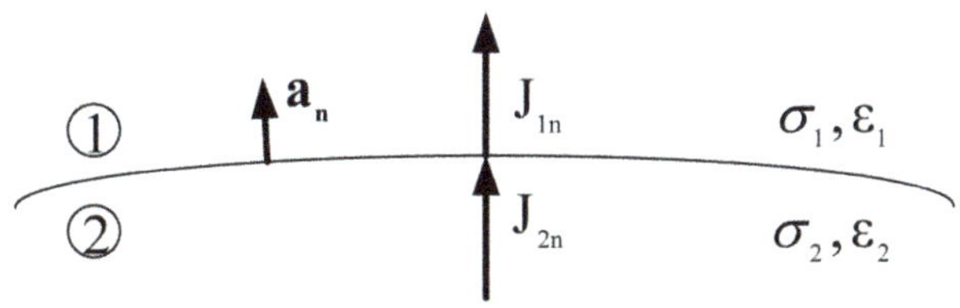

Figure 5-7. Continuity of normal J.

Consider the normal components J_{1n}, J_{2n} of volume current density at the interface. J_{2n} specifies the charge flowing into the interface and J_{1n} determines charge flowing out from the interface. J_{1n}, J_{2n} must be equal for steady currents or else charge would accumulate indefinitely at the interface. Therefore

$$J_{1n} = J_{2n} \text{ for steady currents} \tag{5-14}$$

In summary, the three basic boundary conditions at an interface are:

$E_{1t} = E_{2t}$	Tangential **E** is continuous
$D_{1n} - D_{2n} = \rho_s$	Normal **D** is discontinuous by the amount of free surface charge
$J_{1n} = J_{2n}$	Normal **J** is continuous

EXAMPLE 5-1. DISSIMILAR DIELECTRICS IN A PARALLEL PLATE CAPACITOR

Figure 5-8 shows parallel plates of area A and separation $d_1 + d_2$. The parallel plate region is filled with two different materials of permittivity, conductivity, ε_1, σ_1, and ε_2, σ_2, respectively. The switch is closed and the steady state is eventually reached. Find the current, the total resistance, and the free surface charge at the interface between dissimilar materials.

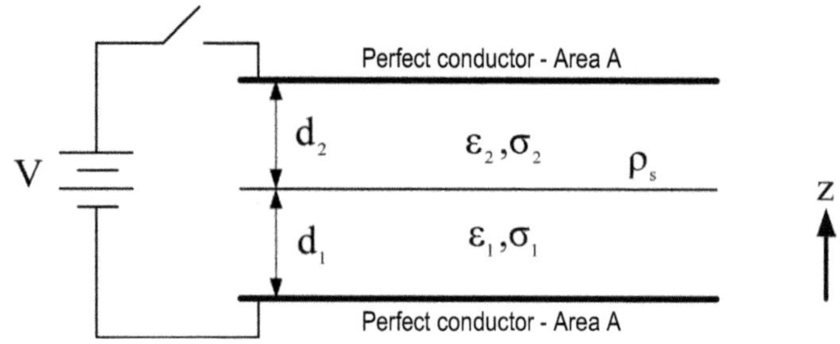

Figure 5-8. Dissimilar dielectrics in a parallel-plate capacitor.

SOLUTION:

In such problems, we concentrate on the current. Because of the continuity of normal **J** in the steady state, we assume a single volume current density for both regions:

$$\mathbf{J} = -\mathbf{a}_z J_o$$

We find the voltage by applying Ohm's law [Eq. (5-9)]:

$$V = -\int \mathbf{E} \cdot d\ell = E_1 d_1 + E_2 d_2 = \frac{J_o d_1}{\sigma_1} + \frac{J_o d_2}{\sigma_2} = J_o\left(\frac{d_1}{\sigma_1} + \frac{d_2}{\sigma_2}\right)$$

and

$$J_o = \frac{V}{\left(\frac{d_1}{\sigma_1} + \frac{d_2}{\sigma_2}\right)}$$

then

$$I = \iint \mathbf{J} \cdot d\mathbf{s} = J_o A = \frac{VA}{\left(\frac{d_1}{\sigma_1} + \frac{d_2}{\sigma_2}\right)}$$

and

$$R = \frac{V}{I} = \frac{d_1}{\sigma_1 A} + \frac{d_2}{\sigma_2 A} \tag{5-15}$$

Note that the potential at the interface $(z = d_1)$ is

$$E_1 d_1 = J_o \frac{d_1}{\sigma_1} \text{ or } V \frac{\frac{d_1}{\sigma_1}}{\frac{d_1}{\sigma_1} + \frac{d_2}{\sigma_2}}$$

with respect to the potential at $z = 0$. It is determined solely by the conductivities σ_1, σ_2.

To find the surface charge density at the interface we use the boundary condition for the normal component of **D**, given by Eq. (3-25).

$$\mathbf{D}_1 = -\mathbf{a}_z \frac{J_o}{\sigma_1}\varepsilon_1, \qquad \mathbf{D}_2 = -\mathbf{a}_z \frac{J_o}{\sigma_2}\varepsilon_2$$

$$\rho_s = D_{2z} - D_{1z} = J_o\left(\frac{\varepsilon_1}{\sigma_1} - \frac{\varepsilon_2}{\sigma_2}\right) \tag{5-16}$$

$$\rho_s \Rightarrow 0 \quad \text{if} \quad \frac{\varepsilon_1}{\sigma_1} = \frac{\varepsilon_2}{\sigma_2}$$

Where did the charge ρ_s come from? Normal **J** is continuous in the steady state and leads to no charge accumulation. We may consider that the charge ρ_s is deposited on the interface during the transient period when normal **J** is not necessarily continuous. A net charge is deposited only if relaxation times are unequal.

5.9. A RELATIONSHIP BETWEEN CAPACITANCE AND RESISTANCE

Figure 5-9 shows two conductors in an infinite homogeneous medium of permittivity ε or conductivity σ or both.

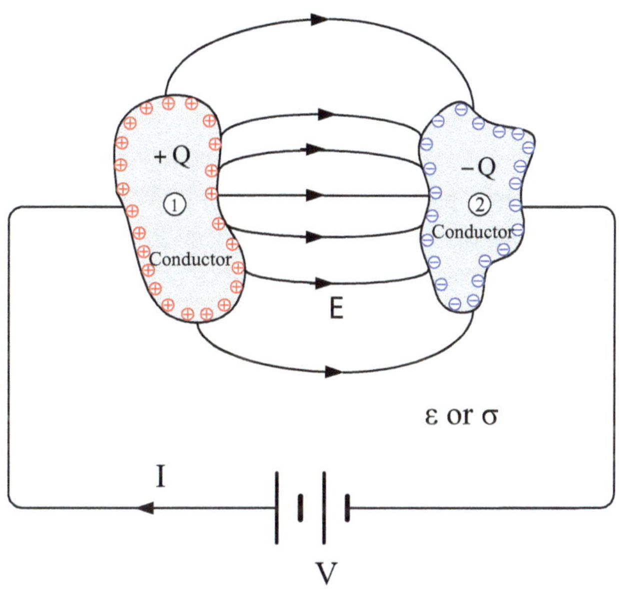

Figure 5-9. A relationship between capacitance and resistance.

Let's consider first the ideal dielectric with permittivity ε. A voltage V is applied between (perfect) conductors, an electric field **E** is established in the dielectric and charges Q, – Q reside on the conductors.

The capacitance is

$$C = \frac{Q}{V} = \frac{\iint_S \mathbf{D} \cdot d\mathbf{s}}{V} = \frac{\varepsilon \iint_S \mathbf{E} \cdot d\mathbf{s}}{V} \tag{5-17}$$

Next consider the resistor with conductivity σ. Voltage V is applied and a steady current I flows in the circuit and through the resistor ($\mathbf{J} = \sigma\,\mathbf{E}$ in the resistor). The resistance is

$$R = \frac{1}{G} = \frac{V}{I} = \frac{V}{\sigma \iint_S \mathbf{E} \cdot d\mathbf{s}} \tag{5-18}$$

since

$$I = \iint_S \mathbf{J} \cdot d\mathbf{s} = \sigma \iint_S \mathbf{E} \cdot d\mathbf{s}$$

G, the inverse of resistance, is called the **conductance** whose unit is mho or siemens. We assume that the voltage V is identical in the two cases. The electric field $\mathbf{E}$ then is also identical in the two cases because of uniqueness and, multiplying Eqs. (5-17) and (5-18), we obtain

$$\boxed{RC = \frac{C}{G} = \frac{\varepsilon}{\sigma}} \tag{5-19}$$

What happens as we change from a perfect to an imperfect dielectric in Figure 5-9? The only thing that changes is that current flows and increases with increasing conductivity σ. V, $\mathbf{E}$, Q, C remain unchanged as σ increases. Only $\mathbf{J}$ and R vary. Thus the resistance and capacitance problems are decoupled.

EXAMPLE 5-2. SHUNT RESISTANCE AND CAPACITANCE OF A COAXIAL LINE

Figure 5-10(a) shows the cross section of a coaxial line with inner, outer radii a, b, respectively and length ℓ. The coaxial line is completely filled with a material of permittivity ε and conductivity σ. A voltage V is applied between inner and outer conductors and a current $\mathbf{I}$ then flows radially outward from inner to outer conductor. Find the shunt resistance and capacitance.

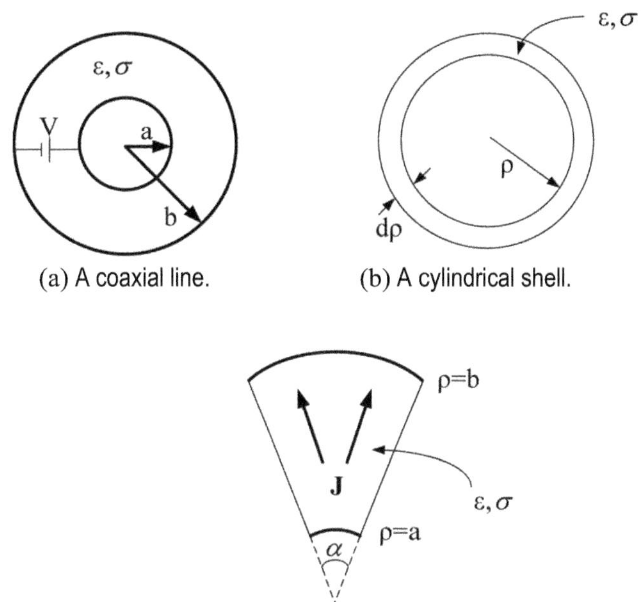

(a) A coaxial line. (b) A cylindrical shell.

(c) A truncated wedge.

Figure 5-10. A coaxial line with inner, outer radii a, b, respectively and length ℓ.

SOLUTION

The volume current density is independent of ϕ and varies as $\left(\frac{1}{\rho}\right)$ so that the total current is independent of ρ:

$$\mathbf{J} = \mathbf{a}_\rho \frac{I}{2\pi\rho\ell} \tag{5-20}$$

An integration over a cylindrical surface of radius ρ and length ℓ shows that the total current is I:

$$\iint \mathbf{J}\cdot d\mathbf{s} = \int_0^{\ell}\int_0^{2\pi} \frac{I}{2\pi\rho\ell}\,\rho\, d\phi\, dz = I$$

Then

$$\mathbf{E} = \frac{\mathbf{J}}{\sigma} = a_\rho \frac{I}{2\pi\rho\sigma\ell}$$

$$V = -\int \mathbf{E}\cdot d\boldsymbol{\ell} = -\int_b^a \frac{I}{2\pi\sigma\ell\rho}\, d\rho = \frac{I}{2\pi\sigma\ell} \ln\left(\frac{b}{a}\right)$$

$$I = \frac{2\pi\sigma\ell V}{\ln(b/a)};\quad \mathbf{E} = \mathbf{a}_\rho \frac{V}{\ln(b/a)}\left(\frac{1}{\rho}\right)$$

and

$$R = \frac{V}{I} = \frac{\ln(b/a)}{2\pi\sigma\ell} \tag{5-21}$$

$$G = \frac{2\pi\sigma\ell}{\ln(b/a)}; \quad \frac{G}{\ell} = \frac{2\pi\sigma}{\ln(b/a)} \tag{5-22}$$

The charge Q in length ℓ (on the inner conductor) is determined as follows:

$$Q = (\text{area})\rho_s = (\text{area})\ D_\rho = (2\pi a \ell)\ \varepsilon \frac{V}{\ln(b/a)} \frac{1}{a} = \frac{2\pi\varepsilon\ell}{\ln(b/a)} V$$

Then

$$C = \frac{Q}{V} = \frac{2\pi\varepsilon\ell}{\ln(b/a)}; \quad \frac{C}{\ell} = \frac{2\pi\varepsilon}{\ln(b/a)} \tag{5-23}$$

Note that R in Eq. (5-21) and C in Eq. (5-23) satisfy $RC = \frac{C}{G} = \frac{\varepsilon}{\sigma}$ as in Eq. (5-19).

The resistance R may also be determined by considering a thin cylindrical shell of radius ρ, length ℓ (Figure 5-10(b)).

$$dR = \frac{d\ell}{\sigma A} = \frac{d\rho}{2\pi\rho\sigma\ell}$$

$$R = \int_a^b dR = \int_a^b \frac{d\rho}{2\pi\sigma\ell\rho} = \frac{\ln(b/a)}{2\pi\sigma\ell}$$

which agrees with Eq. (5-21).

The conductance G may be determined by considering a truncated wedge (Figure 5-10(c)) with radial current flow:

$$\mathbf{J} = \mathbf{a}_\rho \frac{I}{\alpha\rho\ell}; \quad \iint \mathbf{J} \cdot d\mathbf{s} = I$$

$$\mathbf{E} = \mathbf{a}_\rho \frac{I}{\alpha\rho\ell\sigma}; \quad V = -\int \mathbf{E} \cdot d\ell = \frac{I \ln(b/a)}{\alpha\ell\sigma}$$

$$R = \frac{V}{I} = \frac{\ln(b/a)}{\alpha\ell\sigma}; \quad G = \frac{\alpha\ell\sigma}{\ln(b/a)} \quad (\text{wedge}) \tag{5-24}$$

For a thin wedge of angle $d\phi$,

$$dG = \frac{d\phi\,\ell\sigma}{\ln\,(b/a)}$$

and for the whole coaxial line,

$$G = \int dG = \int_0^{2\pi} \frac{\ell \sigma \, d\phi}{\ln(b/a)} = \frac{2\pi\sigma\ell}{\ln(b/a)}$$

which agrees with Eq. (5-22).

The power losses in the coaxial line are determined as follows:

$$P = \iiint \mathbf{E}\cdot\mathbf{J}\, dv = \iiint \sigma\, E_\rho^2\, dv$$
$$= \int_0^{\ell}\int_0^{2\pi}\int_a^{b} \frac{\sigma V^2\, \rho d\rho\, d\phi\, dz}{\ln^2(b/a)\rho^2} = \frac{V^2(2\pi\ell\sigma)}{\ln(b/a)}$$
$$= \frac{V^2}{R} = I^2 R$$

EXAMPLE 5-3. A TRUNCATED CONICAL RESISTOR

Figure 5-11 shows a truncated conical resistor bounded by surfaces $r = a$, $r = b$, and $\theta = 15°$.

The material of the resistor is inhomogeneous:

$$\sigma = \sigma_o \frac{r}{a}$$

A voltage V is applied between surfaces $r = a, b$. Find the electric field and the resistance between the surfaces $r = a, b$.

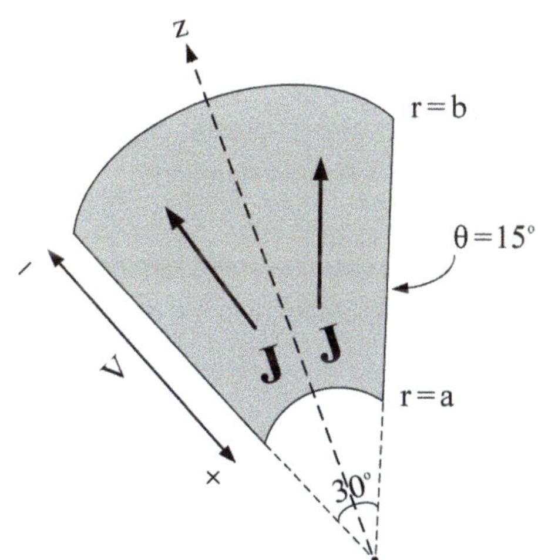

Figure 5-11. A truncated conical resistor.

SOLUTION:

First we assume a radial, inverse square current density which makes the total current independent of r:

$$\mathbf{J} = \mathbf{a}_r \frac{C_1}{r^2}$$

$$I = \iint \mathbf{J} \cdot d\mathbf{s} = \int_0^{2\pi} \int_0^{\pi/12} \frac{C_1 r^2 \sin\theta \, d\theta \, d\phi}{r^2}$$

$$= 2\pi C_1 (1 - \cos 15°)$$

$$\mathbf{E} = \frac{\mathbf{J}}{\sigma} = \mathbf{a}_r \frac{C_1}{r^2} \frac{1}{\sigma_o \frac{r}{a}} = \mathbf{a}_r \frac{C_1 a}{\sigma_o r^3}$$

$$V = -\int \mathbf{E} \cdot d\ell = -\int_b^a E_r \, dr = \frac{C_1 a}{2\sigma_o}\left(\frac{1}{a^2} - \frac{1}{b^2}\right)$$

and

$$C_1 = 2\sigma_o \left(\frac{b^2 a}{b^2 - a^2}\right) V; \quad \mathbf{E} = \mathbf{a}_r \frac{2b^2a^2}{b^2 - a^2}\left(\frac{1}{r^3}\right) V$$

$$R = \frac{V}{I} = \frac{b^2 - a^2}{4\pi\sigma_o ab^2 (1 - \cos 15°)} \qquad (5\text{-}25)$$

PROBLEMS

5-1. A current I_o flows through a conducting wire of circular cross section of radius a (Figure 5-1). Find the volume current density $\mathbf{J}$ in each of the following situations:

(a) The current is uniformly distributed over the cross section, i.e., $\mathbf{J}$ is uniform.

(b) The current density in the wire is proportional to the distance from the axis (say, z-axis) of the wire, i.e., $\mathbf{J} = \mathbf{a}_z k \rho$. Determine k.

5-2. A current I_o flows only on the thin surface of a conducting wire of circular cross section of radius a. This is particularly true for very good conductors at very high frequencies. Assuming that the current is uniformly distributed over the surface of the conductor, find the surface current density $\mathbf{J}_s$.

5-3. A d-c voltage of 0.9 [mV] is applied to the ends of an aluminum wire of 2 [mm] in diameter and 1 [m] in length. The current measured through the wire is 100 [mA]. Find (a) the resistance of the wire, (b) the conductivity of the aluminum, (c) the electric field intensity in the wire, and (d) the power dissipated in the wire using the electric field. Assume that the current is uniformly distributed over the cross section of the wire.

5-4. A copper wire of 1-m length and 1-mm radius has a resistance of 5.5×10^{-3} [Ω] and carries a current of 4 [A]. The number of free electrons per unit volume in copper is 8.43×10^{28} [m^{-3}]. Find (a) the current density, (b) the conductivity of the copper, (c) the power dissipated in the wire, (d) the (average) drift velocity of free electrons, and (e) the mobility of free electrons. Assume that the current is uniformly distributed over the cross section of the wire. Note that the average drift velocity $\mathbf{v}$ of electrons in a conductor is proportional to the electric field $\mathbf{E}$:

$$\mathbf{v} = -\mu \mathbf{E}$$

where μ is called the *mobility* of the charge carriers.

5-5. Lightning strikes a lossy (or conductive) dielectric sphere of $\varepsilon = 2\varepsilon_0$, $\sigma = 5$ [S/m], radius $a = 0.5$ [m] at time $t = 0$, depositing a total charge 1 [mC] in the sphere.

(a) Calculate the time it takes for the charge density in the sphere to diminish to 10% of its initial value.
(b) Find $\mathbf{E}$ and $\mathbf{J}$ for all t both inside and outside the sphere.

[HINT] Assume that the charge is uniformly distributed in the sphere and make use of the result in Example 2-3.

5-6. Two very large planar, parallel plates of metal are separated by the distance $d_1 + d_2$. The space between the plates is filled with two conducting media. The first medium has conductivity σ_1 and thickness d_1, and the second has conductivity σ_2 and thickness d_2. Both plates have the surface area A. Find the resistance between the two metal plates. See Figure 3-24(a) and replace ε_1, ε_2 by σ_1, σ_2.

5-7. Two very large planar, parallel plates of metal are separated by the distance d. The space between the plates is filled with two conducting media next to each other. The first medium has conductivity σ_1 and surface area of $w \times \ell_1$ and the second medium has conductivity σ_2 and surface area of $w \times \ell_2$. Find the resistance between the two metal plates. See Figure 3-24(b) and replace ε_1, ε_2 by σ_1, σ_2.

5-8. Consider Figure 3-13(a) and replace the dielectric by a conductor of conductivity σ. It becomes a *spherical resistor*. A voltage V_0 is applied between inner and outer spherical (perfect) conductors ($\sigma \to \infty$), positive at the inner conductor ($r = a$).

(a) Find the electric field $\mathbf{E}$ and the (volume) current density $\mathbf{J}$ in the region $a \le r \le b$.
(b) Calculate the resistance R.
(c) Calculate the power dissipated in the resistor using Joule's law, Eq. (5-12), and show that $P = V_0^2/R$.

(d) Compare R from (b) above and C from Example 3-6(b) and show that it satisfies Eq. (5-19):

$$RC = \frac{C}{G} = \frac{\varepsilon}{\sigma} \quad \text{(capacitance -conductance analogy)}$$

5-9. Consider Figure 3-13(b) and replace the two dielectrics by two conductors of conductivities σ_1, σ_2. Find the resistance. You can make use of the result of Problem 5-8.

5-10. Consider Figure 3-13(c) and replace the two dielectrics by two conductors of conductivities σ_1, σ_2. Find the resistance.

5-11. Consider Figure 3-13(d) and replace the three dielectrics by three conductors of conductivities σ_1, σ_2, σ_3. Find the resistance.

5-12. Consider Figure 3-13(b) and assume it to be a cross section of a long coaxial line of length ℓ. Replace two dielectrics by two conductors of conductivities σ_1, σ_2. Find the resistance. You can make use of the result of Example 5-2.

5-13. Consider Figure 3-13(c) and assume it to be a cross section of a long coaxial line of length ℓ. Replace two dielectrics by two conductors of conductivities σ_1, σ_2. Find the resistance.

5-14. Consider Figure 3-13(d) and assume it to be a cross section of a long coaxial line of length ℓ. Replace three dielectrics by three conductors of conductivities σ_1, σ_2, σ_3. Find the resistance.

5-15*. A conducting medium of conductivity σ is in the uniform electric field $\mathbf{E} = \mathbf{a}_z E_0$. A spherical cavity of radius a is formed in the medium. Find the electric field inside the cavity $(r < a)$ and the current density in the conducting medium $(r > a)$.

[HINT] Express $\mathbf{E}$ and $\mathbf{J}$ in terms of electric potentials and find the potentials by solving Laplace's equation.

5-16. Kirchhoff current law (KCL) states that the sum of the currents leaving (or entering) a node is zero. Derive KCL from Eq. (5-8), by integrating both sides of Eq. (5-8) over a small volume surrounding a node in the circuit. Discuss the conditions under which KCL is valid.

MAGNETOSTATIC FIELDS

CHAPTER 6

6.1 INTRODUCTION

In the year 1820, a startling discovery was announced. A Danish scientist, Hans Christian Oersted (1777–1851), reported that an electric current flowing in a wire produced a magnetic field. The magnetic field was perpendicular to the wire and consisted of circles about the axis of the wire. His original paper, written in Latin, was published in July 1820 and was immediately translated into several languages. Before Oersted's discovery, most scientists had assumed that electricity and magnetism were unconnected. The two fields had been studied separately for centuries and no one had theorized any relationship between them, although a few researchers had begun to speculate that the two sciences might be connected in some way. Oersted proved that a steady or dc electric current produced a steady or *static* magnetic field. His finding had apparently come about during a classroom demonstration designed to show that electric current in a wire caused the wire to glow.

When the wire was connected to a battery, a compass needle nearby suddenly moved to a position nearly perpendicular to the wire. Eureka! Electricity and magnetism could no longer be viewed as separate phenomena. They were connected in some as yet unknown manner.

The response to Oersted's discovery was striking. Oersted's work was reported to the French Academy in September 1820. New results were then reported by Ampère and Biot almost every week. Ampère published ten papers in 1820; Biot published four papers that year, including the beginnings of the Biot-Savart law. By the end of 1820, they had begun to delineate the relation between electricity and magnetism.

In this chapter, we consider steady (dc) electric currents in free space. The resultant magnetic fields are constant, i.e., they do not vary with time. Thus we are concerned with **static magnetic** or **magnetostatic** fields. In practical situations, we often deal with ac current sources. The magnetostatic analysis is very useful for situations in which fields are varying slowly with time. The study of magnetostatics yields insight into fundamental magnetic phenomena.

6.2 MAGNETIC FIELD

We define the **magnetic induction** or **magnetic flux density** $\mathbf{B}$ by considering the forces on a small test charge q. Suppose the charge q is placed in the region where both the electric field $\mathbf{E}$ and the magnetic field $\mathbf{B}$ exist. If the charge is at rest, it experiences only the electric force $\mathbf{F}_e$ where

$$\mathbf{F}_e = q\,\mathbf{E} \tag{6-1}$$

If the charge is in motion with velocity $\mathbf{v}$, then it experiences an additional force $\mathbf{F}_m$, the *magnetic force,* where

$$\mathbf{F}_m = q\,\mathbf{v} \times \mathbf{B} \tag{6-2}$$

Note that the magnetic force is perpendicular to both the magnetic induction $\mathbf{B}$ and the direction of charge motion ($\mathbf{v}$).

The total force $\mathbf{F}$ on the moving test charge is

$$\boxed{\mathbf{F} = q(\mathbf{E} + \mathbf{v} \times \mathbf{B}) \quad \text{Lorentz Force}} \tag{6-3}$$

Eq. (6-3) is called the **Lorentz force equation**. It can serve as a definition of both $\mathbf{E}$ and $\mathbf{B}$ which can be determined by measurement. For a stationary test charge, $\mathbf{v} = 0$, and $\mathbf{E}$ can be determined by a single measurement of force. Once $\mathbf{E}$ is known, the components of $\mathbf{B}$ may be determined by two different measurements of force $\mathbf{F}$ for two orthogonal velocities $\mathbf{v}$ (see Example 6-1). Although $\mathbf{B}$ is formally named *magnetic flux density* or *magnetic induction*, people often call it simply **magnetic field**. We will also use the latter term. The unit of $\mathbf{B}$ is the tesla [T]. From Eq. (6-2) we see that the unit of $\mathbf{B}$ can be represented as force divided by charge times velocity:

$$1\,\mathrm{T} = \frac{\mathrm{N}}{\mathrm{C}\dfrac{\mathrm{m}}{\mathrm{sec}}} = \frac{\mathrm{N\,sec}}{\mathrm{C\ m}}$$

Note that 1 T is equal to 10^4 gauss and the Earth's magnetic field is about 0.5 gauss.

EXAMPLE 6-1. DETERMINATION OF MAGNETIC FLUX DENSITY B FROM MEASUREMENTS

Determine $\mathbf{B}$ from the Lorentz force equation and measurements.

SOLUTION

Assume that $\mathbf{E}$ has already been determined from a rest velocity measurement ($\mathbf{v} = 0$) of the force and therefore components E_x, E_y, E_z are known.

Consider a velocity $\mathbf{v}$ in the x direction only. We move the charge at velocity $\mathbf{v} = \mathbf{a}_x v_x$ and measure the force $\mathbf{F}$.

$$\mathbf{v} = \mathbf{a}_x v_x;\ \text{Let } \mathbf{B} = \mathbf{a}_x B_x + \mathbf{a}_y B_y + \mathbf{a}_z B_z$$
$$\mathbf{F} = q\,\mathbf{E} + q\,\mathbf{v} \times \mathbf{B} = \mathbf{a}_x F_x + \mathbf{a}_y F_y + \mathbf{a}_z F_z$$
$$\mathbf{F} = q\,\mathbf{E} + q\,(\mathbf{a}_x v_x) \times [\mathbf{a}_x B_x + \mathbf{a}_y B_y + \mathbf{a}_z B_z]$$
$$= \mathbf{a}_x (qE_x) + \mathbf{a}_y (qE_y) + \mathbf{a}_z (qE_z) + \mathbf{a}_z (q\,v_x\,B_y) - \mathbf{a}_y (q\,v_x\,B_z)$$

Taking z components, we obtain $B_y = \dfrac{F_z}{q v_x} - \dfrac{E_z}{v_x}$

Taking y components, we obtain $B_z = \dfrac{-F_y}{q v_x} + \dfrac{E_y}{v_x}$

You can complete the determination of $\mathbf{B}$ by considering a velocity in either the y or z direction. Therefore $\mathbf{E}$, $\mathbf{B}$ can be determined by the Lorentz force equation and three measurements of vector force $\mathbf{F}$.

6.3 BASIC LAWS OF MAGNETOSTATICS

The basic laws of magnetostatics specify the divergence and curl of the static magnetic field $\mathbf{B}$ in free space just as the basic laws of electrostatics specify the divergence and curl of $\mathbf{E}$. The first law, in point form (or differential form) and integral form, is

$$\boxed{\nabla \cdot \mathbf{B} = 0} \qquad \mathbf{B} \text{ is a solenoidal field} \qquad (6\text{-}4)$$

and

$$\boxed{\oint\!\!\oint_S \mathbf{B} \cdot d\mathbf{s} = 0} \qquad (6\text{-}5)$$

where S is any closed surface. The integral form $\oint\!\!\oint_S \mathbf{B} \cdot d\mathbf{s} = 0$ may be derived from the point form by integrating both sides over an arbitrary volume V and applying the divergence theorem: $\iiint_V \nabla \cdot \mathbf{B}\ dv = \oint\!\!\oint_S \mathbf{B} \cdot d\mathbf{s} = 0$ where S bounds V. It is also possible to derive the point form from the integral form; the point and integral forms are thus identical statements. It can also be shown from Eq. (6-5) that the magnetic flux $\iint_S \mathbf{B} \cdot d\mathbf{s}$ through an open surface S is independent of surface as long as the surfaces are bounded by the same contour C. This is a property of a *solenoidal* field just as the line integral is independent of path for conservative fields.

The second law, in point and integral form, is

$$\boxed{\nabla \times \mathbf{B} = \mu_0 \mathbf{J}} \qquad (6\text{-}6)$$

and

$$\boxed{\oint_C \mathbf{B} \cdot d\ell = \mu_0\, I_{enc}} \qquad \text{Ampère's law} \qquad (6\text{-}7)$$

where I_{enc} is the total current enclosed by C and $\mu_0 = 4\pi \times 10^{-7}$ [T · m/A] or [H/m]. μ_0 is a universal constant, called the *permeability of free space*. $\mathbf{J}$ is the volume current density measured in A/m^2. The direction of I_{enc} is related to C by a right-hand rule.

The integral form may be derived from the point form. Integrate both sides of the point form over an arbitrary surface S and apply Stokes' theorem:

$$\iint_S \nabla \times \mathbf{B} \cdot d\mathbf{s} = \oint_C \mathbf{B} \cdot d\ell = \mu_0 \iint_S \mathbf{J} \cdot d\mathbf{s} = \mu_0\, I_{enc}$$

Stokes' theorem specifies a right-hand rule relationship between C and **ds**. $\iint_S \mathbf{J} \cdot d\mathbf{s}$ is the current I_{enc} in the direction $d\mathbf{s}$. Therefore I_{enc} is the total current passing through S (enclosed by C) in a direction $d\mathbf{s}$ related by a right-hand rule to path C. The second law is called **Ampère's law**. The point form of the second law may also be derived from the integral form and thus the point and integral forms are identical statements. The first and second laws may also be derived from the Biot-Savart law in Section 6.5.

6.4 AMPÈRE'S LAW

Ampère's law,

$$\oint_C \mathbf{B} \cdot d\ell = \mu_0 \iint_S \mathbf{J} \cdot d\mathbf{s} = \mu_0 I_{enc}$$

allows us to treat certain magnetostatic problems involving symmetry to obtain the magnetic field **B**, given the current distributions ($\mathbf{J}$, $\mathbf{J}_s$, $\mathbf{I}$). In the expression above, the closed path C bounds the surface S and the directions of C and $d\mathbf{s}$ are related by a right-hand rule, as in Stokes' theorem. I_{enc} is the total current passing through the surface S, in the direction $d\mathbf{s}$, enclosed by the closed path C.

We assume that current **J** is specified and thus the right-hand side of Eq. (6-7) is known or can be determined. **B** is unknown. If **B** were parallel to path C and constant in magnitude over the path chosen, then we could take the unknown outside the integral and solve for **B**. The key to an Ampère's law problem is to choose a closed path or contour C so that it is parallel to the **B** field lines and the magnitude of **B** is constant along the path C. The examples which follow will give us some idea of the applicability of Ampère's law. They will also introduce us to some typical field configurations in magnetostatics. The steps of applying Ampère's law are as follows:

1. Recognize the symmetry.
2. Sketch the magnetic field lines.
3. Choose C parallel to the field lines. Form a closed contour.
4. Solve for B.

Ampère's law plays the role in magnetostatics that Gauss' law assumed in electrostatics.

EXAMPLE 6-2. CURRENT-CARRYING FILAMENTS, WIRES AND CYLINDERS.

(a) *A Current-Carrying Filament*

Figure 6-1 shows a straight current-carrying filament of infinite length which lies along the z axis.

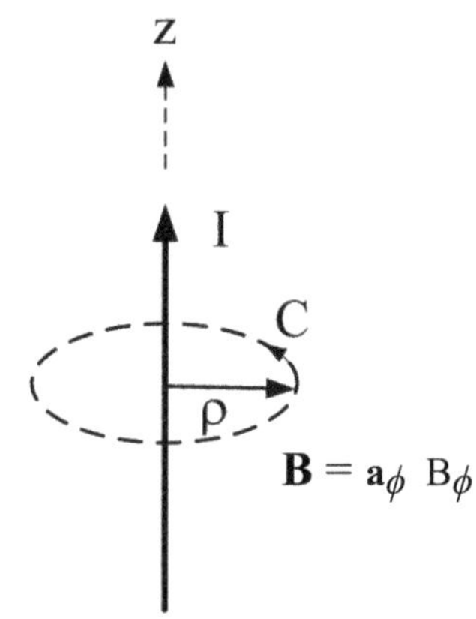

Figure 6-1. A current-carrying filament.

The filament is of zero cross-sectional area and carries a current I. We first note that, by the curl relationship between **B** and **J**, the magnetic field **B** will curl or circulate around the z-directed current source (I); we expect that **B** is ϕ-directed (circumferential), i.e., $\mathbf{B} = \mathbf{a}_\phi B_\phi$. We have also seen a ϕ-directed magnetic field in Oersted's 1820 experiment. Figure 6-1 also shows a circle C of radius ρ, centered on the z axis and lying in a plane z = constant. A similar figure is drawn on the front cover of this book. $\oint_C \mathbf{B} \cdot d\ell$ becomes $\int B_\phi \rho\, d\phi$ and B_ϕ is constant over C by rotational symmetry ($\partial/\partial\phi = 0$). The current I_{enc} enclosed by C, in a direction related to that of C by a right-hand rule, is I. Then application of Ampère's law, Eq. (6-7), leads to

$$\oint_C \mathbf{B} \cdot d\ell = \int_0^{2\pi} B_\phi\, \rho\, d\phi = B_\phi\, 2\pi\rho = \mu_0 I$$

and

$$\boxed{B_\phi = \frac{\mu_0 I}{2\pi\, \rho}} \qquad (6\text{-}8)$$

Note that Eq. (6-8) satisfies both basic laws, i.e., Eqs. (6-4) and (6-6). See Example 1-14.

(b) *A Current-Carrying Wire of Radius a*

Figure 6-2 shows a current-carrying wire of radius a with current I. The current is assumed to be *uniformly* distributed throughout the cross section and thus the volume current density is

$$\mathbf{J} = \mathbf{a}_z \left(\frac{I}{\pi a^2} \right) \ [\text{A/m}^2]$$

This is a two-region problem. **B** is circumferential, i.e., $\mathbf{B} = \mathbf{a}_\phi B_\phi$. Figure 6-2 shows circle C of radius ρ about the axis for the two cases $\rho \leq a$ and $\rho \geq a$. Circles C lie in constant z planes. B_ϕ will again be constant along C.

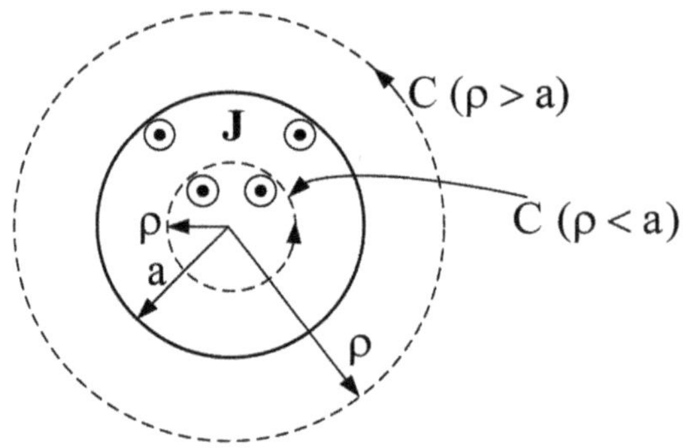

Figure 6-2. A current-carrying wire.

Consider first the field inside the wire $(\rho \le a)$:

$$\oint_C \mathbf{B} \cdot d\ell = B_\phi\, 2\pi\rho = \mu_0 \iint_S \mathbf{J} \cdot d\mathbf{s} = \mu_0 \int_0^{2\pi}\int_0^{\rho} \left(\frac{I}{\pi a^2}\right) \rho d\rho d\phi$$

$$= \mu_0 \left(\frac{I}{\pi a^2}\right) \pi\rho^2 = \frac{\mu_0 I \rho^2}{a^2} \quad (\rho \le a)$$

Then

$$B_\phi = \frac{\mu_0\, I\, \rho}{2\pi a^2} \quad (\rho \le a)$$

Next consider the field outside the wire $(\rho \ge a)$:

$$\oint_C \mathbf{B} \cdot d\ell = B_\phi\, 2\pi\rho = \mu_0 I \; (\rho \ge a) \text{ (since the entire current I is enclosed),}$$

and

$$B_\phi = \frac{\mu_0 I}{2\pi\rho} \quad (\rho \ge a)$$

Collecting expressions for B_ϕ:

$$B_\phi = \frac{\mu_0 I \rho}{2\pi a^2} \quad (\rho \le a) \tag{6-9a}$$

$$= \frac{\mu_0 I}{2\pi\rho} \quad (\rho \ge a) \tag{6-9b}$$

Notice that the field B_ϕ outside $(\rho \ge a)$ is identical to the field of a filament.

(c) *A Current-Carrying Hollow Cylinder*

Figure 6-3 shows a hollow current-carrying cylinder of radius a and infinite length. The cylinder carries a longitudinal (z-directed) current I. The current is uniformly distributed over the surface and thus the surface current density is

$$\mathbf{J}_s = \mathbf{a}_z\left(\frac{I}{2\pi a}\right) \quad \text{[A/m]}$$

This is also a two-region problem. **B** will be circumferential. Now consider circles C of radius ρ for the two cases (ρ < a) and (ρ > a), as shown in Figure 6-3:

$$\oint_C \mathbf{B}\cdot d\boldsymbol{\ell} = B_\phi 2\pi\rho = 0 \qquad (\rho < a) \quad \text{(since no current is enclosed)}$$
$$= \mu_0 I \qquad (\rho > a) \quad \text{(since the total current I is enclosed)}$$

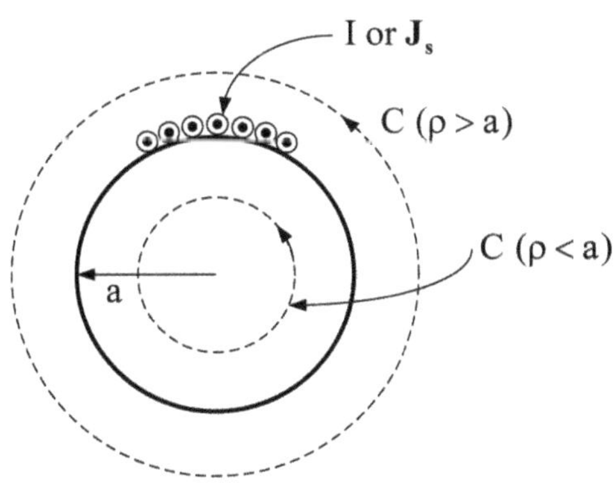

Figure 6-3. A current-carrying hollow cylinder.

Thus

$$B_\phi = 0 \quad (\rho < a) \tag{6-10a}$$

$$= \frac{\mu_0 I}{2\pi\rho} \quad (\rho > a) \tag{6-10b}$$

Note that the magnetic field inside a hollow cylinder with uniform surface current is zero.

(d) *Coaxial Current-Carrying Cylinders (Coaxial Line)*

Figure 6-4 shows coaxial hollow cylinders of infinite length and radii a, b carrying total currents I, –I, respectively, in the z direction. This is the configuration of a coaxial line.

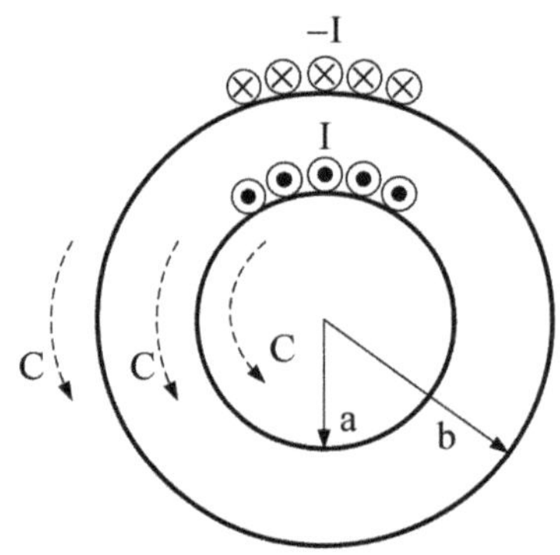

Figure 6-4. Coaxial current-carrying cylinders.

The surface current densities are

$$\mathbf{J}_s = \mathbf{a}_z\left(\frac{I}{2\pi a}\right) \qquad (\rho = a)$$

$$= \mathbf{a}_z\left(\frac{-I}{2\pi b}\right) \qquad (\rho = b)$$

This is a three-region problem.

Consider circles C of radius ρ for the three cases $(\rho < a)$, $(a < \rho < b)$ and $(\rho > b)$:

$$\oint_C \mathbf{B}\cdot d\ell = B_\phi\, 2\pi\rho = 0 \quad (\rho < a)$$

$$= \mu_0 I \quad (a < \rho < b)$$

$$= \mu_0 I + (-\mu_0 I) = 0 \quad (\rho > b)$$

Thus

$$B_\phi = 0 \quad (\rho < a) \qquad (6\text{-}11a)$$

$$= \frac{\mu_0 I}{2\pi\rho} \quad (a < \rho < b) \qquad (6\text{-}11b)$$

$$= 0 \quad (\rho > b) \qquad (6\text{-}11c)$$

Note that, in each of the examples (a)-(d), there is a region with fields identical to that of a current-carrying filament along the z axis.

EXAMPLE 6-3. UNIFORM SURFACE CURRENT SHEET

Figure 6-5 shows a uniform surface current $\mathbf{J}_s = \mathbf{a}_y J_{sy}$ [A/m] flowing over the infinite planar sheet at $z = 0$ (xy plane). Find the magnetic field $\mathbf{B}$ everywhere.

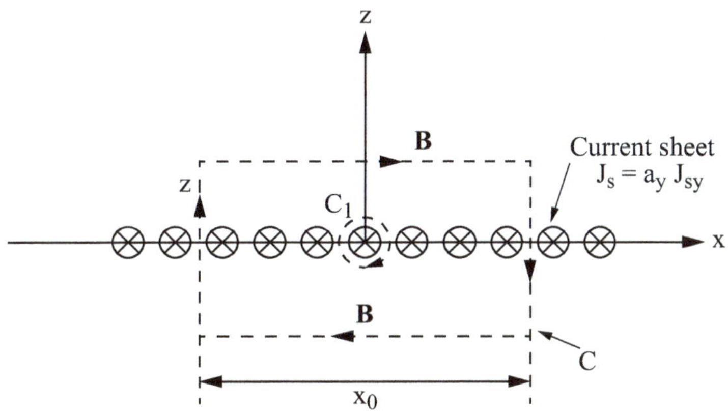

Figure 6-5. Uniform surface current sheet.

SOLUTION

First we need to figure out the direction of $\mathbf{B}$. Breaking the current sheet into a collection of infinitely many thin wires, we observe that each thin wire produces a clockwise circulating magnetic field, shown as C_1 in Figure 6-5. When you consider contributions to $\mathbf{B}$ from all thin wires on the entire infinite sheet, z components are cancelled out and only x components are added up. Thus the resultant magnetic field can be written as follows.

$$\mathbf{B} = \begin{cases} \mathbf{a}_x B_x, & z > 0 \quad \text{(above the plane)} \\ -\mathbf{a}_x B_x, & z < 0 \quad \text{(below the plane)} \end{cases}$$

Due to the planar symmetry, the magnetic fields are equal and opposite above and below the sheet. Construct a closed rectangular loop C with two horizontal lines of length x_0 at distance z from the sheet. Applying Ampère's law,

$$\oint_C \mathbf{B} \cdot d\ell = B_x x_0 + (-B_x)(-x_0) = \mu_0 I_{enc}$$

$$2 B_x x_0 = \mu_0 \int J_s d\ell_\perp = \mu_0 J_{sy} x_0 (d\ell_\perp \text{is along the x axis})$$

$$B_x = \frac{1}{2}\mu_0 J_{sy}$$

Thus

$$\boxed{\begin{aligned} \mathbf{B} &= \mathbf{a}_x \frac{1}{2}\mu_0 J_{sy} \quad (z > 0) \\ &= -\mathbf{a}_x \frac{1}{2}\mu_0 J_{sy} \quad (z < 0) \end{aligned}}$$

(6-12a)

(6-12b)

EXAMPLE 6-4. CURRENT-CARRYING TOROIDS AND INFINITE SOLENOID OF ARBITRARY CROSS-SECTIONS

Figure 6-6(a) shows a toroid of doughnut shape and rectangular cross section, which is closely wound with N turns of wire carrying current I. The toroid has an inner radius a and an outer radius b. Find the magnetic field **B** everywhere.

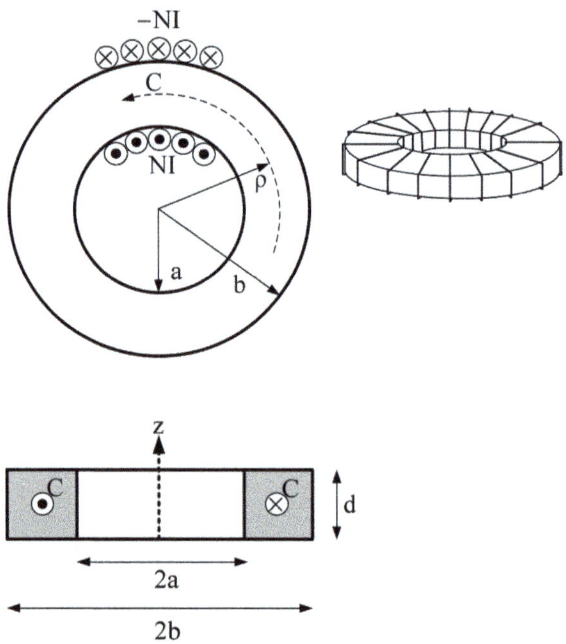

Figure 6-6(a). A toroid of rectangular cross-section.

SOLUTION

The problem has azimuthal symmetry ($\partial/\partial\phi = 0$). This is a good approximation for a large number of turns, assuming that the winding is uniform and tight so that each turn can be considered as a closed loop of current. Then the magnetic fields **B** point perpendicular to the loop cross section and thus are circumferential about the z axis at all points, both inside and outside the toroidal coil. $\mathbf{B} = \mathbf{a}_\phi B_\phi$. Consider circles C of radius ρ for three cases $(\rho < a)$, $(a < \rho < b)$, $(\rho > b)$. We further limit ourselves to the case $0 < z < d$.

Then

$$\oint_C \mathbf{B} \cdot d\boldsymbol{\ell} = B_\phi 2\pi\rho = \mu_0 I_{enc} = 0 \quad (\rho < a)$$
$$= \mu_0 NI \quad (a < \rho < b)$$
$$= \mu_0 (NI - NI) = 0 \quad (\rho > b)$$

Notice that the geometry is very similar to that of Example 6-2(d). For the cases $(z < 0)$ and $(z > d)$ the current enclosed is zero.

Thus

$$B_\phi = \frac{\mu_0 NI}{2\pi\rho} \quad \text{(inside the toroid)} \tag{6-13a}$$

$$= 0 \quad \text{(outside the toroid)} \tag{6-13b}$$

Note that we have not limited our results in any way to the rectangular cross-section. In fact, Eq. (6-13) is valid for a toroid of any arbitrary cross-section including the circular cross-section (doughnut).

Now consider what happens in a toroid of very large radius. Let a, b become very large as (b-a) remains constant. Then the percentage variation of $(1/\rho)$ over the toroid $(a < \rho < b)$ is negligible. The flux density B_ϕ is constant over the toroid. Let N_ℓ denote the number of turns per unit length of the toroid. Then

$$N = N_\ell (2\pi\rho)$$

Substituting this in Eq. (6-13a) yields

$$B_\phi = \frac{\mu_0 N I}{2\pi\rho} = \frac{\mu_0 N_\ell 2\pi\rho I}{2\pi\rho} = \mu_0 N_\ell I \tag{6-14a}$$

Thus the flux density is uniform throughout the toroid. For other toroid cross-section shapes let a, b be minimum and maximum radii, respectively, and the above discussion still holds.

An *infinite solenoid* with its axis along the z axis (Figure 6-6(b)) and with circumferential windings may be regarded as a toroid of infinite radius.

The flux density is uniform throughout the solenoid and z-directed:

$$B_z = \mu_0 N_\ell I \tag{6-14b}$$

The conclusions (Eqs. (6-14a,b)) apply to toroids and infinite solenoids of any cross-sectional shape such as circular, elliptical cross sections. $N_\ell I$ may also be interpreted as surface current density (J_s).

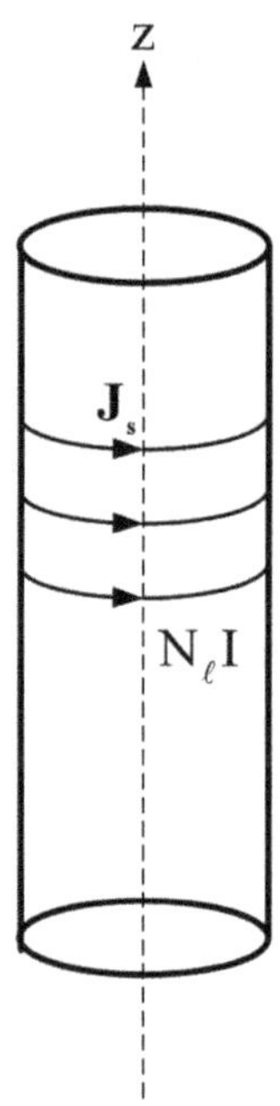

Figure 6-6(b). A solenoid of infinite length.

Table 6-1 summarizes the magnetostatic fields and potentials for certain simple current distributions. The flux density $\mathbf{B}$ is obtained by Ampère's law. Magnetic vector potentials $\mathbf{A}$ are obtained by integrating the differential equation $\nabla \times \mathbf{A} = \mathbf{B}$, as shown in the next section (see Example 6-9). Note the simple dependence on distance for the particular cases shown:

For cylindrical symmetry $(\frac{\partial}{\partial \phi} = \frac{\partial}{\partial z} = 0)$: $\mathbf{B} \sim \frac{1}{\rho}$; $\mathbf{A} \sim \ln \rho$

For planar symmetry $(\frac{\partial}{\partial x} = \frac{\partial}{\partial y} = 0)$: $\mathbf{B}$ is constant; $\mathbf{A} \sim z$

Table 6-1 Basic Magnetostatic Fields and Potentials

Current Distribution	Magnetic Flux Density	Magnetic Vector Potential
Filament $\mathbf{I} = \mathbf{a}_z I$	$B_\phi = \frac{\mu_0 I}{2\pi\rho}$	$A_z = -\frac{\mu_0 I}{2\pi} \ln \frac{\rho}{\rho_0}$ Reference point: $\rho = \rho_0$
Cylindrical Shell $\mathbf{J}_s = \mathbf{a}_z \frac{I}{2\pi a}$ $(\rho = a)$	$B_\phi = \frac{\mu_0 I}{2\pi\rho}$ $(\rho > a)$ $= 0$ $(\rho < a)$	$A_z = -\frac{\mu_0 I}{2\pi} \ln(\frac{\rho}{a})$ $(\rho \geq a)$ Reference point: $\rho = a$

Current Distribution	Magnetic Flux Density	Magnetic Vector Potential
Coaxial Shell $\mathbf{J}_s = \mathbf{a}_z \dfrac{I}{2\pi a} \quad (\rho = a)$ $= -\mathbf{a}_z \dfrac{I}{2\pi b} \quad (\rho = b)$	$B_\phi = \dfrac{\mu_0 I}{2\pi\rho} (a < \rho < b)$ $= 0 \quad (\rho < a, \rho > b)$	$A_z = -\dfrac{\mu_0 I}{2\pi} \ln\left(\dfrac{\rho}{a}\right) \quad (a \le \rho \le b)$ Reference point: $\rho = a$
Toroid (any cross section) N turns Current I	$B_\phi = \dfrac{\mu_0 NI}{2\pi\rho}$ (inside) $= 0$ (outside)	$A_z = -\dfrac{\mu_0 NI}{2\pi} \ln\left(\dfrac{\rho}{\rho_0}\right)$ (inside) $=$ constant (outside) Reference point: $\rho = \rho_0$
Solenoid (any cross section) N_ℓ turns per unit length Current I	$B_z = \mu_0 N_\ell I = \mu_0 J_s$ (inside) $= 0$ (outside)	Circular Cross section (radius a) $A_\phi = \dfrac{\mu_0 N_\ell I}{2} \rho \quad (\rho \le a)$ $= \dfrac{\mu_0 N_\ell I}{2} \dfrac{a^2}{\rho} \quad (\rho \ge a)$
Current Sheet $\mathbf{J}_s = \mathbf{a}_y J_{sy} (z = 0)$	$B_x = \pm \dfrac{\mu_0 J_{sy}}{2} \quad \begin{matrix} +(z>0) \\ -(z<0) \end{matrix}$	$A_y = \mp \dfrac{\mu_0 J_{sy}}{2} z$ Reference point: $z = 0$
Anti-parallel Current Sheets $\mathbf{J}_s = \mathbf{a}_y J_{sy} \quad (z=0)$ $= -\mathbf{a}_y J_{sy} \quad (z=d)$	$B_x = \mu_0 J_{sy} \quad (0 < z < d)$ $= 0 \quad (z < 0, z > d)$	$A_y = -\mu_0 J_{sy} z \quad (0 < z < d)$ Reference point: $z = 0$

6.5 THE MAGNETIC VECTOR POTENTIAL AND THE BIOT-SAVART LAW (THE MAGNETIC FIELDS OF ARBITRARY CURRENT DISTRIBUTIONS)

In this section we consider the magnetic field **B** due to an arbitrary steady current distribution **J**. The methods developed are completely general but are more difficult than the Ampère's law methods of the previous section. We develop two forms for the fields of arbitrary steady currents, one which utilizes a magnetic vector potential **A** and another (the Biot-Savart law) which determines **B** directly. First let us consider the vector potential **A**.

The magnetic field **B** is divergenceless by the first law of Section 6.3; therefore it can be expressed as the curl of a vector **A** which we will call the **magnetic vector potential**.

$$\boxed{\mathbf{B} = \nabla \times \mathbf{A}} \qquad (6\text{-}15)$$

Recall now that a vector is completely defined by its curl and divergence. Therefore Eq. (6-15) only partially defines **A**; we are free to choose $\nabla \cdot \mathbf{A}$ for convenience. Choose $\nabla \cdot \mathbf{A} = 0$; this will turn out to be convenient mathematically. It also corresponds to the choice made in Helmholtz's theorem and is called the **Coulomb condition** or **Coulomb gauge**.

$$\nabla \cdot \mathbf{A} = 0 \tag{6-16}$$

Now take the curl of Eq. (6-15) and utilize the second law (Ampère's law) of Section 6.3 (Eq. (6-6)) and the vector identity $\nabla \times \nabla \times \mathbf{A} = \nabla(\nabla \cdot \mathbf{A}) - \nabla^2\mathbf{A}$ (See Eq. (11) of Table 1-5).

$$\nabla \times \mathbf{B} = \nabla \times \nabla \times \mathbf{A} = \nabla(\nabla \cdot \mathbf{A}) - \nabla^2\mathbf{A} = \mu_0\mathbf{J}$$

Making use of Eq. (6-16), we obtain

$$\boxed{\nabla^2\mathbf{A} = -\mu_0\mathbf{J}} \tag{6-17}$$

or, in rectangular coordinates,

$$\nabla^2 A_x = -\mu_0 J_x; \quad \nabla^2 A_y = -\mu_0 J_y; \quad \nabla^2 A_z = -\mu_0 J_z$$

Eq. (6-17) is similar to Poisson's equation, (4-3), in Chapter 4 and thus it is called **vector Poisson's equation**. The scalar equations encountered here have already been solved in electrostatics:

$$\nabla^2 V = -\rho_v/\varepsilon_0 \Rightarrow V = \frac{1}{4\pi\varepsilon_0}\iiint \frac{\rho_v\,dv'}{R} \tag{2-46a}$$

Hence

$$\nabla^2 A_x = -\mu_0 J_x \Rightarrow A_x = \frac{\mu_0}{4\pi}\iiint \frac{J_x\,dv'}{R}$$

Collecting the solutions for A_x, A_y, A_z:

$$\boxed{\begin{aligned}\mathbf{A}(x,y,z) &= \frac{\mu_0}{4\pi}\iiint \frac{\mathbf{J}(x', y', z')\,dv'}{R}\\ &= \frac{\mu_0}{4\pi}\iiint \frac{\mathbf{J}(x', y', z')\,dv'}{|\mathbf{r}-\mathbf{r}'|}\end{aligned}} \tag{6-18a}$$

where $\mathbf{r}$, $\mathbf{r}'$ are the position vectors of the field and source points, respectively. Refer to Figure 2-6 with the volume charge ρ_v replaced by the volume current $\mathbf{J}$.

For the cases of surface and filamentary currents, $\mathbf{J}\,dv'$ is replaced with $\mathbf{J}_s\,ds'$ and $\mathbf{I}\,d\ell'$, respectively, to obtain

$$\mathbf{A} = \frac{\mu_0}{4\pi}\iint \frac{\mathbf{J}_S\,ds'}{R} \quad \text{for surface current density } \mathbf{J}_s \tag{6-18b}$$

$$= \frac{\mu_0}{4\pi}\int \frac{\mathbf{I}\,d\ell'}{R} \quad \text{for filamentary current } \mathbf{I} \tag{6-18c}$$

Eq. (6-18) can also be obtained directly from Helmholtz's theorem (Eq. (1-49)) since $(\nabla \times \mathbf{F})' = (\nabla \times \mathbf{B})' = \mu_0\,\mathbf{J}(x', y', z')$. Note that $\mathbf{A}$ has $1/R$ dependence just as V has $1/R$ dependence in Eq. (2-46).

Eq. (6-18) allows us to obtain the vector potential $\mathbf{A}$ for an arbitrary current distribution, then $\mathbf{B}$ is obtained by taking the curl of $\mathbf{A}$. Alternatively, we may obtain a direct expression for $\mathbf{B}$ by taking the curl of Eq. (6-18):

$$\mathbf{B} = \nabla \times \mathbf{A} = \frac{\mu_0}{4\pi}\iiint \nabla \times \left(\frac{\mathbf{J}}{R}\right) dv'$$

and, using the vector identity for the curl of a product of a vector and a scalar (see Eq. (1-44))

$$\nabla \times \left(\frac{\mathbf{J}}{R}\right) = \left(\frac{1}{R}\right)\nabla \times \mathbf{J}(x',y',z') + \nabla\left(\frac{1}{R}\right) \times \mathbf{J} = \mathbf{J} \times \frac{\mathbf{R}}{R^3}$$

Note that $\nabla \times \mathbf{J}(x',y',z') = 0$ since $\mathbf{J}$ is a function of primed coordinates. $\nabla(1/R)$ is equal to $-\mathbf{a}_R\left(\frac{1}{R^2}\right) = -\mathbf{R}/R^3$; it may be found by calculating the gradient in spherical coordinates.

Thus

$$\boxed{\begin{aligned}\mathbf{B}(x,y,z) &= \frac{\mu_0}{4\pi}\iiint \frac{\mathbf{J}(x', y', z') \times \mathbf{a}_R\,dv'}{R^2} \\ &= \frac{\mu_0}{4\pi}\iiint \frac{\mathbf{J}(x', y', z') \times \mathbf{R}\,dv'}{R^3} \\ &= \frac{\mu_0}{4\pi}\iiint \frac{\mathbf{J}(x', y', z') \times (\mathbf{r}-\mathbf{r}')\,dv'}{|\mathbf{r}-\mathbf{r}'|^3}\end{aligned}} \tag{6-19a}$$

where $\mathbf{R} = \mathbf{r} - \mathbf{r}'$, $R = |\mathbf{r} - \mathbf{r}'|$.

For the surface and filamentary currents, $\mathbf{J}\,dv'$ is replaced with $\mathbf{J}_s\,ds'$ and $\mathbf{I}\,d\ell'$, respectively, to obtain

$$\mathbf{B} = \frac{\mu_0}{4\pi}\iint \frac{\mathbf{J}_S \times \mathbf{R}\,ds'}{R^3} \quad \text{for surface current density } \mathbf{J}_S \tag{6-19b}$$

$$= \frac{\mu_0}{4\pi}\int \frac{\mathbf{I} \times \mathbf{R}\,d\ell'}{R^3} = \frac{\mu_0 I}{4\pi}\int \frac{d\boldsymbol{\ell}' \times \mathbf{R}}{R^3} \quad \text{for filamentary current } \mathbf{I} \tag{6-19c}$$

($d\boldsymbol{\ell}'$ has the direction of $\mathbf{I}$)

Eq. (6-19) is known as the **Biot-Savart law.** Note that $\mathbf{B}$ has $1/R^2$ dependence just as $\mathbf{E}$ has $1/R^2$ dependence in Eq. (2-27), the generalized Coulomb's law.

Eqs. (6-18) and (6-19) provide two methods for the determination of $\mathbf{B}$ from current distributions:

1. Find $\mathbf{A}$ directly from (6-18), then $\mathbf{B} = \nabla \times \mathbf{A}$.
2. Find $\mathbf{B}$ directly from (6-19) (Biot-Savart law).

Each method has certain advantages. Method 1 has a simpler integral but involves two steps. Method 2 involves a more complex integral but has only one step. One method may have distinct advantages in a particular problem. However, for most magnetostatic problems, we use method 2 (Biot-Savart law).

To find the vector potential $\mathbf{A}$ from the current distribution $\mathbf{J}$, we have a choice of two methods:

1. Find $\mathbf{A}$ directly from (6-18).
2. Find $\mathbf{B}$ first either from (6-19) or by Ampère's law, then integrate the equation $\mathbf{B} = \nabla \times \mathbf{A}$ to find $\mathbf{A}$ (Example 6-9).

EXAMPLE 6-5. ON-AXIS FIELD OF A CURRENT-CARRYING LOOP

(a) *Complete Loop*

Figure 6-7 shows a circular current-carrying loop of radius *a* with current I. Find the magnetic field $\mathbf{B}$ on the loop axis.

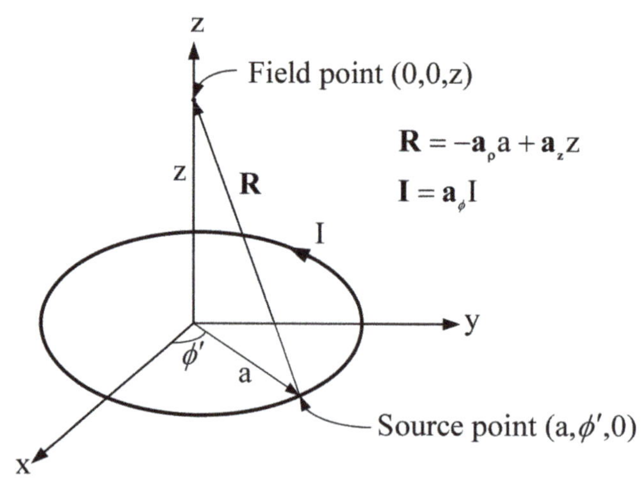

Figure 6-7. A current-carrying loop.

SOLUTION

We use Eq. (6-19c) appropriate to a filamentary current:

$$\mathbf{B} = \frac{\mu_0}{4\pi} \int \frac{\mathbf{I} \times \mathbf{R}\ d\ell'}{R^3}$$

We then identify the fixed field point (0, 0, z) and the variable source point $(\rho', \phi', z') = (a, \phi', 0)$ in cylindrical coordinates.

Then

$$\mathbf{r} = \mathbf{a}_z z;\ \mathbf{r}' = \mathbf{a}_\rho \rho' = \mathbf{a}_\rho a$$

$$\mathbf{R} = \mathbf{r} - \mathbf{r}' = -\mathbf{a}_\rho a + \mathbf{a}_z z \quad (\mathbf{a}_\rho \text{ is a function of } \phi')$$

$$\mathbf{I} = \mathbf{a}_\phi I \quad (\mathbf{a}_\phi \text{ is a function of } \phi')$$

$$R = \sqrt{a^2 + z^2};\quad d\ell' = a\ d\phi'$$

$$\mathbf{I} \times \mathbf{R} = (\mathbf{a}_\phi I) \times (-\mathbf{a}_\rho a + \mathbf{a}_z z) = \mathbf{a}_z (Ia) + \mathbf{a}_\rho (Iz)$$

$$= \mathbf{a}_z (Ia) + \mathbf{a}_x (Iz\cos\phi') + \mathbf{a}_y (Iz\sin\phi')$$

Take the z component of $\mathbf{B}$:

$$B_z = \frac{\mu_0}{4\pi} \int_0^{2\pi} \frac{Ia^2\ d\phi'}{[a^2 + z^2]^{3/2}} = \frac{\mu_0 I a^2}{2[a^2 + z^2]^{3/2}} \tag{6-20}$$

It can be easily shown that $B_x = B_y = 0$. Note that a ϕ-directed current produces a z-directed magnetic field on the axis.

For a distant point ($z >> a$), the loop appears small and the field approaches that of the magnetic dipole which is discussed in the next section:

$$B_z \Rightarrow \frac{\mu_0 (I \pi a^2)}{2\pi z^3} = \frac{\mu_0 m}{2\pi z^3}$$

Compare this result with Eq. (6-31).

(b) *Partial Loop*

Consider a partial loop which is part of a closed circuit:

$$\mathbf{I} = \mathbf{a}_\phi I\ (\rho = a,\ \phi_1 \le \phi \le \phi_2,\ z = 0)$$

Find the magnetic field $\mathbf{B}$ along the z axis.

SOLUTION

$$\mathbf{B} = \frac{\mu_0}{4\pi}\int_{\phi_1}^{\phi_2} \frac{\mathbf{a}_z(\text{I a}) + \mathbf{a}_x\ (\text{I z}\cos\phi') + \mathbf{a}_y\ (\text{I z}\sin\phi')}{[a^2+z^2]^{3/2}}\ \text{a}\ d\phi'$$

$$B_z = \frac{\mu_0\, I\, a^2\,(\phi_2 - \phi_1)}{4\pi\,[a^2+z^2]^{3/2}} \tag{6-21a}$$

$$B_x = \frac{\mu_0\, I\, a\, z}{4\pi\,[a^2+z^2]^{3/2}}\ (\sin\phi_2 - \sin\phi_1) \tag{6-21b}$$

$$B_y = \frac{\mu_0\, I\, a\, z}{4\pi\,[a^2+z^2]^{3/2}}\ (\cos\phi_1 - \cos\phi_2) \tag{6-21c}$$

Compare this example with the charged loop of Problem 2-25.

EXAMPLE 6-6. THE ON-AXIS FIELD OF A SOLENOID OF CIRCULAR CROSS SECTION

Figure 6-8 shows a tightly wound solenoid of length ℓ and radius a. There are N turns of wire carrying a current I. This corresponds to a surface current as follows:

$$\mathbf{J}_s = \mathbf{a}_\phi\ \frac{N\,I}{\ell}\ \left[\frac{A}{m}\right]$$

Find the magnetic field **B** along the z axis.

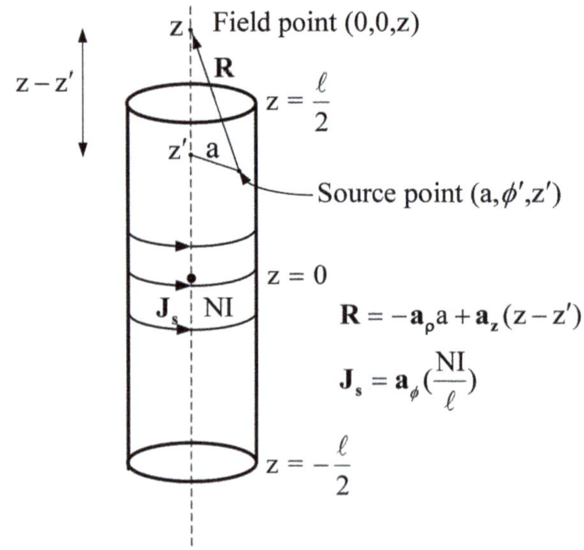

Figure 6-8. A solenoid of circular cross-section.

SOLUTION

We use Eq. (6-19b) appropriate to surface current:

$$\mathbf{B} = \frac{\mu_0}{4\pi} \iint \frac{\mathbf{J}_s \times \mathbf{R}\ ds'}{R^3}$$

The fixed field point is $(0, 0, z)$ and the variable source point is $(\rho', \phi', z') = (a, \phi', z')$.
Then

$$\mathbf{r} = \mathbf{a}_z z; \quad \mathbf{r}' = \mathbf{a}_\rho a + \mathbf{a}_z z'$$

$$\mathbf{R} = -\mathbf{a}_\rho a + \mathbf{a}_z (z - z')$$

$$\mathbf{J}_s \times \mathbf{R} = \left(\mathbf{a}_\phi \frac{NI}{\ell}\right) \times \left(-\mathbf{a}_\rho a + \mathbf{a}_z (z - z')\right)$$

$$= \mathbf{a}_z \left(\frac{NIa}{\ell}\right) + \mathbf{a}_\rho \left(\frac{NI(z-z')}{\ell}\right)$$

$$R = \sqrt{a^2 + (z-z')^2}\ ; \quad ds' = a\, d\phi' dz'$$

Taking the z component of $\mathbf{B}$:

$$B_z = \frac{\mu_0}{4\pi} \int_{-\frac{\ell}{2}}^{\frac{\ell}{2}} \int_0^{2\pi} \frac{NIa^2\, d\phi'\, dz'}{\ell[a^2 + (z-z')^2]^{3/2}} \qquad \text{Let } z'' = z - z', \quad dz'' = -dz'$$

$$= -\frac{\mu_0}{2} \int_{z+\frac{\ell}{2}}^{z-\frac{\ell}{2}} \frac{NIa^2\, dz''}{\ell[a^2 + (z'')^2]^{3/2}} = -\frac{\mu_0 NIa^2}{2\ell} \left[\frac{z''}{a^2\sqrt{a^2 + (z'')^2}}\right]_{z+\frac{\ell}{2}}^{z-\frac{\ell}{2}}$$

$$= \frac{\mu_0 NI}{2\ell} \left[\frac{z + \frac{\ell}{2}}{\sqrt{a^2 + (z + \frac{\ell}{2})^2}} - \frac{z - \frac{\ell}{2}}{\sqrt{a^2 + (z - \frac{\ell}{2})^2}}\right] \tag{6-22a}$$

and $B_x = B_y = 0$ as shown in Example 6-5(a).

At the center $(z = 0)$:

$$B_z = \frac{\mu_0 N I}{2\sqrt{a^2 + \left(\frac{\ell}{2}\right)^2}} \Rightarrow \frac{\mu_0 N I}{\ell} = \mu_0 N_\ell I \quad (\text{when } \ell \gg a) \tag{6-22b}$$

where $N_\ell = N/\ell$ (turns per unit length).

At the end $(z = \ell/2)$:

$$B_z = \frac{\mu_0 NI}{2\sqrt{a^2 + \ell^2}} \Rightarrow \frac{\mu_0 NI}{2\ell} = \frac{\mu_0 N_\ell I}{2} \quad (\text{when } \ell \gg a) \tag{6-22c}$$

Thus for long, thin solenoids ($\ell >> a$), the on-axis field B_z at the ends $(z = \pm\ell/2)$ is half that at the center. Also, compare Eq. (6-22b) with Eq. (6-14b) and note that the on-axis field at the center approaches that of the infinite solenoid.

EXAMPLE 6-7. THE MAGNETIC VECTOR POTENTIAL AND FIELD OF A STRAIGHT CURRENT-CARRYING WIRE.

Figure 6-9 shows a straight section of current-carrying wire which is part of a closed circuit.

The filament carries current I between $z = z_1$ and $z = z_2$. Find the magnetic field $\mathbf{B}$ and the vector potential $\mathbf{A}$ due to this section of a closed circuit.

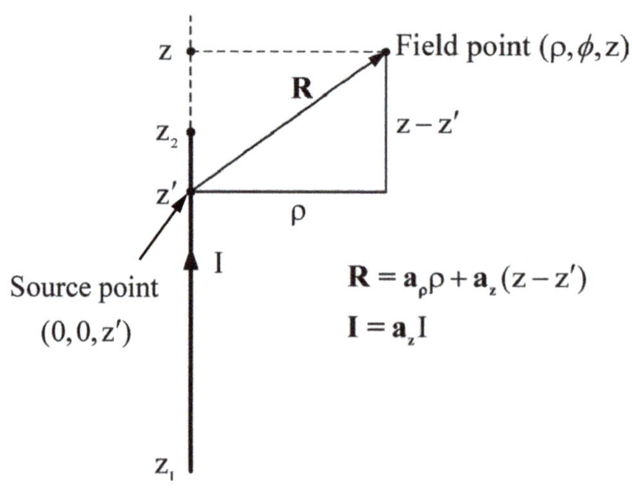

Figure 6-9. A straight current-carrying wire.

SOLUTION

(a) *Magnetic Flux Density* $\mathbf{B}$

We again use

$$\mathbf{B} = \frac{\mu_0}{4\pi} \int \frac{\mathbf{I} \times \mathbf{R} \; d\ell'}{R^3}$$

We identify the field point (ρ, ϕ, z) and the source point $(0, 0, z')$. Then

$$\mathbf{r} = \mathbf{a}_\rho \rho + \mathbf{a}_z z : \quad \mathbf{r}' = \mathbf{a}_z z'$$

$$\mathbf{R} = \mathbf{a}_\rho \rho + \mathbf{a}_z \; (z - z')$$

$$\mathbf{I} = \mathbf{a}_z \, I$$

$$\mathbf{I} \times \mathbf{R} = (\mathbf{a}_z \, I) \times \left(\mathbf{a}_\rho \rho + \mathbf{a}_z (z - z')\right) = \mathbf{a}_\phi \, (I \rho)$$

(A z-directed current produces a ϕ-directed magnetic field)

$$R = \sqrt{\rho^2 + (z - z')^2}\,; \quad d\ell' = dz'$$

Since $\mathbf{a}_\phi$ is independent of z', $\mathbf{B} = \mathbf{a}_\phi \, B_\phi$:

$$B_\phi = \frac{\mu_0}{4\pi} \int_{z_1}^{z_2} \frac{I \; \rho \; dz'}{[\rho^2 + (z - z')^2]^{3/2}} \qquad (\text{Let } z'' = z - z')$$

$$= \frac{-\mu_0}{4\pi} \int_{z-z_1}^{z-z_2} \frac{I \; \rho \; dz''}{[\rho^2 + (z'')^2]^{3/2}} = -\frac{\mu_0 \, I \rho}{4\pi} \left[\frac{z''}{\rho^2 \sqrt{\rho^2 + (z'')^2}} \right]_{z-z_1}^{z-z_2}$$

$$= \frac{\mu_0 \, I}{4\pi\rho} \left[\frac{z - z_1}{\sqrt{\rho^2 + (z - z_1)^2}} - \frac{z - z_2}{\sqrt{\rho^2 + (z - z_2)^2}} \right] \tag{6-23a}$$

Several special cases are of interest here:

For a semi-infinite wire ($z_1 = -\infty$, $z_2 = 0$):

$$B_\phi = \frac{\mu_0 I}{4\pi\rho}\left[1 - \frac{z}{\sqrt{\rho^2+z^2}}\right] \tag{6-23b}$$

$$\Rightarrow \frac{\mu_0 I}{4\pi\rho} \qquad (z=0) \tag{6-23c}$$

Thus in the plane ($z = 0$) perpendicular to an end, the field of a semi-infinite wire is half that of the infinite wire. For an infinite wire ($z_1 = -\infty$, $z_2 = +\infty$):

$$B_\phi = \frac{\mu_0 I}{2\pi\rho}$$

which agrees with Eq. (6-8).

(b) *Magnetic Vector Potential $\mathbf{A}$*

To find the magnetic vector potential $\mathbf{A}$ we use the filamentary form of $\mathbf{A}$:

$$\mathbf{A} = \frac{\mu_0}{4\pi}\int \mathbf{I}\,\frac{d\ell'}{R}$$

Since $\mathbf{I} = \mathbf{a}_z I$, $\mathbf{A} = \mathbf{a}_z A_z$:

$$A_z = \frac{\mu_0}{4\pi}\int_{z_1}^{z_2} \frac{I\,dz'}{\sqrt{\rho^2+(z-z')^2}} \qquad (\text{Let } z'' = z-z')$$

$$= -\frac{\mu_0}{4\pi}\int_{z-z_1}^{z-z_2} \frac{I\,dz''}{\sqrt{\rho^2+(z'')^2}} = -\frac{\mu_0 I}{4\pi}\left[\ln\left(z''+\sqrt{(z'')^2+\rho^2}\right)\right]_{z-z_1}^{z-z_2}$$

$$= \frac{\mu_0 I}{4\pi}\ln\left[\frac{(z-z_1)+\sqrt{\rho^2+(z-z_1)^2}}{(z-z_2)+\sqrt{\rho^2+(z-z_2)^2}}\right] \tag{6-23d}$$

Now the magnetic field $\mathbf{B}$ can also be calculated by $\mathbf{B} = \nabla \times \mathbf{A} = -\mathbf{a}_\phi\, \partial A_z/\partial\rho$. You can obtain the result of Eq. (6-23a). *The results of this example must be used with care. They are valid only when used as a portion of a closed circuit.*

EXAMPLE 6-8. ON-AXIS FIELD OF A SQUARE LOOP

Figure 6-10 shows a square loop of side a and current I. Find the on-axis field $\mathbf{B}$. The loop is in the xy plane, with sides parallel to x, y axes, and is centered at the origin.

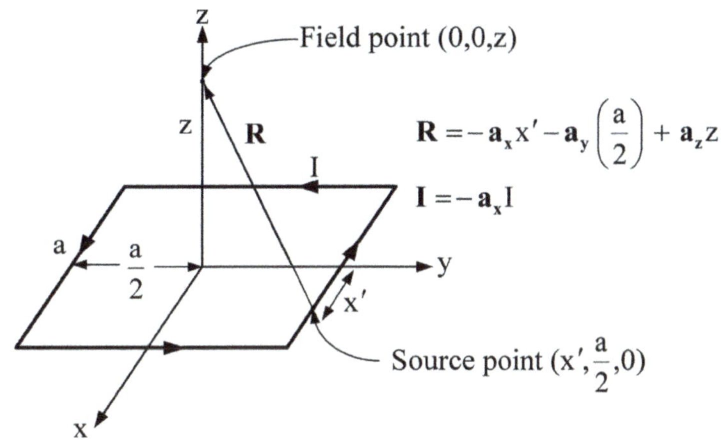

Figure 6-10. A square loop.

SOLUTION

We conclude from a sketch of the field lines that only B_z exists along the z axis. We will calculate only the z component of the contribution of one side of the loop and then multiply by four since all sides are symmetrically disposed with respect to the axis.

$$\mathbf{B} = \frac{\mu_0}{4\pi}\int \frac{(\mathbf{I}\times\mathbf{R})}{R^3}\, d\ell'$$

$$B_z = \frac{\mu_0}{4\pi}\int_{\text{Loop}} \frac{(\mathbf{I}\times\mathbf{R})_z}{R^3}\, d\ell'$$

$$= \frac{4\mu_0}{4\pi}\int_{\text{Side}} \frac{(\mathbf{I}\times\mathbf{R})_z}{R^3}\, d\ell'$$

Choosing one side on the right, we have the source point $(x', \frac{a}{2}, 0)$ and the field point $(0, 0, z)$ as shown in Figure 6-10.

Then

$$\mathbf{r} = \mathbf{a}_z z;\ \mathbf{r}' = \mathbf{a}_x x' + \mathbf{a}_y\left(\frac{a}{2}\right)$$

$$\mathbf{R} = \mathbf{r} - \mathbf{r}' = -\mathbf{a}_x x' - \mathbf{a}_y\left(\frac{a}{2}\right) + \mathbf{a}_z z$$

$$\mathbf{I} = -\mathbf{a}_x I;\ d\ell' = dx'$$

$$\mathbf{I} \times \mathbf{R} = +\mathbf{a}_z\left(\frac{Ia}{2}\right) + \mathbf{a}_y (Iz)$$

$$B_z = \frac{\mu_0}{\pi}\int_{-\frac{a}{2}}^{\frac{a}{2}} \frac{Ia}{2}\frac{dx'}{\left[(x')^2 + (\frac{a}{2})^2 + z^2\right]^{3/2}} \qquad \left(\text{Let } \left(\frac{a}{2}\right)^2 + z^2 = u^2\right)$$

$$= \frac{\mu_0 Ia}{2\pi}\left[\frac{x'}{\sqrt{(x')^2 + (\frac{a}{2})^2 + z^2}}\frac{1}{(\frac{a}{2})^2 + z^2}\right]_{-\frac{a}{2}}^{\frac{a}{2}}$$

$$= \frac{\mu_0 I a}{2\pi}\frac{a}{\sqrt{\frac{a^2}{2} + z^2}}\frac{1}{(\frac{a}{2})^2 + z^2} \qquad (6\text{-}24)$$

Note that as $z \to \infty$:

$$B_z \Rightarrow \frac{\mu_0 I a^2}{2\pi z^3} = \frac{\mu_0 (I a^2)}{2\pi z^3} = \frac{\mu_0 m}{2\pi z^3}$$

in agreement with Eq. (6-31). At distant points $(z \gg a)$, the loop appears small and the field approaches that of a magnetic dipole. The component B_y contributed by this side is cancelled by the side opposite.

EXAMPLE 6-9 MAGNETIC VECTOR POTENTIAL OF A LONG FILAMENT

In this example, we show how the vector potential **A** can be obtained from the magnetic field **B** for an infinitely long current-carrying filament (see Table 6-1). In Example 6-2(a), we obtained **B** using Ampère's law:

$$\mathbf{B} = \mathbf{a}_\phi \frac{\mu_0 I}{2\pi\rho}$$

Note that $B_\rho = B_z = 0$ and B_ϕ depends only on ρ. The vector potential **A** satisfies

$$\nabla \times \mathbf{A} = \mathbf{B}$$

whose scalar equation for the ϕ-component is

$$\frac{\partial A_\rho}{\partial z} - \frac{\partial A_z}{\partial \rho} = B_\phi$$

Noting the rotational symmetry ($\partial/\partial\phi = 0$) and z-independence ($\partial/\partial z = 0$), we conclude that

$$\mathbf{A} = \mathbf{a}_z A_z(\rho)$$

where A_z satisfies

$$-\frac{\partial A_z}{\partial \rho} = B_\phi = \frac{\mu_0 I}{2\pi\rho}$$

Integrating with respect to ρ and taking $\rho = \rho_0$ as a reference point, we obtain

$$A_z = -\int_{\rho_0}^{\rho} \frac{\mu_0 I}{2\pi\rho} d\rho = -\frac{\mu_0 I}{2\pi}\left[\ln \rho\right]_{\rho_0}^{\rho} = -\frac{\mu_0 I}{2\pi} \ln \frac{\rho}{\rho_0}$$

The vector potential of other current distributions in Table 6-1 can be obtained in a similar way.

6.6 THE MAGNETIC DIPOLE*

In this section we consider the magnetic field $\mathbf{B}$ at an arbitrary point (not necessarily along the z-axis) of a small current-carrying loop. Equivalently, we wish to find $\mathbf{B}$ of a loop at far distances ($r >> a$). We will find that the field $\mathbf{B}$ is identical in form to the electric field of an electric dipole. We will also find that many of the characteristics of the loop can be simplified by defining a magnetic dipole moment for the loop.

Figure 6-11(a) shows a circular loop of radius *a* and current I.

*The mathematical detail of this section may be omitted.

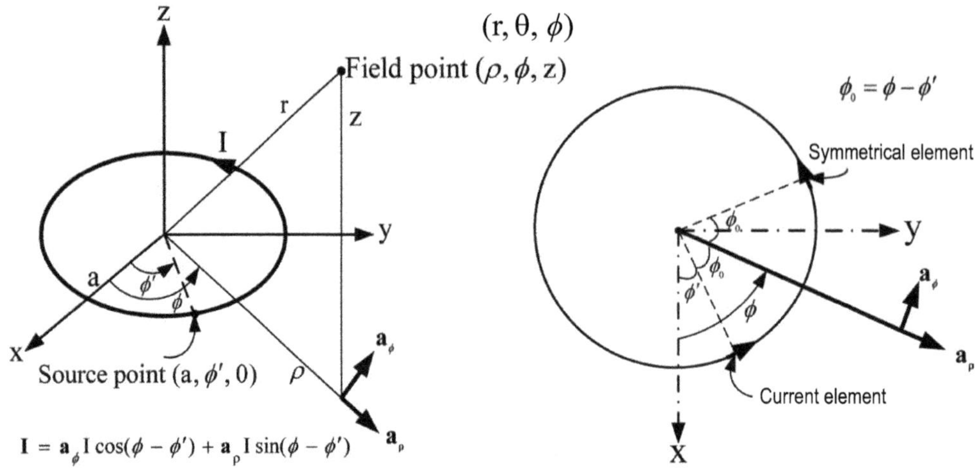

(a) A small circular loop (r >> a). (b) A projection on the *xy* plane.

Figure 6-11. Magnetic Dipole.

The loop lies in the xy plane and is centered at the origin. The source point is (a, ϕ', 0) and the field point is (ρ, ϕ, z). It can be shown that the current vector $\mathbf{I}$ can be written in terms of ϕ', ϕ, $\mathbf{a}_\rho$, $\mathbf{a}_\phi$ as follows:

$$\mathbf{I} = \mathbf{a}_{\phi'}\, I = \mathbf{a}_\phi\, I\cos(\phi-\phi') + \mathbf{a}_\rho\, I\sin(\phi-\phi') \qquad (6\text{-}25)$$

where $\mathbf{a}_{\phi'}$ is a unit vector with respect to the source point and $\mathbf{a}_\rho$, $\mathbf{a}_\phi$ are unit vectors with respect to the field point. See Problem 6-31.

The vector potential $\mathbf{A}$ is

$$\mathbf{A} = \frac{\mu_0}{4\pi}\int \mathbf{I}\,\frac{d\ell'}{R} = \frac{\mu_0}{4\pi}\int_0^{2\pi} \frac{\mathbf{a}_\phi\, I\cos(\phi-\phi') + \mathbf{a}_\rho\, I\sin(\phi-\phi')}{R}\, a\, d\phi'$$

First consider the ϕ component of $\mathbf{A}$:

$$A_\phi = \frac{\mu_0 I a}{4\pi}\int_0^{2\pi} \frac{\cos(\phi-\phi')\, d\phi'}{R}$$

$$R = \sqrt{(x-x')^2+(y-y')^2+(z-z')^2}$$
$$= \sqrt{\rho^2+z^2-2a\rho\cos(\phi-\phi')+a^2}$$
$$= \sqrt{r^2-2a\rho\cos(\phi-\phi')+a^2}$$

$$x = \rho\cos\phi, \quad x' = a\cos\phi'$$
$$y = \rho\sin\phi, \quad y' = a\sin\phi'$$
$$z' = 0$$
$$r^2 = \rho^2+z^2; \quad \frac{\rho}{r} = \sin\theta$$
$$\sqrt{1\pm x} \approx 1 \pm x/2 \quad (|x| \ll 1)$$
$$\frac{1}{1-x} \approx 1+x \quad (|x| \ll 1)$$

At far points ($r >> a$),

$$R = r\sqrt{1 - 2\frac{a}{r}\frac{\rho}{r}\cos(\phi-\phi') + \frac{a^2}{r^2}} \approx r\left[1 - \frac{a}{r}\sin\theta\cos(\phi-\phi')\right] \quad (r>>a)$$

$$\frac{1}{R} \approx \frac{1}{r}\left[1 + \frac{a}{r}\sin\theta\cos(\phi-\phi')\right] \quad (r>>a)$$

$$A_\phi = \frac{\mu_0 I a}{4\pi r}\left[\int_0^{2\pi}\cos(\phi-\phi')d\phi' + \frac{a\sin\theta}{r}\int_0^{2\pi}\cos^2(\phi-\phi')d\phi'\right]$$

$$= \frac{\mu_0 I a}{4\pi r}\left[0 + \frac{a\sin\theta}{r}\frac{2\pi}{2}\right] = \frac{\mu_0 (I\pi a^2)}{4\pi r^2}\sin\theta$$

Similarly, we can show that

$$A_\rho = \frac{\mu_0 I a}{4\pi}\int_0^{2\pi}\frac{\sin(\phi-\phi')\,d\phi'}{R} = 0$$

Substituting our expression for $(1/R)$ in the integral for A_ρ, we obtain integrals of $\sin(\phi-\phi')$ and $\sin(\phi-\phi')\cos(\phi-\phi')$ (which is equal to $1/2\ \sin 2(\phi-\phi')$), each of which yields zero when integrated over a range of 2π. It can also be seen by symmetry that $A_\rho = 0$ since for each current element there is a symmetrically disposed element whose contribution to A_ρ is opposite (see Fig. 6-11(b) which shows a projection on the xy plane.)
Thus

$$\mathbf{A} = \mathbf{a}_\phi \frac{\mu_0 (I\pi a^2)}{4\pi r^2}\sin\theta \tag{6-26}$$

and

$$\mathbf{B} = \nabla\times\mathbf{A} = \frac{\mu_0 (I\pi a^2)}{4\pi r^3}(\mathbf{a}_r\, 2\cos\theta + \mathbf{a}_\theta \sin\theta) \tag{6-27}$$

The fields above are identical in form to the electric fields of an electric dipole (see Eq. (2-68)). Recognizing this identity, we call a small loop of current a **magnetic dipole** and define a *magnetic dipole moment* $\mathbf{m}$ as follows:

$$\mathbf{m} = \mathbf{a}_z m = \mathbf{a}_z (I\pi a^2) = \mathbf{a}_n (IS) \tag{6-28}$$

where S is the surface area of the loop and $\mathbf{a}_n$ is the normal to the plane of the loop with a right-hand rule relationship to the current $\mathbf{I}$. Then the vector potential $\mathbf{A}$ and field $\mathbf{B}$ of the loop of Figure 6-11 may be expressed as follows:

$$\mathbf{A} = \frac{\mu_0\ \mathbf{m}\times\mathbf{a}_r}{4\pi r^2} = \mathbf{a}_\phi \frac{\mu_0\ m}{4\pi r^2}\sin\theta \tag{6-29}$$

$$\mathbf{B} = \frac{\mu_0\ m}{4\pi r^3}(\mathbf{a}_r\, 2\cos\theta + \mathbf{a}_\theta \sin\theta) \tag{6-30}$$

Along the loop axis (z-axis) where $\theta = 0°$, $r = z$ and $\mathbf{a}_r = \mathbf{a}_z$:

$$B_z = \frac{\mu_0\ m}{2\pi z^3} \tag{6-31}$$

which was shown earlier in Example 6-5. In the plane of the loop (xy plane) where $\theta = 90°$:

$$B_\theta = \frac{\mu_0\ m}{4\pi r^3} \tag{6-32}$$

We note that Eqs. (6-29) and (6-30) are valid for any shape of the loop, as long as it is small ($a << r$). We can also view Eq. (6-30) as the field at far points ($r >> a$) of a current-carrying loop.

The magnetic dipole plays an important role in explaining the behavior of magnetic field in matter in Chapter 7.

6.7 FORCES AND TORQUES ON CURRENTS IN MAGNETIC FIELDS

In this section we consider the forces on currents in the presence of magnetic fields. Since the charge in motion experiences a force in the region of $\mathbf{B}$, the currents, which consist of charges in motion, experience a force in the presence of magnetic fields. First let us consider the elementary force on a portion of a volume current distribution $\mathbf{J}$. Figure 6-12(a) shows a volume element of current in the presence of the field $\mathbf{B}$.

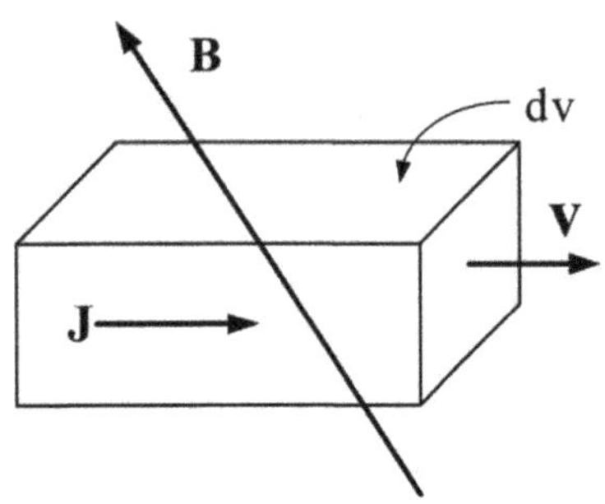

(a) An element of volume current **J** in field **B**.

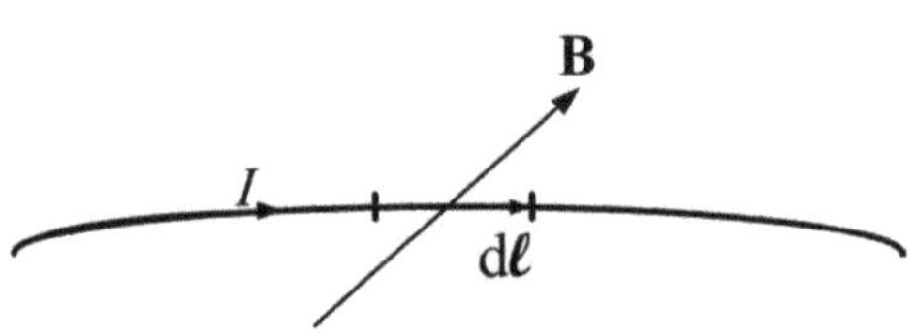

(b) An element of filamentary current in field **B**.

Figure 6-12. A volume current and an element of filamentary current.

The results that follow can be used to find the force acting on a current-carrying wire, a conducting surface or a semiconductor.

The magnetic force $\mathbf{F}_m$ on a charge q moving at velocity $\mathbf{v}$ is

$$\mathbf{F}_m = q\,\mathbf{v} \times \mathbf{B}$$

so that if current $\mathbf{J}$ consists of charges moving at drift velocity $\mathbf{v}$ then the elementary force $d\mathbf{F}_m$ on the volume element dv is

$$d\mathbf{F}_m = dq\ \mathbf{v} \times \mathbf{B} = \rho_v\, dv\ \mathbf{v} \times \mathbf{B}$$

$$\frac{d\mathbf{F}_m}{dv} = (\rho_v \mathbf{v}) \times \mathbf{B} = \mathbf{J} \times \mathbf{B} = \text{force per unit volume}$$

where Eq. (5-4) has been used. For surface current $\mathbf{J}_s$ the force per unit area is $\frac{d\mathbf{F}_m}{ds} = \mathbf{J}_s \times \mathbf{B}$ and, finally, for filamentary current $\mathbf{I}$ (see Figure 6-12(b)), the force per unit length is $\frac{d\mathbf{F}_m}{d\ell} = \mathbf{I} \times \mathbf{B}$. This latter expression may also be written as $d\mathbf{F}_m = I\,(d\boldsymbol{\ell} \times \mathbf{B})$ if $d\boldsymbol{\ell}$ is chosen as a vector in the direction of $\mathbf{I}$. Table 6-2 summarizes the force relationships.

Table 6-2 Magnetic Forces on Currents

$$\frac{d\mathbf{F}_m}{dv} = \mathbf{J} \times \mathbf{B} \quad (6\text{-}33a)$$

$$\frac{d\mathbf{F}_m}{ds} = \mathbf{J}_s \times \mathbf{B} \quad (6\text{-}33b)$$

$$\frac{d\mathbf{F}_m}{d\ell} = \mathbf{I} \times \mathbf{B} \text{ or } d\mathbf{F}_m = I\, d\boldsymbol{\ell} \times \mathbf{B} \quad (6\text{-}33c)$$

To obtain the magnetic force on an entire current distribution we merely integrate over the current distribution

$$\mathbf{F}_m = \iiint \mathbf{J} \times \mathbf{B}\, dv \quad \text{for volume current} \quad (6\text{-}34a)$$

$$= \iint \mathbf{J}_s \times \mathbf{B}\, ds \quad \text{for surface current} \quad (6\text{-}34b)$$

$$= \int \mathbf{I} \times \mathbf{B}\, d\ell = I \oint_C d\boldsymbol{\ell} \times \mathbf{B} \text{ for filamentary current} \quad (6\text{-}34c)$$

Note that there is no net force $\mathbf{F}_m$ on a filamentary loop in a uniform field $\mathbf{B}$ since

$$\mathbf{F}_m = I\oint_C d\ell \times \mathbf{B} = I\left[\oint_C d\ell\right] \times \mathbf{B} = 0 \quad (\text{since } \oint_C d\ell = 0)$$

Now let's consider the torque on current-carrying loops. The torque is important in addition to the force since the torque may exist even though the net force is zero. Recall the definition of torque. Figure 6-13 shows a force $\mathbf{F}$ applied at point $\mathbf{r}$.

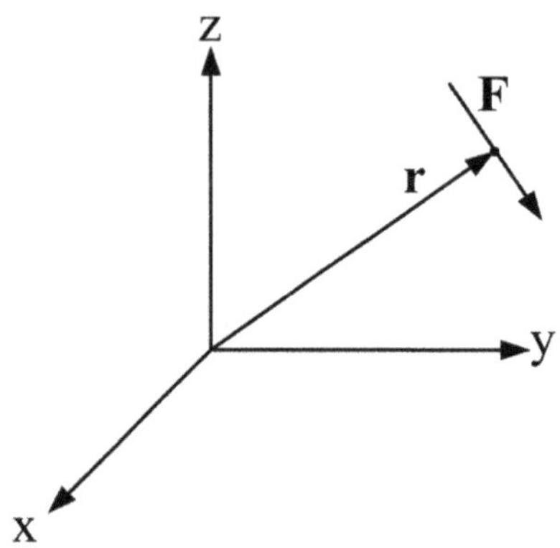

Figure 6-13. The definition of torque T.

The torque $\mathbf{T}$ about a reference point at the origin is defined as

$$\mathbf{T} = \mathbf{r} \times \mathbf{F} \tag{6-35}$$

Note that the torque about any other reference point will, in general, be different. However, if the total force on an object is zero, then the torque is independent of reference.

To calculate the torque on a current-carrying loop we first consider the torque $d\mathbf{T}$ on an elementary portion of the loop:

$$d\mathbf{T} = \mathbf{r} \times d\mathbf{F}_m$$

For a filamentary loop, we integrate over the loop C to obtain

$$\mathbf{T} = I\oint_C \mathbf{r} \times (d\ell \times \mathbf{B}) \tag{6-36}$$

Eq. (6-36) is the general expression for the torque on a filamentary loop. If $\mathbf{B}$ is uniform over the loop, the net force $\mathbf{F}_m$ is zero and the torque (independent of reference) is given by

$$\boxed{\mathbf{T} = \mathbf{m} \times \mathbf{B}} \tag{6-37}$$

where $\mathbf{m}$ is the magnetic dipole moment of the loop. For derivation of Eq. (6-37), see Reitz, Milford and Christy (1993).*

* J. R. Reitz, F. J. Milford, and R. W. Christy, *Foundations of Electromagnetic Theory*, Addison-Wesley, 1993, 4th Ed., Section 8-2.

EXAMPLE 6-10. THE FORCE BETWEEN PARALLEL CURRENT-CARRYING WIRES

Figure 6-14 shows two parallel current-carrying filaments of infinite length, separated by a distance d. Find the force per unit length between wires.

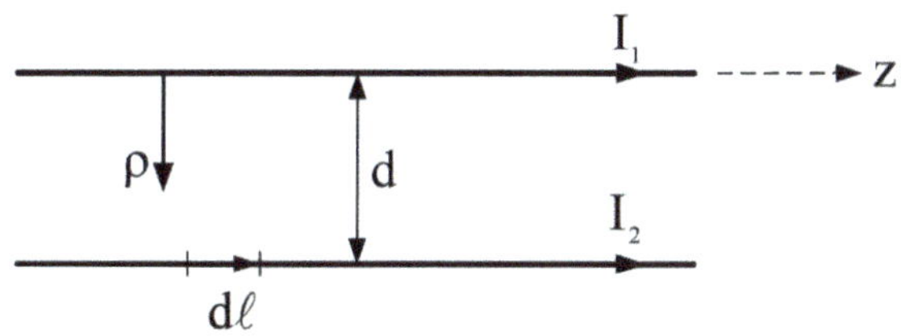

Figure 6-14. Two parallel current-carrying wires.

SOLUTION

We will find the force per unit length on wire 2 with current I_2. This will of course be the negative of the force on the other wire. We treat I_1 as the *source current*, which produces the field $\mathbf{B}$, and I_2 as the *test current*, which experiences a force. We consider the force on an elementary length $d\ell$ of wire 2:

$$d\mathbf{F}_m = I_2\, d\boldsymbol{\ell} \times \mathbf{B}$$

We need to consider only the field due to wire 1. The field due to wire 2 is non-zero, but it cannot exert any net force on itself. Otherwise it could propel itself through space. Therefore we consider only the field at wire 2 due to wire 1, which is $\mathbf{a}_\phi(\mu_0 I_1/2\pi\rho)$.

$$d\mathbf{F}_m = I_2(\mathbf{a}_z\, d\ell)\times(\mathbf{a}_\phi \mu_0 I_1/2\pi d)$$

$$= -\mathbf{a}_\rho \frac{\mu_0 I_1 I_2}{2\pi d}\, d\ell$$

$$\frac{d\mathbf{F}_m}{d\ell} = -\mathbf{a}_\rho \frac{\mu_0 I_1 I_2}{2\pi d} \qquad (6\text{-}38)$$

The force is attractive if currents are parallel (I_1, I_2 both positive or negative) and is repulsive if currents are anti-parallel (one positive, one negative).

EXAMPLE 6-11. THE FORCE BETWEEN A FILAMENT AND A SQUARE LOOP

Figure 6-15(a) shows a filament of infinite length with current I_1. Nearby is a square loop of side ℓ with two sides parallel to the filament. The square loop carries a current I_2 clockwise. The distance between the loop and the filament is d. Find the force between the loop and the filament.

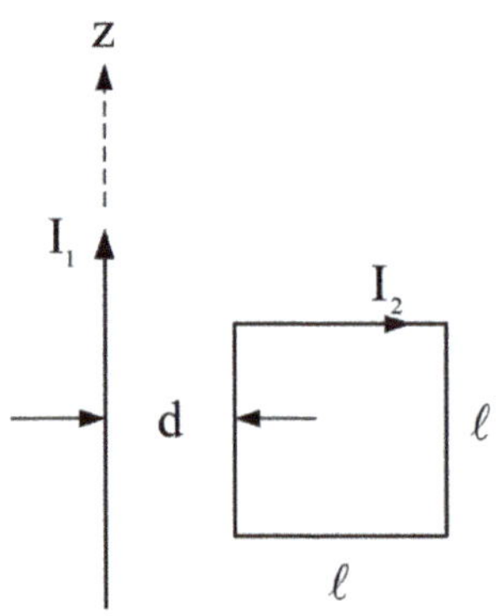

(a) A filament and a square loop.

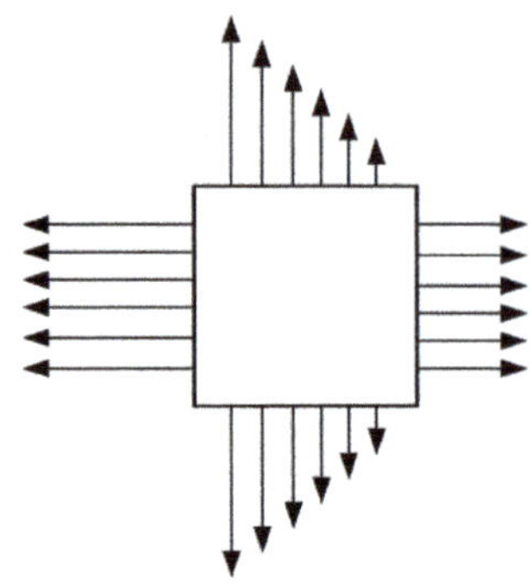

(b) Forces on the square loop.

Figure 6-15. A filament and a square loop.

SOLUTION

We first choose the filament as the source current and the loop as the test current because it is easy to calculate the field due to a filament. To find the force on the loop, we consider the field $\mathbf{B}$ at the loop (due to filamentary current I_1) which is $\mathbf{a}_\phi(\mu_0 I_1/2\pi\rho)$. $\mathbf{B}$ points into the paper. We ignore the self field as before. Figure 6-15(b) shows a sketch of the forces $I_2\, d\boldsymbol{\ell} \times \mathbf{B}$ exerted on different portions of the loop. The z-directed forces cancel but the ρ-directed forces do not because the magnitude of $\mathbf{B}$ decreases as $1/\rho$.

$$
\begin{aligned}
\mathbf{F}_m &= \int_{\text{loop}} \mathbf{I} \times \mathbf{B}\, d\ell \\
&= \int_0^\ell (\mathbf{a}_z I_2) \times \left(\mathbf{a}_\phi \frac{\mu_0 I_1}{2\pi d}\right) dz + \int_0^\ell (-\mathbf{a}_z I_2) \times \left(\mathbf{a}_\phi \frac{\mu_0 I_1}{2\pi(d+\ell)}\right) dz \\
&= -\mathbf{a}_\rho \frac{\mu_0 I_1 I_2 \ell}{2\pi}\left[\frac{1}{d} - \frac{1}{d+\ell}\right] \\
&= -\mathbf{a}_\rho \frac{\mu_0 I_1 (I_2 \ell^2)}{2\pi d(d+\ell)}
\end{aligned}
\tag{6-39}
$$

The force is attractive if I_1, I_2 have the same sign. Note that for this problem it will be very difficult to calculate the force acting on the filament due to the field of the loop because of difficulty in calculating the field of a loop.

Consider the torque on the loop. There is no torque about the center but there is a torque about one of the corners of the loop. Can you calculate it?

6.8 AMPÈRE'S FORCE LAW

A particular application of the theory of Section 6.7 is the calculation of the force on a current-carrying loop due to the field **B** produced by another loop. Figure 6-16 shows two current-carrying loops with currents I_1, I_2.

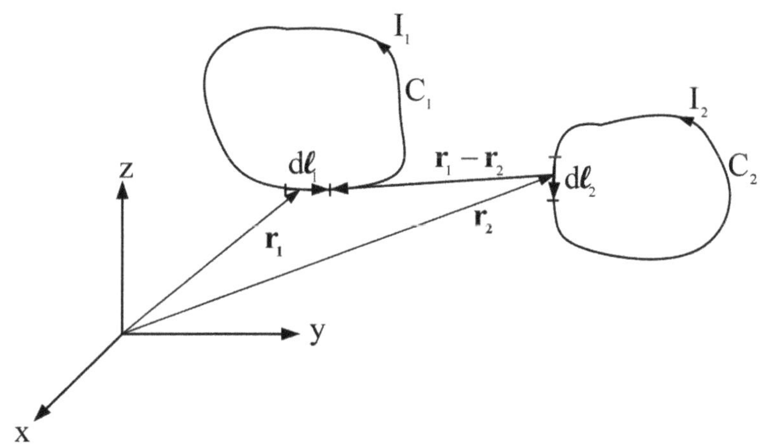

Figure 6-16. Two current-carrying loops.

Typical points on loops (1), (2) are (x_1, y_1, z_1), (x_2, y_2, z_2), respectively, specified by vectors $\mathbf{r}_1$, $\mathbf{r}_2$.

$$\mathbf{F}_1 = \text{Force on loop (1)} = I_1 \oint_{C_1} d\boldsymbol{\ell}_1 \times \mathbf{B} = I_1 \oint_{C_1} d\boldsymbol{\ell}_1 \times (\mathbf{B}_{11} + \mathbf{B}_{12})$$

(where $\mathbf{B}_{11}$, $\mathbf{B}_{12}$ are the fields at loop (1) due to currents I_1, I_2, respectively)

$$\mathbf{F}_1 = I_1 \oint_{C_1} d\boldsymbol{\ell}_1 \times \mathbf{B}_{12} \text{ (since loop (1) exerts no net force on itself)}$$

$$\mathbf{B}_{12} = \frac{\mu_0 I_2}{4\pi} \int_{C_2} d\boldsymbol{\ell}_2 \times \frac{\mathbf{R}_{12}}{R_{12}^3} \text{ (see Eq. (6-19c))}$$

Thus

$$\boxed{\mathbf{F}_1 = \frac{\mu_0 I_1 I_2}{4\pi} \oint_{C_1} \oint_{C_2} \frac{d\boldsymbol{\ell}_1 \times (d\boldsymbol{\ell}_2 \times \mathbf{R}_{12})}{R_{12}^3}} \quad \text{Ampère's force law} \qquad (6\text{-}40)$$

where $\mathbf{R}_{12} = \mathbf{r}_1 - \mathbf{r}_2$.

6.9 ANDRÉ-MARIE AMPÈRE (1775–1836) AND THE MAGNETIC FIELDS OF STEADY CURRENTS

André-Marie Ampère was a French scientist who made contributions to an astonishing number of different fields. He was a mathematician, chemist, physicist, engineer, and philosopher. He possessed encyclopedic knowledge and moved easily from one field to another. He devised a universal language. He wrote a drama about Columbus called *L'Americide*. He invented a table of elements before Mendeleev. His most lasting contribution, however, was his complete exposition of the magnetic fields and forces due to steady electric currents. The chapter which you are reading now is essentially Ampère's chapter. He combined experiment and theory, and amazing insight, in a comprehensive, thorough study unlike any of his other works. Maxwell said of his work: "The whole, theory and experiment, seems as if it had leaped, full grown and full armed, from the brain of the 'Newton of electricity'.* It is perfect in form, and unassailable in accuracy, and it is summed up in a formula from which all the phenomena may be deduced, and which must always remain the cardinal formula of electrodynamics." Ampère coined the word *electrodynamics* to describe his area of study. It is no longer used since his work is limited to steady currents.

A. M. Ampère was born in Lyon in southeastern France into a prosperous middle class family. When Ampère was seven, his father, Jean Jacques Ampère, retired and moved his family to Poleymieux, a small village of several hundred inhabitants perched on a mountainside about 10 kilometers from Lyon. There was no school in Poleymieux and, as a result, Ampère had no formal schooling. He was largely self-educated under the guidance of his father. Ampère described the process as follows: "He never required me to study anything, but he knew how to inspire in me a desire to know." He became interested in mathematics at an early age and eventually was reading the encyclopedia and carrying out the calculations of Laplace's *Celestial Mechanics*.

The French Revolution had erupted in 1789 and it came to Lyon in 1792. Ampère's father found himself caught up in the struggle between two rival revolutionary groups, the moderate republican Girondins and the radical republican Jacobins. Jean Jacques was a justice of the peace when the Girondins took over Lyon. In this position, he certified the arrest of a Jacobin leader who was later guillotined. When the Jacobins took over a few months later after a siege of Lyon, Jean Jacques was arrested, tried, and guillotined. In his final letter home, he pardoned his executioners and wrote, "As for my son, there is nothing that I do not expect of him." Ampère had had a very close relationship with his father and was devastated by the loss. He went into a deep depression for over a year. After he recovered, he resumed his studies, eventually tutored mathematics in Lyon, and taught physics in Bourg. In 1801 the young Ampère presented a lecture in which he rejected action-at-a-distance in electricity: "Bodies which do not touch each other cannot interact." Therefore he rejected Coulomb's law. He retained an aversion to action-at-a-distance throughout much of his life.

He began a short but very happy marriage in 1799. His son, named Jean Jacques, was born in 1800. His wife Julie died in 1803. In 1804 Ampère went to Paris to teach at the Polytechnic School. Over the next ten years he made numerous contributions to mathematics and chemistry, leading to his election to the French Academy of Sciences in 1814.

In the fall of 1820, Ampère first learned of Oersted's discovery, which was reported to the Academy on Sept. 11 by François Arago, who had verified some of the results. This discovery was a shock to Academy members, as it was to scientists all over the world, especially because their highly respected colleague Coulomb had posited many years earlier that there could not be any connection between electricity and magnetism. It would be

* James R. Hofmann, *André-Marie Ampère*, Blackwell Science Biographies, Cambridge, MA., p. 341, 1995.

interesting to know the thoughts of the Academy members as they left that meeting of 11 September 1820. At that time the French Academy included many of the leading scientists in the world. Certainly many of them recognized the historic importance of the discovery and some were determined to play a role in the development of this new theory. All over the world, scientists began to study the interaction of electricity and magnetism. Within a few years the scientific literature on the subject grew to enormous proportions. Ampère and Biot would lead the way in the development of the theory.

Ampère immediately set to work. He repeated Oersted's experiment, but with the terrestrial field cancelled out. In the absence of the earth's field, the compass needle rotated fully to end up perpendicular to the long-wire current and perpendicular to a radial line from wire to field point (compass location). In this final position, the compass needle was either attracted to, or repelled by, the current-carrying wire. If attracted, then it would be repelled if the compass direction were reversed. Ampère also concluded that the earth's magnetic field is caused by electric currents in the earth. These discoveries were made during his first week's work, which was reported on 18 September 1820. Thereafter, he reported to the Academy almost every week for several months. It's not possible to describe here all of his accomplishments but we'll describe a few of the more significant ones. Ampère showed in late September that parallel/antiparallel current-carrying wires attract/repel as shown in Example 6-10 and Eq. (6-38). He showed that helical current-carrying wires (solenoids) act like bar magnets. See Example 6-6 for an approximation to a very tightly-wound solenoid. This work led him to conclude that long bar magnets carry electric currents circumferentially around their periphery as in Figure 6-8. Later he concluded that every small portion of the cross section carried circumferential "Ampèrian" currents as in Figure 7-1(b). These Ampèrian currents would add up to produce circumferential currents around the entire cross section as shown in Figure 7-1(c). These currents then correspond to those of the air-filled solenoid of Figure 6-8.

Since he was able to explain everything in terms of electric currents, Ampère had no need for the various theories of vortices, magnetic fluids, and magnetic charges with which the theory of magnetism was so burdened. He came to the conclusion that magnetic charges did not exist and insisted emphatically on the primacy of *electric currents* in explaining magnetic phenomena. He showed, however, that the magnetic charge model was equivalent mathematically to the electric current model. Once Ampère had become convinced that electric currents were the source of magnetism, he then recognized the primary importance of the force between two current-carrying elements. He set about in a very determined way to find this basic force law. This is a very complex relationship, as we can see by observing the double cross-product of Eq. (6-40), which is obtained by combining the single cross products of Eqs. (6-19) and (6-33). Ampère started with a special case, namely, two short z-directed current elements, both located in the x-y plane. He verified that the force between elements was attractive, inverse square, and of the form $F = K\, I_1\, d\ell_1\, I_2\, d\ell_2 / R_{12}^{\,2}$, where K is a constant and R_{12} is the distance between elements. From this simple result to Eq. (6-40) is a long journey but somehow Ampère was able to get there. Actually his form was equivalent to Eq. (6-40) for the current loops of steady currents. Ampère was handicapped by his belief that the basic force must be radial, along the line connecting the elements. We can see by Eq. (6-40) that this is not necessarily so. The basic force law embedded in Eq. (6-40), or its equivalent, was the fundamental result which was praised so highly by Maxwell in the beginning of this section.

Ampère also showed that an inverse square force law $(1/r^2)$ between elements leads to an inverse force law $(1/\rho)$ between an element and a long current-carrying wire. In other words, if the field of a current element/point charge is of the form $(1/r^2)$, then the field of a long current-carrying wire/long line charge is of the form $(1/\rho)$.

This is shown in Examples 6-2(a), 6-7, Example 2-5, and Problem 2-29(d). Laplace and Biot also showed this principle at about the same time (Laplace was apparently the first). Ampère is also given credit for the force law $d\mathbf{F} = I\, d\ell \times \mathbf{B}$, Eq. (6-33), which gives the force exerted on a current element in a magnetic field $\mathbf{B}$. Ampère was a person of tremendous insight. He proposed the electron shell. He postulated Ampèrian currents. He guessed that the electromagnetic wave was transverse and he invented a telegraphic arrangement, which was first put to use a year after his death.

6.10 THE HALL EFFECT

If a current-carrying conductor is placed in a magnetic field it experiences a separation of charge due to the magnetic force $\mathbf{F}_m$ and a resultant *Hall voltage* V_H appears across the resistor. This effect can be used to measure magnetic fields. Consider a conductor of thickness d, width w, and volume current density $\mathbf{J}$. The current is y-directed (electrons flow in the negative y direction). The magnetic force on an electron is

$$\mathbf{F}_m = q(\mathbf{v} \times \mathbf{B}) = -e(-\mathbf{a}_y\, v_0 \times \mathbf{a}_x\, B_0) = \mathbf{a}_z(-e\, v_o B_0)$$

Thus there is a magnetic force deflecting electrons downwards, leading to a charge separation as shown in Figure 6-17. The charge separation creates an electric field $\mathbf{E}_H$ (the Hall field) and a voltage V_H (the Hall voltage). Associated with $\mathbf{E}_H$ is an electric force $q\mathbf{E}_H$ which opposes the magnetic force.

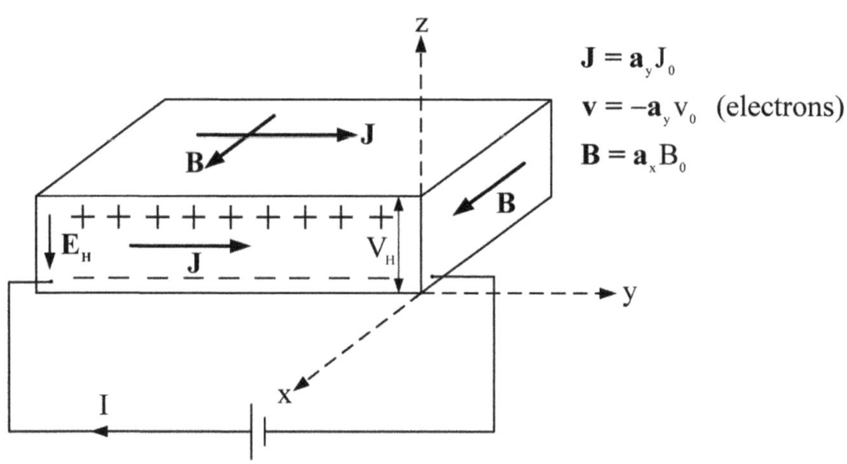

Figure 6-17. The Hall effect.

How long does this charge buildup continue and what is the magnitude of the electric field $\mathbf{E}_H$? Clearly the charge buildup continues until electric and magnetic forces cancel and equilibrium is reached. Equilibrium is reached in about 10^{-14} seconds for copper (relaxation time). At equilibrium the total force on a charge is zero:

$$\mathbf{F} = q(\mathbf{E}_H + \mathbf{v} \times \mathbf{B}) = 0$$

$$\boxed{\mathbf{E}_H = -\mathbf{v} \times \mathbf{B}} \qquad (6\text{-}41)$$

For the example of Figure 6-17:

$$\mathbf{E}_H = \mathbf{a}_z(-v_0 B_0)\,, \qquad V_H = v_0\, B_0\, d$$

(We have assumed a parallel-plate relationship between $\mathbf{E}_H$ and V_H).

If the sample of Figure 6-17 were a semiconductor with both positive and negative charge carriers, the positive charge carriers would produce a charge separation opposite to that of Figure 6-17 with an opposing contribution to $\mathbf{E}_H$, V_H. If the principal charge carriers are positive, then the direction of the Hall voltage in Figure 6-17 is reversed (assuming current direction unchanged). The Hall effect may thus be used to determine whether semiconductors are p-type or n-type. The Hall effect was discovered by Edwin Hall in 1879.

PROBLEMS

6-1. Complete Example 6-1, i.e., show how all components of $\mathbf{E}$, $\mathbf{B}$ can be determined by measurements of force.

6-2. (**Cyclotron**) A particle of charge q and mass m enters a region of uniform magnetic field B_0 (pointing into the page) with uniform velocity v_0. It can be shown that the charge follows a circular trajectory as shown in Figures 6-18.

(a) Is the charge positive or negative?

(b) Find the radius (R) of the trajectory. Make use of the balance between the magnetic force and the centrifugal force.

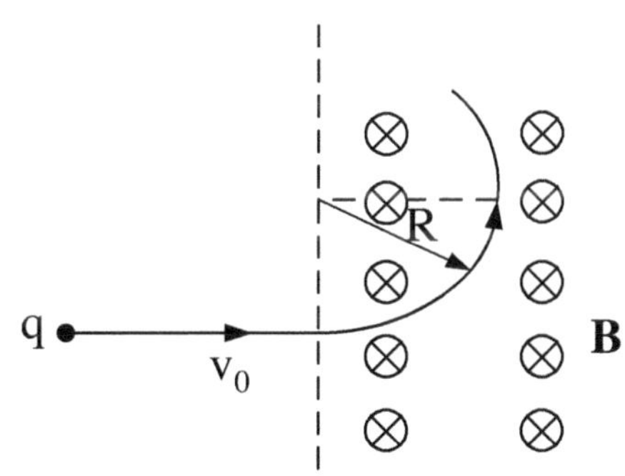

Figure 6-18. Magnetic force on a moving charge.

6-3. Figure 6-19 shows the cross section of a coaxial line with a thick outer conductor of radii b, c surrounding an inner conductor of radius a. Inner, outer conductors carry currents I, –I, respectively; each is uniformly distributed over the appropriate cross section. Find the magnetic field $\mathbf{B}$ everywhere. Confirm that the results for $\mathbf{B}$ satisfy the two basic laws in differential form.

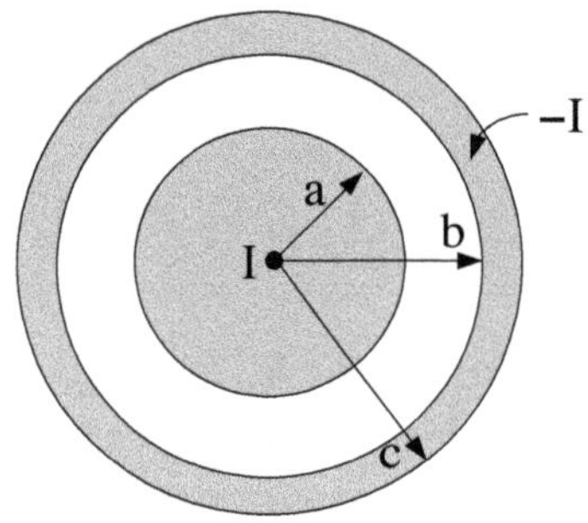

Figure 6-19. A coaxial line.

6-4. A conducting cylinder of radius a and infinite length carries the nonuniform volume current density $\mathbf{J}$ where

$$\mathbf{J} = \mathbf{a}_z J_0 (\frac{\rho}{a}) \qquad (0 < \rho < a)$$

Find the magnetic flux density $\mathbf{B}$ everywhere. Confirm that the results you obtained for $\mathbf{B}$ satisfy the two basic laws: $\nabla \cdot \mathbf{B} = 0$ and $\nabla \times \mathbf{B} = \mu_0 \mathbf{J}$.

6-5. A conducting cylinder of radius a and infinite length carries the following volume current density:

$$\mathbf{J} = \mathbf{a}_z J_0 e^{-\alpha(a-\rho)} \qquad (0 < \rho < a)$$

Find the magnetic induction $\mathbf{B}$ everywhere.

6-6. A hollow cylinder (tube) of radii a, b and infinite length carries the nonuniform volume current density $\mathbf{J}$:

$$\mathbf{J} = \mathbf{a}_z J_0 \left(\frac{a}{\rho}\right) \quad (a < \rho < b)$$

Find the magnetic field $\mathbf{B}$ everywhere.

6-7. A very long conducting cylinder of radius a, carrying uniform volume current $\mathbf{J} = \mathbf{a}_z J_o$, has an off-centered air cavity of radius b along the entire length. The axis of the cavity is offset from the axis of the conducting cylinder by a distance d ($b + d < a$). Find the magnetic field $\mathbf{B}$ in the cavity region.
[HINT] Treat the problem by a superposition of two conducting cylinders of radii a and b with opposite current densities, $\mathbf{J}$ and $-\mathbf{J}$, respectively. You will find that $\mathbf{B}$ is uniform inside the cavity.

6-8. Derive the magnetic fields of two anti-parallel uniform current sheets in Table 6-1. You can make use of the results in Example 6-3.

6-9. Consider four planar current sheets, each of infinite extent (see Figure 6-20).

$$\begin{aligned} \mathbf{J}_s &= \mathbf{a}_x J_{so} && (y = -a/2) \\ &= -\mathbf{a}_x J_{so} && (y = a/2) \\ &= \mathbf{a}_y J_{so} && (x = a/2) \\ &= -\mathbf{a}_y J_{so} && (x = -a/2) \end{aligned}$$

Find the magnetic field $\mathbf{B}$ everywhere in each region of the xy plane.

[HINT] Use superposition.

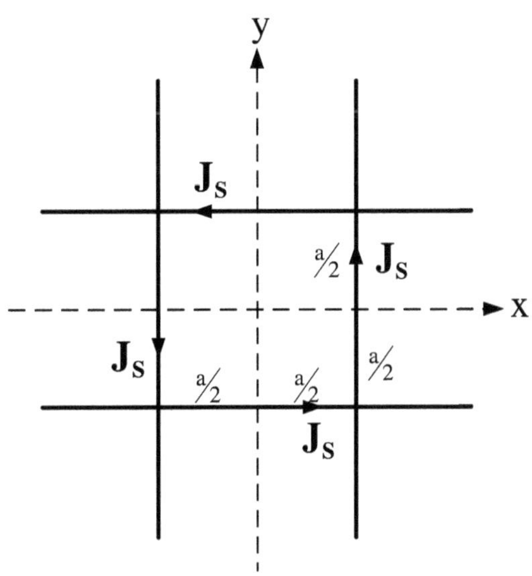

Figure 6-20. Four planar current sheets.

6-10. Consider a slab of uniform volume current density. The slab is of thickness d in the z direction and is of infinite extent in the x, y directions:

$$\mathbf{J} = \mathbf{a}_y J_0 \;\left(-\frac{d}{2} < z < \frac{d}{2}\right)$$

Find the magnetic field $\mathbf{B}$ everywhere. It is expected that the magnetic field is antisymmetric about the xy plane ($B_x(z) = -B_x(-z)$) since the current distribution is even in z.

6-11. Consider an infinite slab of volume current density:

$$\mathbf{J} = \mathbf{a}_y J_0 \qquad (0 < z < d/2)$$
$$= -\mathbf{a}_y J_0 \qquad (-\frac{d}{2} < z < 0)$$

Find the magnetic field $\mathbf{B}$ everywhere. The magnetic field is even since the current distribution is odd in z.

6-12.* Consider an infinite slab of non-uniform volume current density:

$$\mathbf{J} = \mathbf{a}_y J_0 \frac{z}{d} \; (-\frac{d}{2} < z < \frac{d}{2})$$

Find the magnetic field $\mathbf{B}$ everywhere. It is expected that the magnetic field is symmetric (even) about the xy plane ($B_x(z) = B_x(-z)$).

6-13. Consider an infinite solenoid in Figure 6-6(b). Calculate the magnetic field using Ampère's law without using the results of the toroid. Choose the closed loop properly. You can assume that the field is mainly confined in the core of the solenoid and negligible outside.

6-14. Consider the current-carrying circular loop (Example 6-5).

(a) Find the location of the maximum value of the on-axis magnetic field $\mathbf{B}$.
(b) For the half-loop, find values of ϕ_1, ϕ_2 such that $B_x = 0$. Repeat for $B_y = 0$.
(c) Explain the results of (a) and (b) with sketches of magnetic field lines.

6-15. Derive Eq. (6-23a) (field of a straight current-carrying wire) from Eq. (6-23d) utilizing the relationship $\mathbf{B} = \nabla \times \mathbf{A}$.

6-16. Calculate the magnetic vector potential $\mathbf{A}$ for each of the following examples:
(a) Example 6-5. (b) Example 6-6. (c) Example 6-8.

6-17. Find the on-axis magnetic field $\mathbf{B}$ of a current-carrying rectangular loop of sides a, b and current I.

6-18. Find the on-axis magnetic field $\mathbf{B}$ of an equilateral triangular loop of side a and current I.

6-19.* Find the on-axis magnetic field $\mathbf{B}$ of a regular polygonal current-carrying loop of N sides, each of length a and current I.

6-20. Consider two parallel current-carrying circular loops of radius a, separated by a distance d:

$$\mathbf{I} = \mathbf{a}_\phi I_0 \qquad (z = d/2;\; \rho = a)$$
$$= \mathbf{a}_\phi I_0 \qquad (z = -d/2;\; \rho = a)$$

(a) Find the magnetic field $\mathbf{B}$ along the z axis.
(b) Show that the first derivative of the field $\mathbf{B}$ vanishes at the origin. Find the condition for the second derivative to be also zero. Under this condition the field $\mathbf{B}$ is nearly uniform in the region near the origin. The arrangement described above is called a *Helmholtz coil.*

6-21.* A solenoid of square cross section is of length ℓ and sides a. There are N_1 turns per unit length. The current is I.

(a) Find the on-axis magnetic field $\mathbf{B}$.
(b) Find the on-axis magnetic field $\mathbf{B}$ as $\ell \to \infty$ and compare with Eq. (6-14b).

6-22. A thin strip of width w (in the x direction) and infinite length (in the z direction) carries a longitudinal current I. Find the magnetic field $\mathbf{B}$ everywhere. Assume a uniform surface current, i.e., $\mathbf{J}_s = \mathbf{a}_z I/w$ [A/m] $(-w/2 < x < w/2)$, in the xz plane $(y = 0)$.

[HINT] Consider the strip as a collection of thin wires of infinite length and width dx carrying a current I dx/w and make use of Eq. (6-8). The fields at points on the plane of the strip and on the plane bisecting the strip are simpler to obtain.

6-23. Consider two anti-parallel current-carrying strips of width w, infinite length (in the z direction), separation d, and current I, –I.

$$\mathbf{J}_s = \mathbf{a}_z\ (I/w)\ \ (x = \frac{d}{2},\ -\frac{w}{2} < y < \frac{w}{2})$$
$$= -\mathbf{a}_z (I/w)\ \ (x = -\frac{d}{2},\ -\frac{w}{2} < y < \frac{w}{2})$$

(a) Find the magnetic fields everywhere.
(b) Plot the magnetic field along the x axis as a function of x.

6-24. Two anti-parallel, semi-infinite, current-carrying wires of separation d are connected by a semi-circular loop of diameter d (Figure 6-21). The wires lie in the xy plane; the semi-infinite wires are parallel to the x axis. Find the magnetic field $\mathbf{B}$ along the z axis.

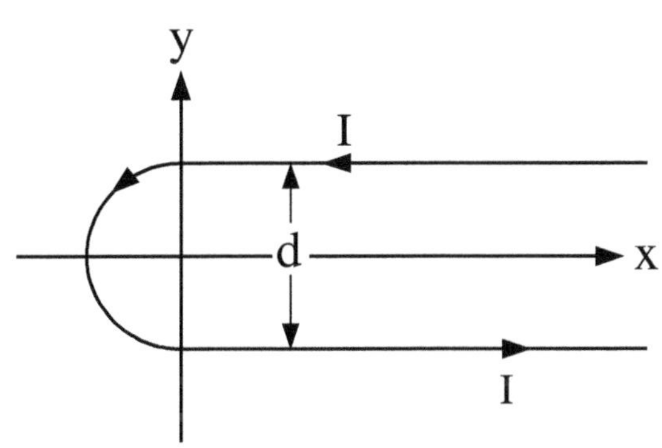

Figure 6-21. Two anti-parallel, semi-infinite current-carrying wires terminated in a semi-circular loop.

6-25. Figure 6-22 shows a planar loop carrying current I. The straight-line portions are connected by circular arcs about point P.

(a) Find the magnetic field **B** at point P.
(b) Find the magnetic moment of the loop.

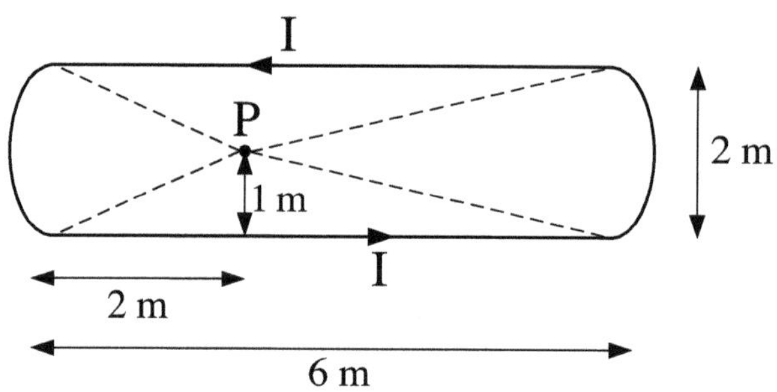

Figure 6-22. A planar current-carrying loop terminated in circular arcs.

6-26. A circular disk of radius a carries a uniform surface current density:

$$\mathbf{J}_s = \mathbf{a}_\phi J_{so} \quad (z = 0; \quad 0 < \rho < a)$$

(a) Find the magnetic field **B** on the z axis.
(b) Find the magnetic field at a distant point ($z >> a$).

6-27. A circular disk of radius a carries a non-uniform surface current density:

$$\mathbf{J}_s = \mathbf{a}_\phi J_{so} \left(1 + \frac{\rho}{a}\right) \quad (z = 0; \quad 0 < \rho < a)$$

Find the magnetic field **B** on the z axis.

6-28. A circular cylinder of radius a and length ℓ carries a uniform volume current density:

$$\mathbf{J} = \mathbf{a}_\phi J_o \quad (-\ell/2 < z < \ell/2; \quad 0 < \rho < a)$$

Find the magnetic field **B** on the z axis.

6-29. Consider the magnetic field **B** due to a semi-infinite strip of width w and current I:

$$\mathbf{J}_s = \mathbf{a}_z \,(I/w) \quad \left(y = 0; \; -\frac{w}{2} < x < \frac{w}{2}; \quad -\infty < z < 0\right)$$

Find the magnetic field **B** (a) in the perpendicular end-plane $z = 0$, and (b) in the plane $y = 0$.

6-30. Find the magnetic field at arbitrary far points $(r \gg a)$ of the square loop of current shown in Example 6-8.

6-31. Derive Eq.(6-25) (current vector of a circular loop).
[NOTE] $\mathbf{I} = \mathbf{a}_{\phi'} I$, where $\mathbf{a}_{\phi'}$ is a unit vector with respect to the source point.

6-32. A uniform surface current flows on the surface of a sphere of radius a:

$$\mathbf{J}_s = \mathbf{a}_\phi J_{so} \qquad (r = a)$$

(a) Find the magnetic field at the center of the sphere.
(b) Find the magnetic field at large distances $(r \gg a)$.
(c) Show that the surface current is approximated by N turns of equally spaced wire with current I where $J_{so} = N\, I/\pi a$.

6-33. A non-uniform surface current flows on the surface of a sphere of radius a:

$$\mathbf{J}_s = \mathbf{a}_\phi J_{so} \sin\theta \qquad (r = a)$$

(a) Find the magnetic field at the center of the sphere.
(b) Find the magnetic field at large distances. $(r \gg a)$

The above distribution is somewhat unique in that it produces a uniform field $\mathbf{B}$ everywhere inside the sphere. Note that the surface current can be approximated by thin wires sinusoidally distributed over the surface or by evenly spaced wires of appropriately varying diameter.

6-34. Consider the helix of Problem 2-40 with a uniform current I. The direction of the current is such that the z-component of current is positive. Find the z component of the magnetic field $\mathbf{B}$ along the z axis.

6-35. A rectangular loop in the xy plane with sides a and b carrying a steady current I clockwise lies in a uniform magnetic field $\mathbf{B} = \mathbf{a}_x B_x + \mathbf{a}_z B_z$. Find the net force $\mathbf{F}_m$ and torque $\mathbf{T}$ on the loop.

6-36. Consider the two anti-parallel current sheets of Table 6-1. Find the force per unit area on the sheet at $z = 0$. Note that one may use Eq. (6-33b) to determine force per unit area but the same problem which we faced in Section 3.8 arises, namely, which value of $\mathbf{B}$ do we use? The elimination of the self field leaves the average field $B/2$, in complete analogy with Section 3.8.

6-37. Consider the four current sheets of Problem 6-9. Find the force per unit area on one complete sheet.

6-38. A square loop of side a and current I lies in the xy plane normal to a uniform magnetic field $\mathbf{B} = \mathbf{a}_z B_0$. Find the forces tending to tear the loop apart about a bisector line.

6-39. A cross section of four parallel, z-directed, current-carrying wires is shown in Figure 6-23. The wires lie at the corners of a square of side a and the current directions alternate around the periphery. Find the vector force per unit length on one wire.

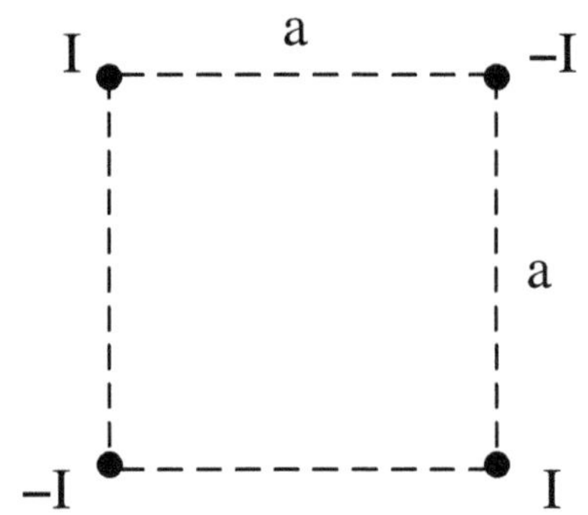

Figure 6-23. Four z-directed, current-carrying wires.

6-40. The upper wires of Figure 6-23 are connected at the ends to form a loop; similarly the lower wires are connected. Find the force and torque per unit length on one loop. One may neglect the connection.

6-41. Figure 6-24(a) shows anti-parallel, infinite-length, current-carrying strip and wire. Find the force per unit length between strip and wire.

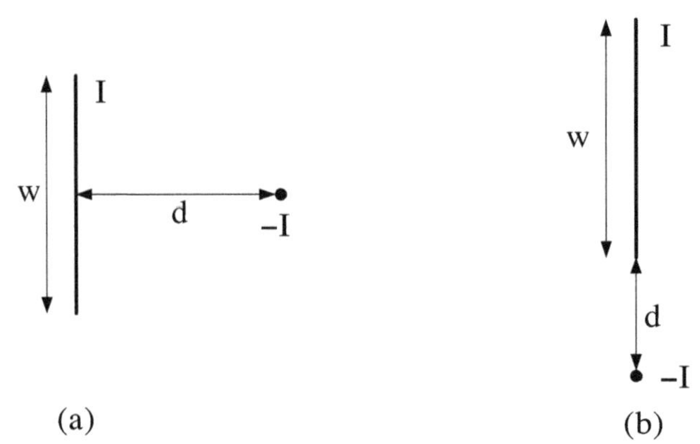

Figure 6-24. Anti-parallel strips and wires (a) wire lies in a perpendicular bisector plane (b) wire lies in plane of strip.

6-42. Figure 6-24(b) shows anti-parallel, infinite-length current-carrying strip and wire. Find the force per unit length between strip and wire.

6-43. Figure 6-25(a) shows two anti-parallel, infinite length strips of width w and vertical separation d. Find the force per unit length between strips.

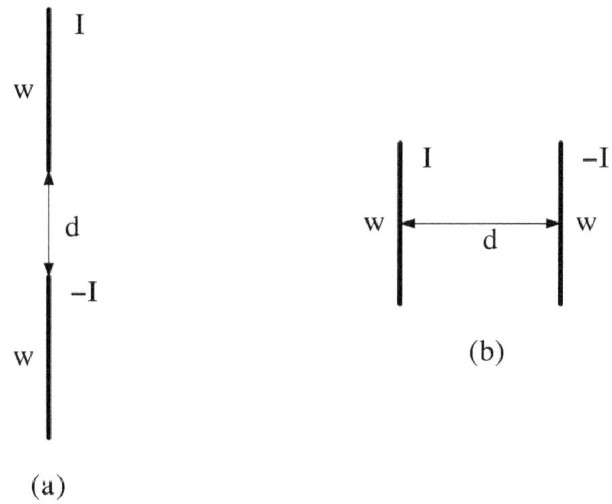

Figure 6-25. Anti-parallel current-carrying strips (a) in the same plane (b) in parallel planes.

6-44. Figure 6-25(b) shows two anti-parallel infinite-length strips of width w and horizontal separation d. Find the force per unit length between strips.

6-45. A two-wire transmission line is connected at the end by a normal straight connection (Figure 6-26). Find the force on the connector.

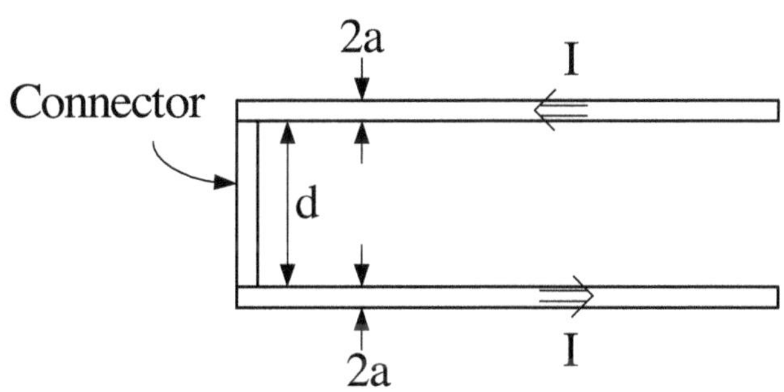

Figure 6-26. A two-wire transmission line terminated by a connector.

6-46. A coaxial line is connected at the end by a thin circular washer (Figure 6-27). Find the force on the washer.

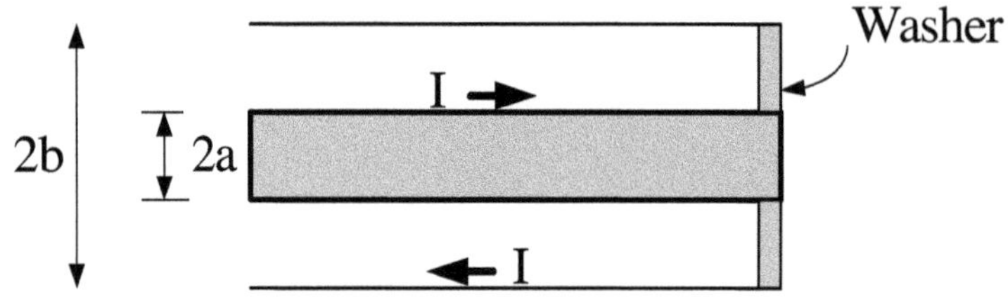

Figure 6-27. A coaxial line terminated by a washer.

6-47.* Figure 6-28 shows a z-directed wire of infinite length and current I_1, and a loop of currents I_2 lying in the xy plane. Find the force and torque (about the origin) on the loop.

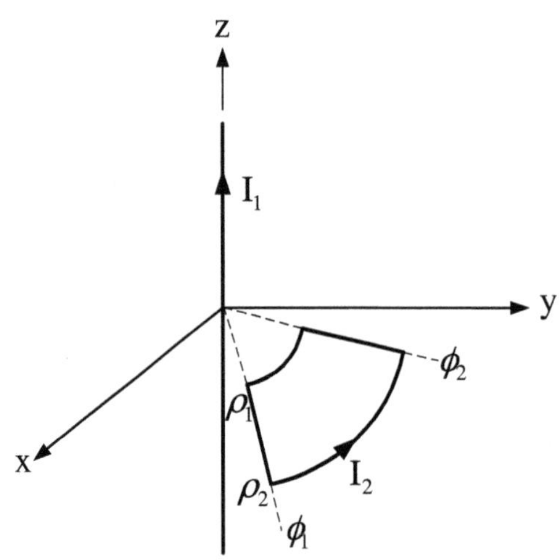

Figure 6-28. An infinite wire and a loop. The loop lies in the xy plane and is bounded by lines $\phi = \phi_1, \phi_2$ and arcs $\rho = \rho_1, \rho_2$.

MAGNETIC FIELDS IN MATTER

CHAPTER 7

7.1 INTRODUCTION

In this chapter, we study some of the characteristics of magnetic materials. We note the close relationships to dielectric materials, but the magnetic effects are much stronger and much more important in everyday life. Whenever we ring a doorbell, start a car, or use a motor or generator, we see firsthand the effects of magnetism.

Shortly after the astounding discoveries of Oersted, Ampère produced a relatively complete formulation for magnetostatic fields, as we have seen. One is amazed at the rapidity of the scientific response and the swift communication between countries. We will see further evidence of this breakthrough in scientific communication and the rapidity of scientific advancement. Within a few years Michael Faraday will discover the effects of time-varying magnetic fields and magnetic induction and will invent the first electric generator (the disc generator). Joseph Henry will precede Faraday in the first discoveries by a few months but will lay aside his work and will not publish immediately. Voilà!! Faraday's law instead of Henry's law. We should not feel unduly sorry for Henry because there will be ample credit for both.

But to return to the years immediately following Oersted's discovery, it is in the 1820s that Ampère has an amazing insight into magnetic effects in material objects. He concludes that there are tiny circulating currents within magnetic materials and that they are the source of the magnetic fields of a magnet. Recall that this is about 70 years before any knowledge of the atom was available. We call the microscopic currents that do indeed exist *Ampèrian* currents in recognition of his significant insight. Ampère also concludes that, if the currents are oriented with parallel magnetic (dipole) moments, a surface current will result. A small loop of current is a magnetic dipole, as discussed in Section 6.6.

Figure 7-1 shows an Ampèrian model of a cylindrical magnet. Here we assume that the Ampèrian currents are identical throughout the body of the material and that their magnetic moments are all z-directed. It is clear from Figure 7-1(b) that adjacent currents cancel each other and that the remaining current is a surface current. This is called a **bound surface current** denoted $\mathbf{J}_{ms}$ and is to be distinguished from *free* surface current $\mathbf{J}_s$. There are no net currents within the volume; the **bound volume current density** $\mathbf{J}_m$ is zero within (Figure 7-1(c)). Therefore, the magnetic material may be represented in terms of the equivalent surface currents $\mathbf{J}_{ms}$. We can ignore the currents within the volume because they cancel.

The model of Figure 7-1 will be useful in describing the magnetic effects that occur in the presence of material objects. In the absence of an applied field $\mathbf{B}$ some materials are unmagnetized either because of an orderly arrangement of magnetic moments which cancel or because of a random arrangement with macroscopic cancellation. In addition, some materials are permanently magnetized. In the presence of an applied field $\mathbf{B}$, most materials show an increase in magnetic moments parallel to the applied field, thus producing a net orderly arrangement of magnetic dipoles such as that of Figure 7-1.

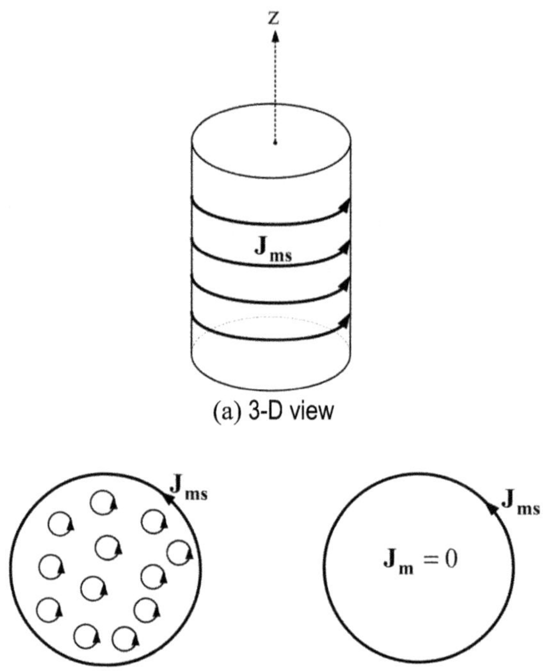

(b) Top view (c) Equivalent currents

Figure 7-1. A magnetized cylinder.

7.2 MAGNETIZATION

We define a *magnetization* $\mathbf{M}$ (x,y,z) to characterize the magnetized state of the material. To calculate $\mathbf{M}$ at a point, we construct a small volume Δv about the point and add vectorially the individual magnetic dipole moments $\mathbf{m}_i$ within the volume Δv. Then

$$\mathbf{M}(x,y,z) = \lim_{\Delta v \to 0} \frac{1}{\Delta v} \sum_i \mathbf{m}_i \tag{7-1}$$

$\mathbf{M}$ is called the **magnetization** or the *magnetic dipole moment per unit volume*. Note that $\mathbf{M}$ is very similar to the polarization $\mathbf{P}$ in dielectric materials for electrostatic problem. By definition it is automatically zero in vacuum (free space). Magnetization $\mathbf{M}$ has the units of magnetic moment per unit volume or $\left(\frac{IS}{\text{volume}}\right)$ which is equal to (amps/m).

Referring to Figure 7-1, we see that our assumption there is that the magnetization is uniform:

$$\mathbf{M} = \mathbf{a}_z M_0$$

7.3 THE MAGNETIC FIELD B OF A MAGNETIZED MATERIAL

We mentioned earlier that a magnetic material becomes magnetized in the presence of an applied field $\mathbf{B}$. An orderly arrangement of net Ampèrian currents and their net dipole moments is created such as that of Figure 7-1. For a uniform magnetization $\mathbf{M}$, there is a uniform surface current $\mathbf{J}_{ms}$. What is the relationship between $\mathbf{M}$ and $\mathbf{J}_{ms}$? If $\mathbf{M}$ is known, what is $\mathbf{J}_{ms}$? To obtain the relationship we require that the magnetized material, and the equivalent currents, have the same distant fields. This is certainly necessary and it will be sufficient to determine the result.

First we consider a uniformly magnetized piece of material. Figure 7-2(a) shows a cylinder of magnetized matter. It is uniformly magnetized with $\mathbf{M} = \mathbf{a}_z M_0$. Figure 7-2(b) shows its equivalent bound currents in free space. Currents cancel macroscopically within the volume and there is a uniform surface current $\mathbf{J}_{ms}$ on the cylindrical surface.

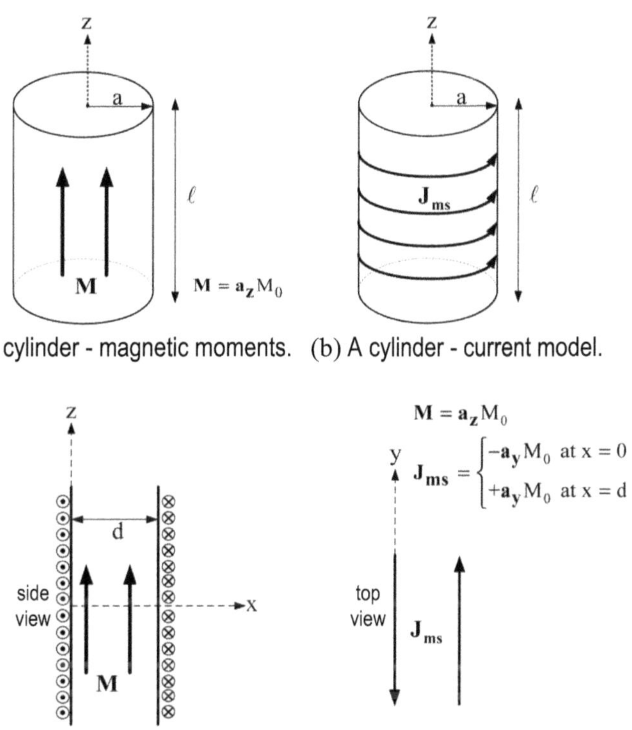

(a) A cylinder - magnetic moments. (b) A cylinder - current model.

(c) An infinite slab of thickness *d*.

Figure 7-2. Uniformly magnetized material.

Now we require that the fields produced by the magnetized material $\mathbf{M}$ and by the equivalent currents $\mathbf{J}_{ms}$ be the same at distant points. This requires that the magnetic dipole moments of $\mathbf{M}$ and $\mathbf{J}_{ms}$ are equal. The dipole moment of the magnetized cylinder is $m = M_0(\text{volume}) = M_0(\pi a^2 \ell)$ (because the magnetization M_0 is the dipole moment per unit volume) and that of the cylindrical surface current is $m = IS = (J_{ms}\ell)(\pi a^2)$. Both are z-directed. Thus

$$(J_{ms}\ell)\pi a^2 = M_0\,(\pi a^2\,\ell)$$

Therefore,

$$J_{ms} = M_0$$

In general, the result is independent of cross-sectional shape and $J_{ms} = M_0$ in all cases. By taking into account the directions of $\mathbf{J}_{ms}$ and $\mathbf{M}$, $J_{ms} = M_0$ can be rewritten, in vector form, as

$$\boxed{\mathbf{J}_{ms} = \mathbf{M} \times \mathbf{a}_n} \tag{7-2}$$

where $\mathbf{a}_n$ is the outward unit vector normal to the surface of the magnetic material. $\mathbf{J}_{ms}$ and $\mathbf{M}$ are related by the right-hand rule. As an example, it can be easily shown from Eq. (7-2) that an infinite slab of uniformly magnetized material ($\mathbf{M} = \mathbf{a}_z M_0$) has two sheets of oppositely directed uniform surface currents $\mathbf{J}_{ms} = \mathbf{a}_y M_0$ as shown in Figure 7-2(c).

Now what happens if the magnetization is not uniform but is a function of position? First we consider a one-dimensional variation $M_z(x)$ as shown in Figure 7-3(a). We can approximate the function $M_z(x)$ by step functions. The first two regions can thus be approximated by two slabs of uniformly magnetized material side by side (Figure 7-3(b)).

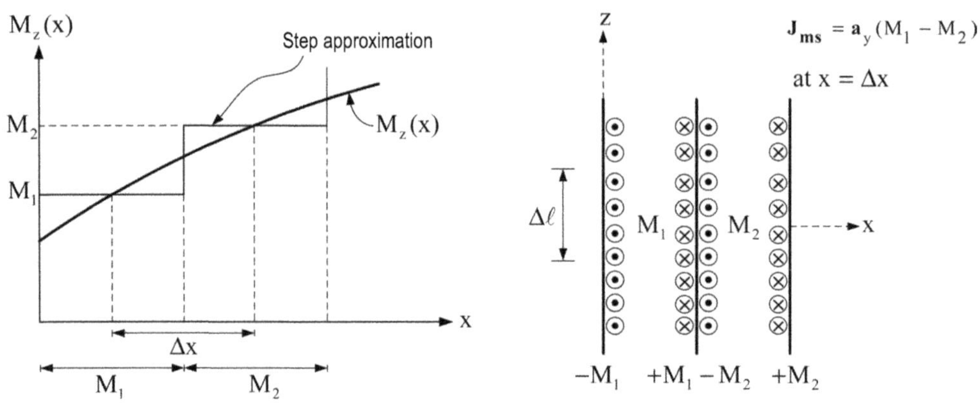

(a) Step approximation $M_z(x)$.

(b) Two–slab model.

Figure 7-3. Non-uniform magnetization $M_z(x)$.

Surface currents on each side of the slabs are proportional to magnetization as in Figure 7-2(c). Since the magnetization of the two slabs differ, there is a net surface current density at $x = \Delta x$. The magnitude of the surface current density at $x = \Delta x$ is equal to the difference $M_1 - M_2$ which is the magnitude of the step discontinuity at $x = \Delta x$. The total current $\Delta\mathbf{I}$ in a length $\Delta\ell$ is

$$\Delta\mathbf{I} = \mathbf{J}_{ms}(\Delta\ell) = \mathbf{a}_y(M_1 - M_2)(\Delta\ell) = -\mathbf{a}_y\left(\frac{\partial M_z}{\partial x}\Delta x\right)\Delta\ell$$

Now let's spread the excess surface current $\Delta\mathbf{I}$ over the region (one half subsection ($\Delta x/2$) to the right and one half subsection to the left) to form the bound volume current $\mathbf{J}_m$. This is equivalent to smaller steps and greater accuracy in the approximation.

$$\Delta\mathbf{I} = \mathbf{J}_m(\Delta x\,\Delta\ell) = -\mathbf{a}_y \frac{\partial M_z}{\partial x}(\Delta x\,\Delta\ell)$$

Thus the z-directed $\mathbf{M}$ nonuniform in x-direction leads to the y-directed bound current:

$$J_{my} = -\frac{\partial M_z}{\partial x}$$

We add an additional magnetization $M_x(z)$ which can be treated in exactly the same way. Then

$$J_{my} = \frac{\partial M_x}{\partial z} - \frac{\partial M_z}{\partial x}$$

Finally, consider all components and variations (these can be treated by coordinate rotations of Figure 7-3(b)) to obtain

$$\boxed{\mathbf{J}_m = \nabla\times\mathbf{M}} \tag{7-3}$$

Table 7-1 gives the equivalent bound current densities for magnetized material. These are *real* bound currents that exist in magnetized material and they are equivalent in the sense that we can obtain the magnetic flux density $\mathbf{B}$ by assuming those currents in free space (so that the formulations of Chapter 6 apply). The bound currents are also called the *magnetization currents*. We observe that the strength of $\mathbf{M}$ determines the bound surface current density $\mathbf{J}_{ms}$ and the spatial rate of change of $\mathbf{M}$ (i.e., $\nabla\times\mathbf{M}$) determines the bound volume current density $\mathbf{J}_m$.

Table 7-1 Equivalent Bound Current Densities

Surface Current [A/m]	Volume Current [A/m²]
$\mathbf{J}_{ms} = \mathbf{M}\times\mathbf{a}_n$	$\mathbf{J}_m = \nabla\times\mathbf{M}$

7.4 THE MAGNETIC INTENSITY H

Free and bound currents are identical in their effect upon magnetic flux density $\mathbf{B}$, as we have seen. The fundamental reason for this is that $\mathbf{B}$ is defined in terms of the Lorentz force law $\mathbf{F} = q(\mathbf{E} + \mathbf{v}\times\mathbf{B})$ and bound currents exert forces in the same way as free currents. Therefore, in the presence of magnetized materials we can replace $\mathbf{J}$ with $(\mathbf{J} + \mathbf{J}_m)$ in Ampère's law (6-6) to obtain

$$\nabla\times\mathbf{B} = \mu_0(\mathbf{J} + \mathbf{J}_m) \tag{7-4a}$$

Rewriting Eq. (7-4a) ,

$$\nabla\times\left(\frac{\mathbf{B}}{\mu_0}\right)=\mathbf{J}+\mathbf{J}_m \tag{7-4b}$$

Substituting $\mathbf{J}_m = \nabla\times\mathbf{M}$,

$$\nabla\times\left(\frac{\mathbf{B}}{\mu_0}-\mathbf{M}\right)=\mathbf{J}$$

The quantity $\mathbf{B}/\mu_0 - \mathbf{M}$ is defined as the *magnetic intensity* $\mathbf{H}$:

$$\frac{\mathbf{B}}{\mu_0}-\mathbf{M}=\mathbf{H}$$

One often calls it also the *magnetic field* $\mathbf{H}$. The magnetic field $\mathbf{B}$ can be written as

$$\boxed{\mathbf{B}=\mu_0\mathbf{H}+\mu_0\mathbf{M}} \tag{7-5}$$

Then Eq. (7-4b) becomes

$$\boxed{\nabla\times\mathbf{H}=\mathbf{J}} \tag{7-6}$$

$\mathbf{H}$ has the units of amps/m. It is a vector which curls around the free current only. Eq. (7-6) can be recognized as a new **Ampère's law** in point form. We can obtain an integral form of Eq. (7-6) by integrating both sides over surface S and applying Stokes' theorem:

$$\iint_S \nabla\times\mathbf{H}\cdot d\mathbf{s}=\oint_C \mathbf{H}\cdot d\boldsymbol{\ell}=\iint_S \mathbf{J}\cdot d\mathbf{s}=I_f$$

$$\boxed{\oint_C \mathbf{H}\cdot d\boldsymbol{\ell}=I_f} \quad \text{(Ampère's law in matter)} \tag{7-7}$$

where I_f is the free current passing through the surface S in direction $d\mathbf{s}$ which is related by a right-hand rule to C. It is thus the current enclosed by C, with a right-hand relationship between the directions of I and C. Eq. (7-7) is a new integral form of Ampère's law. In some cases it is more useful since the free current, in contrast to the bound current, can easily be measured. Ampère's law for $\mathbf{B}$ in the presence of magnetic materials can be obtained by integrating Eq. (7-4a) over the surface S and applying Stokes' theorem:

$$\oint_C \mathbf{B}\cdot d\boldsymbol{\ell}=\mu_0\,(I_f+I_m) \tag{7-4c}$$

where I_f is the free current and I_m is the bound current passing through S bounded by the closed loop C. Table 7-2 summarizes the two forms of Ampère's law for $\mathbf{H}$ and $\mathbf{B}$. In many practical problems where bound currents or the magnetization are not known, Eq. (7-7) helps greatly to find the magnetic field $\mathbf{H}$ first and then the $\mathbf{B}$ field. See Example 7-3.

Table 7-2 Ampère's Law for Magnetic Fields in Matter

Point Form	Integral Form
$\nabla \times \mathbf{H} = \mathbf{J}$	$\oint_C \mathbf{H} \cdot d\ell = I_f$
$\nabla \times \mathbf{B} = \mu_0(\mathbf{J} + \mathbf{J}_m)$	$\oint_C \mathbf{B} \cdot d\ell = \mu_0(I_f + I_m)$

Linear Magnetic Materials

For *linear* magnetic materials, the magnetization $\mathbf{M}$ is proportional to magnetic intensity $\mathbf{H}$:

$$\mathbf{M} = \chi_m \mathbf{H}$$

Then

$$\mathbf{B} = \mu_0\mathbf{H} + \mu_0\mathbf{M} = \mu_0(1+\chi_m)\mathbf{H} = \mu_0\mu_r\,\mathbf{H} = \mu\mathbf{H}$$

χ_m, μ_r, μ are called *magnetic susceptibility*, *relative permeability*, and *permeability*, respectively. They represent three alternative ways of characterizing the linear relationship. We will usually specify the permeability μ. The three parameters (χ_m, μ_r, μ) may be functions of position. Thus $\mu(x, y, z)$ represents an *inhomogeneous*, linear medium and μ = constant represents a *homogeneous*, linear medium.

For a linear medium, we have

$$\boxed{\begin{aligned}\mathbf{B} &= \mu_0(1+\chi_m)\,\mathbf{H} = \mu\mathbf{H}\\ \mathbf{M} &= \left(\frac{\mu-\mu_0}{\mu_0}\right)\mathbf{H}\end{aligned}} \quad \text{for a linear magnetic medium} \qquad (7\text{-}8)$$

Values of μ for a variety of magnetic materials are given in Table 7-4 in Section 7.10.

Linear, Homogeneous Magnetic Materials

If a material is both linear and homogeneous, then

$$\mathbf{B} = \mu\mathbf{H};\ \mu = \text{constant}$$

$$\nabla \times \mathbf{B} = \mu_0(\mathbf{J} + \mathbf{J}_m) = \nabla \times (\mu\mathbf{H}) = \mu\nabla \times \mathbf{H} = \mu\mathbf{J}$$

$$\mathbf{J}_m = \left(\frac{\mu-\mu_0}{\mu_0}\right)\mathbf{J}$$

For a non-conducting magnetic material ($\sigma = 0$, $\mathbf{J} = 0$),

$$\boxed{\mathbf{J}_m = 0} \quad \text{for a linear, homogeneous, non-conducting material.}$$

EXAMPLE 7-1. ON-AXIS FIELDS OF A MAGNETIZED CYLINDER

Consider a uniformly magnetized cylinder of length ℓ, radius a as shown in Figure 7-2(a):

$$\mathbf{M} = \mathbf{a}_z \, M_0 \quad (0 \le \rho \le a, \; -\frac{\ell}{2} \le z \le \frac{\ell}{2})$$

Find the on-axis fields **B**, **H** both inside and outside the cylinder.

SOLUTION

First we find the equivalent bound currents $\mathbf{J}_m$ and $\mathbf{J}_{ms}$:

$$\mathbf{J}_m = \nabla \times \mathbf{M} = 0 \qquad \text{(because } \mathbf{M} \text{ is uniform)}$$
$$\mathbf{J}_{ms} = \mathbf{M} \times \mathbf{a}_n$$

where $\mathbf{a}_n$ is the outward unit vector normal to the cylinder. It is equal to $\pm\,\mathbf{a}_z$ on top and bottom surfaces, respectively, and is equal to $\mathbf{a}_\rho$ on the curved surface $\rho = a$.

$$\mathbf{J}_{ms} = (\mathbf{a}_z \, M_0) \times (\pm\, \mathbf{a}_z) = 0 \qquad \text{on top and bottom surfaces}$$
$$\mathbf{J}_{ms} = (\mathbf{a}_z \, M_0) \times \mathbf{a}_\rho = \mathbf{a}_\phi \, M_0 \qquad \text{on the curved surface } \rho = a.$$

Now the equivalent currents $\mathbf{J}_{ms}$ may be placed in free space (ε_0, μ_0) and the field **B** can then be determined by the Biot-Savart law. To determine the magnetic intensity **H**, we must invoke the general relationship (Eq. (7-5)) between **B**, **M**, **H**.

The equivalent currents $\mathbf{J}_{ms}$ are identical in form to those of the solenoid of Example 6-6 and Figure 6-8; NI/ℓ is replaced everywhere by M_0.

The on-axis field B_z is thus obtained by replacing NI/ℓ in Eq. (6-22a) by M_0:

$$B_z = \frac{\mu_0 M_0}{2}\left[\frac{z+\ell/2}{\sqrt{a^2+(z+\ell/2)^2}} - \frac{z-\ell/2}{\sqrt{a^2+(z-\ell/2)^2}}\right] \tag{7-9}$$

Eq. (7-9) is valid on the z axis both inside and outside the cylinder. To obtain **H** along the z axis, we use Eq. (7-5):

$$\mathbf{B} = \mu_0 \, \mathbf{H} + \mu_0 \mathbf{M}$$

$$\mathbf{H} = \frac{\mathbf{B}}{\mu_0} - \mathbf{M}$$

which gives

$$H_z = \frac{M_0}{2}\left[\frac{z+\ell/2}{\sqrt{a^2+(z+\ell/2)^2}} - \frac{z-\ell/2}{\sqrt{a^2+(z-\ell/2)^2}}\right] - M_0 \left(|z| < \frac{\ell}{2}\right) \quad (7\text{-}10a)$$

$$= \frac{M_0}{2}\left[\frac{z+\ell/2}{\sqrt{a^2+(z+\ell/2)^2}} - \frac{z-\ell/2}{\sqrt{a^2+(z-\ell/2)^2}}\right] \left(|z| > \frac{\ell}{2}\right) \quad (7\text{-}10b)$$

EXAMPLE 7-2. FIELDS AT THE CENTER OF A MAGNETIZED SPHERE

Figure 7-4 shows a magnet sphere of radius a which is uniformly magnetized:

$$\mathbf{M} = \mathbf{a}_z M_0 \quad (r \leq a)$$

Find the fields **B**, **H** at the center of the sphere.

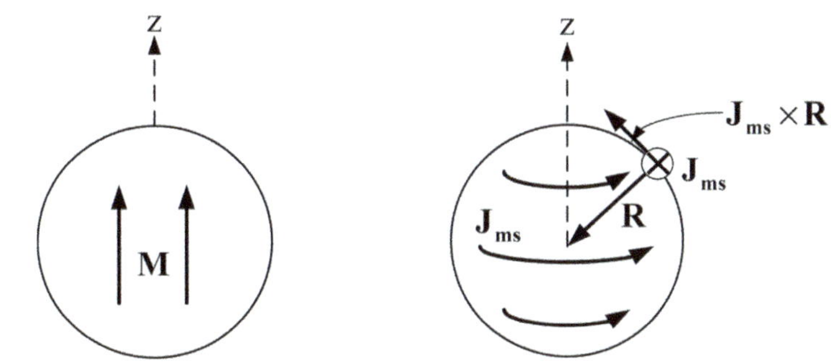

(a) Magnetic dipole model (b) Surface current model

Figure 7-4. A uniformly magnetized sphere.

SOLUTION

First, the equivalent bound currents are determined as follows:

$$\mathbf{J}_m = \nabla \times \mathbf{M} = 0 \qquad \text{(because } \mathbf{M} \text{ is uniform)}$$

$$\mathbf{J}_{ms} = \mathbf{M} \times \mathbf{a}_n = (\mathbf{a}_z M_0) \times \mathbf{a}_r = M_0 (\mathbf{a}_r \cos\theta - \mathbf{a}_\theta \sin\theta) \times \mathbf{a}_r$$

$$= \mathbf{a}_\phi M_0 \sin\theta \qquad (\text{at } r = a)$$

Next we apply the Biot-Savart law:

$$\mathbf{B} = \frac{\mu_0}{4\pi} \iint_s \frac{\mathbf{J}_{ms} \times \mathbf{R}\, ds'}{R^3}$$

The typical source point is a point (r′, θ′, ϕ′) = (a, θ′, ϕ′) on the surface of the sphere. The field point is at the center of the sphere.

Then

$$\mathbf{R} = -\mathbf{a}_r a; \quad R = a \quad (\text{since } \mathbf{r} = 0, \mathbf{r}' = \mathbf{a}_r a)$$

$$\mathbf{J}_{ms} \times \mathbf{R} = (\mathbf{a}_\phi M_0 \sin\theta') \times (-\mathbf{a}_r a)$$
$$= \mathbf{a}_\theta (-M_0 a \sin\theta') = (\mathbf{a}_p \cos\theta' - \mathbf{a}_z \sin\theta')(-M_0 a \sin\theta')$$
$$(\mathbf{J}_{ms} \times \mathbf{R})_z = M_0 a \sin^2\theta'$$

By symmetry, the ρ component of **B** is cancelled out, giving rise to the z-directed **B** field:

$$\mathbf{B} = \frac{\mu_0}{4\pi} \iint \frac{\mathbf{J}_{ms} \times \mathbf{R}\, ds'}{R^3}$$
$$= \mathbf{a}_z \frac{\mu_0}{4\pi} \int_0^{2\pi}\int_0^{\pi} \frac{M_0 a \sin^2\theta' a^2 \sin\theta' d\theta' d\phi'}{a^3}$$

$$B_z = \frac{\mu_0}{2} M_0 \int_0^{\pi} \sin^3\theta \, d\theta = \frac{\mu_0 M_0}{2}\left(\frac{4}{3}\right) = \frac{2\mu_0 M_0}{3} \tag{7-11}$$

and

$$H_z = \frac{B_z}{\mu_0} - M_z = -\frac{M_0}{3} \tag{7-12}$$

Note that B_z and H_z are oppositely directed, a *demagnetizing* effect. It is interesting to note that the magnetic field is given by Eq. (7-11) *at all points* inside the sphere. The fields at arbitrary points both inside and outside the sphere for this problem can be found using the *magnetic scalar potential* approach* and using the solutions of Laplace's equation in Table 4-1.

EXAMPLE 7-3. CURRENT-CARRYING FILAMENTS, WIRES, CYLINDERS AND TOROIDS WITH PERMEABLE MATERIALS

The problems treated by Ampère's law in Examples 6-2 and 6-4 can readily be modified in the presence of permeable magnetic materials. For instance, consider Example 6-2(a) with a filament of infinite length and current I in an unbounded medium of permeability μ. We apply Ampère's law for **H** to the geometry of Figure 6-1, which is modified by replacing free space (μ_0) with material of permeability μ:

$$\oint_C \mathbf{H} \cdot d\ell = \int_0^{2\pi} H_\phi \rho d\phi = H_\phi 2\pi\rho = I_f = I$$

Then

$$H_\phi = \frac{I}{2\pi\rho} \tag{7-13a}$$

* See, for example, D. K. Cheng, *Field and Wave Electromagnetics*, Addison-Wesley, 1989, pp. 242–249.

and

$$B_\phi = \mu H_\phi = \frac{\mu I}{2\pi\rho} \quad (7\text{-}13b)$$

Note that in the presence of highly permeable ($\mu >> \mu_0$) material, the flux density **B** is greatly increased as compared to that of free space.

All of the problems of Examples 6-2 and 6-4 can be treated in this manner. Let the field regions (the regions in which fields are non-zero) be completely filled with material of permeability μ. Equations (6-8), (6-10b), (6-11b), (6-13a), (6-14a), and (6-14b) are modified by replacing μ_0 with μ.

For the wire of finite radius (Example 6-2(b)), let the wire of radius a be of permeability μ_1, and the region outside ($\rho > a$) be of permeability μ_2. Then μ_0 is replaced with μ_1 in Eq. (6-9a) and μ_0 is replaced with μ_2 in Eq. (6-9b).

Polarization Vs. Magnetization

When we review Sections 3.2–3.4 for dielectric polarization and Sections 7.2–7.4 for magnetization, we find a striking parallel between the two. In Table 7-3 we summarize important concepts for electric and magnetic fields in matter by comparison.

Table 7-3. Polarization vs. Magnetization

Electric Fields in Matter	Magnetic Fields in Matter
Electric Dipole: A pair of two opposite charges separated by a small distance (d) Dipole moment $\mathbf{p} = \mathbf{a}_z qd$	Magnetic Dipole: A small loop of current Dipole moment $\mathbf{m} = \mathbf{a}_z I$ (area)
Polarization vector $\mathbf{P} = \lim_{\Delta v \to 0} \frac{1}{\Delta v} \sum_i \mathbf{p}_i$ electric dipole moment per unit volume	Magnetization vector $\mathbf{M} = \lim_{\Delta v \to 0} \frac{1}{\Delta v} \sum_i \mathbf{m}_i$ magnetic dipole moment per unit volume
Bound Charges Surface charge density $\rho_{ps} = \mathbf{P} \cdot \mathbf{a}_n$ [C/m^2] $\mathbf{a}_n$ = unit vector normal outward Volume charge density $\rho_{pv} = -\nabla \cdot \mathbf{P}$ [C/m^3]	Bound Currents Surface current density $\mathbf{J}_{ms} = \mathbf{M} \times \mathbf{a}_n$ [A/m] Volume current density $\mathbf{J}_m = \nabla \times \mathbf{M}$ [A/m^2]
Electric Displacement $\mathbf{D} = \varepsilon_0 \mathbf{E} + \mathbf{P}$	Magnetic Intensity $\mathbf{H} = \frac{\mathbf{B}}{\mu_0} - \mathbf{M}$ or $\mathbf{B} = \mu_0 \mathbf{H} + \mu_0 \mathbf{M}$

(continued)

Gauss' Law in Matter	Ampère's Law in Matter
$\nabla \cdot \mathbf{D} = \rho_v$ $\oiint_S \mathbf{D} \cdot d\mathbf{s} = Q_f$ (free charge enclosed)	$\nabla \times \mathbf{H} = \mathbf{J}$ $\oint_C \mathbf{H} \cdot d\boldsymbol{\ell} = I_f$ (free current)
Linear Dielectric $\mathbf{D} = \varepsilon \mathbf{E}$ ε = permittivity	Linear Magnetic Material $\mathbf{B} = \mu \mathbf{H}$ μ = permeability

7.5 BOUNDARY CONDITIONS

Boundary conditions are fundamental to magnetostatics, as to electrostatics. We need to understand the behavior of the fields **B** and **H** at an interface, i.e., a boundary, between two different media.

First we consider the normal component of **B** using the divergence property that $\nabla \cdot \mathbf{B} = 0$. For this component we could carry out a pillbox analysis identical to that of Section 3-5 for normal **D** (see Figure 3-10). However, let's use the mathematical analogy between electrostatics and magnetostatics and merely substitute **B** for **D**. Consider the divergence law and boundary condition concerning **D**:

$$\nabla \cdot \mathbf{D} = \rho_v, \quad \oiint_S \mathbf{D} \cdot d\mathbf{s} = Q_f \quad \text{and} \quad D_{1n} - D_{2n} = \rho_s$$

Let $\rho_v = 0$ in regions 1, 2:

$$\nabla \cdot \mathbf{D} = 0, \quad \oiint_S \mathbf{D} \cdot d\mathbf{s} = 0 \quad \text{and} \quad D_{1n} = D_{2n}$$

Substituting **B** for **D**:

$$\nabla \cdot \mathbf{B} = 0, \quad \oiint_S \mathbf{B} \cdot d\mathbf{s} = 0 \quad \text{and} \quad B_{1n} = B_{2n}$$

Hence, *normal **B** is continuous.*

$$\boxed{B_{1n} = B_{2n}} \tag{7-14}$$

Next we consider the tangential component of **H** using the curl law $\nabla \times \mathbf{H} = \mathbf{J}$. Figure 7-5 shows an interface between two arbitrary media.

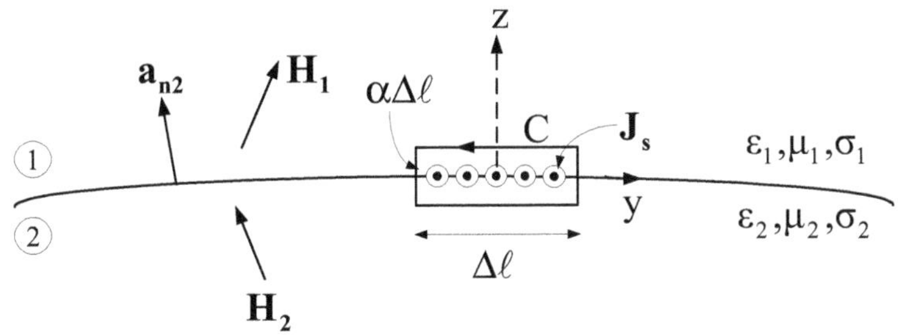

Figure 7-5. Tangential H at a point on a boundary.

We are interested in the behavior of $\mathbf{H}$ at a point on the boundary. At that point we construct a coordinate system (x, y, z) so that the normal is z-directed. The x, y directions lie in the tangent plane. Contour C lies in a plane normal to the interface. $\Delta\ell$ is made small so that $\mathbf{H}_1$, $\mathbf{H}_2$ do not vary over C. α is made arbitrarily small compared to unity so that the end sections of C do not contribute to $\oint_c \mathbf{H} \cdot d\boldsymbol{\ell}$.

Then, orienting length $\Delta\ell$ in the y direction:

$$\nabla \times \mathbf{H} = \mathbf{J}; \ \oint_c \mathbf{H} \cdot d\ell = (H_{2y} - H_{1y})\,\Delta\ell = I_f = J_{sx}\,\Delta\ell$$

if there exists an x-directed free surface current (J_{sx}). Thus,

$$H_{1y} - H_{2y} = -J_{sx}$$

If we orient length $\Delta\ell$ in the x direction, we have

$$H_{1x} - H_{2x} = J_{sy}$$

Combining the above results:

$$\boxed{\mathbf{a}_{n2} \times (\mathbf{H}_1 - \mathbf{H}_2) = \mathbf{J}_s} \qquad (7\text{-}15)$$

where $\mathbf{a}_{n2}$ is a unit vector normal outward from region 2, i.e, it points into region 1. Equation (7-15) indicates that a free surface current on the boundary, in the tangent plane, causes a jump in the tangential component of $\mathbf{H}$, also lying in the tangent plane, orthogonal to $\mathbf{J}_s$.

If there is no $\mathbf{J}_s$ on the surface, then tangential $\mathbf{H}$ is continuous, i.e,

$$\boxed{\mathbf{H}_{1t} = \mathbf{H}_{2t}} \qquad \text{if } \mathbf{J}_s = 0 \qquad (7\text{-}16)$$

Returning to the general boundary condition (Eq. (7-15)) for tangential $\mathbf{H}$, we note that $\mathbf{H}$ determines $\mathbf{J}_s$ but not vice versa. $\mathbf{J}_s$ can be determined by a specification of $\mathbf{H}_1$, $\mathbf{H}_2$.

We might at this point raise the question, "when can (free) surface current $\mathbf{J}_s$ exist at a boundary?" The answer is quite simple. In some cases, we wish to place an ideal source $\mathbf{J}_s$ at the interface to represent, for instance, thin current-carrying wires wound around a magnetic material. In no other cases will $\mathbf{J}_s$ arise except at the surface of a perfect conductor ($\sigma = \infty$). If conductivity σ is finite, volume currents may arise but no surface currents.

Linear Magnetic Materials

For linear magnetic materials, the following relationships are obtained

$$\mathbf{B} = \mu\mathbf{H}; \quad \mathbf{M} = \frac{\mu - \mu_0}{\mu_0}\mathbf{H}$$

$$B_{1n} = B_{2n} \tag{7-17a}$$

$$H_{1n} = \frac{\mu_2}{\mu_1} H_{2n} \tag{7-17b}$$

And if, in addition, $\mathbf{J}_s = 0$, then

$$H_{1t} = H_{2t} \tag{7-18a}$$

$$B_{1t} = B_{2t}\left(\frac{\mu_1}{\mu_2}\right) \tag{7-18b}$$

Behavior of B, H Lines at an Interface

Figure 7-6 shows a vector $\mathbf{B}$ at an interface between two linear magentic media ($\mathbf{B}_1 = \mu_1 \mathbf{H}_1$, $\mathbf{B}_2 = \mu_2 \mathbf{H}_2$).

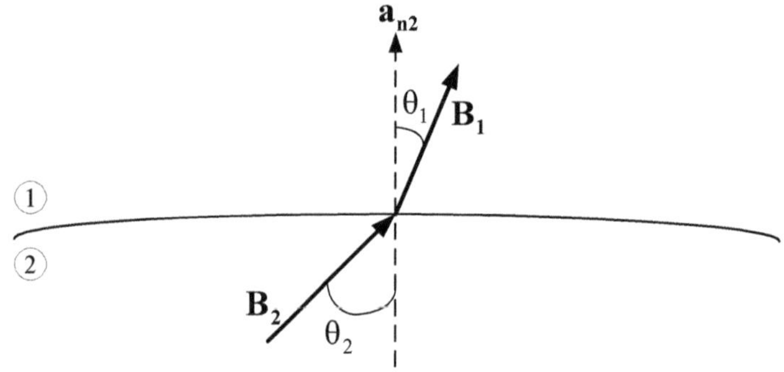

Figure 7-6. B field lines at an interface.

We are interested in the relationship of angles θ_1, θ_2. In other words, how does the direction of the vector change in passing from one medium to another? Let B_1, B_{1n}, B_{1t} represent the magnitudes of total $\mathbf{B}$, normal $\mathbf{B}$, and tangential $\mathbf{B}$, respectively, in region 1 at the interface. Then

$$B_{1n} = B_1 \cos\theta_1$$

$$B_{1t} = B_1 \sin\theta_1$$

and $\tan\theta_1 = \frac{B_{1t}}{B_{1n}}$; similarly, $\tan\theta_2 = \frac{B_{2t}}{B_{2n}}$. Using $H_{1t} = H_{2t}$ and $B_{1n} = B_{2n}$, we obtain

$$\frac{\tan\theta_1}{\tan\theta_2} = \frac{\mu_1}{\mu_2} \tag{7-19}$$

Note that Eq. (7-19) is also valid for the **H** vector since **H** is parallel to **B**. The directions of **B** and **H** are thus identical but the magnitude changes across the interface differ since

$$\frac{B_2}{B_1} = \frac{\mu_2}{\mu_1}\frac{H_2}{H_1}$$

In some cases the ratio μ_2/μ_1 may be very large, for instance, $\mu_r \approx 5000$ for medium silicon steel and $\mu_r \approx 1$ for air. In this case the field lines for **B, H** will be, for the most part, nearly normal to the interface in air and nearly parallel to the interface in the steel. The **B, H** lines passing from steel to air are bent very sharply towards the normal. This fact explains how the magnetic core material of a solenoid or a toroid can guide the magnetic field in the core region even when the winding does not have many turns. Note that Eq. (7-19) is also valid for vectors **D**, **E** at the interface between two isotropic linear dielectric media if we replace μ_1/μ_2 with $\varepsilon_1/\varepsilon_2$.

7.6 INDUCTANCE

In this section, we study the inductance of the current circuit. While a capacitor is a circuit element storing electric energy, an inductor is a circuit element storing magnetic energy. Consider a current-carrying loop with current I (Figure 7-7a).

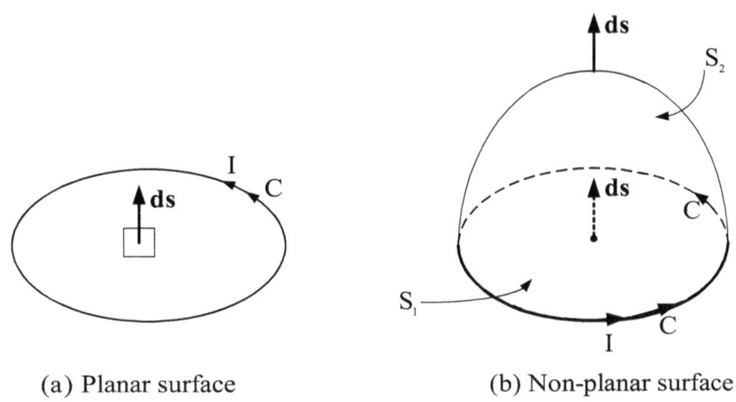

Figure 7-7. Current-carrying loops.

Path C is chosen in the direction of I. The magnetic flux density **B** may be obtained by the Biot-Savart law (Eq. 6-19c)):

$$\mathbf{B} = \frac{\mu_0 I}{4\pi}\oint_C \frac{d\ell \times \mathbf{R}}{R^3}$$

The magnetic flux Φ through the loop (through any surface bounded by the loop*) is defined as

$$\Phi = \iint_S \mathbf{B}\cdot d\mathbf{s}$$

* $\nabla \cdot \mathbf{B} = 0$, therefore $\oiint_S \mathbf{B}\cdot d\mathbf{s} = 0$, and the flux through C is independent of the open surface bounded by C.

where S is bounded by C and ds is related to C by a right-hand rule relationship. Then the **inductance** L (or **self-inductance**) is defined by

$$\boxed{L = \frac{\Phi}{I}} \tag{7-20}$$

Note that we are free to choose any surface S bounded by C. For instance, with a circular loop (Figure 7-7b) we may choose the planar surface S_1, or the hat-shaped surface S_2. The unit of inductance is the henry [H] (named after Joseph Henry). 1 H (henry) is equal to 1 (T · m²/A) or (Webers/A).

There are two steps in the process of finding L. Assuming the current I, we first find **B**. Then we integrate to find the flux Φ and thus Φ/I. The current I will cancel in the process. L is independent of excitation and depends only on geometry and the surrounding medium. L is always positive. It increases as the wire cross section is reduced. For a coil of N turns, the flux Φ linking all turns should be counted:

$$\boxed{\Phi = N\Phi_s} \tag{7-21}$$

where Φ_s is the flux through a surface bounded by a single turn. The above result (Eq. (7-21)) assumes that the same flux links all turns. When would this be a reasonable approximation?

Network Model for Inductance

Figure 7-8 shows the network model for an inductance L.

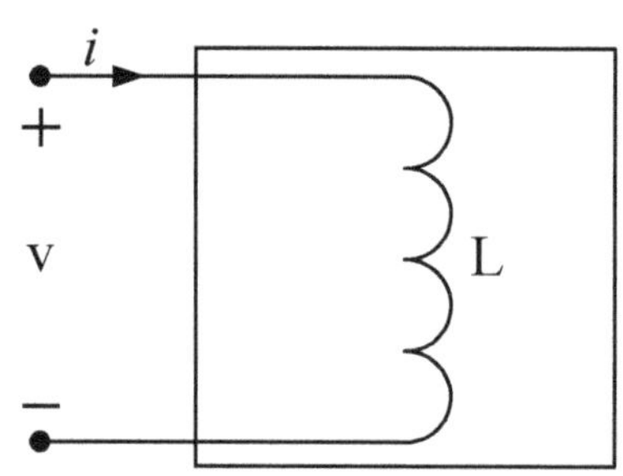

Figure 7-8. Inductance (network model).

The network relationship between v and i across the inductor is:

$$v = \frac{d\Phi}{dt} = \frac{d}{dt}(Li) = L\frac{di}{dt} + i\frac{dL}{dt} \Rightarrow L\frac{di}{dt} \quad \left(\text{if } \frac{dL}{dt} = 0\right)$$

$$\boxed{v = L\frac{di}{dt}} \tag{7-22}$$

Note that $v = \frac{d\Phi}{dt}$ comes from Faraday's law that will be discussed later in Chapter 8.

Mutual Inductance

Consider a set of N current loops with currents $I_1, \ldots, I_N$ as shown in Figure 7-9.

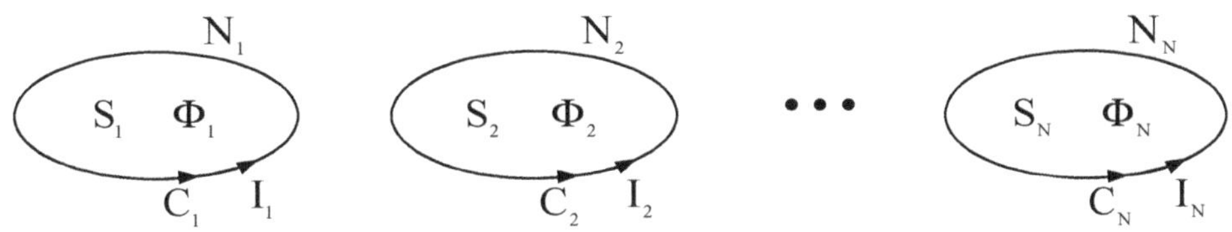

Figure 7-9. Mutual inductance.

Consider the total flux Φ_j through loop j:

$$\Phi_j = \Phi_{j1} + \Phi_{j2} + \ldots + \Phi_{jN} = \sum_{k=1}^{N} \Phi_{jk}$$

where Φ_{jk} is the flux through loop j due to current I_k (all other currents set equal to zero):

$$\Phi_{jk} = \iint_{S_j} \mathbf{B}_k \cdot d\mathbf{s}$$

Then the **mutual inductance** L_{jk} is defined as follows:

$$\boxed{L_{jk} = \frac{\Phi_{jk}}{I_k}} \tag{7-23}$$

If N_j is the number of turns in loop j, then L_{jk} is proportional to $(N_j N_k)$. The inductance L_{kk} is a self inductance, of course. It can be shown that

$$L_{jk} = L_{kj} \tag{7-24}$$

for wires in free space and in general unless a non-reciprocal material such as ferrite is present. Since L_{ij} is usually equal to L_{ji} we can choose to calculate either Φ_{ij} or Φ_{ji}. The calculations may differ considerably.

Note also that $L_{ii} \geq L_{ji}$ since all of the flux generated by loop i passes through loop i but only a portion passes through (or links) loop j. Mutual inductance L_{ij} may be positive or negative. Switching the direction of current I_j will switch the sign of L_{ij}. Similarly, reversing one set of terminals in the network model switches the sign.

Network Model for Coupled Circuits

Figure 7-10 shows the network (two-port) model for a pair of coupled coils.

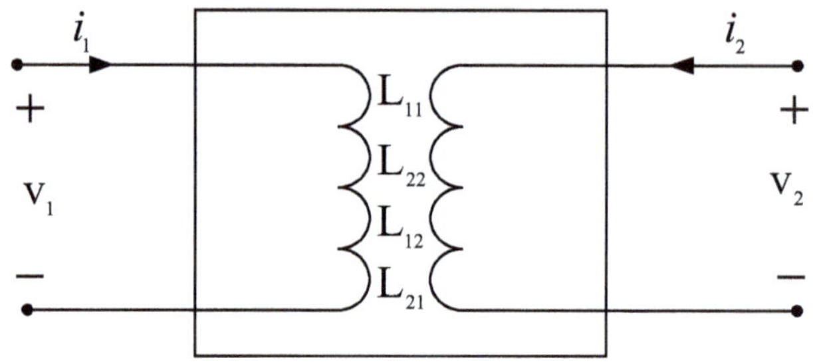

Figure 7-10. Coupled coils (network model).

The network v-i relationships are:

$$v_1 = L_{11}\frac{di_1}{dt} + L_{12}\frac{di_2}{dt} \tag{7-25a}$$

$$v_2 = L_{21}\frac{di_1}{dt} + L_{22}\frac{di_2}{dt} \tag{7-25b}$$

(if L_{11}, L_{12}, L_{21}, L_{22} are all constant in time). Then, for a set of N coupled coils,

$$v_j(t) = \sum_{k=1}^{N} L_{jk}\frac{di_k}{dt} \tag{7-25c}$$

Just as capacitance was increased by the addition of dielectrics, inductance is greatly increased by the addition of magnetic materials.

EXAMPLE 7-4. THE INDUCTANCE OF A TOROID

Consider the toroid of Figure 6-6(a), of rectangular cross section, with N turns and current I. The toroid is completely filled with material of permeability μ. Find the inductance of the toroid.

SOLUTION

As discussed in Example 7-3, the flux density in the toroid with magnetic core is obtained from Eq. (6-13a) by replacing μ_0 by μ:

$$B_\phi = \frac{\mu I N}{2\pi\rho} \quad (a < \rho < b, \quad 0 < z < d) \tag{7-26}$$

Next we find Φ_s, the flux through the surface bounded by a single turn, i.e, the cross-section surface of the toroid:

$$\Phi_s = \int_0^d \int_a^b \frac{\mu N I \, d\rho \, dz}{2\pi\rho} = \frac{I\mu N d}{2\pi} \ln \frac{b}{a}$$

Then the total flux is

$$\Phi = N\Phi_s = \frac{I\mu N^2 d \ \ln(b/a)}{2\pi}$$

Therefore,

$$L = \frac{\Phi}{I} = \frac{\mu N^2 d \ \ln(b/a)}{2\pi} \tag{7-27}$$

Note that the inductance is *proportional* to the permeability (μ) and the *square of turns (N^2)*. Addition of highly permeable materials greatly increases the inductance.

EXAMPLE 7-5 THE INDUCTANCE PER UNIT LENGTH OF AN INFINITE SOLENOID

The inductance per unit length of an infinite solenoid may be obtained readily. Consider the field **B** of an infinite solenoid of arbitrary cross-section as given in Eq. (6-14b):

$$B_z = \mu_0 N_\ell I$$

Now find the flux Φ linked by a unit length of the solenoid with N_ℓ turns:

$$\Phi = N_\ell (\mu_0 N_\ell I) A,$$

where A is the cross-section area of the solenoid. Then the inductance per unit length (L/ℓ) is given by

$$\frac{L}{\ell} = \frac{\Phi}{I} = \mu_0 N_\ell^2 A \tag{7-28}$$

The above result is valid for arbitrary cross-section.

EXAMPLE 7-6. THE MUTUAL INDUCTANCE OF A FILAMENT AND A SQUARE LOOP.

Consider the filament and square loop of Figure 6-15(a). Find the mutual inductance.

SOLUTION

It is simpler to find the flux Φ_{21} through the square loop due to the filament of current I_1 than finding Φ_{12} through the filament due to the loop current:

$$\Phi_{21} = \iint_{S_2} \mathbf{B}_1 \cdot d\mathbf{s} = \int_0^{\ell}\int_d^{d+\ell} \frac{\mu_0 I_1 \, d\rho \, dz}{2\pi\rho} = \frac{\mu_0 I_1 \ell}{2\pi} \ln\left(\frac{d+\ell}{d}\right)$$

where we have used Eq. (6-8) for $\mathbf{B}_1$. Then

$$L_{21} = L_{12} = \frac{\Phi_{21}}{I_1} = \frac{\mu_0 \ell \ln\left(\frac{d+\ell}{d}\right)}{2\pi} \quad (7\text{-}29)$$

EXAMPLE 7-7. THE MUTUAL INDUCTANCE OF A FILAMENT AND A TOROID.

Figure 7-11 shows a filament of infinite length and current I_1 along the axis of a toroid of rectangular cross-section, with current I_2 and N_2 turns. Find the mutual inductance.

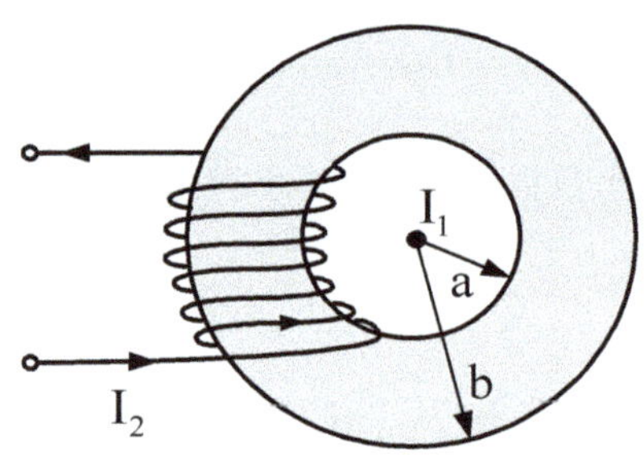

Figure 7-11. A filament and a toroid.

SOLUTION

Once again it is simpler to find the flux through the toroid due to the filament. The $\mathbf{B}$ field is given by Eq. (6-8). The total flux Φ_{21} is N_2 times the flux Φ_{21s} through the surface bounded by a single turn of the toroid, i.e., the cross-section of the toroid (air core):

$$\Phi_{21s} = \int_0^{d}\int_a^{b} \frac{\mu_0 I_1 \, d\rho \, dz}{2\pi\rho} = \frac{\mu_0 I_1 d}{2\pi} \ln(b/a)$$

$$\Phi_{21} = N_2 \Phi_{21s} = \frac{\mu_0 I_1 d N_2 \ \ln(b/a)}{2\pi}$$

$$L_{21} = L_{12} = \frac{\Phi_{21}}{I_1} = \frac{\mu_0 d \ N_2 \ \ln(b/a)}{2\pi} \tag{7-30}$$

EXAMPLE 7-8. THE MUTUAL INDUCTANCE OF TWO LOOPS.

Figure 7-12(a) shows two loops, each centered on the z axis, with radii a, b and currents I_1, I_2, respectively. The loop axis of each is the z axis. The loops are separated by a distance ℓ where $\ell >> a$, $b >> a$. Find the mutual inductance.

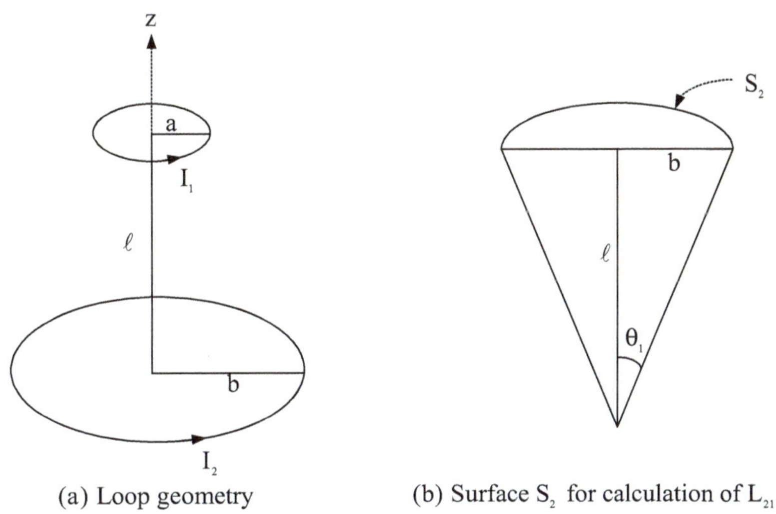

Figure 7-12. Two coaxial loops.

SOLUTION

Method 1—Using the Field $\mathbf{B}_2$ due to I_2:

Since $\ell >> a$ and $b >> a$, we may assume that the field at loop 1 due to current I_2 is approximately equal to the on-axis field of loop 2 (see Example 6-5). The on-axis field B_{2z} of loop 2 at $z = \ell$ is

$$B_{2z} = \frac{\mu_0 I_2 b^2}{2[b^2 + \ell^2]^{3/2}}$$

$$\Phi_{12} = \iint_{S_1} \mathbf{B}_2 \cdot d\mathbf{s} = \frac{\mu_0 I_2 b^2}{2[b^2 + \ell^2]^{3/2}} \ (\pi a^2)$$

$$L_{12} = \frac{\Phi_{12}}{I_2} = \frac{\mu_0 \pi a^2 b^2}{2[b^2 + \ell^2]^{3/2}} \tag{7-31}$$

Method 2—Using The Field $\mathbf{B}_1$ *due to* I_1:

Since $\ell >> a$, we may use the magnetic dipole field developed in Section 6.6 for the field at loop 2 due to loop 1. To find the flux $\Phi_{21,}$ we need to integrate over a surface bounded by loop 2. It is more convenient to use a *spherical* surface of radius $\sqrt{\ell^2+b^2}$ as shown in Figure 7-12(b) rather than a planar surface because of the form of the magnetic dipole fields. The radial component B_{1r} of loop 1 at $r=\sqrt{\ell^2+b^2}$ is, from Eq. (6-30),

$$B_{1r} = \frac{\mu_0 I_1 (\pi a^2)}{4\pi[\ell^2+b^2]^{3/2}}(2\cos\theta)$$

$$\begin{aligned}
\Phi_{21} &= \iint_{S_2} \mathbf{B}_1 \cdot d\mathbf{s} = \iint_{S_2} B_{1r}\, r^2 \sin\theta\, d\theta\, d\phi \\
&= \frac{\mu_0 I_1 (\pi a^2)}{4\pi[\ell^2+b^2]^{3/2}} \int_0^{2\pi}\int_0^{\theta_1} 2\cos\theta\,(\ell^2+b^2)\ \sin\theta\ d\theta\ d\phi \\
&= \frac{\mu_0 I_1 (\pi a^2)}{[\ell^2+b^2]^{3/2}} \int_0^{\theta_1} (\ell^2+b^2)\cos\theta \sin\theta\ d\theta \\
&= \frac{\mu_0 I_1 (\pi a^2)}{[\ell^2+b^2]^{3/2}}(\ell^2+b^2)\left[\frac{\sin^2\theta}{2}\right]_0^{\theta_1} \\
&= \frac{\mu_0 I_1 (\pi a^2)}{[\ell^2+b^2]^{3/2}}\frac{b^2}{2}\left(\text{since } \sin^2\theta_1 = \frac{b^2}{\ell^2+b^2}\right)
\end{aligned}$$

and

$$L_{21} = \frac{\Phi_{21}}{I_1} = \frac{\mu_0 \pi a^2 b^2}{2[\ell^2+b^2]^{3/2}},$$

which agrees with Eq. (7-31). This confirms that Eq. (7-24), the reciprocal property, is satisfied.

7.7 JOSEPH HENRY (1797–1878) AND THE DISCOVERY OF MAGNETIC INDUCTION

In the summer of 1831, Joseph Henry, an instructor at the Albany Academy in Albany, New York, knew that he was on the verge of an important discovery and ready to make a name for himself in the new science of electromagnetism. Henry was already beginning to become known throughout the United States for his practical work with powerful new magnets. He had started by insulating wires from each other by carefully wrapping them in silk (from his wife's petticoat). Thus he was able to use a very large number of turns. He was able to produce electromagnets of great lifting power by (a) insulating wires as described (b) using powerful batteries in series to produce high voltage and (c) using many coils in parallel. These were simple techniques, but clearly the young Henry knew what he was doing. Professor Silliman of Yale requested that Henry construct an electromagnet for his laboratory. Built in 1831, this magnet was able to lift a ton and was the most powerful electromagnet of its day.

Henry, however, had his sights set on bigger fish. For several years he had been experimenting on magnetic induction (which is explained in Chapter 8). His powerful magnets were very useful because they created large magnetic fields which magnified the small effects of induction. By August of 1831, he had completed his work. He understood that inductive effects occurred only when magnetic fields were changing. He was ready to publish but he had not heard anything at all from Europe on the subject and assumed that he was several years ahead of everyone else. He could afford to set aside his work until the following summer and publish later with more complete results. Unfortunately, Michael Faraday, an English chemist and the greatest experimental scientist of his time, was hot on his trail. In ten famous days of experiment in the fall of 1831, he wrapped up the problem very effectively. Faraday read his results on November 1831 and published them in 1832. Henry received the paper in May 1832 and learned that he had been thoroughly beaten.

At first, he was in despair, as he was no longer a young man and had missed his great opportunity. But his friend Professor Silliman encouraged him to publish anyway. Thereby he secured the primacy of his work in self-induction, which would lead eventually to the naming of the unit of inductance as the Henry.

There are several similarities between Henry and Faraday. They were both raised in somewhat impoverished circumstances. Both had minimal schooling, could barely read and, by happenstance almost, both became avid readers. Both became outstanding lecturers who prepared their talks very thoroughly, with dramatic experimental display. Both were interested primarily in scientific principles and would not develop or patent their discoveries. Both worked primarily on their own.

Henry was born in Albany, NY. His father was a day laborer who died young. Joseph Henry had very little schooling and worked as a farm hand and store clerk. He was apprenticed at 14 to a watchmaker and silversmith. He was a member of an amateur theatrical company. At age 16 he became very interested in scientific questions while reading a book on general science. Later in life, he said, "This book, although by no means a profound work, has under providence exerted a remarkable influence on my life. It opened to me a new world of thought and enjoyment, fixed my mind on the study of nature, and caused me to resolve that I would immediately commence to devote my life to the acquisition of knowledge." Attending the Albany Academy, he soon learned enough to become a teacher in a country school. He began his college education at the Albany Academy in 1819 and finished in 1822, studying and teaching simultaneously with a 16-hour schedule.

It was said that the young Henry easily made major decisions but had difficulty with some minor ones. The story is told of a pair of shoes which he ordered. He changed the order from square to round toes and back so many times that the cobbler, out of frustration, finally delivered his shoes with one round and one square toe.

Henry made many additional contributions before retiring from research in 1846. He produced a motor (in 1831), which he regarded as a "scientific toy" and did not patent. It had the basic elements of a D.C. motor including a commutator. He also invented the electric relay and the electric telegraph. He built a telegraph line in 1835 but did not patent it. He also built a transformer capable of stepping the voltage up or down.

Joseph Henry possessed significant talents as an administrator. In 1846, he left Princeton, where he had been a professor since 1832, and became head of the new Smithsonian Institution, where he served ably until his death. He was a key scientific advisor to Abraham Lincoln during the Civil War, and served on many government advisory boards. He played a role in the founding of the National Academy of Sciences, the American Association for the Advancement of Science, the Philosophical Society of Washington, the Lick Observatory, and the U.S. Weather Bureau. In 1893 the International Electrical Congress honored him by naming the unit of inductance the Henry.

7.8 MAGNETIC ENERGY

In Section 3-9 we obtained the electric energy required to assemble a set of charges in the presence of dielectrics. In this section we derive the magnetic energy of a system of current-carrying coils in the presence of magnetic materials. There are many similarities with the analysis of Section 3-9 and it may be helpful to make comparisons. Except for the very first steps, the analysis is limited to the *linear* case ($\Phi = Li$). It is also assumed that all changes are made slowly enough so that the maximum system dimension is small compared to wavelength. This is called the *quasistatic* condition. It is necessary to ensure that radiation does not occur.

First we consider the magnetic energy W_m of a single current-carrying loop (Figure 7-13).

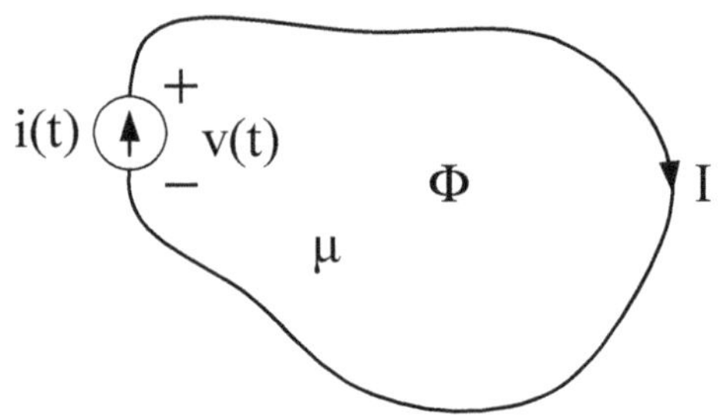

Figure 7-13. A current-carrying loop in the presence of magnetic materials.

Magnetic materials may be present. The current of an ideal current source $i(t)$ is increased slowly from zero to a final value I. Final flux is Φ. You may recall that the current in an inductor does not change readily; work is required. The work W_m done in establishing the current and flux is

$$W_m = \int P(t)\, dt = \int v(t)\, i(t)\, dt = \int \frac{d\Phi}{dt}\, i\, dt = \int i\, d\Phi$$

$$W_m = \int i\, d\Phi \qquad (7\text{-}32)$$

Eq. (7-32) is the general relationship for magnetic energy. It is valid for nonlinear problems, too. Note that $v = d\Phi/dt$ comes from Faraday's law. For linear problems where $\Phi = Li$,

$$W_m = \int i \frac{d\Phi}{dt} dt = \int i \left(L \frac{di}{dt} + i \frac{dL}{dt} \right) dt \Rightarrow \int_0^I L\, i\, di \quad \left(\text{if } \frac{dL}{dt} = 0 \right)$$

$$= \frac{1}{2} L I^2 = \frac{1}{2} I\Phi = \frac{1}{2} \frac{\Phi^2}{L} \quad \left(\text{if } \frac{dL}{dt} = 0 \right)$$

$$\boxed{W_m = \frac{1}{2} LI^2 = \frac{1}{2} I\Phi = \frac{1}{2}\frac{\Phi^2}{L}} \quad \text{Linear case, L constant in time} \quad (7\text{-}33)$$

In the process described above, current and voltage change with time but inductance is constant since there is no distortion of the loop or motion with respect to magnetic materials present. Now consider N coupled coils in the presence of magnetic materials (Figure 7-14).

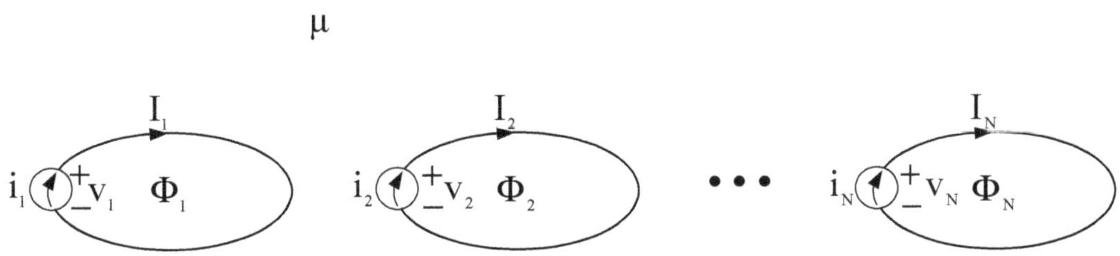

Figure 7-14. N coupled coils in the presence of magnetic materials.

The currents are increased slowly from zero to their final values $I_1, \ldots, I_N$. For simplicity, bring $i_1(t)$ to its final value I_1 with $i_2, \ldots, i_N$ zero. Then bring $i_2(t)$ to its final value I_2 with $i_1 = I_1$ and $i_3, \ldots, i_N$ zero, etc. From our network model (Eq. 7-25) we know that $v_j(t) = \sum_{k=1}^{N} L_{jk}\, di_k/dt$ and that the total work done is

$$W_m = \sum_{j=1}^{N} \int i_j(t)\, v_j(t)\, dt = \sum_{j=1}^{N}\sum_{k=1}^{N} L_{jk} \int i_j(t)\, di_k$$

Consider the contribution of the pair of numbers $(1, 2)$ which is

$$L_{11} \int_0^{I_1} i_1\, di_1 + L_{21} \int_0^{I_1} i_2\, di_1 + L_{22} \int_0^{I_2} i_2\, di_2 + L_{12} \int_0^{I_2} i_1\, di_2$$

Now note that $i_2 = 0$ in the second integral and $i_1 = I_1$ in the fourth integral. The contribution of $(1, 2)$, i.e., the total work for $N = 2$, is

$$\begin{aligned} W_m &= \frac{1}{2} L_{11} I_1^2 + \frac{1}{2} L_{22} I_2^2 + L_{12} I_1 I_2 \\ &= \frac{1}{2} \sum_{j=1}^{2} \sum_{k=1}^{2} L_{jk} I_j I_k \quad \text{(assuming reciprocity, i.e., } L_{12} = L_{21}) \end{aligned} \tag{7-34}$$

For each pair of numbers (j, k) we have four integrals which yield contributions $\frac{1}{2} L_{jj} I_j^2 + \frac{1}{2} L_{kk} I_k^2 + L_{jk} I_j I_k$ as before. One of the integrals is zero because one of the currents (say, i_j) is raised to its final value when the other (say, i_k) is zero. Every pair of numbers has a similar contribution and the final result, assuming reciprocity, is

$$\boxed{W_m = \frac{1}{2} \sum_{j=1}^{N} \sum_{k=1}^{N} L_{jk} I_j I_k} \tag{7-35}$$

Note also that

$$\boxed{W_m = \frac{1}{2} \sum_{j=1}^{N} I_j \Phi_j} \tag{7-36}$$

To see this, substitute $\Phi_j = \sum_{k=1}^{N} L_{jk} I_k$ in Eq. (7-36) to obtain Eq. (7-35).

Magnetic Energy of Volume Current Distributions

Next let's consider a current-carrying loop of finite cross section with volume current density $\mathbf{J}$. Divide the loop into a large number N of loops in parallel. The typical loop has cross section ΔS_k about path C_k, current ΔI_k and flux Φ_k. Then

$$\Phi_k = \iint_{S_k} \mathbf{B} \cdot d\mathbf{s} = \iint_{S_k} (\nabla \times \mathbf{A}) \cdot d\mathbf{s} = \oint_{C_k} \mathbf{A} \cdot d\boldsymbol{\ell} \quad \text{(by Stokes' theorem)}$$

where C_k bounds S_k. Using Eq. (7-36),

$$W_m = \frac{1}{2} \sum_{k=1}^{N} (\Delta I_k) \Phi_k = \frac{1}{2} \sum_{k=1}^{N} \Delta I_k \oint_{C_k} \mathbf{A} \cdot d\boldsymbol{\ell}$$

Since

$$\Delta I_k \, d\boldsymbol{\ell} = J \Delta S_k \, d\boldsymbol{\ell} = \mathbf{J} \, \Delta S_k \, d\ell$$

we have

$$W_m = \frac{1}{2}\sum_{k=1}^{N} \oint_{C_k} \mathbf{A}\cdot\mathbf{J}\,\Delta S_k\, d\ell \Rightarrow \frac{1}{2}\iiint_V \mathbf{A}\cdot\mathbf{J}\,dv \quad \text{as } N\to\infty$$

$$\boxed{W_m = \frac{1}{2}\iiint_V \mathbf{A}\cdot\mathbf{J}\,dv} \tag{7-37}$$

Note that Eq. (7-37) involves an integration over that volume V of space which includes currents since $\mathbf{J} = 0$ elsewhere. The magnetic energy is expressed in terms of the current density and the vector potential distribution just as the electric energy was in terms of the charge density and the scalar potential distribution (See Eq. (3-45)).

Magnetic Energy in Terms of Fields

An alternative form of W_m may be obtained by expanding the range of integration to cover a larger volume V' which includes V.

$$W_m = \frac{1}{2}\iiint_V \mathbf{A}\cdot\mathbf{J}\,dv = \frac{1}{2}\iiint_V \mathbf{A}\cdot(\nabla\times\mathbf{H})\,dv$$

Using the vector identity,

$$\nabla\cdot(\mathbf{A}\times\mathbf{H}) = \mathbf{H}\cdot(\nabla\times\mathbf{A}) - \mathbf{A}\cdot(\nabla\times\mathbf{H}) = \mathbf{H}\cdot\mathbf{B} - \mathbf{A}\cdot(\nabla\times\mathbf{H})$$

Then

$$W_m = \frac{1}{2}\iiint_{V'} \mathbf{B}\cdot\mathbf{H}\,dv - \frac{1}{2}\iiint_{V'} \nabla\cdot(\mathbf{A}\times\mathbf{H})\,dv$$

$$= \frac{1}{2}\iiint_{V'} \mathbf{B}\cdot\mathbf{H}\,dv - \frac{1}{2}\oiint_{S'} (\mathbf{A}\times\mathbf{H})\cdot d\mathbf{s}$$

(by the divergence theorem, S' bounds V')

The above result is valid for any volume V' which includes V. As V' approaches the entire region of space (say, it's a sphere of radius r) then the surface integral vanishes as $r\to\infty$ because $|\mathbf{A}| \sim 1/r$ and $|\mathbf{H}| \sim 1/r^2$ for a current distribution $\mathbf{J}$ of finite extent, and the surface area of the sphere is proportional to r^2. Thus

$$\boxed{W_m = \frac{1}{2}\iiint_{\text{all space}} \mathbf{B}\cdot\mathbf{H}\,dv} \tag{7-38}$$

In Eq. (7-37), the energy is viewed as being stored with the current source, whereas Eq. (7-38) shows the energy being stored with the fields. Both equations yield the same results for a magnetostatic system, but for time-varying sources and fields, Eq. (7-38) makes more sense because the fields move in space, carrying the energy, as will be shown in Chapter 8.

Finally, we note that Eqs. (7-37) and (7-38) can be used to determine the inductance of a circuit, using the relationship $W_m = 1/2\ LI^2$.

$$L = \frac{2W_m}{I^2} = \frac{1}{I^2} \iiint \mathbf{B} \cdot \mathbf{H}\, dv = \frac{1}{I^2} \iiint \mathbf{A} \cdot \mathbf{J}\, dv \qquad (7\text{-}39)$$

We can drop the volume specifications at this point. The integrals exist over the regions of (**B**, **H**) and (**A**, **J**), respectively.

7.9 MAGNETIC FORCES AND TORQUES

Just as the electric force can be calculated from the electric energy for some system as shown in Section 3.9.4, the magnetic force can also be calculated from the magnetic energy for some systems. Consider a solenoid into which is inserted a bar of magnetic material (Figure 7-15).

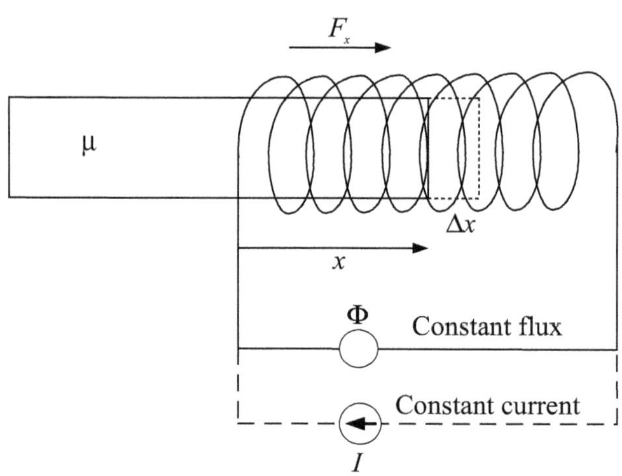

Figure 7-15. The force on a bar of magnetic material in a solenoid.

Its location is specified by the variable x. The source may be either an ideal current source or an ideal flux source. We may think of an ideal flux source as a device which senses flux and varies current to maintain flux constant. Thus it could be merely a variable current or voltage source. With the ideal flux source connected, I varies with x. With the ideal current source connected, Φ varies with x. Magnetic energy may be expressed as $W_m(\Phi, x)$ or $W_m(I, x)$.

Let's consider the two cases shown in Figure 7-15, the constant flux case (shown solid) and the constant current case (shown dotted) separately.

Constant Flux Case:

Consider a small movement dx of the bar. Flux does not change and therefore the flux generator does no work since $\int i(t)\, v(t)\, dt = \int i\, d\Phi = 0$ (upper and lower limits are identical). In general there is a force F_x on the bar. It is either attracted towards or repelled from the solenoid. Mechanical work done is $F_x\, dx$. The mechanical work done may lead to mechanical energy stored in a spring or to kinetic energy. When x varies, the fields **B**, **H** may change, leading to a change dW_m in magnetic energy storage.

Conservation of energy thus requires that

$$F_x\, dx + dW_m = 0$$

$$F_x = -\left.\frac{dW_m}{dx}\right|_{\Phi\ \text{const.}} = -\frac{\partial W_m(\Phi, x)}{\partial x}$$

Similarly, $F_y = -\dfrac{\partial W_m(\Phi, y)}{\partial y}$ and $F_z = -\dfrac{\partial W_m(\Phi, z)}{\partial z}$. In general,

$$\boxed{\mathbf{F} = -\nabla W_m(\Phi, x, y, z)} \quad \text{(constant flux)} \tag{7-40a}$$

If the bar is free to rotate in the θ direction, then work done is (Torque) $d\theta$ and

$$\text{Torque} = -\frac{\partial W_m(\Phi, \theta)}{\partial \theta} \quad \text{(constant flux)} \tag{7-40b}$$

Constant Current Case:

First we disconnect the constant flux source and connect the constant current source. Then consider a small movement dx. Flux changes from Φ to $(\Phi + d\Phi)$. W_m changes from $1/2\ I\Phi$ to $1/2\ I(\Phi + d\Phi)$; dW_m is $1/2\ I\ d\Phi$. Work done by the current source is $\int i(t)\, v(t)\, dt = \int i\, d\Phi = I\, d\Phi = 2\, dW_m$. Therefore energy supplied by the current source is equally divided, half to W_m and half to mechanical energy. Conservation of energy requires that

$$F_x\, dx + dW_m = 2\, dW_m$$

$$F_x = \left.\frac{dW_m}{dx}\right|_{I\ \text{const.}} = \frac{\partial W_m(I, x)}{\partial x}$$

In general,

$$\boxed{\mathbf{F} = \nabla W_m(I, x, y, z)} \quad \text{(constant current)} \tag{7-41a}$$

If the bar is free to rotate in the θ direction,

$$\text{Torque} = \frac{\partial W_m(I,\theta)}{\partial \theta} \quad \text{(constant current)} \tag{7-41b}$$

EXAMPLE 7-9. THE FORCE BETWEEN COUPLED COILS

Consider the force between two coupled coils (see Figure 6-16). Let coil 1 be fixed and let coil 2 be free to move. Then

$$W_m = \frac{1}{2} L_{11} I_1^2 + \frac{1}{2} L_{22} I_2^2 + L_{12} I_1 I_2$$

The force between the two coils is

$$F_x = \frac{\partial W_m}{\partial x} \text{ (constant current)} = I_1 I_2 \frac{\partial L_{12}}{\partial x}$$

$$F_y = I_1 I_2 \frac{\partial L_{12}}{\partial y}$$

and the total force on coil 2 is

$$\boxed{\mathbf{F} = I_1 I_2 \nabla L_{12}} \tag{7-42}$$

7.10 MAGNETIC MATERIALS

Before we close this chapter, we discuss physical characteristics of magnetic materials. There are several types of magnetic materials with distinctly different characteristics. Three of the most common are diamagnetic, paramagnetic, and ferromagnetic materials. We call a material

diamagnetic if $|\chi_m| << 1$, χ_m negative.
paramagnetic if $|\chi_m| << 1$, χ_m positive.
ferromagnetic if $|\chi_m| >> 1$, $\mu \approx \chi_m \mu_0$

Table 7-4 shows typical values of μ_r, χ_m for some magnetic materials.

Table 7-4 Magnetic Susceptibilities of Magnetic Materials

Diamagnetic Materials	
Material	χ_m
Aluminum Oxide	-0.5×10^{-5}
Copper	-0.94×10^{-5}
Gold	-3.6×10^{-5}
Lead	-1.7×10^{-5}
Silver	-2.6×10^{-5}
Bismuth	-1.7×10^{-4}
Sodium Chloride	-1.2×10^{-5}
Paramagnetic Materials	
	χ_m
Aluminum	2.1×10^{-5}
Cr_2O_3	1.7×10^{-3}
Platinum	2.9×10^{-4}
Palladium	8.2×10^{-4}
Oxygen	1.8×10^{-6}
Ferromagnetic Materials	
	$\mu_r \approx \chi_m$
Nickel	250
Cobalt	600
Iron	5×10^3
Permalloy	10^5
Super Malloy	10^6

Since $\mu_r = \chi_m + 1$, the relative permeability of diamagnetic and paramagnetic materials is very close to unity. For ferromagnetic materials μ_r and χ_m are practically equal. Note that $|\chi_m| << 1$ for both diamagnetic and paramagnetic materials. Typical values of χ_m are $\pm 10^{-5}$. Thus the magnetization $\mathbf{M}$ is much smaller in magnitude than the applied field $\mathbf{H}$ since

$$\mathbf{M} = \chi_m \mathbf{H}$$

For instance, consider a long thin solenoid with a current I and N turns per unit length. In the air-filled case, the magnetic flux density is

$$B_1 = \mu_0 H = \mu_0 NI.$$

If the solenoid is completely filled with a diamagnetic or paramagnetic material, then

$$B_2 = \mu_0 H + \mu_0 M = \mu_0 NI\,(1 + \chi_m) = B_1\,(1 + \chi_m)$$

The fractional change in flux density is of the order of 10^{-5}. Thus the effect of the insertion of the cylindrical bar of diamagnetic or paramagnetic material is practically negligible.

Diamagnetic and paramagnetic materials are often called nonmagnetic materials because of the small magnitude of the magnetic effects. For instance, silver and aluminum are often described as nonmagnetic although one is diamagnetic and the other paramagnetic. Michael Faraday discovered diamagnetism and paramagnetism in 1846. He noted that certain materials are attracted towards regions of higher fields and certain materials are repelled. Diamagnetic materials are repelled and all other materials are attracted (Figure 7-16).

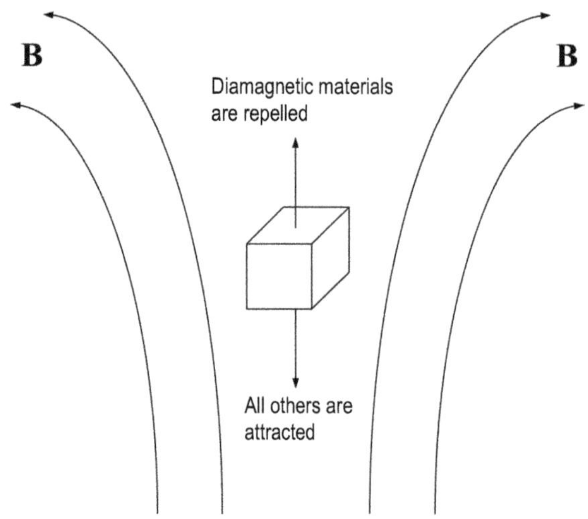

Figure 7-16. Diamagnetic material.

Another example is shown in Figure 7-15. We recall from Example 3-9 that the electric force attracting a dielectric block into a parallel plate capacitor is proportional to ($\varepsilon - \varepsilon_0$ or $\chi_e\,\varepsilon_0$). By analogy, the magnetic force attracting a bar into a solenoid is proportional to ($\mu - \mu_0$ or $\chi_m\,\mu_0$). Thus the force in Figure 7-15 is repulsive if χ_m is negative (diamagnetic). There is no dielectric effect corresponding to diamagnetism because there are no materials with negative χ_e.

Diamagnetic and paramagnetic effects can be explained in terms of the magnetic moments of the molecules involved. There are two principle contributions to the magnetic moment, namely, (a) the orbital motion of electrons about the nucleus and (b) electron spin. (There is also a very small nuclear spin moment).

A diamagnetic material has molecules with zero net magnetic moment due to (a) and (b) in the absence of an applied magnetic field. When a magnetic field is applied, the forces on the moving electron change the orbital velocity to create a net magnetic moment for the magnetic dipole (an electron in an orbital motion constitutes a tiny loop of current). The net magnetic moment **m** and the magnetization **M** per unit volume are such as to oppose the applied field. Thus

$$\mathbf{B} = \mu_0\,\mathbf{H} + \mu_0\,\mathbf{M}$$

where $\mathbf{H}$ is the applied field and $\mathbf{M}$ opposes $\mathbf{H}$. When the applied field $\mathbf{H}$ is removed, the magnetization $\mathbf{M}$ returns to zero, i.e., the effects are reversible.

In diamagnetic materials, there are other effects which occur but the principal one is the effect of the applied field upon the electron orbital motion. Even though the diamagnetic effect is very small, it is always present in all materials. Its effect may be hidden by other, more powerful, effects. The diamagnetic effect is linear as we might expect because of its small, perturbational effect. μ_r and χ_m are independent of temperature for diamagnetic materials. χ_m is of the order of 10^{-5} for most diamagnetic materials. The largest negative magnetic susceptibility χ_m is that of Bismuth, the material used by Faraday in his discovery of diamagnetism.

In paramagnetic materials, the molecules each have a net magnetic moment $\mathbf{m}$, but magnetization $\mathbf{M} = 0$, because of random orientation, in the absence of an applied field. If a magnetic field $\mathbf{H}$ is applied, there are several effects which occur. The principal one is the effect of the applied field upon the electron spin moments, which tend to align parallel to the direction of the applied field. However, random thermal agitation reduces significantly the magnetization $\mathbf{M}$ obtainable by this alignment of spins. Temperature would have to be lowered to a few degrees kelvin to approach the maximum effect. At room temperatures, the random thermal motion almost completely overwhelms the tendency to align leaving only a very small magnetic moment in the direction of the applied field. This moment is, however, strong enough to overcome the diamagnetism which is always present. As we might expect, temperature has a significant effect on χ_m, which decreases rapidly as temperature increases from absolute zero.

In **ferromagnetic** materials such as iron, cobalt, and nickel, the magnetization effect is many orders of magnitude larger than that of paramagnetic and diamagnetic materials. Like paramagnetism, ferromagnetism mainly arises from magnetic dipoles associated with electron spin. One difference is that in ferromagnetic materials, each dipole tends to align in the same direction as its neighboring dipole, forming permanently magnetized "domains." This tendency is known to be a quantum mechanical effect. Each domain contains about 10^{15} atoms or aligned dipoles. In the absence of the magnetic field, these domains are randomly oriented, thus the material as a whole is not magnetized. When the magnetic field is applied, the walls of those domains having dipole moments aligned with the applied field move in such a way that their volumes grow at the expense of other domains. The material becomes magnetized. Because this process is not entirely reversible, when the applied field is removed, there will be some remnant net magnetization. Thus for ferromagnetic materials the magnetization is determined not only by the present field but also by the previous magnetic history of the material. See D. K. Cheng (1989) or D. J. Griffiths (1999) for more detailed discussion.

PROBLEMS

7-1. In each of the cases below, find the equivalent bound surface and volume current densities $\mathbf{J}_{ms}$, $\mathbf{J}_m$.

(a) A uniformly magnetized cylinder of infinite length and radius a:

$$\mathbf{M} = \mathbf{a}_x M_0 \ (0 < \rho < a)$$

(b) A uniformly magnetized infinite slab of thickness d:

$$\mathbf{M} = \mathbf{a}_x M_0 \left(-\frac{d}{2} < z < \frac{d}{2}\right)$$

7-2. In each of the cases below, find the bound surface and volume current densities $\mathbf{J}_{ms}$, $\mathbf{J}_m$:

(a) A magnetized cylinder of length ℓ and radius a:

$$\mathbf{M} = \mathbf{a}_z \left(\frac{\rho}{a}\right) M_0 \quad (0 < \rho < a, -\frac{\ell}{2} < z < \frac{\ell}{2})$$

(b) A uniformly magnetized spherical shell:

$$\mathbf{M} = \mathbf{a}_z M_0 \ (a < r < b)$$

7-3. In each of the cases below, find the bound surface and volume current densities $\mathbf{J}_{ms}$, $\mathbf{J}_m$:

(a) A magnetized cylinder of radius a and infinite length:

$$\mathbf{M} = \mathbf{a}_z \left(\frac{\rho}{a}\right) M_0 \quad (0 < \rho < a)$$

(b) A magnetized sphere of radius a:

$$\mathbf{M} = \mathbf{a}_z \left(\frac{r^2}{a^2}\right) M_0 \qquad (r < a)$$

7-4. (a) Find the magnetic fields **B, H** on the axis of the magnetized cylinder of Problem 7-2(a).
(b) Find the fields **B, H** everywhere of the magnetized cylinder of Problem 7-3(a).

7-5. (a) Find the magnetic field **B** at the center of the magnetized spherical shell of Problem 7-2(b).
(b) Find the magnetic field **B** at the center of the magnetized sphere of Problem 7-3(b).

7-6. Figure 7-17(a) shows a current-carrying z-directed filament of infinite length and current I. Two homogeneous slabs of permeabilities μ_1, μ_2 surround the filament:

$$\begin{aligned} \mu &= \mu_1 \quad (0 < \phi < \pi) \\ &= \mu_2 \quad (\pi < \phi < 2\pi) \end{aligned}$$

Find the fields **M, B, H** everywhere.
[HINT] Apply Ampère's law for **H** and consider fields that satisfy the boundary conditions of Section 7.5.

7-7. Figure 7-17(b) shows a current-carrying z-directed filament of infinite length and current I. Homogeneous regions of permeabilities μ_1, μ_2 surround the filament:

$$\begin{aligned} \mu &= \mu_1 \ (0 < \rho < a) \\ &= \mu_2 \ (\rho > a) \end{aligned}$$

(a) Find the magnetic fields **B, H** everywhere.
(b) Calculate the equivalent bound currents, $\mathbf{J}_m$ and $\mathbf{J}_{ms}$.

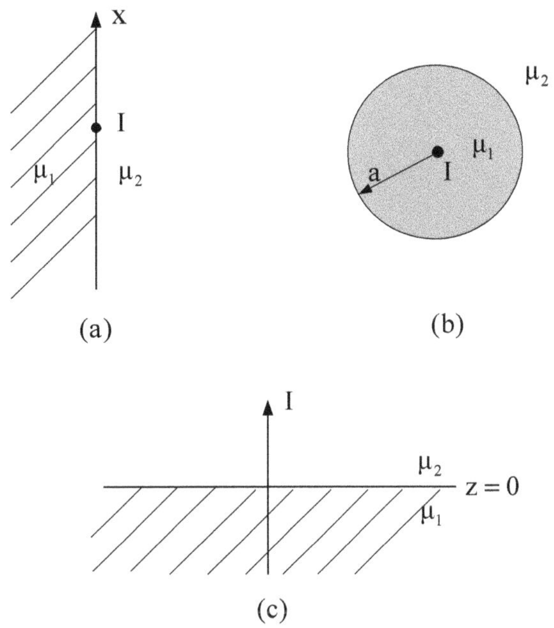

Figure 7-17. Currents in magnetic media.

7-8. Figure 7-17(c) shows a current-carrying z-directed filament of infinite length and current I. Homogeneous regions of permeabilities μ_1, μ_2 surround the filament:

$$\begin{aligned}\mu &= \mu_1 \ (z < 0)\\ &= \mu_2 \ (z > 0)\end{aligned}$$

Find the magnetic fields **B, H** everywhere.

7-9. (a) In Figure 6-4 the region $a < \rho < b$ is filled with a homogeneous material of permeability μ. Find the fields **M, B, H** everywhere.

(b) In Figure 6-4 the inner current is I_1 upward and the outer current is I_2 downward. The regions $a < \rho < b$, $\rho > b$ are filled with materials of permeabilities μ_1, μ_2, respectively. Find the fields **M, B, H** everywhere.

7-10. In Figure 6-6(a) the region inside the toroid is filled with a homogeneous material of permeability μ. Find the fields **M, B, H** everywhere.

7-11. Prove that $\oint_c \mathbf{M} \cdot d\ell = I_m$ where I_m is the bound current enclosed by C, with a right-hand relationship between the directions of I and C.

7-12. Consider the three curl relationships of magnetostatics and the resultant boundary conditions:

Using

$$\nabla \times \mathbf{H} = \mathbf{J},$$

we derived

$$\mathbf{a}_{n2} \times (\mathbf{H}_1 - \mathbf{H}_2) = \mathbf{J}_s \qquad (7\text{-}15)$$

Similarly, using

$$\nabla \times \mathbf{M} = \mathbf{J}_m,$$

show that

$$\mathbf{a}_{n2} \times (\mathbf{M}_1 - \mathbf{M}_2) = \mathbf{J}_{ms}$$

and, using

$$\nabla \times \mathbf{B} = \nabla \times (\mu_0 \mathbf{H} + \mu_0 \mathbf{M}) = \mu_0(\mathbf{J} + \mathbf{J}_m)$$

show that

$$\mathbf{a}_{n2} \times (\mathbf{B}_1 - \mathbf{B}_2) = \mu_0(\mathbf{J}_s + \mathbf{J}_{ms})$$

In particular, the relation

$$\mathbf{a}_{n2} \times (\mathbf{M}_1 - \mathbf{M}_2) = \mathbf{J}_{ms}$$

gives the equivalent surface current at the boundary between two magnetized regions. We may consider it to be the generalization of Eq. (7-2). $\mathbf{a}_{n2}$ is defined to be the outward normal from region 2 as in Figure 7-5. Note that $\mathbf{J}_{ms}$ of Figure 7-3 agrees with the above relation.

7-13. The half-space ($z < 0$) is permanently magnetized:

$$\mathbf{M} = \mathbf{a}_x M_0 \quad (z < 0)$$

The half space $z > 0$ is free space.

(a) Find the equivalent surface and volume (bound) current densities $\mathbf{J}_{ms}$, $\mathbf{J}_m$.
(b) Find the magnetic field $\mathbf{B}$ everywhere, using the equivalent currents.
(c) Find $\mathbf{H}$ everywhere. Check also if the boundary condition (7-15) is satisfied.

7-14. The slab $-\frac{d}{2} < z < \frac{d}{2}$ is permanently magnetized:

$$\mathbf{M} = \mathbf{a}_x M_0 \left(-\frac{d}{2} < z < \frac{d}{2} \right)$$

The regions $|z| > \frac{d}{2}$ are air.

(a) Find the equivalent surface and volume (bound) current densities $\mathbf{J}_{ms}$, $\mathbf{J}_m$.
(b) Find the magnetic field $\mathbf{B}$ everywhere.
(c) Find $\mathbf{H}$ everywhere.

7-15. The half-space regions $z < 0$, $z > 0$ are permanently magnetized:

$$\begin{aligned} \mathbf{M} &= \mathbf{a}_x M_1 \ (z < 0) \\ &= \mathbf{a}_x M_2 \ (z > 0) \end{aligned}$$

(a) Find the equivalent surface and volume (bound) current densities $\mathbf{J}_{ms}$, $\mathbf{J}_m$.
(b) Find the magnetic field $\mathbf{B}$ everywhere.
(c) Find $\mathbf{H}$ everywhere.

7-16*. A circular cylinder of radius a and infinite length carries a uniform surface current density J_{so}. The current is directed as shown in Figure 7-18. This effect could be accomplished by winding the solenoid in the direction shown. Find the magnetic field $\mathbf{B}$ everywhere.

[HINT] Separate the current into longitudinal (z) and circumferential (ϕ) components.

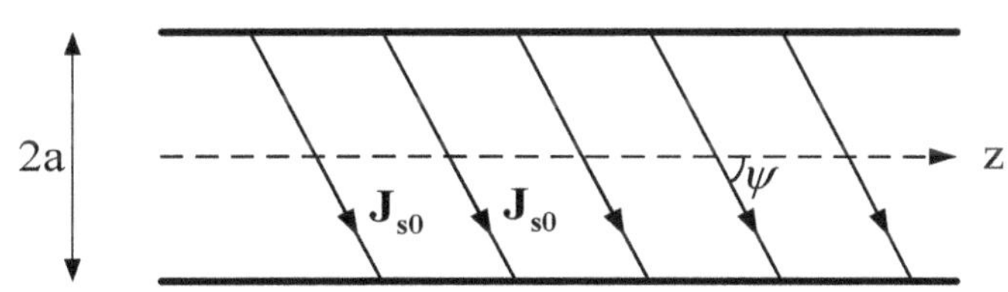

Figure 7-18. A sheath-helical current.

7-17. The current of Figure 7-18 is surrounded by two homogeneous regions of differing permeabilities μ_1, μ_2:

$$\begin{aligned} \mu &= \mu_1 \quad (\rho < a) \\ &= \mu_2 \quad (\rho > a) \end{aligned}$$

(a) Find the magnetic field **B** everywhere.
(b) Check to see that all boundary conditions are satisfied.

7-18. A planar current sheet of infinite extent is surrounded by two regions of different permeabilities:

$$\begin{aligned} \mathbf{J}_s &= \mathbf{a}_x J_{so} \quad (z = 0) \\ \mu &= \mu_1 \quad (z < 0) \\ &= \mu_2 \quad (z > 0) \end{aligned}$$

Find the magnetic field **B** everywhere.

7-19. Two anti-parallel planar current sheets surround a region of permeability μ:

$$\begin{aligned} \mathbf{J}_s &= \mathbf{a}_x J_{so} \quad (z = -d/2) \\ &= -\mathbf{a}_x J_{so} \quad (z = d/2) \end{aligned}$$

$$\text{permeability} = \mu \left(-\frac{d}{2} < z < \frac{d}{2} \right).$$

Find the field **B** everywhere.

7-20. Consider the on-axis fields of a magnetized cylinder (Example 7-1) at large distances $(z \pm \frac{\ell}{2} >> a)$. Show that it agrees with the on-axis field of a loop (Example 6-5) and a magnetic dipole (Eq. (6-31)).

7-21. A semi-infinite cylindrical bar magnet of radius a is uniformly magnetized. Its axis is the z axis:

$$\mathbf{M} = \mathbf{a}_z M_0 \quad (0 < \rho < a, -\infty < z < 0)$$

(a) Find the equivalent surface and volume current densities $\mathbf{J}_{ms}$, $\mathbf{J}_m$.
(b) Find the magnetic field **B** on the axis both inside and outside the bar.
(c) Find **H** on the axis.

7-22. An infinite slab of thickness d is both permeable and conducting. It carries a uniform current J_0 and has a permeability μ:

$$\mathbf{J} = \mathbf{a}_x J_0 \left(-\frac{d}{2} < z < \frac{d}{2}\right)$$

Find the magnetic fields **B, H** everywhere.

7-23. An infinite current-carrying slab of thickness d is characterized by permeability μ. Its current is given as follows:

$$\mathbf{J} = \mathbf{a}_x J_0 \left(\frac{z}{d}\right) \quad (-\frac{d}{2} < z < \frac{d}{2})$$

Find the magnetic fields $\mathbf{B}$, $\mathbf{H}$ everywhere.

7-24. A long straight z-directed current-carrying steel wire of radius a and infinite length has a permeability μ. The volume current density is given as follows:

(a) $\mathbf{J} = \mathbf{a}_z J_0 \quad (\rho < a)$

(b) $\mathbf{J} = \mathbf{a}_z J_0 \left(\frac{\rho}{a}\right) \quad (\rho < a)$

Find the magnetic field $\mathbf{B}$ everywhere in each case.

7-25. Consider a toroidal coil of N turns and circular cross section. The cross-sectional radius is a and the minimum radius of the toroid is b-a as shown in Figure 7-19. The toroid is filled with permeable material of permittivity μ. Find the inductance of the coil.

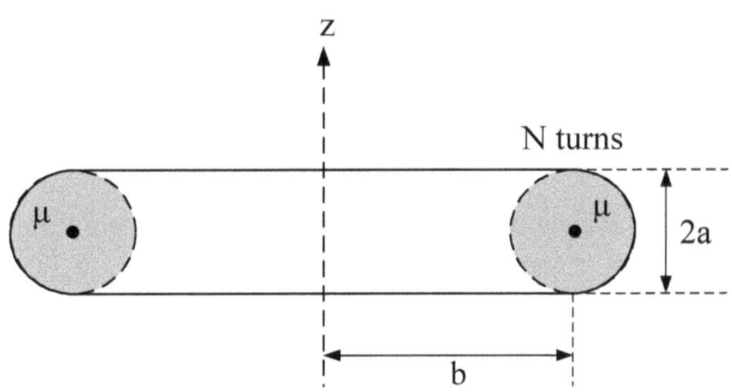

Figure 7-19. A toroidal coil (doughnut) of circular cross section.

7-26. Figure 7-20 shows a toroidal coil of N turns and double-rectangular cross section. The toroid is filled with material of permittivity μ. Find the inductance of the coil.

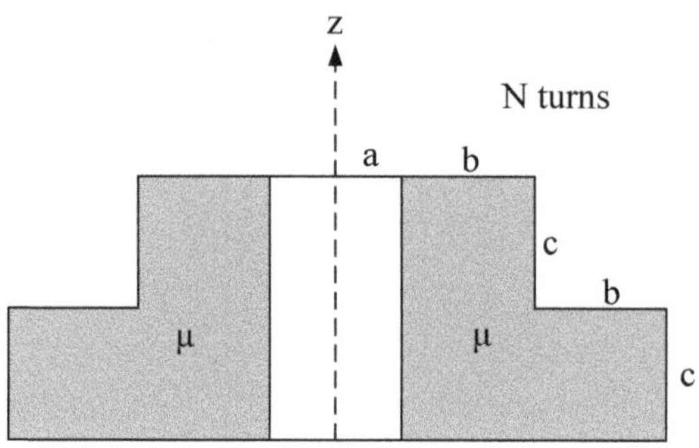

Figure 7-20. A toroidal coil of double-rectangular cross section.

7-27. Consider two coaxial, solenoids of radii a, b (b > a) and infinite length, one contained within the other, each centered on the z axis:

$$\begin{aligned} \mathbf{J}_s &= \mathbf{a}_\phi J_{s1} = \mathbf{a}_\phi N_{\ell 1} I_1 \quad (\rho = a) \\ &= \mathbf{a}_\phi J_{s2} = \mathbf{a}_\phi N_{\ell 2} I_2 \quad (\rho = b) \end{aligned}$$

where $N_{\ell 1}$, $N_{\ell 2}$ represent turns per unit length on inner, outer solenoids and I_1, I_2 represent currents on inner, outer solenoids. Find the mutual inductance per unit length.

7-28. Consider a filament of infinite length and a circular loop of radius a whose center is at a distance d from the filament. See Figure 6-14 and replace the square loop by a circular loop. Find the mutual inductance.

7-29. Consider Example 7-6. Find the force between the filament and the square loop by differentiating the mutual inductance as a function of separation (see Eq. (7-42)). Verify Eq. (6-39). Find the location for maximum force.

7-30. Figure 7-21 shows two loops side by side, of radii a_1, a_2 and currents I_1, I_2. Loop centers are located at $x = 0$, $y = 0$, and $x = d$, $y = 0$. Note that $d \gg a_1$, a_2. Find the approximate mutual inductance and the force between loops. Both loops lie in the x,y plane with the z axis as loop axis.

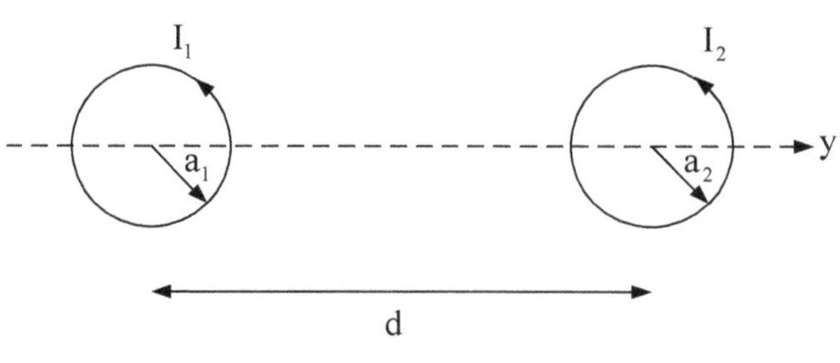

Figure 7-21. Two coplanar current-carrying circular loops.

7-31. Consider two toroids of square cross section, one contained within the other (Figure 7-22). Find the mutual inductance. Find the force between two turns, one on each toroid, both lying in the plane $\phi = 0$. Find the total force between toroids.

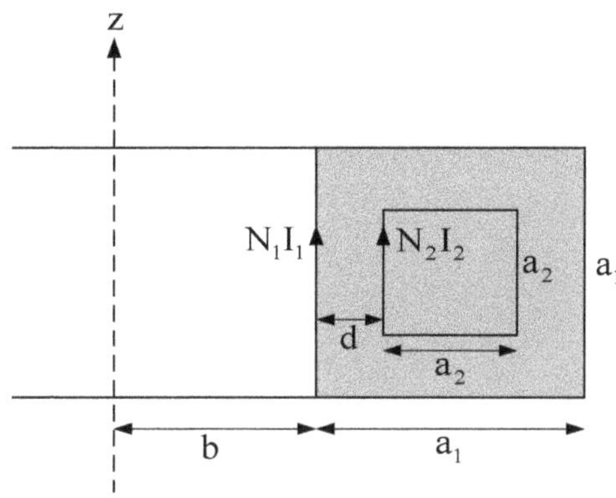

Figure 7-22. Two toroids of square cross section, one contained within the other.

7-32. Figure 7-23 shows a z-directed filament of infinite length and current I_1. An equilateral triangular loop with current I_2 lies in the yz plane. Find the mutual inductance. Find the force on the triangular loop.

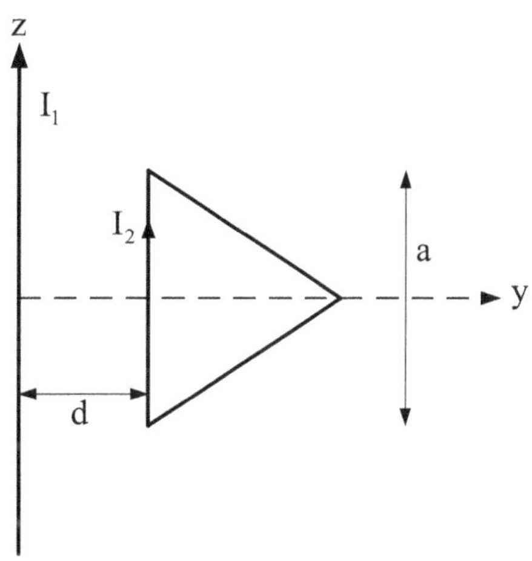

Figure 7-23. An infinite filament and an isosceles triangular loop.

7-33. Consider a toroid of Example 6-4. See Figure 6-6(a). Find the magnetic energy stored in the toroid.

7-34. Consider a coaxial cable of infinite length, inner radius a and outer radius b. The inner and outer currents are as follows:

$$\mathbf{J}_s = \mathbf{a}_z \frac{I}{2\pi a} \quad (\rho = a)$$

$$= -\mathbf{a}_z \frac{I}{2\pi b} \quad (\rho = b)$$

(a) Find the magnetic energy W_m per unit length.
(b) Find the inductance per unit length.

7-35. Consider a coaxial cable similar to that of Problem 7-34, except that the inner conductor is permeable and the inner current is uniformly distributed over the cross section:

$$\mathbf{J} = \mathbf{a}_z \left(\frac{I}{\pi a^2} \right) \quad (0 < \rho < a)$$

$$\mathbf{J}_s = -\mathbf{a}_z \left(\frac{I}{2\pi b} \right) \quad (\rho = b)$$

(a) Find the magnetic energy W_m per unit length.
(b) Find the inductance per unit length. The additional inductance found here is called *internal* inductance.

7-36. A very long solenoid, of radius a, length ℓ and N turns, carries current I and its core is filled with the magnetic material of permeability μ. See Figure 6-6(b). Calculate the magnetic energy stored, using (a) Eq. (7-38) (in terms of fields), (b) Eq. (7-37) (in terms of $\mathbf{A}$), and (c) Eq. (7-33) (in terms of inductance).

7-37. Figure 7-15 shows an iron bar partially inserted into a long solenoid with N turns and length ℓ. The bar and the solenoid have the same cross sectional area S. The solenoid carries current I and the bar has permeability μ.

(a) Find the fields $\mathbf{B}$ and $\mathbf{H}$ in each region.
(b) Find the stored magnetic energy (as a function of x).
(c) Determine the direction and magnitude of the magnetic force acting on the bar.

7-38. A current I in an N-turn coil (Figure 7-24) produces a magnetic flux in the yoke of total mean length ℓ and square cross section $a \times a$. The yoke has an iron core of permeability μ and two air gaps of length x. We neglect fringing and assume that B_x is continuous across the air gap; this assumption is reasonable for very small air gaps.

(a) Find $\mathbf{B}$, $\mathbf{H}$ in the iron core and air gap.
(b) Find the total stored magnetic energy.
(c) Determine the magnitude and direction of the magnetic force acting on the bar.

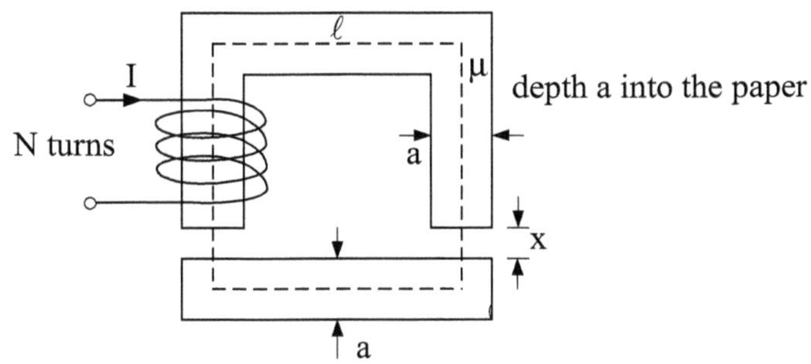

Figure 7-24. A magnetic yoke with airgaps.

7-39*. A linear magnetic sphere of radius a and permeability μ is placed in an initially uniform magnetic field $\mathbf{B}_o = \mathbf{a}_z \mathrm{B}_o$ in air. Refer to Figure 4-14 and replace the dielectric by a magnetic material and apply $\mathbf{B}_o$ instead of $\mathbf{E}_o$. Find the magnetic fields $\mathbf{B}$, $\mathbf{H}$ everywhere (both inside and outside of the sphere).

[HINT] Introduce the *magnetic scalar potential* V_m by letting $\mathbf{H} = -\nabla \mathrm{V}_m$ and show that V_m satisfies Laplace's equation. Make use of the solutions of Laplace's equation in Table 4-1 and apply the boundary conditions in Eqs. (7-14) and (7-16).

7-40*. Consider a uniformly magnetized sphere in Figure 7-4 of Example 7-2. Making use of the magnetic scalar potential described in Problem 7-39, find the magnetic fields $\mathbf{B}$ and $\mathbf{H}$ everywhere (both inside and outside of the sphere). Confirm also that the field you obtained inside the sphere agrees with the result of Example 7-2.

TIME-VARYING FIELDS: FARADAY'S LAW AND MAX-WELL'S EQUATIONS

CHAPTER 8

8.1 INTRODUCTION

In previous chapters, we explored the very rapid progress made in magnetostatics by Oersted, Ampère, Biot-Savart, Arago, and others during the 1820s following Oersted's dramatic discovery of the relationship between electric current and magnetism. By the end of the 1820s the formulation of magnetostatics was nearly complete. Such a rapid development of basic theory was virtually unheard of in those days. James Clerk Maxwell (1831–1879), studying at a later time the development of magnetostatics, said of Ampère's work, "the whole, theory and experiment, seem to have burst forth in full vigor and completely formed from the brain of this Newton of electricity." High praise from Maxwell!

The problems of time-varying fields, however, were much more subtle and much more difficult to crack. Some first steps were made in 1831 when Michael Faraday (1791–1867) carried out in 10 days a famous set of experiments in which he discovered (a) magnetic induction, (b) the similarity of permanent magnets and electromagnets, and (c) the first electric generator (Faraday disk generator). He continued to work on magnetic induction for over twenty years and expressed his conclusions in words equivalent to what we now call Faraday's Law. The first half of the time-varying problem was thus completed by about 1850.

The other half of the time-varying problem was of a more abstract nature. There were no simple experiments which could be carried out to determine the solution. Maxwell, who was 40 years younger than Faraday, made his first step when he translated Faraday's results into mathematical form. The search for the complete formulation of electromagnetics was an international affair. It was widely known that a piece of the puzzle was missing. Virtually all of the leading physicists of the time attempted to solve the problem but with no success. Finally Maxwell, after decades of study, reported his complete theory in 1865, at the end of the American Civil War. He continued to work on electromagnetics for the rest of his life, publishing his great work, *A Treatise on Electricity and Magnetism*, in 1873. He died in 1879. About ten years after his death, Heinrich Hertz (1857–1894) built some of the first antennas and decisively confirmed Maxwell's theory*. About fifteen years after that, Albert Einstein (1879–1955) showed that Maxwell's equations were valid in any moving system and that they were in fact the basis for relativity. Since then, the 20th century has not revealed any flaws in Maxwell's equations. We now recognize them as the crowning achievement of electromagnetics. R. P. Feynman has said,

> "Twenty thousand years from now, Maxwell's equations will be recognized as a pinnacle of 19th century science, while the American Civil War will be regarded as an insignificant brush fire … "

8.2. LAWS OF ELECTROSTATICS AND MAGNETOSTATICS—A SUMMARY

We recall that the basic laws of electrostatics are

$$\boxed{\nabla \times \mathbf{E} = 0} \quad \boxed{\nabla \cdot \mathbf{D} = \rho_v} \tag{8-1}$$

(Electrostatics)

* Brian Santo, "Making waves at 100," *IEEE Spectrum*, May 1988, pp. 58–60.

The first law ($\nabla \times \mathbf{E} = 0$) states the conservative (curl free) nature of the electrostatic field. It implies that the line integral of $\mathbf{E}$ $\left(\oint_C \mathbf{E} \cdot d\boldsymbol{\ell}\right)$ is zero around any closed path C, and that the voltage between any two points is independent of path. It thus implies that the sum of the voltages around a loop is zero, which is known as the Kirchhoff voltage law (KVL) in circuit theory. The second law ($\nabla \cdot \mathbf{D} = \rho_v$), known as Gauss' law, states the divergence relationship between the source ρ_v and the electrostatic field. It implies that the electric flux $\left(\oiint_S \mathbf{D} \cdot d\mathbf{s}\right)$ through any closed surface S is equal to the total charge enclosed within the volume.

The basic laws of magnetostatics are

$$\boxed{\nabla \times \mathbf{H} = \mathbf{J}} \quad \boxed{\nabla \cdot \mathbf{B} = 0} \qquad (8\text{-}2)$$

(Magnetostatics)

The first law ($\nabla \times \mathbf{H} = \mathbf{J}$), known as Ampère's law, states the curl relationship between the source $\mathbf{J}$ and the magnetostatic field. Oersted's discovery was the first indication of this relationship. It implies that the circulation of the magnetic field $\left(\oint_C \mathbf{H} \cdot d\boldsymbol{\ell}\right)$ along any closed contour C is equal to the total current passing through the surface bounded by C. The first law also implies $\nabla \cdot \mathbf{J} = 0$ and thus the sum of the currents into a junction or loose end $\left(\oiint_S \mathbf{J} \cdot d\mathbf{s}\right)$ is zero, which is known as the Kirchhoff current law (KCL) in circuit theory. The second law ($\nabla \cdot \mathbf{B} = 0$) states the solenoidal (divergenceless) nature of the magnetostatic field. It implies that the surface integral $\oiint_S \mathbf{B} \cdot d\mathbf{s}$ over any closed surface S is zero and that the flux $\iint_S \mathbf{B} \cdot d\mathbf{s}$ of the vector $\mathbf{B}$ through any open surface S bounded by C is independent of the surface S.

The laws for electrostatics and magnetostatics are uncoupled. The electrostatic field arises from its divergence-type source ρ_v. The magnetostatic field arises from its curl-type source $\mathbf{J}$. The electric field $\mathbf{E}$ is determined solely by its sources (ρ_v and ρ_{pv}); $\mathbf{B}$ is determined solely by its sources ($\mathbf{J}$ and $\mathbf{J}_m$). Each represents a canonical form. The electrostatic field represents all conservative vector fields; the magnetostatic field represents all solenoidal vector fields.

For time-varying fields, some basic changes are required in the laws above. The two divergence laws, $\nabla \cdot \mathbf{D} = \rho_v$ and $\nabla \cdot \mathbf{B} = 0$, are valid for the time-varying as well as for the static cases, and therefore no changes are necessary for those laws. However, the two curl equations, $\nabla \times \mathbf{E} = 0$ and $\nabla \times \mathbf{H} = \mathbf{J}$, do require drastic revision, which is the subject of Faraday's law and Maxwell's equations, respectively. These revisions will upset all of our previous concepts. You may be surprised to learn that, in the general time-varying case, the sum of the voltages around a loop is not necessarily zero, the voltage is not independent of path, and that the sum of the currents into a junction or loose end is not zero. A completely new ball game! However, despite those drastic changes in the characteristics of the fields, the mathematical changes required are relatively simple. The addition of a single term to each of the two curl equations, $\nabla \times \mathbf{E} = 0$ and $\nabla \times \mathbf{H} = \mathbf{J}$, will be sufficient to complete the formulation for time-varying fields.

8.3. FARADAY'S LAW

Faraday's law appears in various forms and involves some very subtle effects. It is important, therefore, that the different forms and their particular limitations be clearly defined. We will start with the general forms, which are valid in all cases, and then we will look at the special forms which are of particular interest to electrical engineers.

8.3.1. General Forms of Faraday's Law

Faraday's law in differential form is

$$\boxed{\nabla \times \mathbf{E} = -\frac{\partial \mathbf{B}}{\partial t}} \tag{8-3}$$

(Faraday's law in differential form)

We may conclude that a time-varying magnetic field creates an electric field. Referring to Helmholtz's theorem (Section 1.5) we note that $\mathbf{E}$ may arise from $\nabla \times \mathbf{E}$ or $\nabla \cdot \mathbf{E}$. In the electrostatic case, $\nabla \times \mathbf{E} = 0$, and the only possible source is the free and bound charge densities since $\nabla \cdot \mathbf{E} = (\rho_v + \rho_{pv})/\varepsilon_0$. In the time-varying case, however, $\frac{\partial \mathbf{B}}{\partial t}$ acts as an additional source. We note that Eq. (8-3) reduces to the static case, $\nabla \times \mathbf{E} = 0$, if $\frac{\partial \mathbf{B}}{\partial t} = 0$. Equation (8-3) is valid at all times at any point in space in all situations. It is our first general law of electromagnetics since our previous laws were limited to the electrostatic case (charges are nailed down) or to the magnetostatic case (currents are steady).

Integrating both sides of Eq. (8-3) over an open surface S, we obtain the integral form of Faraday's law

$$\iint_S \nabla \times \mathbf{E} \cdot d\mathbf{s} = \oint_C \mathbf{E} \cdot d\boldsymbol{\ell} = -\iint_S \frac{\partial \mathbf{B}}{\partial t} \cdot d\mathbf{s}$$

where the first equality is due to Stokes' theorem. Therefore, we have

$$\boxed{\oint_C \mathbf{E} \cdot d\boldsymbol{\ell} = -\iint_S \frac{\partial \mathbf{B}}{\partial t} \cdot d\mathbf{s}} \tag{8-4}$$

(Faraday's law in integral form)

where the closed path C bounds the surface S. Equations (8-3) and (8-4) are completely general. In case of doubt, we can always return to these general forms.

Modification of Voltage Circuital Law

The line integral of the electric field around the closed loop C in Eq. (8-4) corresponds to the sum of potential drops (or voltages) around the loop. Eq. (8-4) indicates that the sum of the voltages around a loop is not zero, if there exists a time-varying magnetic field, but is equal to the surface integral of the negative rate of change of the magnetic field **B**, which is equivalent to the negative time derivative of the magnetic flux (Φ) passing through the loop as will be shown later in the next Section. The statement of Eq. (8-4) is contrary to the Kirchhoff voltage law (KVL) that you learned in circuit theory. Under the low frequency system for which the time variation is slow, the rate of change of magnetic flux ($-d\Phi/dt$) is considered to be small and thus can be neglected. The KVL in circuit theory is valid only under such circumstances.

In the following example, we illustrate how the time-varying magnetic field induces an electric field.

EXAMPLE 8-1. THE ELECTRIC FIELD INDUCED BY A TIME-VARYING MAGNETIC FIELD

The uniform time-varying magnetic field **B** exists in a cylindrical region of radius a:

$$\mathbf{B} = \mathbf{a}_z B_0 \cos \omega t \quad (\rho < a) \tag{8-5}$$

where B_0 is a constant and ω is an angular frequency. Such a uniform field can be produced by a very long solenoid with time-varying current (see Example 6-4). Find the electric field induced by **B** everywhere.

SOLUTION

First in order to find out the direction of the electric field, we recognize that **E** (effect) and $-\frac{\partial \mathbf{B}}{\partial t}$ (cause) are related through the curl in the same way as **B** (effect) and μ_0 **J** (cause) are related through the curl.

$$\nabla \times \mathbf{E} = -\frac{\partial \mathbf{B}}{\partial t}, \quad \nabla \times \mathbf{B} = \mu_0 \mathbf{J}$$

We know from Example 6-2 by the right-hand rule that the z-directed current (**J**) in a cylindrical region produces a ϕ-directed **B** field. Therefore, we conclude that the z-directed $\frac{\partial \mathbf{B}}{\partial t}$ will induce ϕ-directed E field.

$$\mathbf{E} = \mathbf{a}_\phi E_\phi$$

By cylindrical symmetry, E_ϕ will not vary along the circular loop. Choosing the circle of radius ρ as the closed path C and applying Faraday's law (8-4),

$$\oint_C \mathbf{E}\cdot d\ell = \int_0^{2\pi} E_\phi\, \mathbf{a}_\phi \cdot \mathbf{a}_\phi\, \rho\, d\phi = E_\phi\,(2\,\pi\,\rho) = -\iint_S \frac{\partial \mathbf{B}}{\partial t}\cdot \mathbf{ds}$$

$$= -\int_0^{2\pi}\int_0^{\rho} (-\,\omega B_0 \sin\omega t\,)\,\mathbf{a}_z \cdot \mathbf{a}_z\, \rho\, d\rho\, d\phi = (\omega B_0 \sin\omega t)\,\pi\,\rho^2 \quad (\text{when } \rho \le a)$$

$$= -\int_0^{2\pi}\int_0^{a} (-\,\omega B_0 \sin\omega t\,)\,\mathbf{a}_z \cdot \mathbf{a}_z\, \rho\, d\rho\, d\phi = (\omega B_0 \sin\omega t)\,\pi\, a^2 \quad (\text{when } \rho \ge a)$$

Thus

$$\mathbf{E} = \mathbf{a}_\phi \frac{1}{2}(\omega B_0 \sin\omega t)\rho \quad (\rho \le a)$$

$$= \mathbf{a}_\phi \frac{1}{2}(\omega B_0 \sin\omega t)\frac{a^2}{\rho} \quad (\rho \ge a) \qquad (8\text{-}6)$$

Note that this is not a conservative electrostatic field.

8.3.2. The Effects of Motion Through Magnetic Fields

At this point it is important to recognize that motion has an effect upon the electric and magnetic fields. The stationary observer and the moving observer measure different electric and magnetic fields. For instance, the electric field $\mathbf{E}'$ measured by a moving observer differs from the electric field $\mathbf{E}$ measured by a stationary observer.

In this section we are concerned with the definition of $\mathbf{E}'$, the electric field measured by a moving observer. Consider a perfectly conducting bar* of length ℓ moving through a magnetic field $\mathbf{B}$ at velocity $\mathbf{v}$ (Figure 8-1). For simplicity let $\mathbf{B}$ be uniform, $\mathbf{B} = \mathbf{a}_x B_0$ and let velocity $\mathbf{v}$ be constant, $\mathbf{v} = \mathbf{a}_y v_0$. The free charges on the conducting bar experience a magnetic force q $\mathbf{v} \times \mathbf{B}$ which leads to a separation of charge as indicated in Figure 8-1.

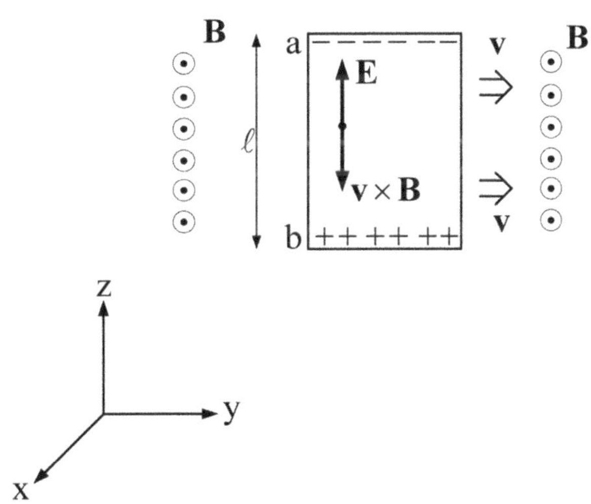

Figure 8-1. A conducting bar moving through a magnetic field.

*See the discussion in Section 8.6 for the definition of a perfect conductor.

See the Hall effect (Section 6.10) for a similar charge-separation process. The separation of charge leads to an electric field $\mathbf{E}$. The charge buildup continues until the electric force $\mathbf{F}_e = q\,\mathbf{E}$ exactly cancels the magnetic force $\mathbf{F}_m = q\,\mathbf{v} \times \mathbf{B}$ and equilibrium is reached. The process is nearly instantaneous. The charge distributes itself in such a manner as to produce a uniform field $\mathbf{E} = -\,\mathbf{v} \times \mathbf{B} = \mathbf{a}_z\,(v_0\,B_0)$ throughout the conductor.

The moving bar therefore acts like a voltage source. If we could make contact with the moving ends of the bar we could tap off a voltage $-B_0\,v_0\,\ell$. This voltage arises from the electric field since

$$V_{ab} = -\int_b^a \mathbf{E} \cdot d\ell = \int_b^a \mathbf{v} \times \mathbf{B} \cdot d\ell = -B_0\,v_0\ell$$

This voltage produced by a conducting bar moving through magnetic field is the basis of electric generators. It may be considered a linear generator. Note that E is not zero inside the moving perfect conductor. This may be somewhat surprising. It is an indication that a new definition of the electric field is required for the moving observer.

Now consider the non-relativistic velocities ($|\mathbf{v}| << c$ where c is the speed of light in vacuum) and let

$$\boxed{\mathbf{E}' = \mathbf{E} + \mathbf{v} \times \mathbf{B}} \qquad (|v| << c) \tag{8-7}$$

Then the field $\mathbf{E}'$ as seen by the moving observer is zero in the conductor. Furthermore, both the stationary and moving observers calculate the same force on a test charge, in the conductor, i.e., $\mathbf{F} = q\,(\mathbf{E} + \mathbf{v} \times \mathbf{B}) = q\,\mathbf{E}' = 0$.

Figure 8-2 shows a system (a black box with terminals a and b) moving at velocity $\mathbf{v}$ with respect to a stationary coordinate system. The voltage V'_{ab} measured by the moving observer is

$$V'_{ab} = -\int_b^a \mathbf{E}' \cdot d\ell \qquad (|\mathbf{v}| << c) \tag{8-8}$$

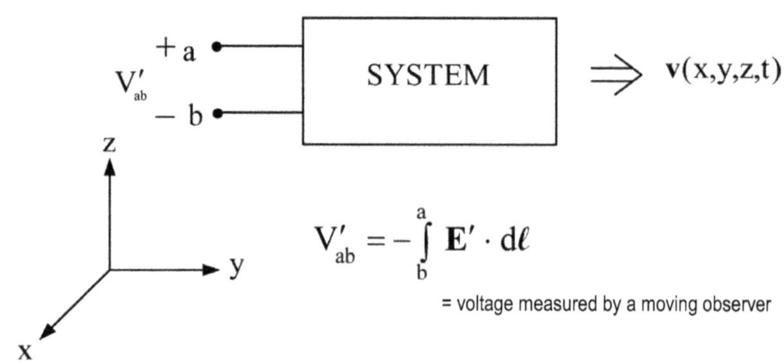

Figure 8-2. Voltage in a moving system.

Equation (8-8) represents the work done in moving a unit charge from b to a while in motion at velocity $\mathbf{v}$. If $\mathbf{v}$ varies throughout the system, then the velocity of the terminals a-b is the velocity of the observer. It is important to point out the difference between $V_{ab} = -\int_b^a \mathbf{E}\cdot d\ell$ and $V'_{ab} = -\int_b^a \mathbf{E}'\cdot d\ell$. Usually we are interested in V'_{ab} which is the open-circuit voltage measured in the system at the terminals a-b. If stationary conducting rails are added to our system as in Figure 8-3, i.e, if the input terminals of the moving part of the system make contact with stationary conducting rails, then we can tap off an induced voltage. In this case the terminals a-b are not moving and V'_{ab} and V_{ab} are identical.

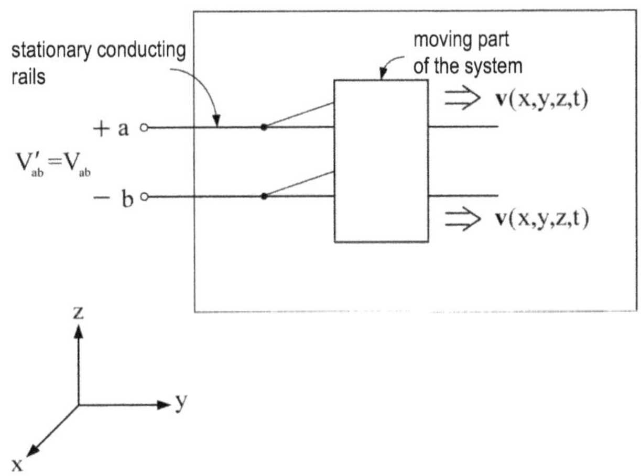

Figure 8-3. How to tap off a voltage.

8.3.3. Non-Relativistic Integral Forms of Faraday's Law

There are several non-relativistic integral forms of Faraday's law which are quite useful in treating problems involving motion of circuits through magnetic fields. These forms are theoretically limited only by the non-relativistic restriction $|\mathbf{v}| << c$, but in order to apply them to network problems of interest we need the *quasistatic* restriction $D_{max} << \lambda$ as well. D_{max} is the maximum dimension of the circuit and λ is the wavelength of the time-varying field. The quasistatic restriction ensures that significant radiation does not occur and that the circuit theory assumptions apply. Thus the results of this section and their application to network problems are limited only by two simple restrictions (a) $|\mathbf{v}| << c$ and (b) $D_{max} << \lambda$.

We now consider the general non-relativistic integral forms of Faraday's law. Substitute $\mathbf{E}' = \mathbf{E} + \mathbf{v} \times \mathbf{B}$ in Eq. (8-4) to obtain the first form:

$$\boxed{\oint_C \mathbf{E}'\cdot d\ell = -\iint_S \frac{\partial \mathbf{B}}{\partial t}\cdot d\mathbf{s} + \oint_C (\mathbf{v}\times\mathbf{B})\cdot d\ell} \qquad (8\text{-}9)$$

$$(|\mathbf{v}| << c)$$

where C bounds S. Equation (8-9) applies to any path C in space whether or not it coincides with a circuit. We may think of $\oint_C \mathbf{E}' \cdot d\ell$ as a voltage around the path C. We call it the *induced voltage*. It is often called an **electromotive force** or **emf**.

The right-hand side of Eq. (8-9) can be manipulated to obtain – dΦ/dt, thus yielding the second form:

$$\boxed{\oint_C \mathbf{E}' \cdot d\ell = -\frac{d\Phi}{dt} = -\frac{d}{dt}\iint_S \mathbf{B} \cdot d\mathbf{s}} \qquad (8\text{-}10)$$

$$(|\mathbf{v}| << c)$$

where C bounds S. The derivation* is lengthy and is omitted here. Equation (8-10) also applies to any path C in space. It states that the induced voltage (emf) around the closed path C is equal to the negative rate of change of magnetic flux (F) passing through the surface S bounded by C.

Now let's apply Eqs. (8-9) and (8-10) to the circuit shown in Fig. 8-4 which includes an open-circuited perfectly conducting loop moving at velocity **v**(x,y,z,t) with respect to a fixed coordinate system (x,y,z). A magnetic field **B**(x,y,z,t) is specified. It is also defined with reference to the fixed (stationary) coordinate system (x,y,z). The velocity **v**(x,y,z,t) may vary from point to point and in time and thus the loop may be distorted as it moves. Some portions of the loop may be completely stationary. There is a small gap in the loop. Path C is completed through the gap. Since the loop is perfectly conducting, **E'** is zero everywhere on C except in the gap region.

$$\oint_C \mathbf{E}' \cdot d\ell = -\int_b^a \mathbf{E}' \cdot d\ell = V'_{ab}(t)$$

The first form, Eq. (8-9), then becomes

$$\boxed{V'_{ab}(t) = -\iint_S \frac{\partial \mathbf{B}}{\partial t} \cdot d\mathbf{s} + \oint_C (\mathbf{v} \times \mathbf{B}) \cdot d\ell} \qquad (8\text{-}11)$$

$$(|\mathbf{v}| << c)$$

*See D. K. Cheng, *Field and Wave Electromagnetics*, Addison-Wesley, 1989, 2nd ed., Chapter 7.

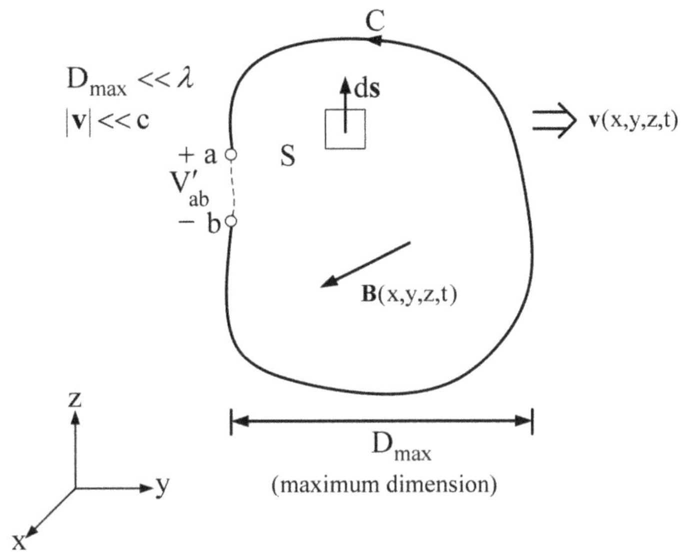

Figure 8-4. A conducting loop C, with a small gap a-b, moving through magnetic fields. Velocities are non-relativistic and the loop is non-radiative. $V'_{ab}(t)$ is the open-circuit voltage as measured by a moving observer. The velocity of the moving observer is that of the terminals a-b. Path C is closed through the gap from a to b.

$V'_{ab}(t)$ is the open-circuit voltage induced across terminals a-b. The open-circuit voltage includes two terms. The first term, $-\iint_S \frac{\partial \mathbf{B}}{\partial t} \cdot d\mathbf{s}$, is called the *transformer emf*. It is the only term present if $\mathbf{v} = 0$. The second term, $\oint_C (\mathbf{v} \times \mathbf{B}) \cdot d\boldsymbol{\ell}$, called the *flux-cutting emf* or the *motional emf*, is the only term present if $\frac{\partial \mathbf{B}}{\partial t} = 0$. In general, both terms are needed and both contributions *must* be considered.

Similarly, the second form, Eq. (8-10), becomes

$$V'_{ab}(t) = -\frac{d\Phi}{dt} = -\frac{d}{dt}\iint_S \mathbf{B} \cdot d\mathbf{s} \qquad (|\mathbf{v}| << c) \qquad (8\text{-}12)$$

(Faraday's law in circuit form)

Equation (8-12) may already be familiar to you. $V'_{ab}(t) = -\frac{d\Phi}{dt}$ is essentially the network model mentioned in Section 7.6. There is one difference in that a negative sign appears in Eq. (8-12). This merely implies that the voltage-current relationships are reversed. We can switch the sign in Eq. (7-22) simply by reversing voltage or current definitions and we can also switch the sign in Eq. (8-12) by reversing the definitions of V'_{ab} or C. For simplicity, the plus sign is used in network theory. In electromagnetic theory we retain the negative sign because it is indicative of a negative or opposing effect as expressed in Lenz's law. Therefore the definition of C is important. It passes through the gap from a (the positive reference for V'_{ab}) to b.

Here the three different forms of Faraday's law are summarized in Table 8-1.

Table 8-1. Faraday's Law

(1) Differential Form: $\nabla \times \mathbf{E} = -\frac{\partial \mathbf{B}}{\partial t}$ • Time-varying magnetic field induces an electric field. • The induced electric field circles around the changing magnetic field. (2) Integral Form: $\oint_C \mathbf{E} \cdot d\ell = -\iint_S \frac{\partial \mathbf{B}}{\partial t} \cdot d\mathbf{s}$ (3) Circuit Form: $V = -\frac{d\Phi}{dt}$ • $V = \oint_C \mathbf{E}' \cdot d\ell =$ induced voltage or emf • $\Phi = \iint_S \mathbf{B} \cdot d\mathbf{s} =$ magnetic flux passing through S

EXAMPLE 8-2. INDUCED VOLTAGE IN A MOVING LOOP

A perfectly conducting square loop of side ℓ with small air gap is at a distance d away from a very long straight wire which carries a steady current I, as shown in Figure 8-5(a).

(a) When you pull the loop away from the straight wire (to the right) at constant speed v, calculate open-circuit voltage V'_{ab} induced across terminals a-b.

(b) Now if you pull the loop upwards at speed v (parallel to the wire), what is the induced voltage V'_{ab}?

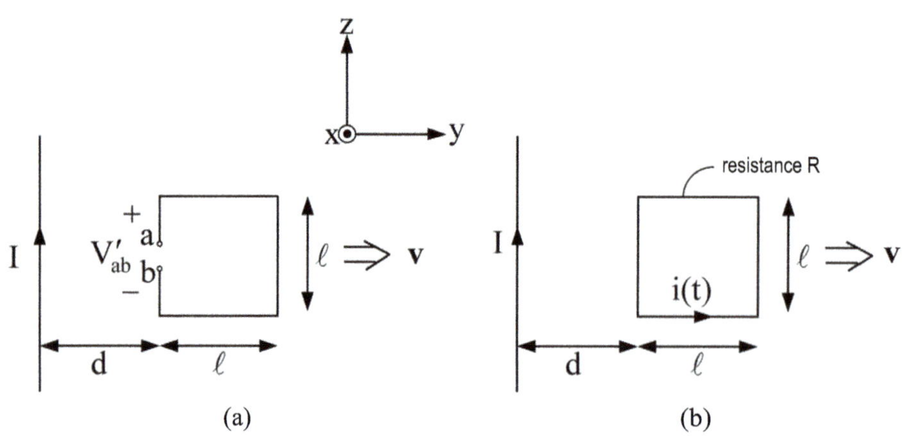

Figure 8-5. Induced voltage and current in a moving loop.

SOLUTION

(a) We can calculate V'_{ab} using Eq. (8-11) or (8-12). Since the magnetic field **B** produced by the steady current I in a straight wire is not time-varying, the first term in Eq. (8-11) doesn't yield any emf, but the second term yields a flux-cutting emf. Assuming the rectangular coordinate system as shown in Fig. 8-5(a),

$$\mathbf{I} = \mathbf{a}_z \, I, \quad \mathbf{v} = \mathbf{a}_y \, v$$

The magnetic field **B** of a filament is given by Eq. (6-8):

$$\mathbf{B} = \mathbf{a}_\phi \frac{\mu_0 I}{2\pi\rho} = -\mathbf{a}_x \frac{\mu_0 I}{2\pi y} \quad \text{(into the paper)}$$

Then the magnetic flux passing through the loop in the direction of $d\mathbf{s}$ (out of the paper because C is chosen counterclockwise) is

$$\Phi = \iint_S \mathbf{B} \cdot d\mathbf{s} = \int_0^{\ell} \int_d^{d+\ell} \left(-\mathbf{a}_x \frac{\mu_0 I}{2\pi y} \right) \cdot \mathbf{a}_x \, dy \, dz$$

$$= -\frac{\mu_0 I}{2\pi} \left[\ln y \right]_d^{d+\ell} (\ell) = -\frac{\mu_0 I}{2\pi} \ell \ln\left(\frac{d+\ell}{d} \right)$$

Now when the loop is pulled to the right with constant velocity v, d should be replaced by $d + vt$. Then

$$\Phi(t) = -\frac{\mu_0 I \ell}{2\pi} \ln\left\{ \frac{(d+vt)+\ell}{(d+vt)} \right\}$$

Thus, from Eq. (8-12), the induced voltage is

$$V'_{ab}(t) = -\frac{d\Phi}{dt} = \frac{\mu_0 I \ell}{2\pi} \frac{d}{dt} \{ \ln(d+vt+\ell) - \ln(d+vt) \}$$

$$= \frac{\mu_0 I \ell}{2\pi} \left\{ \frac{v}{d+vt+\ell} - \frac{v}{d+vt} \right\}$$

$$= \frac{-\mu_0 I \ell^2 v}{2\pi (d+vt+\ell)(d+vt)}$$

(b) When the loop is pulled parallel to the wire, the separation d does not change and consequently the magnetic flux Φ does not change, i.e. $\Phi = \text{constant}$. Thus

$$V'_{ab} = -\frac{d\Phi}{dt} = 0$$

There is no induced voltage.

8.3.4. Lenz's Law

Lenz's law, stated below, is often useful in determining the direction of an induced effect in a magnetic system of conductors, magnetic materials, moving objects, sources, etc. It was deduced in 1834 by Heinrich F. Lenz.

Lenz's Law

> *When there is a change in magnetic flux in a magnetic system, the resulting effect (induced emf) is such as to oppose the change.*

For example, if we try to increase the flux through a closed loop of wire, then the induced emf tends to have currents flow in the loop in such a direction as to decrease the flux. Lenz's law is included already in Faraday's law through a negative sign in Eq. (8-3). However, it is very useful in indicating the sign of a change in current or flux and in checking calculations.

EXAMPLE 8-3. INDUCED CURRENT IN A MOVING LOOP

For the configuration of Example 8-2, consider a *closed* square loop of finite conducting wire of resistance R $[\Omega]$ as shown in Fig. 8-5(b). When the loop is pulled to the right at velocity v, the induced emf causes a current to flow through the loop. Find the induced current $i(t)$ and the direction of the current.

SOLUTION

The induced emf $V(t)$ is obtained in a manner similar to that of Example 8-2 and is related to the current by Ohm's law ($V = Ri$):

$$V(t) = \oint_C \mathbf{E}' \cdot d\ell = -\frac{\mu_0 I \ell^2 v}{2\pi(d+vt+\ell)(d+vt)} = R\ i(t)$$

where the direction of $i(t)$ is taken to be CCW according to the right-hand rule. Therefore,

$$i(t) = -\frac{\mu_0 I \ell^2 v}{2\pi R\,(d+vt+\ell)\,(d+vt)}$$

The negative sign indicates that the current induced flows in a clockwise (CW) direction. We can also check the direction of induced current by Lenz's law. When the loop moves to the right, the magnetic flux passing through the loop (Φ) into the paper decreases because $\mathbf{B}$ decreases as $1/\rho$. According to Lenz's law, the current is induced in such a direction as to oppose the change, i.e., to increase the magnetic flux (into the paper). A clockwise current of the loop would generate the $\mathbf{B}$ field into the paper in the region within the loop (or the surface S). It is assumed in this example that velocities are slow, that frequencies generated are low, and that the resistance of the loop is large compared to its reactance.

8.4. MICHAEL FARADAY'S FAMOUS EXPERIMENTS OF 1831

By 1825, Michael Faraday (1791–1867) had become a very well-known chemist. He had discovered benzene, carried out important research on steel alloys, and been elected a Fellow of the Royal Society. He had also begun research in electromagnetics, repeating many of the basic experiments of Oersted and Ampère. He had also discovered an interesting rotating device described by some as the first motor. Unfortunately, he had been wrongly accused of stealing the idea from his senior colleague Wollaston, an accusation that caused great pain to the honest, upright Faraday.

Faraday was scrupulously honest in his dealings with colleagues. He had a very careful way of raising questions about the research of others that usually led to friendships rather than antagonism. One is reminded of Franklin and his disarming frankness when Faraday writes: "I am by no means decided that there are currents of electricity in the common magnet ... until the presence of electrical currents be proved in the magnet. ... I shall remain in doubt about Ampère's theory." Faraday was a supporter of Ampère's theory in some respects, but he would not accept assumptions until he had seen them proven through experiment. Ampère replied to Faraday at length and the two became friendly correspondents.

Faraday expressed his disappointment at his lack of opportunity for fundamental research when he wrote to Ampère in 1825, "Every letter you write me states how busily you are engaged and I cannot wish it otherwise knowing how well your time is spent. Much of mine is unfortunately occupied in very commonplace employment and this I may offer as an excuse ... for the little I do in original research." Faraday's thoroughness and reliability were too well-known. If Faraday did it, one knew that it would be well done, and Faraday, who worked at the Royal Institution, could not turn down governmental requests.

In 1831, when he carried out his experiments in magnetic induction, Faraday was 40 years old. He had come a long way from the impoverished, uneducated bookbinder's apprentice who, looking for a job in science in 1812, had written to a friend after being turned down: "I am now working at my old trade ... , with respect to the progress of the sciences I know but little and am now likely to know still less. ... I must resign philosophy (science) entirely to those who are more fortunate. ... I am at present in very low spirits ... ". Shortly afterward, Faraday accepted a job as assistant at the Royal Institution and proceeded rapidly in the intervening years to build up his reputation. Now he had come to another similar turning point in his life. He was discouraged about his opportunities for original research, as shown by his letter to Ampère. Now, at 40, was his best work behind him? He would try to answer a basic question: If an electric current causes a magnetic field, will a magnetic field in turn cause an electric current? Perhaps the answer to that question would be as basic and important as Oersted's remarkable discovery of 1820.

In the fall of 1831 Faraday began a series of experiments designed to answer that question. Figure 8-6(a) shows primary and secondary circuits. By closing the switch in the primary circuit, magnetic fields are produced in the vicinity of the secondary. Will a current $i(t)$ be induced in the secondary? Faraday was, in effect, measuring short-circuit current in the secondary. He probably expected that a steady current in the primary would induce a steady current in the secondary. His first experiment took the form shown in Fig. 8-6(b). Both primary and secondary coils were wound about a toroidal iron core of circular cross section. The primary switch was closed suddenly, left closed for some time with a steady current in the primary, then reopened suddenly. A steady current in the primary produced no current in the secondary. It was only upon the closing and opening of the primary switch that any effects were observed in secondary. Small blips occurred upon closing/opening of the switch (Fig. 8-6c).

That was all. The effect which Faraday was looking for did not occur. A secondary current was observed only when the primary current was changing at the closing/opening of the primary switch. What was Faraday to do? Should he abandon his experiments? Was there really any significance in those tiny blips? Fortunately, Faraday was the consummate experimental scientist. In the words of his greatest biographer, "Throughout his life, he followed the same pattern. The unlooked-for result was never ignored or avoided."* He would find out the meaning of those blips, without delay.

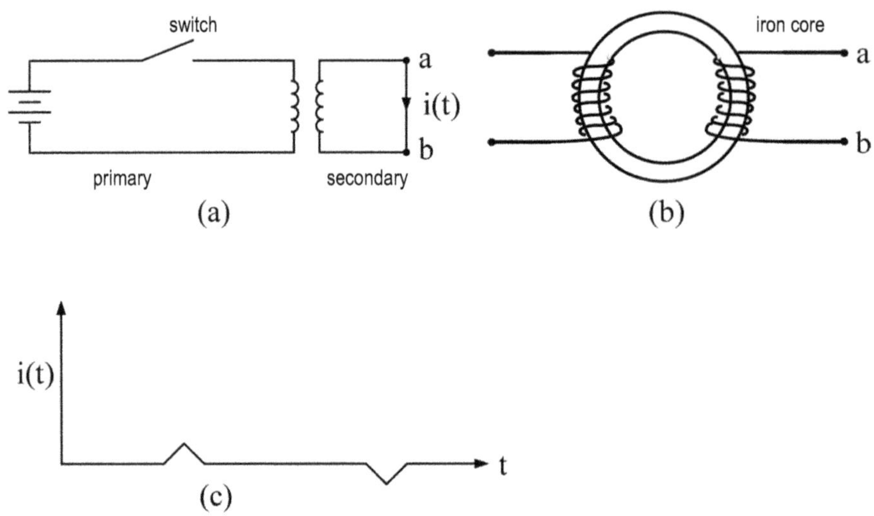

Figure 8-6. Faraday's experiments (a) general form (b) iron core experiment (c) short-circuit current.

Next, Faraday removed the iron. He wound both primary and secondary on a common cylindrical wooden core. At first, no effects were noted. Then, upon increasing the voltage by a factor of ten and increasing the number of turns, the small blips of Figure 8-6(c) were again observed upon closing and opening the switch. Now, what is happening here? A steady primary current produces a steady magnetic field in the vicinity of the secondary and no effect is observed during the steady magnetic field. When the switch is closed, the primary current, and the associated magnetic field, rapidly increase towards their steady values. Similarly, upon opening the switch, the primary current/magnetic field rapidly decrease from their steady values toward zero. Thus, the induced current is obtained only when the magnetic field is changing.

Faraday's next experiment was designed to change the time variation completely. Both primary and secondary were mounted on separate broad boards, the switch was closed, and the primary was moved first towards, then away from the secondary. Thus the primary current would remain steady, or approximately so, but the magnetic field at the vicinity of the secondary would increase/decrease as the primary board approached/receded. When the experiment was performed, the induced secondary current was of one polarity, say positive, as the primary board approached, and negative as the board receded.

*L. Pierce Williams, *Michael Faraday: A Biography*, p. 119, Da Capo Press, NY ,1987, Reprint from Basic Books, NY, 1965.

Now Faraday placed great emphasis on the magnetic lines of force, the magnetic field, associated with the electric current and displayed by the pattern of iron filings, for instance. Doubtless, he could "see" the magnetic field rise up as the switch was thrown, and increase at the secondary as the primary board approached. Thus the common thread of all these experiments was a *changing magnetic field* at the secondary whenever an induced current was observed in the secondary.

Next, Faraday replaced the primary board with a permanent magnet, which was moved towards the secondary circuit board and then moved away from it. The results were identical in form to those obtained with the primary board, the induced secondary current being positive/negative as the magnet approached/receded.

Notice how Faraday moves inexorably toward the conclusion that current is induced in the secondary when the magnetic field in the vicinity of the secondary is *changing*. This conclusion leads to Faraday's law, the first law governing time-varying fields. Faraday knew his abilities and his ambitions, but did he suspect that he was on the verge of such a great discovery during the famous 10 days of fall 1831? Did he suspect that his greatest work was ahead of him, that he would eventually stand in the front rank of English scientists, and that many would rank him, in our own time, as one of the greatest scientists who had ever lived?

During the 10-day period mentioned above, Faraday performed a number of additional experiments. One of these involved the discovery of the first generator, the Faraday disk generator. Faraday also discovered the flux-cutting principle outlined in Section 8.3.3. He read his results on November 24, 1831, to the Royal Society. Later, Faraday would show that induced current was proportional to the area of the secondary loop. He also showed that the induced current was proportional to the conductivity of the wire used in the secondary. These results lead us closer to the mathematical form of Faraday's law given in Eq. (8-12).

8.5. MAXWELL'S EQUATIONS

We have mentioned earlier that Faraday's law was first discovered in 1831. It was expressed in its present form, $\nabla \times \mathbf{E} = -\partial \mathbf{B}/\partial t$, by Maxwell in the 1850s.

There were, at that time, a number of important unanswered questions concerning electromagnetic effects. For instance, Faraday's law describes accurately the voltages and currents induced in a secondary circuit by changes in a primary circuit. However, they do not describe how energy is transferred from primary to secondary or how rapidly. Experiments indicated that the process was nearly instantaneous and instantaneity is implied in the network equations. However, doubt persisted. Is it really instantaneous or is there a finite time delay before the effect on the secondary?

There were also some basic questions at that time concerning the propagation of light. The speed of light was known fairly accurately and with a high degree of confidence since it had been determined by many different methods. It had first been calculated 175 years previously, in 1676, by Ole Roemer (1644–1710), a Danish astronomer. Roemer estimated the speed of light at 2.3×10^8 m/sec, using an astronomical method related to the eclipse of the moons of Jupiter. Later corrections of Roemer's calculations by Isaac Newton in 1704 indicated a velocity of about 2.7×10^8 m/sec. In 1728, J. Bradley (1693–1762), an English astronomer, used an astronomical method related to stellar aberration to estimate the speed of light at 3.05×10^8 m/sec. Finally H. L. Fizeau and L. Foucault made direct measurements of the speed of light by two different methods in 1849 and 1852, yielding values of

3.15×10^8 m/sec and 2.98×10^8 m/sec, respectively. Although the speed of light was fairly well established, the nature of light and the mechanism of propagation were unknown. However, the propagation of sound waves was well understood, having been investigated by many researchers. In particular, D'Alembert had shown in 1750 that sound waves satisfied a scalar wave equation,

$$\nabla^2 \psi = \frac{1}{v_s^2} \frac{\partial^2 \psi}{\partial t^2} \tag{8-13}$$

where ψ and v_s are sound pressure and sound velocity, respectively.

Another question which would naturally have been raised during the 1850s and 1860s was the following: Faraday's law implies that a changing magnetic field produces an electric field. Is the inverse true, i.e., does a changing electric field produce a magnetic field?

8.5.1. Displacement Current

To answer the question just raised we consider the curl equations of electrostatics and magnetostatics. The first curl equation, $\nabla \times \mathbf{E} = 0$, becomes $\nabla \times \mathbf{E} = -\partial \mathbf{B}/\partial t$ in the time-varying case. What, if anything, happens to the second curl equation, $\nabla \times \mathbf{H} = \mathbf{J}$, in the time-varying case?

We recall the equation of continuity,

$$\nabla \cdot \mathbf{J} = -\frac{\partial \rho_v}{\partial t} \tag{5-7}$$

which relates $\mathbf{J}$ and ρ_v in general, including non-steady (time-varying) currents. For magnetostatics

$$\nabla \times \mathbf{H} = \mathbf{J} \tag{8-2a}$$

We take the divergence of Eq. (8-2a) above

$$\nabla \cdot (\nabla \times \mathbf{H}) = \nabla \cdot \mathbf{J} = -\frac{\partial \rho_v}{\partial t} \tag{8-14}$$

The left-hand side, the divergence of the curl, is always zero, whereas the right-hand side clearly is non-zero in the general time-varying case. Therefore the curl equation $\nabla \times \mathbf{H} = \mathbf{J}$ is incorrect for the time-varying case and requires modification at least. How can we modify the curl equation so that there is no contradiction? One simple resolution would be to modify Eq. (8-14) by adding $\frac{\partial \rho_v}{\partial t}$ to the right-hand side.

$$\nabla \cdot (\nabla \times \mathbf{H}) = \nabla \cdot \mathbf{J} + \frac{\partial \rho_v}{\partial t} = 0$$

To see how this changes the curl equation, we substitute $\rho_v = \nabla \cdot \mathbf{D}$ to obtain

$$\nabla \cdot (\nabla \times \mathbf{H}) = \nabla \cdot \left(\mathbf{J} + \frac{\partial \mathbf{D}}{\partial t} \right) = 0 \tag{8-15}$$

which is satisfied by

$$\boxed{\nabla \times \mathbf{H} = \mathbf{J} + \frac{\partial \mathbf{D}}{\partial t}} \tag{8-16}$$

One could also add a divergenceless (solenoidal) vector to the right-hand side of Eq. (8-16), but the result would not yield the proper static form in Eq. (8-2a) when $\partial/\partial t = 0$. The term $\frac{\partial \mathbf{D}}{\partial t} \equiv \mathbf{J}_d$ added to the second curl equation has the units of volume current density (amps/m^2). It is called **displacement current density** even though it may not actually represent current flow. This is a major contribution of Maxwell. He introduced this term on a purely theoretical basis. Maxwell's theory was confirmed about 20 years later by Heinrich Hertz's experiments on electromagnetic radiation in 1888. As will be shown later, inclusion of $\frac{\partial \mathbf{D}}{\partial t}$ in the curl equation is crucial in predicting the propagation of electromagnetic waves, without which we would not live in a "communication" age today.

Integrating both sides of Eq. (8-16) over an open surface S and applying Stokes' theorem, we obtain the integral form:

$$\boxed{\oint_C \mathbf{H} \cdot d\boldsymbol{\ell} = \iint_S \left(\mathbf{J} + \frac{\partial \mathbf{D}}{\partial t} \right) \cdot d\mathbf{s}} \tag{8-17}$$

Eqs. (8-16) and (8-17) are now known as the **generalized Ampère's law**.

EXAMPLE 8-4 DISPLACEMENT CURRENT IN A PARALLEL-PLATE CAPACITOR

A parallel-plate capacitor consists of two circular plates of radius a separated by a distance d (assume that $a >> d$). The region between the plates is filled with an ideal dielectric (perfect insulator) of permittivity ε. The capacitor is charged to a potential difference V(t) (Figure 8-7). Find the magnetic field $\mathbf{H}$ inside the capacitor due to displacement current.

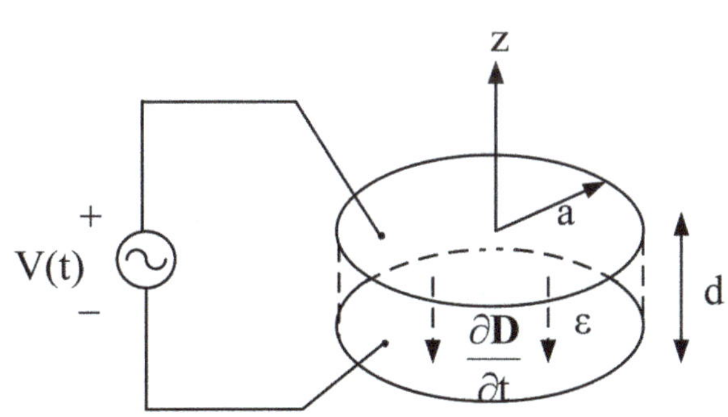

Figure 8-7. Displacement current and induced magnetic field in a parallel-plate capacitor.

SOLUTION

Since $a >> d$, the electric field between the plates is uniform and z-directed:

$$\mathbf{E} = -\mathbf{a}_z E = -\mathbf{a}_z \frac{V(t)}{d}$$

Then the displacement current density is given by

$$\frac{\partial \mathbf{D}}{\partial t} = \frac{\partial}{\partial t}(\varepsilon \mathbf{E}) = -\mathbf{a}_z \frac{\varepsilon}{d} \frac{\partial V}{\partial t}$$

Since the dielectric is a perfect insulator, there is no conduction current, i.e. $\mathbf{J} = 0$. Thus Eq. (8-16) reduces to

$$\nabla \times \mathbf{H} = \frac{\partial \mathbf{D}}{\partial t}$$

and its integral form is

$$\iint_S (\nabla \times \mathbf{H}) \cdot d\mathbf{s} = \oint_C \mathbf{H} \cdot d\boldsymbol{\ell} = \iint_S \frac{\partial \mathbf{D}}{\partial t} \cdot d\mathbf{s} \qquad \text{(8-17a)}$$

We recognize that the z-directed displacement current $\left(\frac{\partial \mathbf{D}}{\partial t}\right)$ induces the ϕ-directed magnetic field $\mathbf{H}$ by the right-hand rule:

$$\mathbf{H} = \mathbf{a}_\phi H_\phi$$

Applying Eq. (8-17a) over the circle of radius ρ:

$$\oint_C \mathbf{H} \cdot d\boldsymbol{\ell} = H_\phi (2\pi\rho) = \iint_S \frac{\partial \mathbf{D}}{\partial t} \cdot d\mathbf{s} = \left(\frac{\partial D_z}{\partial t}\right)(\pi\rho^2)$$

$$H_\phi = \frac{\partial D_z}{\partial t} \cdot \frac{\rho}{2} = -\frac{\varepsilon}{d}\frac{\partial V}{\partial t}\frac{\rho}{2} \quad (\rho \le a)$$

Thus a magnetic field H_ϕ, proportional to $\frac{\partial V}{\partial t}$ and increasing with ρ, is induced within the capacitor.

Modification of Current Circuital Law

The vector $(\mathbf{J} + \partial \mathbf{D}/\partial t)$ is divergenceless, i.e., $\nabla \cdot (\mathbf{J} + \partial \mathbf{D}/\partial t) = 0$ from Eq. (8-15). Integrating Eq. (8-15) over volume V we obtain

$$\iiint_V \nabla \cdot \left(\mathbf{J} + \frac{\partial \mathbf{D}}{\partial t}\right) dv = \oiint_S \left(\mathbf{J} + \frac{\partial \mathbf{D}}{\partial t}\right) \cdot d\mathbf{s} = 0 \qquad \text{(8-18)}$$

Thus

$$-\oiint_S \mathbf{J}\cdot d\mathbf{s} = \oiint_S \frac{\partial \mathbf{D}}{\partial t}\cdot d\mathbf{s} \tag{8-19}$$

Equation (8-18) states that the sum of current and displacement current flowing into (or out of) a closed surface is zero. Thus $\mathbf{J} + \partial\mathbf{D}/\partial t$ plays the same role in time-varying problems as did $\mathbf{J}$ in magnetostatic problems ($\nabla \cdot \mathbf{J} = 0$ for the steady currents of magnetostatics). Equation (8-19), on the other hand, indicates that the current $I = -\oiint_S \mathbf{J}\cdot d\mathbf{s}$ flowing into a surface S is not zero but is equal to the displacement current $\oiint_S \frac{\partial \mathbf{D}}{\partial t}\cdot d\mathbf{s}$ flowing out of S. Thus, the sum of the currents flowing into a junction is not zero but is equal to the displacement current flowing out of the junction. This is contrary to the Kirchhoff current law (KCL) you learned in circuit theory.

Figure 8-8(a) shows conduction current and displacement current at the open end of a conducting wire of radius a. The current at the end of the wire does not go to zero but terminates in displacement current. Consider the closed surface S. The current $I = -\oiint_S \mathbf{J}\cdot d\mathbf{s}$ flowing in from the left equals the displacement current flowing out on the right.

Figure 8-8(b) shows two fat open-ended wires close together, separated by a distance d, forming a capacitor. Current $\mathbf{J}$ flows in from the left and out on the right in the wires; only displacement current flows in the gap region.

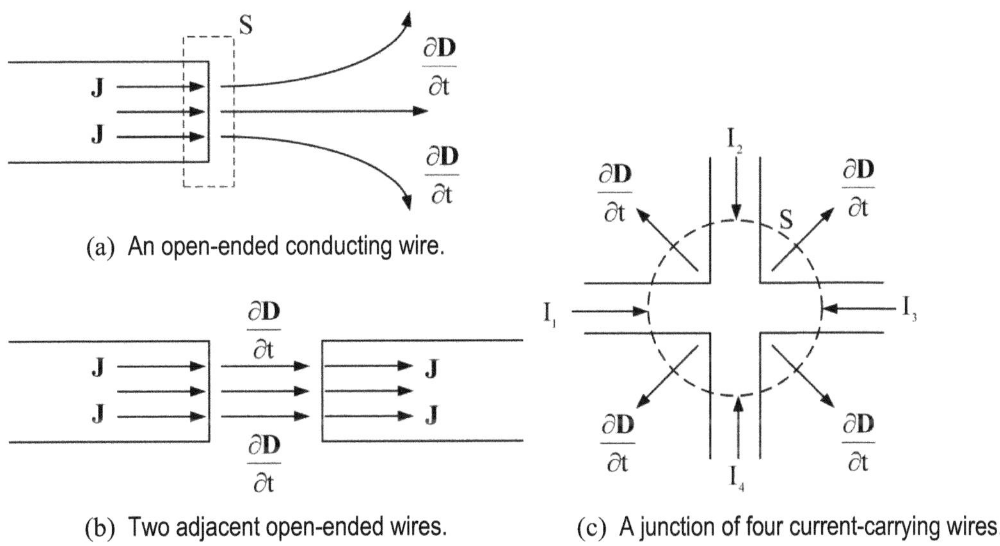

Figure 8-8. Conduction current and displacement current.

Figure 8-8(c) shows a junction of four current-carrying wires. The sum of the currents flowing into the junction is not zero but is equal to the displacement current flowing out. So if we consider any closed surface S such as that of Fig. 8-8(c), the current $I = I_1 + I_2 + I_3 + I_4 = -\oiint_S \mathbf{J} \cdot d\mathbf{s}$ flowing into S equals the displacement current $\oiint_S \frac{\partial \mathbf{D}}{\partial t} \cdot d\mathbf{s}$ flowing out.

The effects of displacement current depend on frequency and geometry. For instance, consider the isolated open-ended wire with radius a in Fig. 8-8(a). For $a << \lambda$ (corresponding to low frequency because wavelength is inversely proportional to frequency) the current at the open end (and the displacement current) is very small and the effect of displacement current is not significant at low frequencies. If, however, another conductor is brought into close proximity, as shown in Fig. 8-8(b), then the effects of displacement current may become significant. In Fig. 8-8(b) for any a/λ, no matter how small, we can choose d (gap) such that the capacitive reactance is small and the circuit looks more like a short circuit rather than an open circuit. In Fig. 8-8(c), we can assume that displacement current is negligible if $a << \lambda$ and the junction is isolated. If however, other conductors, for instance, another junction, are brought into close proximity, then the displacement current may become significant even if $a << \lambda$.

It may be necessary then to take into account the so-called *stray capacitance* to the nearby object. Including the stray capacitance allows an additional path for current flow in the circuit and permits the currents to be balanced. Thus displacement current upsets the Kirchhoff current law that the sum of currents into a junction is zero. The addition of stray or mutual capacitances helps but makes the circuit more complex. There is, however, no end to the various stray capacitances which can be taken into account. Eventually the circuit representation breaks down as frequency increases.

8.5.2. Maxwell's Equations

As discussed earlier Eqs. (8-3) and (8-16) are generalizations of the two curl laws in Eqs. (8-1) and (8-2). While the latter two are valid only for static fields, the former two are valid for general time-varying fields. The two divergence laws in Eqs. (8-1) and (8-2) are still valid for time-varying fields. We collect the four equations, (8-3), (8-16), and the two divergence equations in (8-1), (8-2), for general time-varying fields:

$$\nabla \times \mathbf{E} = -\frac{\partial \mathbf{B}}{\partial t} \quad \text{(Faraday's law)} \tag{8-20a}$$

$$\nabla \times \mathbf{H} = \mathbf{J} + \frac{\partial \mathbf{D}}{\partial t} \quad \text{(Generalized Ampère's law)} \tag{8-20b}$$

$$\nabla \cdot \mathbf{D} = \rho_v \quad \text{(Gauss' law for electric field)} \tag{8-20c}$$

$$\nabla \cdot \mathbf{B} = 0 \quad \text{(Gauss' law for magnetic field)} \tag{8-20d}$$

(Maxwell's equations in differential form)

The four basic equations above are called **Maxwell's equations** and are the basis of electromagnetics. They apply in *any* and *all* macroscopic situations and cover the general time-varying case. They are also valid in any and all moving systems as was shown by Einstein. They are valid in any media including non-linear, anisotropic and non-reciprocal media.

For completeness we need to add the additional relationships among the field quantities of Maxwell's equations. For instance, $\mathbf{J}$ and ρ_v are not independent but are related by the equation of continuity,

$$\nabla \cdot \mathbf{J} = -\frac{\partial \rho_v}{\partial t}$$

For *linear* isotropic media one can add the *constitutive* relations to express $\mathbf{D}$ in terms of $\mathbf{E}$ and $\mathbf{B}$ in terms of $\mathbf{H}$:

$$\mathbf{D} = \varepsilon\,\mathbf{E}, \quad \mathbf{B} = \mu\,\mathbf{H} \tag{8-21}$$

For other media, one merely adds the appropriate relations among the field quantities:

$$\mathbf{D} = \varepsilon_0\,\mathbf{E} + \mathbf{P}, \quad \mathbf{B} = \mu_0\,\mathbf{H} + \mu_0\,\mathbf{M} \tag{8-22}$$

Finally, one should add the Lorentz force law

$$\boxed{\mathbf{F} = q\,(\mathbf{E} + \mathbf{v} \times \mathbf{B})} \tag{8-23}$$

for calculation of electric and magnetic forces. It is important to notice that Maxwell's equations describe the action of charges on fields (how charges produce fields), and the Lorentz force law explains the action of fields on charges (how fields affect charges). Their combination summarizes the entire set of electromagnetic laws.

Note also that the two divergence relations, $\nabla \cdot \mathbf{D} = \rho_v$ and $\nabla \cdot \mathbf{B} = 0$ do not have to be established separately because they can be deduced from the two curl equations (8-20a), (8-20b) and the equation of continuity. This is done merely by taking the divergence of the curl equations and using the equation of continuity.

Maxwell's equations may also be stated in integral form. The first two integral forms may be derived by integrating the two curl equations (8-20 a,b) over surfaces and applying Stokes' theorem. The last two forms are obtained by integrating the two divergence equations (8-20 c,d) over volume V and applying the divergence theorem. The results are:

$$\oint_C \mathbf{E} \cdot d\boldsymbol{\ell} = -\iint_S \frac{\partial \mathbf{B}}{\partial t} \cdot d\mathbf{s} \tag{8-24a}$$

$$\oint_C \mathbf{H} \cdot d\boldsymbol{\ell} = \iint_S \left(\mathbf{J} + \frac{\partial \mathbf{D}}{\partial t} \right) \cdot d\mathbf{s}$$

$$= I + \iint_S \frac{\partial \mathbf{D}}{\partial t} \cdot d\mathbf{s} \tag{8-24b}$$

$$\oint\!\!\oint_S \mathbf{D} \cdot d\mathbf{s} = \iiint_V \rho_v \, dv = Q \tag{8-24c}$$

$$\oint\!\!\oint_S \mathbf{B} \cdot d\mathbf{s} = 0 \tag{8-24d}$$

(Maxwell's equations in integral form)

Comparing the equations with the static forms, we note that there are only two changes, i.e., the additions of the terms $-\iint_S \frac{\partial \mathbf{B}}{\partial t}\cdot d\mathbf{s}$, $\iint_S \frac{\partial \mathbf{D}}{\partial t}\cdot d\mathbf{s}$ to the first, second equations, respectively. However, all vectors and scalars are now functions of time, i.e., $\mathbf{E}(x,y,z,t)$, $\mathbf{B}(x,y,z,t)$, $Q(t)$, etc.

8.5.3. The Wave Equation—Electromagnetic Waves!

The addition of the displacement current term ($\partial\mathbf{D}/\partial t$) ties in electromagnetics with wave action and light. It also answers a number of questions raised in the beginning of this section. Consider the electric and magnetic fields in *free space* (or open air), which require

$$\rho_v = \mathbf{J} = 0 \quad \text{(in a source-free region)}$$

$$\mathbf{D} = \varepsilon_0 \mathbf{E}, \quad \mathbf{B} = \mu_0 \mathbf{H} \quad \text{(in free space or air)}$$

Then Maxwell's equations (8-20) reduce to

$$\nabla \times \mathbf{E} = -\mu_0 \frac{\partial \mathbf{H}}{\partial t} \tag{8-25a}$$

$$\nabla \times \mathbf{H} = \varepsilon_0 \frac{\partial \mathbf{E}}{\partial t} \tag{8-25b}$$

$$\nabla \cdot (\varepsilon_0 \mathbf{E}) = 0 \quad \text{or} \quad \nabla \cdot \mathbf{E} = 0 \tag{8-25c}$$

$$\nabla \cdot (\mu_0 \mathbf{H}) = 0 \quad \text{or} \quad \nabla \cdot \mathbf{H} = 0 \tag{8-25d}$$

Taking the curl of Eq. (8-25a) and using the vector identity (11) in Table 1-5,

$$\nabla\times\nabla\times\mathbf{E} = \nabla(\nabla\cdot\mathbf{E}) - \nabla^2\mathbf{E} = -\nabla\times\left(\mu_0\frac{\partial \mathbf{H}}{\partial t}\right) = -\mu_0\frac{\partial}{\partial t}(\nabla\times\mathbf{H})$$

Using Eqs. (8-25b) and (8-25c),

$$-\nabla^2\mathbf{E} = -\mu_0\frac{\partial}{\partial t}\left\{\varepsilon_0\frac{\partial \mathbf{E}}{\partial t}\right\} = -\mu_0\varepsilon_0\frac{\partial^2\mathbf{E}}{\partial t^2}$$

Therefore, we have

$$\boxed{\nabla^2\mathbf{E} = \mu_0\varepsilon_0\frac{\partial^2\mathbf{E}}{\partial t^2}} \tag{8-26}$$

(Helmholtz's wave equation)

Equation (8-26) is a vector form of the scalar wave equation (8-13) mentioned earlier. We can also replace **E** with **H** in Eq. (8-26). Thus the components of **E**, **H** satisfy the scalar wave equation, which indicates wave propagation at velocity c:

$$c = \frac{1}{\sqrt{\mu_0 \varepsilon_0}} = \frac{1}{(4\pi \times 10^{-7} \times 8.85 \times 10^{-12})^{1/2}} \approx 3 \times 10^8 \text{ [m/sec]}$$

Maxwell could thus predict the velocity of electromagnetic waves. His calculation, with some error in μ_0, ε_0, yielded a velocity of 3.11×10^8 m/sec, which was very close to the measured speed of light. Later more precise measurements showed that the speed of light c is 2.998×10^8 m/sec. So Maxwell was able to conclude that there are electromagnetic waves with a given velocity of propagation in free space, that all components of **E**, **H** propagate at that velocity, and that light is an electromagnetic phenomenon, which was speculated by Faraday 20 years earlier. All of this was possible only because of the addition of the displacement current term.

EXAMPLE 8-5 FIELD SOLUTIONS OF MAXWELL'S EQUATIONS

The electric field of an electromagnetic wave in air, free of sources, is given by

$$\mathbf{E} = \mathbf{a}_x E_o \cos(\omega t - kz)$$

where ω and k are constants.

(a) Find the corresponding magnetic field **H** of this electromagnetic wave.
(b) Confirm that **E** satisfies Gauss' law, Eq. (8-20c).

SOLUTION

(a) If you consider this problem as one in which the electric field (or equivalently the displacement current) induces a magnetic field, you would be tempted to use the integral form of Ampère's law to find the magnetic field. However, the difficulty lies in the prediction of how the magnetic field (**H**) will behave in terms of its direction and spatial dependence. For a source-free problem like this, it is much easier to use the differential form of Maxwell's equations. Knowing **E**, we can calculate **H**, by using the differential form of Faraday's law, Eq. (8-20a), and integrating with respect to time.

$$-\frac{\partial \mathbf{B}}{\partial t} = \nabla \times \mathbf{E} = \begin{vmatrix} \mathbf{a}_x & \mathbf{a}_y & \mathbf{a}_z \\ \frac{\partial}{\partial x} & \frac{\partial}{\partial y} & \frac{\partial}{\partial z} \\ E_x & E_y & E_z \end{vmatrix} \qquad \text{Note that} \quad \begin{aligned} \frac{\partial}{\partial x} &= \frac{\partial}{\partial y} = 0 \\ E_y &= E_z = 0 \end{aligned}$$

$$= \mathbf{a}_y \frac{\partial E_x}{\partial z} = \mathbf{a}_y \frac{\partial}{\partial z}\{E_0 \cos(\omega t - kz)\}$$

$$= \mathbf{a}_y E_0 \, k \sin(\omega t - kz)$$

Since $\mathbf{B} = \mu_o \mathbf{H}$ in air,

$$\mathbf{H} = \int \frac{1}{\mu_0}\frac{\partial \mathbf{B}}{\partial t}\,dt = -\frac{1}{\mu_0}\int \mathbf{a}_y E_0 k \sin(\omega t - kz)\,dt$$

$$= \mathbf{a}_y E_0 \frac{k}{\mu_0 \omega} \cos(\omega t - kz)$$

Note that $\mathbf{E}$ and $\mathbf{H}$ should also satisfy Ampère's law if they represent *electromagnetic* fields. (see Problem 8-18).

(b) In air, free of sources, Gauss' law becomes

$$\nabla \cdot \mathbf{D} = 0 \quad \text{and} \quad \mathbf{D} = \varepsilon_0 \mathbf{E}$$

$$\nabla \cdot \mathbf{D} = \varepsilon_0 \left\{ \frac{\partial E_x}{\partial x} + \frac{\partial E_y}{\partial y} + \frac{\partial E_z}{\partial z} \right\}$$

$$= \varepsilon_0 \frac{\partial}{\partial x} \{E_0 \cos(\omega t - kz)\} = 0$$

Thus Gauss' law is satisfied. One can also show that Gauss' law for $\mathbf{B}$, Eq. (8-20d), is satisfied.

8.5.4. James Clerk Maxwell (1831–1879)

In the previous sections of this Chapter, we outlined some of Maxwell's and Faraday's contributions to time-varying electromagnetic fields. Now let's go back to Maxwell's origins and describe briefly his upbringing, education, and his scientific contributions leading up to his complete formulation of electromagnetic theory.

Maxwell was born in 1831. His only sibling, Elizabeth, died in infancy before James was born. James spent most of his childhood on the family estate called Glenlair, which is 3 miles south of the village Corsock and about 50 miles west of Carlisle, in southwestern Scotland. His parents were John Clerk Maxwell and Francis Cay Maxwell, both of whom took a serious interest in his upbringing and education. His mother took up his education and soon he was reading very widely, including a large number of works in literature and history. It would have been difficult to find a better background for his early upbringing and education. James, like many a bright young lad, took a great interest in the world around him. He was especially interested in learning exactly how things worked and often wanted to perform complex tasks without help.

James's mother died when he was 8. At the age of 10, he was sent to live with his aunt Jane and attend Edinburgh Academy. For completeness, let's list here the schools and colleges which James attended:

Edinburgh Academy (1841–1847)
Edinburgh University (1847–1850)
Cambridge University (1850–1856)

and those at which he taught:

Marischal College (1856–1860)
King's College, London (1860–1865)
Cambridge University (1871–1879)

After a period of adjustment, James made rapid progress in his studies at Edinburgh Academy. He published a mathematical paper at age 15. He was becoming a person of a very kind and generous spirit; he loved animals and rode horses but did not go hunting. He was strong and athletic but did not participate in schoolboy games. He did a lot of walking, swimming, and exercising. He could defend himself but did not pick fights. He could be critical, usually in a whimsical way. He had his own world of his projects and simple toys, often of his own making, and his ideas which he shared with a few friends. He wrote poetry, mostly of a whimsical nature.

At Edinburgh University, James took to the very broad education offered in subjects such as philosophy and logic. He also had access to Professor Edward Forbes's laboratory. He became interested in polarized light and the strain patterns which it revealed; this led to an excellent paper on elastic solids.*

James found Cambridge University to be a very convivial place. He made many friends and joined in discussions on every subject imaginable. Often he could offer his peers a new approach to a subject because of his wide reading. He had no interest whatsoever in institutional discussions of the conflict between science and religion, categorizing this issue as of a personal nature. In his senior year (1854) there were two important mathematical exams: the tripos and the Smith's prize exam. James took second in the tripos and shared first in the Smith's prize.

He would stay on for two more years at Cambridge, working on electromagnetics among other things and publishing a paper entitled "On Faraday's Lines of Force," in which he developed a fluid analogy for static electric and magnetic problems.

Maxwell reminds us of Benjamin Franklin and Michael Faraday in his careful and generous way of discussing the research of others. We have already seen how generous Maxwell was with Gibbs's, Coulomb's, and Ampère's work. James praises Faraday's work in particular because he carefully delineates all steps he has taken. Maxwell particularly admired Faraday's "lines of force." Faraday had sensed something of great importance. It was the beginning of field theory, fields that may occupy all space. Faraday was criticized by some for "fuzzy ideas" but received support from Maxwell. Maxwell was a great contributor to the concept of field theory but Faraday also deserves credit.

In 1856, Maxwell was appointed Professor at Marischal College in Aberdeen. He was there for 4 years, was married to Katherine Mary Dewar in 1858, and then was appointed Professor at King's College, London, in 1860.

In 1861–62, James published a paper called "On Physical Lines of Force." This paper used a very complex mechanical model with springy, spinning cells, and much else. The springiness led to the existence of displacement current in the electromagnetic system and electromagnetic waves (as shown in Section 8.5.3).

James had a complete model for electromagnetics, but he was not at all content. He wanted a more transparent system, namely a set of equations of motion which could then be solved by LaGrange's methods, a set of mathematical equations rather than the spinning cells. He worked for several more years on this project and

*Basil Mahon, *The Man Who Changed Everything: The Life of James Clerk Maxwell*, Chapter 3, John Wiley and Sons, Hoboken, NJ, 2003.

published it in 1865, under the title, "A Dynamical Theory of the Electromagnetic Field." This was also a complex system, consisting of 20 equations in 20 unknowns, which was reduced to the four equations with which we are familiar, namely, equations (8-20a) to (8-20d). A few years after Maxwell's death the reduction was completed by Oliver Heaviside*, using the vector notation system of Josiah Willard Gibbs.

So now Maxwell had two separate solutions to electromagnetic problems, but he was still not satisfied. He had benefitted from the respective models in developing his own understanding but he did not want the models to play a dominant role. They were sufficient but not necessary. It was time to downplay the models and so he wrote, "I have on a former occasion attempted to describe a particular kind of strain.... In the present paper I avoid any hypothesis of this kind."** Thus Maxwell focused attention on the mathematical steps and the final results. This was a difficult but wise choice. Nowadays we set aside the springy cells as well as LaGrange's system and start with Maxwell's equations. But in order to understand how Maxwell arrived at his results, we need to recall the scaffolding and the steps he took. Yet, in 1865, Maxwell's results were just too deep and too complicated to be fully accepted. They were, in a sense, held in abeyance. Maxwell went back to work and in 1873 published his monumental, 1,000-page] *A Treatise on Electricity and Magnetism.* Eight years after Maxwell died, Heinrich Hertz observed electromagnetic waves in 1887 and verified Maxwell's theory.

Einstein said, "One scientific epoch ended and another began with James Clerk Maxwell."

8.6. BOUNDARY CONDITIONS FOR TIME-VARYING FIELDS

The boundary conditions for static electric fields were derived in Section 3.5 and those for static magnetic fields were derived in Section 7.5. The boundary conditions for time-varying fields are shown to be the same as those for static fields and they are given below:

$$\mathbf{E}_{1t} = \mathbf{E}_{2t} \tag{8-27a}$$

$$\mathbf{a}_{n2} \times (\mathbf{H}_1 - \mathbf{H}_2) = \mathbf{J}_s \tag{8-27b}$$

$$D_{1n} - D_{2n} = \rho_s \tag{8-27c}$$

$$B_{1n} = B_{2n} \tag{8-27d}$$

The subscripts t and n denote the *tangential* and *normal* components, respectively. Note that $\mathbf{a}_{n2}$ is a unit vector normal to the boundary interface, pointing from region 2 into region 1. The boundary conditions at the interface between two different media are derived using procedures quite similar to those used for static problems (see Figure 8-9).

* Bruce J. Hunt, "Oliver Heaviside: A first-rate oddity," *Physics Today*, pp. 48–54, November 2012.

** R.A.R. Tricker, *The Contributions of Faraday and Maxwell to Electrical Science*, p. 266, Pergamon Press, NY, 1966.

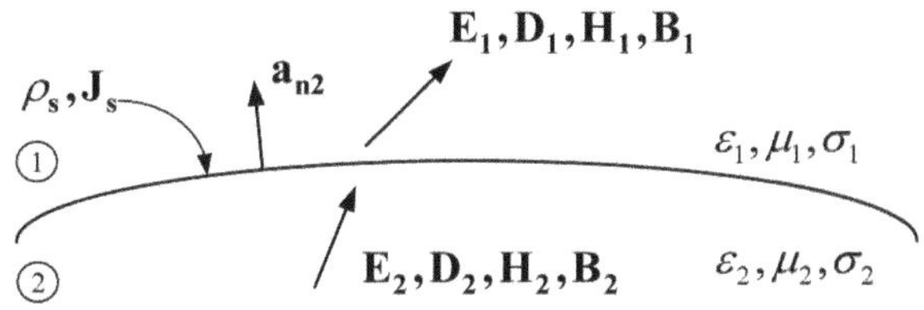

Figure 8-9. Boundary conditions at an interface between two different media.

We start with Maxwell's equations in integral form, Eqs. (8-24a)-(8-24d), and apply them to the closed contour C and the Gaussian pillbox S as shown in Sections 3.5 and 7.5. The integral forms, Eqs. (8-24c) and (8-24d), are identical to the static forms and so the process is identical, yielding the boundary conditions, Eqs. (8-27c) and (8-27d), which are identical to Eqs. (3-25) and (7-14). The integral forms, Eqs. (8-24a) and (8-24b), have the additional terms, $-\iint_S \frac{\partial \mathbf{B}}{\partial t} \cdot d\mathbf{s}$ and $\iint_S \frac{\partial \mathbf{D}}{\partial t} \cdot d\mathbf{s}$, respectively, which are not present in the static case. The corresponding additional terms are integrated over a surface whose area vanishes in the limit ($\alpha \to 0$), so these additional terms do not contribute to the boundary conditions, Eqs. (3-24) and (7-15). Hence we again have the same boundary conditions, Eqs. (8-27a) and (8-27b), as before. Therefore, four boundary conditions are identical in form to the static boundary conditions.

There are two special cases for boundary conditions which will prove to be useful in later chapters.

Case 1: Dielectric—Perfect Conductor Interface

When the medium 2 is a perfect conductor which has an infinite conductivity ($\sigma = \infty$), in order to support a finite current ($\mathbf{J} = \sigma\mathbf{E}$) the electric field $\mathbf{E}$ inside the perfect conductor must approach zero. Thus $\mathbf{E}_2 = 0$, which also leads to $\mathbf{H}_2 = 0$ for the time-varying fields, because $\nabla \times \mathbf{E} = -\mu \frac{\partial \mathbf{H}}{\partial t} = 0$ and $\mathbf{H}$ is assumed to be not a static field. Inside a perfect conductor ($\sigma = \infty$), all time-varying fields are zero: $\mathbf{E} = \mathbf{H} = 0$. The first two boundary conditions, Eqs. (8-27a) and (8-27b), reduce to

$$\boxed{\begin{aligned} &\mathbf{E}_{1t} = 0 \\ &\mathbf{a}_{n2} \times \mathbf{H}_1 = \mathbf{J}_s \end{aligned}} \tag{8-28}$$

at a dielectric-perfect conductor interface.

These boundary conditions will be used in the waveguide problems in Chapter 11. Here we have written only the boundary conditions for the tangential components because for time-varying field problems it is sufficient to use these two conditions.

Case 2: Dielectric—Dielectric Interface

When both media are not perfect conductors, the current can not flow on the surface, i.e., $\mathbf{J}_s = 0$. If the media have finite conductivity, a volume current $\mathbf{J}$ can flow inside the media but not on the surface. In this case, the first two boundary conditions become

$$\boxed{\begin{aligned} \mathbf{E}_{1t} &= \mathbf{E}_{2t} \\ \mathbf{H}_{1t} &= \mathbf{H}_{2t} \end{aligned}} \qquad (8\text{-}29)$$

at dielectric-dielectric interface.

These boundary conditions will be used in the wave reflection and transmission problems in Chapter 10.

Finally, the boundary condition for the current density $\mathbf{J}$ at an interface differs in form from that of the steady-current case, Eq. (5-14). It can be derived directly from the equation of continuity whose integral form is

$$\iiint_V \nabla \cdot \mathbf{J}\, dv = \oiint_S \mathbf{J} \cdot d\mathbf{s} = -\iiint_V \frac{\partial \rho_v}{\partial t}\, dv$$

When it is applied to the pillbox S, the result is

$$\boxed{J_{1n} - J_{2n} = -\frac{\partial \rho_S}{\partial t}} \qquad (8\text{-}30)$$

It is useful to note the boundary condition for normal $\mathbf{J}$. It is not necessary to include it in our basic set, Eq. (8-27), since $\mathbf{J}$ can be determined from $\mathbf{E}$ and $\rho_s(t)$ can be determined from $\mathbf{D}$.

8.7 FLOW OF ELECTROMAGNETIC POWER: POYNTING'S THEOREM

In this last section we derive the theorem that illustrates conservation of electromagnetic energy and also shows the direction of electromagnetic power flow. We start with Maxwell's equations. If we dot-multiply Eq. (8-20a) with $\mathbf{H}$ and Eq. (8-20b) with $\mathbf{E}$ and subtract the second equation from the first equation, we have

$$\begin{aligned} &\mathbf{H} \cdot (\nabla \times \mathbf{E}) - \mathbf{E} \cdot (\nabla \times \mathbf{H}) \\ &= \nabla \cdot (\mathbf{E} \times \mathbf{H}) = -\mathbf{H} \cdot \frac{\partial \mathbf{B}}{\partial t} - \mathbf{E} \cdot \frac{\partial \mathbf{D}}{\partial t} - \mathbf{E} \cdot \mathbf{J} \end{aligned} \qquad (8\text{-}31)$$

Here we have used a vector identity (6) in Table 1-5. If the medium is linear (permittivity ε and permeability μ) and ε and μ do not depend on time (such a medium is called non-dispersive), then

$$\mathbf{E}\cdot\frac{\partial \mathbf{D}}{\partial t}=\mathbf{E}\cdot\frac{\partial}{\partial t}(\varepsilon\mathbf{E})=\varepsilon\mathbf{E}\cdot\frac{\partial \mathbf{E}}{\partial t}=\frac{1}{2}\varepsilon\frac{\partial}{\partial t}\{\mathbf{E}\cdot\mathbf{E}\}$$
$$=\frac{\partial}{\partial t}\left\{\frac{1}{2}\mathbf{E}\cdot(\varepsilon\mathbf{E})\right\}=\frac{\partial}{\partial t}\left\{\frac{1}{2}\mathbf{E}\cdot\mathbf{D}\right\} \quad (8\text{-}32a)$$

Similarly,

$$\mathbf{H}\cdot\frac{\partial \mathbf{B}}{\partial t}=\mathbf{H}\cdot\frac{\partial}{\partial t}(\mu\,\mathbf{H})=\frac{1}{2}\mu\frac{\partial}{\partial t}\{\mathbf{H}\cdot\mathbf{H}\}=\frac{\partial}{\partial t}\left\{\frac{1}{2}\mathbf{H}\cdot\mathbf{B}\right\} \quad (8\text{-}32b)$$

Substituting Eq. (8-32) in Eq. (8-31), we obtain

$$\boxed{\nabla\cdot(\mathbf{E}\times\mathbf{H})=-\frac{\partial}{\partial t}\left\{\frac{1}{2}(\mathbf{E}\cdot\mathbf{D}+\mathbf{H}\cdot\mathbf{B})\right\}-\mathbf{J}\cdot\mathbf{E}} \quad (8\text{-}33)$$

Equation (8-33) is called **Poynting's theorem** in differential form. The first term on the right side is the negative time rate of change of the sum of electric and magnetic energy densities (or energy per unit volume). See Eqs. (3-46) and (7-38). The second term is, if $\mathbf{J}=\sigma\mathbf{E}$, the negative of the Joule heat power per unit volume (see Eq. (5-12)). The quantity on the left is in question. In order to interpret this equation in terms of power flow, we integrate both sides of Eq. (8-33) over volume V bounded by closed surface S:

$$\iiint_V \nabla\cdot(\mathbf{E}\times\mathbf{H})\,dv=-\frac{d}{dt}\iiint_V \frac{1}{2}(\mathbf{E}\cdot\mathbf{D}+\mathbf{H}\cdot\mathbf{B})\,dv-\iiint_V \mathbf{J}\cdot\mathbf{E}\,dv$$

Applying the divergence theorem to the left side:

$$\oiint_S (\mathbf{E}\times\mathbf{H})\cdot d\mathbf{s}=-\frac{d}{dt}\iiint_V \frac{1}{2}(\mathbf{E}\cdot\mathbf{D}+\mathbf{H}\cdot\mathbf{B})\,dv-\iiint_V \mathbf{J}\cdot\mathbf{E}\,dv$$

Rewriting:

$$\boxed{-\iiint_V (\mathbf{J}\cdot\mathbf{E})\,dv=\frac{d}{dt}\iiint_V \frac{1}{2}(\mathbf{E}\cdot\mathbf{D}+\mathbf{H}\cdot\mathbf{B})\,dv+\oiint_S (\mathbf{E}\times\mathbf{H})\cdot d\mathbf{s}} \quad (8\text{-}34)$$

This is Poynting's theorem in integral form. Let us interpret each term.

Case 1: If there are no sources of emf in the volume V, then $-\iiint_V \mathbf{J} \cdot \mathbf{E}\, dv$ is the negative of the Joule heat or power dissipation in V (see Section 5.6). Note that when $\mathbf{J}$ is a conduction current ($\mathbf{J}_c = \sigma\mathbf{E}$), $\mathbf{J}_c \cdot \mathbf{E} = \sigma\mathbf{E} \cdot \mathbf{E} = \sigma|\mathbf{E}|^2$ is positive.

Case 2: If there is a source, then this term becomes positive and it is equal to the power generated by the source in V. Note that when $\mathbf{J}$ is a source current ($\mathbf{J}_i$), it is anti-parallel to $\mathbf{E}$ and $\mathbf{J}_i \cdot \mathbf{E}$ is negative.

The first term on the right side of Eq. (8-34) is the rate of change (increase) of electromagnetic energy stored in V, i.e., $\frac{d}{dt}(W_e + W_m)$ (see Sections 3.9.3 and 7.8). The second term on the right can then be interpreted as the power flowing out of the volume V through the surface S, carried by the electromagnetic fields. In other words, in the case of a source present in V (Case 2), Poynting's theorem, Eq. (8-34), states that

> *The net power generated by the source in a certain volume equals the sum of the rate of increase of electromagnetic energy stored and the electromagnetic power flowing out of the volume.*

This represents the conservation of electromagnetic energy in V. If the medium in V is dissipative ($\mathbf{J} = \sigma\mathbf{E}$ is present), the interpretation is different but electromagnetic energy is still conserved. Thus the special quantity

$$\boxed{\mathbf{S} = \mathbf{E} \times \mathbf{H}} \tag{8-35}$$

denotes the *power* (or energy per unit time) *per unit area* carried by the electric and magnetic fields ($\mathbf{E}$, $\mathbf{H}$). It is called the **Poynting vector** which gives us the direction of electromagnetic energy flow. We will see many applications of the Poynting vector later.

EXAMPLE 8-6 POWER FLOW IN A PARALLEL-PLATE CAPACITOR

Consider the parallel-plate capacitor in Example 8-4.

(a) Find the magnitude and direction of the Poynting vector $\mathbf{S}$ at a point on the cylindrical surface ($\rho = a$) of the dielectric.
(b) Integrate $\mathbf{S}$ over the cylindrical surface of the dielectric and show that it is equal to the rate of change of the stored electrostatic energy.

SOLUTION

(a) In Example 8-4, we have obtained

$$\mathbf{E} = -\mathbf{a}_z \frac{V(t)}{d}, \quad \mathbf{H} = -\mathbf{a}_\phi \frac{\varepsilon}{d} \frac{dV}{dt} \frac{a}{2} \qquad (\text{at } \rho = a)$$

Thus

$$\mathbf{S} = \mathbf{E} \times \mathbf{H} = \mathbf{a}_z \times \mathbf{a}_\phi \frac{V(t)}{d} \frac{\varepsilon a}{2d} \frac{dV}{dt} = -\mathbf{a}_\rho \frac{\varepsilon a}{2d^2} V \frac{dV}{dt}$$

The electromagnetic power flows in a radial direction.

(b) Noting that $d\mathbf{s} = \mathbf{a}_\rho\, a\, d\phi\, dz$, the power flowing *out* of the capacitor is

$$\oint\!\!\oint_S (\mathbf{E} \times \mathbf{H}) \cdot d\mathbf{s} = \int_0^d \int_0^{2\pi} \left\{ -\frac{\varepsilon a}{2d^2} V \frac{dV}{dt} \right\} \mathbf{a}_\rho \cdot \mathbf{a}_\rho \, a\, d\phi\, dz$$

$$= -\frac{\varepsilon a}{2d^2} V \frac{dV}{dt} \cdot (2\pi a) d = -\frac{\pi a^2 \varepsilon}{d} V \frac{dV}{dt}$$

The power flowing into the capacitor is $\frac{\pi a^2 \varepsilon}{d} V \frac{dV}{dt}$. The electrostatic energy stored in the capacitor is

$$W_e = \iiint_V \frac{1}{2} \mathbf{E} \cdot \mathbf{D}\, dv = \iiint_V \frac{1}{2} \varepsilon |\mathbf{E}|^2 dv \qquad (\mathbf{E} \text{ is uniform in } V)$$

$$= \frac{1}{2} \varepsilon \left\{ \frac{V(t)}{d} \right\}^2 (\pi a^2 d) = \frac{\pi a^2 \varepsilon}{2d} V^2(t)$$

and

$$\frac{dW_e}{dt} = \frac{\pi a^2 \varepsilon}{2d} \frac{d}{dt} \left\{ V^2(t) \right\} = \frac{\pi a^2 \varepsilon}{d} V \frac{dV}{dt}$$

Therefore, the power flowing into the capacitor is equal to the rate of increase of the stored electrostatic energy. This is the illustration of Poynting's theorem. For this problem, there is no source between the plates (in volume V). The medium is a perfect dielectric; thus there is no power dissipation. Then Eq. (8-34) reduces to

$$-\oint\!\!\oint_S (\mathbf{E} \times \mathbf{H}) \cdot d\mathbf{s} = \frac{d}{dt} \iiint_V \frac{1}{2} (\mathbf{E} \cdot \mathbf{D} + \mathbf{H} \cdot \mathbf{B}) dv = \frac{d}{dt}(W_e + W_m)$$

Assuming low frequency, the magnetic energy associated with $\mathbf{H}$ is negligible compared to the electric energy. To describe the energy conservation more accurately, we should recalculate $\mathbf{E}$ and $\mathbf{H}$, including all time-varying effects.

PROBLEMS

8-1. A non-uniform time-varying magnetic field $\mathbf{B}$ exists in a cylindrical region of radius a:

$$\mathbf{B} = \mathbf{a}_z B_0 \frac{\rho}{a} \cos\omega t \quad (\rho < a)$$

where ω is an angular frequency and B_0 is a constant.

(a) Find the electric field $\mathbf{E}$ induced by $\mathbf{B}$ everywhere ($\rho < a$ and $\rho > a$).
(b) Now you put a circular conducting loop of radius b ($b < a$) and resistance $R[\Omega]$ inside the cylindrical region. Find the induced *emf* $v(t)$ and current $i(t)$ on the loop.
(c) Determine the direction of current flow using Lenz's law. Consider when $\omega t = \pi/4$.

8-2. Repeat the same questions in Problem 8-1 when the time-varying magnetic field is given by

$$\mathbf{B} = \mathbf{a}_z B_0 \left(\frac{\rho}{a}\right)^2 \sin\omega t \ (\rho < a)$$

8-3. Consider a coaxial cable with an outer conductor of radii b, c ($c > b$) surrounding an inner conductor of radius a. An alternating current $I(t) = I_0 \cos\omega t$ flows down the inner wire and returns back along the outer conducting cylinder. Assume that the current flows only on the surface ($\rho = a$) of the inner conductor and on the inner surface ($\rho = b$) of the outer conductor. See Figure 6-19 of Problem 6-3.

(a) The time-varying magnetic field produced by currents will induce an electric field by Faraday's law. In what direction does the induced electric field point?
(b) Determine $\mathbf{E}$ as a function of ρ in each region:
(i) $\rho < a$, (ii) $a < \rho < b$, and (iii) $\rho > b$

8-4. Consider a very long solenoid, of radius a, length ℓ ($\ell >> a$), and number of turns N, driven by an alternating current:

$$i(t) = I_0 \sin(2\pi\, 10^6\, t) \quad [A]$$

(a) Find the magnetic field $B(t)$ by applying Ampère's law. Include the direction of $\mathbf{B}(t)$ in your answer.
(b) According to what law is the electric field induced? In what direction does the induced electric field point (radial, circumferential, or longitudinal)?
(c) Find the electric field $\mathbf{E}$ at a distance ρ from the axis of the solenoid ($\rho < a$).
(d) Now you place a small solenoid, of radius b ($b < a$), length L and number of turns N_s, inside the long solenoid, coaxial. Find the voltage v(t) induced in the small coil.

8-5. **(Induction Heating)**
Consider Example 8-1. Suppose a long circular cylindrical shell of length ℓ, radius b ($b < a$) and conductivity σ is placed coaxially in the cylindrical region.

(a) Find the power dissipated in the conducting shell, by making use of Joule's law in Eq. (5-12). Assume that the thickness of the shell is d and $d << b$. This is the principle behind the *induction heating* in which the metal can be heated or melted using the electromagnetic induction. The induction furnaces can reach a much higher temperature than ordinary furnaces that use coal or oil.

(b) Show that $P = V(t)\, i(t)$.

8-6. A rectangular loop of sides $a \times b$ [m²] is placed in the xy plane as shown in Figure 8-10. The following ac magnetic field is applied:

$$\mathbf{H} = \mathbf{a}_x H_x \cos \omega t + \mathbf{a}_z H_z \sin \omega t \text{ [A/m]}$$

(a) Find the induced voltage $V(t)$ [V] with the polarity as shown.

(b) Now close the loop and assume the resistance of the loop R [Ω]. Find the induced current $i(t)$ [A] in the loop. What is the direction of the current flow at $\omega t = \pi/4$, assuming $H_x > 0$, $H_z > 0$?

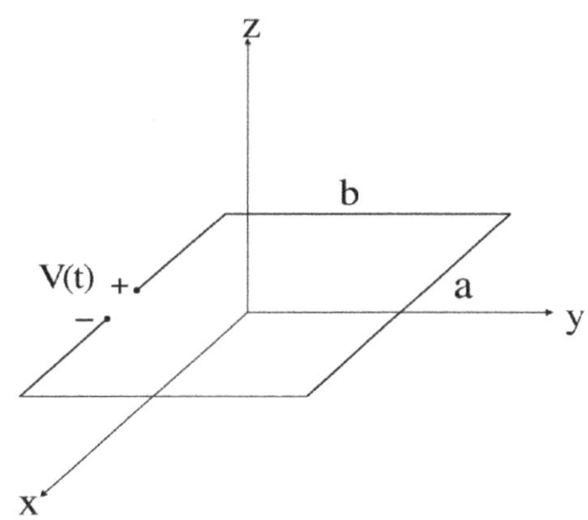

Figure 8-10. A rectangular loop in a time-varying magnetic field.

8-7. The rectangular conducting loop of $a \times b$ [m²] is placed in the xy plane in a non-uniform time-varying magnetic field

$$\mathbf{B} = \mathbf{a}_z B_0 \cos\left(\frac{\pi}{b} y\right) \sin(2\pi f t) \quad [\text{T}]$$

where $f = 200$ [kHz] and $B_0 = 0.1$ [T]. Assume that the loop is centered at the origin and the resistance of the loop is $R = 20$ [kΩ]. Let $a = 0.5$ [m], $b = 0.7$ [m].

(a) Find the induced *emf* $V(t)$ and current $i(t)$ and calculate their peak values.

(b) Determine the direction of current flow at $t = 1$ μsec using (i) Lenz's law and (ii) Faraday's law.

8-8. Consider a very long wire and a square loop of sides $\ell \times \ell$ [m^2], separated by a distance d [m], as shown in Figure 6-15(a).

(a) The long wire carries an alternating current $i_1(t) = I_1 \cos \omega t$. Find the induced voltage $V_2(t)$ in the loop.

(b) With the result of mutual inductance from Example 7-6, show that V_2 and i_1 satisfy $V_2 = L_{21} \frac{di_1}{dt}$.

8-9. An equilateral triangular loop, with side length a and resistance R [Ω], is at a distance d from a very long straight wire which carries an alternating current

$$i_1(t) = I_0 \sin \omega t \text{ [A]}$$

(See Figure 7-23). Find the current $i_2(t)$ induced in the triangular loop.

8-10. An inductor is formed by winding 20 turns of a thin wire around a wooden dowel rod that has a diameter of 1 cm. If a uniform, sinusoidal magnetic field with an rms magnitude of 0.01 Wb/m^2 and a frequency of 30 kHz is directed along the axis of the dowel rod, determine the voltage induced between the two ends of the wire, assuming that the two ends are close together.

8-11. (a) A pair of conducting rails has a shorting metal bar moving at velocity v along the rails, as shown in Figure 8-11. A uniform ac magnetic field $\mathbf{B} = \mathbf{a}_z B_0 \cos \omega t$ [T] is directed out of the page. Determine the voltage V(t) induced across the small gap a-b as shown. Note that your result should consist of both the *transformer emf* and *motional emf*. Assume that the bar is at $x = d$ at $t = 0$.

(b) This circuit can be used as a **rail gun** in which a very strong current can accelerate the metal bar or projectile to a very high velocity. Now assume that **B** is a dc magnetic field, a voltage source V is *applied* across the terminals a-b and the total resistance of the rails and the bar is R [Ω]. Calculate the force acting on the metal bar. What should the polarity of V be, i.e., positive or negative?

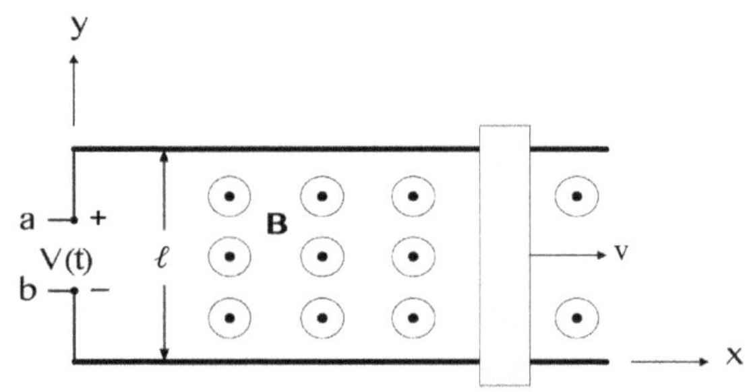

Figure 8-11. A metal bar moving along conducting rails.

8-12. **(AC generator)**
Consider a loop of perfectly conducting wire with side dimensions of ℓ and d rotating with the angular velocity ω [radians/sec] in a constant dc magnetic field $\mathbf{B} = \mathbf{a}_y B_0$, as shown in Figure 8-12. According to Faraday's law, the emf or voltage is induced in a small gap in the loop. Assume that the plane of the loop starts in the

xz plane at $t = 0$ and find the induced voltage $V(t)$. This is a simple *ac generator*. Suppose the loop is closed. Then there will be an induced current flow $i(t)$. Using Lenz's law, find the direction of i at $t = 0^+$.

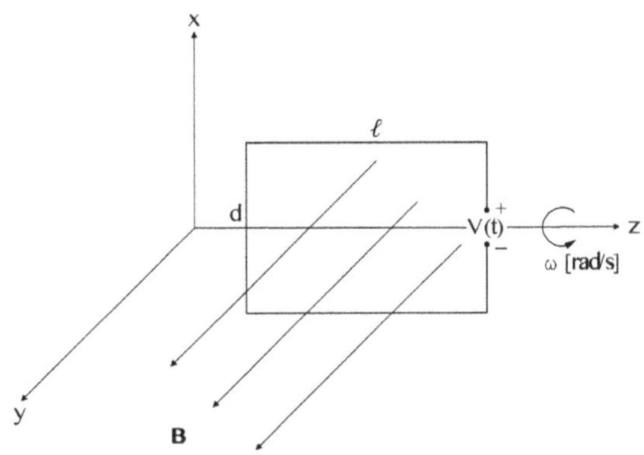

Figure 8-12. A rotating loop in a magnetic field.

8-13. Consider Example 8-2. Assume that $I(t) = I_0 \cos \omega t$ and repeat the same questions (a), (b). Note that your result should consist of both the *transformer emf* and *motional emf*.

8-14. **(Faraday disk generator)**
A circular metal disk of radius a rotates around a vertical axis with a velocity of N revolutions per second through a uniform dc magnetic field $\mathbf{B}$ parallel to the axis of rotation. Brush contacts are attached to the axle and the outer rim of the disk, as shown in Figure 8-13. Determine the voltage $V(t)$ induced between the two brushes.

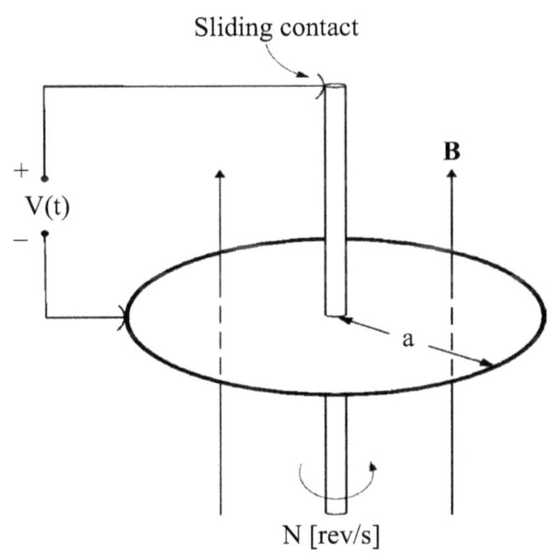

Figure 8-13. Faraday disk generator.

8-15. A long cylindrical magnet of length ℓ and radius a ($\ell >> a$) carries a uniform magnetization M_0 parallel to its axis. Suppose this magnet passes through a circular conducting loop of radius b ($b > a$) with velocity v_0. The magnet starts at a distance from the loop, passes through the loop, and moves away from the loop.

(a) Find the approximate magnetic field inside the magnet. Note that $\ell >> a$.
(b) Describe the emf or voltage induced in the loop, as the magnet moves, as a function of time.

8-16. Consider the capacitor problem in Example 8-4.

(a) Find the magnetic field $\mathbf{H}$, produced by the displacement current, outside the capacitor ($\rho > a$). Neglect the fringing electric field.
(b) Find the relationship between the current entering the top plate of the capacitor $I(t)$ and the voltage $V(t)$ by making use of generalized Ampère's law. Does it agree with the V-I relationship for the capacitor that you learned in circuit theory?

8-17. (Displacement current vs. conduction current)
A semiconductor (silicon) has $\mu = \mu_0$, $\varepsilon = 11.7\,\varepsilon_0$ and $\sigma = 4.4 \times 10^{-4}$ S/m. An electric field $\mathbf{E} = \mathbf{a}_z\, 3 \cos \omega t$ [V/m] exists in the semiconductor.

(a) Find the conduction current density J_c and displacement current density J_d in the semiconductor.
(b) Under what frequency range is the displacement current dominant over the conduction current?
(c) Are the conduction current density and the displacement current density *in phase*? If not, what is the difference in phase angle?

8-18. Consider Example 8-5.

(a) The given electric field $\mathbf{E}$ and the magnetic field $\mathbf{H}$ obtained in the example should also satisfy Ampère's law if they represent electromagnetic fields. Then ω (angular frequency) and k (known as *wave number*) have a certain relationship. Find this relationship, using the differential form of Ampère's law. Note that the medium is air and free of sources.
(b) Confirm that $\mathbf{H}$ satisfies the last Maxwell's equation, Eq. (8-20d).

8-19. (Field solutions of Maxwell's equations)
A magnetic field in a *lossless* dielectric medium (*free of sources*) is given by

$$\mathbf{H} = \mathbf{a}_x H_m \sin(\omega t - kz)$$

where ω and k are constants. The medium has permittivity ε and permeability μ.

(a) Find the corresponding electric field $\mathbf{E}$, using the differential form of Ampère's law.
(b) $\mathbf{H}$ given above and $\mathbf{E}$ you obtained in (a) should also satisfy Faraday's law if they represent electromagnetic fields. Then ω (angular frequency) and k (wave number) have a certain relationship. Find this relationship, using the differential form of Faraday's law. Note that the medium is lossless and free of sources. The relationship is known as the "dispersion relation."
(c) Confirm that the remaining Maxwell's equations are satisfied.

8-20. An electric field in *air* (free of sources) is given by

$$\mathbf{E} = \mathbf{a}_y E_m \sin(\alpha x) \cos(\omega t - \beta z)$$

Find the corresponding magnetic field from Faraday's law. Under what conditions on (α, β, ω) do these fields satisfy all of Maxwell's equations?

8-21. The electric field of an AM radio signal propagating in air is given by

$$\mathbf{E}(x, t) = \mathbf{a}_y 3 \sin(\omega t - 0.02\, x) \ [\text{V/m}]$$

(a) Find the corresponding magnetic field $\mathbf{H}(x, t)$, using one of Maxwell's equations.
(b) Determine the frequency of the AM signal f [Hz], using another one of Maxwell's equations.

8-22. **(Boundary conditions on dielectric interface)**
An interface between two media lies in the xy plane at $z = 0$. Medium 1 has permittivity ε_1, permeability μ_1 and conductivity σ_1, and medium 2 has parameters ε_2, μ_2, and σ_2. The electric field in region 1 at the interface ($z = 0$) is given by

$$\mathbf{E}_1 = \mathbf{a}_x E_{1x} + \mathbf{a}_y E_{1y} + \mathbf{a}_z E_{1z}$$

Determine $\mathbf{E}_2$ at $z = 0$ if both regions are non-conducting, that is, $\sigma_1 = \sigma_2 = 0$.

8-23. **(Boundary conditions on perfect conductor)**
An interface between two media lies in the xz plane at $y = 0$. Medium 1 is air and medium 2 is a perfect conductor ($\sigma_2 = \infty$). The electric and magnetic fields in region 1 at the interface ($y = 0$) are to be determined:

$$\mathbf{E}_1 = \mathbf{a}_x E_{1x} + \mathbf{a}_y E_{1y} + \mathbf{a}_z E_{1z} \ [\text{V/m}]$$
$$\mathbf{H}_1 = \mathbf{a}_x H_{1x} + \mathbf{a}_y H_{1y} + \mathbf{a}_z H_{1z} \ [\text{A/m}]$$

There exist the surface charge ρ_s [C/m^2] and the surface current $\mathbf{J}_s = \mathbf{a}_x J_{so}$ [A/m] on the surface of the perfect conductor. Determine the values of E_{1x}, E_{1y}, E_{1z}, H_{1x}, H_{1y} and H_{1z}.

8-24. Derive the boundary conditions, Eq. (8-27), for time-varying electric and magnetic fields.

8-25. (Poynting's theorem)
Consider Examples 8-4 and 8-6. Let $V(t) = V_0 \sin \omega t$.

(a) Assume that the dielectric between the plates is not a perfect insulator and has non-zero conductivity σ [S/m]. Including both the displacement current and conduction current, recalculate the magnetic field $\mathbf{H}$ inside the capacitor ($\rho \le a$).
(b) Find the Poynting vector $\mathbf{S}$ at a point on the cylindrical surface ($\rho = a$) of the lossy dielectric. What is the direction of power flow?
(c) Check if Poynting's theorem, Eq. (8-34), is satisfied for the volume V enclosed by the two plates and the cylindrical surface ($\rho = a$). Note that there is no source current in the volume V and for low frequency (when ω is small) the magnetic energy associated with $\mathbf{H}$ is negligible (or much smaller) compared to the electric energy associated with $\mathbf{E}$.

8-26. (Poynting vector)
Consider the electromagnetic fields shown in Example 8-5. Determine the Poynting vector $\mathbf{S}$ and calculate the instantaneous power crossing the surface area bounded by $x = 0$, $x = a$ [m], $y = 0$, and $y = b$ [m]. What is the direction of power flow?

8-27. (Poynting vector in a solenoid)
A very long solenoid with n turns per unit length and radius a carries a current i that is increasing at a constant rate K, i.e., $i(t) = Kt$.

(a) Calculate the magnetic field $\mathbf{B}$ and the induced electric field $\mathbf{E}$ at a point inside the solenoid ($\rho \le a$).
(b) Find the Poynting vector $\mathbf{S}$ at a point on the cylindrical surface ($\rho = a$) of the solenoid. What is the direction of power flow?
(c) Consider a cylindrical solenoid volume of radius a and length ℓ. Find the power flowing into the solenoid and the magnetic energy stored in the volume. Confirm that Poynting's theorem, Eq. (8-34), is satisfied. Here the electric energy associated with $\mathbf{E}$ is much smaller than the magnetic energy.

8-28. (Power flow in a coaxial cable)
Consider a coaxial cable of Example 4-2 (see Figure 4-2). A voltage $V(t)$ is applied between inner and outer conductors. Assume also that the inner and outer conductors carry the currents $I(t)$ and $-I(t)$ (see Example 6-2(d) and Figure 6-4). Calculate the Poynting vector $\mathbf{S}$ from the electric and magnetic fields and the power transported down the cable.

ELECTROMAGNETIC WAVES

CHAPTER 9

9.1. INTRODUCTION

In the previous chapter we learned the two fundamental laws for time-varying electric and magnetic fields, namely, Faraday's law and Maxwell's correction on Ampère's law. Combined with the laws for static fields, a set of four laws, known as Maxwell's equations, govern all electromagnetic phenomena involving both static and dynamic (time-varying) electric and magnetic fields. In particular, we showed briefly in Section 8.5.3 that a combination of Faraday's law (time-varying magnetic field induces an electric field) and Maxwell's displacement current effect (time-varying electric field induces a magnetic field) leads to the wave equation, whose solution will represent the "electromagnetic wave."

In this chapter and next chapter we study the electromagnetic waves as solutions of Maxwell's equations and their properties in detail—propagation, attenuation, dispersion, polarization, reflection, and transmission. We begin by deriving the wave equation from Maxwell's equations in a source-free region and show that the solutions represent a "wave," called an electromagnetic wave. Then we introduce the complex phasor to treat the time-harmonic (or sinusoidal) electromagnetic fields.

First, we consider the propagation of electromagnetic waves in an unbounded, lossless medium. We study the solutions for electric and magnetic fields, the direction of propagation, the time-average power and other various properties for a so-called uniform plane wave. Next we consider what happens when the wave propagates in a lossy or conducting medium. It will be shown that the wave will attenuate as it propagates through a lossy medium. We then discuss the dispersion and polarization of waves. The polarization is a unique property that the electromagnetic wave has but the acoustic wave does not.

9.2. THE WAVE EQUATION IN A SOURCE-FREE REGION

We first derive the wave equation that governs the propagation of electromagnetic waves and find the solutions for the simplest one-dimensional wave. Let's consider the fields in a *source-free* region where there is no free charge or no free current, i.e., $\rho_v = 0$ and $\mathbf{J} = 0$. Why do we consider such a situation? It's because in many practical problems of interest such as reflection, transmission, guidance, and resonance of waves we deal with the wave solutions in a region free of sources. When we later consider the radiation of waves in the antenna problem, we will need to include the sources, ρ_v and $\mathbf{J}$.

Electromagnetic fields in a source-free region where $\rho_v = \mathbf{J} = 0$ satisfy the following Maxwell's equations from Eq. (8-20):

$$\nabla \times \mathbf{E} = -\frac{\partial \mathbf{B}}{\partial t} \qquad \nabla \cdot \mathbf{D} = 0$$
$$\nabla \times \mathbf{H} = \frac{\partial \mathbf{D}}{\partial t} \qquad \nabla \cdot \mathbf{B} = 0 \tag{9-1}$$

If the medium is *linear, isotropic**, *homogeneous* (see Section 3.4), and *nonconducting* (also called *lossless*),

$$\mathbf{D} = \varepsilon\mathbf{E}, \quad \mathbf{B} = \mu\mathbf{H}, \quad \mathbf{J}_c = \sigma\mathbf{E} = 0 \tag{9-2}$$

where $\mathbf{J}_c$ is the conduction current density. In such medium, Maxwell's equations (9-1) reduce to

$$\nabla\times\mathbf{E} = -\mu\frac{\partial\mathbf{H}}{\partial t} \tag{9-3a}$$

$$\nabla\times\mathbf{H} = \varepsilon\frac{\partial\mathbf{E}}{\partial t} \tag{9-3b}$$

$$\nabla\cdot\mathbf{E} = 0 \tag{9-3c}$$

$$\nabla\cdot\mathbf{H} = 0 \tag{9-3d}$$

Note that we have written Maxwell's equations in terms of only $\mathbf{E}$ and $\mathbf{H}$. In the dynamic problem, we often deal with $\mathbf{E}$ and $\mathbf{H}$ instead of $\mathbf{E}$ and $\mathbf{B}$ because a pair of $\mathbf{E}$ and $\mathbf{H}$ have certain symmetry and duality. Eqs. (9-3) constitute a set of *coupled*, first-order partial differential equations (PDEs) for $\mathbf{E}$ and $\mathbf{H}$. They can be *decoupled* or separated into a second-order PDE for $\mathbf{E}$ or $\mathbf{H}$ alone as follows. To eliminate $\mathbf{H}$, we take the curl of Eq. (9-3a), interchange the curl and time derivative, and make use of Eq. (9-3b):

$$\nabla\times(\nabla\times\mathbf{E}) = \nabla\times\left\{-\mu\frac{\partial\mathbf{H}}{\partial t}\right\} = -\mu\frac{\partial}{\partial t}(\nabla\times\mathbf{H}) = -\mu\varepsilon\frac{\partial^2\mathbf{E}}{\partial t^2}$$

Making use of the vector identity $\nabla\times(\nabla\times\mathbf{A}) = \nabla(\nabla\cdot\mathbf{A}) - \nabla^2\mathbf{A}$ (see Table 1-5) and Eq.(9-3c), we obtain the following **wave equation** for the electric field $\mathbf{E}$:

$$\boxed{\nabla^2\mathbf{E} - \mu\varepsilon\frac{\partial^2\mathbf{E}}{\partial t^2} = 0} \tag{9-4}$$

where

$$\nabla^2 = \frac{\partial^2}{\partial x^2} + \frac{\partial^2}{\partial y^2} + \frac{\partial^2}{\partial z^2}$$

is a Laplacian operator. Similarly, by eliminating $\mathbf{E}$ in Eq. (9-3), we can derive the same wave equation for the magnetic field $\mathbf{H}$:

$$\boxed{\nabla^2\mathbf{H} - \mu\varepsilon\frac{\partial^2\mathbf{H}}{\partial t^2} = 0} \tag{9-5}$$

Eqs. (9-4) and (9-5) are called the vector wave equations because their solutions represent "waves."

**Isotropic* means that ε and μ do not depend on the orientation or direction of the fields, $\mathbf{E}$ and $\mathbf{H}$. In an isotropic medium, ε and μ are scalars. In an *anisotropic* medium, ε and μ become tensors.

9.2.1. One-Dimensional Wave Solutions

Let's first consider the wave equation when the fields vary only in one spatial direction, say, in the z-direction; thus they are independent of x and y $\left(\frac{\partial}{\partial x}=\frac{\partial}{\partial y}=0\right)$. Assume that $\mathbf{E}$ has only an x component:

$$\mathbf{E}=\mathbf{a}_x E_x(z,t)$$

Then from Eq. (9-4), E_x satisfies the following one-dimensional scalar wave equation:

$$\frac{\partial^2 E_x}{\partial z^2}-\mu\varepsilon\frac{\partial^2 E_x}{\partial t^2}=0 \tag{9-6}$$

Now we try the solutions for $E_x(z,t)$ of the following form:

$$E_x(z,t)=f(z\pm vt) \tag{9-7}$$

Letting $u=z\pm vt$ and making use of the chain rule,

$$\begin{aligned}\frac{\partial E_x}{\partial z}&=\frac{\partial E_x}{\partial u}\frac{\partial u}{\partial z}=f'(u)=f'(z\pm vt)\\ \frac{\partial^2 E_x}{\partial z^2}&=\frac{\partial u}{\partial z}\frac{\partial}{\partial u}\left\{\frac{\partial E_x}{\partial z}\right\}=f''(u)=f''(z\pm vt)\end{aligned} \tag{9-8a}$$

Similarly,

$$\begin{aligned}\frac{\partial E_x}{\partial t}&=\frac{\partial E_x}{\partial u}\frac{\partial u}{\partial t}=f'(u)\,(\pm v)=\pm vf'(z\pm vt)\\ \frac{\partial^2 E_x}{\partial t^2}&=\frac{\partial u}{\partial t}\frac{\partial}{\partial u}\left\{\frac{\partial E_x}{\partial t}\right\}=(\pm v)^2 f''(u)=v^2 f''(z\pm vt)\end{aligned} \tag{9-8b}$$

Substituting Eqs. (9-8a) and (9-8b) into Eq. (9-6), it is shown that the wave equation (9-6) is satisfied if

$$v=\frac{1}{\sqrt{\mu\varepsilon}} \tag{9-9}$$

Therefore, any function whose argument is of the form $z\pm vt$ or $z\pm\frac{1}{\sqrt{\mu\varepsilon}}t$ satisfies Eq. (9-6).

Physically $E_x(z,t)=f(z-vt)$ represents a "wave" of constant shape $f(u)$ traveling in the positive z direction with velocity v. Consider $f(z)$ being a Gaussian pulse with a peak at the origin. If we plot $E_x(z,t)$ as a function of z at two successive instants, $t=0$ and $t=t_o$ ($t_o>0$), then it is given by $f(z)$ and $f(z-vt_o)$ as shown in Figure 9-1.

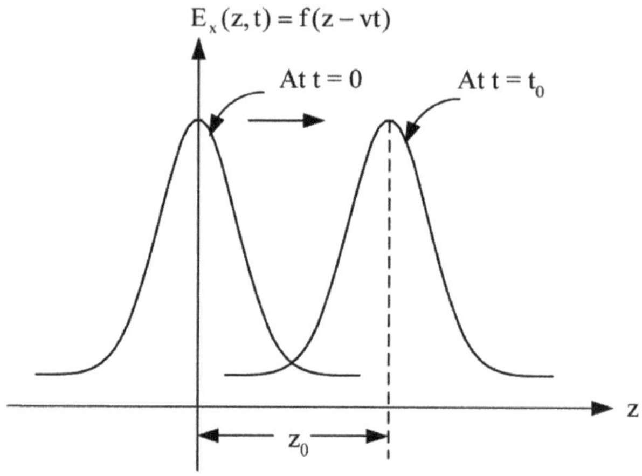

Figure 9-1. Gaussian pulse moving with velocity v.

The Gaussian pulse has traveled a distance $vt_o \equiv z_o$ in the $+z$ direction as time progresses. The velocity of travel is of course given by $\frac{z_o}{t_o} = v = \frac{1}{\sqrt{\mu\varepsilon}}$. It is also clear that $E_x(z,t) = f(z + vt)$ represents a wave traveling in the negative z direction with velocity v. In summary,

$$\mathbf{E} = \mathbf{a}_x\, f(z \pm vt) = \mathbf{a}_x\, f(z \pm \frac{1}{\sqrt{\mu\varepsilon}}\, t) \tag{9-10}$$

is the simple form of a one-dimensional wave. Electromagnetic waves propagate through a linear homogeneous medium (μ,ε) with a velocity v:

$$v = \frac{1}{\sqrt{\mu\varepsilon}}$$

In free space or vacuum,

$$v = \frac{1}{\sqrt{\mu_o \varepsilon_o}} \approx \frac{1}{\sqrt{4\pi\times10^{-7}\times\frac{10^{-9}}{36\pi}}} = 3\times10^{8}\left[\frac{m}{sec}\right]$$

which is precisely the speed of light in vacuum. Note that we have used a useful approximate number:

$\varepsilon_0 = 8.854\times10^{-12} \approx \frac{10^{-9}}{36\pi}\left[\frac{F}{m}\right]$. This implies that *light is an electromagnetic wave*. Moreover, electromagnetic waves at any frequency propagate with the same velocity. See Example 9-2(c) in the next section for an example of an electromagnetic wave.

9.3. TIME-HARMONIC ELECTROMAGNETIC FIELDS

In Section 9.2, we have shown that any function whose argument is of the form $z \pm vt$ satisfies the wave equation. Among all possible functions for sources and fields, the sinusoidal function is the most popular. Therefore, we will mainly deal with the fields that are sinusoidal in time. We often call them **time-harmonic** fields.

There are a few good reasons why we emphasize the time-harmonic fields. First of all, almost all the electric sources that produce charges and currents are sinusoidal (known as ac sources) with a single frequency. Secondly, when the charges and currents are sinusoidal with frequency f in hertz, the fields and waves that they produce are also sinusoidal with the exact same frequency f in Hz in the steady state because a set of Maxwell's equations are linear: for example, when ρ_v is increased twice, Gauss' law confirms that $\mathbf{E}$ is also increased twice. This fact will greatly simplify the mathematical complexity, using the concept of complex phasor as will be shown later. Thirdly, even when the sources are not sinusoidal, any function of time can be expanded as a linear combination of harmonic sinusoidal functions according to Fourier analysis. Thus knowing the solutions for time-harmonic electromagnetic fields will help find the solutions for non-sinusoidal fields.

9.3.1 Phasor Representation of Time-Harmonic Fields

You may recall that in the analysis of AC circuits where voltages and currents are sinusoidal, the **complex phasor** is introduced to represent the sinusoidal or ac voltages and currents. We will do the same here. Suppose the volume charge density ρ_v is a sinusoidal function of the following form:

$$\rho_v(\mathbf{r}, t) = \rho_o(\mathbf{r}) \cos(\omega t + \phi) \tag{9-11}$$

where $\rho_o(\mathbf{r})$ is the amplitude, ω is the angular frequency in radian(s) per second and ϕ is called the phase. ω is related to the frequency f in Hz by

$$\omega = 2\pi f \tag{9-12}$$

Using Euler's formula $e^{j\theta} = \cos\theta + j \sin\theta$, ρ_v in (9-11) can be written as

$$\rho_v(\mathbf{r}, t) = \text{Re}\{\rho_o(\mathbf{r})\ e^{j(\omega t + \phi)}\} = \text{Re}\{\rho_o(\mathbf{r})\, e^{j\phi} e^{j\omega t}\}$$

where Re{ } means taking the real part of the quantity in the brackets { }.

Eliminating the complex exponential time function $e^{j\omega t}$, we define

$$\underline{\rho}_v(\mathbf{r}) \equiv \rho_o(\mathbf{r})\ e^{j\phi} \tag{9-13}$$

as the complex phasor of $\rho_v(\mathbf{r}, t)$. Thus the real time $\rho_v(\mathbf{r}, t)$ and its complex phasor $\underline{\rho}_v(\mathbf{r})$ are related by

$$\rho_v(\mathbf{r}, t) = \text{Re}\{\underline{\rho}_v(\mathbf{r})\ e^{j\omega t}\} \tag{9-14}$$

which represents a sinusoidal charge density. We can similarly define the complex phasors for all of the scalar and vector field quantities that are time-harmonic. For example, the time-harmonic electric field can be written as

$$\mathbf{E}(\mathbf{r},t) = \mathrm{Re}\left\{\underline{\mathbf{E}}(\mathbf{r})\ e^{j\omega t}\right\} \tag{9-15}$$

Note that $\mathbf{E}(\mathbf{r},t)$ is a real function of t, whereas $\underline{\mathbf{E}}(\mathbf{r})$ is a complex phasor and is no longer a function of t. Now consider the time derivative of the electric field in (9-15).

$$\frac{\partial}{\partial t}\{\mathbf{E}(\mathbf{r},t)\} = \frac{\partial}{\partial t}\mathrm{Re}\left\{\underline{\mathbf{E}}(\mathbf{r})\, e^{j\omega t}\right\} = \mathrm{Re}\left\{\frac{\partial}{\partial t}\underline{\mathbf{E}}(\mathbf{r})\, e^{j\omega t}\right\} = \mathrm{Re}\left\{j\omega\ \underline{\mathbf{E}}(\mathbf{r})\, e^{j\omega t}\right\} \tag{9-16}$$

Therefore, the complex phasor of $\frac{\partial \mathbf{E}}{\partial t}$ is $j\omega\underline{\mathbf{E}}(\mathbf{r})$, which is an algebraic multiplication of $\underline{\mathbf{E}}(\mathbf{r})$, the phasor of $\mathbf{E}(\mathbf{r},t)$, by a simple factor $j\omega$. In other words, $j\omega$ can replace the time derivative $\frac{\partial}{\partial t}$ in the phasor representation of time-harmonic quantities:

$$\boxed{\frac{\partial}{\partial t}\mathbf{E}(\mathbf{r},t) \leftrightarrow j\omega\ \underline{\mathbf{E}}(\mathbf{r})} \tag{9-17}$$

(time domain) (phasor domain)

This identity simplifies Maxwell's equations for time-harmonic fields because the time derivatives can be eliminated.

EXAMPLE 9-1. PHASOR REPRESENTATION OF TIME-HARMONIC QUANTITIES

Find the complex phasors of the following quantities that are sinusoidal in time.

(a) $\rho_v(t) = \rho_o \sin(10^6 t)$

(b) $\mathbf{E}(z,t) = \mathbf{a}_x E_1 \cos(10^9 t - 4z + \pi/4) + \mathbf{a}_y E_2 \cos(10^9 t - 4z + 3\pi/4)$

SOLUTION

(a)
$$\begin{aligned}\rho_v(t) &= \rho_o \sin(10^6 t) = \rho_o \cos(10^6 t - \pi/2) = \mathrm{Re}\left\{\rho_o\ e^{j(10^6 t - \pi/2)}\right\}\\ &= \mathrm{Re}\left\{\rho_o\ e^{-j\pi/2}\ e^{j10^6 t}\right\} = \mathrm{Re}\left\{\underline{\rho}_v\ e^{j10^6 t}\right\}\end{aligned}$$

Thus, the phasor of $\rho_v(t)$ is given by $\underline{\rho}_v = \rho_o\, e^{-j\pi/2}$ with $\omega = 10^6$ rad/s .

(b)
$$\begin{aligned}\mathbf{E}(z,t) &= \mathrm{Re}\left\{\mathbf{a}_x E_1\ e^{j(10^9 t - 4z + \pi/4)} + \mathbf{a}_y\ E_2\ e^{j(10^9 t - 4z + 3\pi/4)}\right\}\\ &= \mathrm{Re}\left\{\left(\mathbf{a}_x\ E_1\, e^{-j4z}\ e^{j\pi/4} + \mathbf{a}_y\ E_2\, e^{-j4z}\ e^{j3\pi/4}\right) e^{j10^9 t}\right\} = \mathrm{Re}\left\{\underline{\mathbf{E}}(z)\, e^{j10^9 t}\right\}\end{aligned}$$

Thus the phasor of $\mathbf{E}(z,t)$ is given by

$$\underline{\mathbf{E}}(z) = \left\{ \mathbf{a}_x E_1 e^{j\pi/4} + \mathbf{a}_y E_2 e^{j3\pi/4} \right\} e^{-j4z} \text{ with } \omega = 10^9 \text{ rad/s} .$$

9.3.2 Maxwell's Equations for Time-Harmonic Fields

When the sources—charges and currents—are sinusoidal with the angular frequency ω, Gauss' law and Ampère's law guarantee that all the electric and magnetic field quantities in a linear medium become sinusoidal or time-harmonic with the same angular frequency ω. Thus, if we define complex phasors for all scalar and vector field quantities that appear in Maxwell's equations (8-20) as shown in Eqs. (9-14) and (9-15), then using the identity (9-17), Maxwell's equations (8-20) in the most general form reduce to

$$\nabla \times \underline{\mathbf{E}} = -j\omega \underline{\mathbf{B}} \quad \text{(9-18a)}$$

$$\nabla \times \underline{\mathbf{H}} = \underline{\mathbf{J}} + j\omega \underline{\mathbf{D}} \quad \text{(9-18b)}$$

$$\nabla \cdot \underline{\mathbf{D}} = \underline{\rho}_v \quad \text{(9-18c)}$$

$$\nabla \cdot \underline{\mathbf{B}} = 0 \quad \text{(9-18d)}$$

(Maxwell's equations for time-harmonic fields)

where $\underline{\mathbf{E}}$, $\underline{\mathbf{H}}$, $\underline{\mathbf{D}}$, $\underline{\mathbf{B}}$, $\underline{\mathbf{J}}$, $\underline{\rho}_v$ are complex phasor representations of their corresponding quantities and they now depend only on spatial variables. When the medium is linear, isotropic, homogeneous and lossless as characterized by Eq. (9-2), time-harmonic electromagnetic fields in the source-free region where $\underline{\rho}_v = \underline{\mathbf{J}} = 0$, satisfy the following Maxwell's equations.

$$\nabla \times \underline{\mathbf{E}} = -j\omega\mu \underline{\mathbf{H}} \quad \text{(9-19a)}$$

$$\nabla \times \underline{\mathbf{H}} = j\omega\varepsilon \underline{\mathbf{E}} \quad \text{(9-19b)}$$

$$\nabla \cdot \underline{\mathbf{E}} = 0 \quad \text{(9-19c)}$$

$$\nabla \cdot \underline{\mathbf{H}} = 0 \quad \text{(9-19d)}$$

(Maxwell's equations for time-harmonic fields
in simple medium in the source-free region)

Note that Eq. (9-19) is the time-harmonic version of Eq. (9-3). In Maxwell's equations, the time derivatives have been eliminated and replaced by an algebraic multiplication of jω. This will greatly simplify finding solutions for the fields from Maxwell's equations.

9.3.3 Complex Poynting Theorem—Real Power Flow

There is a complex version of Poynting's theorem that was presented in Section 8.7, which illustrates the flow of real electromagnetic power in terms of complex phasor fields. If we dot-multiply Eq. (9-18a) with $\underline{\mathbf{H}}^*$ and the complex conjugate of Eq. (9-18b) with $\underline{\mathbf{E}}$, and subtract the second term from the first term, we obtain

$$\begin{aligned}&\underline{\mathbf{H}}^*\cdot(\nabla\times\underline{\mathbf{E}})-\underline{\mathbf{E}}\cdot(\nabla\times\underline{\mathbf{H}}^*)\\&=\nabla\cdot(\underline{\mathbf{E}}\times\underline{\mathbf{H}}^*)=-j\omega\ (\underline{\mathbf{B}}\cdot\underline{\mathbf{H}}^*-\underline{\mathbf{E}}\cdot\underline{\mathbf{D}}^*)-\underline{\mathbf{E}}\cdot\underline{\mathbf{J}}^*\end{aligned} \tag{9-20}$$

where we have used the vector identity (6) in Table 1-5. Eq. (9-20) is the complex Poynting's theorem in differential form and is the complex version of Eq. (8-31). If we follow the same procedure as in Section 8.7, i.e., integrate both sides of Eq. (9-20) over the volume V bounded by closed surface S and apply the divergence theorem to the left side, we obtain the complex Poynting's theorem in integral form:

$$\oiint_S (\underline{\mathbf{E}}\times\underline{\mathbf{H}}^*)\cdot d\mathbf{s}=j\omega\iiint_V(\underline{\mathbf{E}}\cdot\underline{\mathbf{D}}^*-\underline{\mathbf{H}}^*\cdot\underline{\mathbf{B}})\,dv-\iiint_V \underline{\mathbf{E}}\cdot\underline{\mathbf{J}}^*dv \tag{9-21}$$

If the medium is linear and μ, ε are assumed to be real, the first term of the right side of Eq. (9-21) becomes purely imaginary. Suppose $\mathbf{J}$ consists of the source current $\underline{\mathbf{J}}_i$ (often called the *impressed* current) and the conduction current $\underline{\mathbf{J}}_c$, i.e.,

$$\underline{\mathbf{J}}=\underline{\mathbf{J}}_i+\underline{\mathbf{J}}_c=\underline{\mathbf{J}}_i+\sigma\underline{\mathbf{E}} \tag{9-22}$$

Taking the real part of Eq. (9-21), dividing by 2, and rearranging the terms, we obtain

$$\iiint_V \frac{1}{2}\,\mathrm{Re}\,\{-\underline{\mathbf{E}}\cdot\underline{\mathbf{J}}_i^*\}\,dv=\oiint_S\frac{1}{2}\ \mathrm{Re}\ \{\underline{\mathbf{E}}\times\underline{\mathbf{H}}^*\}\cdot d\mathbf{s}+\iiint_V\frac{1}{2}\,\underline{\mathbf{E}}\cdot\underline{\mathbf{J}}_c^*\,dv \tag{9-23}$$

where the last term is shown to be real because $\underline{\mathbf{E}}\cdot\underline{\mathbf{J}}_c^*=\underline{\mathbf{E}}\cdot\sigma\underline{\mathbf{E}}^*=\sigma\left|\underline{\mathbf{E}}\right|^2$.

The physical interpretation of each term is given as follows. The term on the left side is the *real* or *time-average* power supplied by the sinusoidal source current $\mathbf{J}_i$ in the volume V. The first term on the right side is the *real* or *time-average* power flowing out of the volume V and the second term is the *real* or *time-average* power dissipated as the Joule heat in the volume V as discussed in Section 5.6. Thus Eq. (9-23) represents the conservation of real electromagnetic power in V. If we define the complex quantity

$$\underline{\mathbf{S}}\equiv\underline{\mathbf{E}}\times\underline{\mathbf{H}}^* \tag{9-24}$$

half its *real* part gives the time-average power per unit area carried by the electric and magnetic fields $(\underline{\mathbf{E}}, \underline{\mathbf{H}})$. $\underline{\mathbf{S}}$ is called the *complex Poynting vector*. In summary, it is important to recognize that

$$\iint_S \frac{1}{2}\,\mathrm{Re}\{\underline{\mathbf{E}}\times\underline{\mathbf{H}}^*\}\cdot d\mathbf{s} = \begin{pmatrix}\text{Time-average power flowing}\\ \text{through the surface S}\end{pmatrix} \tag{9-25}$$

and

$$\iiint_V \frac{1}{2}\,\mathrm{Re}\{\underline{\mathbf{E}}\cdot\underline{\mathbf{J}}_c^*\}\ dv = \iiint_V \frac{1}{2}\sigma\left|\underline{\mathbf{E}}\right|^2 dv = \begin{pmatrix}\text{Time-average power}\\ \text{dissipated in the volume V}\end{pmatrix} \tag{9-26}$$

Finally, we wish to explain why the real part of $\frac{1}{2}\left(\underline{\mathbf{E}}\times\underline{\mathbf{H}}^*\right)$ gives the time-average of $\mathbf{E}(t)\times\mathbf{H}(t)$ for the time-harmonic (or sinusoidal) fields:

$$\boxed{\mathbf{S}_{av} \equiv \left\langle \mathbf{E}(t)\times\mathbf{H}(t)\right\rangle = \frac{1}{2}\,\mathrm{Re}\left\{\underline{\mathbf{E}}\times\underline{\mathbf{H}}^*\right\}} \tag{9-27}$$

where < > denotes the time average. Note that for time-harmonic fields, the Poynting vector $\mathbf{S} = \mathbf{E}(t) \times \mathbf{H}(t)$ represents an instantaneous power density. In practice, we are more interested in time-average power rather than an instantaneous power. The relationship (9-27) is analogous to how the average power is calculated from the complex phasor voltage and current in the analysis of ac circuits in the sinusoidal steady state, that is,

$$P_{av} \equiv \left\langle V(t)\ I(t)\right\rangle = \frac{1}{2}\mathrm{Re}\left\{\underline{V}\ \underline{I}^*\right\} \tag{9-28}$$

where $\underline{V}$ and $\underline{I}$ are complex phasors corresponding to $V(t)$ and $I(t)$, respectively. In order to prove Eq. (9-27), we write

$$\underline{\mathbf{E}} = \mathbf{E}_R + j\mathbf{E}_I, \qquad \underline{\mathbf{H}} = \mathbf{H}_R + j\mathbf{H}_I \tag{9-29}$$

where $\mathbf{E}_R$, $\mathbf{H}_R$ are real parts and $\mathbf{E}_I$, $\mathbf{H}_I$ are imaginary parts of $\underline{\mathbf{E}}, \underline{\mathbf{H}}$. Then

$$\begin{aligned}
&\mathbf{E}(t) = \mathrm{Re}\ \{\ \underline{\mathbf{E}}\ e^{j\omega t}\} = \mathbf{E}_R \cos \omega t - \mathbf{E}_I \sin \omega t\\
&\mathbf{H}(t) = \mathrm{Re}\ \{\underline{\mathbf{H}}\ e^{j\omega t}\} = \mathbf{H}_R \cos \omega t - \mathbf{H}_I \sin \omega t\\
&\mathbf{E}(t) \times \mathbf{H}(t) = \mathbf{E}_R \times \mathbf{H}_R \cos^2 \omega t + \mathbf{E}_I \times \mathbf{H}_I \sin^2 \omega t - (\mathbf{E}_R \times \mathbf{H}_I + \mathbf{E}_I \times \mathbf{H}_R) \cos \omega t \sin \omega t
\end{aligned} \tag{9-30}$$

Making use of the following identities on the time-averages of the sinusoidal functions with angular frequency $\omega = 2\pi f = \frac{2\pi}{T}$ (T is a period):

$$\left\langle \cos^2\omega t \right\rangle \equiv \frac{1}{T}\int_0^T \cos^2\omega t \, dt = \frac{1}{T}\int_0^T \frac{1}{2}\left(1+\cos 2\omega t\right) dt$$

$$= \frac{1}{2T}\left[t + \frac{1}{2\omega}\sin(2\omega t) \right]_0^T = \frac{1}{2}$$

$$\left\langle \sin^2\omega t \right\rangle = \frac{1}{T}\int_0^T \frac{1}{2}\left(1-\cos 2\omega t\right) dt = \frac{1}{2}$$

$$\left\langle \sin\omega t \cos \omega t \right\rangle = \frac{1}{T}\int_0^T \frac{1}{2}\sin 2\omega t \, dt = 0,$$

time-averaging both sides of Eq. (9-30) leads to

$$\left\langle \mathbf{E}(t)\times\mathbf{H}(t) \right\rangle = \frac{1}{2}\left(\mathbf{E}_R \times \mathbf{H}_R + \mathbf{E}_I \times \mathbf{H}_I\right) \tag{9-31}$$

From Eq. (9-29),

$$\begin{aligned} \underline{\mathbf{E}}\times\underline{\mathbf{H}}^* &= \left\{\mathbf{E}_R + j\mathbf{E}_I\right\}\times\left\{\mathbf{H}_R - j\mathbf{H}_I\right\} \\ &= \left(\mathbf{E}_R \times \mathbf{H}_R + \mathbf{E}_I \times \mathbf{H}_I\right) + j\left(\mathbf{E}_I \times \mathbf{H}_R - \mathbf{E}_R \times \mathbf{H}_I\right) \end{aligned} \tag{9-32}$$

Taking the real part of Eq. (9-32) and comparing with Eq. (9-31) proves the time-average relationship in Eq. (9-27). Note also that a factor of ½ in Eq. (9-26) signifies the *time average* of the dissipated power.

EXAMPLE 9-2 CALCULATION OF TIME-AVERAGE POWER

Consider an electromagnetic wave propagating in air (free of sources) which has the following time-harmonic electric field

$$\mathbf{E} = \mathbf{a}_x \, E_0 \cos\left(9\times 10^8 t - 3z\right)$$

where $E_o = 10$ [V/m].

(a) Find the magnetic field $\mathbf{H}$ of this wave, using the complex phasors.
(b) Find the time-average power density in watts/m^2 carried by the electromagnetic wave.
(c) Show that $\mathbf{E}$ satisfies the wave equation, Eq. (9-4).

SOLUTION

(a) First, we find the corresponding complex phasor of $\mathbf{E}$:

$$\mathbf{E}(z,t) = \text{Re}\left\{\mathbf{a}_x E_0\, e^{j(9\times10^8 t - 3z)}\right\} = \text{Re}\left\{\mathbf{a}_x E_0\, e^{-j\beta z}\, e^{j\omega t}\right\}$$

where $\omega = 9 \times 10^8$ and $\beta = 3$. Then the complex electric field is given by

$$\underline{\mathbf{E}} = \mathbf{a}_x E_0 e^{-j\beta z}$$

The complex magnetic field $\underline{\mathbf{H}}$ can be calculated from Faraday's law for time-harmonic fields, Eq. (9-18a):

$$\nabla\times\underline{\mathbf{E}} = -j\omega\underline{\mathbf{B}} = -j\omega\mu_0\underline{\mathbf{H}} \quad [\text{In air}, \mu = \mu_0]$$

$$\underline{\mathbf{H}} = \frac{\nabla\times\underline{\mathbf{E}}}{-j\omega\mu_0} = \frac{1}{-j\omega\mu_0}\mathbf{a}_y\frac{\partial \underline{E}_x}{\partial z} = \mathbf{a}_y\frac{(-j\beta)}{-j\omega\mu_0}E_0\, e^{-j\beta z} = \mathbf{a}_y\frac{\beta}{\omega\mu_0}E_0\, e^{-j\beta z}$$

The magnetic field in real time is given by

$$\mathbf{H}(z,t) = \text{Re}\left\{\underline{\mathbf{H}}\, e^{j\omega t}\right\} = \text{Re}\left\{\mathbf{a}_y\frac{\beta}{\omega\mu_0}E_0\, e^{-j\beta z}\, e^{j\omega t}\right\}$$

$$= \mathbf{a}_y\frac{\beta}{\omega\mu_0}E_0\cos(\omega t - \beta z)$$

$$= \mathbf{a}_y\frac{3\cdot 10}{9\times10^8\cdot 4\pi\cdot 10^{-7}}\cos(9\times10^8 t - 3z) = \mathbf{a}_y 2.65\times10^{-2}\cos(9\times10^8 t - 3z)\ [\text{A/m}]$$

(b) The time-average Poynting vector is given by

$$\mathbf{S}_{av} = \frac{1}{2}\text{Re}\{\underline{\mathbf{E}}\times\underline{\mathbf{H}}^*\} = \frac{1}{2}\text{Re}\left\{\mathbf{a}_x E_0\, e^{-j\beta z}\times\left[\mathbf{a}_y\frac{\beta}{\omega\mu_0}E_0\, e^{-j\beta z}\right]^*\right\}$$

$$= \mathbf{a}_z\frac{1}{2}\text{Re}\left\{E_0\left(\frac{\beta}{\omega\mu_0}\right)E_0^*\right\} = \mathbf{a}_z\frac{1}{2}\frac{\beta}{\omega\mu_0}|E_0|^2$$

The time-average power density is

$$\frac{1}{2}\frac{\beta}{\omega\mu_0}E_0^2 = \frac{1}{2}\frac{3}{9\cdot10^8\cdot4\pi\cdot10^{-7}}(10)^2 = 0.133\,[\text{W/m}^2]$$

(c) Let $\mathbf{E} = \mathbf{a}_x E_o \cos(\omega t - \beta z) \equiv \mathbf{a}_x E_x$

where $E_o = 10$, $\omega = 9\times10^8$, and $\beta = 3$. Note that $\frac{\partial}{\partial x} = \frac{\partial}{\partial y} = 0$.

$$\frac{\partial E_x}{\partial z} = \beta E_o \sin(\omega t - \beta z), \quad \frac{\partial^2 E_x}{\partial z^2} = -\beta^2 E_o \cos(\omega t - \beta z)$$

$$\frac{\partial E_x}{\partial t} = -\omega E_o \sin(\omega t - \beta z), \quad \frac{\partial^2 E_x}{\partial t^2} = -\omega^2 E_o \cos(\omega t - \beta z).$$

then

$$\nabla^2 E_x - \mu\varepsilon \frac{\partial^2 E_x}{\partial t^2} = \frac{\partial^2 E_x}{\partial z^2} - \mu_o \varepsilon_o \frac{\partial^2 E_x}{\partial t^2} \quad \text{(in air)}$$

$$= -\beta^2 E_o \cos(\omega t - \beta z) + \omega^2 \mu_o \varepsilon_o E_o \cos(\omega t - \beta z)$$

$$= (\omega^2 \mu_o \varepsilon_o - \beta^2) E_o \cos(\omega t - \beta z) = 0$$

because

$$\omega^2 \mu_o \varepsilon_o \approx (9\times10^8)^2 4\pi\times10^{-7} \cdot \frac{10^{-9}}{36\pi} = 9 = \beta^2.$$

Thus, the wave equation (9-4) is satisfied.

9.4. UNIFORM PLANE WAVES IN LOSSLESS MEDIA

We first consider an electromagnetic wave propagating in an unbounded lossless medium free of sources. Assume that the medium has permittivity ε and permeability μ and is lossless ($\sigma = 0$) and consider time-harmonic fields. In order to find solutions for the time-harmonic fields we again start from Maxwell's equations (9-19) and derive the differential equation for $\mathbf{E}$ or $\mathbf{H}$ separately as we have done in Section 9.2. To eliminate $\mathbf{H}$, we take the curl of Eq. (9-19a) and make use of Eq. (9-19b):

$$\nabla\times(\nabla\times\mathbf{E}) = -j\omega\mu(\nabla\times\mathbf{H}) = -j\omega\mu(j\omega\varepsilon\mathbf{E}) = \omega^2\mu\varepsilon\mathbf{E}$$

Making use of the vector identity $\nabla\times(\nabla\times\mathbf{E}) = \nabla(\nabla\cdot\mathbf{E}) - \nabla^2\mathbf{E}$ (see Table 1-5) and Eq. (9-19c), we obtain the following second-order partial differential equation (PDE) for the electric field $\mathbf{E}$:

$$\boxed{\nabla^2\mathbf{E} + k^2\mathbf{E} = 0} \tag{9-33}$$

where

$$k^2 = \omega^2 \mu \varepsilon \tag{9-34}$$

or

$$\boxed{k = \omega\sqrt{\mu\varepsilon}} \tag{9-35}$$

Eq. (9-33) is called the **Helmholtz equation** or the wave equation for a (time-harmonic) complex field. Eq. (9-35) is called the **dispersion relation** and k is called the **propagation constant**.

Similarly, by eliminating $\mathbf{E}$ in Eq. (9-19), we can derive the same PDE for the magnetic field $\mathbf{H}$:

$$\boxed{\nabla^2 \mathbf{H} + k^2 \mathbf{H} = 0} \tag{9-36}$$

A final note of caution is in order. In the equations above, we have used $\mathbf{E}$ and $\mathbf{H}$ without underbar although they represent complex phasor fields. From now on we will not use the separate notations (underbars) for complex phasor fields, because we will mainly consider time-harmonic fields and the reader is to understand that they are phasors from the absence of time derivatives in the equations.

Let's consider a simple solution for time-harmonic electric field $\mathbf{E}$ which points in the x-direction and varies only in the z-direction, thus independent of x and y. The complex field $\mathbf{E}$ can be written as

$$\mathbf{E} = \mathbf{a}_x E_x(z) \tag{9-37}$$

Substituting Eq. (9-37) into Eq. (9-33) and removing the unit vector, we obtain

$$\nabla^2 E_x + k^2 E_x = \frac{\partial^2 E_x}{\partial z^2} + k^2 E_x = 0 \tag{9-38}$$

This is a second-order constant-coefficient ordinary differential equation whose two independent solutions can be written as either sinusoidal functions $(\sin, \cos)$ or complex exponential functions as follows.

$$E_x(z) = A \cos(kz) + B \sin(kz) \tag{9-39}$$

or

$$E_x(z) = E_0^+ e^{-jkz} + E_0^- e^{+jkz} \tag{9-40}$$

where A, B, E_0^+, E_0^- are arbitrary coefficients.

A and B are related to E_0^+ and E_0^- by using Euler's formula (see Problem 9-14). We will take the form (9-40) for our discussion because each term in Eq. (9-40) has a clear physical interpretation and is easier to manipulate mathematically. The first and second terms represent waves propagating in the positive and negative z direction, respectively. Let's consider the first term and let

$$\mathbf{E} = \mathbf{a}_x E_0\, e^{-jkz} \tag{9-41}$$

Assuming that the constant E_0 is real (E_0 could be complex in general), the electric field in real time is given as

$$\mathbf{E}(z,t) = \mathrm{Re}\left\{\mathbf{E}\, e^{j\omega t}\right\} = \mathrm{Re}\left\{\mathbf{a}_x E_0\, e^{j(\omega t - kz)}\right\} = \mathbf{a}_x E_0 \cos(\omega t - kz) \tag{9-42}$$

Rewriting Eq. (9-42),

$$\mathbf{E}(z,t) = \mathbf{a}_x E_0 \cos\left\{(-k)\left(z - \frac{\omega}{k}t\right)\right\} = \mathbf{a}_x E_0 \cos\left\{k\left(z - \frac{1}{\sqrt{\mu\varepsilon}}t\right)\right\} \tag{9-43}$$

where we have used Eq. (9-35). Since the field takes the form of $f\left(z - \frac{1}{\sqrt{\mu\varepsilon}}t\right)$ where f is the cosine function for this case, Eq. (9-42), or alternatively Eq. (9-41), represents an electromagnetic wave traveling in the positive z direction with velocity $v = \frac{1}{\sqrt{\mu\varepsilon}}$, as we have discussed in Section 9.2.1. In fact, we can see this by plotting the x component of the electric field in Eq. (9-42) as a function of space (z) at two successive time instants, $t = 0$ and $t = \frac{\pi}{2\omega}$ as shown in Figure 9-2 (a), (b).

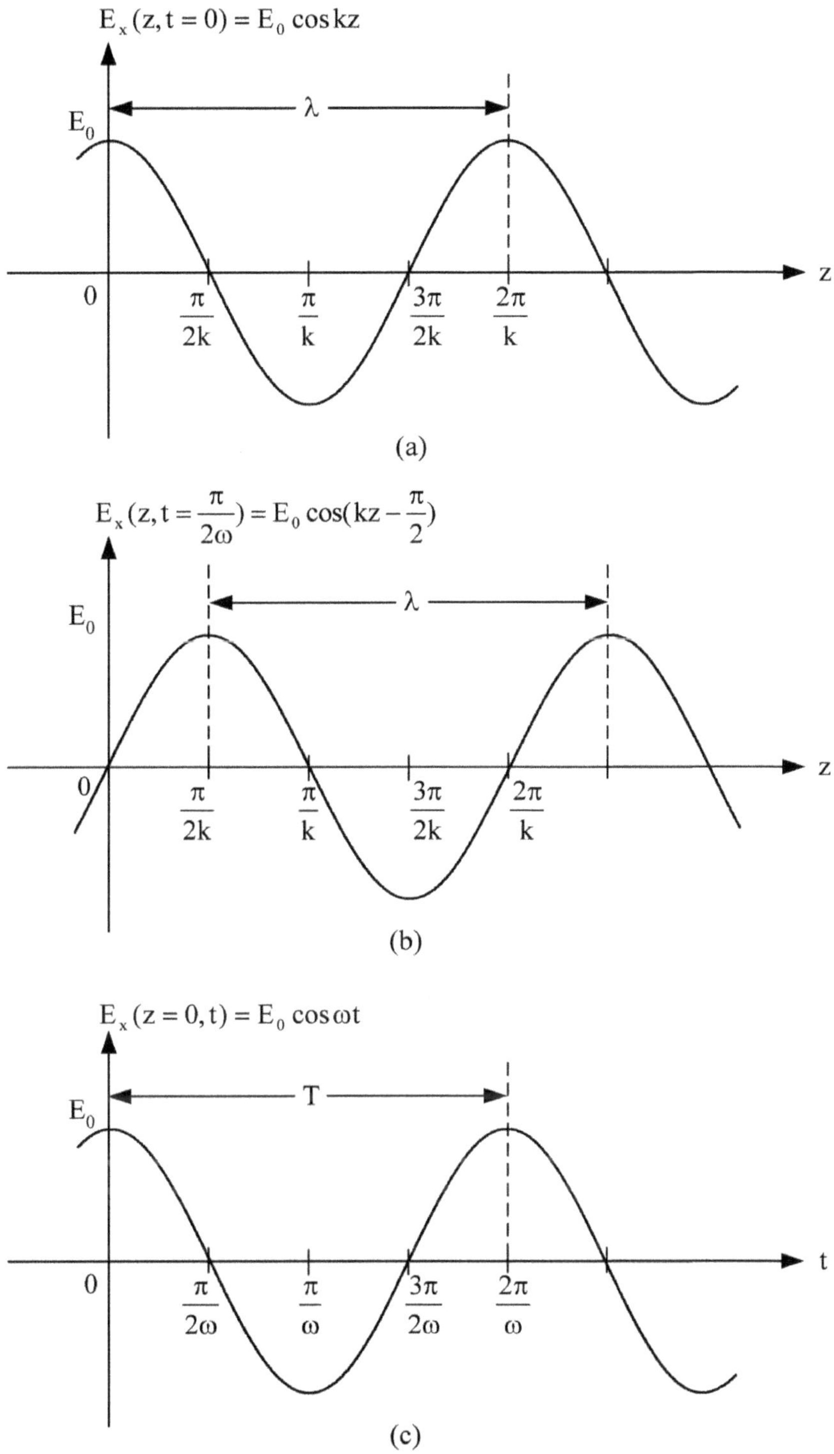

Figure 9-2. Time-harmonic electromagnetic field.

We observe that the sinusoidal wave has traveled a distance $z_0 = \pi/2k$ in the $+z$ direction over the period of time $t_0 = \pi/2\omega$. The velocity of the wave is given by

$$v = \frac{z_0}{t_0} = \frac{\omega}{k} = \frac{1}{\sqrt{\mu\varepsilon}} \tag{9-44}$$

as expected. The velocity v is often called the **phase velocity** because it measures the velocity of the phase change (kz) in Eq. (9-41). It is striking to observe that when the fields are sinusoidal in time (see Figure 9-2(c)) the fields are also sinusoidal in space (see Figure 9-2(a), (b)), thus periodic as a function of space. This spatial period of the wave is called the **wavelength** and is given by

$$\lambda = \frac{2\pi}{k} = \frac{v}{f} \tag{9-45}$$

where $v = \frac{\omega}{k} = \frac{2\pi f}{k}$ has been used in the second equality. Note that the period in time of the wave is $T = \frac{1}{f} = \frac{2\pi}{\omega}$ [Figure 9-2(c)]. For example, if the wave has the frequency $f = 1$ GHz, the wavelength in free space is $\lambda = \frac{v}{f} = \frac{3\times10^8}{1\times10^9} = 30\text{cm}$. On the other hand, for an AM signal of $f = 1000$ kHz, $\lambda = \frac{3\times10^8}{1\times10^6} = 300\,\text{m}$. The propagation constant k is related to the wavelength λ by

$$k = \frac{2\pi}{\lambda} \tag{9-46}$$

indicating that k gives the number of wavelengths in a spatial distance of 2π. Thus, k is also called the **wavenumber**.

Given the electric field in Eq. (9-41), the complex magnetic field $\mathbf{H}$ of the wave can be calculated by using Faraday's law (9-19a):

$$\mathbf{H} = \frac{\nabla\times\mathbf{E}}{-j\omega\mu} = \frac{1}{-j\omega\mu}\mathbf{a}_y\frac{\partial E_x}{\partial z} = \mathbf{a}_y\frac{(-jk)}{(-j\omega\mu)}E_0 e^{-jkz} = \mathbf{a}_y\sqrt{\frac{\varepsilon}{\mu}}\,E_0\,e^{-jkz} \tag{9-47}$$

where $\frac{k}{\omega\mu} = \frac{\sqrt{\mu\varepsilon}}{\mu} = \sqrt{\frac{\varepsilon}{\mu}}$ has been used. The magnetic field in real time is given by

$$\mathbf{H}(z,t) = \text{Re}\left\{\mathbf{H}\,e^{j\omega t}\right\} = \mathbf{a}_y\sqrt{\frac{\varepsilon}{\mu}}\,E_0\cos(\omega t - kz) \tag{9-48}$$

We note that the magnetic field takes the same form as the electric field both in time and space, except that **H** points in the y direction. It is interesting to find that the ratio of the amplitudes of **E** and **H** is given by

$$\frac{E_x}{H_y} = \frac{E_0}{\sqrt{\frac{\varepsilon}{\mu}}\,E_0} = \sqrt{\frac{\mu}{\varepsilon}} \equiv \eta \tag{9-49}$$

Thus the ratio depends only on the medium parameters. This quantity η is called the **intrinsic impedance** of the medium. In free space or air for which $\varepsilon_0 = 8.854 \times 10^{-12} \approx 1/36\pi \times 10^{-9}$, $\mu_0 = 4\pi \times 10^{-7}$, the phase velocity, wavelength, intrinsic impedance and wavenumber are given by

$$\begin{aligned}
&v = \frac{1}{\sqrt{\mu_0\varepsilon_0}} = 3\times10^{8}\left[\frac{m}{s}\right] \equiv c \quad \text{in free space or air}\\
&\lambda = \frac{c}{f} = \lambda_0 = \text{free-space wavelength [m]}\\
&\eta = \sqrt{\frac{\mu_0}{\varepsilon_0}} = 377 \approx 120\pi\,[\Omega] \equiv \eta_0\\
&k = \omega\sqrt{\mu_0\varepsilon_0} = \frac{\omega}{c} = \frac{2\pi}{\lambda_0} \equiv k_0\ [m^{-1}]
\end{aligned} \tag{9-50}$$

In the medium of permittivity $\varepsilon = \varepsilon_r\varepsilon_0$ and permeability $\mu = \mu_r\mu_0$,

$$\begin{aligned}
&v = \frac{1}{\sqrt{\mu\varepsilon}} = \frac{c}{\sqrt{\mu_r\varepsilon_r}} = \frac{c}{n}, \quad \lambda = \frac{v}{f} = \frac{\lambda_0}{n},\\
&\eta = \sqrt{\frac{\mu}{\varepsilon}} = \eta_0\sqrt{\frac{\mu_r}{\varepsilon_r}}, \quad k = \omega\sqrt{\mu\varepsilon} = \omega\sqrt{\mu_0\varepsilon_0}\sqrt{\frac{\mu\varepsilon}{\mu_0\varepsilon_0}} = nk_0
\end{aligned} \tag{9-51}$$

The wave slows down by a factor of n,

$$\boxed{n \equiv \frac{c}{v} = \sqrt{\frac{\mu\varepsilon}{\mu_0\varepsilon_0}} = \sqrt{\mu_r\varepsilon_r}} \tag{9-52}$$

called the **index of refraction** or the **refractive index**, which will play an important role in the reflection and transmission (or refraction) of the wave. The instantaneous Poynting vector and the time-average Poynting vector can be calculated as

$$\mathbf{S} = \mathbf{E}(z,t) \times \mathbf{H}(z,t) = \mathbf{a}_z \frac{E_0^2}{\eta} \cos^2(\omega t - kz) \tag{9-53a}$$

$$\mathbf{S}_{av} = \frac{1}{2} \text{Re}\{\mathbf{E} \times \mathbf{H}^*\} = \mathbf{a}_z \frac{1}{2} \text{Re}\left\{E_0 \left(\frac{E_0}{\eta}\right)^*\right\} = \mathbf{a}_z \frac{|E_0|^2}{2\eta} \tag{9-53b}$$

As expected, $\mathbf{S}$ points in the direction of travel ($+z$). We can also show that the time-average of Eq. (9-53a) gives Eq. (9-53b). In summary, the wave has the electric field $\mathbf{E}$ in the x direction and the magnetic field $\mathbf{H}$ in the y direction, and propagates in the z direction.

$$\mathbf{E} = \mathbf{a}_x E_0 e^{-jkz}, \quad \mathbf{H} = \mathbf{a}_y \frac{E_0}{\eta} e^{-jkz}, \quad \mathbf{S}_{av} = \mathbf{a}_z \frac{|E_0|^2}{2\eta} \tag{9-54}$$

It is important to note that

$$\mathbf{E} \perp \mathbf{H}, \quad \mathbf{E} \perp \mathbf{S}, \quad \mathbf{H} \perp \mathbf{S}$$

and ($\mathbf{E}$, $\mathbf{H}$, $\mathbf{S}$) make a right-handed orthogonal coordinate system. Using Eqs. (9-42) and (9-48), the electric and magnetic fields as a function of space at a fixed time (say, $t = 0$) are plotted in Figure 9-3.

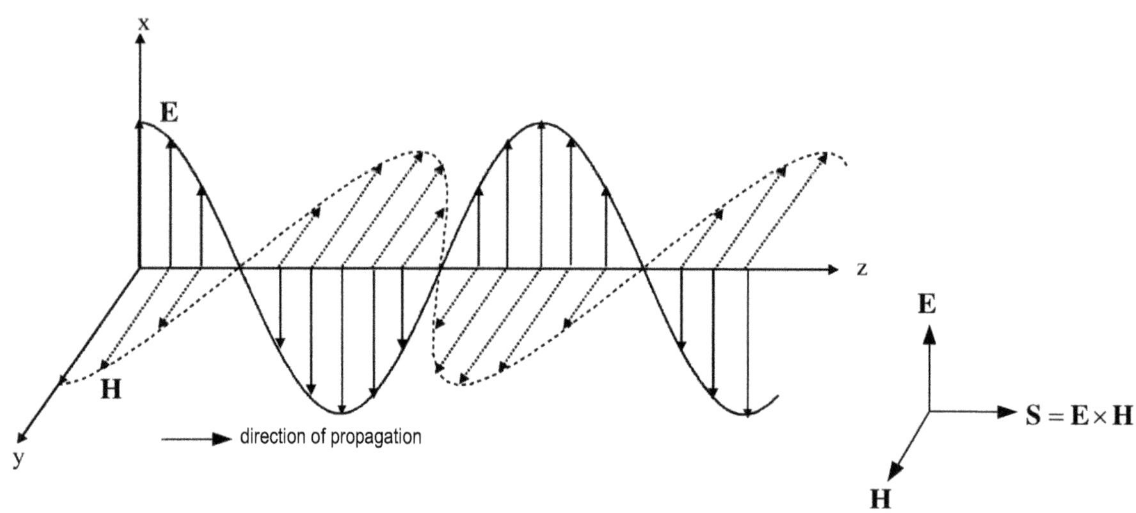

Figure 9-3. Electric and magnetic fields as a function of space at fixed time.

The wave we have just described is called a **uniform plane wave**. The term *plane* indicates that the surface on which the phase of the wave is constant (called the constant-phase front) is a plane. For the wave shown in Figure 9-3, the constant-phase front is given by

$$kz = \text{const } (\equiv C)$$

Then $z = \frac{C}{k}$ denotes a plane parallel to the xy plane ($z = 0$). The term *uniform* implies that amplitudes of the wave fields on the constant-phase fronts (planes) are uniform, or independent of the position. For the wave described above, at any given constant-phase plane, amplitudes of $\mathbf{E}$ and $\mathbf{H}$, E_0 and E_0/η, are independent of x and y, i.e., uniform. When the wave is generated from a localized source, it will start as a spherical wave, but after it propagates some distance, the wave will behave like a plane wave at points far from the source. Thus, the uniform plane wave represents the simplest and most useful form of wave propagation.

As a final note, for the second term of the $\mathbf{E}$ field solution in Eq. (9-40), we can follow the similar procedure and obtain

$$\mathbf{E} = \mathbf{a}_x\, E_0^- e^{jkz} \tag{9-55a}$$

$$\mathbf{E}(z,t) = \mathbf{a}_x\, E_0^- \cos(\omega t + kz) \tag{9-55b}$$

$$\mathbf{H} = -\mathbf{a}_y \frac{E_0}{\eta} e^{jkz} \tag{9-56}$$

$$\mathbf{S}_{av} = -\mathbf{a}_z \frac{|E_0|^2}{2\eta} \tag{9-57}$$

The wave has the electric field in the positive x direction and the magnetic field in the negative y direction, and propagates in the *negative* z direction. The relationship of ($\mathbf{E}$, $\mathbf{H}$, $\mathbf{S}$) still holds. It is important to note that the complex field e^{-jkz} represents a wave traveling in the $+z$ direction and e^{jkz} represents a wave traveling in the $-z$ direction.

EXAMPLE 9-3 UNIFORM PLANE WAVE IN A LOSSLESS MEDIUM

A uniform plane wave is traveling in the positive x direction in a *lossless* medium, with the 10 V/m electric field in the z direction. The wavelength of the wave is 20 cm and the velocity of wave propagation is 2×10^8 m/s.

(a) Determine the frequency in Hz of the wave and the permittivity of the medium, assuming the medium is non-magnetic, i.e., $\mu = \mu_o$.

(b) Write the complete complex phasor expressions for the electric and magnetic field vectors. Discuss the relationship between the direction of $\mathbf{E}$, $\mathbf{H}$ and $\mathbf{S}$.

SOLUTION

(a) The frequency can be obtained from the relationship (9-45):

$$f = \frac{v}{\lambda} = \frac{2\times10^8}{0.2} = 1\times10^9\,[\text{Hz}]$$

If we let the dielectric constant of the medium be ε_r, then

$$v = \frac{1}{\sqrt{\mu\varepsilon}} = \frac{1}{\sqrt{\mu_0\varepsilon_0\varepsilon_r}} = \frac{c}{\sqrt{\varepsilon_r}} = 2\times10^8$$

$$\varepsilon_r = \left[\frac{c}{2\times10^8}\right]^2 = \left[\frac{3\times10^8}{2\times10^8}\right]^2 = 2.25 \;\therefore\; \varepsilon = 2.25\varepsilon_0$$

(b) According to the description of the uniform plane wave, the complex electric field can be written as follows:

$$\mathbf{E} = \mathbf{a}_z E_0\, e^{-jkx}$$

where $E_0 = 10$ [V/m] and the propagation constant k is determined by

$$k = \omega\sqrt{\mu\varepsilon} = 2\pi f\sqrt{\mu_0(2.25\varepsilon_0)} = 2\pi\cdot1\times10^9\cdot\frac{1.5}{3\times10^8} = 10\pi\;[\text{m}^{-1}]\quad.$$

The magnetic field can be obtained by Faraday's law:

$$\mathbf{H} = \frac{\nabla\times\mathbf{E}}{-j\omega\mu} = \frac{1}{-j\omega\mu}\left\{-\mathbf{a}_y\frac{\partial E_z}{\partial x}\right\} = \mathbf{a}_y\frac{E_0}{j\omega\mu_0}(-jk)\,e^{-jkx} = -\mathbf{a}_y H_0 e^{-jkx}$$

where $H_0 = \dfrac{k}{\omega\mu_0}E_0 = \dfrac{10\pi\cdot10}{2\pi\cdot1\times10^9\cdot4\pi\times10^{-7}} = 0.0398\,[\text{A/m}]$.

Thus,

$$\mathbf{E} = \mathbf{a}_z 10\, e^{-j10\pi x},\quad \mathbf{H} = -\mathbf{a}_y 0.0398\, e^{-j10\pi x}$$

The fields in real time are given by

$$\mathbf{E}(x,t) = \mathbf{a}_z 10\cos(2\pi\,10^9 t - 10\pi x)$$

$$\mathbf{H}(x,t) = -\mathbf{a}_y\, 0.0398\cos(2\pi 10^9 t - 10\pi x)$$

We note that the electric field points in $+z$ direction, the magnetic field points in $-y$ direction, and the wave propagates in $+x$ direction. It is clear that $(\mathbf{E}, \mathbf{H}, \mathbf{S})$ make a right-handed orthogonal coordinate system. Making use of this fact and knowing that the ratio of amplitudes of $\mathbf{E}$ and $\mathbf{H}$ is equal to the intrinsic impedance (η) of the medium, we can simply calculate the magnetic field, without using Faraday's law, as follows:

$$\mathbf{H} = \frac{1}{\eta}\mathbf{a}_x \times \mathbf{E} = \sqrt{\frac{2.25\varepsilon_0}{\mu_0}}\,\mathbf{a}_x \times \mathbf{a}_z\, E_0\, e^{-jkx} = -\mathbf{a}_y \frac{1.5}{\eta_0} \cdot 10\, e^{-jkx}$$

$$= -\mathbf{a}_y \frac{15}{377} e^{-jkx} = -\mathbf{a}_y 0.0398\, e^{-j\,10\,\pi\,x}$$

9.4.1 Uniform Plane Waves Propagating in Arbitrary Direction

So far we have considered the wave in a simple form, that is, a uniform plane wave propagating in a direction with one particular coordinate (z), so that the fields vary only as a function of one coordinate. Suppose the wave propagates in an arbitrary direction which involves all three coordinates. The fields of such a plane wave can be expressed in the following general form:

$$\mathbf{E}(\mathbf{r}) = \mathbf{e}E_0\, e^{-jk_x x}\, e^{-jk_y y}\, e^{-jk_z z} = \mathbf{e}\,E_0\, e^{-j(k_x x + k_y y + k_z z)} \tag{9-58}$$

where $\mathbf{e}$ is a unit vector in the direction of the electric field and E_0 is a (complex) constant for a uniform plane wave. If we define a vector

$$\mathbf{k} = \mathbf{a}_x k_x + \mathbf{a}_y k_y + \mathbf{a}_z k_z \tag{9-59}$$

and make use of a position vector $\mathbf{r} = \mathbf{a}_x x + \mathbf{a}_y y + \mathbf{a}_z z$, Eq. (9-58) can be rewritten as

$$\mathbf{E}(\mathbf{r}) = \mathbf{e}\,E_0\, e^{-j\mathbf{k}\cdot\mathbf{r}} \tag{9-60}$$

We note that $e^{-j\mathbf{k}\cdot\mathbf{r}}$ is the general form of the uniform plane wave that propagates in an arbitrary direction. The vector $\mathbf{k}$ gives the direction of propagation, thus we will call $\mathbf{k}$ the **propagation vector** or the **wave vector**. Its magnitude gives the propagation constant k and is related to the angular frequency ω by Eq. (9-35), the dispersion relation. Substituting Eq. (9-58) into Eq. (9-33) and making use of

$$\frac{\partial}{\partial u} e^{-jk_u u} = (-jk_u)e^{-jk_u u}, \quad u = x, y, z,$$

we obtain

$$\left\{(-jk_x)^2 + (-jk_y)^2 + (-jk_z)^2 + k^2\right\}\mathbf{E} = 0$$

$$|\mathbf{k}|^2 = k_x^2 + k_y^2 + k_z^2 = k^2 = \omega^2\mu\varepsilon \tag{9-61}$$

The phase of the wave will be constant when

$$\mathbf{k} \cdot \mathbf{r} = \text{constant}$$

This defines a *constant-phase front*, which is a plane perpendicular to $\mathbf{k}$ as shown in Figure 9-4. Note that for all observation points $\mathbf{r}$ on this phase front (which is a plane), $\mathbf{k} \cdot \mathbf{r}$ is the same, i.e., is a constant.

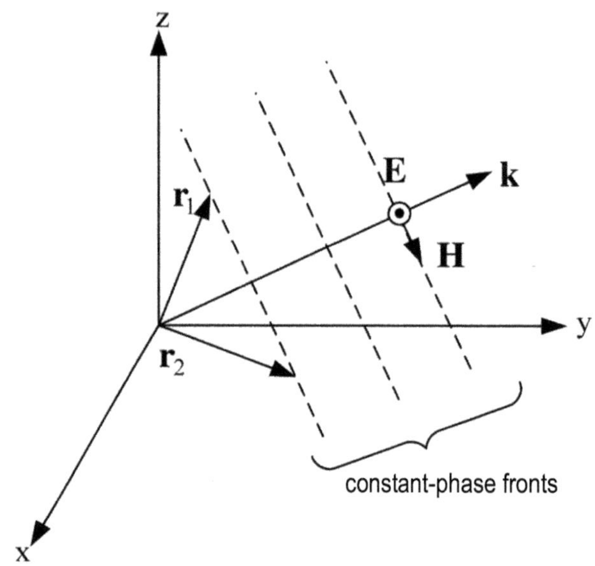

Figure 9-4. Uniform plane wave propagating in an arbitrary direction.

Now we write the corresponding magnetic field in a similar form:

$$\mathbf{H}(\mathbf{r}) = \mathbf{h}H_0 e^{-j\mathbf{k}\cdot\mathbf{r}} = \mathbf{h}H_0 e^{-j(k_x x + k_y y + k_z z)} \tag{9-62}$$

If we substitute Eqs. (9-58) and (9-62) into Maxwell's equations (9-19), making use of the algebraic identity for plane waves

$$\begin{aligned} \nabla &= \mathbf{a}_x \frac{\partial}{\partial x} + \mathbf{a}_y \frac{\partial}{\partial y} + \mathbf{a}_z \frac{\partial}{\partial z} = \mathbf{a}_x(-jk_x) + \mathbf{a}_y(-jk_y) + \mathbf{a}_z(-jk_z) \\ &= -j(\mathbf{a}_x k_x + \mathbf{a}_y k_y + \mathbf{a}_z k_z) = -j\mathbf{k}, \end{aligned} \tag{9-63}$$

we obtain

$$\begin{aligned} \mathbf{k} \times \mathbf{E} &= \omega\mu\mathbf{H} \qquad \mathbf{k} \cdot \mathbf{E} = 0 \\ \mathbf{k} \times \mathbf{H} &= -\omega\varepsilon\mathbf{E} \qquad \mathbf{k} \cdot \mathbf{H} = 0 \end{aligned} \tag{9-64}$$

From Eq. (9-64) we conclude that for uniform plane wave, ($\mathbf{E}$, $\mathbf{H}$, $\mathbf{k}$) are perpendicular to each other and they make a right-handed orthogonal coordinate system:

$$\mathbf{E} \perp \mathbf{H}, \quad \mathbf{E} \perp \mathbf{k}, \quad \mathbf{H} \perp \mathbf{k}, \quad \mathbf{E} \times \mathbf{H} // \mathbf{k}$$

Thus for a uniform plane wave, knowing $\mathbf{E}$ and the direction of propagation, we can easily obtain magnetic field $\mathbf{H}$ without using one of Maxwell's equations as follows:

$$\boxed{\mathbf{H} = \frac{1}{\omega\mu}\mathbf{k} \times \mathbf{E} = \frac{1}{\eta}\mathbf{a}_k \times \mathbf{E}} \tag{9-65}$$

where $\mathbf{a}_k = \frac{\mathbf{k}}{k}$ is a unit vector in the direction of the propagation vector $\mathbf{k}$ and η is the intrinsic impedance of the medium. Similarly, we can also calculate $\mathbf{E}$ from $\mathbf{H}$ for a uniform plane wave by

$$\boxed{\mathbf{E} = \frac{1}{-\omega\varepsilon}\mathbf{k} \times \mathbf{H} = \eta \mathbf{H} \times \mathbf{a}_k} \tag{9-66}$$

9.5. UNIFORM PLANE WAVES IN LOSSY MEDIA

In Section 9.4, we have considered the field solutions for the wave in a lossless or nonconducting medium for which $\sigma = 0$. What happens to the wave when propagating in a lossy or conducting medium? The lossy medium could be either a good conductor or a dielectric with small conductivity. As discussed in Section 5.6 (Joule's law) and Section 9.3.3 (Poynting's theorem), there will be power dissipation or loss due to conduction current $\mathbf{J}_c = \sigma\mathbf{E}$. Since the power is carried by the wave, we expect that the wave will attenuate in its amplitude as it propagates through a lossy medium.

9.5.1 Attenuation of Waves

In a lossy medium with permittivity ε, permeability μ and conductivity σ, we need to correct Ampère's law (9-3b) and (9-19b) by adding the conduction current density $\mathbf{J}_c$:

$$\mathbf{J}_c = \sigma\mathbf{E}, \quad \mathbf{J}_i = 0 \quad \text{(source-free)}$$

Ampère's law now reads

$$\nabla \times \mathbf{H} = \mathbf{J}_c + \varepsilon\frac{\partial \mathbf{E}}{\partial t} \quad \text{(in real time)} \tag{9-67a}$$

$$\nabla \times \mathbf{H} = \mathbf{J}_c + j\omega\varepsilon\mathbf{E} = (\sigma + j\omega\varepsilon)\mathbf{E} \tag{9-67b}$$

(for time-harmonic fields)

Making use of Eq. (9-67a) along with Eq. (9-3a), the wave equation (9-4) in the time domain is modified to

$$\nabla^2\mathbf{E} - \mu\sigma\frac{\partial \mathbf{E}}{\partial t} - \mu\varepsilon\frac{\partial^2 \mathbf{E}}{\partial t^2} = 0 \tag{9-68}$$

Making use of Eq. (9-67b) along with Eq. (9-13a), the Helmholtz equation (9-33) or the wave equation for a complex field is modified to

$$\nabla^2\mathbf{E} - j\omega\mu(\sigma + j\omega\varepsilon)\,\mathbf{E} = 0 \tag{9-69}$$

Without resolving the new PDE (9-68), we can find solutions of Eq. (9-69) from the solutions of Eq. (9-33) by redefining the propagation constant k as the complex propagation constant such that

$$k^2 = -j\omega\mu\,(\sigma + j\omega\varepsilon) = \omega^2\mu\,\{\varepsilon - j\frac{\sigma}{\omega}\} \tag{9-70a}$$

$$k = \sqrt{-j\omega\mu\,(\sigma + j\omega\varepsilon)} = \omega\sqrt{\mu\{\varepsilon - j\frac{\sigma}{\omega}\}} \tag{9-70b}$$

Then Eq. (9-69) takes the same form as Eq. (9-33) for lossless media and we can make use of the field solutions that we obtained in Section 9.4 with the understanding of k being a complex number. Eq. (9-70) can be simplified by introducing an **effective permittivity**

$$\boxed{\varepsilon_{eff} = \varepsilon - j\frac{\sigma}{\omega}} \tag{9-71}$$

for the lossy medium which has permittivity ε and conductivity σ. Note that ε_{eff} is complex. Then we have

$$k^2 = \omega^2\mu\varepsilon_{eff} \tag{9-72a}$$

$$k = \omega\sqrt{\mu\varepsilon_{eff}} \tag{9-72b}$$

Ampère's law (9-67) reduces to a regular form

$$\nabla\times\mathbf{H} = j\omega\varepsilon_{eff}\,\mathbf{E} \tag{9-73}$$

which is the same as Eq. (9-19b) with ε being replaced by ε_{eff}. This implies that if we take the solutions of Section 9.4 and replace ε by ε_{eff}, we find the solutions for the wave in lossy medium.

Now we again consider a simple solution for time harmonic field $\mathbf{E}$ which points in the x direction and varies only in the z direction, as formulated in Eq. (9-37). Then the solution for $E_x(z)$ will be given by Eq. (9-40) except that k now becomes a *complex propagation constant.*

If we define

$$k = \omega\sqrt{\mu\left(\varepsilon - j\frac{\sigma}{\omega}\right)} \equiv \beta - j\alpha \tag{9-74}$$

where β, α are real, for a wave traveling in the positive z direction, the complex electric field $\mathbf{E}$ is written from Eq. (9-41) as

$$\mathbf{E} = \mathbf{a}_x E_0 e^{-jkz} = \mathbf{a}_x E_0 e^{-j(\beta - j\alpha)z} = \mathbf{a}_x E_0 e^{-\alpha z} e^{-j\beta z} \tag{9-75}$$

Since $e^{-\alpha z}$ represents an exponential decay in $+z$ direction, the wave attenuates in a lossy medium as it travels in $+z$ direction. If we assume E_o is real (if it is complex, we can write $E_o = Ae^{j\phi}$), the electric field in real time is given by

$$\mathbf{E}(z,t) = \mathrm{Re}\{\mathbf{a}_x E_0\, e^{-\alpha z}\, e^{-j\beta z}\, e^{j\omega t}\} = \mathbf{a}_x\, E_0\, e^{-\alpha z} \cos(\omega t - \beta z) \tag{9-76}$$

At fixed time, say $t = 0$, the electric field is plotted as a function of space z in Figure 9-5.

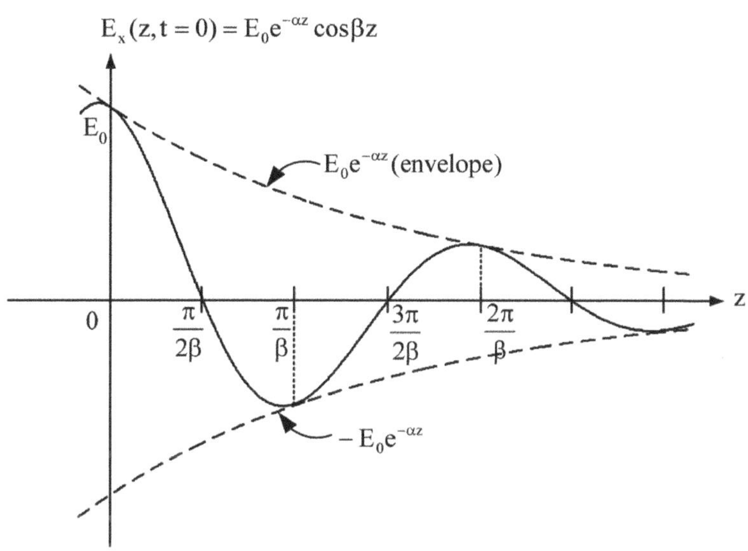

Figure 9-5. Time-harmonic electromagnetic field with attenuation.

We observe that β plays the same role as k of the wave in lossless medium and α describes the rate of attenuation. Thus we call β the **phase constant** and α the **attenuation constant**. The unit of β is radians per meter and that of α is called nepers per meter, in short, Np/m. The wavelength and phase velocity are given by

$$\lambda = \frac{2\pi}{\beta} = \frac{v}{f} \tag{9-77}$$

$$v = \frac{\omega}{\beta} \neq \frac{1}{\sqrt{\mu\varepsilon}} \tag{9-78}$$

Since β depends on the frequency (ω) as seen from Eq. (9-74), the phase velocity of the wave in a lossy medium is not a constant, but *dependent on the frequency* of the wave. Such a medium is called **dispersive**.

The complex magnetic field $\mathbf{H}$ can be calculated using Faraday's law (9-19a):

$$\mathbf{H} = \frac{\nabla \times \mathbf{E}}{-j\omega\mu} = \mathbf{a}_y \frac{k}{\omega\mu} E_0\, e^{-jkz} \equiv \mathbf{a}_y \frac{E_0}{\hat{\eta}} e^{-\alpha z}\, e^{-j\beta z} \tag{9-79}$$

where

$$\frac{E_x}{H_y} = \hat{\eta} = \frac{\omega\mu}{k} = \sqrt{\frac{\mu}{\varepsilon_{eff}}} = \sqrt{\frac{\mu}{\varepsilon - j\sigma/\omega}} \equiv |\hat{\eta}|\, e^{j\theta_\eta} \tag{9-80}$$

The intrinsic impedance $\hat{\eta}$ becomes *complex* for lossy media. The "hat" symbol signifies that η is complex. The complex magnetic field can be rewritten as

$$\mathbf{H} = \mathbf{a}_y \frac{E_0}{|\hat{\eta}|} e^{-\alpha z}\, e^{-j(\beta z + \theta_\eta)} \tag{9-81}$$

and the magnetic field in real time is given by

$$\mathbf{H}(z,t) = \mathbf{a}_y \frac{E_0}{|\hat{\eta}|} e^{-\alpha z} \cos(\omega t - \beta z - \theta_\eta) \tag{9-82}$$

This implies that the electric and magnetic field in lossy media are *out of phase* by the phase angle θ_η. From Eqs. (9-75) and (9-79) or (9-81), the time-average Poynting vector is given by

$$\begin{aligned}\mathbf{S}_{av} &= \frac{1}{2}\mathrm{Re}\{\mathbf{E} \times \mathbf{H}^*\} = \frac{1}{2}\mathrm{Re}\left\{\mathbf{a}_z\, E_0\, e^{-\alpha z}\, e^{-j\beta z} \frac{E_0^*}{|\hat{\eta}|} e^{-\alpha z}\, e^{j(\beta z + \theta_\eta)}\right\} \\ &= \mathbf{a}_z \frac{|E_0|^2}{2|\hat{\eta}|} \cos\theta_\eta\, e^{-2\alpha z}\end{aligned} \tag{9-83}$$

The power density of the wave attenuates like $(e^{-\alpha z})^2 = e^{-2\alpha z}$ as it travels in $+z$ direction.

For a wave traveling in the negative z direction, we follow a similar procedure to obtain

$$\mathbf{E} = \mathbf{a}_x\, E_0^-\, e^{+jkz} = \mathbf{a}_x\, E_0^-\, e^{\alpha z}\, e^{j\beta z} \tag{9-84a}$$

$$\mathbf{E}(z,t) = \mathbf{a}_x\, E_0^-\, e^{\alpha z} \cos(\omega t + \beta z) \tag{9-84b}$$

$$\mathbf{H} = -\mathbf{a}_y \frac{E_0^-}{|\hat{\eta}|} e^{\alpha z} \cos(\omega t + \beta z - \theta_\eta) \tag{9-84c}$$

$$\mathbf{S}_{av} = -a_z \frac{|E_0^-|^2}{2|\hat{\eta}|} \cos\theta_\eta\, e^{2\alpha z} \tag{9-84d}$$

How to find α, β, $|\hat{\eta}|$, θ_η

The values of the attenuation constant, the phase constant and the complex intrinsic impedance can be obtained by solving the two complex equations for k and $\hat{\eta}$, given by Eqs. (9-70) and (9-80). For example, substituting Eq. (9-74) into Eq. (9-70a), we have

$$k^2 = (\beta - j\alpha)^2 = (\beta^2 - \alpha^2) - j2\alpha\beta = \omega^2\mu\varepsilon - j\omega\mu\sigma \qquad (9\text{-}85)$$

Matching the real and imaginary parts on both sides and solving two equations for α and β (see Problem 9-23), we obtain the expressions for α and β in terms of ω, μ, ε, σ as follows.

$$\alpha = \omega\sqrt{\frac{\mu\varepsilon}{2}}\left[\sqrt{1+\left(\frac{\sigma}{\omega\varepsilon}\right)^2}-1\right]^{\frac{1}{2}} \left[\frac{\text{Np}}{\text{m}}\right] \qquad (9\text{-}86a)$$

$$\beta = \omega\sqrt{\frac{\mu\varepsilon}{2}}\left[\sqrt{1+\left(\frac{\sigma}{\omega\varepsilon}\right)^2}+1\right]^{\frac{1}{2}} \left[\frac{\text{rad}}{\text{m}}\right] \qquad (9\text{-}86b)$$

Although Eqs. (9-86) can be used for any lossy medium with (μ, ε, σ), in the cases of good conductors and good dielectric media with small loss, we can obtain much simpler approximate expressions for α, β, $\hat{\eta}$. Another important parameter that is often used is the **penetration depth** d_p. It is the distance over which the wave amplitude is attenuated by a factor of $\frac{1}{e} = e^{-1}$, which is equivalent to 36.8%.

$$\frac{E_x(z = d_p)}{E_x(z = 0)} = \frac{E_0\, e^{-\alpha d_p}}{E_0} = e^{-\alpha d_p} = e^{-1} \rightarrow \alpha d_p = 1$$

Thus the penetration depth is the inverse of the attenuation constant:

$$d_p = \frac{1}{\alpha} \qquad (9\text{-}87)$$

The penetration depth measures how far the wave can penetrate into the lossy medium. When the medium is very lossy, d_p will be very short since the wave will attenuate fast. When the medium is slightly lossy, d_p will be long and the wave can penetrate deep into the medium without much attenuation.

9.5.2 Good Dielectric vs. Good Conductor

Whether the medium is a good dielectric (with small loss) or a good conductor (highly conducting) is determined by comparing the real (ε) and imaginary (σ/ω) parts of the effective permittivity given in Eq. (9-71). If $\varepsilon \gg \sigma/\omega$, the medium is a good dielectric. If $\varepsilon \ll \sigma/\omega$, the medium is highly conducting or very lossy. There is a parameter, which determines this criterion, called the **loss tangent**. It is the ratio of the imaginary and real parts of ε_{eff} and also the ratio of amplitudes of the conduction current density ($\mathbf{J}_c = \sigma\mathbf{E}$) and the displacement current density ($\mathbf{J}_d = j\omega\varepsilon\mathbf{E}$).

$$\text{Loss tangent (L.T.)} = \frac{\text{Im}(\varepsilon_{eff})}{\text{Re}(\varepsilon_{eff})} = \frac{\sigma}{\omega\varepsilon} = \frac{|\mathbf{J}_c|}{|\mathbf{J}_d|} \tag{9-88}$$

The loss tangent is a measure of how lossy the medium is and helps to classify the media between good conductors and good dielectrics:

$$\text{L.T.} = \frac{\sigma}{\omega\varepsilon} << 1, \quad \sigma << \omega\varepsilon, \quad |\mathbf{J}_c| << |\mathbf{J}_d| \rightarrow \text{ Good dielectric}$$

$$\text{L.T.} = \frac{\sigma}{\omega\varepsilon} >> 1, \quad \sigma >> \omega\varepsilon, \quad |\mathbf{J}_c| >> |\mathbf{J}_d| \rightarrow \text{ Good conductor}$$

A. Waves in Slightly Lossy Dielectric (L.T. << 1)

In a good dielectric with small loss, $\sigma << \omega\varepsilon$ or L.T. $<< 1$. The complex propagation constant can be approximated by

$$k = \omega\sqrt{\mu\left(\varepsilon - j\frac{\sigma}{\omega}\right)} = \omega\sqrt{\mu\varepsilon}\left[1 - j\frac{\sigma}{\omega\varepsilon}\right]^{\frac{1}{2}} \approx \omega\sqrt{\mu\varepsilon}\left[1 - j\frac{\sigma}{2\omega\varepsilon}\right]$$

where the last equality can be shown by using the Taylor expansion of the square root function or the binomial expansion (see Eq. (2-32b)) with $\frac{\sigma}{\omega\varepsilon}$ being a small parameter. Then

$$k \approx \omega\sqrt{\mu\varepsilon} - j\frac{\sigma}{2}\sqrt{\frac{\mu}{\varepsilon}} = \beta - j\alpha$$

and

$$\hat{\eta} = \sqrt{\frac{\mu}{\varepsilon - j\frac{\sigma}{\omega}}} \approx \sqrt{\frac{\mu}{\varepsilon}} \quad \text{since } \frac{\sigma}{\omega} << \varepsilon.$$

Thus, for a slightly lossy dielectric, we have

$$\begin{aligned} &\beta = \omega\sqrt{\mu\varepsilon}, \quad \alpha = \frac{\sigma}{2}\sqrt{\frac{\mu}{\varepsilon}} \\ &d_p = \frac{2}{\sigma}\sqrt{\frac{\varepsilon}{\mu}}, \quad \hat{\eta} = \sqrt{\frac{\mu}{\varepsilon}} \end{aligned} \quad \text{(Low-loss dielectric) } \omega\varepsilon >> \sigma \tag{9-89}$$

Note that the phase constant and intrinsic impedance have not been affected much due to small loss.

EXAMPLE 9-4 WAVE IN A SLIGHTLY LOSSY MEDIUM

An electromagnetic wave of 10 MHz propagates in a lossy medium of $\varepsilon = 2\varepsilon_0$, $\mu = \mu_0$ and $\sigma = 5 \times 10^{-5}$ [S/m]. Determine the attenuation constant and the phase constant using both the approximate and exact expressions and compare the results.

SOLUTION

First, we calculate the loss tangent to determine whether the medium is slightly lossy or highly conducting or neither at the given frequency.

$$\text{L.T.} = \frac{\sigma}{\omega\varepsilon} = \frac{\sigma}{2\pi f \cdot 2\varepsilon_0} = \frac{5\times10^{-5}}{2\pi\cdot 10\times10^{6}\cdot 2\times\dfrac{10^{-9}}{36\pi}} = 0.045 \ll 1$$

Thus the medium is slightly lossy and we can use the approximate result of Eq. (9-89):

$$\alpha = \frac{\sigma}{2}\sqrt{\frac{\mu}{\varepsilon}} = \frac{\sigma}{2}\sqrt{\frac{\mu_0}{2\varepsilon_0}} = \frac{\sigma}{2\sqrt{2}}\eta_0 = \frac{5\times10^{-5}}{2\sqrt{2}}\times 377 = 6.66\times10^{-3}\left[\frac{\text{Np}}{\text{m}}\right]$$

$$\beta = \omega\sqrt{\mu\varepsilon} = 2\pi f\sqrt{\mu_0\cdot 2\varepsilon_0} = 2\pi f\frac{\sqrt{2}}{c} = \frac{2\pi\times10^{7}\cdot\sqrt{2}}{3\times10^{8}} = 0.296\left[\frac{\text{rad}}{\text{m}}\right]$$

If we use the exact expressions in Eq. (9-86), we obtain

$$\alpha = \frac{0.296}{\sqrt{2}}\left[\sqrt{1+(0.045)^2}-1\right]^{1/2} = 6.61\times10^{-3}\left[\frac{\text{Np}}{\text{m}}\right]$$

$$\beta = \frac{0.296}{\sqrt{2}}\left[\sqrt{1+(0.045)^2}+1\right]^{1/2} = 0.296\left[\frac{\text{rad}}{\text{m}}\right]$$

We observe that the approximate results agree well with the exact results; there is very little error.

B. Waves in Good Conductor (L.T. >> 1)

In a highly conducting medium, $\sigma \gg \omega\varepsilon$ or L.T. >> 1. In this case, the complex propagation constant can be approximated by dropping the real part of ε_{eff}:

$$k = \omega\sqrt{\mu\left(\varepsilon - j\frac{\sigma}{\omega}\right)} \approx \omega\sqrt{\mu\left(-j\frac{\sigma}{\omega}\right)} = \sqrt{\omega\mu\sigma}\left(\frac{1}{\sqrt{2}} - j\frac{1}{\sqrt{2}}\right) \equiv \beta - j\alpha$$

The complex intrinsic impedance is approximated by

$$\hat{\eta} = \sqrt{\frac{\mu}{\varepsilon - j\dfrac{\sigma}{\omega}}} \approx \sqrt{\frac{\mu}{-j\sigma/\omega}} = \sqrt{\frac{\omega\mu}{\sigma}}\left(\frac{1}{\sqrt{2}} + j\frac{1}{\sqrt{2}}\right) = \sqrt{\frac{\omega\mu}{\sigma}}\,e^{j\frac{\pi}{4}}$$

Thus, for a highly conducting medium, we have

$$\beta=\alpha=\sqrt{\frac{\omega\mu\sigma}{2}}, \quad d_p=\sqrt{\frac{2}{\omega\mu\sigma}}$$
$$\hat{\eta}=\sqrt{\frac{\omega\mu}{2\sigma}}(1+j)=\sqrt{\frac{\omega\mu}{\sigma}}e^{j\frac{\pi}{4}} \qquad \text{(Good conductor) } \sigma \gg \omega\varepsilon \tag{9-90}$$

We observe that in very good conductors, the attenuation constant is approximately equal to the phase constant and $\mathbf{E}$ and $\mathbf{H}$ are out of phase by $45°$. The penetration depth d_p will be very small for good conductors, in particular, at high frequencies where ω is very large, so d_p is often called the **skin depth** for good conductors. Since the field cannot penetrate much into the medium, most of the conduction current $\mathbf{J}_c$ concentrates on the surface of the conductor and flows very little inside. This phenomenon is known as the *skin effect*. For example, for copper which has $\sigma = 5.8\times10^7$ S/m, $\varepsilon \approx \varepsilon_o$, $\mu \approx \mu_o$, the penetration depth is 8.53 mm at $f = 60$ Hz and 2.09 μm at 1 GHz, which looks like a skin. We can use the skin effect in effectively shielding the electronic devices from interference caused by external fields, by highly conducting enclosures whose wall thickness is greater than a few skin depths since the external interfering field attenuates to a negligible value after a skin depth. The exact and approximate expressions for α, β, d_p and $\hat{\eta}$ are summarized in Table 9-1.

Table 9-1. Waves in Lossy Media

	Exact	Approximate: Low-loss Dielectric ($\omega\varepsilon \gg \sigma$)	Approximate: Good Conductor ($\omega\varepsilon \ll \sigma$)
α attenuation constant	$\omega\sqrt{\frac{\mu\varepsilon}{2}}\left[\sqrt{1+\left(\frac{\sigma}{\omega\varepsilon}\right)^2}-1\right]^{\frac{1}{2}}$	$\frac{\sigma}{2}\sqrt{\frac{\mu}{\varepsilon}}$	$\sqrt{\frac{\omega\mu\sigma}{2}}$
β phase constant	$\omega\sqrt{\frac{\mu\varepsilon}{2}}\left[\sqrt{1+\left(\frac{\sigma}{\omega\varepsilon}\right)^2}+1\right]^{\frac{1}{2}}$	$\omega\sqrt{\mu\varepsilon}$	$\sqrt{\frac{\omega\mu\sigma}{2}}$
d_p penetration depth	$\frac{1}{\alpha}$	$\frac{2}{\sigma}\sqrt{\frac{\varepsilon}{\mu}}$	$\sqrt{\frac{2}{\omega\mu\sigma}}$
$\hat{\eta}$ intrinsic impedance	$\sqrt{\frac{\mu}{\varepsilon_{eff}}}=\sqrt{\frac{\mu}{\varepsilon-j\sigma/\omega}}$	$\sqrt{\frac{\mu}{\varepsilon}}$	$\sqrt{\frac{\omega\mu}{2\sigma}}(1+j)$

EXAMPLE 9-5 COMMUNICATION WITH A SUBMARINE

A ship at the ocean surface is supposed to communicate with a submarine at the depth of 100 meters below the surface, using a ULF (ultra low frequency) electromagnetic signal at 1 kHz. How much attenuation in dB does the signal suffer at the location of the submarine when the sea water has permittivity $\varepsilon = 81\ \varepsilon_0$, permeability $\mu = \mu_0$, and conductivity $\sigma = 4$ [S/m] at this frequency?

SOLUTION

Since ocean water is a lossy medium, when the electromagnetic wave travels from the ship at the surface to the submarine at the depth z, its field amplitude will attenuate exponentially as follows:

$$|E(z)| = E_0\, e^{-\alpha z}$$

where E_0 is the amplitude at the surface ($z = 0$) and α is the attenuation constant. Thus we need to calculate α. First we calculate the loss tangent at $f = 1$ kHz:

$$\text{L.T.} = \frac{\sigma}{\omega\varepsilon} = \frac{4}{2\pi \cdot 1\times 10^3 \cdot 81 \times \dfrac{10^{-9}}{36\pi}} = 8.9\times 10^5 \gg 1$$

Seawater is a highly lossy medium at ULF, thus we can use Eq. (9-90):

$$\alpha \approx \sqrt{\frac{\omega\mu\sigma}{2}} = \sqrt{\frac{2\pi\cdot 10^3 \cdot 4\pi\times 10^{-7}\cdot 4}{2}} = 4\pi\times 10^{-2} = 0.1257$$

The attenuation in dB at $z = 100$ [m] can be calculated from the field as follows:

$$20\ \log\left|\frac{E(z=100)}{E_0}\right| = 20\ \log[e^{-\alpha z}] = 20\ \log[e^{-0.126\times 100}] = -109\ \text{dB}$$

Note that the lower the frequency, the less the attenuation.

Dielectric Loss

The loss can be caused not only by the dc conduction current but also by the high frequency ac current in phase with the applied electric field. When the dielectric is placed in an appreciable alternating field, the polarized atoms and molecules go through the continuous process of switching their dipole orientation, which results in a damping effect.* To account for such *dielectric loss*, often permittivity of the lossy dielectric is described by the **complex permittivity** whose imaginary part accounts for the loss:

$$\hat{\varepsilon} = \varepsilon' - j\varepsilon'' \qquad (9\text{-}91)$$

* see D. J. Griffiths (1999), *Introduction to Electrodynamics*, pp. 398-404, or J. D. Kraus (1992), *Electromagnetics*, pp. 442-445 for further discussion.

Comparing Eq. (9-91) with the effective permittivity in Eq. (9-71), ε' plays the role of ε and ε'' plays the role of σ/ω, or equivalently, $\sigma = \omega\varepsilon''$ is the equivalent conductivity. The loss tangent is given by $\text{L.T.} = \frac{\varepsilon''}{\varepsilon'}$.

9.6 DISPERSION OF WAVES – GROUP VELOCITY

In Section 9.4, we have seen that the velocity of propagation of an electromagnetic wave in a lossless medium of permittivity ε and permeability μ is given by the phase velocity

$$\upsilon_p = \frac{\omega}{k} = \frac{1}{\sqrt{\mu\varepsilon}} \tag{9-44}$$

For most media, μ and ε are constants and do not depend on the frequency. The electromagnetic waves at different frequencies propagate with the same velocity. However, for some media μ or ε or the index of refraction (n) is not constant and depends on the frequency, or equivalently, the wavelength. For example, the refractive index of typical glass is dependent on the wavelength at optical frequencies. When the white light consisting of different frequencies enters a prism, the refracted light gives a rainbow of colors because the glass prism has different indices of refraction at different color wavelengths. The phenomenon is called **dispersion**. When the medium parameters (ε and/or μ) depend on frequency, the phase velocity of the wave in that medium depends on the frequency and the medium is called **dispersive**. A lossy medium we studied in Section 9.5 exhibits a *dispersive* behavior because the phase velocity in a lossy medium is given by

$$\upsilon_p = \frac{\omega}{\beta} \tag{9-92}$$

where for a highly conducting medium, $\beta = \sqrt{\frac{\omega\mu\sigma}{2}}$, as shown in Eq. (9-90), and υ_p will depend on frequency. We will also find dispersive behavior when the waves propagate through a waveguide, as will be shown in Chapter 11.

In transmitting information such as speech, image, digital data, etc., in a "baseband" signal which ranges from dc to some kHz or MHz, we translate the low-frequency baseband signal to higher "carrier" frequency because higher frequencies will permit more reasonable sizes of waveguides and antennas. Hence a composite signal consists of a multitude of many different frequencies. If the medium or the system is not dispersive, all different frequency components travel at the same velocity and the received signal in time domain will not be distorted. This is the case for a uniform plane wave in unbounded lossless medium. On the other hand, if the medium is dispersive, the individual frequency components travel at different phase velocities and the received signal will be *distorted*. Usually an information-bearing composite signal has a small spread of frequencies (narrow band signals) around a high carrier frequency. Such a signal forms a wave packet and changes shape as it propagates. While each individual frequency component travels at the phase velocity, the wave packet as a whole (the envelope) travels at a different velocity, called the **group velocity**.

In order to observe the behavior of the wave packet, let's consider a simple case of a wave packet that consists of two travelling waves with the same amplitude and slightly different frequencies, $\omega_1 = \omega_c - \Delta\omega$ and $\omega_2 = \omega_c + \Delta\omega$ where $\Delta\omega << \omega_c$. The propagation constants (k) in lossless media or the phase constants (β) in lossy media that correspond to waves of two different frequencies will also be slightly different. We let $k_1 = k_c - \Delta k$ and $k_2 = k_c + \Delta k$, respectively. Again $\Delta k << k_c$. Then a composite wave of two frequencies traveling in $+z$ direction can be described by

$$\begin{aligned}
E(z,t) &= E_0 \cos(\omega_1 t - k_1 z) + E_0 \cos(\omega_2 t - k_2 z) \\
&= E_0 2\cos\left[\frac{1}{2}\{(\omega_2 t - k_2 z) - (\omega_1 t - k_1 z)\}\right] \\
&\quad \cos\left[\frac{1}{2}\{(\omega_2 t - k_2 z) + (\omega_1 t - k_1 z)\}\right] \\
&= 2E_0 \cos\left[\frac{\omega_2 - \omega_1}{2} t - \frac{k_2 - k_1}{2} z\right] \cos\left[\frac{\omega_2 + \omega_1}{2} t - \frac{k_2 + k_1}{2} z\right] \\
&= 2E_0 \underbrace{\cos(\Delta\omega t - \Delta k z)}_{\text{envelope}} \underbrace{\cos(\omega_c t - k_c z)}_{\text{carrier}}
\end{aligned} \tag{9-93}$$

It is seen that the composite wave consists of a carrier signal with high frequency ω_c and an envelope with low frequency $\Delta\omega$ (slowly varying in time) as shown in Figure 9-6. Each part travels with different velocity, namely, the carrier travels with the velocity $\upsilon_p = \frac{\omega_c}{k_c}$, which is the phase velocity, and the envelope travels with the velocity $\upsilon_g = \frac{\Delta\omega}{\Delta k}$, which we call the *group velocity*. Therefore, the group velocity of the wave in a dispersive medium is given by

$$\boxed{\upsilon_g = \frac{d\omega}{dk} = \frac{1}{\left(\frac{dk}{d\omega}\right)} \quad \text{or} \quad \frac{d\omega}{d\beta} = \frac{1}{\left(\frac{d\beta}{d\omega}\right)}} \tag{9-94}$$

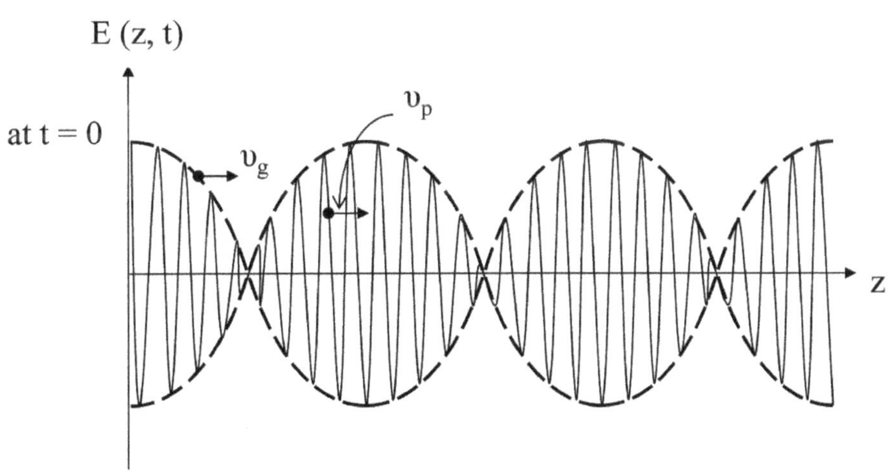

Figure 9-6. Group velocity of a wave packet.

In the case of an AM signal, υ_p is the velocity of a carrier signal and υ_g gives the velocity of a modulation envelope. Eq. (9-93) can be rewritten as

$$E(z,t) = 2E_0 \cos\left[\Delta k(z - \upsilon_g t)\right]\cos\left[k_c(z - \upsilon_p t)\right] \tag{9-95}$$

EXAMPLE 9-6 DISPERSION IN A LOSSY MEDIUM

Find the group velocity of a wave in a lossy medium (ε, μ, σ). Consider both cases: (a) slightly lossy dielectric and (b) highly conducting medium. Assume that ε, μ, σ are constants, i.e., independent of frequency.

SOLUTION

As shown in Section 9.5.2, for waves in a slightly lossy medium,

$$\beta = \omega\sqrt{\mu\varepsilon} \tag{9-89}$$

The group velocity is given by

$$\upsilon_g = \frac{d\omega}{d\beta} = \frac{1}{\frac{d\beta}{d\omega}} = \frac{1}{\sqrt{\mu\varepsilon}}$$

which is the same as the phase velocity, $\upsilon_p = \frac{\omega}{\beta} = \frac{1}{\sqrt{\mu\varepsilon}}$ and no dispersion occurs.

For waves in a highly conducting medium,

$$\beta = \sqrt{\frac{\omega\mu\sigma}{2}} \tag{9-90}$$

The group velocity is given by

$$\upsilon_g = \frac{d\omega}{d\beta} = \frac{1}{\frac{d}{d\omega}\left\{\sqrt{\frac{\omega\mu\sigma}{2}}\right\}} = \frac{1}{\frac{1}{2}\omega^{-1/2}\sqrt{\frac{\mu\sigma}{2}}} = 2\sqrt{\frac{2\omega}{\mu\sigma}}$$

which is different from the phase velocity, $\upsilon_p = \frac{\omega}{\beta} = \sqrt{\frac{2\omega}{\mu\sigma}}$, and dispersion occurs.

9.7 POLARIZATION OF WAVES

Polarization (to be distinguished from dielectric polarization) is a unique feature of an electromagnetic wave that the longitudinal acoustic wave does not have. At a fixed point in space, the $\mathbf{E}$ and $\mathbf{H}$ field vectors of a time-harmonic electromagnetic wave vary sinusoidally in time. Given the direction of propagation, the field vectors are in a plane perpendicular to the direction of propagation as discussed in Section 9.4. The orientation of these field vectors is an important characteristic of the electromagnetic wave. The **wave polarization** is defined or described by the orientation of the electric field vector $\mathbf{E}$ as a function of time, at a fixed point in space. Specifying the orientation for $\mathbf{E}$ is sufficient because the magnetic field $\mathbf{H}$ can be readily obtained from $\mathbf{E}$ by one of Maxwell's equations or Eq. (9-65) for uniform plane wave. When the electric field vectors lie along a line as time progresses at any given point in space, the wave is said to be **linearly polarized**. The wave whose electric field is given by Eq. (9-41) is an example of linearly polarized wave. When the tip of $\mathbf{E}$ follows a circle as time progresses, the wave is called **circularly polarized**. If the tip follows an ellipse, the wave is **elliptically polarized**. When the $\mathbf{E}$ vector does not follow any of these but moves around randomly, the wave is said to be **unpolarized**. Sunlight and light emitted from fluorescent lamps are unpolarized. The polarization of the wave becomes a very important characteristic, for example, when the wave is incident upon a different medium at an oblique angle. The reflection and transmission of the wave depends upon the polarization of the incident wave as will be discussed in Chapter 10.

Let's consider a time-harmonic uniform plane wave that propagates in the positive z direction in a lossless medium. Since the electric field has to lie in a plane perpendicular to the z axis, the electric field has only x and y components and, in general, the complex electric field can take the following form:

$$\mathbf{E} = (\mathbf{a}_x E_x + \mathbf{a}_y E_y) e^{-jkz} \tag{9-96}$$

where E_x and E_y are complex constants. The wave polarization depends on the amplitude ratio and phase difference of E_x, E_y. For simplicity let

$$E_x = E_{x0} \text{ (real)}$$
$$E_y = E_{y0}\, e^{j\phi} \equiv E_{x0} A\, e^{j\phi} \quad (A, \phi \text{ real})$$

so that

$$\frac{E_y}{E_x} = A\, e^{j\phi} \tag{9-97}$$

where $A = \left|\frac{E_y}{E_x}\right|$ is the amplitude ratio and ϕ is the phase difference between E_x and E_y.

The electric field in real time is given by

$$\begin{aligned}\mathbf{E}(z,t) &= \mathrm{Re}\left\{(\mathbf{a}_x E_{x0} + \mathbf{a}_y E_{x0} A\, e^{j\phi})\, e^{-jkz}\, e^{j\omega t}\right\} \\ &= \mathbf{a}_x E_{x0} \cos(\omega t - kz) + \mathbf{a}_y E_{x0} A \cos(\omega t - kz + \phi)\end{aligned} \tag{9-98}$$

At a fixed point in space, say $z = 0$,

$$\mathbf{E}(z=0,t) = \mathbf{a}_x E_{x0} \cos\omega t + \mathbf{a}_y E_{x0} A\cos(\omega t + \phi) \tag{9-99}$$

Let us examine different cases for orientations of the electric field.

Case 1 $A = 0$ $[E_y = 0]$

$$\mathbf{E} = \mathbf{a}_x E_{x0} e^{-jkz}$$
$$\mathbf{E}(z,t) = \mathbf{a}_x E_{x0}\cos(\omega t - kz)$$

At $z = 0$,

$$\mathbf{E}(0,t) = \mathbf{a}_x E_{x0} \cos \omega t.$$

As time progresses, the electric field oscillates on a straight line—the horizontal axis—if we plot $\mathbf{E}(z = 0, t)$ on the E_x–E_y plane, as shown in Figure 9-7(a). So the wave is *linearly polarized* (LP)—*x-polarized* when $E_y = 0$ and *y-polarized* when $E_x = 0$.

Case 2 $A = 2$, $\phi = 0$ [E_x and E_y are in phase]

$$\mathbf{E} = (\mathbf{a}_x + \mathbf{a}_y 2) E_{x0} e^{-jkz}$$

At $z = 0$,

$$\mathbf{E}(0,t) = (\mathbf{a}_x + \mathbf{a}_y 2) E_{x0} \cos \omega t$$

Again the $\mathbf{E}$ vector lies on a straight line as time progresses (see Figure 9-7(b)). Thus the wave is *linearly polarized*. Note that the wave is linearly polarized when $\phi = 0$ or $180°$.

Case 3 $A = 1$, $\phi = -90°$ [$E_{x0} = E_{y0}$, $90°$ out of phase]

$$\mathbf{E} = \{ \mathbf{a}_x E_{x0} + \mathbf{a}_y E_{x0} e^{-j\pi/2}\} e^{-jkz} = (\mathbf{a}_x - j\mathbf{a}_y) E_{x0} e^{-jkz} \tag{9-100}$$

At $z = 0$,

$$\begin{aligned}\mathbf{E}(0,t) &= \mathbf{a}_x E_{x0} \cos \omega t + \mathbf{a}_y E_{x0} \cos(\omega t - 90°)\\ &= \mathbf{a}_x E_{x0} \cos \omega t + \mathbf{a}_y E_{x0} \sin \omega t \equiv \mathbf{a}_x E_x(t) + \mathbf{a}_y E_y(t)\end{aligned}$$

When we plot $\mathbf{E}$ in the E_x–E_y plane as a function of time, we find that the tip of the $\mathbf{E}$ vector traces a circle, rotating in the counterclockwise direction, as shown in Figure 9-7(c). The wave is said to be *circularly polarized*. The circularly polarized wave has a sense of rotation, or handedness. According to the definition of the IEEE standards, the wave just shown is right-handed because if we place the fingers of the right hand in the direction of rotation, the thumb will point in the direction of propagation (+z in this case). It is then clear that when $E_x = E_y$ and $\phi = 90°$,

$$\mathbf{E} = (\mathbf{a}_x + j\mathbf{a}_y)\, E_{x0}\, e^{-jkz}$$

and the wave will be *left-hand circularly polarized.* For both cases, we can show that $E_x(t)$ and $E_y(t)$ satisfies the equation of a circle:

$$E_x^2 + E_y^2 = E_{x0}^2 \quad \text{or} \quad \left(\frac{E_x}{E_{x0}}\right)^2 + \left(\frac{E_y}{E_{x0}}\right)^2 = 1$$

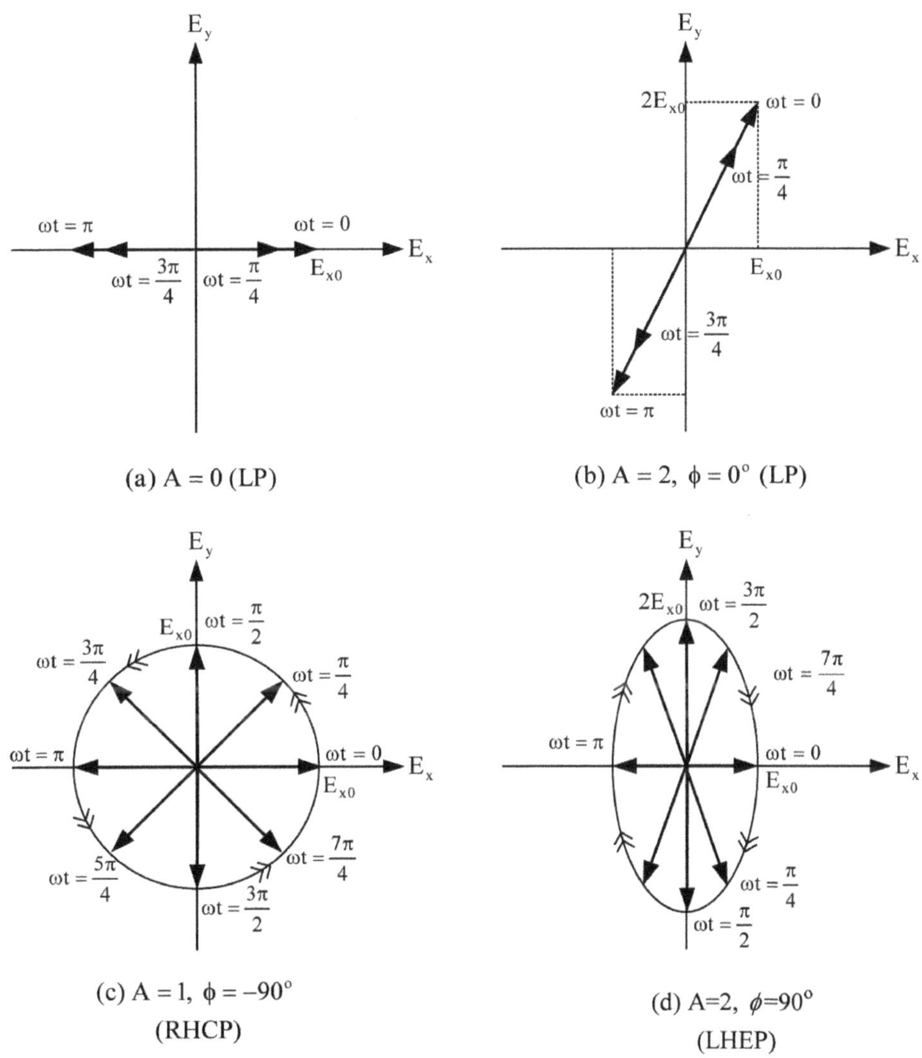

Figure 9-7. Wave polarization

Case 4 $A = 2, \phi = 90°$ [$E_{x0} \neq E_{y0}$, 90° out of phase]

$$\mathbf{E} = \{ \mathbf{a}_x E_{x0} + \mathbf{a}_y 2E_{xo} e^{j\pi/2} \} e^{-jkz} = (\mathbf{a}_x + \mathbf{a}_y 2j) E_{xo} e^{-jkz} \tag{9-101}$$

At $z = 0$,

$$\mathbf{E}(0,t) = \mathbf{a}_x E_{x0} \cos \omega t - \mathbf{a}_y 2E_{x0} \sin \omega t$$

In this case, we find that as time progresses the $\mathbf{E}$ vector traces an ellipse and rotates in the clockwise direction (see Figure 9-7(d)). Hence the wave is said to be *left-hand elliptically polarized.* $E_x(t)$ and $E_y(t)$ satisfies the equation of an ellipse:

$$\left(\frac{E_x}{E_{x0}}\right)^2 + \left(\frac{E_y}{2E_{x0}}\right)^2 = 1$$

Note that the wave is elliptically polarized when $E_{x0} \neq E_{y0}$, $\phi = \pm 90°$ or $\phi \neq 0°, \pm 90°, 180°$ in which case the polarization ellipse will be tilted. All cases of wave polarization depending on A and ϕ are summarized in Table 9-2. It is important to recognize that any wave having some general elliptical polarization can be decomposed into the superposition of two orthogonal, linearly polarized waves—for example, the x-polarized and y-polarized waves as shown in Eq. (9-101). Therefore, the analysis of a uniform plane wave having general polarization can be done by decomposing it into two orthogonal, linearly polarized waves, analyzing each linearly polarized wave separately and combining the two solutions. This will greatly simplify the analysis of many complex wave problems. One immediate example will be shown in studying reflection and transmission of the wave in the next chapter.

Table 9-2. Wave Polarization

$\mathbf{E} = (\mathbf{a}_x E_x + \mathbf{a}_y E_y) e^{-jkz}$	$\frac{E_y}{E_x} = A e^{j\phi}$
Linear Polarization	(i) $E_x = 0$, (ii) $E_y = 0$ (iii) $\phi = 0°$ or $180°$ (E_x and E_y are in phase)
Circular Polarization	$A = 1, \phi = \pm 90° \begin{cases} + \text{left-handed} \\ - \text{right-handed} \end{cases}$ when $\mathbf{k} \mathbin{//} +\mathbf{a}_z$ $(\lvert E_x\rvert = \lvert E_y\rvert,\ \pm 90°$ out of phase$)$
Elliptical Polarization	(i) $\phi \neq 0°, \pm 90°, 180°$ (ii) $\phi = \pm 90°, A \neq 1$ $(\lvert E_x\rvert \neq \lvert E_y\rvert)$ $\left.\begin{cases} \phi > 0 \to \text{left-handed} \\ \phi < 0 \to \text{right-handed} \end{cases}\right\}$ when $\mathbf{k} \mathbin{//} +\mathbf{a}_z$

Sunlight is *unpolarized* or not polarized because although its electric field has two orthogonal components, each component is randomly varying and E_x and E_y would not have any deterministic relationship. However, when the sunlight is reflected from the road surface or snow on the ground, the reflected light known as the *glare* is *partially polarized* because one linearly polarized component is reflected more while the other component is mostly transmitted into the ground (see Figure 10-6 and discussion in Section 10.5). A partially polarized wave can be viewed as a mixture of polarized waves and unpolarized waves. Skylight is another example of a partially polarized wave. Some species of ants and horseshoe crabs are known to be sensitive to polarized light and they use it for navigation.*

Linearly polarized waves are used in AM radio and TV broadcasting systems. AM broadcast stations operate at relatively low frequencies ($f = 535–1605$ kHz) and require large antenna towers in order to produce longer wavelengths ($\lambda = 187–561$m). These tall antennas generate *vertically polarized* waves, with the $\mathbf{E}$ field perpendicular to the ground and parallel to the antenna. For maximum reception of AM signals, receiving antennas of the radio must be oriented parallel to the electric field, thus perpendicular (vertical) to the ground. On the other hand, most TV broadcasting signals are *horizontally polarized*, with the $\mathbf{E}$ fields parallel to the ground, so for good reception, rooftop TV antennas should be oriented horizontally and also perpendicular to the direction from which the signal comes. Most of the FM radio stations ($f = 88–108$ MHz) in the United States utilize *circular polarization*, which has the advantage that any orientation of an FM receiving antenna can detect a signal as long as the antenna is on the plane perpendicular to the direction of the signal propagation because a circularly polarized wave has two orthogonal, linearly polarized components.

PROBLEMS

9-1. Consider the electromagnetic wave shown in Example 8-5.

(a) Confirm that $\mathbf{E}(z, t)$ satisfies the wave equation, Eq. (9-4), under certain condition.

(b) Write the electric field in the form of $f(z - vt)$ and determine the direction and the velocity of wave propagation.

(c) Does the direction of wave propagation agree with the direction of electromagnetic power flow which is given by the Poynting vector? Does the velocity agree with the expression in Eq. (9-9)?

9-2. An electric field is given by

$$\mathbf{E} = \mathbf{a}_z E_0 \cos(\omega t - kz)$$

Does this field satisfy the wave equation? Does this represent an electromagnetic wave? To represent an *electromagnetic* wave, all of Maxwell's equations should be satisfied.

9-3. A light propagates through a glass which has permittivity $\varepsilon = 2.56\varepsilon_0$ and permeability $\mu = \mu_0$. What is the velocity of propagation? Compare the result with the velocity of light propagating in air.

*See T. H. Waterman, "Polarized light and animal navigation", *Scientific American*, July 1955, pp. 88.

9-4. (a) Although the average distance between the Earth and Mars is 50 million miles, the Red Planet came very close to the Earth, 34.6 million miles away, on August 27, 2003. How long does it take light to travel from Mars to the Earth?

(b) The star Vegas is 26 light years from the Earth. How far is Vegas from the Earth in miles? A light year is a unit of length representing the distance light travels in one year.

9-5. Find the complex phasor representation of the following quantities that are sinusoidal in time.

(a) $\rho_v(r,t)=\rho_0 e^{-\frac{r}{r_0}}\cos(10^9 t-\pi/4)$

(b) $\mathbf{J}(\rho,t)=\mathbf{a}_z J_0\dfrac{\rho}{a}\sin(2\pi 10^6 t-\pi/3)$

(c) $\mathbf{E}(z,t)=\mathbf{a}_x E_0\cos(\omega t-2z)+\mathbf{a}_y E_0\sin(\omega t-2z)$

(d) $\mathbf{H}(x,t)=\mathbf{a}_y H_0\sin(\omega t-5x)+\mathbf{a}_z H_0\cos(\omega t+5x)$

Assume that ρ_0, r_0, J_0, a, E_0 and H_0 are real constants.

9-6. Find the time-harmonic fields in real time for the following complex phasor fields. Assume that the angular frequency is ω [rad/s].

(a) $\mathbf{E}=(\mathbf{a}_x+\mathbf{a}_y 2j)\,e^{-j2z}$

(b) $\mathbf{E}=\left[\mathbf{a}_y(1+j)+\mathbf{a}_z(2-j)\right]\,e^{-j5x}$

(c) $\mathbf{H}=\mathbf{a}_x(1-j)e^{-j2z}+\mathbf{a}_y\left(1-j\sqrt{3}\right)e^{+j2z}$

9-7. By taking the imaginary part of the complex Poynting theorem, (9-21), derive the following relationship between the reactive power and the stored electromagnetic energy:

$$\text{Im}(P_s)=\text{Im}(P_f)+2\omega(W_m-W_e)$$

where

$$P_s=\iiint_V\left(-\underline{\mathbf{E}}\cdot\underline{\mathbf{J}}_i^*\right)dv=\begin{pmatrix}\text{complex power supplied by}\\ \text{the source current } \underline{\mathbf{J}}_i\end{pmatrix}$$

$$P_f=\oint\!\!\oint_S \underline{\mathbf{E}}\times\underline{\mathbf{H}}^*\cdot ds=\begin{pmatrix}\text{complex power flowing out}\\ \text{through the closed surface S}\end{pmatrix}$$

$$W_m = \iiint_V \frac{1}{2}\mu\left|\underline{\mathbf{H}}\right|^2 dv = (\text{magnetic energy stored in the volume})$$

$$W_e = \iiint_V \frac{1}{2}\varepsilon\left|\underline{\mathbf{E}}\right|^2 dv = (\text{electric energy stored in the volume})$$

Assume that μ and ε are real.

9-8. Show that for an electromagnetic wave the amplitudes of the electric field (E) and the magnetic field (B) are related by

$$E = v\,B$$

where v is the velocity of the wave in the medium.

9-9. Show that for a time-harmonic electromagnetic wave, with peak electric field amplitude E_0, traveling in a lossless medium of (μ, ε), the time-average power density carried by the wave is given by

$$S_{av} = \frac{\left|E_0\right|^2}{2\eta} \quad \left[\frac{W}{m^2}\right]$$

where $\eta = \sqrt{\dfrac{\mu}{\varepsilon}}$ is the intrinsic impedance of the medium.

9-10. At a distance of 10 km from an AM radio station antenna, the electric field has a peak amplitude $0.5\ \mathrm{V/m}$.

(a) What is the peak amplitude of the magnetic field $(\mathbf{H})$ at this point?
(b) Assuming that the antenna radiates equally in all directions above the ground, what is the total power in watts of the station?
(c) Assuming that the total power is maintained at any distance, calculate the strength of the electric field at the AM radio receiver, $50\ \mathrm{km}$ away from the station.

9-11. **(Radiation pressure)**
Just as the electromagnetic wave transports energy, it carries *momentum*. When the electromagnetic wave hits the surface of an object, the momentum is transferred to the object. The rate at which momentum is transferred is the *force* on the surface. The average force per unit area due to the wave is called the *radiation pressure*.

(a) Show that the radiation pressure that an electromagnetic plane wave exerts normally on a totally absorbing surface is given by

$$p_{rad} = \sqrt{\mu\varepsilon}\; S_{av} = \frac{S_{av}}{v} \quad \left[\frac{N}{m^2}\right]$$

where S_{av} is the time-average Poynting power density. When the wave is totally reflected (e.g., when the object is a perfect conductor), the momentum change is twice as compared to the perfect absorber and the radiation pressure will be twice the above amount.

(b) The direct sunlight carries the average power density of $1.4\ \mathrm{kW/m^2}$ when it passes through the earth's atmosphere. Suppose an earth-orbiting satellite has a solar panel of total area $8\ \mathrm{m}^2$. Assuming that the solar panel is a perfect absorber, calculate the total force exerted on the panel by the sunlight.

9-12. Consider an electromagnetic wave propagating in air (free of sources) which has the following time-harmonic magnetic field

$$\mathbf{H} = \mathbf{a}_x H_0 \cos\left(2\pi\, 10^9 t - \frac{20\pi}{3} z\right)$$

where $H_0 = 10$ [mA/m].

(a) Find the electric field $\mathbf{E}$ (z,t) of this wave, using complex phasors.

(b) Find the time-average power density in watts/m^2 carried by the electromagnetic wave.

9-13. A time-harmonic electromagnetic wave with the electric field

$$\mathbf{E} = \mathbf{a}_z\, 5\sin\left(\omega t - 0.5x\right)\left[\frac{\mathrm{V}}{\mathrm{m}}\right]$$

propagates in a lossless dielectric medium of $\varepsilon = 2.0\,\varepsilon_0$, $\mu = \mu_0$ (free of sources).

(a) Find the complex phasor representation $\underline{\mathbf{E}}$ of the electric field.

(b) Find the complex magnetic field $\underline{\mathbf{H}}$, using Faraday's law for time-harmonic fields. Write down the magnetic field in real time, $\mathbf{H}(x,t)$.

(c) The complex fields $\underline{\mathbf{E}}$ and $\underline{\mathbf{H}}$ should also satisfy Ampère's law for time-harmonic fields. Determine the value of ω in rad/s.

(d) Find the time-average Poynting vector $\langle \mathbf{S}(x, t)\rangle \equiv \mathbf{S}_{av}$ using (i) the fields in real time, and (ii) the complex fields. Show that they agree with each other.

9-14. Derive the relationships between (A, B) and $\left(E_0^+, E_0^-\right)$ shown in Eqs. (9-39) and (9-40), using Euler's formula, $e^{j\theta} = \cos\theta + j\sin\theta$.

9-15. For an electromagnetic wave in air of the following frequency, calculate the wavelength in meters.

(a) $f = 60$ Hz (household electricity)

(b) $f = 300$ kHz (AM radio)

(c) $f = 2.45$ GHz (microwave oven)

9-16. For an electromagnetic wave in air of the following wavelength, calculate the frequency in Hz.

(a) $\lambda = 3.3$ m (FM radio)
(b) $\lambda = 5.2$ cm (cellular phone)
(c) $\lambda = 460$ nm (blue light)

9-17. **(The wave equation in a source-free region)**
Consider the magnetic field $\mathbf{H}(z,t)$ given in Problem 9-12.

(a) Confirm that $\mathbf{H}(z,t)$ satisfies the wave equation, Eq. (8-26) or (9-5).
(b) Express $\mathbf{H}(z,t)$ in the form of general solutions of the wave equation, i.e., $f(z-vt)$, and determine the velocity v of the electromagnetic wave. What is the direction of propagation of this wave?
(c) Determine the Poynting vector $\mathbf{S}$ and confirm that the direction of power flow is the same as the direction of wave propagation you obtained in (b).
(d) Calculate the total *time-average power* crossing the surface area bounded by $x = 0$, $x = a$ [m], $y = 0$, and $y = b$ [m], (i) by using the fields or the Poynting vector in real time, and (ii) by using the complex fields or the complex Poynting vector.

9-18. The magnetic field of a TV broadcasting signal propagating in air is given by

$$\mathbf{H} = \mathbf{a}_z\, 2 \sin(\omega t - 10x) \text{ [mA/m]}$$

(a) Find the corresponding electric field $\mathbf{E}(x,t)$. You can use one of Maxwell's equations in either time domain or complex (phasor) domain. If you choose the latter approach, you should first convert $\mathbf{H}(x,t)$ to the complex phasor $\mathbf{H}(x)$ *before* you apply Maxwell's equations.
(b) Determine the frequency of the TV signal f [Hz], using another one of Maxwell's equations or using the dispersion relation.

9-19. The electric field of an FM radio signal propagating in air is given by

$$\mathbf{E}(x,t) = \mathbf{a}_y\, 5 \cos(2\pi\, 10^8 t - k_0 x) \text{ [V/m]}$$

(a) Does this field represent a *uniform plane wave*?
(b) Determine the propagation constant or the wave number k_0 of the wave.
(c) Determine the time-average Poynting vector $\langle \mathbf{S} \rangle$. Discuss the relationship between the directions of $\mathbf{E}$, $\mathbf{H}$ and $\mathbf{S}$.
(d) Find the wavelength in [m], phase velocity and the intrinsic impedance.

9-20. **(Uniform plane wave in lossless medium)**
Suppose a uniform plane wave is traveling in the negative y direction in a *lossless* medium, with the 2 V/m (rms) electric field in the z direction. The wavelength of the wave is 10 cm and the velocity of wave propagation is 2.5×10^8 m/s.

(a) Determine the frequency in Hz of the wave and the permittivity of the medium, assuming the medium is non-magnetic, i.e., $\mu = \mu_0$.
(b) What is the refractive index of the medium?
(c) Write complete complex phasor expressions for the electric and magnetic field vectors. Determine the rms value (in A/m) of the magnetic field **H**.
(d) What is the ratio of amplitudes of **E** and **H** along with the unit? Did you obtain the expected result that was discussed in Section 9.4?

9-21. Suppose a uniform plane wave is traveling in the negative x direction in a *lossless* medium of $\mu = \mu_0$, $\varepsilon = 9\,\varepsilon_0$, with the 5 V/m (rms) electric field in the +y direction. The wavelength of the wave is 2 m.

(a) Determine the frequency in Hz and velocity in m/s of the wave, and the refractive index of the medium.
(b) Write complete time-domain expressions of the electric and magnetic field vectors. Determine the rms value (in A/m) of the magnetic field **H**.
(c) Calculate the ratio of the amplitudes of **E** and **H** along with the unit and confirm that it is equal to the intrinsic impedance of the medium.

9-22. **(Wave propagating in an arbitrary direction)**
Consider a time-harmonic electromagnetic plane wave of electric field amplitude 5 [V/m], frequency 1 GHz, and phase angle zero that is traveling in air in the direction from the origin to the point (1, 2, 3), with polarization (or the direction of the electric field) *parallel to the xy plane*.

(a) Write down the explicit expression for the propagation vector or wave vector **k** in Cartesian coordinates. Consider both the direction and the magnitude.
(b) Find the explicit expression for the real or complex electric field **E**.
(c) Find the real or complex magnetic field **H**, using the relationship, Eq. (9-65).
(d) Calculate **H** using Faraday's law and show that it agrees with the result of (c).

9-23. Derive the exact expressions for the attenuation constant (α) and the phase constant (β) for the uniform plane wave in lossy media shown in Eq. (9-86).

$$\alpha = \frac{k_u}{\sqrt{2}}\left[\sqrt{1+(LT)^2}-1\right]^{1/2}$$

$$\beta = \frac{k_u}{\sqrt{2}}\left[\sqrt{1+(LT)^2}+1\right]^{1/2}$$

where $k_u = \omega\sqrt{\mu\varepsilon}$, $LT = \frac{\sigma}{\omega\varepsilon}$.

9-24. Consider a 1.9 GHz electromagnetic wave, received by a mobile telephone, penetrating into the human head. Assume that a human brain tissue has the permittivity $50\ \varepsilon_0$, conductivity $\sigma = 1.4$ [S/m] and permeability $\mu \approx \mu_0$ at 1.9 GHz.

(a) Calculate the attenuation constant α and the penetration depth of the wave inside the brain tissue.

(b) Assuming that the peak amplitude of the electric field at the point of entry inside the brain tissue is about 10 [V/m], calculate the peak amplitude of the electric field at the depth of 2 cm inside the brain tissue.

9-25. (**Microwave oven**)
At the operating frequency (2.45 GHz) of a microwave oven, the permittivity of sirloin steak is about $\varepsilon = 40\ \varepsilon_0$, the permeability is μ_0, and the conductivity is $\sigma = 0.5$ [S/m].

(a) Calculate the attenuation constant α and the penetration depth d_p exactly.

(b) Calculate the loss tangent, determine whether the steak is *slightly lossy* or *highly conducting* at the given frequency, and then use the approximate expressions to calculate α and d_p. Compare the results of (a) and (b) and discuss the errors.

(c) Now consider polystyrene foam which has the permittivity $\varepsilon = 1.03\varepsilon_0$, the permeability μ_0, and the conductivity $\sigma = 10^{-5}$ [S/m]. Calculate the attenuation constant and the penetration depth of the foam and compare with the results of the steak. Is it OK to put the steak inside the foam cover (container)?

9-26. Calculate the attenuation constant, penetration depth and intrinsic impedance of earth for an electromagnetic wave of 15 MHz. Assume that the earth has $\varepsilon = 3\ \varepsilon_0$, $\mu = \mu_0$, $\sigma = 10^{-4}$ [S/m]. Calculate also the same parameters when the earth is wet. Assume that the wet earth has $\varepsilon = 12\ \varepsilon_0$, $\mu = \mu_0$, $\sigma = 10^{-2}$ [S/m]. Compare the two results.

9-27. A highly conducting material can be used to shield an area against the electromagnetic interference. Suppose you want to shield a room from radio interference by enclosing a layer of copper. The radio frequencies range from 10 kHz to 100 MHz. You need to make the copper layer be at least five times the skin depth of the copper. Assuming that copper has $\varepsilon = \varepsilon_0$, $\mu = \mu_0$, $\sigma = 5.8 \times 10^7$ [S/m], find the thickness of the copper layer that is required to shield the radio interference.

9-28. The effective complex permittivity of walls in building should be known for wireless communication applications. Suppose the complex permittivity of the concrete wall of a building is found to be $(6.5 - j0.3)\,\varepsilon_0$ at 1.8 GHz. Assuming that the electric field strength is 1 V/m when the signal enters the building and the thickness of the wall is 50 cm, what is the electric field strength in V/m when you answer the cell phone inside the building? You can calculate α directly from ε_{eff} or by using the relationship between ε'', σ and ω.

9-29*. Consider a uniform plane wave of electric field amplitude $10\ \mathrm{V/m}$ and frequency $1\ \mathrm{GHz}$ propagating into freshly fallen snow which has the dielectric constant 1.2, the loss tangent 3×10^{-2} and permeability μ_0.

(a) Find the conductivity, the attenuation constant, and the penetration depth of the snow.
(b) Calculate the average power dissipated in a surface block of snow having surface area of $100\ \mathrm{cm}^2$ as the wave travels a depth of $10\ \mathrm{cm}$.

9-30. As discussed in Example 9-5, sea water becomes very lossy at high frequencies. Consider an airplane flying over the surface of the ocean and communicating with a submarine underneath the water. The airplane transmits a signal of $1\ \mathrm{MHz}$ using a trailing, long-wire antenna. The electric field amplitude of the signal is $10\ \mathrm{V/m}$ upon entering the ocean surface. If the submarine requires a minimum signal level of $10\ \mu\mathrm{V/m}$, determine the maximum communication depth of the submarine.

9-31. **(Dispersion in conducting medium)**
An electromagnetic wave of a carrier frequency $1\ \mathrm{MHz}$ with a narrow band voice signal propagates through sea water that has $\varepsilon = 81\ \varepsilon_0$, $\mu = \mu_0$, $\sigma = 4\ [\mathrm{S/m}]$. Find the phase velocity and the group velocity of this composite signal.

9-32. **(Dispersion in a plasma medium)**
A plasma medium is a neutral ionized gas consisting of free electrons and positive charged ions. The permittivity of the plasma is known to be*

$$\varepsilon = \varepsilon_0\left[1-\left(\frac{\omega_P}{\omega}\right)^2\right]$$

where $\omega_p = \sqrt{\dfrac{Ne^2}{m\varepsilon_0}} \approx 56.4\sqrt{N}$ is the plasma frequency. N is the number of electrons per unit volume, e and m are the charge and mass of an electron. Find the phase velocity and the group velocity of the wave in plasma.

9-33. Determine the (wave) polarization of the electromagnetic wave with the following electric fields by plotting the electric field vectors as a function of time at a fixed point in space, like Figure 9-7. When the polarization is circular or elliptical, determine also *handedness* of the wave polarization.

(a) $\mathbf{E}(z,t) = \mathbf{a}_x\, 2\cos(\omega t + kz) - \mathbf{a}_y\, 3\cos(\omega t + kz)$
(b) $\mathbf{E}(z,t) = \mathbf{a}_x\, 2\cos(\omega t - kz) - \mathbf{a}_y\, 2\sin(\omega t - kz)$
(c) $\mathbf{E}(z,t) = \mathbf{a}_x\, 2\cos(\omega t + kz) - \mathbf{a}_y\, 3\sin(\omega t + kz)$
(d) $\mathbf{E}(x,t) = \mathbf{a}_y\, 2\cos(\omega t + kx) + \mathbf{a}_z\, 2\sin(\omega t + kx)$
(e) $\mathbf{E}(z,t) = \mathbf{a}_x\, 2\cos(\omega t - kz) + \mathbf{a}_y\, 2\cos(\omega t - kz + 45°)$

* See J. A. Kong, *Electromagnetic Wave Theory* (2008), pp. 271–274 or D. H. Staelin et al., *Electromagnetic Waves* (1994), pp. 99–101.

9-34. Repeat Problem 9-33 when the wave has the following complex electric fields.

(a) $\mathbf{E} = (\mathbf{a}_x 3 + \mathbf{a}_y 2)\, e^{-jkz}$
(b) $\mathbf{E} = (\mathbf{a}_x 3 - \mathbf{a}_y j3)\, e^{+jkz}$
(c) $\mathbf{E} = (\mathbf{a}_x 3 - \mathbf{a}_y j2)\, e^{-jkz}$
(d) $\mathbf{E} = (\mathbf{a}_x 3 + \mathbf{a}_z j3)\, e^{-jky}$
(e) $\mathbf{E} = [\mathbf{a}_x j3 + \mathbf{a}_y (3 + j3)]\, e^{-jkz}$

9-35. Prove that a circularly polarized wave can be decomposed into a sum of two orthogonal, linearly polarized waves.

9-36. Prove that a linearly polarized wave can be decomposed into a sum of a right-handed circularly polarized wave and a left-handed circularly polarized wave. Take, for example, $\mathbf{E} = \mathbf{a}_x E_0\, e^{-jkz}$.

REFLECTION AND TRANSMISSION OF WAVES

CHAPTER 10

10.1 INTRODUCTION

In Chapter 9 we considered a wave propagating in a single unbounded medium which can be lossless or lossy, along with its propagation characteristics, dispersion, and polarization. We will now consider what happens when an electromagnetic wave hits the boundary of another medium. We expect that a part of the wave will be reflected from the surface and a part of it will transmit through the second medium. Thus we study the reflection and transmission of the wave at a boundary separating two different media. In each medium the fields satisfy Maxwell's equations with appropriate medium parameters or constitutive relations. At the interface, the fields on both sides must satisfy the boundary conditions that are given by Eq. (8-27). Depending on the types of the media, the boundary conditions may take different forms. Here we solve the boundary value problems for time-varying fields in bounded media. First we consider the problem of a wave incident normally upon a dielectric interface, which is simpler to solve. Then we solve the problem of oblique incidence with two different distinct polarizations—perpendicular and parallel. The interpretation of the results on the solutions for the fields is given. Interesting phenomena such as total internal reflection and Brewster's angle effect are discussed. Lastly, the reflection (and no transmission) of waves from a perfectly conducting surface is studied along with the concept of standing waves.

10.2 NORMAL INCIDENCE AT A DIELECTRIC BOUNDARY

Let's consider a uniform plane wave, propagating in the $+x$ direction in medium 1 having material constants ε_1, μ_1. The wave is incident normally upon medium 2 having material constants ε_2, μ_2. The interface is at $x = 0$ (yz plane) as shown in Figure 10-1. We assume now that both media are lossless dielectrics ($\sigma_1 = \sigma_2 = 0$), but each of them can be lossy in our formulation, in which case ε_1 and/or ε_2 can be simply replaced by complex effective permittivities, making our results still valid for lossy media.

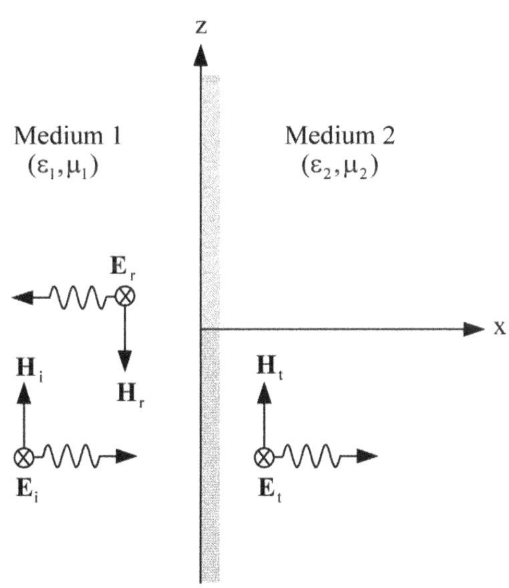

Figure 10-1. Reflection and transmission of waves at normal incidence.

Assuming the incident wave is linearly polarized in the y direction (we can assume z-polarization, which will give the same result), since the wave propagates in the +x direction, its electric and magnetic fields can be written as

$$\mathbf{E}_i = \mathbf{a}_y E_0 e^{-jk_1x} \tag{10-1a}$$

$$\mathbf{H}_i = \frac{1}{\eta_1}\mathbf{a}_x \times \mathbf{E}_i = \mathbf{a}_z \frac{E_0}{\eta_1} e^{-jk_1x} \tag{10-1b}$$

where

$$k_1 = \omega\sqrt{\mu_1\varepsilon_1} \tag{10-2a}$$

$$\eta_1 = \sqrt{\frac{\mu_1}{\varepsilon_1}} \tag{10-2b}$$

k_1 is the propagation constant of the incident wave and η_1 the intrinsic impedance of medium 1. We have used the property of the uniform plane wave in deriving Eq. (10-1b), instead of using Faraday's law. The presence of the boundary gives rise to a reflected wave in medium 1 and a transmitted wave in medium 2. Since the reflected wave travels in the –x direction in medium 1, its electric and magnetic fields can be written in the following form:

$$\mathbf{E}_r = \mathbf{a}_y E_r e^{+jk_1x} \equiv \mathbf{a}_y R E_0 e^{jk_1x} \tag{10-3a}$$

$$\mathbf{H}_r = \frac{1}{\eta_1}(-\mathbf{a}_x) \times \mathbf{E}_r = -\mathbf{a}_z \frac{RE_0}{\eta_1} e^{jk_1x} \tag{10-3b}$$

where

$$R \equiv \frac{E_r}{E_0} \tag{10-4}$$

is the ratio of $\mathbf{E}$ field amplitudes of the reflected vs. incident waves, called the **reflection coefficient**. Note that the reflected wave has the same k and η as the incident wave but its spatial dependence is changed to $\exp(+jk_1x)$. The transmitted wave propagates in +x direction in medium 2 and its fields are written as

$$\mathbf{E}_t = \mathbf{a}_y E_t e^{-jk_2x} \equiv \mathbf{a}_y T E_0 e^{-jk_2x} \tag{10-5a}$$

$$\mathbf{H}_t = \frac{1}{\eta_2}\mathbf{a}_x \times \mathbf{E}_t = \mathbf{a}_z \frac{TE_0}{\eta_2} e^{-jk_2x} \tag{10-5b}$$

where

$$k_2 = \omega\sqrt{\mu_2\varepsilon_2} \tag{10-6a}$$

$$\eta_2 = \sqrt{\frac{\mu_2}{\varepsilon_2}} \tag{10-6b}$$

k_2 is the propagation constant of the transmitted wave and η_2 is the intrinsic impedance of medium 2.

$$T \equiv \frac{E_t}{E_0} \tag{10-7}$$

is the ratio of $\mathbf{E}$ field amplitudes of the transmitted vs. incident waves, called the **transmission coefficient**.

All the parameters are known except the two coefficients, R and T. They are determined by applying the boundary conditions for the fields at the boundary ($x = 0$). Although we have four boundary conditions in Eq. (8-27), we only need to use two of them on tangential components, i.e., Eqs. (8-27a) and (8-27b). $\mathbf{E}_1$, $\mathbf{H}_1$ consist of the incident and reflected fields and $\mathbf{E}_2$, $\mathbf{H}_2$ have only transmitted fields. Since medium 2 is not a perfect conductor for the given problem, there exists no surface current at the boundary, i.e., $\mathbf{J}_s = 0$. Thus we have the following two boundary conditions at the dielectric interface:

$$\mathbf{E}_{1t} = \mathbf{E}_{2t} \tag{10-8a}$$

$$\mathbf{H}_{1t} = \mathbf{H}_{2t} \tag{10-8b}$$

where the subscript "t" means the tangential component. Note that Eq. (10-8) is the same as Eq. (8-29).

For this problem, we have

$$\mathbf{E}_i + \mathbf{E}_r = \mathbf{E}_t \quad \text{at } x = 0 \tag{10-9a}$$

$$\mathbf{H}_i + \mathbf{H}_r = \mathbf{H}_t \quad \text{at } x = 0 \tag{10-9b}$$

Substituting $x = 0$ in Eqs. (10-1), (10-3), (10-5) and applying Eq. (10-9), we obtain

$$1 + R = T \tag{10-10a}$$

$$\frac{1}{\eta_1}(1 - R) = \frac{1}{\eta_2} T \tag{10-10b}$$

Solving Eq. (10-10) for R and T, we obtain

$$R = \frac{\eta_2 - \eta_1}{\eta_2 + \eta_1} \quad \text{(Reflection coefficient for normal incidence)} \tag{10-11a}$$

$$T = \frac{2\eta_2}{\eta_2 + \eta_1} \quad \text{(Transmission coefficient for normal incidence)} \tag{10-11b}$$

Note that $|R| \leq 1$, although the magnitude of T may exceed 1. R and T are real when both media are lossless, and they become complex when either one of the two is lossy. As will be seen later in Section 11.7, the expression of R is very similar to that for the reflection coefficient of the voltage wave along the transmission line.

Let's calculate the Poynting vectors of the incident, reflected and transmitted waves and find the relationships regarding the power conservation. The time-average Poynting vectors are given by (assuming lossless media)

$$\mathbf{S}_{av,i} = \frac{1}{2}\mathrm{Re}\left(\mathbf{E}_i \times \mathbf{H}_i^*\right) = \mathbf{a}_x \frac{|E_0|^2}{2\eta_1} \tag{10-12a}$$

$$\mathbf{S}_{av,r} = \frac{1}{2}\mathrm{Re}\left(\mathbf{E}_r \times \mathbf{H}_r^*\right) = -\mathbf{a}_x |R|^2 \frac{|E_0|^2}{2\eta_1} \tag{10-12b}$$

$$\mathbf{S}_{av,t} = \frac{1}{2}\mathrm{Re}\left(\mathbf{E}_t \times \mathbf{H}_t^*\right) = \mathbf{a}_x |T|^2 \frac{|E_0|^2}{2\eta_2} \tag{10-12c}$$

According to the conservation of power, it should hold that

$$P_i = P_r + P_t \tag{10-13a}$$

or

$$\left|\mathbf{S}_{av,i}\right| = \left|\mathbf{S}_{av,r}\right| + \left|\mathbf{S}_{av,t}\right| \tag{10-13b}$$

where P_i, P_r, P_t are the powers of the incident, reflected and transmitted waves, respectively. Eq. (10-13b) reduces to

$$1 = |R|^2 + \frac{\eta_1}{\eta_2}|T|^2 \tag{10-14}$$

Eq. (10-14) can be easily verified by substituting Eq. (10-11). The fraction or percentage of power reflected from the boundary is given by

$$\frac{P_r}{P_i} = \frac{\left|\mathbf{S}_{av,r}\right|}{\left|\mathbf{S}_{av,i}\right|} = |R|^2 \tag{10-15}$$

and the fraction of power transmitted through medium 2 is given by

$$\frac{P_t}{P_i} = \frac{\left|\mathbf{S}_{av,t}\right|}{\left|\mathbf{S}_{av,i}\right|} = |T|^2 \frac{\eta_1}{\eta_2} = 1 - |R|^2 \tag{10-16}$$

Note that the fraction of power transmitted is not equal to $|T|^2$.

EXAMPLE 10-1 WAVE NORMALLY INCIDENT ON THE LAKE

An electromagnetic plane wave of 1 MHz is incident normally on Onondaga Lake. Assume that the lake water has $\varepsilon = 81\,\varepsilon_0$, $\mu = \mu_0$, and $\sigma = 10^{-4}$ [S/m].

(a) Calculate the reflection and transmission coefficients of the incident wave, ignoring the loss ($\sigma = 0$).

(b) Calculate the reflection and transmission coefficients, including the loss ($\sigma = 10^{-4}$). Write down the complete expression for the electric field of the transmitted wave.

SOLUTIONS:

(a) When the wave is incident from air upon the lake, η_1 and η_2 are given by

$$\eta_1 = \sqrt{\frac{\mu_0}{\varepsilon_0}} = \eta_0, \quad \eta_2 = \sqrt{\frac{\mu_2}{\varepsilon_2}} = \sqrt{\frac{\mu_0}{81\varepsilon_0}} = \frac{1}{9}\eta_0$$

R and T are given by Eq. (10-11):

$$R = \frac{\eta_2 - \eta_1}{\eta_2 + \eta_1} = \frac{\frac{1}{9}\eta_0 - \eta_0}{\frac{1}{9}\eta_0 + \eta_0} = \frac{\frac{1}{9} - 1}{\frac{1}{9} + 1} = -0.8$$

$$T = \frac{2\eta_2}{\eta_2 + \eta_1} = \frac{2 \cdot \frac{1}{9}\eta_0}{\frac{1}{9}\eta_0 + \eta_0} = \frac{\frac{2}{9}}{\frac{1}{9} + 1} = 0.2$$

Note that R and T satisfy the relationship $1 + R = T$.

(b) When the lake is lossy, η_2 becomes complex and $\hat{\eta}_2 = \sqrt{\frac{\mu_2}{\varepsilon_{2,\text{eff}}}} = \sqrt{\frac{\mu_0}{81\varepsilon_0 - j\frac{\sigma}{\omega}}}$.

$$\text{At } f = 1 \text{ MHz, L.T.} = \frac{\sigma}{\omega\varepsilon} = \frac{10^{-4}}{2\pi \cdot 10^6 \times 81 \times \frac{10^{-9}}{36\pi}} = 0.022 << 1$$

The lake is *slightly lossy* at $f = 1$ MHz, so $\frac{\sigma}{\omega} << \varepsilon = 81\,\varepsilon_0$ and $\hat{\eta}_2 \approx \sqrt{\frac{\mu_0}{81\varepsilon_0}} = \frac{1}{9}\eta_0$, which is approximately the same as η_2 when the lake is lossless. Thus,

$$R = \frac{\hat{\eta}_2 - \eta_1}{\hat{\eta}_2 + \eta_1} \approx -0.8, \quad T = \frac{2\hat{\eta}_2}{\hat{\eta}_2 + \eta_1} \approx 0.2$$

You can calculate $\hat{\eta}_2$ exactly in complex numbers and find more accurate results for R and T (see Problem 10–3). Since the lake is lossy, the transmitted wave will propagate with some attenuation and the electric field can be written as

$$\mathbf{E}_t = \mathbf{a}_y \, \mathrm{T}\mathrm{E}_0 \, e^{-j(\beta - j\alpha)x} = \mathbf{a}_y \, 0.2\,\mathrm{E}_0 \, e^{-j\beta x} \, e^{-\alpha x}$$

where E_0 is the amplitude of the incident electric field and β and α can be calculated from Eq. (9-89) for slightly lossy media:

$$\beta \approx \omega\sqrt{\mu\varepsilon} = 2\pi f \sqrt{\mu_0 \cdot 81\varepsilon_0} = 2\pi \cdot 10^6 \cdot 9 \cdot \frac{1}{3\times 10^8} = 0.188 \left[\frac{\text{rad}}{\text{m}}\right]$$

$$\alpha \approx \frac{\sigma}{2}\sqrt{\frac{\mu}{\varepsilon}} = \frac{10^{-4}}{2}\sqrt{\frac{\mu_0}{81\varepsilon_0}} = \frac{10^{-4}}{2} \cdot \frac{377}{9} = 2.09\times 10^{-3} \left[\frac{\text{Np}}{\text{m}}\right]$$

10.3 OBLIQUE INCIDENCE AT A DIELECTRIC BOUNDARY

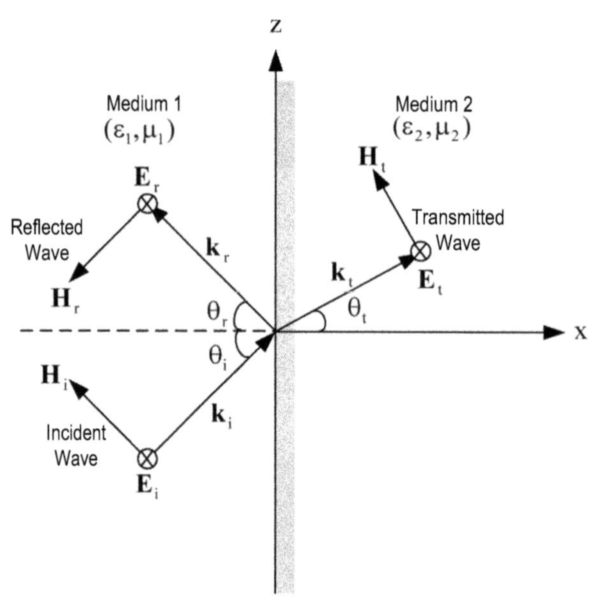

Figure 10-2. Reflection and transmission of waves with perpendicular polarization at oblique incidence

We now consider the case when a uniform plane wave is incident from medium 1 at an oblique angle upon medium 2 as shown in Figure 10-2. The interface is at $x = 0$. The angle between the incident wave vector $\mathbf{k}_i$ and the normal of the boundary (x axis) is θ_i, called the *angle of incidence*. Let the angle between the reflected wave vector $\mathbf{k}_r$ and the normal be θ_r, called the *angle of reflection* and the angle between the transmitted wave vector $\mathbf{k}_t$ and the normal be θ_t, called the *angle of transmission*. The objective of the problem is to determine the fields of the reflected and transmitted waves along with θ_r, θ_t, given the fields of the incident wave with angular frequency ω at an angle θ_i. In the case of oblique incidence, the polarization of the incident wave affects how much the wave is reflected and transmitted. Thus we decompose the fields of the incident wave into two *linearly polarized* waves, orthogonal to each other. The decomposition is done with respect to a plane, known as the *plane of incidence*, which is formed by the incident wave vector ($\mathbf{k}_i$) and the normal of the boundary (x axis). In Figure 10-2, the xz-plane or the plane of the paper is the plane of incidence. We decompose the electric field into a component perpendicular to the plane of incidence and one parallel to the plane of incidence, and solve each problem separately. The former is said to have **perpendicular polarization**, *horizontal* polarization, s-polarization (s for "senkrecht", the German word for perpendicular), or *TE* (transverse electric) polarization. The latter is said to have **parallel polarization**, *vertical* polarization, *p*-polarization (*p* for "parallel"), or *TM* (transverse magnetic) polarization.

10.3.1. Perpendicular Polarization

First we assume the incident wave is perpendicularly polarized. Since the electric field $\mathbf{E}_i$ is perpendicular to the plane of incidence, it has only the y component (see Figure 10-2). Since the incident wave propagates in the positive x and positive z directions, the wave vector can be written as (see Section 9.4.1)

$$\mathbf{k}_i = \mathbf{a}_x k_x + \mathbf{a}_z k_z \tag{10-17}$$

The electric and magnetic fields of the incident wave are written as

$$\mathbf{E}_i = \mathbf{a}_y E_0 e^{-j\mathbf{k}_i \cdot \mathbf{r}} = \mathbf{a}_y E_0 e^{-jk_x x - jk_z z} \tag{10-18a}$$

$$\mathbf{H}_i = \frac{\nabla \times \mathbf{E}_i}{-j\omega\mu_1} = (-\mathbf{a}_x k_z + \mathbf{a}_z k_x)\frac{E_0}{\omega\mu_1} e^{-jk_x x - jk_z z} \tag{10-18b}$$

Note that the magnetic field is parallel to the plane of incidence for a perpendicularly polarized wave and Eq. (10-18b) can also be obtained by $\mathbf{H}_i = \dfrac{\mathbf{k}_i \times \mathbf{E}_i}{\omega\mu_1}$ from Eq. (9-65). The components of the wave vector, k_x and k_z, can be expressed in terms of medium constants, the frequency of the wave, and the angle of incidence as follows:

$$k_x = |\mathbf{k}_i| \cos\theta_i = k_1 \cos\theta_i \tag{10-19a}$$

$$k_z = |\mathbf{k}_i| \sin\theta_i = k_1 \sin\theta_i \tag{10-19b}$$

$$k_x^2 + k_z^2 = k_1^2 = \omega^2 \mu_1 \varepsilon_1 \tag{10-19c}$$

Since the reflected wave propagates in the negative x and positive z directions, its wave vector and fields can be written as

$$\mathbf{k}_r = -\mathbf{a}_x k_{rx} + \mathbf{a}_z k_{rz} \tag{10-20}$$

$$\mathbf{E}_r = \mathbf{a}_y E_r e^{-j\mathbf{k}_r \cdot \mathbf{r}} = \mathbf{a}_y R_\perp E_o\, e^{+jk_{rx}x - jk_{rz}z} \tag{10-21a}$$

$$\mathbf{H}_r = \frac{\nabla \times \mathbf{E}_r}{-j\omega\mu_1} = (-\mathbf{a}_x k_{rz} - \mathbf{a}_z k_{rx}) \frac{R_\perp E_0}{\omega\mu_1}\, e^{jk_{rx}x - jk_{rz}z} \tag{10-21b}$$

where $R_\perp$ is the reflection coefficient for the perpendicularly polarized wave. Because of the choice of signs in Eq. (10-20), k_{rx} and k_{rz} are considered to be positive and they can be expressed in terms of the angle of reflection θ_r.

$$k_{rx} = |\mathbf{k}_r| \cos\theta_r = k_1 \cos\theta_r \tag{10-22a}$$

$$k_{rz} = |\mathbf{k}_r| \sin\theta_r = k_1 \sin\theta_r \tag{10-22b}$$

$$k_{rx}^2 + k_{rz}^2 = k_1^2 = \omega^2 \mu_1 \varepsilon_1 \tag{10-22c}$$

Note that $|\mathbf{k}_r| = k_1 = |\mathbf{k}_i|$ because the reflected wave propagates in the same medium as the incident wave. For the transmitted wave, we can write

$$\mathbf{k}_t = \mathbf{a}_x k_{tx} + \mathbf{a}_z k_{tz} \tag{10-23}$$

$$\mathbf{E}_t = \mathbf{a}_y E_t e^{-j\mathbf{k}_t \cdot \mathbf{r}} = \mathbf{a}_y T_\perp E_0 e^{-jk_{tx}x - jk_{tz}z} \tag{10-24a}$$

$$\mathbf{H}_t = \frac{\nabla \times \mathbf{E}_t}{-j\omega\mu_2} = (-\mathbf{a}_x k_{tz} + \mathbf{a}_z k_{tx}) \frac{T_\perp E_0}{\omega\mu_2}\, e^{-jk_{tx}x - jk_{tz}z} \tag{10-24b}$$

where $T_\perp$ is the transmission coefficient for the perpendicularly polarized wave. Similarly, k_{tx} and k_{tz} satisfy

$$k_{tx} = |\mathbf{k}_t| \cos\theta_t = k_2 \cos\theta_t \tag{10-25a}$$

$$k_{tz} = |\mathbf{k}_t| \sin\theta_t = k_2 \sin\theta_t \tag{10-25b}$$

$$k_{tx}^2 + k_{tz}^2 = k_2^2 = \omega^2 \mu_2 \varepsilon_2 \tag{10-25c}$$

Now the problem reduces to determining θ_r, θ_t, $R_\perp$, $T_\perp$, given θ_i, ω, (ε_1, μ_1), (ε_2, μ_2). Once we obtain θ_r, θ_t, $R_\perp$, $T_\perp$, we have a full knowledge on the reflected and transmitted waves. As we have done in Section 10.2, we apply the boundary conditions, Eq. (10-8), to determine these unknowns. For this problem, the tangential component of $\mathbf{H}$ is only its z component. Substituting $x = 0$ in Eqs. (10-18), (10-21), (10-24) and applying Eq. (10-8), we obtain

$$E_y:\quad e^{-jk_z z} + R_\perp\, e^{-jk_{rz} z} = T_\perp\, e^{-jk_{tz} z} \tag{10-26a}$$

$$H_z:\quad \frac{k_x}{\omega\mu_1} e^{-jk_z z} - \frac{k_{rx}}{\omega\mu_1} R_\perp\, e^{-jk_{rz} z} = \frac{k_{tx}}{\omega\mu_2} T_\perp\, e^{-jk_{tz} z} \tag{10-26b}$$

In order that Eq. (10-26) be satisfied at all points on the boundary ($x = 0$), i.e., for *all values of z*, the following condition must be met first:

$$k_z = k_{rz} = k_{tz} \tag{10-27}$$

so that the exponential functions match on both sides. Eq. (10-27) implies that the tangential components of three wave vectors $\mathbf{k}_i$, $\mathbf{k}_r$, $\mathbf{k}_t$ are equal. The condition is known as the *phase matching condition*. Substituting Eqs. (10-19b), (10-22b), (10-25b) into Eq. (10-27) gives rise to

$$k_1 \sin\theta_i = k_1 \sin\theta_r = k_2 \sin\theta_t \tag{10-28}$$

The first equality gives

$$\boxed{\theta_i = \theta_r} \quad \text{(Law of reflection)} \tag{10-29}$$

It says that the angle of reflection is equal to the angle of incidence. Making use of Eqs. (10-19c) and (10-25c), the second equality gives

$$\sqrt{\mu_1\varepsilon_1}\,\sin\theta_i = \sqrt{\mu_2\varepsilon_2}\,\sin\theta_t$$

or

$$\boxed{n_1 \sin\theta_i = n_2 \sin\theta_t} \quad \text{(Snell's law)} \tag{10-30}$$

where n_1, n_2 are the indices of refraction of media 1, 2, respectively (see Eq. (9-52) for its definition). Eq. (10-30) is known as **Snell's law** or *law of refraction*, from which θ_t can be expressed in terms of θ_i as

$$\theta_t = \sin^{-1}\left(\frac{n_1}{n_2}\sin\theta_i\right) \tag{10-31}$$

Eqs. (10-29) and (10-30) are well-known fundamental laws of geometrical optics.

Making use of Eq. (10-27), Eq. (10-26) becomes

$$1 + R_\perp = T_\perp \tag{10-32a}$$

$$\frac{k_x}{\mu_1}\left(1-R_\perp\right)=\frac{k_{tx}}{\mu_2}\,T_\perp \tag{10-32b}$$

where we have used

$$k_{rx}=k_x=k_1\cos\theta_i \tag{10-33}$$

because of Eq. (10-29). Now solving Eq. (10-32) for $R_\perp$ and $T_\perp$, we obtain

$$R_\perp=\frac{\mu_2 k_x-\mu_1 k_{tx}}{\mu_2 k_x+\mu_1 k_{tx}}\quad\left(\begin{array}{l}\text{Reflection coefficient for}\\ \text{perpendicular polarization}\end{array}\right) \tag{10-34a}$$

$$T_\perp=\frac{2\mu_2 k_x}{\mu_2 k_x+\mu_1 k_{tx}}\quad\left(\begin{array}{l}\text{Transmission coefficient}\\ \text{for perpendicular polarization}\end{array}\right) \tag{10-34b}$$

Making use of Eqs. (10-19), (10-25), (10-2), and (10-6), Eq. (10-34) can be written as

$$R_\perp=\frac{\eta_2\cos\theta_i-\eta_1\cos\theta_t}{\eta_2\cos\theta_i+\eta_1\cos\theta_t} \tag{10-35a}$$

$$T_\perp=\frac{2\eta_2\cos\theta_i}{\eta_2\cos\theta_i+\eta_1\cos\theta_t} \tag{10-35b}$$

where η_1, η_2 are intrinsic impedances of the two media. These are known as the **Fresnel equations**. We note that at normal incidence ($\theta_i=\theta_t=0$), Eq. (10-35) reduces to Eq. (10-11). When both media are non-magnetic, i.e., $\mu_1=\mu_2=\mu_0$, Eq. (10-35) reduces to

$$R_\perp=\frac{n_1\cos\theta_i-n_2\cos\theta_t}{n_1\cos\theta_i+n_2\cos\theta_t},\qquad T_\perp=\frac{2n_1\cos\theta_i}{n_1\cos\theta_i+n_2\cos\theta_t}\quad\left(\begin{array}{l}\text{Fresnel coefficients for}\\ \text{non-magnetic media}\\ \text{for }\perp\text{ polarization}\end{array}\right) \tag{10-36}$$

where n_1, n_2 are refractive indices of the two media.

EXAMPLE 10-2. REFLECTION AND TRANSMISSION OF PERPENDICULARLY POLARIZED WAVE

A 1 GHz uniform plane wave with 2 V/m is incident obliquely at $\theta_i = 60°$ from air upon a lossless dielectric (glass) of $\varepsilon = 1.96\,\varepsilon_0$ and $\mu = \mu_0$, whose surface is at $x = 0$ (see Figure 10-2). The incident wave has perpendicular polarization.

(a) Find the complete expression for the electric field $\mathbf{E}_i$ of the incident wave.
(b) Find the electric field $\mathbf{E}_r$ of the reflected wave.
(c) Find the magnetic field $\mathbf{H}_t$ of the transmitted wave.

SOLUTIONS:

(a) Assuming that the plane of incidence is the xz plane, the electric field of the perpendicularly polarized wave is given by

$$\mathbf{E}_i = \mathbf{a}_y E_0 e^{-j(k_x x + k_z z)}$$

$$\text{where } E_0 = 2 \text{ [V/m]},\ k_1 = \frac{\omega}{c} \text{ (in air)} = \frac{2\pi \times 10^9}{3 \times 10^8} = \frac{20\pi}{3} \left[\text{rad/m}\right]$$

$$k_x = k_1 \cos\theta_i = \frac{20\pi}{3}\cos 60° = \frac{20\pi}{3} \cdot \frac{1}{2} = \frac{10\pi}{3} = 10.47$$

$$k_z = k_1 \sin\theta_i = \frac{20\pi}{3}\sin 60° = \frac{20\pi}{3} \cdot \frac{\sqrt{3}}{2} = \frac{10\pi}{3}\sqrt{3} = 18.14$$

$$\therefore\ \mathbf{E}_i = \mathbf{a}_y 2 e^{-j\frac{10\pi}{3}(x+\sqrt{3}z)} = \mathbf{a}_y 2 e^{-j(10.5x + 18.1z)} \text{ [V/m]}$$

(b) First, we need to find θ_r and θ_t to calculate R.
By the law of reflection, $\theta_r = \theta_i = 60°$.

$$\text{By Snell's law, } \theta_t = \sin^{-1}\left(\frac{n_1}{n_2}\sin\theta_i\right) = \sin^{-1}\left(\frac{1}{\sqrt{1.96}}\sin 60°\right) = 38.2°$$

$$\text{where } n_2 = \sqrt{\frac{\mu_2\varepsilon_2}{\mu_0\varepsilon_0}} = \sqrt{\frac{\mu_0 \cdot 1.96\varepsilon_0}{\mu_0\varepsilon_0}} = \sqrt{1.96} = 1.4.$$

The electric field of the reflected wave is given by

$$\mathbf{E}_r = \mathbf{a}_y R_\perp E_0 e^{+jk_x x - jk_z z} = \mathbf{a}_y (-0.75)\, e^{+j(10.5x - 18.1z)} \ [\text{V/m}]$$

$$\text{where } R_\perp = \frac{n_1 \cos\theta_i - n_2 \cos\theta_t}{n_1 \cos\theta_i + n_2 \cos\theta_t} = \frac{1 \cos 60° - 1.4\cos 38.2°}{1 \cos 60° + 1.4\cos 38.2°} = -0.375$$

Note that the sign of the normal (x-) component of the k vector is changed.

(c) The magnetic field of the transmitted wave is given by

$$\mathbf{H}_t = (-\mathbf{a}_x k_{tz} + \mathbf{a}_z k_{tx}) \frac{T_\perp E_o}{\omega\mu_2} e^{-jk_{tx}x - jk_{tz}z}$$

where

$$T_\perp = \frac{2n_1 \cos\theta_i}{n_1 \cos\theta_i + n_2 \cos\theta_t} = 1 + R_\perp = 0.625$$

$$k_2 = \omega\sqrt{\mu_2\varepsilon_2} = \omega\sqrt{\mu_0 1.96\varepsilon_0} = 1.4k_1 = 1.4\times\frac{20\pi}{3} = 29.32$$

$$k_{tz} = k_z = \frac{10\pi}{3}\sqrt{3} = 18.14, \quad k_{tx} = k_2 \cos\theta_t = 1.4\times\frac{20\pi}{3}\cos 38.2° = 23.04$$

$$\omega\mu_2 = 2\pi\times10^9\times4\pi\times10^{-7} = 7.90\times10^3$$

$$\therefore \mathbf{H}_t = (-\mathbf{a}_x 18.1 + \mathbf{a}_z 23.0)\, 1.58\times10^{-4}\, e^{-j(23.0x + 18.1z)} \ [\text{A/m}]$$

Note that the tangential (z-) component of the **k** vector is unchanged (due to the *phase matching* condition) and the normal (x-) component is changed.

10.3.2. Parallel Polarization

In the second case, we assume the incident wave is *parallel polarized*. Since the electric field is parallel with the plane of incidence, the magnetic field **H** is perpendicular to the plane of incidence as shown in Figure 10-3. In this case it is advantageous to formulate the magnetic field first.

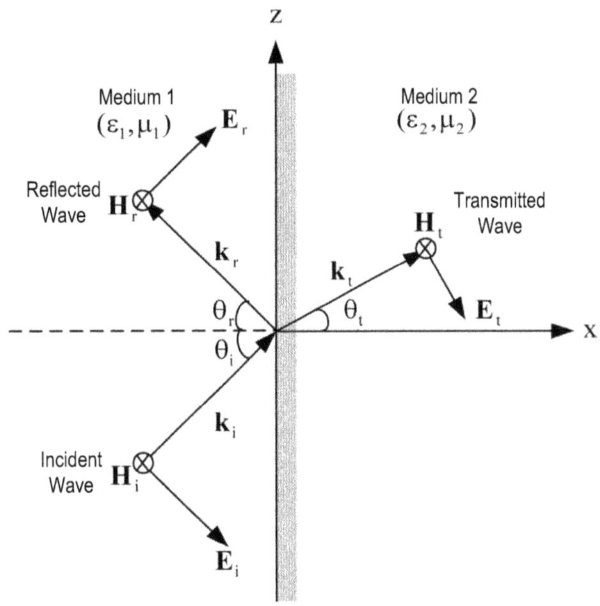

Figure 10-3. Reflection and transmission of waves with parallel polarization at oblique incidence

Considering the direction of the incident wave vector as shown, the magnetic and electric fields of the incident wave are written as follows:

$$\mathbf{H}_i = \mathbf{a}_y \, H_0 \, e^{-j\mathbf{k}_i \cdot \mathbf{r}} = \mathbf{a}_y \, H_0 \, e^{-jk_x x - jk_z z} \tag{10-37a}$$

$$\mathbf{E}_i = \frac{\nabla \times \mathbf{H}_i}{j\omega\varepsilon_1} = (\mathbf{a}_x \, k_z - \mathbf{a}_z \, k_x) \, \frac{H_0}{\omega\varepsilon_1} \, e^{-jk_x x - jk_z z} \tag{10-37b}$$

where H_0 is the amplitude of the incident magnetic field and k_x, k_z are given in Eq. (10-19). $\mathbf{E}_i$ can also be obtained by $\mathbf{E}_i = \frac{\mathbf{k}_i \times \mathbf{H}_i}{-\omega\varepsilon_1}$ from Eq. (9-66). The fields of the reflected and transmitted waves can be formulated in the similar way as done in Section 10.3.1.

$$\mathbf{H}_r = \mathbf{a}_y \, H_r \, e^{-j\mathbf{k}_r \cdot \mathbf{r}} = \mathbf{a}_y \, R_{\parallel} \, H_0 \, e^{+jk_{rx} x - jk_{rz} z} \tag{10-38a}$$

$$\mathbf{E}_r = \frac{\nabla \times \mathbf{H}_r}{j\omega\varepsilon_1} = (\mathbf{a}_x \, k_{rz} + \mathbf{a}_z \, k_{rx}) \, \frac{R_{\parallel} \, H_0}{\omega\varepsilon_1} \, e^{jk_{rx} x - jk_{rz} z} \tag{10-38b}$$

$$\mathbf{H}_t = \mathbf{a}_y \, H_t \, e^{-j\mathbf{k}_t \cdot \mathbf{r}} = \mathbf{a}_y \, T_{\parallel} \, H_0 \, e^{-jk_{tx} x - jk_{tz} z} \tag{10-39a}$$

$$\mathbf{E}_t = \frac{\nabla \times \mathbf{H}_t}{j\omega\varepsilon_2} = (\mathbf{a}_x \, k_{tz} - \mathbf{a}_z \, k_{tx}) \, \frac{T_{\parallel} \, H_0}{\omega\varepsilon_2} \, e^{-jk_{tx} x - jk_{tz} z} \tag{10-39b}$$

where $R_{\|}$ and $T_{\|}$ are, respectively, the reflection and transmission coefficients for the magnetic field for the parallel polarized wave. k_{rx}, k_{rz}, k_{tx}, k_{tz} are given in Eqs. (10-22) and (10-25). Now the problem again reduces to determining θ_r, θ_t, $R_{\|}$, $T_{\|}$, given θ_i, ω, (ε_1, μ_1), (ε_2, μ_2). Applying the boundary conditions—continuity of tangential components of **H** (y component) and **E** (z component)—at $x = 0$, we obtain

$$H_y: \; e^{-jk_z z} + R_{\|} \; e^{-jk_{rz} z} = T_{\|} \; e^{-jk_{tz} z} \tag{10-40a}$$

$$E_z: \; -\frac{k_x}{\omega\varepsilon_1} e^{-jk_z z} + \frac{k_{rx}}{\omega\varepsilon_1} R_{\|} \; e^{-jk_{rz} z} = -\frac{k_{tx}}{\omega\varepsilon_2} T_{\|} \; e^{-jk_{tz} z} \tag{10-40b}$$

Following the same procedures done in Section 10.3.1 in solving Eq. (10-40), we find the same law of reflection and Snell's law given by Eqs. (10-29) and (10-30) and $R_{\|}$, $T_{\|}$ are obtained as follows.

$$R_{\|} = \frac{\varepsilon_2 k_x - \varepsilon_1 k_{tx}}{\varepsilon_2 k_x + \varepsilon_1 k_{tx}} \quad \begin{pmatrix} \text{Reflection coefficient for } \mathbf{H} \\ \text{for parallel polarization} \end{pmatrix} \tag{10-41a}$$

$$T_{\|} = \frac{2\varepsilon_2 k_x}{\varepsilon_2 k_x + \varepsilon_1 k_{tx}} \quad \begin{pmatrix} \text{Transmission coefficient for } \mathbf{H} \\ \text{for parallel polarization} \end{pmatrix} \tag{10-41b}$$

The results in Eq. (10-41) can also be obtained simply by replacing μ by ε in Eq. (10-34) for perpendicular polarization, because the mathematical expressions for **E** and **H** for parallel polarization, given by Eqs. (10-37), (10-38), (10-39), are identical to those for perpendicular polarization, given by Eqs. (10-18), (10-21), (10-24), if we replace **E** by **H**, **H** by –**E**, ε by μ, and μ by ε. This is known as *duality* in electromagnetic theory. Making use of Eqs. (10-19), (10-25), (10-21), and (10-6), Eq. (10-41) can be rewritten in terms of angles of incidence and transmission.

$$R_{\|} = \frac{\eta_1 \cos\theta_i - \eta_2 \cos\theta_t}{\eta_1 \cos\theta_i + \eta_2 \cos\theta_t} \tag{10-42a}$$

$$T_{\|} = \frac{2\eta_1 \cos\theta_i}{\eta_1 \cos\theta_i + \eta_2 \cos\theta_t} \tag{10-42b}$$

We note that at normal incidence ($\theta_i = \theta_t = 0$), Eq. (10-42) doesn't reduce to Eq. (10-11) because Eqs. (10-41) and (10-42) are the ratio of magnetic fields. If we take the ratios of the z components of the electric fields, we find the reflection and transmission coefficients for **E** for parallel polarization as follows.

$$R_{\|}^{E} = \frac{E_{rz}}{E_{iz}} = -R_{\|} = \frac{\eta_2 \cos\theta_t - \eta_1 \cos\theta_i}{\eta_2 \cos\theta_t + \eta_1 \cos\theta_i} \quad \left(\begin{array}{l}\text{Reflection coefficient for } \mathbf{E}\\ \text{for parallel polarization}\end{array}\right) \tag{10-43a}$$

$$T_{\|}^{E} = \frac{E_{tz}}{E_{iz}} = T_{\|}\frac{k_{tx}\varepsilon_1}{k_x\varepsilon_2} = \frac{2\eta_2 \cos\theta_t}{\eta_2 \cos\theta_t + \eta_1 \cos\theta_i} \quad \left(\begin{array}{l}\text{Transmission coefficient for } \mathbf{E}\\ \text{for parallel polarization}\end{array}\right) \tag{10-43b}$$

Now at normal incidence, Eq. (10-43) reduces to Eq. (10-11) as expected. When both media are non-magnetic, i.e., $\mu_1 = \mu_2 = \mu_o$, Eqs. (10-42) and (10-43) reduce to

$$R_{\|} = \frac{n_2 \cos\theta_i - n_1 \cos\theta_t}{n_2 \cos\theta_i + n_1 \cos\theta_t} \tag{10-44a}$$

$$T_{\|} = \frac{2n_2\cos\theta_i}{n_2\cos\theta_i + n_1\cos\theta_t} \tag{10-44b}$$

(Fresnel coefficients for non-magnetic media for $\|$ polarization)

$$R_{\|}^{E} = \frac{n_1 \cos\theta_t - n_2 \cos\theta_i}{n_1 \cos\theta_t + n_2 \cos\theta_i} = -R_{\|}$$

$$T_{\|}^{E} = \frac{2n_1 \cos\theta_t}{n_1 \cos\theta_t + n_2 \cos\theta_i} \tag{10-45}$$

Power Conservation

In order to observe power conservation, the time-average Poynting vectors of the incident, reflected and transmitted waves for the perpendicular polarization are calculated from the fields in Eqs. (10-18), (10-21), (10-24).

$$\mathbf{S}_{av}^{i} = \frac{1}{2}\text{Re}\left(\mathbf{E}_i \times \mathbf{H}_i^*\right) = \frac{|E_0|^2}{2\omega\mu_1}\left\{\mathbf{a}_x k_x + \mathbf{a}_z k_z\right\} \tag{10-46a}$$

$$\mathbf{S}_{av}^{r} = \frac{1}{2}\text{Re}\left(\mathbf{E}_r \times \mathbf{H}_r^*\right) = \frac{|E_0|^2}{2\omega\mu_1}|R_\perp|^2\left\{-\mathbf{a}_x k_x + \mathbf{a}_z k_z\right\} \tag{10-46b}$$

$$\mathbf{S}_{av}^{t} = \frac{1}{2}\text{Re}\left(\mathbf{E}_t \times \mathbf{H}_t^*\right) = \frac{|E_0|^2}{2\omega\mu_2}|T_\perp|^2\,\text{Re}\left\{\mathbf{a}_x k_{tx}^* + \mathbf{a}_z k_z\right\} \tag{10-46c}$$

where Eqs. (10-27) and (10-33) have been used. Note that k_{tx} can become complex in some instances as will be discussed later. Power conservation holds for the x components of the power. The fraction of power reflected is given by

$$\frac{P_r}{P_i} = \frac{S_{av,x}^{r}}{S_{av,x}^{i}} = |R_\perp|^2 \tag{10-47a}$$

The fraction of power transmitted is given by

$$\frac{P_t}{P_i}=\frac{S^t_{av,x}}{S^i_{av,x}}=\left|T_\perp\right|^2\frac{\mu_1}{\mu_2}\frac{\operatorname{Re}(k^*_{tx})}{k_x} \tag{10-47b}$$

Substituting Eq. (10-34) into Eq. (10-47), one can show that

$$\frac{P_r}{P_i}+\frac{P_t}{P_i}=1 \quad \text{or} \quad P_i=P_r+P_t \tag{10-48}$$

We can derive the same power conservation relationship (10-48) for parallel polarization (see Problem 10-16). Eq. (10-47b) should be adjusted for parallel polarization. The reflection and transmission coefficients for perpendicular and parallel polarized waves are summarized in Table 10-1.

Table 10-1. Reflection and Transmission Coefficients

1. **Perpendicular (TE) Polarization**

$$R_\perp=\frac{E_r}{E_i}=\frac{\eta_2\cos\theta_i-\eta_1\cos\theta_t}{\eta_2\cos\theta_i+\eta_1\cos\theta_t} \qquad n_1\sin\theta_i=n_2\sin\theta_t,\ \ n_i=\sqrt{\frac{\mu_i\varepsilon_i}{\mu_o\varepsilon_o}},\ i=1,\ 2$$

$$T_\perp=\frac{E_t}{E_i}=\frac{2\eta_2\cos\theta_i}{\eta_2\cos\theta_i+\eta_1\cos\theta_t} \qquad \eta_i=\sqrt{\frac{\mu_i}{\varepsilon_i}},\ i=1,\ 2$$

For non-magnetic media ($\mu_1=\mu_2=\mu_0$),

$$R_\perp=\frac{n_1\cos\theta_i-n_2\cos\theta_t}{n_1\cos\theta_i+n_2\cos\theta_t},\quad T_\perp=\frac{2n_1\cos\theta_i}{n_1\cos\theta_i+n_2\cos\theta_t}$$

2. **Parallel (TM) Polarization**

$$R_\parallel=\frac{H_r}{H_i}=\frac{\eta_1\cos\theta_i-\eta_2\cos\theta_t}{\eta_1\cos\theta_i+\eta_2\cos\theta_t}$$

$$T_\parallel=\frac{H_t}{H_i}=\frac{2\eta_1\cos\theta_i}{\eta_1\cos\theta_i+\eta_2\cos\theta_t}$$

For non-magnetic media ($\mu_1=\mu_2=\mu_0$),

$$R_\parallel=\frac{n_2\cos\theta_i-n_1\cos\theta_t}{n_2\cos\theta_i+n_1\cos\theta_t},\quad T_\parallel=\frac{2n_2\cos\theta_i}{n_2\cos\theta_i+n_1\cos\theta_t}$$

EXAMPLE 10-3. REFLECTION AND TRANSMISSION OF PARALLEL POLARIZED WAVE

A uniform plane wave whose electric field is given by

$$\mathbf{E}_i = (\mathbf{a}_x \sqrt{3} - \mathbf{a}_z)\ e^{-j(2x+2\sqrt{3}z)}\ [\text{V/m}]$$

is incident from air upon a lossless dielectric (glass) of $\varepsilon = 1.96\ \varepsilon_0$ and $\mu = \mu_0$, whose surface is at $x = 0$ (see Figure 10-3).

(a) Determine the angles of incidence, reflection and transmission.
(b) Find the electric field of the reflected wave.
(c) Find the magnetic field of the transmitted wave.
(d) Calculate the fraction of power reflected from the dielectric surface.

SOLUTION

(a) Noting that $k_x = 2,\ k_z = 2\sqrt{3}$, we find

$$k_1 = \sqrt{k_x^2 + k_z^2} = \sqrt{(2)^2 + (2\sqrt{3})^2} = 4$$

$$k_z = k_1 \sin\theta_i \rightarrow \theta_i = \sin^{-1}\left(\frac{k_z}{k_1}\right) = \sin^{-1}\left(\frac{2\sqrt{3}}{4}\right) = 60^\circ$$

By the law of reflection, $\theta_r = \theta_i = 60^\circ$

By Snell's law, $\theta_t = \sin^{-1}\left(\frac{n_1}{n_2}\sin\theta_i\right) = \sin^{-1}\left(\frac{1}{\sqrt{1.96}}\sin 60^\circ\right) = 38.2^\circ$

(b) Since the incident electric field $\mathbf{E}$ is parallel to the plane of incidence (xz plane), the incident wave is *parallel polarized.* Because the reflection $(R_{||})$ and transmission $(T_{||})$ coefficients for parallel polarized are defined in terms of ratio of magnetic fields, we calculate the magnetic field $\mathbf{H}_i$ of the incident wave

$$\mathbf{H}_i = \frac{\nabla \times \mathbf{E}_i}{-j\omega\mu_1} = \frac{\mathbf{a}_{ki} \times \mathbf{E}_i}{\eta_1} = \frac{1}{377}\left(\mathbf{a}_x \frac{1}{2} + \mathbf{a}_z \frac{\sqrt{3}}{2}\right) \times \left(\mathbf{a}_x\ \sqrt{3} - \mathbf{a}_z\right) e^{-j(2x+2\sqrt{3}z)}$$

$$= \mathbf{a}_y \frac{2}{377}\ e^{-j(2x+2\sqrt{3}z)}\ [\text{A/m}]$$

where $\mathbf{a}_{ki} = \frac{\mathbf{k}_i}{k_1} = \frac{\mathbf{a}_x 2 + \mathbf{a}_z 2\sqrt{3}}{4} = \mathbf{a}_x \frac{1}{2} + \mathbf{a}_z \frac{\sqrt{3}}{2}$

As expected, $\mathbf{H}_i$ is perpendicular to the plane of incidence. Now we calculate $R_{\parallel}$ and $T_{\parallel}$:

$$R_{\parallel} = \frac{n_2 \cos\theta_i - n_1 \cos\theta_t}{n_2 \cos\theta_i + n_1 \cos\theta_t} = \frac{1.4\cos 60° - 1\cos 38.2°}{1.4\cos 60° + 1\cos 38.2°} = -0.058$$

$$T_{\parallel} = 1 + R_{\parallel} = 0.942$$

The fields of the reflected wave are given by

$$\mathbf{H}_r = \mathbf{a}_y R_{\parallel} \frac{2}{377} e^{+j(2x-2\sqrt{3}z)} = -\mathbf{a}_y 3.08\times10^{-4}\, e^{+j(2x-2\sqrt{3}z)} \text{ [A/m]}$$

$$\mathbf{E}_r = \eta_1 \mathbf{H}_r \times \mathbf{a}_{k_r} = R_{\parallel}\, 2\, \mathbf{a}_y \times \left(-\mathbf{a}_x \frac{1}{2} + \mathbf{a}_z \frac{\sqrt{3}}{2}\right) e^{+j(2x-2\sqrt{3}z)}$$
$$= R_{\parallel}\left(\mathbf{a}_x \sqrt{3} + \mathbf{a}_z\right) e^{+j(2x-2\sqrt{3}z)} = -0.058\left(\mathbf{a}_x \sqrt{3} + \mathbf{a}_z\right) e^{+j(2x-2\sqrt{3}z)} \text{ [V/m]}$$

Note that one component of the reflected electric field is changed in sign because the sign of the x component of $\mathbf{k}_r$ is changed.

(c) The magnetic field of the transmitted wave is given by

$$\mathbf{H}_t = \mathbf{a}_y T_{\parallel} \frac{2}{377} e^{-j(k_{tx}x+2\sqrt{3}z)} = \mathbf{a}_y\, 5.0\times10^{-3}\, e^{-j(4.40x+2\sqrt{3}z)} \text{ [A/m]}$$

where

$$k_{tx} = k_2 \cos\theta_t = \omega\sqrt{\mu_o 1.96\varepsilon_o}\, \cos\theta_t = 1.4k_1 \cos 38.2° = 4.40.$$

(d) The fraction (or percentage) of the power reflected is given by

$$\frac{P_r}{P_i} = \left|R_{\parallel}\right|^2 = \left|-0.058\right|^2 = 0.0034$$

Thus, 0.34% of the incident power is reflected at $\theta_i = 60°$.

10.4 TOTAL INTERNAL REFLECTION

Let us examine Snell's law in Eq. (10-30) or equivalently, the phase matching condition in Eq. (10-27) or (10-28). We first note:

$$\text{If } n_1 < n_2, \text{ then } \theta_i > \theta_t$$

$$\text{If } n_1 > n_2, \text{ then } \theta_i < \theta_t$$

When the wave is incident from a less dense medium to a denser medium, for example, from air to glass, the transmitted wave is bent toward the normal. When the wave is incident from a denser medium to a less dense medium, the transmitted wave is bent toward the surface. This can be illustrated graphically by the k_x-k_z diagram shown in Figure 10-4, (a) and (b). Note that Eqs. (10-19c), (10-22c), and (10-25c) represent circles in the k_x-k_z plane. The radius of the semicircle represents the wave number (k) for each of the incident, reflected and transmitted waves (k_1 and k_2). Since $k_i = \omega\sqrt{\mu_i \varepsilon_i} = n_i \omega\sqrt{\mu_o \varepsilon_o}$, the less dense medium has smaller semicircle. The arrows are the wave vectors, $\mathbf{k}_i$, $\mathbf{k}_r$ and $\mathbf{k}_t$. The phase matching condition in Eq. (10-27) or (10-28) requires that the z component of $\mathbf{k}$ or the component parallel to the boundary should match. In (a), since $k_1 < k_2$, we have $\theta_t < \theta_i$. In (b), since $k_1 > k_2$, we have $\theta_t > \theta_i$.

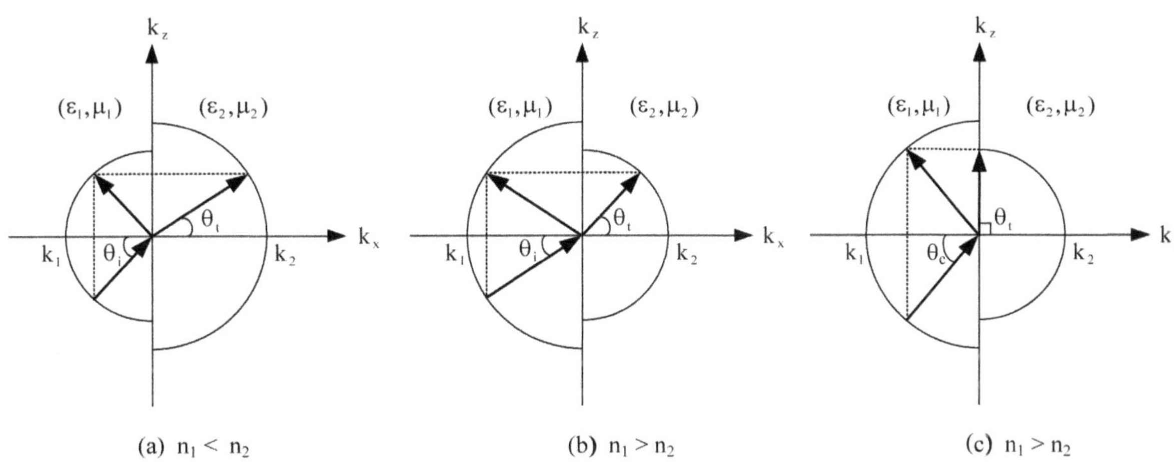

Figure 10-4. Snell's law and total internal reflection.

Now in the case of $n_1 > n_2$, there exists an angle of incidence ($\theta_i = \theta_c$) such that the angle of transmission becomes 90°, as shown in Figure 10-4(c).

$$n_1 \sin\theta_c = n_2 \sin 90° = n_2$$

or

$$\theta_c = \sin^{-1}\frac{n_2}{n_1} \qquad \text{(critical angle)} \tag{10-49}$$

$$= \sin^{-1}\sqrt{\frac{\varepsilon_2}{\varepsilon_1}} \qquad \text{(for non-magnetic media)}$$

θ_c is called the **critical angle**. It is interesting to see what happens when the angle of incidence exceeds the critical angle ($\theta_i > \theta_c$). For discussion, let us repeat the dispersion relations and Snell's law here.

$$k_x^2 + k_z^2 = k_1^2 = \omega^2\mu_1\varepsilon_1 = \left(\frac{\omega}{c}n_1\right)^2 \tag{10-19c}$$

$$k_{tx}^2 + k_{tz}^2 = k_2^2 = \omega^2\mu_2\varepsilon_2 = \left(\frac{\omega}{c}n_2\right)^2 \tag{10-25c}$$

$$k_z = k_{tz} \text{ or } k_1 \sin\theta_i = k_2 \sin\theta_t \tag{10-28}$$

When $\theta_i > \theta_c$,

$$k_z = k_1 \sin\theta_i > k_1 \sin\theta_c = k_2 \sin 90° = k_2$$

Then, from Eq. (10-25c) and Eq. (10-27),

$$k_{tx}^2 = k_2^2 - k_{tz}^2 = k_2^2 - k_z^2 < 0$$

Thus k_{tx} has to become *purely imaginary*, so we let

$$k_{tx} = -j\alpha_2 \tag{10-50}$$

where

$$\alpha_2 = \sqrt{k_z^2 - k_2^2} = \sqrt{k_1^2 \sin^2\theta_i - k_2^2} \tag{10-51}$$

is real. Note that we choose $-j\alpha_2$ instead of $j\alpha_2$ so that $\mathbf{E}_t$ does not grow, i.e., $j\alpha_2$ gives a physically unacceptable solution. The transmitted electric field in Eq. (10-24) for perpendicular polarization becomes

$$\mathbf{E}_t = \mathbf{a}_y T_\perp E_0 e^{-\alpha_2 x} e^{-jk_{tz}z} \tag{10-52}$$

The transmitted wave *propagates in the* z *direction* (parallel to the surface) with the phase constant $k_{tz} = k_z = k_1 \sin\theta_i$ and *attenuates exponentially in the* x *direction*. It is a *nonuniform* plane wave propagating along the boundary because $|\mathbf{E}_t| = |T_\perp E_o| e^{-\alpha_2 x}$ is not constant over a plane perpendicular to the direction of propagation (xy plane) but is attenuated away from the boundary. Such a wave is called the **surface wave**. The wave can be tightly bound to the surface of the interface if α_2 is large. More importantly, this wave carries no real power into medium 2. We show this by examining the reflection coefficient for perpendicular polarization in Eq. (10-34a).

$$R_\perp = \frac{\mu_2 k_x - \mu_1 k_{tx}}{\mu_2 k_x + \mu_1 k_{tx}} = \frac{\mu_2 k_x + j\mu_1\alpha_2}{\mu_2 k_x - j\mu_1\alpha_2} = \frac{1 + j\dfrac{\mu_1\alpha_2}{\mu_2 k_x}}{1 - j\dfrac{\mu_1\alpha_2}{\mu_2 k_x}} \tag{10-53}$$

The fraction of power reflected is given by

$$\frac{P_r}{P_i} = |R_\perp|^2 = \frac{1 + \left(\dfrac{\mu_1\alpha_2}{\mu_2 k_x}\right)^2}{1 + \left(\dfrac{\mu_1\alpha_2}{\mu_2 k_x}\right)^2} = 1 \tag{10-54a}$$

or

$$P_r = P_i, \; P_t = P_i - P_r = 0 \tag{10-54b}$$

The wave is totally reflected and there is no time-average (or real) power transmitted in the x direction or into medium 2. Hence we call this phenomenon **total internal reflection**. This can also be shown by substituting Eq. (10-50) into Eq. (10-46c). The total internal reflection is the major principle of the dielectric waveguide such as optical fiber which guides the light beam with very little loss because the metal is not used.

In summary, there are two conditions for total internal reflection (TIR) to occur:

(i) $n_1 > n_2$ (ii) $\theta_i > \theta_c = \sin^{-1}\dfrac{n_2}{n_1}$

The consequences of TIR are:

(i) $k_{tx} = -j\alpha_2 \rightarrow \mathbf{E}_t$ becomes a surface wave.

(ii) $|R_\perp| = 1$. The wave is totally reflected.

As a final note, although we illustrated TIR for perpendicular polarization, it also applies to parallel polarized wave (see Problem 10-22).

EXAMPLE 10-4. GUIDANCE OF A LIGHT WAVE ALONG OPTICAL FIBER

An optical fiber is a dielectric pipe made of two concentric glass cylinders as shown in Figure 10-5. The inner cylinder, called the *core*, has the refractive index n_1 and the outer cylinder, called the *cladding*, has the index n_2, which is slightly smaller than n_1, i.e., $n_2 < n_1$. Light waves traveling in the core are totally reflected from the cladding and thus guided if their angle of incidence is greater than the critical angle.

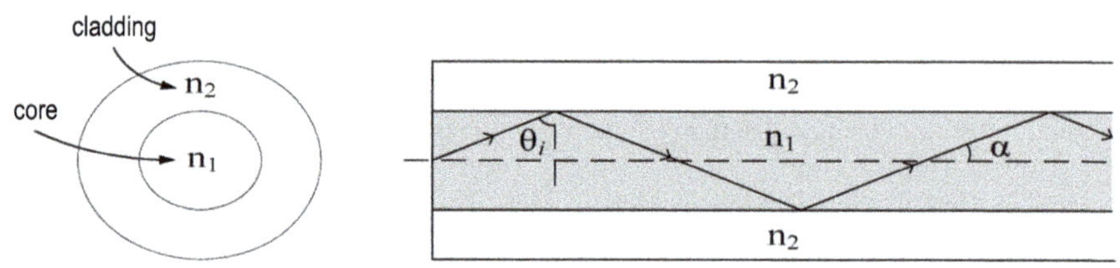

Figure 10-5. Optical fiber.

(a) A silica glass fiber has $n_1 = 1.475$ and $n_2 = 1.460$. What should the angle between the fiber axis and a light ray be in order that the light will propagate along the fiber?

(b) When the cladding is removed, i.e., when the core is surrounded by air, what should that angle be?

SOLUTION

(a) In order for the TIR to occur at the interface between the core and the cladding, θ_i must be greater than the critical angle (θ_c).

$$\theta_i > \theta_c = \sin^{-1}\frac{n_2}{n_1} = \sin^{-1}\frac{1.460}{1.475} = 81.8°$$

Then the angle between the fiber axis and the ray (α) should be

$$\alpha = 90° - \theta_i < 90° - \theta_c = 8.2°$$

(b) When the core is surrounded by air, $n_2 = 1$.

$$\theta_i > \theta_c = \sin^{-1}\frac{1}{1.475} = 42.7°$$

Then $\alpha < 90° - \theta_c = 47.3°$

10.5 BREWSTER ANGLE EFFECT

Let us now investigate how the reflection coefficients vary as a function of incidence angle θ_i. The expressions for $R_\perp$ and $R_\parallel$ for electric fields are given in Eqs. (10-36) and (10-45) in the case of non-magnetic media. Consider the case when the wave is incident from air ($n_1 = 1$) upon glass with $\varepsilon_2 = 4\varepsilon_o$. The glass has the index of refraction $n_2 = \sqrt{\dfrac{\mu_o\varepsilon_2}{\mu_o\varepsilon_o}} = 2$.

Figure 10-6(a) shows the plot of $|R_\perp|$ and $|R_\parallel|$ versus θ_i when $n_1 = 1$, $n_2 = 2$. It is observed that as θ_i varies from 0° (normal incidence) to 90° (grazing), $|R_\perp|$ increases monotonically, reaching a maximum value of 1 at 90°, while $|R_\parallel|$ decreases initially, reaches a minimum value of 0 at some angle $\theta_i = \theta_B$ and then increases up to 1 at 90°.

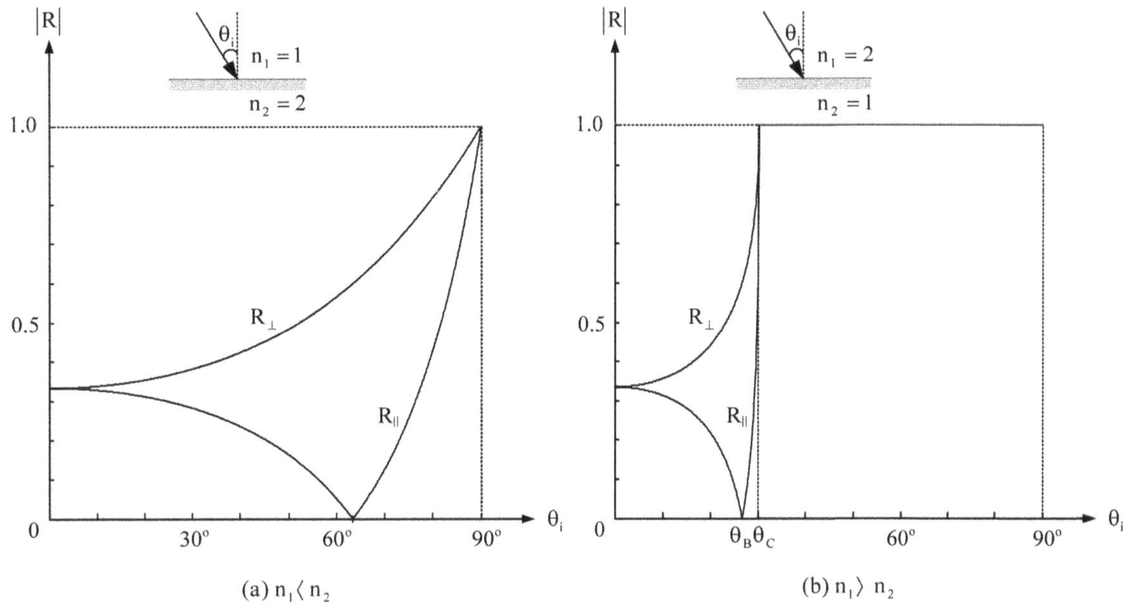

Figure 10-6. Reflection coefficients vs. incidence angle.

Thus we have

(i) $|R_\perp| \geq |R_\parallel|$ for all angles θ_i

(ii) $R_\parallel = 0$ at $\theta_i = \theta_B$

The perpendicularly polarized wave is reflected more than is the parallel polarized wave. There exists an angle of incidence $\theta_i = \theta_B$ at which $R_\parallel = 0$, i.e., there is *no reflection*, thus *total transmission* for parallel polarized wave. Such an angle θ_B is called **Brewster's angle.** θ_B can be calculated by letting $R_\parallel = 0$ in Eq. (10-45) for non-magnetic media, with the use of Snell's law. Letting $R_\parallel = 0$ leads to

$$n_1 \cos\theta_t = n_2 \cos\theta_B \qquad (10\text{-}55)$$

From Snell's law, Eq. (10-30), we have

$$n_2 \sin\theta_t = n_1 \sin\theta_B \qquad (10\text{-}56)$$

Eliminating θ_t from Eqs. (10-55) and (10-56), we derive Brewster's angle as follows (see Problem 10-28):

$$\boxed{\theta_B = \tan^{-1}\frac{n_2}{n_1}} \qquad \left(\begin{array}{c}\text{Brewster's angle for} \\ \text{non-magnetic media}\end{array}\right) \qquad (10\text{-}57)$$

When $n_1 = 1$, $n_2 = 2$, Brewster's angle occurs or $\theta_B = 63.4°$. When the wave with elliptical polarization or the unpolarized wave is incident on the dielectric surface exactly at Brewster's angle, the reflected wave will contain only the perpendicularly polarized component because the parallel polarized component of the incident wave is totally transmitted into the dielectric. The reflected wave becomes *linearly polarized.* **Polaroid sunglasses** reduce glares based on this principle. When the unpolarized sunlight is reflected from the ground such as the asphalt and snow surface, the reflected light, that is, the glare becomes *partially polarized*, having mostly the perpendicularly polarized component for which the electric field is horizontal with respect to the reflecting surface. The material used in sunglasses is *anisotropic* such that it passes the vertical component of the electric field but blocks (or absorbs) the horizontal component (perpendicularly polarized component).

In Figure 10-6(b), $|R_\perp|$ and $|R_\parallel|$ versus θ_i are plotted when the wave is incident from glass (denser, $n_1 = 2$) upon air (less dense, $n_2 = 1$). We find similar characteristics except that $|R_\perp| = |R_\parallel| = 1$ for $\theta_i \geq \theta_c = \sin^{-1} n_2/n_1 = 30°$ because of the total internal reflection, as discussed in Section 10.4. Brewster's angle occurs at $\theta_B = \tan^{-1} n_2/n_1 = 26.6°$. Finally, we note that Brewster's angle exists only for parallel polarized wave when both media are non-magnetic ($\mu_1 = \mu_2 = \mu_o$). There can exist Brewster's angle for perpendicularly polarized wave, when two media have different magnetic properties ($\mu_1 \neq \mu_2$) (see Problem 10-30).

10.6 REFLECTION FROM PERFECT CONDUCTOR – STANDING WAVES

In this section we consider the situation when the wave is incident upon a perfectly conducting medium which has an infinite conductivity. Referring to Figure 10-2, we consider the oblique incidence of the perpendicularly polarized wave whose fields are given by Eq. (10-18). Now the medium 2 has μ_2, ε_2 and $\sigma_2 = \infty$. First we note that inside a perfect conductor the fields are zero, therefore, we have

$$\mathbf{E}_t = \mathbf{H}_t = 0 \text{ in medium 2} \qquad (10\text{-}58)$$

and there will be no transmitted wave, i.e., $T_\perp = 0$. The electric and magnetic fields of the reflected wave can be written as in Eq. (10-21). The boundary conditions at the surface of the perfect conductor ($x = 0$) are given from Eqs. (8-27a,b) by letting $\mathbf{E}_2 = \mathbf{H}_2 = 0$ or from Eq. (8-28),

$$\mathbf{E}_{1t} = 0 \tag{10-59a}$$

$$\mathbf{a}_{n2} \times \mathbf{H}_1 = \mathbf{J}_s \tag{10-59b}$$

where $\mathbf{J}_s$ is the (free) surface current density. Substituting $x = 0$ in Eqs. (10-18a), (10-21a) and applying the boundary condition, Eq. (10-59a), we obtain

$$e^{-jk_z z} + R_\perp e^{-jk_{rz} z} = 0 \tag{10-60}$$

Following a similar procedure shown in Section 10.3, we have

$$k_z = k_{rz} \text{ or } \theta_r = \theta_i$$

and

$$1 + R_\perp = 0$$

Thus we have

$$R_\perp = -1, \qquad T_\perp = 0 \tag{10-61}$$

At the surface of the perfectly conducting medium, the wave is *totally reflected*, i.e., the perfect reflection occurs, and there will be no transmission. This fact becomes a major principle of metallic waveguides for which the wave bounces off the conducting walls by perfect reflection and is guided in a certain direction. We can also obtain the results of Eq. (10-61) from the results of a dielectric boundary, Eq. (10-35), by treating the medium 2 as a lossy dielectric with infinite conductivity,

$$\varepsilon_{2,\text{eff}} = \varepsilon_2 - j\frac{\sigma_2}{\omega}, \quad \sigma_2 \to \infty$$

$$\eta_2 = \sqrt{\frac{\mu_2}{\varepsilon_2 - j\frac{\sigma_2}{\omega}}} \to 0 \quad \text{as } \sigma_2 \to \infty$$

Letting $\eta_2 = 0$ in Eq. (10-35) leads to Eq. (10-61).

In the case of a parallel polarized wave, substituting $x = 0$ in Eqs. (10-37b), (10-38b) and applying Eq. (10-59a), we obtain

$$R_\parallel = 1, \quad T_\parallel = 0 \tag{10-62a}$$

$$R_\parallel^E = -1, \quad T_\parallel^E = 0 \tag{10-62b}$$

We note that the reflection coefficient is $+1$ for the magnetic field and -1 for the electric field. For both polarizations, we find that

$$P_r = P_i, \qquad P_t = 0 \tag{10-63}$$

Standing Waves

Let us now look at the total fields in medium 1 for the perpendicular polarization and observe the field pattern as a function of distance from the surface of the perfect conductor (or medium 2).

$$\begin{aligned}\mathbf{E}_1 &= \mathbf{E}_i + \mathbf{E}_r = \mathbf{a}_y E_0\, e^{-jk_x x - jk_z z} - \mathbf{a}_y E_0\, e^{jk_x x - jk_z z} \\ &= \mathbf{a}_y E_0\, e^{-jk_z z}\left\{e^{-jk_x x} - e^{jk_x x}\right\} = \mathbf{a}_y E_m\, e^{-jk_z z} \sin(k_x x)\end{aligned} \tag{10-64a}$$

where

$$E_m = -j2E_0\,.$$

Similarly, we can obtain

$$\mathbf{H}_1 = \mathbf{H}_i + \mathbf{H}_r = \frac{2E_0}{\omega\mu_1} e^{-jk_z z}\left\{\mathbf{a}_x jk_z \sin(k_x x) + \mathbf{a}_z k_x \cos(k_x x)\right\} \tag{10-64b}$$

Assuming $E_m = A\, e^{j\phi}$, the electric field in real time is given by

$$\mathbf{E}_1(\mathbf{r},t) = \mathrm{Re}\{\mathbf{E}_1 e^{j\omega t}\} = \mathbf{a}_y A \sin(k_x x)\cos(\omega t - k_z z + \phi) \tag{10-65}$$

The wave represented by Eqs. (10-64) and (10-65) propagates in the $+z$ direction but does not propagate or travel in the x direction. Such a wave is called a **standing wave** in the x direction while $e^{-jk_x x}$ and $e^{jk_x x}$ represent traveling waves. The amplitude of the electric field is given by

$$|\mathbf{E}_1| = |E_m|\,|\sin(k_x x)| \tag{10-66}$$

and plotted as a function of x in Figure 10-7. It gives a periodic pattern for which the minima (with zero value) occur at $x = -n\pi/k_x$ $(n = 0, 1, 2, \ldots.)$ and the maxima (with the value of $2|E_0|$) at $x = -\left(n + \frac{1}{2}\right)\pi/k_x$ $(n = 0, 1, 2, \ldots.)$.

The period of the standing wave pattern is

$$\frac{\pi}{k_x} = \frac{\pi}{k_1 \cos\theta_i} = \frac{\lambda_1}{2}\frac{1}{\cos\theta_i}\,. \tag{10-67}$$

Note that at normal incidence ($\theta_i = 0$), the period is half the wavelength ($\lambda/2$).

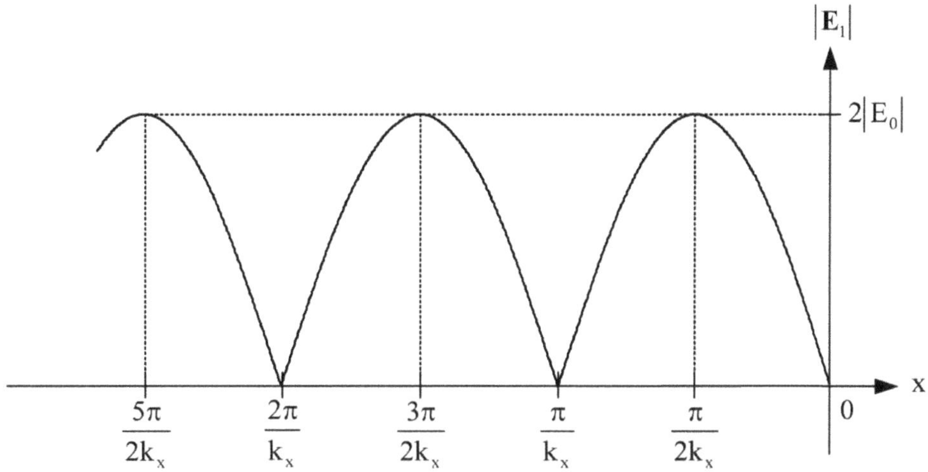

Figure 10-7. Standing wave pattern

Noting that

$$\mathbf{E}_1 = 0 \text{ at } x = -n\frac{\pi}{k_x} = -n\frac{\lambda}{2}\frac{1}{\cos\theta_i}, \quad n = 0, 1, 2, \ldots ,$$

perfectly conducting plates can be placed at these points without affecting the solution between the original and added conductors. This interpretation will be quite useful in understanding the field solutions of the metallic waveguide. In fact, the total fields we just discussed are proved to be the TE (transverse electric) mode solutions for the parallel-plate waveguide as we shall see in Chapter 11.

The time-average power density or Poynting vector of the total wave is given by

$$\mathbf{S}_{av} = \frac{1}{2}\text{Re}\{\mathbf{E}_1 \times \mathbf{H}_1^*\} = \mathbf{a}_z \frac{2|E_0|^2}{\omega\mu_1} k_z \sin^2(k_x x) = \mathbf{a}_z \frac{2|E_0|^2}{\eta_1} \sin\theta_i \sin^2(k_x x) \quad (10\text{-}68)$$

Thus, the time-average power flow is in the $+z$ direction. Note that the x component of $\mathbf{S}_{av}$ is zero because E_y and H_z are 90° out of phase in time. The phase velocity in the z direction is given by

$$v_p = \frac{\omega}{k_z} = \frac{\omega}{k_1 \sin\theta_i} = \frac{v_1}{\sin\theta_i}$$

where v_1 is the phase velocity of a wave in an unbounded medium 1.

Now, if we use the second boundary condition given in Eq. (10-59b), we can obtain the surface current $\mathbf{J}_s$ that flows on the surface of the perfect conductor from the magnetic field as follows.

$$\mathbf{J}_s = (-\mathbf{a}_x) \times \mathbf{H}_1\Big|_{x=0} = \mathbf{a}_y H_z\Big|_{x=0} = \mathbf{a}_y \frac{2E_0}{\omega\mu_1} k_x e^{-jk_z z} \quad (10\text{-}69)$$

If the conducting plate is replaced by a grating of parallel conducting wires arranged in the direction of the surface current flow (y direction in this case), these wires can serve as a good reflector as effectively as a solid conducting plate. The gratings are used often as reflecting antennas because they reduce weight, save material cost, and reduce wind loading.

Normal Incidence

As a special case, when the wave is incident normal to the boundary, letting $\theta_i = 0$, we obtain the following results:

$$k_x = k_1 \cos\theta_i = k_1,\ k_z = k_1 \sin\theta_i = 0$$

$$\mathbf{E}_1 = \mathbf{a}_y E_0 \left\{ e^{-jk_1 x} - e^{jk_1 x} \right\} = \mathbf{a}_y (-j2E_0)\ \sin(k_1 x)$$

$$\mathbf{H}_1 = \mathbf{a}_z \frac{E_0}{\eta_1} \left\{ e^{-jk_1 x} + e^{jk_1 x} \right\} = \mathbf{a}_z \frac{2E_0}{\eta_1} \cos(k_1 x)$$

$$\mathbf{E}_1 = 0 \text{ at } x = -n\frac{\pi}{k_1} = -n\frac{\lambda}{2}, \quad n = 0, 1, 2 \ldots.$$

$$\mathbf{S}_{av} = 0$$

$$\mathbf{J}_s = \mathbf{a}_y \frac{2E_0}{\eta_1}$$

We can also obtain similar results and interpretations when the parallel polarized wave is incident upon a perfectly conducting medium (see Problem 10-36).

EXAMPLE 10-5. STANDING WAVE IN FRONT OF A PERFECT CONDUCTOR

A uniform plane wave of 1 GHz is incident normally on a large copper plate at $x = 0$ (see Figure 10-1). The electric field is z-polarized and has an amplitude of 10 V/m.

(a) Find the total electric field and the locations of its first maximum and minimum in front of the plate.
(b) Find the total magnetic field and the locations of its first maximum and minimum.

SOLUTIONS

(a) The electric fields of the incident and reflected waves can be written as

$$\mathbf{E}_i = \mathbf{a}_z 10e^{-jkx}, \quad k = \frac{\omega}{c} = \frac{2\pi\times10^9}{3\times10^8} = \frac{20\pi}{3}, \quad \lambda = \frac{2\pi}{k} = 30 \text{ [cm]}$$

$$\mathbf{E}_r = -\mathbf{a}_z 10e^{+jkx} \quad \text{Note that } R = -1 \text{ for } \mathbf{E}.$$

The total electric field is given by

$$\mathbf{E}_1 = \mathbf{E}_i + \mathbf{E}_r = \mathbf{a}_z(-j20)\sin(kx).$$

The first maximum occurs at $kx = \frac{2\pi}{\lambda}x = -\frac{\pi}{2}, \quad x_{max} = -\frac{\lambda}{4} = -7.5$ [cm]

The first minimum (excluding $x = 0$) occurs at $kx = \frac{2\pi}{\lambda}x = -\pi, \quad x_{min} = -\frac{\lambda}{2} = -15$ [cm].

(b) The magnetic fields are given by

$$\mathbf{H}_i = \frac{1}{\eta_0}\mathbf{a}_x \times \mathbf{E}_i = -\mathbf{a}_y \frac{10}{377} e^{-jkx}$$

$$\mathbf{H}_r = \frac{1}{\eta_0}(-\mathbf{a}_x)\times\mathbf{E}_r = -\mathbf{a}_y \frac{10}{377} e^{+jkx} \quad \text{Note that } R = +1 \text{ for } \mathbf{H}.$$

$$\mathbf{H}_1 = \mathbf{H}_i + \mathbf{H}_r = -\mathbf{a}_y \frac{20}{377}\cos(kx)$$

The first maximum (excluding $x = 0$) occurs at $kx = -\pi, \; x_{max} = -\frac{\lambda}{2} = -15$ [cm]

The first minimum occurs at $kx = -\frac{\pi}{2}, \; x_{min} = -\frac{\lambda}{4} = -7.5$ [cm]

Note that the electric and magnetic fields alternate their maxima and minima. This behavior is similar to that of the transmission line with short-circuited load for which the voltage and current alternate their maxima and minima, as will be shown in Section 11-7.

PROBLEMS

10-1. Show that the reflection and transmission coefficients at normal incidence, when both media are non-magnetic ($\mu_1 = \mu_2 = \mu_0$) and lossless, are given by

$$R = \frac{n_1 - n_2}{n_1 + n_2}$$

$$T = \frac{2n_1}{n_1 + n_2}$$

where n_1, n_2 are the refractive indices of the two media.

10-2. Show that the power conservation relationship, Eq. (10-14), holds, by substituting the reflection and transmission coefficients in Eq. (10-11). Assume that the media are lossless.

10-3. Consider the wave incident on the lake in Example 10-1. Including the loss in the lake water, calculate the reflection and transmission coefficients and the fields of the transmitted wave exactly and compare it with the approximate results. Discuss the errors.

10-4. A light is incident normally on a glass which has a refractive index 1.5. Find the reflection and transmission coefficients. The glass is non-magnetic and lossless.

10-5. A microwave with $f = 1$ GHz is incident normally on alumina ($A\ell_2 O_3$), which has the dielectric constant 9.7 and the loss tangent 0.002 at 1 GHz. Calculate the reflection and transmission coefficients. Assume that alumina is non-magnetic. Write down the expression for the electric field of the wave transmitted through alumina and determine the parameters that appear in the field expression. Let the electric field amplitude of the incident wave be E_0.

10-6. Derive $R_\perp$ and $T_\perp$ in Eq. (10-34) by solving the two equations in Eq. (10-32).

10-7. Derive Eq. (10-35) ($R_\perp$ and $T_\perp$ in terms of angles of incidence and transmission) and Eq. (10-36) ($R_\perp$ and $T_\perp$ for non-magnetic media) from Eq. (10-34).

10-8. **(A uniform plane wave propagating in an arbitrary direction)**

Consider the time-harmonic electromagnetic wave propagating in air whose complex electric field is given by

$$\mathbf{E} = \mathbf{a}_y\, 5\, e^{-j(3x + 4z)}\ [\mathrm{V/m}]$$

(a) Is this a uniform plane wave? What are the constant-phase fronts?

(b) Write down the wave vector or the propagation vector $\mathbf{k}$. What is the direction of propagation and the state of polarization of this wave?

(c) Find the frequency f in Hz of this wave.
(d) Find the associated magnetic field $\mathbf{H}$ and the time-average power density in W/m^2 and the direction of power flow.

10-9. **(Reflection and transmission of perpendicularly polarized wave)**
Suppose the wave in Problem 10-8 is incident upon a lossless dielectric (e.g., glass) surface at $z = 0$. Assume that the glass has a dielectric constant 2.25 and is non-magnetic. See Figure 10-2 but let the interface be at $z = 0$, i.e., the z axis is normal to the surface.

(a) What is the angle of incidence θ_i?
(b) Find the angles of reflection θ_r and transmission θ_t.
(c) Find the electric field $\mathbf{E}_r$ and the magnetic field $\mathbf{H}_r$ of the reflected wave.
(d) Find the electric field $\mathbf{E}_t$ and the magnetic field $\mathbf{H}_t$ of the transmitted wave.
(e) Find the fraction of power reflected from the glass surface.
(f) Find the fraction of power transmitted to the glass surface.

10-10. **(Reflection and transmission of parallel polarized wave)**
A uniform plane wave whose electric field is given by

$$\mathbf{E} = (\mathbf{a}_x\, 4 - \mathbf{a}_z\, 3)\, e^{-j(3x+4z)}\ [V/m]$$

is incident from air upon a lossless dielectric (e.g., glass) surface at $z = 0$. Assume that the glass has dielectric constant 2.25 and is non-magnetic. See Figure 10-3 but let the surface be at $z = 0$, i.e., the z axis is normal to the surface. Repeat the same questions in Problem 10-9.

10-11. Derive $R_{||}$ and $T_{||}$ in Eq. (10-41) by solving the two equations in Eq. (10-40).

10-12. Derive Eq. (10-42) ($R_{||}$ and $T_{||}$ in terms of angles of incidence and transmission) and Eq. (10-44) ($R_{||}$ and $T_{||}$ for non-magnetic media) from Eq. (10-41).

10-13. Derive Eqs. (10-43) and (10-45), the reflection and transmission coefficients for the *electric field* of the parallel polarized wave.

10-14. A uniform plane wave whose electric field is given by

$$\mathbf{E} = \mathbf{a}_y\, 5\, e^{-j(5x+12z)}\ [V/m]$$

is incident from air upon a lossless dielectric-magnetic surface at $x = 0$ (see Figure 10-2). The second medium has $\varepsilon = 4\,\varepsilon_0$, $\mu = 9\,\mu_0$. Repeat the same questions in Problem 10-9.

10-15. A uniform plane wave whose magnetic field is given by

$$\mathbf{H} = \mathbf{a}_y\ 5\ e^{-j(5x+12z)}\ [\text{mA/m}]$$

is incident from air upon a lossless dielectric-magnetic surface at $x = 0$ (see Figure 10-3). The second medium has $\varepsilon = 4\,\varepsilon_0$, $\mu = 9\,\mu_0$. Repeat the same questions in Problem 10-9.

10-16. Confirm that the power conservation relationship in Eq. (10-48) holds for the perpendicularly polarized wave.

10-17. Derive the expression for the fraction of power transmitted for the parallel polarized wave, similar to Eq. (10-47b), and confirm that the power conservation relationship in Eq. (10-48) holds for the parallel polarized wave.

10-18. The index of refraction of a material plays an important role in optics applications involving reflection and refraction (or transmission) of a light ray. When the red light in air enters the human eye, the wavelength of the light is changed from 633 nm to 474 nm. Find the index of refraction of the aqueous humor in the eye.

10-19. A uniform plane wave propagating in air has the following time-harmonic electric field:

$$\mathbf{E}_i = \left\{\mathbf{a}_x\sqrt{2} + \mathbf{a}_y\, j\sqrt{3} - \mathbf{a}_z 1\right\}\ e^{-j(2x+2\sqrt{2}z)}\ [\text{V/m}]$$

The wave is incident on the planar surface (located at $x = 0$) of a dielectric with $\varepsilon = 4\,\varepsilon_0$, $\mu = \mu_0$. See the geometry in either Figure 10-2 or 10-3.

(a) Determine the frequency in Hz and wavelength in m of this wave.
(b) Determine the angle of incidence and the angle of transmission.
(c) What is the polarization of the incident wave—LP, CP or EP? Discuss also its handedness, if applicable.
(d) Find the complete expression for the electric field of the reflected wave.
(e) Find the complete expression for the magnetic field of the transmitted wave.

10-20. The wave is incident from air upon snow surface. Assume that the snow medium has permittivity $(1.0 + 2.0\rho)\,\varepsilon_0$ $(0.1 \le \rho \le 0.5)$ and neglect its small conductivity. ρ is the relative density of dry snow (compared to water).

(a) Choose the value of ρ and plot the fraction of the power reflected as a function of incident angle θ_i for both perpendicular and parallel polarizations, using a computer program.
(b) Calculate the Brewster angle θ_B.

Now suppose the wave is incident from snow upon air.

(c) Plot the fraction of the power reflected as a function of incident angle θ_i for both perpendicular and parallel polarizations.
(d) Calculate the Brewster angle θ_B and the critical angle θ_c.

10-21. The wave is incident from air upon ice surface. Assume that ice has permittivity $(3.2 + \alpha)\,\varepsilon_0$ $(0.01 \le \alpha \le 0.09)$ and neglect its conductivity. Choose the value of α and repeat the same questions in Problem 10-20.

10-22. Rewrite the reflection coefficient ($R_{\parallel}$) for parallel polarization when the total internal reflection occurs and find the expression like the one shown in Eq. (10-53). Show that the fraction of power reflected is equal to 1 when total internal reflection occurs for the parallel polarized wave.

10-23. Prove that when total internal reflection occurs the reflection coefficients for perpendicular and parallel polarized waves can be written as

$$R_{\perp} = 1e^{j2\phi_{\perp}}$$
$$R_{\parallel} = 1e^{j2\phi_{\parallel}}$$

where $\phi_{\perp} = \tan^{-1}\left(\dfrac{\mu_1\alpha_2}{\mu_2 k_x}\right)$, $\phi_{\parallel} = \tan^{-1}\left(\dfrac{\varepsilon_1\alpha_2}{\varepsilon_2 k_x}\right)$. This implies that there occurs a phase shift in the reflected wave at the total internal reflection. $2\phi_{\perp}$ and $2\phi_{\parallel}$ are known as the *Goos-Hänchen* shift.

10-24. A light source is submerged under the water to illuminate a surface of a body of water. The refractive index of water is 1.33 at optical frequencies. When the light source is located at a depth of 4 m, find the area of water surface illuminated by the light.

10-25. (**Numerical aperture of an optical fiber**)
One end of an optical fiber is illuminated by light source such as a light-emitting diode as shown in Figure 10-8. Show that the maximum value (θ_a) of the angle of incidence which guarantees the guidance of light in the fiber core is given by

$$\theta_a = \sin^{-1}\sqrt{n_1^2 - n_2^2}$$

where n_1 and n_2 are the refractive indices of the core and the cladding, respectively. $\sin\theta_a = \sqrt{n_1^2 - n_2^2}$ is known as the *numerical aperture* of the fiber.

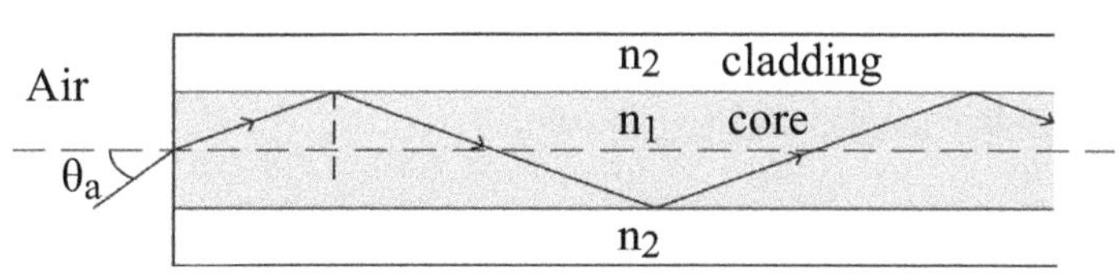

Figure 10-8. Acceptance angle of an optical fiber.

10-26*. Consider the geometry of Figure 10-2. Assume the medium 1 is glass with $\varepsilon_1 = 1.96\,\varepsilon_0$, $\mu_1 = \mu_0$ and the medium 2 is air. A 1 GHz uniform plane wave with perpendicular polarization and electric field amplitude 2 V/m is incident at $\theta_i = 60°$ from glass upon air.

(a) Find the complete expression for the electric field of the incident wave.
(b) Find the fraction of power reflected from the air interface.
(c) Find the complete expression for the electric field of the transmitted wave.

10-27. Repeat the same questions in Problem 10-26 when the incident wave is parallel polarized.

10-28. Derive the expression for Brewster's angle, Eq. (10-57), when both media are non-magnetic.

10-29. Show that Brewster's angle ($\theta_{B\parallel}$) for parallel polarized wave when $\mu_1 \neq \mu_2$ is given by

$$\sin^2\theta_{B\parallel} = \frac{1-\dfrac{\mu_2\varepsilon_1}{\mu_1\varepsilon_2}}{1-\left(\dfrac{\varepsilon_1}{\varepsilon_2}\right)^2}$$

10-30. When two media have different magnetic properties, i.e., $\mu_1 \neq \mu_2$, there can exist Brewster's angle ($\theta_{B\perp}$) for perpendicularly polarized wave at which $R_\perp = 0$. Show that $\theta_{B\perp}$ is given by

$$\sin^2\theta_{B\perp} = \frac{1-\dfrac{\mu_1\varepsilon_2}{\mu_2\varepsilon_1}}{1-\left(\dfrac{\mu_1}{\mu_2}\right)^2}$$

10-31. Sunlit glares caused by reflections of the sun light from the road or snow surface are "partially" linearly polarized.

(a) What is the polarization of the dominant component of the glare? Perpendicular (TE) or parallel (TM)?
(b) Your Polaroid sunglasses are made up of *anisotropic* material so that they absorb one linearly polarized component of the incident wave and pass the other linearly polarized component. To minimize sun glare, what component, *perpendicular* or *parallel*, should reach your eyes after passing through the sunglasses? Explain why.

10-32. (Brewster's window in a gas laser)*

A gas laser is composed of a tube containing gas fitted with a Brewster-angle window and external mirrors as shown in Figure 10-9. Two mirrors reflect the light back and forth in such a manner that a standing wave is sustained in the "cavity". The cavity is said to be at resonance. Part of the light is transmitted through mirrors and becomes the output of the laser. Now, if the angle (θ) of the window is at a certain angle, the output of the laser becomes almost completely *linearly polarized*, because of the Brewster angle effect.

(a) Explain why. What would be the polarization of the output wave, i.e., is $\mathbf{E}$ perpendicular or parallel to the plane of incidence?
(b) Suppose the window is made of quartz of refractive index 1.54. Find the angle θ.

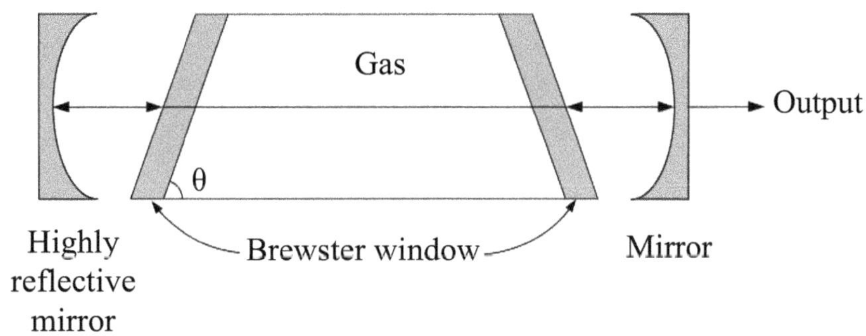

Figure 10-9. A gas laser with Brewster windows.

10-33. Show that when the wave is incident at Brewster's angle ($\theta_i = \theta_B$), the angle of transmission (θ_t) is given by $\theta_t = 90° - \theta_B$.

10-34. Consider Problem 10-19 and let $\varepsilon_2 = 2\,\varepsilon_0$.

(a) Recalculate the angle of transmission.
(b) Find the complete expression for the electric field of the reflected wave.
(c) What is the polarization of the reflected wave?

10-35. (Reflection from a Perfect Conductor)

Suppose the wave in Problem 10-8 is incident from air upon a perfectly conducting plane at $x = 0$. Refer to Figure 10-2 for the geometry.

(a) Find the electric field $\mathbf{E}_r$ and the magnetic field $\mathbf{H}_r$ of the reflected wave.
(b) Find the fraction of the power reflected from the conducting surface.

* L.C. Shen and J.A. Kong, *Applied Electromagnetism*, PWS Publishing Company, 1995, 3rd ed.

10-36. Consider the problem of a parallel polarized wave incident at oblique angle (θ_i) upon a perfectly conducting medium ($\sigma_2 = \infty$). Refer to Figure 10-3 for the geometry.

(a) Find the total fields $\mathbf{H}_1$ and $\mathbf{E}_1$ as a function of distance from the surface of the perfect conductor.
(b) Plot the standing wave pattern for the amplitude of the magnetic field $|\mathbf{H}_1|$.
(c) Find the location of points at which the magnetic field amplitude is zero ($\mathbf{H}_1 = 0$).
(d) Calculate the surface current density $\mathbf{J}_s$ on the surface of the perfect conductor.

10-37. Consider the perpendicularly polarized wave in Problem 10-8 incident upon a perfect conductor surface at $z = 0$.

(a) Find the electric field $\mathbf{E}_r$ of the reflected wave and the total electric field $\mathbf{E}_1$ in front of the conductor.
(a) Plot the amplitude of the total electric field as a function of distance (z) from the conductor surface.
(b) Determine the location of points at which the electric field is minimum (zero).

10-38. A dipole antenna is placed in front of a large conducting plate. The signal received by the antenna coming from the normal direction is twice as strong as the incident field, when it is placed at a certain distance (say, 1 m at the closest) in front of the plate. Can you determine the frequency of the signal? What would the strength of the received signal be if the frequency of the signal is changed to 50 MHz?

10-39*. **(Standing wave in front of a lossy dielectric)**
Consider the perpendicularly polarized wave in Example 10-2 incident upon a lossy dielectric medium of $\varepsilon = 1.96\,\varepsilon_0$, $\mu = \mu_0$, $\sigma = 0.1$ S/m, whose surface is at $x = 0$.

(a) Find the electric field of the reflected wave. Note that $R_\perp$ becomes complex.
(b) Plot the amplitude of the total electric field $|\mathbf{E}_1|$ in air (medium 1) as a function of the distance (x) from the dielectric interface. You will find the standing wave pattern.
(c) Find the electric field of the transmitted wave and plot its amplitude $|\mathbf{E}_t|$ as a function of the distance (x) from the surface. Note that there will be an attenuation.

WAVEGUIDES, RESONATORS, AND TRANSMISSION LINES

CHAPTER 11

11.1 INTRODUCTION

In the previous chapters we studied propagation, polarization, reflection, and transmission of electromagnetic waves. In this chapter, we study how the electromagnetic wave is "guided" from one point to another, using certain transmission structures called *waveguides*. There are two ways to guide the wave in a bounded system, i.e., use either highly conducting materials or dielectric materials. When very good (so-called perfect) conductors such as metals are used, the wave is guided due to total reflection of waves on a perfectly conducting surface and the system is called the *metallic waveguide*. When dielectric material is used, the wave is guided based on the principle of total internal reflection and the system is called *dielectric waveguide*.

There are various forms of metallic waveguides. The simplest structure would be to use a pair of good conductors where the waves bounce off these conducting walls and are guided along the axis of these conducting structures. The examples are the parallel plate waveguide, the two-wire transmission line and the coaxial cable as shown in Figure 11-1(a)-(c). All of these waveguides support the transverse electromagnetic (TEM) mode which has no cutoff frequency. They are also called transmission lines. The TEM mode can propagate along these waveguides from dc up to some high frequency where the cross-section of the waveguide becomes electrically large. Hence they are useful in guiding signals with relatively low frequencies.

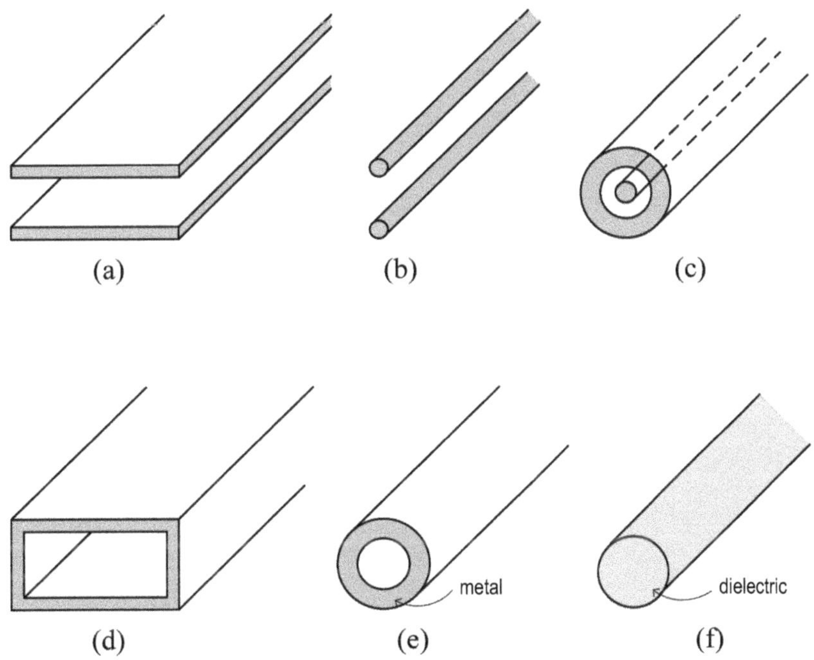

Figure 11-1. Various waveguides. (a) Parallel-plate waveguide, (b) two-wire transmission line, (c) coaxial cable, (d) rectangular waveguide, (e) circular waveguide, (f) optical fiber.

At high frequencies such as microwaves, the use of a two-conductor guiding system becomes impractical because the attenuation of TEM waves along the waveguide due to conductor loss (because the conductor is imperfect) increases with frequency. So at microwave frequencies, other forms of transmission system are used. The single-conductor hollow waveguides [see Figure 11-1(d), (e)] are able to transmit or guide waves in the GHz range over long distances without incurring large losses. Since metal pipes are single conductors with large surface areas, we expect that the attenuation due to conductor loss would be smaller. As will be shown, the hollow waveguides do not support the TEM mode but higher order modes—the TM (transverse magnetic) and TE (transverse electric) modes.

At optical frequencies higher than 1 THz, the metallic waveguides become impractical again due to increased conductor loss at higher frequencies, but the dielectric waveguides such as optical fibers shown in Figure 11-1(f) can provide guidance of optical signals with very small loss.

In this chapter we first discuss the general solution methods for waveguides with uniform cross section and introduce the concepts of TE, TM and TEM modes. We consider the following waveguides and resonators and find the solutions for the electric and magnetic fields and their applications.

1. Parallel plate waveguide
2. Rectangular waveguide
3. Coaxial cable
4. Rectangular cavity resonator

In the last section, we study characteristics of the transmission lines and their TEM mode analysis using the concepts of the voltage, current, and impedance.

11.2 SOLUTION METHODS FOR UNIFORM WAVEGUIDES

For the waveguiding structure whose cross section is uniform (or invariant) along a certain direction (say, z-axis), the following method of analysis is generally applicable. We assume that the metallic walls are perfect conductors ($\sigma = \infty$) and the medium inside the waveguide is lossless ($\sigma = 0$) and has the real material constants μ, ε, as shown in Figure 11-2. The medium could be air.

In the medium within the guiding structure, **E** and **H** satisfy the Helmholtz wave equation [see Eqs. (9-33) and (9-36)]:

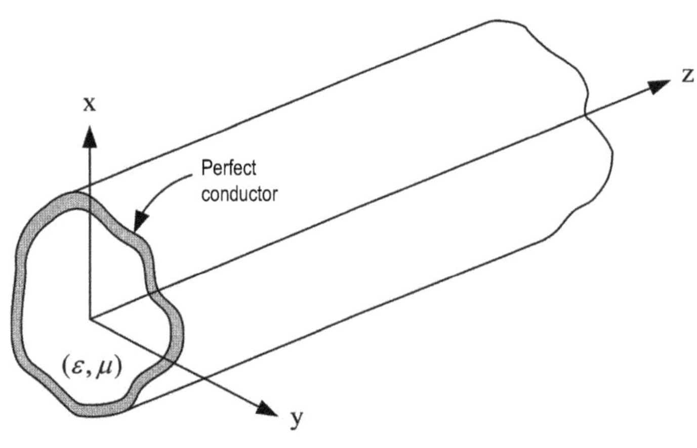

Figure 11-2. Uniform Waveguide

$$(\nabla^2 + k^2)\,\mathbf{E} = 0 \tag{11-1a}$$

$$(\nabla^2 + k^2)\,\mathbf{H} = 0 \tag{11-1b}$$

where

$$k^2 = \omega^2\,\mu\varepsilon \tag{11-2}$$

Let us assume that we wish to *guide* the wave in the $+z$ *direction*. Then the field solutions are assumed to take the following form:

$$\mathbf{E}(x,y,z) = \hat{\mathbf{E}}(x,y)\; e^{-jk_z z} \tag{11-3a}$$

$$\mathbf{H}(x,y,z) = \hat{\mathbf{H}}(x,y)\; e^{-jk_z z} \tag{11-3b}$$

because $e^{-jk_z z}$ represents a wave traveling in the $+z$ direction.

Noting that

$$\frac{\partial}{\partial z}\begin{Bmatrix}\mathbf{E}\\ \mathbf{H}\end{Bmatrix} = -jk_z\begin{Bmatrix}\mathbf{E}\\ \mathbf{H}\end{Bmatrix}, \tag{11-4}$$

substitution of Eq. (11-3) into Eq. (11-1) yields

$$\left(\frac{\partial^2}{\partial x^2} + \frac{\partial^2}{\partial y^2} - k_z^2 + k^2\right)\begin{Bmatrix}\mathbf{E}\\ \mathbf{H}\end{Bmatrix} = 0 \tag{11-5}$$

Making use of the identity, Eq. (11-4), from the two Maxwell's vector curl equations

$$\nabla \times \mathbf{E} = -\,j\omega\mu\,\mathbf{H} \tag{11-6a}$$

$$\nabla \times \mathbf{H} = j\omega\varepsilon\,\mathbf{E} \tag{11-6b}$$

we obtain the following six scalar equations:

$$\frac{\partial E_z}{\partial y} + jk_z E_y = -j\omega\mu\,H_x \tag{11-7a}$$

$$-jk_z E_x - \frac{\partial E_z}{\partial x} = -j\omega\mu\,H_y \tag{11-7b}$$

$$\frac{\partial E_y}{\partial x}-\frac{\partial E_x}{\partial y}=-j\omega\mu H_z \tag{11-7c}$$

and

$$\frac{\partial H_z}{\partial y}+jk_z H_y=j\omega\varepsilon E_x \tag{11-8a}$$

$$-jk_z H_x-\frac{\partial H_z}{\partial x}=j\omega\varepsilon E_y \tag{11-8b}$$

$$\frac{\partial H_y}{\partial x}-\frac{\partial H_x}{\partial y}=j\omega\varepsilon E_z \tag{11-8c}$$

Because of the assumed form of the z dependence, from Eqs. (11-7a,b) and (11-8a,b) we can express E_x, E_y, H_x, H_y in terms of E_z and H_z. For example, from Eq. (11-7b), we find

$$H_y=\frac{k_z}{\omega\mu}E_x+\frac{1}{j\omega\mu}\frac{\partial E_z}{\partial x} \tag{11-9}$$

Substituting Eq. (11-9) into Eq. (11-8a) and solving for E_x, we obtain

$$E_x=\frac{-1}{k^2-k_z^2}\left\{jk_z\frac{\partial E_z}{\partial x}+j\omega\mu\frac{\partial H_z}{\partial y}\right\} \tag{11-10a}$$

Similarly, we can derive (see Problem 11-1)

$$E_y=\frac{1}{k^2-k_z^2}\left\{-jk_z\frac{\partial E_z}{\partial y}+j\omega\mu\frac{\partial H_z}{\partial x}\right\} \tag{11-10b}$$

$$H_x=\frac{1}{k^2-k_z^2}\left\{j\omega\varepsilon\frac{\partial E_z}{\partial y}-jk_z\frac{\partial H_z}{\partial x}\right\} \tag{11-10c}$$

$$H_y=\frac{-1}{k^2-k_z^2}\left\{j\omega\varepsilon\frac{\partial E_z}{\partial x}+jk_z\frac{\partial H_z}{\partial y}\right\} \tag{11-10d}$$

Note that E_x and H_y are derived from Eqs. (11-7b) and (11-8a), while E_y and H_x are derived from Eqs. (11-7a) and (11-8b). Here we have expressed the transverse components (E_x, E_y, H_x, H_y) of the fields in terms of the longitudinal components (E_z, H_z) of the fields. We mean "transverse" with respect to the direction of guidance (z axis in this case). Thus we only need to solve for the z components (E_z, H_z) of the fields, which satisfy the following wave equation from Eq. (11-5).

$$\frac{\partial^2 E_z}{\partial x^2}+\frac{\partial^2 E_z}{\partial y^2}+(k^2-k_z^2)\,E_z=0 \tag{11-11a}$$

$$\frac{\partial^2 H_z}{\partial x^2}+\frac{\partial^2 H_z}{\partial y^2}+(k^2-k_z^2)\,H_z=0 \tag{11-11b}$$

Eqs. (11-11) are the second-order partial differential equations which are similar to Laplace's equation [see Eq. (4-39)] we treated in Chapter 4 except for the existence of the third term. Thus the general solutions can be obtained by using the same technique, that is, the *method of separation of variables*. As will be shown in Section 11.4, the general solutions of Eq. (11-11) for E_z (or H_z) take the following form.

$$E_z(x, y, z) = \{A \sin k_x x + B \cos k_x x\} \{C \sin k_y y + D \cos k_y y\} e^{-jk_z z} \tag{11-12}$$

for which k_x, k_y, k_z should satisfy the dispersion relation:

$$k_x^2 + k_y^2 + k_z^2 = k^2 = \omega^2 \mu \varepsilon \tag{11-13}$$

The allowed or acceptable values for k_x, k_y and some of the coefficients (A, B, C, D) are determined by applying the appropriate boundary conditions at the surface of the conducting walls, i.e., $\mathbf{E}_{tan} = 0$, in the case of metallic waveguide.

It is convenient to classify the propagating waves in a waveguiding system into three types according to whether E_z or H_z exists.

(i) $E_z = 0$, $H_z \neq 0$: Transverse electric (TE) mode
(ii) $H_z = 0$, $E_z \neq 0$: Transverse magnetic (TM) mode
(iii) $E_z = H_z = 0$: Transverse electromagnetic (TEM) mode

It will be easier to find solutions for each type, separately. This also reduces the complexity of finding general waveguide solutions. For TE modes, the electric field has only transverse (x- and y-) components while the magnetic field may have both transverse and longitudinal (z-) components. Thus in order to obtain the TE mode solutions we only need to determine H_z; then all the other components of $\mathbf{E}$ and $\mathbf{H}$ can be obtained through Eq. (11-10). For TM modes, the magnetic field has only transverse components while the electric field may have both components. Thus for the TM mode solutions we seek to determine E_z first, from which all the other components can be calculated. In the case of the TEM mode, one has to find the transverse component directly. It is known that there exists a TEM mode for a two-conductor transmission system such as a two-wire transmission line, a parallel-plate waveguide and a coaxial cable. The TEM wave does not exist for a one-conductor system such as rectangular waveguide as will be shown later.

In summary, to obtain the field solutions for metallic waveguides whose cross section is uniform in the z direction, we take the following steps:

Step 1. Solve for E_z (for TM modes) or H_z (for TE modes) that satisfy Eq. (11-11), subject to the boundary conditions on the perfectly conducting surface [$\mathbf{E}_{tan} = 0$].

Step 2. Determine the transverse components (E_x, E_y, H_x, H_y) from E_z or H_z, using Eq. (11-10).

We will start with the simplest waveguide, the parallel-plate waveguide, and then consider the rectangular waveguide and also the rectangular cavity resonator, whose solutions are a simple extension of the rectangular waveguide solutions.

11.3. PARALLEL-PLATE WAVEGUIDE

A parallel-plate waveguide consists of two perfectly conducting plates separated by a distance a and filled with a medium having material constants μ, ε as shown in Figure 11-3. The width w of the conducting plate is assumed to be much greater than the separation a. For the purpose of finding simple and analytical solutions we assume that the cross section is infinite in extent in the y direction, i.e., $w \to \infty$. As mentioned in the previous section, we look for waves propagating in the $+z$ direction.

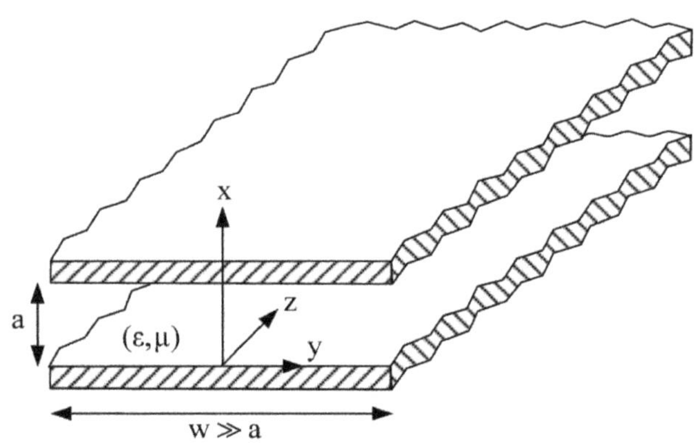

Figure 11-3. A parallel-plate waveguide.

First of all, because the plates are infinite in extent in the y direction and edge effects are neglected, we assume that all fields do not vary in the y direction. They are independent of y and all derivatives with respect to y are zero, i.e., $\frac{\partial}{\partial y} = 0$.

Then Eqs. (11-10) reduce to

$$E_x = -\frac{jk_z}{k_x^2}\frac{\partial E_z}{\partial x}, \quad E_y = -\frac{j\omega\mu}{k_x^2}\frac{\partial H_z}{\partial x}$$
$$H_x = -\frac{jk_z}{k_x^2}\frac{\partial H_z}{\partial x}, \quad H_y = -\frac{j\omega\varepsilon}{k_x^2}\frac{\partial E_z}{\partial x} \qquad (11\text{-}14)$$

where

$$k_x^2 = k^2 - k_z^2 \qquad (11\text{-}15)$$

Eqs. (11-11) reduce to

$$\frac{\partial^2 E_z}{\partial x^2} + k_x^2 E_z = 0, \quad \frac{\partial^2 H_z}{\partial x^2} + k_x^2 H_z = 0 \tag{11-16}$$

Eq. (11-16) has two independent solutions, sin $k_x x$ and cos $k_x x$. We now consider two types of solutions separately.

11.3.1 TM Mode Solutions

For TM modes, $H_z = 0$ and we seek the solutions for $E_z(x,z)$ which satisfy Eq. (11-16). The solutions for E_z can be written as

$$E_z(x,z) = \{A \sin k_x x + B \cos k_x x\} \quad e^{-jk_z z} \tag{11-17}$$

where A and B are arbitrary constants. We now apply the boundary conditions (BC's) on the surface of the perfect conductor:

$$\mathbf{E}_{\tan} = 0 \text{ or } E_y = E_z = 0 \text{ at } x = 0 \text{ and } x = a$$

Since we have formulated E_z, we can apply the BC's on E_z.

BC (i) $E_z = 0$ at $x = 0$
BC (ii) $E_z = 0$ at $x = a$

BC (i) leads to $B = 0$, thus

$$E_z = A \sin(k_x x) \quad e^{-jk_z z} \tag{11-18}$$

BC (ii) gives the following condition and determines the allowed values of k_x.

$$\sin(k_x a) = 0 \quad \rightarrow k_x a = m\pi, \quad m = \text{integar}$$

$$\boxed{k_x = \frac{m\pi}{a}}, \quad m = 1, 2, 3, \ldots \tag{11-19}$$

The values obtained for k_x are known as the *eigenvalues* or the *characteristic values*. Making use of Eq. (11-15) and Eq. (11-2), we obtain the following dispersion relation:

$$k_x^2 + k_z^2 = k^2 \quad \text{or} \quad \left(\frac{m\pi}{a}\right)^2 + k_z^2 = \omega^2 \mu \varepsilon \tag{11-20}$$

Therefore, the final solutions for E_z for TM modes are

$$\boxed{E_z = E_0 \sin\left(\frac{m\pi}{a}x\right)e^{-jk_z z}, \quad TM_m \text{ mode}} \tag{11-21}$$

where

$$k_z = \sqrt{\omega^2\mu\varepsilon - \left(\frac{m\pi}{a}\right)^2}, \quad m = 1, 2, 3 \ldots \tag{11-22}$$

We have let $A = E_o$ in Eq. (11-18). Eq. (11-22) is obtained from Eq. (11-20). It is to be noted that the field distribution and the propagation constant k_z depend on the value of m; thus given m we call it the *TM_m mode.*

The transverse components of the fields are obtained from Eq. (11-14):

$$\begin{aligned} E_x &= -j\frac{k_z}{m\pi/a} E_0 \cos\left(\frac{m\pi}{a}x\right) e^{-jk_z z} \\ H_y &= -j\frac{\omega\varepsilon}{m\pi/a} E_0 \cos\left(\frac{m\pi}{a}x\right) e^{-jk_z z} \\ E_y &= H_x = 0 \end{aligned} \tag{11-23}$$

Propagating Mode vs. Cutoff

First of all, as seen in Eq. (11-21), we note that in order for the wave to be guided in the z direction without attenuation the propagation constant k_z must be real. From Eq. (11-22), we observe that

(i) When $\omega\sqrt{\mu\varepsilon} > \frac{m\pi}{a}$, k_z becomes real.

$$k_z \equiv \beta = \sqrt{\omega^2\mu\varepsilon - \left(\frac{m\pi}{a}\right)^2} \tag{11-24}$$

then the wave propagates.

(ii) When $\omega\sqrt{\mu\varepsilon} < \frac{m\pi}{a}$, k_z becomes purely imaginary because $k_z^2 < 0$.

$$k_z = -j\alpha = -j\sqrt{\left(\frac{m\pi}{a}\right)^2 - \omega^2\mu\varepsilon} \rightarrow e^{-jk_z z} = e^{-\alpha z} \tag{11-25}$$

then the wave attenuates rapidly with z and is called an *evanescent* wave. We say that the wave is in the *cutoff* region. The frequency at which cutoff occurs, i.e., where the propagation constant k_z changes from β to $-j\alpha$, is called the **cutoff frequency** and it is obtained by letting $k_z = 0$ in Eq. (11-24).

$$(2\pi f_c)\sqrt{\mu\varepsilon} = \frac{m\pi}{a} \rightarrow \boxed{f_c = \frac{1}{\sqrt{\mu\varepsilon}}\frac{m}{2a} = \frac{v}{2a}m} \tag{11-26}$$

The cutoff frequency depends on the mode number m. Eq. (11-26) gives the cutoff frequency for the TM_m mode. Then the propagating mode vs. cutoff are determined by whether the operating frequency f is greater than the cutoff frequency f_c.

(i) If $f > f_{c,m} = \frac{v}{2a}m$, then the TM_m mode propagates.

(ii) If $f < f_{c,m} = \frac{v}{2a}m$, then the TM_m mode is cutoff.

Dominant Mode

The mode that has the lowest cutoff frequency is called the **dominant mode**. From Eq. (11-26), when $m = 0$, $f_c = 0$, thus the lowest possible value for the cutoff frequency. Therefore, the $\mathbf{TM_0}$ **mode** is the dominant mode among the TM modes of the parallel-plate waveguide. When $m = 0$, it is seen from Eqs. (11-21) and (11-23) that

$$E_z = 0, \quad \boxed{E_x = E_0' e^{-jkz}, \; H_y = \frac{E_0'}{\eta} e^{-jkz} \text{ for } TM_0 \text{ mode}} \tag{11-27}$$

where we have redefined $E_0' = -j\frac{k_z}{m\pi/a}E_0$ and used $k_z = k$ and $\frac{k_z}{\omega\varepsilon} = \eta$ when $m = 0$.

In this case, $E_z = 0$ as well as $H_z = 0$. Both electric and magnetic fields are transverse to the direction of guidance (z). Thus the TM_0 mode is a **TEM** (transverse electromagnetic) mode. The mode that has the next higher cutoff frequency is the TM_1 mode for which

$$f_{c,1} = \frac{v}{2a} \tag{11-28}$$

and the corresponding cutoff wavelength is

$$\lambda_{c,1} = \frac{v}{f_{c,1}} = 2a \tag{11-29}$$

If we choose the operating frequency such that $f_{c,o} < f < f_{c,1}$ or $0 < f < \frac{v}{2a}$ or $\lambda > 2a$, only the dominant TEM mode will propagate and all the higher-order modes will be cutoff, because $f < f_{c,m} = \frac{v}{2a}m \quad (m = 1, 2, \ldots.)$. Such choice is known as the *single mode operation*. In most practical applications, the waveguide operates this way because the presence of multiple modes creates a dispersion problem where each mode propagates with different velocity, (see Eq. (11-32)), leading to the distortion of the signal.

Guide Wavelength, Phase Velocity and Group Velocity

When the wave is in the propagating mode, the phase constant is given by

$$k_z \equiv \beta = \omega\sqrt{\mu\varepsilon}\,\sqrt{1-\left(\frac{v}{f}\frac{m}{2a}\right)^2} = k\sqrt{1-\left(\frac{f_c}{f}\right)^2} \tag{11-30}$$

The corresponding wavelength in the waveguide, called the **guide wavelength**, is given by

$$\lambda_g \equiv \frac{2\pi}{k_z} = \frac{2\pi}{k\sqrt{1-(f_c/f)^2}} = \frac{\lambda}{\sqrt{1-\left(\frac{f_c}{f}\right)^2}}, \quad f > f_c \tag{11-31}$$

where λ is the wavelength of the wave in an unbounded medium. We note that $\lambda_g > \lambda$.
The phase velocity of the wave in the waveguide is given by

$$v_p \equiv \frac{\omega}{k_z} = \frac{\omega}{k\sqrt{1-\left(\frac{f_c}{f}\right)^2}} = \frac{v}{\sqrt{1-\left(\frac{f_c}{f}\right)^2}}, \quad f > f_c \tag{11-32}$$

where v is the phase velocity of the wave in an unbounded medium. We note that $v_p > v$ and v_p depends on the frequency and the mode number. Thus *the waveguide is a dispersive transmission system*, similar to the dispersion in a lossy medium as discussed in Section 9.6. The group velocity will be different from the phase velocity and is given by

$$v_g = \frac{\partial\omega}{\partial k_z} = \frac{1}{\left(\frac{\partial k_z}{\partial\omega}\right)} = v\sqrt{1-\left(\frac{f_c}{f}\right)^2} \tag{11-33}$$

We note that $v_g < v$, i.e., the group velocity cannot exceed v because the velocity of the energy transport in a lossless waveguide is equal to the group velocity*. The derivation of Eq. (11-33) is assigned as Problem 11-3.

Wave Impedance

The wave impedance defined as the ratio of the electric and magnetic fields in their transverse components is given by

$$Z_{TM} \equiv \frac{E_x}{H_y} = \frac{k_z}{\omega\varepsilon} = \eta\sqrt{1-\left(\frac{f_c}{f}\right)^2}, \quad f > f_c \tag{11-34}$$

*D. K. Cheng, *Field and Wave Electromagnetics*, pp. 541–543, Addison-Wesley Publishing Co., 1989, 2nd ed.

It is to be noted that while the wave impedance is real in the propagating mode (when $f > f_c$), it becomes purely imaginary or reactive when the wave is cutoff ($f < f_c$) because $k_z = -j\alpha$.

Time-Average Power Flow

The time-average Poynting vector for the TM_m mode is given by

$$\begin{aligned}\mathbf{S}_{av} &= \frac{1}{2}\mathrm{Re}\{\mathbf{E}\times\mathbf{H}^*\} = \frac{1}{2}\mathrm{Re}\{(\mathbf{a}_x E_x + \mathbf{a}_z E_z)\times\mathbf{a}_y H_y^*\}\\ &= \frac{1}{2}\mathrm{Re}\{-\mathbf{a}_x E_z H_y^* + \mathbf{a}_z E_x H_y^*\}\\ &= \mathbf{a}_z \frac{1}{2} E_x H_y^* = \mathbf{a}_z \frac{1}{2} k_z\ \omega\varepsilon \left(\frac{a}{m\pi}\right)^2 |E_0|^2 \cos^2\left(\frac{m\pi}{a}x\right), \quad f > f_c\end{aligned} \tag{11-35}$$

when the wave is in the propagating mode. $E_z H_y^*$ is purely imaginary. When the wave is cutoff, k_z becomes purely imaginary and $\mathbf{S}_{av} = 0$, i.e., the wave will not propagate as expected. The time-average power flowing through the waveguide can be easily calculated by integrating $\mathbf{S}_{av}$ over the cross section of the waveguide.

Finally we interpret the propagation of the wave within the waveguide as represented by the field solutions. The solutions given in Eqs. (11-21) and (11-23) represent a standing wave in the x direction and a traveling wave in the z direction. The standing wave can be written as a superposition of two oppositely traveling waves:

$$\sin\left(\frac{m\pi}{a}x\right) = \frac{1}{2j}\left(e^{j\frac{m\pi}{a}x} - e^{-j\frac{m\pi}{a}x}\right)$$

Thus

$$E_z = \frac{(-E_0)}{2j}\left(e^{-j\frac{m\pi}{a}x}\, e^{-jk_z z} - e^{j\frac{m\pi}{a}x}\, e^{-jk_z z}\right) \tag{11-36}$$

The first term represents a plane wave traveling in the $+x$ and $+z$ directions and the second term represents a wave traveling in the $-x$ and $+z$ directions. These two waves can be illustrated in Figure 11-4 where the upward traveling wave corresponds to the first term and the downward traveling wave corresponds to the second term, which is due to the perfect reflection from the upper conducting boundary. The ratio of the amplitudes of the two waves should be a reflection coefficient for an electric field at the perfectly conducting surface, $R = -1$. Eq. (11-36) is consistent with this interpretation. When $m \neq 0$, the wave doesn't travel straight down within the waveguide but bounces between the two conducting walls as it is guided.

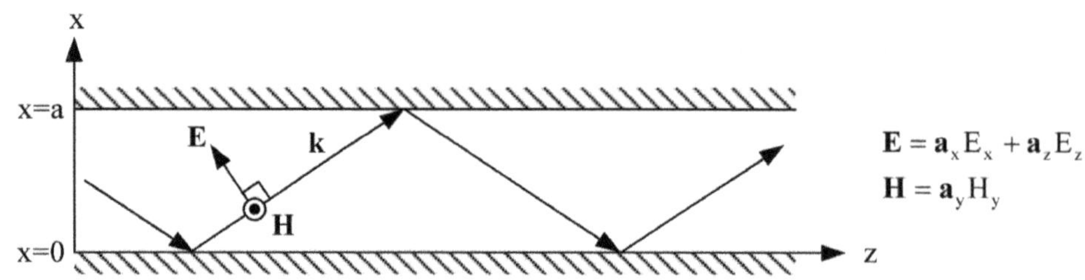

Figure 11-4. Guidance of the TM wave in a parallel-plate waveguide.

11.3.2 TE Mode Solutions

For the second type of solutions, the TE modes, $E_z = 0$ and we seek the solutions for $H_z(x, z)$ which satisfy Eq. (11-16). The solutions for H_z can be similarly written as

$$H_z(x,z) = \{C \sin k_x x + D \cos k_x x\}\ e^{-jk_z z} \qquad (11\text{-}37)$$

where C and D are arbitrary constants. The boundary conditions (BC's) on the conducting plates are

$$E_y = \frac{j\omega\mu}{k_x^2}\frac{\partial H_z}{\partial x} = 0 \qquad \text{at} \qquad x = 0 \ \text{ and } \ x = a$$

The first BC, $E_y \propto \dfrac{\partial H_z}{\partial x} = 0$ at $x = 0$, leads to $C = 0$, thus

$$H_z = D\cos(k_x x)\ e^{-jk_z z} \qquad (11\text{-}38)$$

The second BC, $E_y \propto \dfrac{\partial H_z}{\partial x} = 0$ at $x = a$, determines the allowed values of k_x.

$$\sin(k_x a) = 0 \rightarrow k_x = \frac{m\pi}{a}, \ m = 1,2,3,... \qquad (11\text{-}19)$$

which is the same result as that of the TM modes, Eq. (11-19). k_x and k_z will also satisfy the same dispersion relation, Eq. (11-20).

The final solutions for H_z for the TE modes are

$$\boxed{H_z = H_0 \cos\left(\frac{m\pi}{a}x\right)\ e^{-jk_z z}, \quad TE_m \text{ mode}} \qquad (11\text{-}39)$$

where

$$k_z = \sqrt{\omega^2\mu\varepsilon - \left(\frac{m\pi}{a}\right)^2}, \quad m = 1,2,3,... \qquad (11\text{-}22)$$

We have let $D = H_o$ in Eq. (11-38). We again notice that the field distribution and the propagation constant k_z depend on the mode number m. The transverse components of the fields are obtained from Eq. (11-14) as follows.

$$\begin{aligned} E_y &= -j\frac{\omega\mu}{m\pi/a} H_0 \sin\left(\frac{m\pi}{a}x\right) e^{-jk_z z} \\ H_x &= j\frac{k_z}{m\pi/a} H_0 \sin\left(\frac{m\pi}{a}x\right) e^{-jk_z z} \\ E_x &= H_y = 0 \end{aligned} \qquad (11\text{-}40)$$

The TM and TE solutions for the parallel-plate waveguide are summarized in Table 11-1.

Table 11-1 Field Solutions for Parallel-Plate Waveguide

TM_m mode	TE_m mode
$\mathbf{E} = \mathbf{a}_x E_x + \mathbf{a}_z E_z$	$\mathbf{E} = \mathbf{a}_y E_y$
$\mathbf{H} = \mathbf{a}_y H_y$	$\mathbf{H} = \mathbf{a}_x H_x + \mathbf{a}_z H_z$
$E_z = E_o \sin\left(\frac{m\pi}{a}x\right) e^{-jk_z z}$	$H_z = H_o \cos\left(\frac{m\pi}{a}x\right) e^{-jk_z z}$
$E_x = -\frac{jk_z}{m\pi/a} E_o \cos\left(\frac{m\pi}{a}x\right) e^{-jk_z z}$	$H_x = \frac{jk_z}{m\pi/a} H_o \sin\left(\frac{m\pi}{a}x\right) e^{-jk_z z}$
$H_y = -\frac{j\omega\varepsilon}{m\pi/a} E_o \cos\left(\frac{m\pi}{a}x\right) e^{-jk_z z}$	$E_y = -\frac{j\omega\mu}{m\pi/a} H_o \sin\left(\frac{m\pi}{a}x\right) e^{-jk_z z}$
Phase Constant	$k_z = \sqrt{\omega^2\mu\varepsilon - \left(\frac{m\pi}{a}\right)^2} = k\sqrt{1-\left(\frac{f_c}{f}\right)^2}$
Cutoff Frequency	$f_{c,m} = \frac{v}{2a} m = \frac{1}{\sqrt{\mu\varepsilon}} \frac{m}{2a}$
Guide Wavelength	$\lambda_g = \frac{2\pi}{k_z} = \frac{\lambda}{\sqrt{1-(f_c/f)^2}}$
Phase Velocity	$v_p = \frac{\omega}{k_z} = \frac{v}{\sqrt{1-(f_c/f)^2}}$
Wave Impedance	$Z_{TM} = \eta\sqrt{1-\left(\frac{f_c}{f}\right)^2}, \quad Z_{TE} = \frac{\eta}{\sqrt{1-(f_c/f)^2}}$
Dominant Mode: TM_o or TEM mode $\mathbf{E} = \mathbf{a}_x E_o e^{-jkz}, \quad \mathbf{H} = \mathbf{a}_y \frac{E_o}{\eta} e^{-jkz}$ $k = \omega\sqrt{\mu\varepsilon}, \quad f_{c,0} = 0$ (no cutoff frequency)	

All the discussions about the cutoff frequency, guide wavelength, phase velocity, group velocity, wave impedance and the standing wave interpretation, presented in the previous section, are valid also for the TE modes except for the following differences.

(i) The dominant TE mode which has the lowest cutoff frequency among the TE modes is not the $m = 0$ mode but the TE_1 mode for which $f_{c,1} = v/2a$.
When $m = 0$, $E_y = H_x = 0$; such a mode does not exist. Overall, the TM_0 or TEM mode is the dominant mode for parallel plate waveguide. The field lines for the TE_1 mode and the TM_1 mode are shown in Figure 11-5.

(ii) The wave impedance for the TE mode is given by

$$Z_{TE} = -\frac{E_y}{H_x} = \frac{\omega\mu}{k_z} = \frac{\eta}{\sqrt{1-\left(\frac{f_c}{f}\right)^2}}, \qquad f > f_c \tag{11-41}$$

We note that $Z_{TM} < \eta$ and $Z_{TE} > \eta$ for the propagating modes and both Z_{TM} and Z_{TE} approach η (the intrinsic impedance of the medium) as $f \to \infty$.

(iii) The time-average Poynting vector for the TE_m mode is given by

$$\begin{aligned}\mathbf{S}_{av} &= \frac{1}{2}\,\mathrm{Re}\{\mathbf{E}\times\mathbf{H}^*\} = \frac{1}{2}\,\mathrm{Re}\left\{\mathbf{a}_y E_y \times \left(\mathbf{a}_x H_x + \mathbf{a}_z H_z\right)^*\right\} \\ &= \frac{1}{2}\,\mathrm{Re}\{\mathbf{a}_x\, E_y H_z^* - \mathbf{a}_z E_y\, H_x^*\} \\ &= \mathbf{a}_z\left\{-\frac{1}{2}\,E_y\,H_x^*\right\} = \mathbf{a}_z\,\frac{1}{2}\,k_z\;\omega\mu\left(\frac{a}{m\,\pi}\right)^2 |H_0|^2\,\sin^2\!\left(\frac{m\pi}{a}x\right), \quad f > f_c\end{aligned}$$

Note that when the wave is in the propagating mode, $E_y H_z^*$ is purely imaginary.

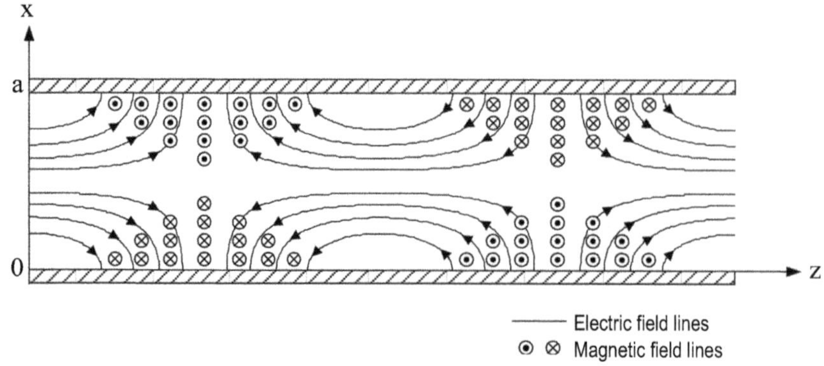

(a) TM_1 mode

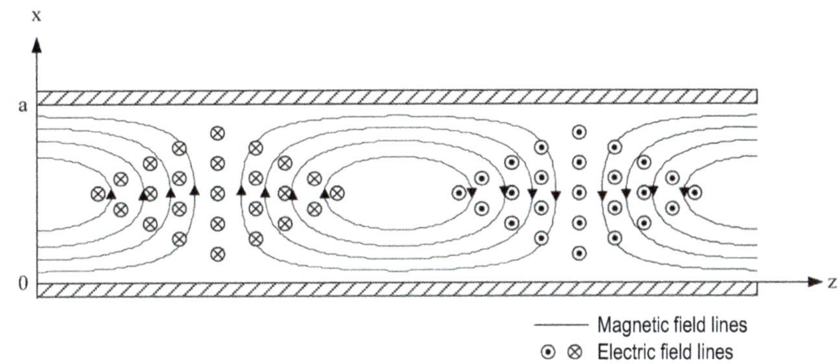

(b) TE_1 mode

Figure 11-5. Field lines for the TM_1 and TE_1 modes in a parallel-plate waveguide.

EXAMPLE 11-1. MICROSTRIP LINE

Microstrip line is one of the most commonly used planar transmission line structures in microwave integrated circuits because it can be easily fabricated by printed circuit techniques. It consists of a metallic strip of width w and a large conducting ground plane separated by a thin layer of dielectric substrate of thickness d as shown in Figure 11-6.

The complete analysis of the electromagnetic fields that propagate in microstrip line is very complicated. The fields exist inside the dielectric layer as well as in the upper air region. However, if the width of the strip is much larger than the thickness of the layer, then the microstrip line can be modeled as a parallel-plate waveguide and the fields of the dominant mode can be approximated by the TEM mode (or the TM_0 mode) discussed earlier.

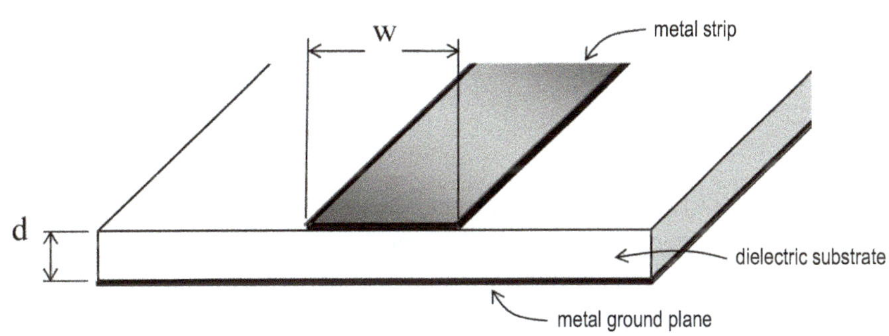

Figure 11-6: Microstrip transmission line

Suppose the microstrip line has $w = 0.7$ cm and $d = 1.4\ \mathrm{mm}$, and the substrate has $\varepsilon = 1.96\,\varepsilon_o$, $\mu = \mu_o$, $\sigma = 0$.

(a) What is the range of frequencies for which only the dominant mode propagates?

(b) Calculate the time-average power for the dominant mode that is transmitted by the line when the amplitude of the electric field is $7\ \mathrm{kV/m}$.

SOLUTIONS

(a) If we model the microstrip line as parallel-plate waveguide, the dominant mode is the TEM mode which has no cutoff frequency (or zero cutoff frequency $f_{c,0} = 0$) and the next higher mode is either TM_1 or TE_1 mode whose cutoff frequency is given by

$$f_{c,1} = \frac{v}{2d} = \frac{1}{\sqrt{\mu\varepsilon}}\frac{1}{2d} = \frac{1}{\sqrt{\mu_o 1.96\varepsilon_o}}\frac{1}{2\times1.4\times10^{-3}} = \frac{3\times10^8}{1.4\times2.8\times10^{-3}} = 7.65\times10^{10}\,\mathrm{Hz}$$

Thus the range of frequencies for which only the dominant mode propagates is

$$0 < f < 76.5\ \mathrm{GHz}$$

(b) For the TEM mode, the fields are given by

$$\mathbf{E} = \mathbf{a}_x E_o\ e^{-jkz},\quad \mathbf{H} = \mathbf{a}_y \frac{E_o}{\eta}\ e^{-jkz}$$

The time-average Poynting vector is given by

$$\mathbf{S}_{av} = \frac{1}{2}\,\mathrm{Re}(\mathbf{E}\times\mathbf{H}^*) = \mathbf{a}_z\frac{1}{2}\,\mathrm{Re}\left\{E_o\left(\frac{E_o}{\eta}\right)^*\right\} = \mathbf{a}_z\frac{|E_o|^2}{2\eta}$$

The time-average power for the TEM mode transmitted by the line is approximately obtained by integrating over the dielectric cross section of width w and thickness d:

$$P_{av} = \iint \mathbf{S}_{av}\cdot d\mathbf{s} = \int_0^w\!\!\int_0^d \frac{|E_o|^2}{2\eta}\,dx\,dy = \frac{|E_o|^2\,wd}{2\sqrt{\dfrac{\mu_o}{1.96\varepsilon_o}}} = \frac{(7\times10^3)^2\cdot7\times10^{-3}\cdot1.4\times10^{-3}}{2\times\dfrac{377}{1.4}} = 0.89\,[\mathrm{W}]$$

Note that we have ignored the field that exists in the dielectric beyond the width w.

EXAMPLE 11-2. DIELECTRIC BREAKDOWN

A dielectric substrate in a waveguide breaks down when a strong electric field is applied. The maximum electric field magnitude that a dielectric material can withstand without breakdown is called the **dielectric strength** of the material. Suppose you design a parallel-plate waveguide that has $w = 20\ d$, filled with the dielectric (polystyrene) whose dielectric constant is 2.6 and dielectric strength is $20 \times 10^6\ V/m$. What is the smallest thickness of the dielectric that can propagate $1\ kW$ of power?

SOLUTIONS

Assuming that the dominant (TEM) mode propagates in the waveguide, the time-average power transmitted (see Example 11-1) is given by

$$P_{av} = \frac{|E_o|^2}{2\eta} wd = \frac{|E_o|^2}{2\sqrt{\frac{\mu_o}{2.6\varepsilon_o}}} 20d^2 = \frac{|E_o|^2}{2\times 233.8} 20d^2 = 10^3\ [W]$$

Since the dielectric strength is $20\times10^6\ V/m$, $|E_o| \le 2\times10^7\ [V/m]$.

The thickness of the dielectric should be

$$d = \sqrt{\frac{10^3\times2\times233.8}{20|E_o|^2}} \ge \sqrt{\frac{10^2\times233.8}{\left(2\times10^7\right)^2}} = 7.65\times10^{-6}\ [m].$$

Attenuation in the Waveguide

In this analysis, we have assumed that the conducting plates are perfect conductors ($\sigma = \infty$). In practice, the metals that are used in the waveguide are *not perfect* and they have high, but *finite*, conductivities. Thus the wave can penetrate into the conducting walls as it propagates along the waveguide. Some of the power leaks into the conducting surfaces and is dissipated as heat. Then the signal in the waveguide is attenuated as it propagates. The attenuation constant (α) can be obtained by calculating the power flowing along the waveguide and the power dissipated on the conducting walls (see Problem 11-16).

Additional Notes on Solution Method

In our approach we have formulated the longitudinal components first, namely, E_z for TM modes and H_z for TE modes and then we obtained the transverse components, E_x, H_y for the TM modes and E_y, H_x for TE modes. There is another approach taken by others [Shen and Kong (1995)*] for the parallel-plate waveguide. They formulate the transverse components first, namely, H_y for TM modes and E_y for TE modes and then calculate

* L. C. Shen and J. A. Kong, *Applied Electromagnetism*, PWS Publishing Company, 1995, 3rd Ed.

the remaining components of the fields using Maxwell's curl equations. This approach works for the parallel-plate waveguide but doesn't work for the rectangular waveguide because the fields depend on both x and y. For the parallel-plate waveguide for which $\frac{\partial}{\partial y}=0$, the fields for the TE and TM modes are obtained as follows.

TE (Transverse electric) modes:

$$\mathbf{E}=\mathbf{a}_y E_y, \quad \mathbf{H}=\mathbf{a}_x H_x+\mathbf{a}_z H_z$$

$$H_x=\frac{1}{j\omega\mu}\frac{\partial E_y}{\partial z}, \quad H_z=-\frac{1}{j\omega\mu}\frac{\partial E_y}{\partial x} \tag{11-42}$$

TM (Transverse magnetic) modes:

$$\mathbf{H}=\mathbf{a}_y H_y, \quad \mathbf{E}=\mathbf{a}_x E_x+\mathbf{a}_z E_z$$

$$E_x=-\frac{1}{j\omega\varepsilon}\frac{\partial H_y}{\partial z}, \quad E_z=\frac{1}{j\omega\varepsilon}\frac{\partial H_y}{\partial x} \tag{11-43}$$

11.4. RECTANGULAR WAVEGUIDE

In practice, we cannot construct an "ideal" parallel-plate waveguide that is infinite in extent in one (y) direction. Since the width w in the y direction is finite, we will have some fringing fields at both ends. This leads to some loss of energy and imperfect guidance. Such a problem can be avoided by placing another two parallel conducting walls at both ends in the y direction. Then it becomes a rectangular waveguide. Now the waves inside the waveguide bounce off the side walls as well as the top and bottom conductors as they are guided within the closed structure. Since all four conducting walls are connected, they form a *one-conductor* transmission system. The geometry of a rectangular waveguide is shown in Figure 11-7. We again assume that the metallic walls are perfect conductors ($\sigma=\infty$) and the medium inside the waveguide is lossless ($\sigma=0$) and has the real material constants μ, ε. The x and y dimensions of the cross section are a and b, respectively.

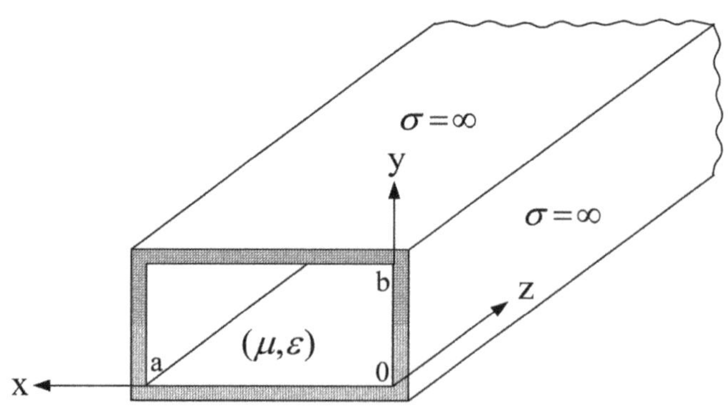

Figure 11-7. A rectangular waveguide.

11.4.1 General Field Solutions

Since the cross section of the waveguiding structure is uniform (it is a rectangle), the method of analysis described in Section 11.2 is applicable. We will first find the general form of field solutions for the longitudinal components E_z, H_z by solving Eq. (11-11) with the method of separation of variables. The unknown coefficients that appear in Eq. (11-12) are determined separately for the TM and TE modes, by applying the boundary conditions. Finally, the transverse components are obtained by Eq. (11-10).

First we rewrite the wave equation for uniform waveguides that E_z and H_z should satisfy.

$$\frac{\partial^2 E_z}{\partial x^2}+\frac{\partial^2 E_z}{\partial y^2}+(k^2-k_z^2)\, E_z=0 \tag{11-11a}$$

In order to separate the two variables, we assume the solution of Eq. (11-11a) for $E_z(x,y,z)$ in the following form:

$$E_z(x,y,z)=X(x)\, Y(y)\, e^{-jk_z Z} \tag{11-44}$$

where $X(x)$ and $Y(y)$ are functions of only x and y, respectively. Substitution of Eq. (11-44) into Eq. (11-11a) yields

$$Y(y)\frac{\partial^2 X(x)}{\partial x^2}+X(x)\frac{\partial^2 Y(y)}{\partial y^2}+(k^2-k_z^2)\, X(x)\,Y(y)=0 \tag{11-45}$$

Dividing Eq. (11-45) by $X(x)\, Y(y)$, we obtain

$$\frac{1}{X(x)}\frac{d^2 X(x)}{dx^2}+\frac{1}{Y(y)}\frac{d^2 Y(y)}{dy^2}+(k^2-k_z^2)=0 \tag{11-46}$$

Note that the partial derivatives are replaced by the ordinary derivatives because $X(x)$ and $Y(y)$ are functions of one variable. In Eq. (11-46), we observe that the first term is a function of x only, the second term is a function of y only, the third term is a constant, and the sum has to be zero. In order for Eq. (11-46) to be satisfied for *all values of x and y*, each term must be a constant which is independent of x and y.

Thus we let

$$\frac{1}{X(x)}\frac{d^2 X(x)}{dx^2}=C_1\equiv -k_x^2 \rightarrow \frac{d^2 X(x)}{dx^2}+k_x^2\, X(x)=0 \tag{11-47a}$$

$$\frac{1}{Y(y)}\frac{d^2 Y(y)}{dy^2}=C_2\equiv -k_y^2 \rightarrow \frac{d^2 Y(y)}{dy^2}+k_y^2\, Y(y)=0 \tag{11-47b}$$

A choice of separation constants in a particular form allows the solutions of Eq. (11-47) to be written in a physically meaningful way. As discussed in Section 4.4, the general solutions of Eqs. (11-47a) and (11-47b) are, when $k_x \neq 0$, $k_y \neq 0$, respectively,

$$X(x)=A\sin(k_x x)+B\cos(k_x x) \tag{11-48a}$$

$$Y(y)=C\sin(k_y y)+D\cos(k_y y) \tag{11-48b}$$

Therefore, the general solution for $E_z(x, y, z)$ or $H_z(x, y, z)$ can be written as

$$F(x, y, z) = \{A \sin k_x x + B \cos k_x x\} \{C \sin k_y y + D \cos k_y y\} \quad e^{-jk_z z}$$
$$F(x, y, z) = E_z(x, y, z) \text{ or } H_z(x, y, z) \tag{11-49}$$

We recognize that the fields in Eq. (11-49) represent a standing wave in both the x and y directions and a traveling wave in the z direction, as expected. The values of k_x and k_y are not arbitrary but related to the propagation constant k_z. The relationship, known as the *dispersion relation*, is obtained by substituting Eq. (11-47) into Eq. (11-46).

$$\boxed{k_x^2 + k_y^2 + k_z^2 = k^2 = \omega^2 \mu \varepsilon} \tag{11-50}$$

As was done in Section 11.3, we will seek field solutions for the TM and TE modes, separately.

11.4.2 TM Mode Solutions

For TM modes, $H_z = 0$ and we seek the solutions for $E_z(x,y,z)$ in the form of Eq. (11-49). The boundary conditions are

$$\mathbf{E}_{tan} = 0 \quad \text{on four conducting walls}$$

Thus we have

$$\text{(i) } E_z = 0, E_y = 0 \text{ at } x = 0 \tag{11-51a}$$
$$\text{(ii) } E_z = 0, E_y = 0 \text{ at } x = a \tag{11-51b}$$
$$\text{(iii) } E_z = 0, E_x = 0 \text{ at } y = 0 \tag{11-51c}$$
$$\text{(iv) } E_z = 0, E_x = 0 \text{ at } y = b \tag{11-51d}$$

Since we have formulated E_z, we can use the boundary condition (BC), $E_z = 0$ directly.
We follow the similar procedures which were used in the parallel-plate waveguide.

$$\text{BC (i): } E_z = 0 \rightarrow X(x) = 0 \text{ at } x = 0 \rightarrow X(x) = A \sin(k_x x)$$
$$\text{BC (iii): } E_z = 0 \rightarrow Y(y) = 0 \text{ at } y = 0 \rightarrow Y(y) = C \sin(k_y y)$$

Then $E_z(x,y,z)$ can be written as

$$E_z(x, y, z) = E_0 \sin(k_x x) \sin(k_y y) \; e^{-jk_z z} \tag{11-52}$$

where $E_o = AC$. The separation constants k_x, k_y are determined by applying the BC's, (ii) and (iv).

BC (ii): $E_z = 0 \rightarrow X(x) = 0$ at $x = a \rightarrow \sin k_x a = 0$

$$k_x a = m\pi \quad \text{or} \quad \boxed{k_x = \frac{m\pi}{a}}, \quad m = 1, 2, 3, \ldots \tag{11-53a}$$

BC (iv): $E_z = 0 \rightarrow Y(y) = 0$ at $y = b \rightarrow \sin k_y b = 0$

$$k_y b = n\pi \quad \text{or} \quad \boxed{k_y = \frac{n\pi}{b}}, \quad n = 1, 2, 3, \ldots \tag{11-53b}$$

Note that when m or $n = 0$, the field vanishes. Each choice of the integers m and n yields the TM_{mn} mode. The final solution for E_z for the TM modes is then given by

$$\boxed{E_z(x,y,z) = E_0 \sin\left(\frac{m\pi}{a}x\right) \sin\left(\frac{n\pi}{b}y\right) e^{-jk_z z}, \quad TM_{mn} \text{ mode}} \tag{11-54}$$

The propagation constant (or phase constant) k_z is obtained from Eq. (11-50).

$$\boxed{k_z = \sqrt{k^2 - (k_x^2 + k_y^2)} = \sqrt{\omega^2 \mu \varepsilon - \left[\left(\frac{m\pi}{a}\right)^2 + \left(\frac{n\pi}{b}\right)^2\right]}} \tag{11-55}$$

The transverse components of the fields can be calculated by using Eq. (11-10) with $H_z = 0$. They are listed in Table 11-2. Unlike the parallel-plate waveguide, we have all four components of the fields, E_x, E_y, H_x, H_y.

Propagating Mode vs. Cutoff

As explained in Section 11.3, in order for the wave to be guided in the z direction without attenuation the propagation constant k_z must be real. From Eq. (11-55), we observe that

(i) When $\omega\sqrt{\mu\varepsilon} > \sqrt{\left(\frac{m\pi}{a}\right)^2 + \left(\frac{n\pi}{b}\right)^2}$, k_z becomes real.

$$k_z \equiv \beta = \sqrt{\omega^2 \mu \varepsilon - \left[\left(\frac{m\pi}{a}\right)^2 + \left(\frac{n\pi}{b}\right)^2\right]}, \tag{11-56}$$

then the wave propagates.

Table 11-2 Field Solutions for Rectangular Waveguide

	TM_{mn} mode	TE_{mn} mode
E_z	$E_o \sin\left(\frac{m\pi}{a}x\right)\sin\left(\frac{n\pi}{b}y\right)e^{-jk_z z}$	0
H_z	0	$H_o \cos\left(\frac{m\pi}{a}x\right)\cos\left(\frac{n\pi}{b}y\right)e^{-jk_z z}$
E_x	$-\frac{jk_z}{k_\rho^2}\left(\frac{m\pi}{a}\right)E_o \cos\left(\frac{m\pi}{a}x\right)\sin\left(\frac{n\pi}{b}y\right)e^{-jk_z z}$	$\frac{j\omega\mu}{k_\rho^2}\left(\frac{n\pi}{b}\right)H_o \cos\left(\frac{m\pi}{a}x\right)\sin\left(\frac{n\pi}{b}y\right)e^{-jk_z z}$
E_y	$-\frac{jk_z}{k_\rho^2}\left(\frac{n\pi}{b}\right)E_o \sin\left(\frac{m\pi}{a}x\right)\cos\left(\frac{n\pi}{b}y\right)e^{-jk_z z}$	$-\frac{j\omega\mu}{k_\rho^2}\left(\frac{m\pi}{a}\right)H_o \sin\left(\frac{m\pi}{a}x\right)\cos\left(\frac{n\pi}{b}y\right)e^{-jk_z z}$
H_x	$\frac{j\omega\varepsilon}{k_\rho^2}\left(\frac{n\pi}{b}\right)E_o \sin\left(\frac{m\pi}{a}x\right)\cos\left(\frac{n\pi}{b}y\right)e^{-jk_z z}$	$\frac{jk_z}{k_\rho^2}\left(\frac{m\pi}{a}\right)H_o \sin\left(\frac{m\pi}{a}x\right)\cos\left(\frac{n\pi}{b}y\right)e^{-jk_z z}$
H_y	$-\frac{j\omega\varepsilon}{k_\rho^2}\left(\frac{m\pi}{a}\right)E_o \cos\left(\frac{m\pi}{a}x\right)\sin\left(\frac{n\pi}{b}y\right)e^{-jk_z z}$	$\frac{jk_z}{k_\rho^2}\left(\frac{n\pi}{b}\right)H_o \cos\left(\frac{m\pi}{a}x\right)\sin\left(\frac{n\pi}{b}y\right)e^{-jk_z z}$
$k_\rho^2 = \left(\frac{m\pi}{a}\right)^2 + \left(\frac{n\pi}{b}\right)^2 = \omega^2\mu\varepsilon - k_z^2$		
Phase Constant	$k_z = \sqrt{\omega^2\mu\varepsilon - \left[\left(\frac{m\pi}{a}\right)^2 + \left(\frac{n\pi}{b}\right)^2\right]}$	
Cutoff Frequency	$f_{c,mn} = \frac{v}{2}\sqrt{\left(\frac{m}{a}\right)^2 + \left(\frac{n}{b}\right)^2}$	

Dominant Mode: TE_{10} mode $(a > b)$

$$\mathbf{E} = \mathbf{a}_y E_y, \quad \mathbf{H} = \mathbf{a}_x H_x + \mathbf{a}_z H_z$$

$$H_z = H_o \cos\left(\frac{\pi}{a}x\right)e^{-jk_z z} \text{(same as the } TE_1 \text{ mode of the parallel-plate waveguide)}$$

$$f_{c,10} = \frac{v}{2a} = \frac{1}{\sqrt{\mu\varepsilon}}\frac{1}{2a}, \qquad \lambda_{c,10} = 2a$$

(ii) When $\omega\sqrt{\mu\varepsilon} < \sqrt{\left(\frac{m\pi}{a}\right)^2 + \left(\frac{n\pi}{b}\right)^2}$, k_z becomes purely imaginary.

$$k_z = -j\alpha = -j\sqrt{\left(\frac{m\pi}{a}\right)^2 + \left(\frac{n\pi}{b}\right)^2 - \omega^2\mu\varepsilon} \rightarrow e^{-jk_z z} = e^{-\alpha z} \tag{11-57}$$

then the wave attenuates rapidly with z and becomes an *evanescent* wave. The wave is in the *cutoff* region. The frequency at which cutoff occurs, the *cutoff frequency,* is obtained by letting $k_z = 0$ in Eq. (11-56).

$$(2\pi f_c)\sqrt{\mu\varepsilon} = \sqrt{\left(\frac{m\pi}{a}\right)^2 + \left(\frac{n\pi}{b}\right)^2} \rightarrow \boxed{f_c = \frac{1}{2\sqrt{\mu\varepsilon}}\sqrt{\left(\frac{m}{a}\right)^2 + \left(\frac{n}{b}\right)^2}} \tag{11-58}$$

The cutoff frequency depends on the mode numbers m and n. Eq. (11-58) gives the cutoff frequency for the TM_{mn} mode. Then the propagating mode vs. cutoff are determined by whether the operating frequency f is greater than the cutoff frequency f_c.

(i) If $f > f_{c,mn} = \frac{v}{2}\sqrt{\left(\frac{m}{a}\right)^2 + \left(\frac{n}{b}\right)^2}$, then the TM_{mn} mode propagates.

(ii) If $f < f_{c,mn} = \frac{v}{2}\sqrt{\left(\frac{m}{a}\right)^2 + \left(\frac{n}{b}\right)^2}$, then the TM_{mn} mode is cutoff.

Dominant Mode

For TM modes, m and n cannot be zero because $m = 0$ or $n = 0$ leads to no fields. The mode that has the lowest cutoff frequency occurs when $m = 1$, $n = 1$. Thus the TM_{11} mode is the dominant mode among TM modes of the rectangular waveguide. Its cutoff frequency is given by

$$f_{c,11} = \frac{v}{2}\sqrt{\left(\frac{1}{a}\right)^2 + \left(\frac{1}{b}\right)^2} \tag{11-59}$$

The field lines of the TM_{11} mode are plotted in Figure 11-8.

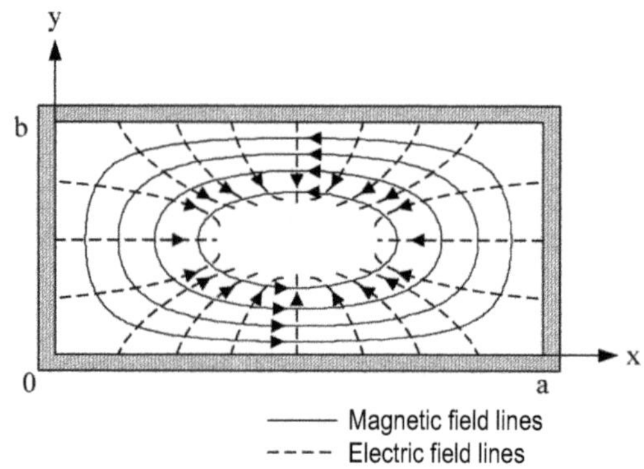

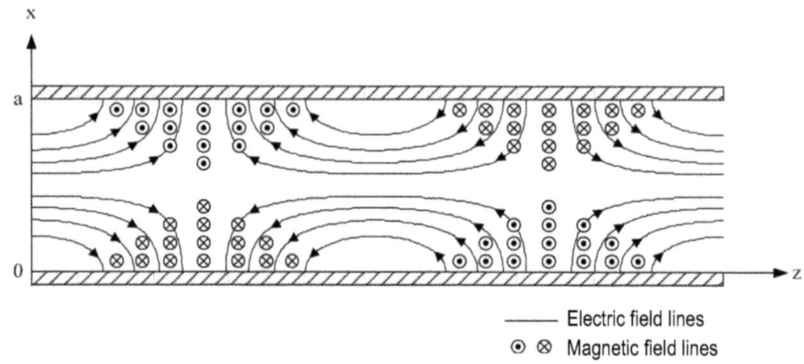

Figure 11-8. Field lines of the TM_{11} mode in a rectangular waveguide.

When the wave is in the propagating mode, the phase constant ($k_z = \beta$), the guide wavelength (λ_g), the phase velocity (v_p) and the group velocity (v_g) are given by the same expressions obtained for the parallel-plate waveguide, namely, Eqs. (11-30), (11-31), (11-32), and (11-33), respectively, except that f_c is given by Eq. (11-58). If we take the ratios of the transverse components of the electric and magnetic fields, we obtain the wave impedance as follows.

$$Z_{TM} = \frac{E_x}{H_y} = -\frac{E_y}{H_x} = \frac{k_z}{\omega\varepsilon} = \eta\sqrt{1-\left(\frac{f_c}{f}\right)^2}, \qquad f > f_c \tag{11-60}$$

Again the wave impedance is real in the propagating mode and it becomes purely imaginary or reactive when the wave is cutoff ($f < f_c$). The minus sign in the second ratio is due to the wave traveling in the $+z$ direction.

The time-average Poynting vector for the TM_{mn} mode of the rectangular waveguide consists of two terms:

$$\mathbf{S}_{av} = \mathbf{a}_z \frac{1}{2}\{E_x H_y^* - E_y H_x^*\} = \mathbf{a}_z \frac{1}{2Z_{TM}}\left\{|E_x|^2 + |E_y|^2\right\} \tag{11-61}$$

where we have used the following relationships:

$$E_x = Z_{TM} H_y\,,\; E_y = -Z_{TM} H_x \tag{11-62}$$

The time-average power flowing through the waveguide can be calculated by integrating $\mathbf{S}_{av}$ over the cross section of the waveguide ($0 < x < a$, $0 < y < b$).

We can interpret the propagation of the wave within the rectangular waveguide as represented by the field solution in Eq. (11-54). $\sin\left(\frac{m\pi}{a}x\right)$ represents a standing wave in the x direction which consists of two oppositely traveling (in $\pm x$) waves as described in Section 11.3.1. The wave bounces off two conducting walls at $x = 0$ and $x = a$ as it is guided. $\sin\left(\frac{n\pi}{b}y\right)$ similarly represents a standing wave in the y direction, which consists of two oppositely traveling (in $\pm y$) waves. The wave bounces off two conducting walls at $y = 0$ and $y = b$. Overall the wave doesn't travel straight down within the waveguide but bounces between the four conducting walls and travels zigzag as it is guided.

11.4.3 TE Mode Solutions

For transverse electric (TE) modes, we let $E_z = 0$ and seek the solutions for $H_z(x,y,z)$ in the form of Eq. (11-49).

$$H_z(x,y,z) = \{A\sin k_x x + B\cos k_x x\}\{C\sin k_y y + D\cos k_y y\}\, e^{-jk_z z} \tag{11-49}$$

The boundary conditions are given by Eqs. (11-51). From Eq. (11-10) with $E_z = 0$, we have

$$E_x = \frac{-j\omega\mu}{k^2 - k_z^2}\frac{\partial H_z}{\partial y}, \quad E_y = \frac{j\omega\mu}{k^2 - k_z^2}\frac{\partial H_z}{\partial x} \tag{11-63}$$

We now apply the four boundary conditions (BC's).

$$\text{BC (i): } E_y = 0 \rightarrow \frac{\partial H_z}{\partial x} = 0 \ \text{ at } x = 0 \rightarrow X(x) = B\cos(k_x x)$$

$$\text{BC (iii): } E_x = 0 \rightarrow \frac{\partial H_z}{\partial y} = 0 \ \text{ at } y = 0 \rightarrow Y(y) = D\cos(k_y y)$$

Then $H_z(x,y,z)$ can be written as

$$H_z(x,y,z) = H_0 \cos(k_x x)\cos(k_y y)\, e^{-jk_z z} \tag{11-64}$$

where $H_0 = BD$. Note that H_z for TE modes has the cosine dependence in x, y directions whereas E_z for TM modes has the sine dependence in x, y directions. The separation constants k_x, k_y are again determined by the remaining BC's.

$$\text{BC (ii): } E_y = 0 \rightarrow \frac{\partial H_z}{\partial x} = 0 \ \text{ at } x = a \rightarrow \boxed{k_x = \frac{m\pi}{a}}, \ m = 0, 1, 2, \ldots \tag{11-65a}$$

$$\text{BC (iv): } E_x = 0 \rightarrow \frac{\partial H_z}{\partial y} = 0 \ \text{ at } y = b \rightarrow \boxed{k_y = \frac{n\pi}{b}}, \ n = 0, 1, 2, \ldots \tag{11-65b}$$

Note that the results in Eqs. (11-65) are identical to Eqs. (11-53) for TM modes except that *m or n can be zero* for TE modes. Each choice of the integers m and n defines the TE_{mn} mode. The final solution for H_z for the TE modes is then given by

$$\boxed{H_z(x,y,z) = H_0 \cos\left(\frac{m\pi}{a}x\right)\cos\left(\frac{n\pi}{b}y\right)e^{-jk_z z}, \quad TE_{mn}\ \text{mode}} \tag{11-66}$$

The propagation constant (or phase constant) k_z is again given by Eq. (11-55). The transverse components of the fields can be calculated by using Eq. (11-10) and $E_z = 0$. They are listed in Table 11-2.

All the discussions about the cutoff frequency, guide wavelength, phase velocity, wave impedance, time-average Poynting vector and the standing wave interpretation, presented in the previous section, are valid also for the TE modes except for the following differences.

(i) The wave impedance for the TE mode is given by

$$Z_{TE} = \frac{E_x}{H_y} = -\frac{E_y}{H_x} = \frac{\omega\mu}{k_z} = \frac{\eta}{\sqrt{1-\left(\frac{f_c}{f}\right)^2}}, \qquad f > f_c \tag{11-67}$$

From Eqs. (11-60) and (11-67), we find that $Z_{TM} < \eta$ and $Z_{TE} > \eta$ for the propagating modes and both Z_{TM} and Z_{TE} approach the intrinsic impedance of the medium as $f \to \infty$. We also note that Eqs. (11-61) and (11-62) for the time-average Poynting vector are valid for TE modes when Z_{TM} is replaced by Z_{TE}.

(ii) **Dominant Modes**

One of the most important issues in any waveguide system is the question of what the dominant mode is because waveguides are often designed so as to support the lowest-order mode and suppress all the higher-order modes. Recall that the lowest-order (dominant) mode for TM modes is the TM_{11} mode because when $m = 0$ or $n = 0$ fields vanish. However, for TE modes in rectangular waveguides, *either m or n can be zero, but not both.* If $m = n = 0$, then all the transverse components of the fields vanish and H_z alone would not satisfy Maxwell's equations. Thus the lowest-order mode for TE modes can be either the TE_{10} mode ($m = 1, n = 0$) or the TE_{01} mode ($m = 0, n = 1$). Which mode gives the lowest cutoff frequency depends on the sizes a, b. If $a > b$ (we will assume this throughout the book), the TE_{10} mode is dominant. If $b > a$ (recall that a is the x dimension and b is the y dimension), the TE_{01} mode is dominant. Assuming $a > b$, the $\mathbf{TE_{10}}$ **mode** is the *dominant mode overall*, considering all TM and TE modes in rectangular waveguide, because it gives the lowest cutoff frequency which is obtained from Eq. (11-58) by letting $m = 1, n = 0$.

$$f_{c,10} = \frac{1}{2\sqrt{\mu\varepsilon}}\frac{1}{a} = \frac{v}{2a} \tag{11-68}$$

The corresponding cutoff wavelength is given by

$$\lambda_{c,10} = \frac{v}{f_{c,10}} = 2a \tag{11-69}$$

Eq. (11-69) is easy to remember. The cutoff wavelength is twice the longer (x) dimension of the rectangular waveguide. The TE_{10} mode is of particular importance because it also gives the lowest attenuation among all modes of the rectangular waveguide. It is useful to know the field distributions of the dominant TE_{10} mode, which are given by

$$\begin{aligned} H_z &= H_0 \cos\left(\frac{\pi}{a}x\right)e^{-jk_z z} \\ H_x &= j\frac{k_z}{\pi/a} H_0 \sin\left(\frac{\pi}{a}x\right)e^{-jk_z z} \\ E_y &= -j\frac{\omega\mu}{\pi/a} H_0 \sin\left(\frac{\pi}{a}x\right)e^{-jk_z z} \end{aligned} \tag{11-70}$$

The electric field of the TE_{10} mode is polarized only in one (y) direction everywhere. The field distributions are identical to those of the TE_1 mode of the parallel-plate waveguide. The field lines of the TE_{10} mode are plotted in Figure 11-9. It is to be emphasized that the TEM mode does not exist in the rectangular waveguide. In general, *a single conductor waveguiding system does not support a TEM mode.*

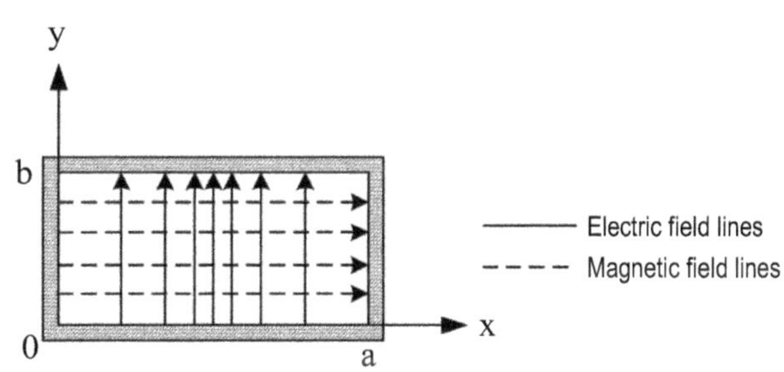

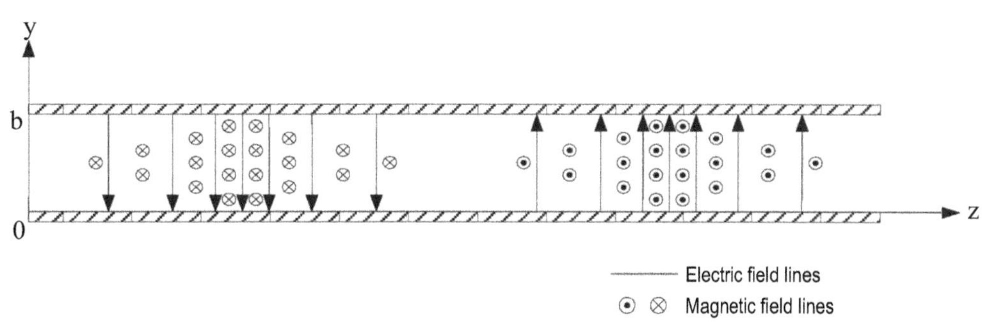

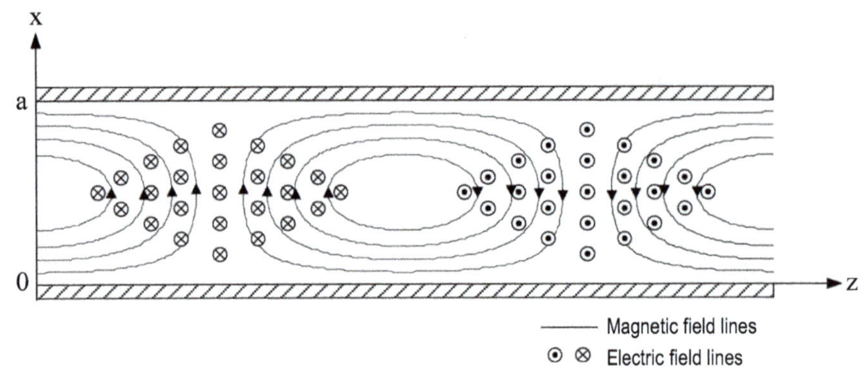

Figure 11-9. Field lines of the TE_{10} mode in a rectangular waveguide.

EXAMPLE 11-3. X-BAND WAVEGUIDE

Consider an air-filled rectangular waveguide having dimensions of $a = 0.9$ in. (2.286 cm) and $b = 0.4$ in (1.016 cm).

(a) Suppose only one mode (i.e., the dominant mode) should be transmitted. What is the range of frequencies that can be used?

(b) When the operating frequency is 18 GHz, which TE and TM modes can propagate in the waveguide?

SOLUTIONS

(a) The dominant mode of the rectangular waveguide is TE_{10} mode and its cutoff frequency is given by

$$f_{c,10} = \frac{v}{2a} = \frac{3\times10^8}{2\times2.286\times10^{-2}} = 6.56\times10^9\ [\text{Hz}]$$

The next possible higher-order mode is either TE_{01} or TE_{20} mode:

$$f_{c,01} = \frac{v}{2b} = \frac{3\times10^8}{2\times1.016\times10^{-2}} = 14.76\times10^9\ [\text{Hz}]$$

$$f_{c,20} = \frac{v}{2a}\cdot 2 = \frac{3\times10^8}{2.286\times10^{-2}} = 13.12\times10^9\ [\text{Hz}] < f_{c,01}$$

Thus, the range of frequencies for single-mode operation is

$$6.56\ \text{GHz} < f < 13.12\ \text{GHz}$$

Note that this waveguide is used for radar applications at X-band frequencies (8 – 12.4 GHz).

(b) Since $f = 18$ GHz $> f_{c,10}, f_{c,01}, f_{c,20}$, all the above three modes will propagate. The next higher-order modes to be considered are 11, 21 and 30 modes:

$$f_{c,11} = \frac{v}{2}\sqrt{\frac{1}{a^2}+\frac{1}{b^2}} = 16.16 \text{ GHz} < f = 18 \text{ GHz}$$

$$f_{c,21} = \frac{v}{2}\sqrt{\left(\frac{2}{a}\right)^2+\frac{1}{b^2}} = 19.74 \text{ GHz} > f$$

$$f_{c,30} = \frac{v}{2a}\cdot 3 = 19.68 \text{ GHz} > f$$

Therefore, TE_{10}, TE_{20}, TE_{01}, TE_{11} and TM_{11} modes can propagate.

EXAMPLE 11-4 WR-137 WAVEGUIDE

A rectangular waveguide having dimensions of $a = 3.484$ cm and $b = 1.58$ cm (WR-137 waveguide) is filled with polyethylene ($\varepsilon = 2.25\,\varepsilon_o$, $\mu = \mu_o$). The operating frequency is 4 GHz.

(a) Find the phase constant (β), the guide wavelength (λ_g), the phase velocity (υ_p), the group velocity (υ_g), and the wave impedance (Z) of the dominant mode.

(b) Compute the propagation constant when the operating frequency is 2 GHz. How much does the wave attenuate over the distance of 2 cm?

SOLUTIONS

(a) The cutoff frequency of the dominant (TE_{10}) mode is

$$f_{c,10} = \frac{v}{2a} = \frac{c}{\sqrt{2.25}}\frac{1}{2a} = \frac{2\times10^8}{2\times3.484\times10^{-2}} = 2.87\times10^9\,[\text{Hz}]$$

Since most of the above quantities can be written in terms of $\sqrt{1-\left(\frac{f_c}{f}\right)^2}$, we calculate this factor first:

$$F \equiv \sqrt{1-\left(\frac{f_c}{f}\right)^2} = \sqrt{1-\left(\frac{2.87\times10^9}{4\times10^9}\right)^2} = 0.697$$

Using Eqs. (11-30) – (11-33) and (11-41), we obtain

$$k_z = \beta = \omega\sqrt{\mu\varepsilon}\ F = 2\pi\times4\times10^9\times\frac{1}{2\times10^8}\times0.697 = 87.5\left[\frac{\text{rad}}{\text{m}}\right]$$

$$\lambda_g = \frac{2\pi}{k_z} = \frac{2\pi}{87.5} = 7.17\ [\text{cm}]$$

$$v_p = \frac{v}{F} = \frac{2\times10^8}{0.697} = 2.87\times10^8\left[\frac{\text{m}}{\text{s}}\right]$$

$$v_g = v\,F = 2\times10^8\times0.697 = 1.39\times10^8\left[\frac{\text{m}}{\text{s}}\right]$$

$$Z_{TE} = \frac{\eta}{F} = \sqrt{\frac{\mu_o}{2.25\varepsilon_o}}\frac{1}{F} = \frac{377}{1.5}\times\frac{1}{0.697} = 360.6\ [\Omega]$$

(b) When $f = 2$ GHz, since $f < f_{c,10}$ the wave is in the cutoff region (evanescent wave) and k_z becomes purely imaginary:

$$k_z = -j\alpha = -j\sqrt{\left(\frac{\pi}{a}\right)^2 - \omega^2\mu\varepsilon} = -j\omega\sqrt{\mu\varepsilon}\ \sqrt{\left(\frac{f_c}{f}\right)^2 - 1}$$

$$= -j\frac{2\pi\times2\times10^9}{2\times10^8}\sqrt{\left(\frac{2.87}{2.0}\right)^2 - 1} = -j64.7\left[\frac{\text{rad}}{\text{m}}\right]$$

The field attenuates as $e^{-\alpha z}$ and over 2 cm propagation the field amplitude is reduced by a factor of $e^{-64.7\times0.02} = 0.274$, which is a quite rapid attenuation. This demonstrates that the signal attenuates very rapidly with distance when its operating frequency is below the cutoff frequency.

11.5 RECTANGULAR CAVITY RESONATOR

In the rectangular waveguide, if we place two perfectly conducting walls normal to the z axis (direction of guidance), then the fields will form a standing wave in the z direction, too. The structure becomes a **resonator**. The electromagnetic fields exist in a source-free region enclosed by a perfect conductor, i.e., in a *cavity* and they exist only at specific frequencies, called *resonant frequencies*, which will depend on the geometry of the resonator. This cavity resonator behaves like an LC resonator of low frequencies. At high frequencies (microwaves), ordinary lumped circuit elements such as capacitors and inductors become impractical due to their sizes comparable to operating wavelength and their large losses at high frequencies. The losses come from radiation and high resistance. The cavity resonator having an enclosure with a large area of conducting surface can eliminate radiation and resistive losses, yielding a very high quality factor (Q).

Figure 11-10 shows the geometry of a rectangular cavity resonator. It consists of six perfectly conducting ($\sigma = \infty$) walls at $x = 0$, $x = a$, $y = 0$, $y = b$, $z = 0$, and $z = d$. The medium inside the resonator is assumed to be lossless ($\sigma = 0$) and has permittivity ε and permeability μ.

Since the resonator is constructed of the rectangular waveguide with two additional perfect conductors at $z = 0$ and $z = d$, the field solutions for the resonator can be obtained by taking the solutions of the rectangular waveguide and applying two additional boundary conditions at the added two walls. Thus we will again seek the field solutions for the TM and TE modes, separately. It is to be noted that here the TM and TE modes are with respect to the z direction, although they can also be formulated with respect to the x or y direction.

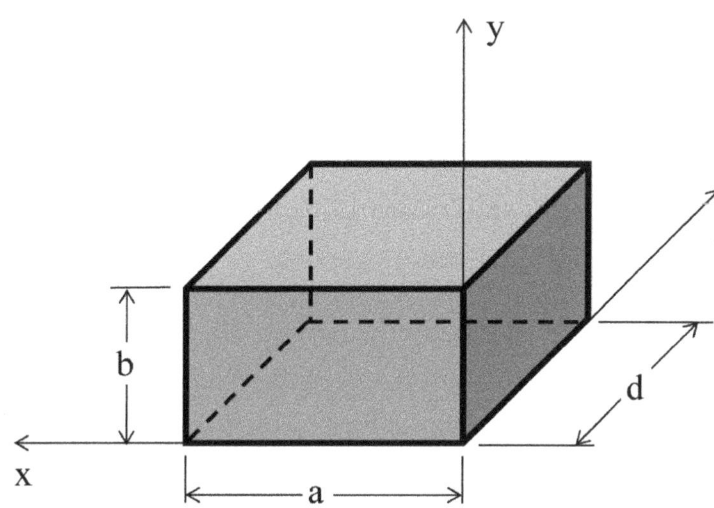

Figure 11-10. A rectangular cavity resonator.

11.5.1 TM Mode Solutions

For TM modes, $H_z = 0$ and we seek solutions for $E_z(x,y,z)$. In addition to four boundary conditions at $x = 0$, $x = a$, $y = 0$, and $y = b$, given by Eq. (11-51), the fields must also satisfy

$$\text{(v)}\quad E_x = 0, \quad E_y = 0 \text{ at } z = 0 \tag{11-71a}$$

$$\text{(vi)}\quad E_x = 0, \quad E_y = 0 \text{ at } z = d \tag{11-71b}$$

The TM_{mn} mode of the rectangular waveguide, shown in Eq. (11-54), represents a guided wave traveling in +z direction. In the rectangular cavity resonator, there will be a combination of a wave traveling in +z direction

($e^{-jk_z z}$) and a wave traveling in $-z$ direction ($e^{jk_z z}$) due to reflections from the conducting walls at $z = 0$ and $z = d$, forming a *standing wave in the z direction*. Therefore, the z dependence of the field can be written as a linear combination of $e^{-jk_z z}$ and $e^{jk_z z}$ or a linear combination of $\sin(k_z z)$ and $\cos(k_z z)$. Hence the solution for E_z for the TM (to z) modes can be written as

$$E_z(x,y,z) = \sin\left(\frac{m\pi}{a}x\right)\sin\left(\frac{n\pi}{b}y\right)\left\{F\,e^{-jk_z z} + G\,e^{+jk_z z}\right\} \qquad (11\text{-}72)$$

From Eq. (11-10) with $H_z = 0$, we have

$$E_x = \frac{-jk_z}{k^2 - k_z^2}\frac{\partial E_z}{\partial x} \qquad E_y = \frac{-jk_z}{k^2 - k_z^2}\frac{\partial E_z}{\partial y} \qquad (11\text{-}73)$$

for the wave traveling in $+z$ direction ($e^{-jk_z z}$). Similarly, for the wave traveling in $-z$ direction ($e^{jk_z z}$), we will have

$$E_x = \frac{jk_z}{k^2 - k_z^2}\frac{\partial E_z}{\partial x} \qquad E_y = \frac{jk_z}{k^2 - k_z^2}\frac{\partial E_z}{\partial y} \qquad (11\text{-}74)$$

Now we apply the two boundary conditions (BCs):

$$\text{BC(v)}: E_x = 0 \rightarrow -F\,e^{-jk_z z} + G\,e^{jk_z z} = 0 \text{ at } z = 0 \rightarrow F = G$$

$$\text{BC(vi)}: E_x = 0 \rightarrow -F\,e^{-jk_z z} + G\,e^{jk_z z} = 0 \text{ at } z = d$$

$$F\left(e^{jk_z d} - e^{-jk_z d}\right) = F(2j)\sin(k_z d) = 0$$

$$k_z d = p\pi \text{ or } \boxed{k_z = \frac{p\pi}{d}}, \quad p = 0, 1, 2, \ldots \qquad (11\text{-}75)$$

We can also rewrite

$$F e^{-jk_z z} + G\,e^{jk_z z} = F\left\{e^{-jk_z z} + e^{jk_z z}\right\} = 2F\cos(k_z z)$$

The final solution for E_z for the TM mode is then given by

$$\boxed{E_z(x,y,z) = E_o \sin\left(\frac{m\pi}{a}x\right)\sin\left(\frac{n\pi}{b}y\right)\cos\left(\frac{p\pi}{d}z\right), \ \text{TM}_{mnp} \text{ mode}} \qquad (11\text{-}76)$$

$$m = 1, 2, 3, \ldots;\ n = 1, 2, 3 \ldots;\ p = 0, 1, 2, \ldots$$

Each choice of the integers m, n, p defines a particular mode (TM_{mnp}). $k_x = \frac{m\pi}{a}$, $k_y = \frac{n\pi}{b}$ and $k_z = \frac{p\pi}{d}$ should satisfy the following dispersion relation (see Eq. (11-50)):

$$\left(\frac{m\pi}{a}\right)^2 + \left(\frac{n\pi}{b}\right)^2 + \left(\frac{p\pi}{d}\right)^2 = k^2 = \omega^2\mu\varepsilon \qquad (11\text{-}77)$$

It means that each mode exists only at a certain single frequency that satisfies Eq. (11-77):

$$\omega = 2\pi f = \frac{1}{\sqrt{\mu\varepsilon}}\sqrt{\left(\frac{m\pi}{a}\right)^2+\left(\frac{n\pi}{b}\right)^2+\left(\frac{p\pi}{d}\right)^2}$$

$$\boxed{f = \frac{1}{2\sqrt{\mu\varepsilon}}\sqrt{\left(\frac{m}{a}\right)^2+\left(\frac{n}{b}\right)^2+\left(\frac{p}{d}\right)^2} \equiv f_{r,mnp}} \tag{11-78}$$

This frequency is called the **resonant frequency** of the TM_{mnp} mode. The mode that has the lowest resonant frequency among TM modes is the TM_{110} mode whose f_r is given by

$$f_{r,110} = \frac{1}{2\sqrt{\mu\varepsilon}}\sqrt{\left(\frac{1}{a}\right)^2+\left(\frac{1}{b}\right)^2} = \frac{v}{2}\sqrt{\frac{1}{a^2}+\frac{1}{b^2}} \tag{11-79}$$

The remaining field components (E_x, E_y, H_x, H_y) can be obtained by using Eqs. (11-10) with substitution of $-jk_z$ by $\frac{\partial}{\partial z}$ and $k^2-k_z^2$ by $k_x^2+k_y^2=\left(\frac{m\pi}{a}\right)^2+\left(\frac{n\pi}{b}\right)^2$. The fields of the dominant TM_{110} mode are given by

$$\mathbf{E} = \mathbf{a}_z E_o \sin\left(\frac{\pi}{a}x\right)\sin\left(\frac{\pi}{b}y\right), \quad TM_{110} \text{ mode}$$

$$\mathbf{H} = \frac{jE_o}{\omega\mu}\left\{\mathbf{a}_x \frac{\pi}{b}\sin\left(\frac{\pi}{a}x\right)\cos\left(\frac{\pi}{b}y\right) - \mathbf{a}_y \frac{\pi}{a}\cos\left(\frac{\pi}{a}x\right)\sin\left(\frac{\pi}{b}y\right)\right\} \tag{11-80}$$

11.5.2. TE Mode Solutions

For TE modes, we let $E_z = 0$ and seek the solutions for $H_z(x,y,z)$. The boundary conditions are still given by Eq. (11-71). We again take the TE_{mn} mode solutions of the rectangular waveguide, shown in Eq. (11-66), and consider the standing wave (in z direction) nature of the resonator as discussed earlier. The solution for H_z for the TE (to z) modes can be written as

$$H_z(x,y,z) = \cos\left(\frac{m\pi}{a}x\right)\cos\left(\frac{n\pi}{b}y\right)\left\{F\,e^{-jk_z z} + G\,e^{jk_z z}\right\} \tag{11-81}$$

From Eq. (11-10) with $E_z = 0$, we have

$$E_x = \frac{-j\omega\mu}{k^2-k_z^2}\frac{\partial H_z}{\partial y}, \quad E_y = \frac{j\omega\mu}{k^2-k_z^2}\frac{\partial H_z}{\partial x} \tag{11-82}$$

This time Eq. (11-82) is valid for the wave traveling in either $+z$ or $-z$ direction. Now we apply two BC's at $z = 0$ and $z = d$:

$$\mathrm{BC(v)}: E_x = 0 \text{ or } \frac{\partial H_z}{\partial y} = 0 \rightarrow F\, e^{-jk_z z} + G\, e^{jk_z z} = 0 \text{ at } z = 0 \rightarrow F = -G$$

$$\mathrm{BC(vi)}: E_x = 0 \text{ or } \frac{\partial H_z}{\partial y} = 0 \rightarrow F\, e^{-jk_z z} + G\, e^{jk_z z} = 0 \text{ at } z = d$$

$$F\left(e^{-jk_z d} - e^{jk_z d}\right) = F(-j2)\sin(k_z d) = 0$$

$$k_z d = p\pi \text{ or } \boxed{k_z = \frac{p\pi}{d}},\ p = 1, 2, 3, \ldots \tag{11-83}$$

We can also rewrite,

$$F e^{-jk_z z} + G e^{jk_z z} = F\left(e^{-jk_z z} - e^{jk_z z}\right) = (-j2)\, F \sin(k_z z)$$

The final solution for H_z for the TE modes is then given by

$$\boxed{H_z(x,y,z) = H_0 \cos\left(\frac{m\pi}{a}x\right)\cos\left(\frac{n\pi}{b}y\right)\sin\left(\frac{p\pi}{d}z\right),\ TE_{mnp} \text{ mode}} \tag{11-84}$$

$m = 0, 1, 2, \ldots ; n = 0, 1, 2, \ldots ; p = 1, 2, 3, \ldots$
$m = n = 0$ is excluded

Note that p cannot be zero because it leads to no field. Each choice of the integers m,n,p defines a particular mode (TE_{mnp}). Again $\frac{m\pi}{a}, \frac{n\pi}{b}, \frac{p\pi}{d}$ satisfies the dispersion relation, Eq. (11-77), which gives the same expression, Eq. (11-78), for the resonant frequency of the TE_{mnp} mode. The mode that has the lowest resonant frequency among TE modes is the TE_{101} mode (when $a > b$) or the TE_{011} mode (when $b > a$). Assuming $a > b$, the resonant frequency of the dominant TE mode is given by

$$f_{r,101} = \frac{1}{2\sqrt{\mu\varepsilon}}\sqrt{\left(\frac{1}{a}\right)^2 + \left(\frac{1}{d}\right)^2} = \frac{v}{2}\sqrt{\frac{1}{a^2} + \frac{1}{d^2}} \tag{11-85}$$

Comparing Eq. (11-85) with Eq. (11-79), the overall dominant mode (including both TE and TM modes) depends upon $d > b$ or $b > d$. The dominant mode of a rectangular cavity resonator is

TE_{101} mode when $a \geq d > b$ or $d > a > b$
TM_{110} mode when $a > b > d$

The remaining field components (E_x, E_y, H_x, H_y) can again be obtained by using Eqs. (11-10) with appropriate substitutions. The fields of the dominant TE_{101} mode are given by

$$\mathbf{E} = \mathbf{a}_y(-j\omega\mu)\frac{H_o}{\pi/a}\sin\left(\frac{\pi}{a}x\right)\sin\left(\frac{\pi}{d}z\right), \quad TE_{101}\text{ mode}$$
$$\mathbf{H} = H_o\left\{\mathbf{a}_z\cos\left(\frac{\pi}{a}x\right)\sin\left(\frac{\pi}{d}z\right) - \mathbf{a}_x\frac{a}{d}\sin\left(\frac{\pi}{a}x\right)\cos\left(\frac{\pi}{d}z\right)\right\} \tag{11-86}$$

11.5.3. Quality Factor of the Cavity Resonator

In practice, the conducting enclosure is not perfect (finite σ) and there will be power dissipated in the walls. The quality factor (or Q) of the resonator measures the bandwidth of the resonator and the degree of power loss. As in the case of lumped resonant circuits, the quality factor is defined by

$$Q = \frac{\omega W}{P_d} = 2\pi f_r\frac{(\text{time-average energy stored})}{(\text{time-average power dissipated})} \tag{11-87}$$

An alternative definition of Q is

$$Q = \frac{\omega}{\Delta\omega} = \frac{f_r}{B} \tag{11-88}$$

where f_r is the resonant frequency and B is the bandwidth or the difference in frequencies between two half-power points where the oscillation amplitude is reduced from its maximum value at f_r by 3 dB. Q or B indicates the sharpness of resonance; a high Q means narrower bandwidth and smaller loss.

EXAMPLE 11-5 QUALITY FACTOR FOR TE101 MODE

Find the quality factor for the TE_{101} mode of the rectangular cavity resonator with dimensions $a \times b \times d$.

SOLUTION

Letting $(-j\omega\mu)\, H_o\frac{a}{\pi} = E_o$ in Eq. (11-86), the *time-average* electric energy stored in the resonator volume is given by

$$W_e = \frac{1}{2}\text{Re}\left\{\iiint_V \frac{1}{2}\mathbf{E}\cdot\mathbf{D}^* dv\right\} = \frac{1}{4}\iiint_V \varepsilon|\mathbf{E}|^2 dv \quad \text{(see Problem 9-7)} \tag{11-89}$$

where a factor of ½ is due to the time average.

$$W_e = \frac{\varepsilon}{4}\int_0^d\int_0^b\int_0^a |E_o|^2 \sin^2\left(\frac{\pi}{a}x\right)\sin^2\left(\frac{\pi}{d}z\right)dx\,dy\,dz = \frac{\varepsilon}{4}|E_o|^2\frac{abd}{4}$$

The time-average magnetic energy stored in the resonator is

$$W_m = \frac{1}{2}\,\mathrm{Re}\left\{\iiint_V \frac{1}{2}\mathbf{H}^*\cdot\mathbf{B}\,dv\right\} = \frac{1}{4}\iiint_V \mu\,|\mathbf{H}|^2\,dv \tag{11-90}$$

$$= \frac{\mu}{4}\int_0^d\int_0^b\int_0^a |H_o|^2\left\{\cos^2\left(\frac{\pi}{a}x\right)\sin^2\left(\frac{\pi}{d}z\right)\right.$$

$$\left. + \left(\frac{a}{d}\right)^2\sin^2\left(\frac{\pi}{a}x\right)\cos^2\left(\frac{\pi}{d}z\right)\right\}dx\,dy\,dz = \frac{\mu}{4}|H_o|^2\frac{abd}{4}\left\{1+\left(\frac{a}{d}\right)^2\right\}$$

It can be shown that at the resonant frequencies ($f = f_r$ or $\omega = 2\pi f_r$), $W_e = W_m$ (see Problem 11-36). Thus, the total time-average stored energy is

$$W = W_e + W_m = 2W_m = \frac{1}{8}\mu|H_o|^2\,a\,b\,d\left\{1+\left(\frac{a}{d}\right)^2\right\} \tag{11-91}$$

The time-average power dissipated or power loss *per unit area* of a conducting wall is given by

$$p_d = \frac{1}{2}|H_{tan}|^2 R_s \quad [\mathrm{W/m^2}] \qquad \text{(see Problem 11-33)} \tag{11-92}$$

where H_{tan} is the tangential component of the magnetic field on the conducting surface and R_s is the surface resistance of the conducting wall. Recognizing that the power losses on two opposite sides of the walls are equal to each other, the time-average power dissipated on six walls is calculated by

$$P_d = \frac{1}{2}R_s\left\{2\int_0^b\int_0^a |H_x|^2_{z=0}\,dxdy + 2\int_0^d\int_0^b |H_z|^2_{x=0}\,dy\,dz\right.$$

$$\left. +2\int_0^d\int_0^a \left(|H_x|^2+|H_z|^2\right)_{y=0} dxdz\right\} \tag{11-93}$$

$$= R_s|H_o|^2\left\{\left(\frac{a}{d}\right)^2\frac{ab}{2}+\frac{bd}{2}+\frac{ad}{4}\left[\left(\frac{a}{d}\right)^2+1\right]\right\} \quad [\mathrm{W}]$$

Substituting Eqs. (11-91) and (11-93) into Eq. (11-87), we obtain Q for the TE_{101} mode:

$$Q = \frac{\pi f_{r,101}\,\mu\,a\,b\,d(a^2+d^2)}{R_s(2a^3b+2bd^3+a^3d+ad^3)} \tag{11-94}$$

For an air-filled cavity with $a = d = 2b = 2$ cm, the resonant frequency of the dominant mode is 10.6 GHz and the quality factor (Q) is approximately 10^4 when the cavity is made of copper walls (see Problem 11-34). Q is considerably larger than that of lumped-circuit resonator at much lower frequencies.

11.6. COAXIAL CABLE

The coaxial line, commonly known as the coaxial cable, is probably one of the most commonly used two-conductor waveguiding structures. The coaxial line can be viewed as a circular cylindrical version of a parallel plate waveguide. It consists of an inner conductor of radius a and an outer conductor of radius b, as shown in Figure 11-11(a). The space in between is normally filled with the dielectric of permittivity ε and permeability μ. Both conductors are usually made with many strands of thin copper wire in order to provide flexibility. Polyethylene or Teflon is commonly used for dielectric filling in coaxial cable.

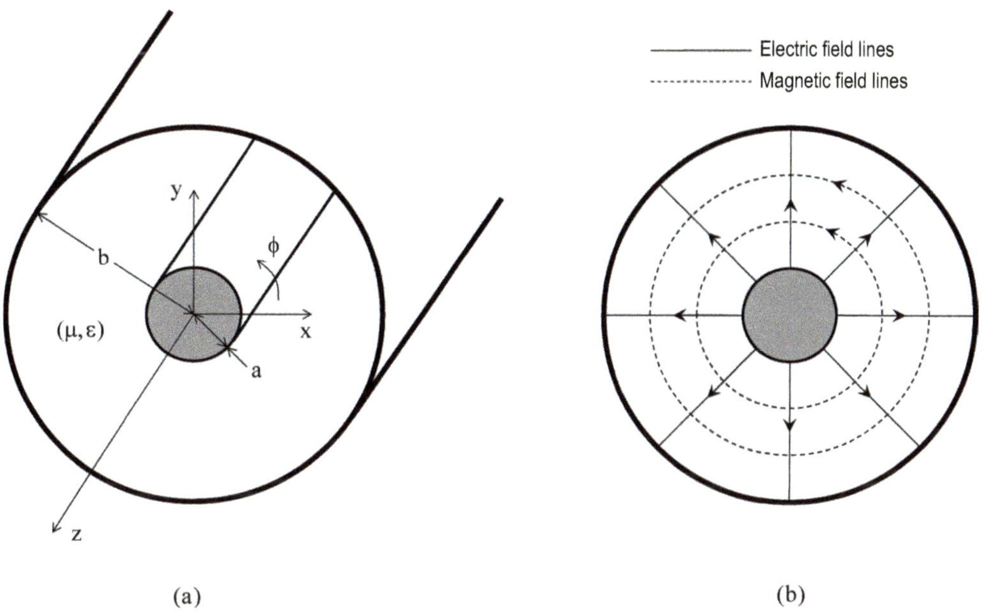

Figure 11-11. Coaxial cable: (a) geometry (b) field lines of a TEM mode.

One major advantage of the coaxial line is that the fields are completely confined within the dielectric region while in parallel plate waveguide there will be fringing or leakage fields at the edges of parallel plates.

The field solutions for the coaxial line include the TEM mode as the dominant mode (because it has two conductors) as well as higher-order TE and TM modes, just as in the case of the parallel plate waveguide. The general solution for the fields can be obtained by solving the Helmholtz wave equation, Eq. (11-1), in cylindrical coordinates, which involves the Bessel functions. This is beyond the scope of this book, so we will not consider general higher-order TE and TM solutions. However, the dominant TEM mode solution can be obtained without advanced mathematics.

11.6.1. TEM Mode Solution

First of all, for the TEM mode, $E_z = H_z = 0$ since both electric and magnetic fields are transverse to the direction of guidance (z). Since the coaxial line can be viewed as a cylindrical version of the parallel plate waveguide (PPW), we expect that the TEM mode fields of the coax (short word for coaxial line) will behave much like those of the PPW. From Eq. (11-27), we re-write here the fields of the TEM mode of the PPW of Figure 11-3.

$$\mathbf{E} = \mathbf{a}_x E_o e^{-jkz}, \quad \mathbf{H} = \mathbf{a}_y \frac{E_o}{\eta} e^{-jkz} \tag{11-95}$$

where

$$k(=k_z) = \omega\sqrt{\mu\varepsilon}, \quad \eta = \sqrt{\frac{\mu}{\varepsilon}} \tag{11-96}$$

Noting that for the PPW, **E** has only a component normal to the surface of the conductors and **H** has only a component tangential to the conductor surface, we assume that for the coaxial line the **E** and **H** fields take the following form:

$$\mathbf{E} = \mathbf{a}_\rho \tilde{E}_\rho(\rho) e^{-jkz}, \quad \mathbf{H} = \mathbf{a}_\phi \tilde{H}_\phi(\rho) e^{-jkz} \tag{11-97}$$

whose field lines are sketched in Figure 11-11(b). Note that **E** is normal to the conductor surface and **H** is tangential to the conductor surface. Note also that no ϕ dependence is assumed due to azimuthal symmetry of the geometry.

Substituting Eq. (11-97) into $\nabla \cdot \mathbf{E} = 0$ and $\nabla \times \mathbf{E} = -j\omega\mu\mathbf{H}$ (see Eqs. (9-19)), we obtain

$$\frac{1}{\rho}\frac{\partial}{\partial\rho}(\rho\tilde{E}_\rho) = 0 \rightarrow \tilde{E}_\rho = \frac{K_o}{\rho}$$

$$\frac{\partial E_\rho}{\partial z} = -jk\tilde{E}_\rho e^{-jkz} = -j\omega\mu H_\phi = -j\omega\mu \tilde{H}_\phi e^{-jkz}$$

$$\rightarrow \tilde{H}_\phi = \tilde{E}_\rho \frac{k}{\omega\mu} = \frac{\tilde{E}_\rho}{\eta} = \frac{K_o}{\eta\rho}$$

where K_o is an arbitrary constant. Note that $E_\rho = \tilde{E}_\rho e^{-jkz}$.

Therefore, the field solutions for the TEM mode in a coaxial line are

$$\boxed{\mathbf{E} = \mathbf{a}_\rho \frac{K_o}{\rho} e^{-jkz}, \quad \mathbf{H} = \mathbf{a}_\phi \frac{K_o}{\eta\rho} e^{-jkz}} \tag{11-98}$$

where k and η are given by Eq. (11-96). It can be shown that these fields satisfy also $\nabla \cdot \mathbf{H} = 0$ and $\nabla \times \mathbf{H} = j\omega\varepsilon\,\mathbf{E}$ (see Problem 11-38). We observe that $\tilde{E}_\rho = \frac{K_o}{\rho}$ behaves much like a *static electric field* of an infinitely long line charge (see Eq. (2-22)) and of a coaxial cable with opposite charges (see Eq. (3-21a)). $\tilde{H}_\varphi = \frac{K_o}{\eta\rho}$ behaves much like a *static magnetic field* of an infinitely long current-carrying wire (see Eq. (6-8)) and of two concentric cylinders carrying opposite currents (see Eq. (6-11b)). Thus, the TEM mode solutions can be obtained by multiplying the static field solutions by e^{-jkz} (wave characteristic). The same argument can also be applied to the TEM mode solutions of the PP waveguide. It is to be emphasized that the TEM mode of the coax has no cutoff frequency, i.e., its cutoff frequency is zero just like the TEM mode of the PP line. The coaxial cable is typically used at lower frequencies because its loss due to imperfect conductors increases dramatically at higher frequencies.

For example, the RG6 75Ω coaxial cable, made by CommScope, has the attenuation 0.70 dB/100 ft. (or 2.30 dB/100 m), 2.01 dB/100 ft, and 7.45 dB/100 ft at 10 MHz, 100 MHz, and 1 GHz, respectively. Today coaxial cables are widely used in the broadband cable TV industry. The subscribers are served by an HFC (Hybrid Fiber Coax) architecture which consists of optical fiber cables, transmitters and receivers' amplifiers in the transport segment and coaxial cables, amplifiers and splitters in the distribution segment that directly serves homes.

EXAMPLE 11-6

Find the total time-average power transmitted along the coaxial line for the TEM mode, assuming that the voltage between the two conductors is V_o in its amplitude.

SOLUTION

For the TEM mode fields, given by Eq. (11-98), the voltage or potential difference between the two conductors is obtained by

$$V = -\int \mathbf{E}\cdot d\ell = -\int_b^a \frac{K_o}{\rho} e^{-jkz}\, \mathbf{a}_\rho \cdot \mathbf{a}_\rho\, d\rho = K_o \ln\frac{b}{a} e^{-jkz}$$

$$= V_o e^{-jkz} \rightarrow K_o = \frac{V_o}{\ln(b/a)} \tag{11-99}$$

Thus

$$\mathbf{E} = \mathbf{a}_\rho \frac{V_o}{\ln(b/a)} \frac{1}{\rho} e^{-jkz} \tag{11-100}$$

The time-average Poynting vector is given by

$$\mathbf{S}_{av} = \frac{1}{2}\mathrm{Re}\left(\mathbf{E}\times\mathbf{H}^*\right) = \frac{1}{2}\mathbf{a}_\rho \times \mathbf{a}_\phi \frac{|K_o|^2}{\eta}\frac{1}{\rho^2} = \mathbf{a}_z \frac{1}{2\eta}\frac{|V_o|^2}{[\ln(b/a)]^2}\frac{1}{\rho^2}$$

The time-average power transmitted is

$$P = \iint \mathbf{S}_{av}\cdot d\mathbf{s} = \int_0^{2\pi}\int_a^b \frac{1}{2\eta}\frac{|V_o|^2}{[\ln(b/a)]^2}\frac{1}{\rho^2}\,\rho\, d\rho\, d\phi = \frac{\pi |V_o|^2}{\eta \ln(b/a)}$$

When you calculate the current flowing through the conductors from the magnetic field and define its amplitude to be I_o, it can be shown that the time-average power can be obtained from $\frac{1}{2}\mathrm{Re}(V_o I_o^*)$ (see Problem 11-39).

11.7 TRANSMISSION LINES

In previous sections we learned that a two-conductor waveguiding system such as parallel-plate waveguide and coaxial cable can support a *TEM wave* for which both electric and magnetic fields are *transverse* to the direction of propagation or guidance. Such a two-conductor transmission system is often called a **transmission line**. Examples, as shown in Figure 11-1(a)-(c), include the parallel-plate waveguide (also called parallel-plate transmission line), the two-wire transmission line, and the coaxial line. The parallel-plate (PP) line is fabricated in microstrip circuits at microwave frequencies, the two-wire transmission line is seen in overhead power and telephone lines, and the coaxial line is used in telephone, TV and internet cables.

When we deal with only the TEM wave in transmission lines, it is very convenient to introduce the concepts of the voltage, current, and impedance in the analysis and application of the transmission lines, since you are familiar with these concepts from circuit theory. We can use the voltage $V(z)$ and current $I(z)$, instead of the electric and magnetic fields, to describe the propagation and reflection of the wave. They satisfy a pair of coupled differential equations, known as the transmission-line equations, which are equivalent to the two Maxwell's curl equations. These equations will include the circuit parameters such as inductance, capacitance, resistance and conductance. Many characteristics of a TEM wave guided by transmission lines are very similar to the characteristics of a uniform plane wave in an unbounded lossless or lossy medium that you learned in Chapters 9-10.

11.7.1 The Transmission-Line Equations—Lumped-Circuit Model

Let's consider a section of a two-conductor transmission line shown in Figure 11-12(a). We assume that $V(z)$ is the voltage across the conductors at $z = z$ and $I(z)$ is the current flowing through one conductor section at $z = z$, implying that the same current flows through the other conductor section in the opposite direction. Assuming that the section length (Δz) is much smaller than the operating wavelength, the equivalent circuit can be drawn in Figure 11-12(b).

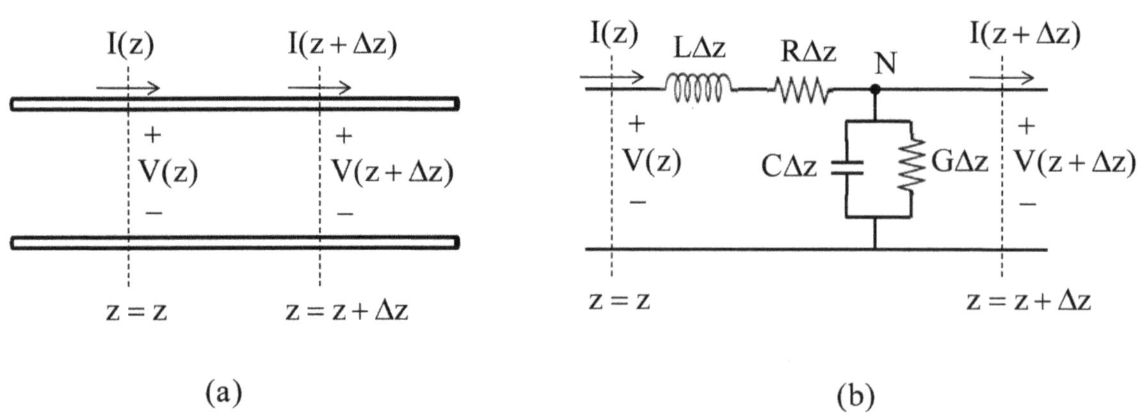

Figure 11-12. A transmission line and its lumped-circuit model.

A differential section (Δz) of a transmission line (T.L.) is described by four circuit parameters:

L = inductance per unit length in H/m

R = resistance per unit length in Ω/m

C = capacitance per unit length in F/m

G = conductance per unit length in S/m

When the T.L. is lossless, R and G are zero. R comes from the loss due to the imperfect conductors and G comes from the imperfect dielectric filling the T.L., i.e., the dielectric loss. Note that L and R are series elements and C and G are shunt elements.

First, we apply the Kirchhoff voltage law (KVL) around the loop and obtain:

$$V(z) = (j\omega L + R)\ \Delta z\ I(z) + V(z + \Delta z)$$

$$\rightarrow \frac{V(z+\Delta z) - V(z)}{\Delta z} = -(j\omega L + R)\ I(z)$$

Thus we have, as $\Delta z \rightarrow 0$ (incrementally small),

$$\boxed{\frac{dV}{dz} = -(j\omega L + R)I = -ZI} \quad (11\text{-}101)$$

where $Z = j\omega L + R$ is the series impedance per unit length in Ω/m of the T.L.

Secondly, we apply the Kirchhoff current law (KCL) at the node N and obtain:

$$I(z) = (j\omega C + G)\ \Delta z\ V(z+\Delta z) + I(z + \Delta z)$$

$$\rightarrow \frac{I(z+\Delta z) - I(z)}{\Delta z} = -(j\omega C + G)\ V(z + \Delta z)$$

Thus we have, as $\Delta z \rightarrow 0$ (incrementally small),

$$\boxed{\frac{dI}{dz} = -(j\omega C + G)V = -YV} \quad (11\text{-}102)$$

where $Y = j\omega C + G$ is the shunt admittance per unit length in S/m of the T.L. Eqs. (11-101) and (11-102) are a pair of coupled, first-order differential equations for $V(z)$ and $I(z)$, which are called the **transmission-line equations**.

Differentiating Eq. (11-101) with respect to z and substituting Eq. (11-102), we obtain

$$\frac{d^2V}{dz^2} = ZY\ V(z) \ \text{ or } \boxed{\frac{d^2V}{dz^2} + k^2\ V(z) = 0} \quad (11\text{-}103)$$

where

$$k^2 = -ZY = -(j\omega L + R)(j\omega C + G) \tag{11-104}$$

Eq. (11-103) is the *wave equation* for the voltage $V(z)$. Similarly, we can show that $I(z)$ satisfies the same wave equation:

$$\frac{d^2 I}{dz^2} = ZY\, I(z) \quad \text{or} \quad \boxed{\frac{d^2 I}{dz^2} + k^2 I(z) = 0} \tag{11-105}$$

$V(z)$ and $I(z)$ can now be seen as the *voltage wave* and the *current wave*, respectively. Eq. (11-103) is very similar to Eq. (9-38), the wave equation for the $\mathbf{E}$ field in the source-free region.

Lossless Transmission Line

In the absence of loss, $R = G = 0$ and

$$k^2 = -(j\omega L)(j\omega C) = \omega^2 LC \rightarrow \boxed{k = \omega\sqrt{LC}} \tag{11-106}$$

The general solution to Eq. (11-103) can be written as

$$V(z) = V_o^+\, e^{-jkz} + V_o^-\, e^{+jkz} \tag{11-107}$$

just like the field solutions in Eq. (9-40). The first and second terms represent the voltage waves traveling in $+z$ and $-z$ directions, respectively. k plays the same role as k of the uniform plane wave, so k is called the *propagation constant*. Substituting Eq. (11-107) into Eq. (11-101), we obtain

$$I(z) = \frac{1}{Z_o}(V_o^+\, e^{-jkz} - V_o^-\, e^{+jkz}) \tag{11-108}$$

where

$$\boxed{Z_o = \frac{\omega L}{k} = \sqrt{\frac{L}{C}}\ [\Omega]} \tag{11-109}$$

Z_o, known as the **characteristic impedance** of the T.L., is the ratio of the voltage amplitude vs. the current amplitude and plays the same role as the wave impedance of a uniform plane wave—ratio of electric and magnetic field amplitudes—which is equal to the intrinsic impedance of the medium $(\eta = \sqrt{\mu/\varepsilon})$ (see Eq. (9-49).

In the presence of conductor loss (R) and/or dielectric loss (G), k in Eq. (11-104) becomes complex ($\beta - j\alpha$) and the voltage and current waves will experience attenuation as they propagate along the T.L.

11.7.2. The Transmission-Line Equations from Field Theory

The T.L. equations shown in Eqs. (11-101) and Eq. (11-102) can also be derived from Maxwell's equations by relating the voltage $V(z)$ and the current $I(z)$ to the electric field $\mathbf{E}$ and the magnetic field $\mathbf{H}$ of a TEM wave in a two-conductor T.L. For illustration, we will consider the lossless parallel-plate waveguide. The following procedure gives us the physical insight as to how field concepts are converted to circuit concepts.

As shown in Section 11.3.1, the dominant TEM (or TM_o) mode fields of a parallel-plate transmission line of Figure 11-3 are described by

$$\mathbf{E} = \mathbf{a}_x E_x(z)$$
$$\mathbf{H} = \mathbf{a}_y H_y(z) \qquad (11\text{-}110)$$

Substituting Eq. (11-110) into $\nabla \times \mathbf{E} = -j\omega\mu\mathbf{H}$ and $\nabla \times \mathbf{H} = j\omega\varepsilon\mathbf{E}$, we obtain

$$\frac{\partial E_x}{\partial z} = -j\omega\mu H_y \qquad (11\text{-}111a)$$

$$\frac{\partial H_y}{\partial z} = -j\omega\varepsilon E_x \qquad (11\text{-}111b)$$

Now express the voltage $V(z)$ and the current $I(z)$ in terms of the fields using the electrostatic and magnetostatic theory. The voltage across the two conductors is related to $\mathbf{E}$ by

$$V(z) = -\int_{C_1} \mathbf{E} \cdot d\boldsymbol{\ell} \qquad (11\text{-}112a)$$

where C_1 is the integration path from one conductor to the other. The current flowing through each conductor is related to $\mathbf{H}$ by Ampère's law:

$$I(z) = \oint_{C_2} \mathbf{H} \cdot d\boldsymbol{\ell} \qquad (11\text{-}112b)$$

where C_2 is the closed loop that encloses one conductor. The current can also be related to $\mathbf{H}$ by using the boundary condition at the conductor surface, $\mathbf{a}_n \times \mathbf{H} = \mathbf{J}_s$. The integration paths are shown in Figure 11-13:

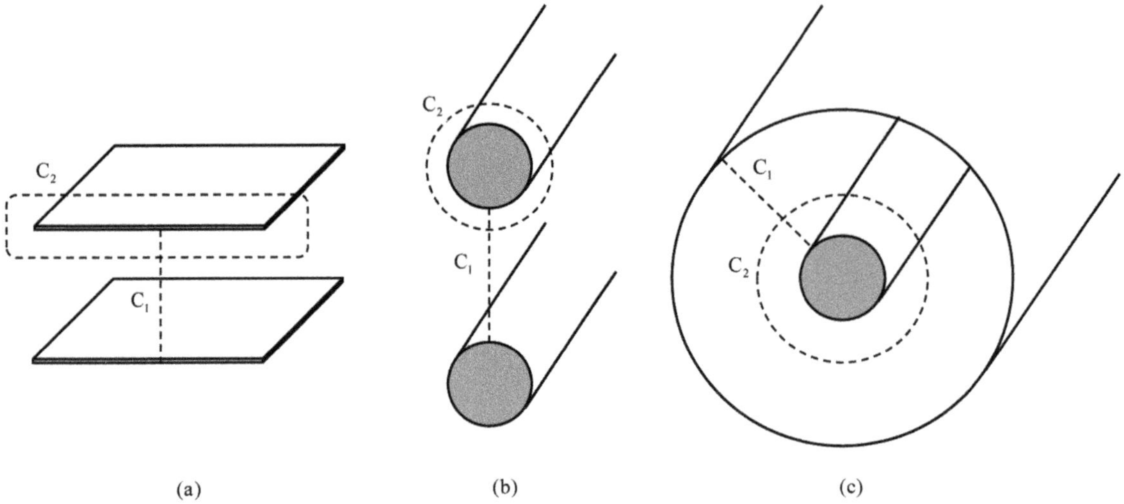

Figure 11-13. Integration paths defining the voltage and the current.

In the case of the PP line, Eqs. (11-112) gives

$$V(z) = E_x(z)\, a \tag{11-113a}$$

$$I(z) = H_y(z)\, w \tag{11-113b}$$

Making use of $V - E_x$ and $I - H_y$ relationships above, Eqs. (11-111) reduce to

$$\frac{dV}{dz} = -j\omega\mu \frac{a}{w}\, I(z) = -j\omega L\, I \tag{11-114a}$$

$$\frac{dI}{dz} = -j\omega\varepsilon \frac{w}{a}\, V(z) = -j\omega C\, V \tag{11-114b}$$

where we define

$$L = \mu \frac{a}{w} = \text{inductance per unit length of a PP line} \tag{11-115a}$$

$$C = \varepsilon \frac{w}{a} = \text{capacitance per unit length of a PP line} \tag{11-115b}$$

Eqs. (11-114) are, in fact, the transmission-line equations for a lossless ($R = G = 0$) T.L., as shown earlier in Eqs. (11-101) and (11-102).

Eliminating H_y in Eqs. (11-111), E_x satisfies

$$\frac{\partial^2 E_x}{\partial z^2} + k^2 E_x = 0, \quad k^2 = \omega^2 \mu \varepsilon \tag{11-116}$$

Eliminating I in Eqs. (11-114), V satisfies

$$\frac{\partial^2 V}{\partial z^2} + k^2 V = 0, \quad k^2 = \omega^2 LC \tag{11-117}$$

From Eq. (11-115), it is seen that the following relationship holds:

$$\boxed{LC = \mu\varepsilon} \tag{11-118}$$

For the wave traveling in $+z$ direction, we have

$$\mathbf{E} = \mathbf{a}_x E_o^+ e^{-jkz}, \quad \mathbf{H} = \mathbf{a}_y \frac{E_o^+}{\eta} e^{-jkz} \tag{11-119}$$

$$V = V_o^+ e^{-jkz}, \quad I = \frac{V_o^+}{Z_o} e^{-jkz} \tag{11-120}$$

The time-average power along the PP transmission line can be obtained by

$$P_{av} = \iint \frac{1}{2} \operatorname{Re}(\mathbf{E} \times \mathbf{H}^*) \cdot \mathbf{a}_z \, dx\, dy = \frac{|E_o^+|^2}{2\eta} \text{ wa} \tag{11-121}$$

or

$$P_{av} = \frac{1}{2} \operatorname{Re}\{V(z)\, I(z)^*\} = \frac{|V_o^+|^2}{2Z_o} \tag{11-122}$$

It can be easily shown that Eq. (11-121) and Eq. (11-122) are equal, using

$$Z_o = \sqrt{\frac{L}{C}} = \sqrt{\frac{\mu}{\varepsilon}} \left(\frac{a}{w}\right) \quad \text{for a PP line} \tag{11-123}$$

The analogy between the characteristics of the voltage and the current on the T.L. and those of the electric and magnetic fields of a uniform plane wave or a TEM mode on the T.L. is summarized in Table 11-3.

Table 11-3. Analogy between Transmission-Line Waves and Uniform Plane Waves (or TEM Waves)

Transmission Line (Lossless)	Uniform Plane Wave or TEM Wave
$V(z) = -\int_{C_1} \mathbf{E} \cdot d\ell$	$\mathbf{E} = \mathbf{a}_x E_x(z)$
$I(z) = \oint_{C_2} \mathbf{H} \cdot d\ell$	$\mathbf{H} = \mathbf{a}_y H_y(z)$

$\frac{dV}{dz} = -(j\omega L)\, I$ $\frac{dI}{dz} = -(j\omega C)\, V$	$\frac{\partial E_x}{\partial z} = -j\omega\mu\, H_y$ $\frac{\partial H_y}{\partial z} = -j\omega\varepsilon\, E_x$
$\frac{d^2V}{dz^2} + k^2 V = 0$ $\frac{d^2I}{dz^2} + k^2 I = 0$ $k^2 = \omega^2 LC$	$\frac{\partial^2 E_x}{\partial z^2} + k^2 E_x = 0$ $\frac{\partial^2 H_y}{\partial z^2} + k^2 H_y = 0$ $k^2 = \omega^2 \mu\varepsilon$
$V(z) = V_o^+ e^{-jkz} + V_o^- e^{jkz}$ $I(z) = \frac{1}{Z_o}\left(V_o^+ e^{-jkz} - V_o^- e^{jkz}\right)$ $Z_o = \frac{V_o^+}{I_o^+} = \sqrt{\frac{L}{C}}$ (characteristic impedance)	$E_x(z) = E_o^+ e^{-jkz} + E_o^- e^{jkz}$ $H_y(z) = \frac{1}{\eta}\left(E_o^+ e^{-jkz} - E_o^- e^{jkz}\right)$ $\eta = \frac{E_o^+}{H_o^+} = \sqrt{\frac{\mu}{\varepsilon}}$ (intrinsic impedance)
$P_{av} = \frac{1}{2}\,\mathrm{Re}\,(VI^*)$	$\mathbf{S}_{av} = \frac{1}{2}\,\mathrm{Re}\left(\mathbf{E}\times\mathbf{H}^*\right)$

11.7.3. Transmission-Line Circuit Parameters

As shown in the transmission-line equations, the properties of a T.L. are characterized by its four circuit parameters L, C, R and G. We would like to learn how to calculate these parameters for a given T.L. from the electromagnetic analysis. These parameters depend on the geometry and the medium parameters of the T.L.

The capacitance (C) and inductance (L) per unit length are calculated from the electrostatic and magnetostatic analysis as they are defined by

$$C = \frac{Q/\ell}{V} = \frac{(\text{Charge per unit length})}{(\text{Potential difference})} \tag{3-26}$$

$$L = \frac{\Phi/\ell}{I} = \frac{(\text{Magnetic flux per unit length})}{(\text{Current of closed loop})} \tag{7-20}$$

Since L and C are related to $\mu\varepsilon$ through Eq. (11-118), if L is known for a T.L., C can be easily obtained, and vice versa.

The shunt conductance (G) per unit length is due to dielectric loss when the dielectric medium (of permittivity ε) in the T.L. has a small conductivity σ. In Section 5.9, we have shown a relationship between capacitance and conductance when the dielectric medium surrounding two conductors has ε and σ as follows:

$$\frac{C}{G} = \frac{\varepsilon}{\sigma} \tag{5-19}$$

Hence once C of a T.L. is known, G can be calculated easily by

$$G = C\frac{\sigma}{\varepsilon} \qquad (11\text{-}124)$$

Finally, the series resistance (R) per unit length is due to conductor loss when the conductors of the line are not perfectly conducting, i.e., σ_c (conductivity) is finite. For good conductors at high frequencies, the current mainly flows or penetrates over the skin depth as discussed in Section 9.5.2. Thus R can be approximately calculated by assuming that the total currents in the conductors are uniformly distributed over the skin depth d_p. The resistance of a homogeneous resistor of length ℓ and uniform cross sections S is obtained in Section 5.4 as follows.

$$R = \frac{\ell}{\sigma S} \qquad (5\text{-}11)$$

For the transmission line of unit length that has two conductors, the series resistance per unit length (R) is given by

$$R = R_1 + R_2 = \frac{1}{\sigma_c}\left(\frac{1}{S_1} + \frac{1}{S_2}\right) = \frac{1}{\sigma_c d_p}\left(\frac{1}{C_1} + \frac{1}{C_2}\right) \qquad (11\text{-}125)$$

where C_1 and C_2 are the widths or the circumferences of the two conductors. The skin depth d_p of the good conductor is given by Eq. (9-90):

$$d_p = \sqrt{\frac{2}{\omega\mu_c\sigma_c}} = \sqrt{\frac{1}{\pi f \mu_c \sigma_c}} \qquad (11\text{-}126)$$

where μ_c is the permeability of the conductor. Note that $\frac{1}{\sigma_c d_p} = \sqrt{\frac{\omega\mu_c}{2\sigma_c}}$ is known as the *surface resistance* (R_s) of the conductor (see Problem 11-33). For a parallel plate line,

$$R = 2R_1 = \frac{2}{\sigma_c d_p w} = R_s \frac{2}{w} \qquad (11\text{-}127)$$

EXAMPLE 11-7 TRANSMISSION LINE PARAMETERS OF A COAXIAL LINE

Find L, C, R, G and Z_o of a coaxial transmission line of Figure 11-11.

SOLUTION

The capacitance per unit length of a coaxial line is obtained from the capacitance of a concentric cylindrical capacitor, which was solved in Example 3-7. Eq. (3-36) gives

$$C = \frac{2\pi\varepsilon}{\ln\left(\frac{b}{a}\right)} \qquad (11\text{-}128)$$

Making use of the relationship, $LC = \mu\varepsilon$, we obtain the inductance per unit length:

$$L = \frac{\mu\varepsilon}{C} = \frac{\mu}{2\pi} \ln\left(\frac{b}{a}\right) \tag{11-129}$$

which is also treated in Problem 7-34. Applying the C-G relationship in Eq. (11-124), we obtain

$$G = \frac{2\pi\sigma}{\ln\left(\frac{b}{a}\right)} \tag{11-130}$$

when σ is the conductivity of the dielectric filling. The series resistance per unit length is obtained from Eq. (11-125) by substituting $C_1 = 2\pi a$, $C_2 = 2\pi b$:

$$R = \frac{1}{\sigma_2 d_p}\left(\frac{1}{2\pi a} + \frac{1}{2\pi b}\right) = \frac{R_s}{2\pi}\left(\frac{1}{a} + \frac{1}{b}\right) \tag{11-131}$$

The characteristic impedance is obtained by

$$Z_o = \sqrt{\frac{L}{C}} = \sqrt{\frac{\mu}{\varepsilon}}\frac{1}{2\pi} \ln\left(\frac{b}{a}\right) = \frac{\eta}{2\pi} \ln\frac{b}{a} \tag{11-132}$$

The transmission line parameters (L, C, R, G and Z_o) are summarized in Table 11-4 for three types of transmission lines—parallel-plate line, coaxial line and two-wire line.

Table 11-4. Transmission Line Parameters (L, C, R, G, Z_o)

Parameters	Parallel-Plate Line	Coaxial Line	Two-Wire Line
L	$\mu\frac{a}{w}$	$\frac{\mu}{2\pi}\ln\frac{b}{a}$	$\frac{\mu}{\pi}\cosh^{-1}\left(\frac{D}{2a}\right)$, $D = \text{separation}$
C	$\varepsilon\frac{w}{a}$	$\frac{2\pi\varepsilon}{\ln(b/a)}$	$\frac{\pi\varepsilon}{\cosh^{-1}(D/2a)}$
R	$R_s\frac{2}{w}$	$\frac{R_s}{2\pi}\left(\frac{1}{a}+\frac{1}{b}\right)$	$\frac{R_s}{\pi a}$, $R_s = \frac{1}{\sigma_c d_p} = \sqrt{\frac{\omega\mu_c}{2\sigma_c}}$
G	$\sigma\frac{w}{a}$	$\frac{2\pi\sigma}{\ln(b/a)}$	$\frac{\pi\sigma}{\cosh^{-1}(D/2a)}$
Z_o	$\eta\frac{a}{w}$	$\frac{\eta}{2\pi}\ln\frac{b}{a}$	$\frac{\eta}{\pi}\cosh^{-1}\left(\frac{D}{2a}\right)$

11.7.4. Finite Transmission Line with Load

Let's consider a section of transmission line of characteristic impedance Z_o terminated with a load impedance Z_L as shown in Figure 11-14. We investigate the wave characteristics, in particular, how the wave reflected from the load is related to the wave traveling to the right at any point on the T.L.

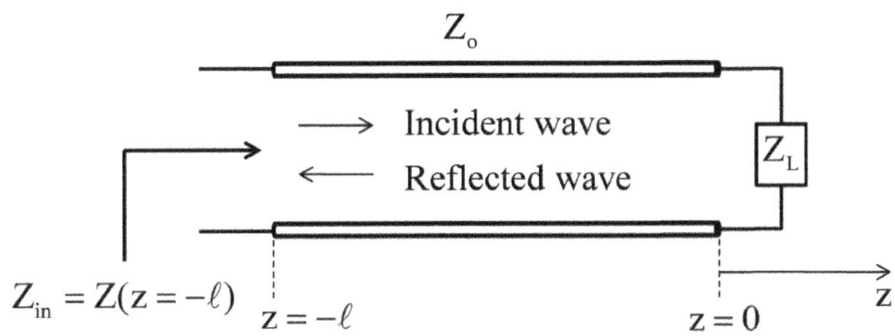

Figure 11-14. A finite transmission line terminated with load.

The general solution for the voltage wave on the line can be written as

$$V(z) = \underbrace{V_o^+ e^{-jkz}}_{\substack{\text{forward traveling} \\ \text{incident wave}}} + \underbrace{V_o^- e^{jkz}}_{\substack{\text{backward traveling} \\ \text{reflected wave}}} \tag{11-107}$$

The current wave on the line is similarly given by

$$I(z) = \frac{V_o^+}{Z_o} e^{-jkz} - \frac{V_o^-}{Z_o} e^{jkz} \tag{11-108}$$

At the location of the load $(z = 0)$,

$$\begin{aligned} V(z=0) &= V_L = V_o^+ + V_o^- \\ I(z=0) &= I_L = \frac{1}{Z_o}\left(V_o^+ - V_o^-\right) \end{aligned} \tag{11-133}$$

If we define the ratio of V_o^- (reflected amplitude) and V_o^+ (incident amplitude) to be the reflection coefficient Γ_L (at the load)

$$\Gamma_L \equiv \frac{V_o^-}{V_o^+} \tag{11-134}$$

We can express Γ_L in terms of Z_o and Z_L using the following relationship:

$$\frac{V_L}{I_L} = Z_L = \frac{V_o^+ + V_o^-}{\frac{1}{Z_o}\left(V_o^+ - V_o^-\right)} = Z_o \frac{V_o^+(1+\Gamma_L)}{V_o^+(1-\Gamma_L)} = Z_o \frac{1+\Gamma_L}{1-\Gamma_L}$$

which leads to

$$\boxed{\Gamma_L = \frac{Z_L - Z_o}{Z_L + Z_o}} \tag{11-135}$$

Note that the reflection coefficient in Eq. (11-135) is very similar to the reflection coefficient of a uniform plane wave normally incident from the medium of η_1 upon the medium of η_2, as shown in Section 10.2:

$$R = \frac{\eta_2 - \eta_1}{\eta_2 + \eta_1} \tag{10-11a}$$

We observe the following three special cases:

(i) Short circuit ($Z_L = 0$) : $\Gamma_L = \frac{V_o^-}{V_o^+} = -1$. This is equivalent to having a perfect conductor in the second medium for the uniform plane wave problem.

(ii) Open circuit ($Z_L = \infty$) : $\Gamma_L = \frac{V_o^-}{V_o^+} = 1$. In this case, $I(z = 0) = 0$.

(iii) Matched load ($Z_L = Z_o$) : $\Gamma_L = 0$ or $V_o^- = 0$. There is no reflected wave.

The impedance $Z(z)$ at any point on the line is defined as

$$Z(z) = \frac{V(z)}{I(z)} = Z_o \frac{V_o^+\left(e^{-jkz} + \Gamma_L\, e^{jkz}\right)}{V_o^+\left(e^{-jkz} - \Gamma_L\, e^{jkz}\right)} \tag{11-136}$$

At the source end ($z = -\ell$) of the line, looking into the line, we see an input impedance $Z_{in} = Z(z = -\ell)$:

$$\begin{aligned} Z(z=-\ell) &= Z_o \frac{e^{jk\ell} + \Gamma_L\, e^{-jk\ell}}{e^{jk\ell} - \Gamma_L\, e^{-jk\ell}} = Z_o \frac{(Z_L + Z_o)\, e^{jk\ell} + (Z_L - Z_o)\, e^{-jk\ell}}{(Z_L + Z_o)\, e^{jk\ell} - (Z_L - Z_o)\, e^{-jk\ell}} \\ &= Z_o \frac{Z_L\, 2\cos(k\ell) + Z_o\, 2\,j\sin(k\ell)}{Z_L\, 2\,j\sin(k\ell) + Z_o\, 2\cos(k\ell)} = Z_o \frac{Z_L + jZ_o \tan(k\ell)}{Z_o + jZ_L \tan(k\ell)} \end{aligned} \tag{11-137}$$

Thus, we can evaluate the input impedance as a function of line length ℓ, given the characteristic impedance of the line and load impedance. $Z(-\ell)$ will vary periodically with period $k\ell = (2\pi/\lambda)\ell = \pi$ or $\ell = \lambda/2$. There are a few special cases for discussion:

(i) Short circuit ($Z_L = 0$) : $Z(-\ell) = jZ_o \tan(k\ell)$. (11-138)
Z_{in} is reactive. For a short section ($k\ell << 1$), $Z(-\ell) \approx jZ_o\, k\ell$. The short-circuited line behaves like an inductor.

(ii) Open circuit ($Z_L = \infty$): $Z(-\ell) = \dfrac{Z_o}{j\tan(k\ell)} = -jZ_o \cot(k\ell)$ (11-139)

Z_{in} is also reactive. For a short section ($k\ell << 1$), $Z(-\ell) \approx -j\frac{Z_o}{k\ell}$. The open-circuited line behaves like a capacitor.

(iii) Matched load ($Z_L = Z_o$) : $Z(-\ell) = Z_o$. The input impedance is equal to the characteristic impedance at all points because there is no reflected wave.

(iv) Quarter-wave section ($\ell = \lambda/4$): $k\ell = \pi/2$, $\tan k\ell \to \infty$

$$Z_{in} = \frac{Z_o^2}{Z_L} \quad \text{or} \quad \boxed{\frac{Z_{in}}{Z_o} = \left(\frac{Z_L}{Z_o}\right)^{-1}} \tag{11-140}$$

A quarter-wave lossless transmission line transforms the normalized load impedance (normalized by Z_o) to its inverse. It is called the **quarter wave transformer**. For example, an open-circuited, quarter wave line appears as a short circuit.

(v) Half-wave section ($\ell = \lambda/2$): $k\ell = \pi$, $\tan k\ell = 0$, $Z_{in} = Z_L$, as expected. When the length of a line is an integer multiplier of $\lambda/2$, Z_{in} does not change.

Analysis of generalized reflection coefficient and the impedance of the transmission line involves tedious manipulations of complex numbers. A graphical chart, known as the **Smith Chart**, greatly simplifies the analysis. See D.K. Cheng (1993) or Shen and Kong (1995) for detailed analysis using the Smith Chart.

Standing Wave Pattern

Whenever the load is mismatched ($Z_L \neq Z_o$), $\Gamma_L \neq 0$ and $V_o^- \neq 0$. Then both forward and backward waves propagate on the line. The total voltages and currents are the superposition of these two and form a **standing wave** just like the standing wave formed by a uniform plane wave incident upon a perfect conductor as discussed in Section 10.6. Voltage maxima occur where the two (incident and reflected) waves are in phase; minima where they are 180° out of phase. Similarly, there are maxima and minima for current along the line. There are local accumulations of electric energy at voltage maxima and of magnetic energy at current maxima, which may reduce the power handling capacity of the line.

Defining the generalized reflection coefficient (at z)

$$\Gamma(z) = \frac{V_o^- \; e^{jkz}}{V_o^+ \; e^{-jkz}} = \Gamma_L \; e^{j2kz} \;, \tag{11-141}$$

the voltage wave from Eq. (11-107) can be re-written as

$$V(z) = V_o^+ \, e^{-jkz} \left\{1 + \Gamma(z)\right\} \tag{11-142}$$

The current wave is similarly given by

$$I(z) = \frac{V_o^+}{Z_o} \; e^{-jkz} \left\{1 - \Gamma(z)\right\} \tag{11-143}$$

It is observed from Eq. (11-141) that for a lossless line

$$|\Gamma(z)| = |\Gamma_L|$$

Since $|\Gamma(z)|$ is constant while its phase varies with z, the amplitude of $V(z)$, i.e., $|V(z)|$, will be a maximum when $\Gamma(z) = |\Gamma_L|$, and it will be a minimum when $\Gamma(z) = -|\Gamma_L|$. But the current amplitude $|I(z)|$ is a minimum at the maximum of the voltage and vice versa. The ratio of the voltage maximum to the voltage minimum is defined as **(voltage) standing wave ratio** or VSWR:

$$\text{VSWR} = \frac{|V(z)|_{max}}{|V(z)|_{min}} = \frac{1+|\Gamma_L|}{1-|\Gamma_L|} \tag{11-144}$$

$|\Gamma_L|$ can also be written in terms of VSWR:

$$|\Gamma_L| = \frac{\text{VSWR}-1}{\text{VSWR}+1} \tag{11-145}$$

Note that since the distance between $\Gamma(z) = |\Gamma_L|$ and $\Gamma(z) = -|\Gamma_L|$ is $\lambda/4$ $(2kd = \pi)$, the distance between two maxima is $\lambda/2$. Thus, although $V(z)$ and $I(z)$ are periodic with λ, $|V(z)|$ and $|I(z)|$ are periodic with $\lambda/2$. Two special cases are for discussion.

(i) Matched load $(Z_L = Z_o) : \Gamma_L = 0$, $\text{VSWR} = 1$, $V(z) = V_o^+ e^{-jkz}$. There is no standing wave.

(ii) Short circuit $(Z_L = 0) : \Gamma_L = -1$, $\text{VSWR} \to \infty$. $|V(z)|$ will be minimum (zero) and $|I(z)|$ will be maximum at the load. They will switch at $z = -\lambda/4$.

The standing wave patterns for $|V(z)|$ and $|I(z)|$ of the above two cases are plotted in Figure 11-15. A finite transmission line with open-circuited load is considered in Problem 11-50.

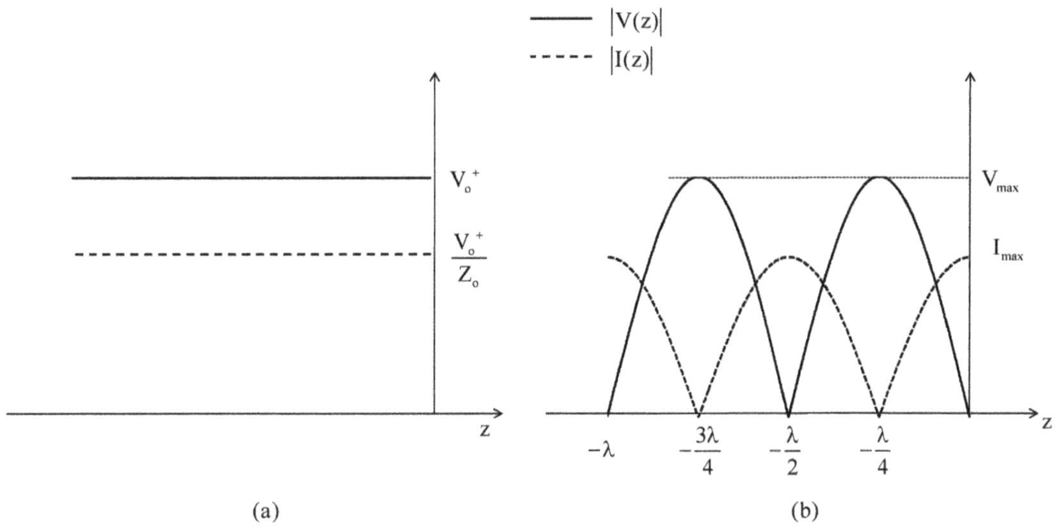

Figure 11-15. Standing wave patterns on a transmission line with (a) matched load and (b) short-circuited load.

PROBLEMS

11-1. Derive the equations for the transverse components in terms of the longitudinal components (E_z, H_z) in the uniform waveguide, given by Eq. (11-10a) through (11-10d).

11-2. In Section 11.3 we derived the dominant (TEM) mode of the parallel-plate waveguide from the TM_m mode by letting $m = 0$. Its fields are given by Eq. (11-27).

(a) Find the surface charge density ρ_s and the surface current density $\mathbf{J}_s$ on the surfaces of both conducting plates.

(b) Show that the electric field is very similar to the static field between a pair of parallel plates carrying opposite charges (see Eq. (2-23c)) and the magnetic field is very similar to the static field between a pair of anti-parallel current sheets (see Table 6-1).

11-3. Derive the expression for the group velocity in the waveguide, given by Eq. (11-33). Show also that v_p and v_g satisfy $v_p v_g = v^2$.

11-4. Derive the expressions for the phase constant, guide wavelength, phase velocity, group velocity, and the wave impedance for the TE_m mode of a parallel-plate waveguide.

11-5. Consider a parallel-plate waveguide filled with a dielectric of $\varepsilon = 6.25\ \varepsilon_o$. The separation between the plates is 1.0 cm.

(a) Find the cutoff frequencies of the first five lowest TM modes.
(b) Suppose the operating frequency is 20 GHz; which TM modes can propagate?

11-6. Consider the microstrip line shown in Example 11-1. Calculate the time-average power that is transmitted by the line for the TM_1 and TE_1 modes. Let $f = 100$ GHz.

11-7. Consider a parallel-plate waveguide filled with a dielectric of $\varepsilon = 4\varepsilon_o$. The separation between the plates is 3 cm.

(a) Find the cutoff frequencies of the first three lowest TE modes.
(b) Determine the range of the frequencies for which only the dominant mode propagates.
(c) Suppose the operating frequency is 6 GHz. List all the modes that can propagate without being cutoff.

11-8. Consider the parallel-plate waveguide of Problem 11-5. Calculate the phase constant, guide wavelength, phase velocity and the wave impedance of the TM_2 mode when $f = 15$ GHz. Find also the time-average Poynting vector of the TM_2 mode.

11-9. Consider the parallel-plate waveguide of Problem 11-7. (a) Calculate the propagation constant and wave impedance of the TE_1 mode when $f = 4$ GHz. Find also the time-average Poynting vector of the TE_1 mode. (b) Repeat the same questions for the TE_2 mode.

11-10. The field solutions that we obtained for the parallel-plate waveguide represent a wave, traveling in the +z direction and standing in the x direction. We can express these fields as a combination of two oppositely traveling waves.

(a) Write the magnetic field H_z for TE modes in a parallel-plate waveguide as a combination of two traveling waves.
(b) Interpret each term.
(c) Explain the ratio of amplitudes of the two traveling waves in terms of the boundary condition (or the reflection coefficient) at the surface of the perfect conductor.

11-11. **(Ionosphere-Earth Waveguide)**
The ionosphere consists of several layers of plasma, with the lowest layer approximately 80 km above the surface of the Earth. At low frequencies in the MHz range both the Earth and the ionosphere are good conductors. Thus, the ionosphere and the Earth form a natural parallel-plate waveguide.

(a) What is the range of frequencies for which only the dominant mode (overall) propagates?
(b) What is the range of frequencies for which only the dominant mode among TE modes propagates?
(c) If the operating frequency is 1 MHz (AM), what modes can propagate?

11-12. The exact fields of a microstrip line shown in Figure 11-6 (see Example 11-1) are neither TM nor TE modes but a hybrid TM-TE mode. However, in practice, the dielectric substrate is electrically very thin ($d << \lambda$) and the fields of the dominant mode are quasi-TEM. The phase constant of the quasi-TEM wave can be expressed as

$$k_z = \omega\sqrt{\mu_o \varepsilon_o}\ \sqrt{\varepsilon_e}$$

where ε_e is called the *effective dielectric constant* of the microstrip line, which depends on the actual dielectric constant ε_r and thickness d of the substrate, and the strip width w. The approximate formula for ε_e is known to be

$$\varepsilon_e = \frac{\varepsilon_r + 1}{2} + \frac{\varepsilon_r - 1}{2}\frac{1}{\sqrt{1+12d/w}}$$

[Pozar (1998)*]. Find the effective dielectric constant of the microstrip line shown in Example 11-1.

11-13. Calculate the surface charge (ρ_s) and the surface current ($\mathbf{J}_s$) on the top and bottom plates of a parallel-plate waveguide for the TM_m mode.

11-14. Repeat Problem 11-13 for the TE_m mode.

11-15. Sketch the (electric and magnetic) field lines of the TE_2 and the TM_2 modes, similar to Figure 11-5.

11-16* **(Attenuation in the parallel-plate waveguide)**
The conducting plates or walls that are used in the metallic waveguide are not perfect conductors; thus the wave can penetrate into the conductors as it propagates along the waveguide. This leads to loss of electromagnetic power and attenuation of the wave. It will also modify the electric and magnetic fields within the guide, but the changes are minimal because the losses are usually small. The attenuation constant α (so that the field decays as $e^{-\alpha z}$ and the total propagating power decays as $e^{-2\alpha z}$) can be calculated, using a perturbation approach [Kong (2008)†], as follows.

$$\alpha = \frac{P_d}{2P_f}$$

where P_f is the time-average power flowing along the waveguide and P_d is the power dissipated (or power penetrated into the conducting walls) per unit guide length (in z direction). Show that the attenuation constant for the TEM mode of parallel-plate waveguide is given by

$$\alpha^{TEM} = \frac{1}{a}\sqrt{\frac{\omega\varepsilon}{2\sigma_c}}$$

where σ_c is the conductivity of the conducting plates.

* D.M. Pozar, *Microwave Engineering*, John Wiley & Sons, 1998, 2nd Ed., p. 162.
†J.A. Kong, *Electromagnetic Wave Theory*, EMW Publishing, 2008, p. 411.

11-17*. Study Problem 11-16 and show that the attenuation constants for the TE_m and TM_m modes of the parallel-plate waveguide are given by

$$\alpha^{TE} = \frac{1}{a}\sqrt{\frac{\omega\varepsilon}{2\sigma_c}}\frac{2(f_{c,m}/f)^2}{\sqrt{1-(f_{c,m}/f)^2}}$$

$$\alpha^{TM} = \frac{1}{a}\sqrt{\frac{\omega\varepsilon}{2\sigma_c}}\frac{2}{\sqrt{1-(f_{c,m}/f)^3}}, \quad m = 1, 2, 3, \ldots$$

11-18. **(Power-handling capacity of a parallel-plate waveguide)**
A parallel-plate air-filled waveguide with the plate separation of 2.0 cm is operated at $f = 10$ GHz. Find the maximum time-average power per unit guide width (w = 1 cm) in watts/cm that can be carried by (a) TEM mode, (b) TM_1 mode and (c) TE_1 mode. Assume that the dielectric strength of air in atmospheric pressure is 3×10^6 V/m.

11-19. Consider a rectangular waveguide having dimensions $a = 3.484$ cm and $b = 1.58$ cm (WR-137 rectangular waveguide). It is filled with a dielectric of $\varepsilon = 2.25\ \varepsilon_o$.

(a) When the operating frequency is 8 GHz, which TM and TE modes can propagate in this waveguide?
(b) Find the phase constant k_z, the cutoff frequency f_c, the phase velocity v_p, the guide wavelength λ_g, and the wave impedance Z_{TM} of the TM_{11} mode.
(c) Suppose only one mode (i.e., the dominant mode) should be transmitted (or guided). What is the range of operating frequencies that can be used?
(d) What is the range of corresponding wavelengths in free space?

11-20. Find the time-average power flowing through the rectangular waveguide for the TM_{mn} mode by integrating $\mathbf{S}_{av}$ (see Eq. (11-61)) over the cross section of the waveguide.

11-21. Consider a rectangular waveguide having dimensions $a = 1.067$ cm and $b = 0.432$ cm (WR-42 rectangular waveguide). It is filled with a dielectric of $\varepsilon = 1.44\ \varepsilon_o$.

(a) When the operating frequency is 35 GHz, which TM and TE modes can propagate in this waveguide?
(b) Find the phase constant k_z, the cutoff frequency f_c, the phase velocity v_p, the group velocity v_g, the guide wavelength λ_g, and the wave impedance Z_{TE} of the TE_{11} mode.
(c) What is the range of frequencies for which only the dominant mode can propagate?

11-22. Find the expression for the time-average Poynting vector $\mathbf{S}_{av}$ for the TE_{mn} mode of the rectangular waveguide and calculate the time-average power flowing through the waveguide for the TE_{mn} mode by integrating $\mathbf{S}_{av}$ over the cross section.

11-23. The field solutions that we obtained for the rectangular waveguide represent a wave, traveling in $+z$ direction and standing in x and y directions, as discussed in Section 11.4.2.

(a) Express the electric field solution E_z for the TM_{mn} mode in a rectangular waveguide as a combination of four traveling waves (in $\pm x$ and $\pm y$ directions).
(b) Interpret each term.

11-24. Repeat Problem 11-23 for the magnetic field solution H_z for the TE_{mn} mode in a rectangular waveguide.

11-25. Calculate the surface charge (ρ_s) and the surface current (J_s) on each of four conducting walls of rectangular waveguide for the TM_{mn} mode.

11-26. Repeat Problem 11-25 for the TE_{mn} mode.

11-27. **(Reception of radio signals inside a tunnel)**
Suppose you are driving inside the Lehigh Tunnel on the Northeast Extension of the Pennsylvania Turnpike. Assume that each of two tunnels is 22 ft wide and 13.5 ft tall.

(a) Can you receive an AM radio signal in an automobile? Explain why.
(b) Also determine whether FM signals can be received inside the tunnel.

Note that the tunnel can be approximately modeled as a rectangular waveguide.

11-28. An air-filled rectangular waveguide ($b < a < 2b$) is to be designed to operate at 6 GHz (C band) in the dominant mode. It is desired that the operating frequency be at least 25% higher than the cutoff frequency of the dominant mode and also at least 25% lower than the cutoff frequency of the next higher-order mode. Determine the typical design for the dimensions a and b.

11-29. The dominant mode of a rectangular waveguide ($a > b$) is the TE_{10} mode. The next higher-order mode is either the TE_{01} or TE_{20} mode, depending on whether b is greater than $a/2$ (or equivalently, whether a is smaller than $2b$). Determine in which case ($b < a/2$ vs. $b > a/2$) we obtain the maximum operating frequency range (band). Show also that when $b = a/2$, the time-average power being transmitted in the waveguide is maximized without sacrificing the operating frequency bandwidth.

11-30. Design an air-filled rectangular waveguide ($a = 2b$) that will be operated at 15 GHz (Ku band). This frequency should be at the middle of the frequency range for single mode operation. Find also the maximum power the waveguide can transmit, assuming that air breaks down at $E = 3\times10^6$ V/m. Use a safety factor of 2.

11-31. Prove that the TEM mode cannot exist in a single-conductor waveguide such as a rectangular waveguide.

11-32. Consider an air-filled L-band rectangular waveguide having the dimensions $a = 10.922$ cm and $b = 5.461$ cm (WR-430 waveguide). Compute the propagation constant when the operating frequency is one-half the cutoff frequency of the dominant mode. How much does the wave attenuate in dB over the distance of 2 cm?

11-33* **(Attenuation in a rectangular waveguide)**
The attenuation due to conductor loss in a metallic waveguide is explained in Problem 11-16.

(a) Show that the power dissipated per unit guide length on the side wall (at $x = 0$ or $x = a$ in Figure 11-7) for the TE_{10} mode is given by

$$P_d = \frac{b}{2} H_o^2 R_s \left[\frac{W}{m}\right]$$

where

$$R_s = \sqrt{\frac{\omega\mu_c}{2\sigma_c}} = \frac{1}{\sigma_c d_p} [\Omega]$$

is the real part of the complex intrinsic impedance of the conductor wall, known as the *surface resistance*, σ_c is its conductivity and d_p is the penetration depth or skin depth.

(b) Show that the power dissipated per unit guide length on top or bottom wall (at $y = 0$ and $y = b$) is given by

$$P_d = \frac{a}{4}\left[1 + \left(\frac{k_z a}{\pi}\right)^2\right] H_0^2 R_s = \frac{a}{4}\left(\frac{f}{f_{c,10}}\right)^2 H_o^2 R_s \left[\frac{W}{m}\right]$$

where Eq. (11-30) has been used.

(b) Show that the attenuation constant for the TE_{10} mode in a rectangular waveguide is given by

$$\alpha = \frac{\left[b + \frac{a}{2}\left(\frac{f}{f_c}\right)^2\right] R_s}{\omega\mu k_z \frac{ab}{2}\left(\frac{a}{\pi}\right)^2} = \frac{R_s\left[1 + \frac{2b}{a}\left(\frac{f_c}{f}\right)^2\right]}{\eta b \sqrt{1 - \left(\frac{f_c}{f}\right)^2}} \left[\frac{Np}{m}\right]$$

11-34. Consider an air-filled rectangular cavity resonator having dimensions $a = d = 2$ cm, $b = 1$ cm and copper walls. Calculate the resonant frequency and Q of the dominant mode.

11-35. Design a square-base $(a = d)$ rectangular cavity resonator with height one-half the width of the base $(b = a/2)$ to resonate with the dominant mode at 1 GHz (a) when it is air filled and (b) when it is filled with polystyrene ($\varepsilon = 2.55\,\varepsilon_o$).

11-36. Prove that for the rectangular cavity resonator $W_e = W_m$ at resonant frequencies.

11-37. Calculate the first five higher-order resonant frequencies for the rectangular cavity of Problem 11-34.

11-38. Confirm that the electric and magnetic fields for the TEM mode of a coaxial cable, shown in Eq. (11-98), satisfy all four Maxwell's equations in cylindrical coordinates.

11-39. Study Example 11-6. From the magnetic field of a coaxial line, calculate the surface current density $\mathbf{J}_s$ on the surfaces of the inner and outer conductors (at $\rho = a$ and $\rho = b$), and total currents flowing along each conductor line. Assuming that the current has amplitude I_o, show that

(a) $I_o = \dfrac{V_o}{Z_o}$ where $Z_o = \dfrac{\eta}{2\pi} \ln \dfrac{b}{a}$

(b) $P = \dfrac{1}{2}\mathrm{Re}(V_o I_o^*) = \dfrac{1}{2Z_o}|V_o|^2$

Z_o is called the *characteristic impedance* of the coaxial line.

11-40. **(Coaxial resonator)**

Just as the rectangular cavity resonator is constructed of the rectangular waveguide by placing two conducting walls normal to the z axis, we can construct a coaxial resonator by taking a section of a coaxial cable and placing perfect conductors at both ends (at $z = 0$ and $z = d$). Combining waves traveling in both $+z$ and $-z$ directions and applying additional boundary conditions at two ends, show that the TEM field solutions are given by

$$\mathbf{E} = \mathbf{a}_\rho \frac{K_o}{\rho} \sin(kz)$$
$$\mathbf{H} = \mathbf{a}_\phi \frac{jK_o}{\eta\rho} \cos(kz)$$

where K_o is an arbitrary constant and

$$k = \omega\sqrt{\mu\varepsilon} = \frac{m\pi}{d}, \quad m = 1, 2, 3, \ldots$$

Show that the resonant frequencies are given by

$$f_{r,m} = \frac{1}{\sqrt{\mu\varepsilon}} \frac{m}{2d} = \frac{v}{2d} m, \quad m = 1, 2, 3, \ldots$$

For the dominant mode ($m = 1$), $f_{r,1} = v/(2d) = v/\lambda$ and $d = \lambda/2$, i.e., the length of the resonator should be half wavelength.

11-41* **(Attenuation in a coaxial cable)**

The attenuation due to conductor loss in a metallic waveguide is explained in Problem 11-16. Following the same procedure, show that the attenuation constant for the TEM mode of the coaxial line is given by

$$\alpha = \frac{R_s}{2\eta \ln(b/a)}\left[\frac{1}{a}+\frac{1}{b}\right]$$

where $R_s = \sqrt{\dfrac{\omega\mu_c}{2\sigma_c}}$ is the surface resistance of the conductor (see Problem 11-33).

11-42. Consider a low-loss transmission line for which $R \neq 0$, $G \neq 0$ but $R << \omega L$, $G << \omega C$. The propagation constant becomes complex in a lossy T.L.

$$k = \sqrt{-(j\omega L + R)\,(j\omega C + G)} = \omega\sqrt{LC}\sqrt{\left(1+\frac{R}{j\omega L}\right)\left(1+\frac{G}{j\omega C}\right)} = \beta - j\alpha$$

Show that for a low-loss T.L. the approximate expressions for β and α are

$$\beta \approx \omega\sqrt{LC}$$
$$\alpha \approx \frac{1}{2}\left(R\sqrt{\frac{C}{L}} + G\sqrt{\frac{L}{C}}\right)$$

11-43. **(Distortionless Line)**

For a lossy transmission line, the phase velocity of the wave depends on the frequency, leading to *distortion* of a signal when it consists of a band of frequencies. However, we can avoid this by making the T.L. so that the following condition is satisfied:

$$\frac{L}{R} = \frac{C}{G}$$

Show that under this condition, α, β, and v_p (phase velocity) are given by

$$\beta = \omega\sqrt{LC} \quad \text{(a linear function of f)}$$

$$\alpha = R\sqrt{\frac{C}{L}} = \sqrt{RG} \quad \text{(independent of f)}$$

$$v_p = \frac{\omega}{\beta} = \frac{1}{\sqrt{LC}} \quad \text{(independent of f)}$$

Such a transmission line is a *distortionless* line.

11-44. Calculate L, C, R, G and Z_o of a two-wire transmission line of radius a (for each conductor) and separation D (between two conductors). For calculation of C, see Example 4-4 and Problem 4-14. The results are in Table 11-4.

11-45. Design a microstrip line (modeled as a parallel-plate line) which has the characteristic impedance $75\ \Omega$ and the maximum time-average power allowed 100 kW. The dielectric material to be used is a polystyrene which has $\varepsilon = 2.55\ \varepsilon_o$, $\mu = \mu_o$, $\sigma \approx 0$ and the dielectric strength 2×10^7 V/m. Use a safety factor for 10 so that the maximum electric field does not exceed 2×10^6 V/m. Determine the width (w) of the metal strip and the thickness (a) of the substrate. Calculate also L and C.

11-46. The coaxial cable RG-213/U is filled with a dielectric of $\varepsilon = 2.56\ \varepsilon_o$, $\mu = \mu_o$. The capacitance per unit length of the cable is $C = 50$ [pF/m]. Find the velocity of wave propagation, the inductance per unit length, and the characteristic impedance. Suppose the dielectric is lossy and its conductivity is $\sigma = 10^{-14}$ [S/m]. Find the conductance per unit length of the cable.

11-47. Consider a finite transmission line terminated with load. When $Z_o = 50\ \Omega$, $Z_L = 100\ \Omega$ and $\ell = 2.1\ \lambda$, calculate Γ_L and Z_{in} (at $z = -\ell$).

11-48. For a finite transmission line of length $\ell = \lambda/8$ and $Z_o = 50\ \Omega$ terminated with load, find the load impedance Z_L that will provide $Z_{in} = 100 + j150\ \Omega$.

11-49. Show that for short-circuited or open-circuited lossless line, the input impedance repeats itself as inductive and capacitive alternately for every $\ell = \lambda/4$, by plotting Eqs. (11-138) and (11-139) as a function of $k\ell$.

11-50. Consider a finite lossless transmission line with open-circuited load. Find VSWR and sketch the standing wave patterns for $|V(z)|$ and $|I(z)|$.

11-51. Repeat Problem 11-50 when $Z_L = 100\ \Omega$, $Z_o = 50\ \Omega$.

11-52. In a laboratory experiment conducted on a lossless $100\ \Omega$ transmission line terminated with an unknown load, the VSWR is measured to be 2.0. The first and second voltage minima occur at 3 cm and 23 cm from the load, respectively. Determine the reflection coefficient (Γ_L) at the load and the load impedance (Z_L). Where would the first voltage maximum be located if the load were replaced by a short circuit?

RADIATION AND ANTENNAS

CHAPTER 12

12.1 INTRODUCTION

We have so far studied propagation, reflection, transmission, guidance, and resonance of the electromagnetic plane waves in the source-free region, that is, at points away from the source, without knowing how the waves were generated by the sources. In this chapter, we consider how the time-varying current and charge sources generate the waves and radiate the electromagnetic energy. It is to be noted that a charge at rest does not generate electromagnetic waves, nor does a steady (dc) electric current. It takes accelerating charges or time-varying currents to produce electromagnetic waves. An antenna is a device or structure that is designed to launch or radiate electromagnetic waves efficiently into space and to focus or direct these waves in a certain direction. The antenna is also used to receive an electromagnetic signal.

We first present the potential formulation where the electric and magnetic fields are expressed in terms of the scalar and vector potentials. The retarded potentials are found for both arbitrary and time-harmonic charge and current sources. Then we consider the elementary (Hertzian) dipole antenna and the linear antenna with sinusoidal current distribution. Finally we study a linear array of identical, parallel antennas with different phases and amplitudes. Several special cases are considered.

12.2 POTENTIAL FORMULATION—METHOD OF SOLUTIONS FOR RADIATION PROBLEMS

The antenna (or radiation) problem consists of solving for the fields that are generated by an impressed current distribution – $\mathbf{J}$, $\mathbf{J}_s$ or $\mathbf{I}$, depending on the type of the antenna. In many practical problems the current distribution is not known in advance, but it is obtained during the solution process. How to obtain the current distribution is beyond the scope of this book. For the moment, assuming that the current distribution $\mathbf{J}$ is known, we wish to determine the electric field $\mathbf{E}$ and the magnetic field $\mathbf{H}$. The typical approach is the use of the scalar electric potential V and the (magnetic) vector potential $\mathbf{A}$, known as the *potential formulation*. Since we deal with time-varying fields, we expect that some of the results we obtained in the static problems will not be valid here.

We again start with Maxwell's equations that are re-written here.

$$\nabla \times \mathbf{E} = -\frac{\partial \mathbf{B}}{\partial t} \tag{8-20a}$$

$$\nabla \times \mathbf{H} = \mathbf{J} + \frac{\partial \mathbf{D}}{\partial t} \tag{8-20b}$$

$$\nabla \cdot \mathbf{D} = \rho_v \tag{8-20c}$$

$$\nabla \cdot \mathbf{B} = 0 \tag{8-20d}$$

First, since the magnetic field $\mathbf{B}$ is divergenceless from Eq. (8-20d), it can be expressed as the curl of another vector function $\mathbf{A}$, as this was done in magnetostatics (Section 6.5).

$$\boxed{\mathbf{B} = \nabla \times \mathbf{A}} \tag{12-1}$$

$\mathbf{A}$ is called the **magnetic vector potential** or simply **vector potential**. The magnetic field $\mathbf{H}$ is obtained as $\mathbf{H} = \frac{1}{\mu}\mathbf{B} = \frac{1}{\mu}\nabla \times \mathbf{A}$. Substituting Eq. (12-1) into Faraday's law, Eq. (8-20a), we obtain

$$\nabla \times \mathbf{E} = -\frac{\partial}{\partial t}(\nabla \times \mathbf{A})$$

or

$$\nabla \times \left\{\mathbf{E} + \frac{\partial \mathbf{A}}{\partial t}\right\} = 0 \tag{12-2}$$

Since the sum of two quantities inside the bracket is curl-free, it can be expressed as the gradient of a scalar function V (see Section 1.5), as this was done in electrostatics (Section 2.7).

$$\mathbf{E} + \frac{\partial \mathbf{A}}{\partial t} = -\nabla V \tag{12-3}$$

from which we obtain

$$\boxed{\mathbf{E} = -\nabla V - \frac{\partial \mathbf{A}}{\partial t}} \tag{12-4}$$

V is called the **scalar electric potential** or simply **scalar potential**. Note that in the static case, $\frac{\partial \mathbf{A}}{\partial t} = 0$, thus Eq. (12-4) reduces to $\mathbf{E} = -\nabla V$, Eq. (2-41). For time-varying fields, the electric field depends on both V and $\mathbf{A}$. Now the $\mathbf{E}$ and $\mathbf{H}$ fields are expressed in terms of the potential functions, V and $\mathbf{A}$. If we find V and $\mathbf{A}$, we can obtain easily the fields.

In order to find the solutions for the potentials, we derive the partial differential equations that will be satisfied by V and $\mathbf{A}$ as follows. Let's substitute Eqs. (12-1) and (12-4) into Eq. (8-20b) and make use of the constitutive relations:

$$\mathbf{D} = \varepsilon \mathbf{E}, \ \mathbf{B} = \mu \mathbf{H} \tag{8-21}$$

We have

$$\nabla \times \left(\frac{1}{\mu}\nabla \times \mathbf{A}\right) = \mathbf{J} + \varepsilon \frac{\partial}{\partial t}\left\{-\nabla V - \frac{\partial \mathbf{A}}{\partial t}\right\} \tag{12-5}$$

Using the vector identity, $\nabla \times \nabla \times \mathbf{A} = \nabla(\nabla \cdot \mathbf{A}) - \nabla^2 \mathbf{A}$ (see Table 1-5), we obtain

$$\nabla(\nabla \cdot \mathbf{A}) - \nabla^2 \mathbf{A} = \mu \mathbf{J} - \nabla\left(\mu\varepsilon \frac{\partial V}{\partial t}\right) - \mu\varepsilon \frac{\partial^2 \mathbf{A}}{\partial t^2}$$

or

$$\nabla^2 \mathbf{A} - \mu\varepsilon \frac{\partial^2 \mathbf{A}}{\partial t^2} = -\mu \mathbf{J} + \nabla\left\{\nabla \cdot \mathbf{A} + \mu\varepsilon \frac{\partial V}{\partial t}\right\} \tag{12-6}$$

Now, according to Helmholtz's theorem discussed in Section 1.5, a vector field is completely determined if its divergence and curl are specified everywhere. For the vector potential $\mathbf{A}$, its curl is given as $\mathbf{B}$ in Eq. (12-1) but its divergence is not specified. Thus we are free to choose $\nabla \cdot \mathbf{A}$ for convenience. We let

$$\nabla \cdot \mathbf{A} = -\mu\varepsilon \frac{\partial V}{\partial t} \tag{12-7}$$

which makes the second term in Eq. (12-6) disappear and reduces Eq. (12-6) to a simple partial differential equation (PDE). The relation between $\mathbf{A}$ and V in Eq. (12-7) is called the **Lorenz condition** or **Lorenz gauge** for potentials. For static fields, it reduces to $\nabla \cdot \mathbf{A} = 0$, which is known as the Coulomb gauge, for magnetostatic fields, as discussed in Section 6.5. Applying Eq. (12-7), we have the following PDE for $\mathbf{A}$:

$$\boxed{\nabla^2 \mathbf{A} - \mu\varepsilon \frac{\partial^2 \mathbf{A}}{\partial t^2} = -\mu \mathbf{J}} \tag{12-8}$$

Eq. (12-8) is called the **inhomogeneous wave equation for the vector potential $\mathbf{A}$** in the presence of the current source $\mathbf{J}$. In the source-free region, Eq. (12-8) reduces to the wave equation we have seen earlier for $\mathbf{E}$ and $\mathbf{H}$ in Eqs. (9-4) and (9-5). Similarly, we can also derive the PDE for the scalar potential V, by substituting Eq. (12-4) into Eq. (8-20c) and making use of Eqs. (8-21) and (12-7), as follows.

$$\boxed{\nabla^2 V - \mu\varepsilon \frac{\partial^2 V}{\partial t^2} = -\frac{\rho_v}{\varepsilon}} \tag{12-9}$$

which is called the **inhomogeneous wave equation for the scalar potential** V in the presence of the charge source ρ_v. The Lorenz condition made it possible to decouple the wave equations for $\mathbf{A}$ and V. We find striking similarities between the two wave equations. They are identical when the following substitutions are made:

$$\mathbf{A} \to V, \quad \mu \mathbf{J} \to \frac{\rho_v}{\varepsilon}$$

Thus we expect that the solutions for $\mathbf{A}$ and V, given $\mathbf{J}$ and ρ_v, respectively, will take similar forms. However, we don't need to find both $\mathbf{A}$ and V in order to determine the $\mathbf{E}$ and $\mathbf{H}$ fields because V is related to $\mathbf{A}$ through the Lorenz condition, Eq. (12-7). This is also expected through the relationship between the current and charge densities ($\mathbf{J}$ and ρ_v), given by the equation of continuity:

$$\nabla \cdot \mathbf{J} = -\frac{\partial \rho_v}{\partial t} \tag{5-7}$$

Therefore, for radiation problems, we try to determine only the vector potential $\mathbf{A}$ by solving Eq. (12-8) and calculate the $\mathbf{E}$ and $\mathbf{H}$ fields from $\mathbf{A}$ using Eqs. (12-1) and (12-4) along with Eq. (12-7).

12.2.1 Retarded Potentials

Now our task is to find the solutions of Eqs. (12-8) and (12-9) for $\mathbf{A}(\mathbf{r}, t)$ and $V(\mathbf{r}, t)$, given the sources $\mathbf{J}(\mathbf{r}, t)$ and $\rho_v(\mathbf{r}, t)$. The solutions to these equations for time-varying sources can be obtained by making use of the potential solutions for electrostatic and magnetostatic problems. In the static limit where $\frac{\partial}{\partial t} = 0$, Eqs. (12-8) and (12-9) reduce to the vector and scalar Poisson's equations, (6-17) and (4-3), which are repeated here:

$$\nabla^2 \mathbf{A} = -\mu \mathbf{J} \qquad \text{(6-17) or (12-10a)}$$

$$\nabla^2 V = -\frac{\rho_v}{\varepsilon} \qquad \text{(4-3) or (12-10b)}$$

The solutions to Poisson's equations for $\mathbf{A}$ and V were obtained in Sections 6.5 and 2.8 as follows:

$$\mathbf{A}(\mathbf{r}) = \iiint \frac{\mu \mathbf{J}(r')dv'}{4\pi R} \qquad \text{(6-18a) or (12-11a)}$$

$$V(\mathbf{r}) = \iiint \frac{\rho_v(r')dv'}{4\pi\varepsilon R} \qquad \text{(2-46a) or (12-11b)}$$

where $R = |\mathbf{r} - \mathbf{r}'|$ is the distance from the source point $\mathbf{r}'$ to the field point $\mathbf{r}$. The integrals in Eqs. (12-11) show how the static sources, $\mathbf{J}$ and ρ_v at the source point $\mathbf{r}'$, generate the potentials, $\mathbf{A}$ and V at the field point $\mathbf{r}$. Knowing these relationships, we expect that the time-varying sources, $\mathbf{J}$ and ρ_v at the source point $\mathbf{r}'$, would generate the time-varying potentials, $\mathbf{A}(\mathbf{r}, t)$ and $V(\mathbf{r}, t)$ at the field point $\mathbf{r}$, in a similar fashion except for some time delay because it takes time for temporal variation of the sources to reach the field point. Using these arguments, without mathematical proof we present the solutions to Eqs. (12-8) and (12-9) as follows.

$$\mathbf{A}(\mathbf{r},t) = \iiint \frac{\mu \mathbf{J}(\mathbf{r}',t')}{4\pi R}\, dv' \tag{12-12}$$

$$V(\mathbf{r},t) = \iiint \frac{\rho_v(\mathbf{r}',t')}{4\pi\varepsilon R}\, dv' \tag{12-13}$$

where

$$R = |\mathbf{r} - \mathbf{r}'|$$
$$t' = t - \frac{R}{v} = t - \sqrt{\mu\varepsilon}\,|\mathbf{r} - \mathbf{r}'| \tag{12-14}$$

$\mathbf{A}(\mathbf{r}, t)$ and $V(\mathbf{r}, t)$ given in Eqs. (12-12) and (12-13) are called the **retarded vector potential** and the **retarded scalar potential**, respectively. t' in Eq. (12-14), called the **retarded time**, is a time *earlier* than t by $\Delta t = \frac{R}{v}$, which is how long it takes the electromagnetic wave to travel a distance R. The physical interpretation of the retarded potentials is given as follows. As shown in Eqs. (12-12) and (12-13), at any given point $\mathbf{r}$ and given time t, the potentials ($\mathbf{A}$ and V) are determined by the current and charge distributions that existed at other points $\mathbf{r}'$ (source point) at earlier time t'. The time difference between t and t' is the time required for the electromagnetic "news," due to temporal change of $\mathbf{J}$ and ρ_v, to travel from the source point $\mathbf{r}'$ to the field point $\mathbf{r}$. In other words, any temporal change that occurs at the sources will be felt *later* to the observer at some distance away.

The mathematical proof that the retarded potentials in Eqs. (12-12) and (12-13) satisfy the wave equations, Eqs. (12-8) and (12-9), can be done by evaluating the Laplacian $\nabla^2 V = \nabla \cdot (\nabla V)$ or $\nabla^2 \mathbf{A}$ (each component), recognizing that t' has a spatial dependence in it, and making use of the static solutions of Poisson's equation. See Griffiths (1999)* for proof of the retarded scalar potential. The integral expressions for the retarded potentials can also be derived by considering the point sources at the origin, solving the wave equation in spherical coordinates with the use of spherical symmetry, and lastly applying the superposition principle for the distributed sources. See, for example, Reitz, Milford and Christy (1993)† for this alternative approach.

One final note is that the potential integrals in Eqs. (12-12) and (12-13) with $t' = t + \frac{R}{v}$ (advanced time) also satisfy Eqs. (12-8) and (12-9). However, we discard these solutions because they represent potentials that depend on the sources at future time. These are physically unacceptable solutions and they violate the principle of *causality*.

12.2.2 Retarded Vector Potential for Time-Harmonic Fields

When the current sources are sinusoidal with single frequency, i.e., time-harmonic, the expressions for the fields and potentials can be simplified. As discussed in Section 9.3.1, for time-harmonic problems the time derivative $\frac{\partial}{\partial t}$ can be replaced by the algebraic product of $j\omega$. Thus the electric field in Eq. (12-4) can be written as

$$\mathbf{E} = -j\omega\,\mathbf{A} - \nabla V \tag{12-15}$$

and the scalar potential is related to the vector potential from Eq. (12-7) (the Lorenz gauge) as follows:

$$\nabla \cdot \mathbf{A} = -j\omega\mu\varepsilon V \tag{12-16}$$

*D. J. Griffiths, *Introduction to Electrodynamics*, Prentice-Hall, 3rd Ed., 1999.
†J. R. Reitz, F. J. Milford and R. W. Christy, *Foundations of Electromagnetic Theory*, Addison-Wesley, 4th Ed., 1993.

or

$$V = \frac{-\nabla \cdot \mathbf{A}}{j\omega\mu\varepsilon} \tag{12-17}$$

Substituting Eq. (12-17) into Eq. (12-15) leads to

$$\boxed{\mathbf{E} = -j\omega\mathbf{A} + \frac{\nabla(\nabla \cdot \mathbf{A})}{j\omega\mu\varepsilon}} \tag{12-18}$$

Therefore, the electric field can be completely determined once the vector potential $\mathbf{A}$ is obtained. The vector potential integral in Eq. (12-12) for time-harmonic sources can be obtained as follows. Assume the time-harmonic current density to be of the form:

$$\mathbf{J}(\mathbf{r}, t) = \mathrm{Re}\left\{\mathbf{J}(\mathbf{r})e^{j\omega t}\right\} \tag{12-19}$$

in which $\mathbf{J}(\mathbf{r})$ is interpreted as the complex phasor representation of $J(\mathbf{r},t)$. The current density that appears in the retarded vector potential [see Eq. (12-12)] is then given by

$$\begin{aligned} \mathbf{J}(\mathbf{r}', t') &= \mathbf{J}\left(\mathbf{r}', t - \frac{R}{v}\right) = \mathrm{Re}\left\{\mathbf{J}(\mathbf{r}')e^{j\omega\left(t - \frac{R}{v}\right)}\right\} \\ &= \mathrm{Re}\left\{\mathbf{J}(\mathbf{r}')e^{-j\frac{\omega}{v}R}\, e^{j\omega t}\right\} = \mathrm{Re}\left\{\mathbf{J}(\mathbf{r}')e^{-jkR}\, e^{j\omega t}\right\} \end{aligned} \tag{12-20}$$

where $\frac{\omega}{v} = \omega\sqrt{\mu\varepsilon} = k$ has been used. In Eq. (12-20), we observe that $J(\mathbf{r}')e^{-jkR}$ is the complex phasor representation of $\mathbf{J}(\mathbf{r}', t')$. Therefore, the retarded vector potential for time-harmonic fields is given by

$$\boxed{\begin{aligned} \mathbf{A}(\mathbf{r}) &= \iiint_V \frac{\mu\,\mathbf{J}(\mathbf{r}')e^{-jkR}}{4\pi R}\,dv' \\ &= \iiint_V \frac{\mu\,\mathbf{J}(\mathbf{r}')e^{-jk|\mathbf{r}-\mathbf{r}'|}}{4\pi\,|\mathbf{r}-\mathbf{r}'|}\,dv' \end{aligned}} \tag{12-21}$$

where $\mathbf{r}$ indicates the position of the field point where $\mathbf{A}$ is to be evaluated, $\mathbf{r}'$ indicates the position of the source point where the current source $\mathbf{J}$ is located and $R = |\mathbf{r} - \mathbf{r}'|$ is the distance between the source point $\mathbf{r}'$ and the field point $\mathbf{r}$, as shown in Figure 12-1.

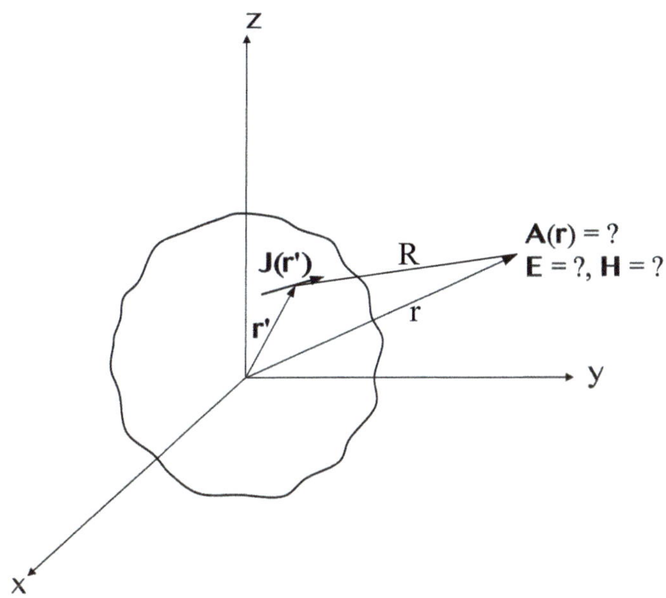

Figure 12-1. The vector potential A at r due to the current source J at r′.

12.2.3. Finding Antenna Fields

Given the time-harmonic current density $\mathbf{J}$ distributed over the antenna source, the procedures of finding the electric and magnetic fields produced by this antenna can be summarized in Table 12-1.

Table 12-1. Road Map for Finding Antenna Fields

Step 1. Calculate the retarded vector potential $\mathbf{A}$: $$\mathbf{A}(\mathbf{r}) = \iiint_V \frac{\mu \mathbf{J}(\mathbf{r}') \; e^{-jk\lvert\mathbf{r}-\mathbf{r}'\rvert}}{4\pi \lvert \mathbf{r}-\mathbf{r}' \rvert} dv'$$
Step 2. Find the magnetic field $\mathbf{H}$: $$\mathbf{H} = \frac{1}{\mu} \nabla \times \mathbf{A}$$
Step 3. Find the electric field $\mathbf{E}$: $$\mathbf{E} = \frac{\nabla \times \mathbf{H}}{j\omega\varepsilon}$$

Note that in Step 3 the electric field can also be calculated directly from the vector potential $\mathbf{A}$, using Eq. (12-18). Normally this requires more computation, so Ampère's law is more often used to calculate $\mathbf{E}$ from $\mathbf{H}$. As a final note, although Eq. (12-21) gives the vector potential for the volume current distribution $\mathbf{J}$, we can easily write the potential integral for the surface current distribution $\mathbf{J}_s$ (for example, dish antenna), simply by replacing

$\mathbf{J}(\mathbf{r}')\,dv'$ by $\mathbf{J}_s(\mathbf{r}')\,ds'$ and changing the volume integral to the surface integral. Similarly, the vector potential integral for the line current distribution $\mathbf{I}$ (for example, wire antenna) is obtained by replacing $\mathbf{J}(\mathbf{r}')\,dv'$ by $\mathbf{I}(\mathbf{r}')\,d\ell'$ as follows:

$$\mathbf{A}(\mathbf{r}) = \int_L \frac{\mu \mathbf{I}(\mathbf{r}')\, e^{-jk|\mathbf{r}-\mathbf{r}'|}}{4\pi|\mathbf{r}-\mathbf{r}'|}\, d\ell' \tag{12-22}$$

(Potential integral for wire antenna)

12.3 ELEMENTARY DIPOLE ANTENNA—HERTZIAN DIPOLE

The first antenna configuration to consider is the so-called **Hertzian dipole**, the simplest and most fundamental antenna. It is a thin wire piece of *uniform amplitude* current of *infinitesimal length* or of very small finite length much shorter than wavelength, i.e., $\Delta\ell << \lambda$. It does not exist by itself in practice, but it may be considered to be a piece of an actual wire antenna. Understanding fully the radiation characteristics of the Hertzian dipole antenna is very important because the radiation properties of the actual wire antenna result from the superposition of many infinitesimal (Hertzian) dipoles. The Hertzian dipole also gives a good approximation of an electrically small ($\Delta\ell << \lambda$) dipole antenna.

We assume that the antenna has an infinitesimal length $\Delta\ell$, and the current (phasor) flowing through the antenna has amplitude I, pointing in the z direction, i.e., the antenna is placed along the z axis (Figure 12-2). Then the line current source is given by

$$\mathbf{I} = \mathbf{a}_z I, \quad -\frac{\Delta\ell}{2} \le z \le \frac{\Delta\ell}{2} \tag{12-23}$$

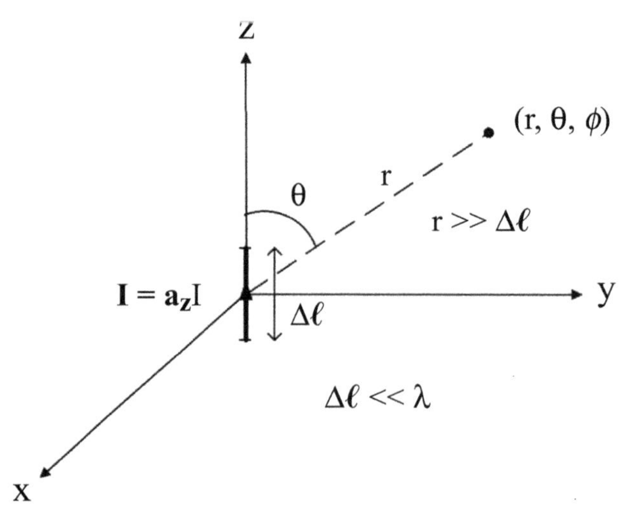

Figure 12-2. Hertzian dipole antenna.

We will use Eq. (12-22) to calculate the vector potential $\mathbf{A}$. As mentioned above, there are three conditions that define a Hertzian dipole:

(i) *I is constant*, independent of the position, i.e., the wire has a uniform current distribution.

(ii) $\Delta\ell << r = |\mathbf{r}|$ (r is the distance to the field point), i.e., we observe the field at points far from the source. As $\Delta\ell \to 0$, fields we will obtain are valid *everywhere*.

(iii) $\Delta\ell << \lambda$, λ is the free-space wavelength of the time-harmonic current source and $\lambda = \frac{v}{f} = \frac{2\pi}{k}$.

Note that the second condition (ii) applies to all antenna problems. In other words, we don't observe the fields at points close to the antenna because the main purpose of the antenna is to radiate the electromagnetic energy to far distances. Since we observe the field at far points, it is more convenient to work the problem in spherical coordinates (r, θ, ϕ) than rectangular or cylindrical coordinates.

Now in order to find the $\mathbf{E}$ and $\mathbf{H}$ fields we follow the procedures outlined in Table 12-1.

Step 1. Calculate the retarded vector potential $\mathbf{A}$ first.

Substituting Eq. (12-23) into Eq. (12-22), we obtain

$$\mathbf{A}(\mathbf{r}) = \int_{-\frac{\Delta\ell}{2}}^{\frac{\Delta\ell}{2}} \frac{\mathbf{a}_z \mu I e^{-jk|\mathbf{r}-\mathbf{r}'|}}{4\pi |\mathbf{r}-\mathbf{r}'|} dz' = \mathbf{a}_z \frac{\mu I}{4\pi} \int_{-\frac{\Delta\ell}{2}}^{\frac{\Delta\ell}{2}} \frac{e^{-jk|\mathbf{r}-\mathbf{r}'|}}{|\mathbf{r}-\mathbf{r}'|} dz' \tag{12-24}$$

where $\mathbf{r} = \mathbf{a}_r r$ and $\mathbf{r}' = \mathbf{a}_z z'$. Noting that $|\mathbf{r}'|_{max} = (z')_{max} = \frac{\Delta\ell}{2}$ and using the condition (ii), $r >> \Delta\ell$, we can approximate

$$\frac{1}{|\mathbf{r}-\mathbf{r}'|} \approx \frac{1}{r} \tag{12-25a}$$

Now consider the effect of $\mathbf{r}'$ on $e^{-jk|\mathbf{r}-\mathbf{r}'|}$. We write

$$\begin{aligned} |\mathbf{r}-\mathbf{r}'| &= \sqrt{|\mathbf{r}-\mathbf{r}'|^2} = \sqrt{(\mathbf{r}-\mathbf{r}')\cdot(\mathbf{r}-\mathbf{r}')} \\ &= \left[\mathbf{r}\cdot\mathbf{r} - 2\mathbf{r}\cdot\mathbf{r}' + \mathbf{r}'\cdot\mathbf{r}'\right]^{1/2} = \left[|\mathbf{r}|^2 - 2r\,\mathbf{a}_r\cdot\mathbf{r}' + |\mathbf{r}'|^2\right]^{1/2} \\ &\approx \left[r^2 - 2r\,\mathbf{a}_r\cdot\mathbf{r}'\right]^{1/2} \quad \text{when} \quad r \gg |\mathbf{r}'| = z' \\ &\approx r - z'\cos\theta \end{aligned} \tag{12-25b}$$

where θ is the angle between two vectors $\mathbf{r}$ and $\mathbf{r}'$ (z-axis). Note that $\mathbf{a}_r \cdot \mathbf{r}' = z'\cos\theta$.

The complex exponential term in Eq. (12-24) is given by

$$e^{-jk|\mathbf{r}-\mathbf{r}'|} \approx e^{-jk(r - z'\cos\theta)} = e^{-jkr}\, e^{jkz'\cos\theta} \tag{12-26}$$

Under the third condition (iii), $\Delta\ell << \lambda$,

$$kz' = \frac{2\pi}{\lambda} z' \ll 1 \qquad \text{(becomes very small)}$$

Therefore, the correction term in Eq. (12-26) is small and we can approximate the complex exponential, for the Hertzian dipole, as follows:

$$e^{-jk|\mathbf{r}-\mathbf{r}'|} \approx e^{-jkr} \tag{12-27}$$

Making use of the two approximations, Eqs., (12-25) and (12-27), we obtain finally

$$\mathbf{A}(\mathbf{r}) = \mathbf{a}_z \frac{\mu I}{4\pi} \int_{-\frac{\Delta\ell}{2}}^{\frac{\Delta\ell}{2}} \frac{e^{-jkr}}{r} dz' = \mathbf{a}_z \frac{\mu I \Delta\ell}{4\pi r} e^{-jkr} \equiv \mathbf{a}_z A_z \tag{12-28}$$

<u>Step 2</u>. Find the magnetic field $\mathbf{H}$ by taking the curl of $\mathbf{A}$.

We note that $\mathbf{A}$ in Eq. (12-28) is written in mixed coordinates—$\mathbf{a}_z$ in rectangular or cylindrical coordinate and the component A_z in terms of spherical coordinate (r). In order to calculate $\nabla \times \mathbf{A}$, we need to write $\mathbf{A}$ in one coordinate system. Using

$$\mathbf{a}_z = \mathbf{a}_r \cos\theta - \mathbf{a}_\theta \sin\theta$$

from Table 1-3, $\mathbf{A}$ can be rewritten as

$$\begin{aligned}\mathbf{A} &= (\mathbf{a}_r \cos\theta - \mathbf{a}_\theta \sin\theta)\, A_z \\ &= \mathbf{a}_r (A_z \cos\theta) + \mathbf{a}_\theta(-A_z \sin\theta) \equiv \mathbf{a}_r A_r + \mathbf{a}_\theta A_\theta\end{aligned} \tag{12-29}$$

Noting that $A_\phi = 0$ and $\frac{\partial}{\partial\phi} = 0$ (A_r and A_θ have no ϕ dependence), and referring to Table 1-4, we calculate $\mathbf{H}$ as follows.

$$\begin{aligned}\mathbf{H} &= \frac{1}{\mu}\nabla\times\mathbf{A} = \frac{1}{\mu}\mathbf{a}_\phi \frac{1}{r}\left[\frac{\partial}{\partial r}(rA_\theta) - \frac{\partial A_r}{\partial\theta}\right] \\ &= \mathbf{a}_\phi \frac{1}{\mu}\frac{1}{r}\left\{\frac{\partial}{\partial r}\left[\frac{\mu}{4\pi} I\Delta\ell\, e^{-jkr}(-\sin\theta)\right] - \frac{\partial}{\partial\theta}\left[\frac{\mu}{4\pi} I\Delta\ell \frac{e^{-jkr}}{r}\cos\theta\right]\right\} \\ &= \mathbf{a}_\phi \frac{I\Delta\ell}{4\pi} jk \frac{e^{-jkr}}{r}\left[1 + \frac{1}{jkr}\right]\sin\theta \equiv \mathbf{a}_\phi H_\phi\end{aligned} \tag{12-30}$$

Step 3. Find the electric field $\mathbf{E}$ by taking the curl of $\mathbf{H}$.

Noting that $H_r = H_\theta = 0$ and $\frac{\partial}{\partial \phi} = 0$, we calculate $\mathbf{E}$ (see Table 1-4):

$$\begin{aligned}\mathbf{E} &= \frac{1}{j\omega\varepsilon}\nabla\times\mathbf{H} = \frac{1}{j\omega\varepsilon}\left[\mathbf{a}_r \frac{1}{r\sin\theta}\frac{\partial}{\partial\theta}\left(H_\phi \sin\theta\right) - \mathbf{a}_\theta \frac{1}{r}\frac{\partial}{\partial r}\left(rH_\phi\right)\right] \\ &= \frac{I\Delta\ell}{4\pi}\, j\omega\mu\, \frac{e^{-jkr}}{r}\left\{\mathbf{a}_r\left[\frac{1}{jkr}+\frac{1}{(jkr)^2}\right]2\cos\theta + \mathbf{a}_\theta\left[1+\frac{1}{jkr}+\frac{1}{(jkr)^2}\right]\sin\theta\right\} \\ &\equiv \mathbf{a}_r E_r + \mathbf{a}_\theta E_\theta\end{aligned} \tag{12-31}$$

where $k^2 = \omega^2\mu\varepsilon$ has been used. Thus, Eqs. (12-30) and (12-31) provide a complete expression of $\mathbf{E}$ and $\mathbf{H}$ for the Hertzian dipole antenna. We observe that the magnetic field has only $\mathbf{a}_\phi$ component, circulating around the axis of the dipole, and the electric field has both $\mathbf{a}_r$ and $\mathbf{a}_\theta$ components that are perpendicular to $\mathbf{H}$. In the following we will discuss the properties of these fields in two different ranges—(i) radiation fields in the far-field region and (ii) the near field behavior. We define the *far-field region* where the observation distance is much greater than the wavelength ($r >> \lambda$) and the *near-field* region where the field point is closer to the antenna relative to the wavelength ($r << \lambda$).

12.3.1 Radiation Fields of a Hertzian Dipole

In the far-field region where the field point is at great distances from the antenna as compared to wavelength of the signal ($r >> \lambda$), the parameter jkr that appears in Eqs. (12-30) and (12-31) becomes very large, i.e.,

$$kr = \frac{2\pi}{\lambda} r >> 1$$

Among the terms of three different orders of magnitude

$$1 >> \frac{1}{jkr} >> \frac{1}{(jkr)^2}$$

the first term is dominant. The fields in the far-field region, called the **radiation fields**, are given by

$$\boxed{\begin{aligned}\mathbf{E} &= \mathbf{a}_\theta \frac{I\Delta\ell}{4\pi}\, j\omega\mu\, \frac{e^{-jkr}}{r}\sin\theta \\ \mathbf{H} &= \mathbf{a}_\phi \frac{I\Delta\ell}{4\pi}\, jk\, \frac{e^{-jkr}}{r}\sin\theta\end{aligned}} \tag{12-32}$$

The interpretation of these results is as follows.

(i) The dependence on the distance r:

The radiation fields are proportional to $\frac{e^{-jkr}}{r}$, which is known to be true for all types of antennas. First and foremost, $\frac{1}{r}$ dependence guarantees the fact that the electromagnetic signal can be sent and detected at far distances without wire ("wireless") because the fields decrease much slower as r increases, as compared to the static fields (electric or magnetic) that behave like $\frac{1}{r^2}$ or $\frac{1}{r^3}$ at best, as shown in Chapters 2 and 6. Secondly, the e^{-jkr} term implies that the constant phase front, given by $kr = \text{constant}$, is a spherical surface. Thus, we have a *spherical wave.* $\frac{e^{-jkr}}{r}$ typically represents a spherical wave.

(ii) The plane wave characteristics:

At far distances, a spherical surface with large radius looks more like a plane, so we expect to see some aspects of the plane wave characteristics. First, we note that

$$\mathbf{E} = \mathbf{a}_\theta E_\theta,\ \mathbf{H} = \mathbf{a}_\phi H_\phi,\ \mathbf{k} = \mathbf{a}_r k$$

E, **H** and **k** are perpendicular to each other and they make a right-handed orthogonal coordinate system as they do for a uniform plane wave. The directions of these vectors are depicted in Figure 12-3.

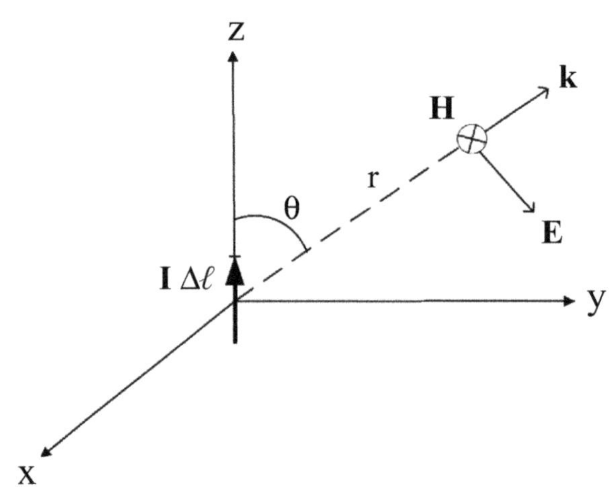

Figure 12-3. Radiation fields of a Hertzian dipole.

Secondly, the ratio of the electric and magnetic field amplitudes is given by

$$\frac{E_\theta}{H_\phi} = \frac{\omega\mu}{k} = \frac{\omega\mu}{\omega\sqrt{\mu\varepsilon}} = \sqrt{\frac{\mu}{\varepsilon}} = \eta$$

which is the intrinsic impedance of the medium. This is also one of the properties of a uniform plane wave. Due to this property, we don't have to calculate both $\mathbf{E}$ and $\mathbf{H}$ if we are interested only in radiation fields. You just need to calculate one of the two and obtain one from the other. It makes sense to see that the electromagnetic wave leaves the antenna as a spherical wave and eventually becomes at very far distances something that looks like a plane wave locally while retaining the $1/r$ dependence of a spherical wave.

(iii) The dependence on observation angle (θ)—Radiation pattern:
From Eq. (12-32), we observe that both E_θ and H_ϕ have the $\sin\theta$ dependence:

$$|E_\theta| \text{ or } |H_\phi| \propto \sin\theta$$

Thus, the field amplitude is maximum at $\theta = 90°$, which corresponds to observation points in the xy plane, perpendicular to the axis of the dipole. $|E_\theta|$ is zero at $\theta = 0°$ and $\theta = 180°$, which correspond to the direction along the axis of the dipole. The Hertzian dipole antenna gives maximum radiation in the horizontal plane and no radiation along its axis. The field amplitude versus the observation angle θ can be plotted in Figure 12-4, which is called the **radiation pattern** of the antenna. In general, the radiation pattern of the antenna depends on both θ and ϕ, one needs to draw it in three-dimensional space or draw two separate plots—one vs. θ on the constant-ϕ plane and one vs. ϕ on the constant-θ surface. The radiation pattern of a Hertzian dipole is independent of ϕ.

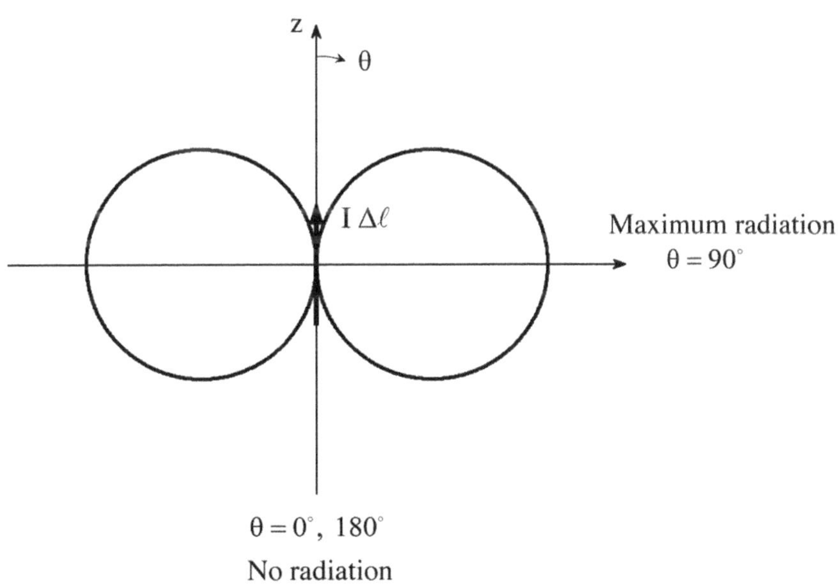

Figure 12-4. Radiation pattern of a Hertzian dipole.

12.3.2 Near Fields of a Hertzian Dipole

In the near-field region where the field point is at close distances from the antenna as compared to wavelength ($r << \lambda$, or equivalently, $kr << 1$, but still $r >> \Delta\ell$) we have,

$$1 << \frac{1}{jkr} << \frac{1}{(jkr)^2}$$

and the third term becomes dominant. From Eqs. (12-30) and (12-31), the dominant fields in the near-field region are given by

$$\begin{aligned} \mathbf{E} &\approx \frac{I\Delta\ell}{4\pi} j\omega\mu \frac{e^{-jkr}}{r} \frac{1}{(jkr)^2} \{\mathbf{a}_r 2\cos\theta + \mathbf{a}_\theta \sin\theta\} \\ &= \frac{I\Delta\ell\omega\mu}{j4\pi k^2} \frac{1}{r^3} (\mathbf{a}_r 2\cos\theta + \mathbf{a}_\theta \sin\theta) \\ \mathbf{H} &\approx \mathbf{a}_\phi \frac{I\Delta\ell}{4\pi} \frac{1}{r^2} \sin\theta \end{aligned} \quad (12\text{-}33)$$

It is interesting to note that the electric field resembles the *static* electric field of a (static) electric dipole as shown in Eq. (2-68). If we calculate $\frac{1}{2}\mathbf{E}\times\mathbf{H}^*$ from near fields in Eq. (12-33), it becomes purely imaginary, indicating that no real power is carried by the near field terms ($1/r^3$ and $1/r^2$) in Eq. (12-33); instead it is carried by the far field ($1/r$) terms in Eq. (12-32). In the near fields of the Hertzian dipole there are both electric and magnetic stored energy but the electric stored energy is much greater than the magnetic stored energy (see Problem 12-11).

EXAMPLE 12-1. RADIATED POWER OF A HERTZIAN DIPOLE

Calculate the total power radiated by a Hertzian dipole antenna of $I\,\Delta\ell$ at the frequency f [Hz] in open air.

SOLUTIONS:

The total radiated power can be calculated by integrating the time-average Poynting vector over the entire spherical surface with radius r, using the radiation fields we obtained in Eq. (12-32). It is expected that the total power should be independent of the distance r since the air medium is lossless.

The time-average Poynting vector is given by (see Eq. (9-27))

$$\begin{aligned} \mathbf{S}_{av} &= \frac{1}{2}\mathrm{Re}\{\mathbf{E}\times\mathbf{H}^*\} = \mathbf{a}_\theta\times\mathbf{a}_\phi \frac{1}{2}\mathrm{Re}\{E_\theta H_\phi^*\} = \mathbf{a}_r \frac{1}{2}\mathrm{Re}\left\{E_\theta\left(\frac{E_\theta}{\eta}\right)^*\right\} \\ &= \mathbf{a}_r \frac{|E_\theta|^2}{2\eta} = \mathbf{a}_r \frac{|I\Delta\ell|^2}{32\pi^2} \frac{(\omega\mu)^2}{\eta} \frac{\sin^2\theta}{r^2} \quad \left[\frac{W}{m^2}\right] \end{aligned} \quad (12\text{-}34)$$

The total radiated power is given by

$$P_r = \iint_{\text{Sphere}} \mathbf{S}_{av} \cdot \mathbf{a}_r \, ds = \int_0^{2\pi} \int_0^{\pi} \frac{|I|^2}{32\pi^2} (\Delta\ell)^2 \frac{(k\eta)^2}{\eta} \frac{\sin^2\theta}{r^2} r^2 \sin\theta \, d\theta \, d\phi$$

$$= \frac{|I|^2}{32\pi^2} (\Delta\ell)^2 \left(\frac{2\pi}{\lambda_o}\right)^2 \eta \, (2\pi) \int_0^{\pi} \sin^3\theta \, d\theta = \frac{\pi\eta_o}{3} \left(\frac{\Delta\ell}{\lambda_o}\right)^2 |I|^2 \; [W] \tag{12-35}$$

where $\eta = \eta_o = 377$ [Ω] in air and $\lambda_o = c/f$ is the wavelength in air.

Radiation Resistance

A useful measure of the amount of power radiated by an antenna is the quantity called the **radiation resistance**. The radiation resistance R_r of an antenna is defined by the following equation:

$$P_r = \frac{1}{2} I_A^2 \, R_r \qquad \text{or} \qquad R_r = \frac{2P_r}{I_A^2} \tag{12-36}$$

where I_A is the current at the input terminals of an antenna or the maximum current along the antenna. Thus the radiation resistance is the value of a resistance that would dissipate an amount of power equal to the radiated power P_r when the current in the resistance is equal to the input or maximum current of the antenna.

The radiation resistance of a Hertzian dipole is obtained from Eq. (12-35):

$$R_r = \frac{2P_r}{|I|^2} = \frac{2\pi}{3}\eta_o \left(\frac{\Delta\ell}{\lambda_o}\right)^2 = 80\pi^2 \left(\frac{\Delta\ell}{\lambda_o}\right)^2 \; [\Omega] \tag{12-37}$$

For example, an elementary dipole of $\Delta\ell = 0.1\lambda_o$ has $R_r = 7.9$ [Ω], which is small for an antenna. As will be shown later, a more practical straight wire antenna that is half-wavelength long has $R_r = 73$ [Ω].

Note the rapid change of R_r with *electrical length* ($\Delta\ell/\lambda_o$). If we reduce the length $\Delta\ell$ or the frequency by a factor of 10, then R_r is reduced by a factor of 100. This is our first indication of the basic limitations of "electrically small antennas" of any shape. As the electrical length is decreased, the efficiency and bandwidth is lowered very rapidly. Half wavelength antennas can be very efficient and of reasonable bandwidth. See Problem 12-23 for discussion of *radiation efficiency* of an antenna.

EXAMPLE 12-2. DIRECTIVE GAIN OF AN ANTENNA

One very important characteristic of an antenna is how much the antenna sends energy in a certain direction in preference to radiation in other directions. A parameter that measures such characteristic is called the **directive gain**, $D(\theta, \phi)$, which is defined by

$$D(\theta,\phi) = \frac{S_{av}(\theta,\phi)}{P_r/4\pi r^2} \qquad (12\text{-}38)$$

where S_{av} is the time-average Poynting power density in a given direction denoted by the observation angles (θ, ϕ) and P_r is the total power radiated over the entire space. Thus, the directive gain measures a ratio of the power density in a certain direction at given r (distance) to the average power density at that range. Find the directive gain and its maximum value for a Hertzian dipole antenna.

SOLUTIONS:

In Example 12-1, we have calculated S_{av} and P_r of the Hertzian dipole:

$$S_{av} = \frac{|I\Delta\ell|}{32\pi^2}\frac{(\omega\mu)^2}{\eta}\frac{\sin^2\theta}{r^2} \equiv K\frac{\sin^2\theta}{r^2}$$

$$P_r = \iint S_{av}\, r^2 \sin\theta\, d\theta\, d\phi = K\int_0^{2\pi}\int_0^{\pi} \sin^3\theta\; d\theta\, d\phi = K(2\pi)\frac{4}{3}$$

The directive gain of the Hertzian dipole is

$$D(\theta,\phi) = \frac{S_{av}}{P_r/4\pi r^2} = \frac{K\sin^2\theta}{K\frac{8\pi}{3}/4\pi} = \frac{3}{2}\sin^2\theta \qquad (12\text{-}39)$$

The plot of $D(\theta, \phi)$ vs. observation angles (θ, ϕ) is basically the radiation pattern of the antenna shown in Figure 12-4. In a direction $\theta = 90°$, the gain is maximum with the value of 1.5 whereas at $\theta = 0°$ and $180°$, the gain is zero. The value of the directive gain in the direction of its maximum value is called the **directivity** of an antenna. Thus the directivity of a Hertzian dipole is

$$D_{max} = 1.5 \text{ for Hertzian dipole} \qquad (12\text{-}40)$$

As will be shown in the next section, the directivity of a half-wavelength dipole antenna is 1.67.

12.4 LINEAR ANTENNA—LONG DIPOLE

In the previous section we considered an elementary dipole antenna, which is electrically short ($\Delta\ell << \lambda$) and is assumed to have uniform current distribution. However, such a short antenna has a very small radiation resistance and is not a good radiator. A more practical and efficient antenna would be a straight wire whose length is comparable to a wavelength, known as the **linear dipole antenna**. A household TV antenna is one example. If the current distribution along the wire is known we can easily incorporate it in the vector potential integral and find the radiation fields. However, the exact current distribution on the thin wire antenna is not known and its approximate distribution can be obtained through the use of advanced numerical techniques such as the method of moments. We will not deal with such sophisticated techniques and assume that the current distribution on the wire is *sinusoidal in space* along the wire, with zero at the ends. This functional dependence on z is known to be a good approximation to actual current distribution.

Figure 12-5 shows a straight thin wire antenna of length ℓ excited by a source in the middle, whose current distribution is assumed to take the following form:

$$I(z) = I_o \sin\left[k\left(\frac{\ell}{2} - |z|\right)\right], \quad |z| < \frac{\ell}{2} \qquad (12\text{-}41)$$

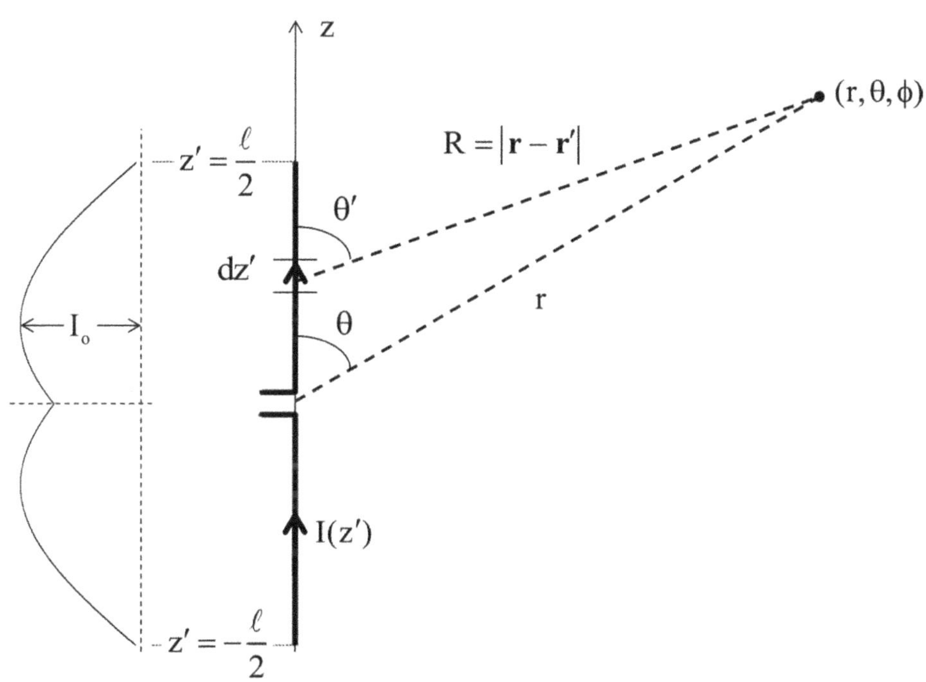

Figure 12-5. Long dipole antenna with sinusoidal current distribution.

where $k = 2\pi/\lambda$ is a wavenumber corresponding to an excitation frequency. The current is symmetric about $z = 0$ (center-driven) and goes to zero smoothly at the ends ($z = \pm\ell/2$). I_o is the maximum value of the current on the wire and its position depends on the length of the antenna.

In order to obtain the fields radiated by the antenna we would normally calculate the vector potential first. However, since we are primarily interested in the radiation (far) fields and we know the radiation fields of an elementary dipole, we can make use of the result of Eq. (12-32) and apply superposition to calculate directly the radiation fields of the long dipole.

Consider a differential element $I(z')\,dz'$ located at $z = z'$. Its radiation electric field can be written from Eq. (12-32) as follows:

$$d\mathbf{E} = \mathbf{a}_\theta \frac{I(z')dz'}{4\pi} j\omega\mu \frac{e^{-jkR}}{R} \sin\theta' \tag{12-42}$$

where $R = |\mathbf{r} - \mathbf{r}'|$ is the distance from the source point ($z = z'$) to the field point (r, θ, ϕ) and θ' is the angle between the wire axis (z-axis) and the vector $\mathbf{R} = \mathbf{r} - \mathbf{r}$. At far points ($r >> z'$), the following approximations can be made (see Eq. (12-25)).

$$R = |\mathbf{r} - \mathbf{r}'| \approx r - z'\cos\theta$$

$$\text{when } r >> |\mathbf{r}'| = z'$$

$$\frac{1}{R} \approx \frac{1}{r}, \qquad \sin\theta' \approx \sin\theta$$

Then Eq. (12-42) reduces to

$$d\mathbf{E} \approx \mathbf{a}_\theta \frac{I(z')dz'}{4\pi} j\omega\mu \frac{e^{-jkr}}{r} e^{jkz'\cos\theta} \sin\theta \tag{12-43}$$

The radiation electric field due to the entire wire can be obtained by integrating $d\mathbf{E}$ over the source distribution.

$$\mathbf{E} = \int d\mathbf{E} = \mathbf{a}_\theta \frac{j\omega\mu}{4\pi} \frac{e^{-jkr}}{r} \sin\theta \int_{-\ell/2}^{\ell/2} I(z') e^{jkz'\cos\theta}\, dz' \tag{12-44}$$

$$\begin{aligned} E_\theta &= \frac{j\omega\mu}{4\pi} \frac{e^{-jkr}}{r} \sin\theta \left\{ \int_0^{\ell/2} I_o \sin\left[k\left(\frac{\ell}{2} - z\right)\right] e^{jkz'\cos\theta}\, dz' \right. \\ &\quad \left. + \int_{-\ell/2}^{0} I_o \sin\left[k\left(\frac{\ell}{2} + z\right)\right] e^{jkz'\cos\theta}\, dz' \right\} \\ &= j\eta 2 I_o \frac{e^{-jkr}}{4\pi r} \frac{\cos\left(\frac{k\ell}{2}\cos\theta\right) - \cos\left(\frac{k\ell}{2}\right)}{\sin\theta} \end{aligned} \tag{12-45}$$

The derivation of Eq. (12-45) from the integrals is assigned as Problem 12-15. The radiation magnetic field is simply given by

$$\mathbf{H} = \mathbf{a}_\phi \, H_\phi = \mathbf{a}_\phi \frac{E_\theta}{\eta} \tag{12-46}$$

due to its plane wave characteristic at far distances. The radiation pattern of the long dipole antenna is given by

$$F(\theta) = \frac{\cos\left(\frac{k\ell}{2}\cos\theta\right) - \cos\left(\frac{k\ell}{2}\right)}{\sin\theta} \tag{12-47}$$

Note that it is independent of ϕ (because the current is z-directed) and depends on the length of the dipole. Making use of $k = \frac{2\pi}{\lambda}$, we have $\frac{k\ell}{2} = \pi\frac{\ell}{\lambda}$. For example,

$$\ell = \frac{\lambda}{2} \rightarrow F(\theta) = \frac{\cos\left(\frac{\pi}{2}\cos\theta\right)}{\sin\theta} \tag{12-48a}$$

$$\ell = \lambda \rightarrow F(\theta) = \frac{\cos\left(\pi\cos\theta\right) + 1}{\sin\theta} \tag{12-48b}$$

$$\ell = \frac{3\lambda}{2} \rightarrow F(\theta) = \frac{\cos\left(\frac{3}{2}\pi\cos\theta\right)}{\sin\theta} \tag{12-48c}$$

$$\ell = 2\lambda \rightarrow F(\theta) = \frac{\cos\left(2\pi\cos\theta\right) - 1}{\sin\theta} \tag{12-48d}$$

The radiation patterns of the long dipole for four different lengths, $\ell = \frac{\lambda}{2}, \lambda, \frac{3}{2}\lambda$ and 2λ, are plotted in Figure 12-6. It is observed that a full-wavelength dipole ($\ell = \lambda$) has a more directive beam along $\theta = 90°$ than a half-wave dipole $\left(\ell = \frac{\lambda}{2}\right)$. As the length of the dipole increases further, the main beam (or maximum radiation) shifts away from $\theta = 90°$ and nulls (zero radiation) occur in certain directions. For example, $\ell = \frac{3}{2}\lambda$ dipole has a null at $\theta = 70.5°$ and $\ell = 2\lambda$ dipole has a null at $\theta = 90°$.

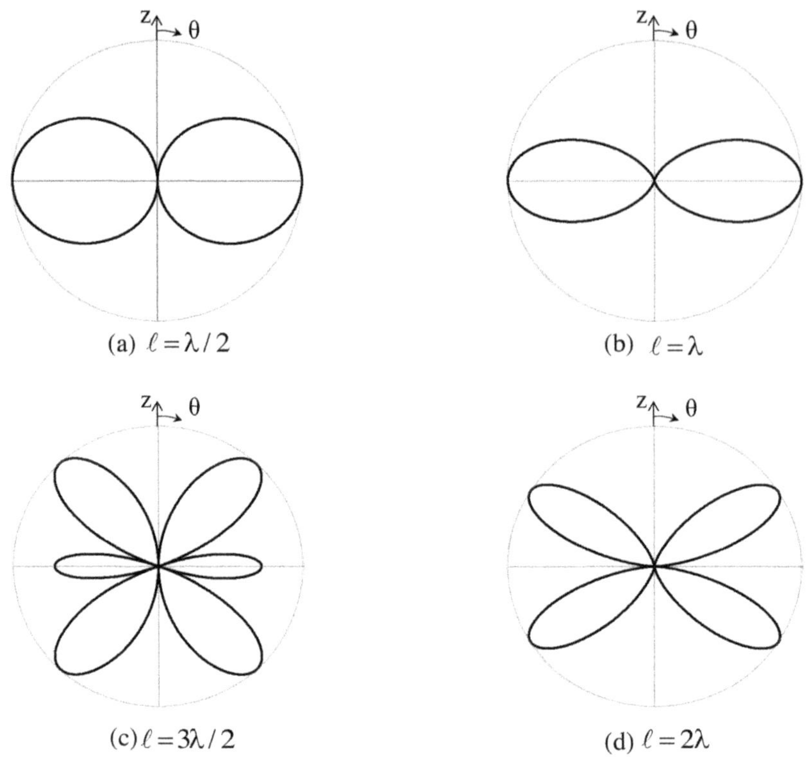

Figure 12-6. Radiation patterns of long dipole antennas.
(a) $\ell = \lambda/2$, (b) $\ell = \lambda$, (c) $\ell = 3\lambda/2$, (d) $\ell = 2\lambda$

12.4.1 Half-Wave Dipole

A linear antenna having a length $\ell = \lambda/2$ (half wavelength), known as the **half-wave dipole**, is very widely used and is of practical importance because of its impedance characteristic. The half-wave dipole has a current distribution that is one-half of a sine wave with maximum at the center.

$$I(z) = I_o \sin\left[k\left(\frac{\lambda}{4} - |z|\right)\right] = I_o \sin\left[\frac{\pi}{2} - \frac{2\pi}{\lambda}|z|\right],\ |z| < \frac{\lambda}{4} \tag{12-49}$$

A unique advantage of a half-wave dipole is that it can be made to *resonate* and to have zero reactance components, making it easy to tune. To make the dipole resonant its physical length must be a little bit shorter than a free-space half wavelength.*

The radiation fields of the half-wave dipole are given by

$$E_\theta = j\eta 2 I_o \frac{e^{-jkr}}{4\pi r} \frac{\cos\left(\frac{\pi}{2}\cos\theta\right)}{\sin\theta},\ H_\phi = \frac{E_\theta}{\eta} \tag{12-50}$$

*W.L. Stutzman and G.A. Thiele, *Antenna Theory and Design*, John Wiley & Sons, 1998, 2nd Ed., Section 5.1.

whose pattern is shown in Figure 12-6(a). The pattern is very similar to that of a Hertzian dipole, i.e., it has a maximum radiation along $\theta = 90°$ and no radiation along its axis ($\theta = 0°$). However, the beam width is a little bit sharper.

EXAMPLE 12-3. RADIATION RESISTANCE OF A HALF-WAVE DIPOLE.

Calculate the radiation resistance and directivity of a half-wave dipole antenna.

SOLUTIONS:

The total radiated power of a half-wave dipole is given by

$$P_r = \iint \mathbf{S}_{av} \cdot \mathbf{a}_r \, ds = \iint \frac{|E_\theta|^2}{2\eta} r^2 \sin\theta \, d\theta \, d\phi$$

$$= \int_0^{2\pi}\int_0^{\pi} \frac{\eta}{2}\left(\frac{2I_o}{4\pi}\right)^2 \frac{\left[\cos\left(\frac{\pi}{2}\cos\theta\right)\right]^2}{\sin\theta} d\theta\, d\phi = \frac{\eta}{4\pi} I_o^2 \int_0^{\pi} \frac{\left[\cos\left(\frac{\pi}{2}\cos\theta\right)\right]^2}{\sin\theta} d\theta \qquad (12\text{-}51)$$

The integral can be written in terms of the cosine integral function or can be evaluated numerically to give a value of 1.22.

Then

$$P_r = \frac{\eta}{4\pi} I_o^2 \,(1.22) = 36.6\, I_o^2 \;\; [\mathrm{W}] \qquad (12\text{-}52)$$

The radiation resistance is given by (see Eq. (12-36)).

$$R_r = \frac{2P_r}{I_o^2} = 73\,[\Omega] \qquad (12\text{-}53)$$

The directive gain of the dipole is given by

$$D(\theta,\phi) = \frac{S_{av}}{P_r/4\pi r^2} = \frac{|E_\theta|^2/2\eta}{P_r/4\pi r^2} = \frac{\frac{\eta}{2}\left(\frac{2I_o}{4\pi}\right)^2}{\eta\left(\frac{I_o}{4\pi}\right)^2 (1.22)} \left[\frac{\cos\left(\frac{\pi}{2}\cos\theta\right)}{\sin\theta}\right]^2 \qquad (12\text{-}54)$$

$$= 1.64\,[F(\theta)]^2$$

As shown in Figure 12-6(a), F(θ) has a maximum value 1 when $\theta = 90°$, so the directivity of a half-wave dipole is

$$D_{max} = 1.64 \text{ for half-wave dipole} \tag{12-55}$$

This is a little bit higher than D_{max} $(= 1.5)$ of a Hertzian dipole, which results in a slightly sharper main beam along $\theta = 90°$. A full-wave dipole ($\ell = \lambda$) gives a more directive pattern, as shown in Figure 12-6(b), and it has the directivity of 2.41.

12.5 ANTENNA ARRAYS

In the previous sections we considered the radiation of a single-element antenna—the Hertzian dipole, the small loop antenna (Problem 12-12), the long dipole and the monopole antenna (Problem 12-17). We observe that the radiation patterns of all these antennas are omnidirectional (independent of ϕ) in the plane perpendicular to the antenna and they have low directivity due to a broad main beam. Then how do we send energy in certain directions or areas and avoid radiation in other directions? We can achieve this by using an **array** of several antennas in different configurations—on a line, circle or plane, etc. By changing the spacing between antennas and the phase differences between excitation currents of antenna elements, we can create a more directional pattern of the array. For example, many broadcast antenna stations use an array of multiple vertical antennas above ground (monopoles). An array of many small antennas can also be used to obtain a performance similar to that of a single large antenna. We will first treat the simplest array with two-element antennas and then extend to an array of multiple antennas.

12.5.1 Two-Element Array

Let's consider two identical antennas separated by a distance d on the x axis as shown in Figure 12-7. For example, they could be a pair of vertical half-wave dipoles or quarter-wave monopoles above ground. The current excitations are I_1 and I_2; they have the same amplitude and the phase difference of α. For simplicity, let

$$I_1 = I_o, \qquad I_2 = I_o\, e^{j\alpha} \tag{12-56}$$

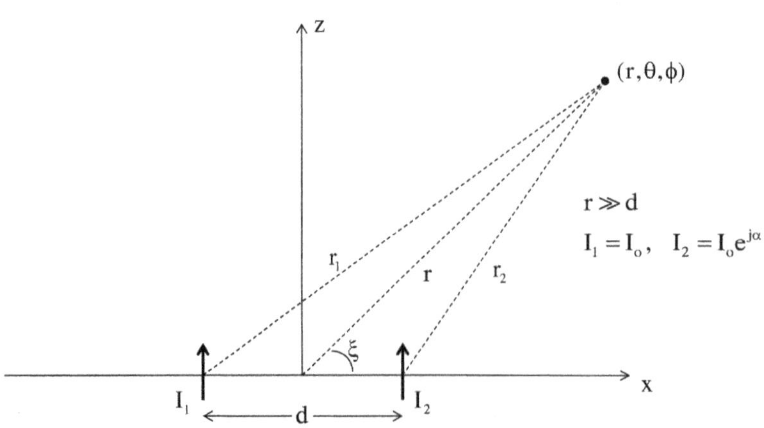

Figure 12-7. A two-element array.

The radiation field at far point $P(r, \theta, \phi)$ $(r >> d)$ is a superposition of radiation fields due to two antennas. Since two antennas are identical, they will have the same radiation pattern but with different current excitations. Hence we can write

$$E_1 = I_1 K F(\theta,\phi)\frac{e^{-jkr_1}}{r_1} \tag{12-57}$$

$$E_2 = I_2 K F(\theta,\phi)\frac{e^{-jkr_2}}{r_2} \tag{12-58}$$

where $F(\theta, \phi)$ is the radiation pattern function of each individual antenna and K is a multiplying constant. Substituting Eq. (12-56) into E_1 and E_2, the total radiation field is given by

$$E = E_1 + E_2 = I_o K F(\theta,\phi)\left\{\frac{e^{-jkr_1}}{r_1} + \frac{e^{-jkr_2}\, e^{j\alpha}}{r_2}\right\} \tag{12-59}$$

When the field (or observation) point P is in the far distances $(r >> d)$,

$$r_1 \approx r + \frac{d}{2}\cos\xi \tag{12-60a}$$

$$r_2 \approx r - \frac{d}{2}\cos\xi \tag{12-60b}$$

where ξ is the angle between the direction of the observation ($\mathbf{a}_r$) and the axis of the array ($\mathbf{a}_x$) and its cosine is given by

$$\cos\xi = \mathbf{a}_r \cdot \mathbf{a}_x = \sin\theta\cos\phi$$

The radiation field can be approximated by

$$\begin{aligned} E &\approx I_o K F(\theta,\phi)\left\{\frac{e^{-jk\left(r+\frac{d}{2}\cos\xi\right)}}{r_1} + \frac{e^{-jk\left(r-\frac{d}{2}\cos\xi\right)}\, e^{j\alpha}}{r_2}\right\} \\ &\approx I_o K F(\theta,\phi)\frac{e^{-jkr}}{r} e^{j\frac{\alpha}{2}}\left\{e^{-j\left(\frac{1}{2}kd\cos\xi + \frac{\alpha}{2}\right)} + e^{j\left(\frac{1}{2}kd\cos\xi + \frac{\alpha}{2}\right)}\right\} \\ &= I_o K\, e^{j\frac{\alpha}{2}}\, \frac{e^{-jkr}}{r} F(\theta,\phi)\cdot 2\cos\left[\frac{1}{2}(kd\cos\xi + \alpha)\right] \end{aligned} \tag{12-61}$$

The magnitude of the electric field of the array is

$$|E| = \frac{|2I_o K|}{r} |F(\theta,\phi)| \left|\cos\frac{\psi}{2}\right| \tag{12-62}$$

where

$$\psi = kd\cos\xi + \alpha = \frac{2\pi d}{\lambda}\sin\theta\cos\phi + \alpha \tag{12-63}$$

Note that $|F(\theta,\phi)|$ represents the radiation pattern of an individual antenna, called the **element pattern**, and $\left|\cos\frac{\psi}{2}\right|$ is the additional pattern factor due to the array configuration, called the **array factor** (AF). The array factor depends on the array geometry and the phase difference between current excitations of the antenna elements. From Eq. (12-62), we observe that the radiation pattern of an array of identical antennas can be obtained by a product of the element pattern and the array factor. Such property is known as the **principle of pattern multiplication**. For a more detailed discussion, see, for example, Stutzman and Thiele (1998).

EXAMPLE 12-4. RADIATION PATTERN OF A TWO-ELEMENT ARRAY

Sketch the electric field pattern of an array of two vertical (z-directed) half-wave dipoles in the xy plane $\left(\theta = \frac{\pi}{2}\right)$ for the following four cases: (i) $d = \lambda/2$, $\alpha = 0$, (ii) $d = \lambda/2$, $\alpha = \pi$, (iii) $d = \lambda/4$, $\alpha = -\pi/2$, (iv) $d = \lambda$, $\alpha = 0$.

SOLUTION:

Since we observe the pattern in the xy plane ($\theta = 90°$), the element pattern of a half-wave dipole is $F(\theta = \pi/2, \phi) = 1$. The radiation pattern of the array is simply given by the array factor:

$$AF = \left|\cos\frac{\psi}{2}\right| = \left|\cos\left[\frac{\pi d}{\lambda}\cos\phi + \frac{\alpha}{2}\right]\right| \tag{12-64}$$

Case 1: $d = \lambda/2$, $\alpha = 0$

$$AF = \left|\cos\left(\frac{\pi}{2}\cos\phi\right)\right| \tag{12-65}$$

The pattern is sketched in Figure 12-8(a). It has the maximum radiation along $\phi = 90°$ (perpendicular to the array axis) and no radiation along the axis of the array ($\phi = 0°$, $180°$). Because two antennas are identical and in phase, in the direction along $\phi = 90°$ the radiation fields from the two antennas add, whereas in the direction along $\phi = 0°$ or $180°$ the radiation fields from the two antennas are out of phase by $kd = (2\pi/\lambda) \cdot \lambda/2 = \pi$ and they cancel each other. This type of array is known as the *broadside array*.

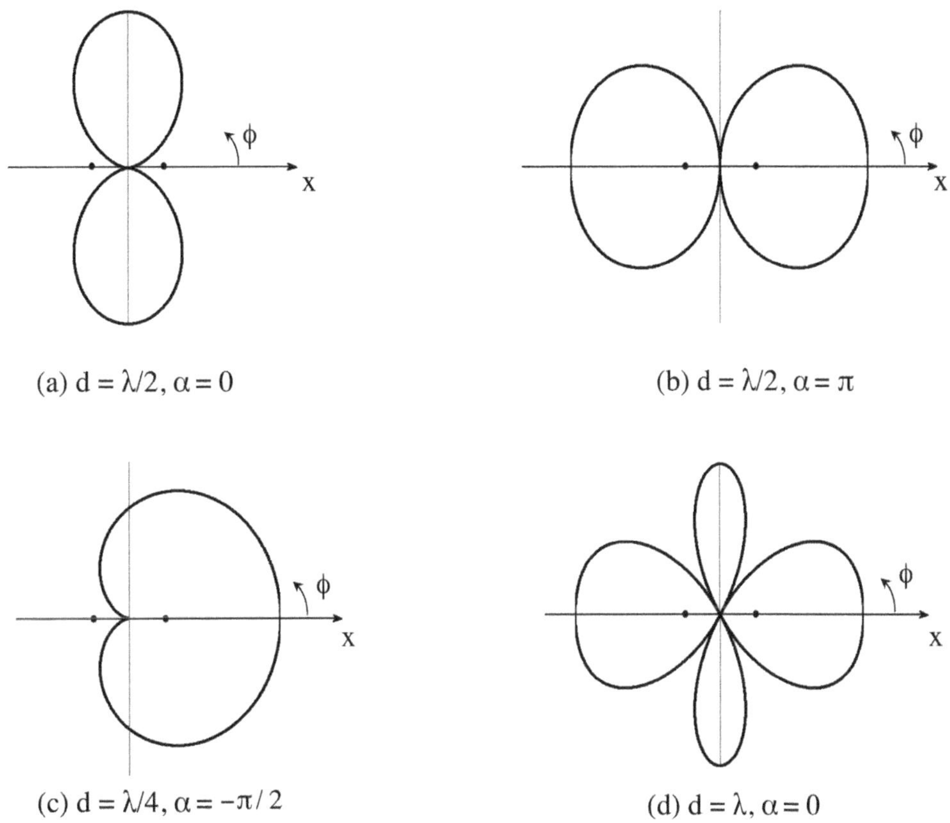

Figure 12-8. Radiation pattern of a two-element array.
(a) $d = \lambda/2, \alpha = 0$, **(b)** $d = \lambda/2, \alpha = \pi$, **(c)** $d = \lambda/4, \alpha = -\pi/2$, **(d)** $d = \lambda, \alpha = 0$.

Case 2: $d = \lambda/2, \alpha = \pi$

$$AF = \left|\cos\left(\frac{\pi}{2}\cos\phi + \frac{\pi}{2}\right)\right| = \left|\sin\left(\frac{\pi}{2}\cos\phi\right)\right| \tag{12-66}$$

The pattern, shown in Figure 12-8(b) is completely opposite to that of Case 1. It has the maximum radiation along the array axis ($\phi = 0°$, $180°$) and no radiation (nulls) along $\phi = 90°$. This time two antennas are $180°$ out of phase, thus the radiation fields from the two cancel each other along $\phi = 90°$ and add along $\phi = 0°$ because the total phase difference is $kd + \alpha = \pi + \pi = 2\pi$ in the direction along the array axis. An array which has the maximum in the direction along the array axis is called an *end fire array*.

Case 3: $d = \lambda/4, \alpha = -\pi/2$

$$AF = \left|\cos\left(\frac{\pi}{4}\cos\phi - \frac{\pi}{4}\right)\right| \tag{12-67}$$

The radiation pattern is sketched in Figure 12-8(c). It has a maximum along $\phi = 0°$ and null at $\phi = 180°$, known as the *cardioid pattern*. This array can be used when two tall antennas need to be placed near the sea. You want to send signal in the residential area but no signal in the direction of the sea side. This array is also frequently used in acoustics for microphone patterns.

<u>**Case 4**</u>: $d = \lambda,\ \alpha = 0$

$$AF = \left|\cos\left(\pi\cos\phi\right)\right| \tag{12-68}$$

The pattern, shown in Figure 12-8(d), has maximum radiation in four directions ($\phi = 0°, 90°, 180°, 270°$) and nulls in between ($\phi = \pm 60°, \pm 120°$). Nulls (perfect cancellation) occur when the path length difference from two antennas is one-half wavelength, which is equivalent to a phase difference of π, i.e., $d\cos\phi = \lambda\cos\phi = \lambda/2$ or $\cos\phi = 1/2$.

12.5.2 Uniform Linear Array

An array of multiple (more than two) elements can be analyzed in a similar manner. Among various types of an array, we consider an array of equally spaced identical antennas that have current excitations with the same amplitude and *uniform progressive phase shift* because its analysis is straightforward. Such an array is called a **uniform linear array**. Suppose N identical antennas are placed along the x axis. They are separated by d and the phase difference between two neighboring elements is α, i.e., the current excitations can be written as

$$I_1 = I_o,\ I_2 = I_o e^{j\alpha},\ I_3 = I_o e^{j2\alpha},\ \ldots,\ I_N = I_o e^{j(N-1)\alpha}$$

The configuration of the array is shown in Figure 12-9. At far points ($r >> d$),

$$r_1 = r, \quad r_2 \approx r - d\cos\xi, \quad r_3 = r - 2d\cos\xi,\ \ldots,\ r_N = r - (N-1)\,d\cos\xi$$

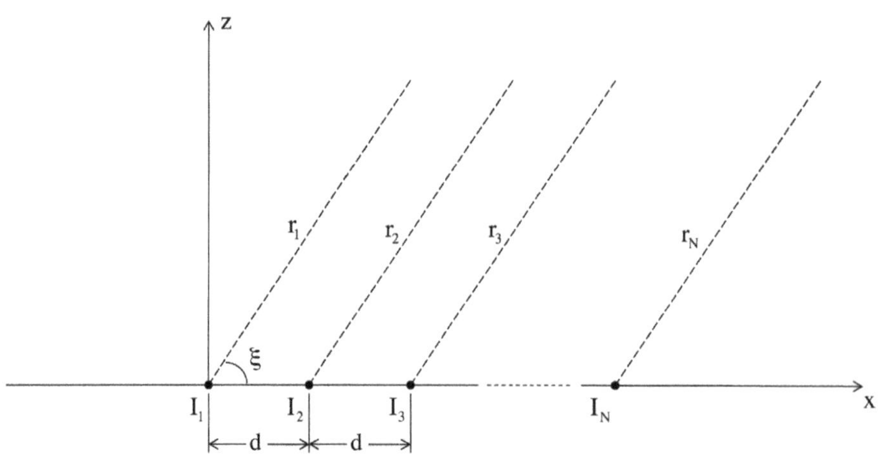

Figure 12-9. A uniform linear array.

Following a similar analysis of the two-element array (see Eq. (12-61)), the array factor of the uniform linear array is given by

$$AF = 1 + e^{j(kd\cos\xi)} e^{j\alpha} + e^{j(2kd\cos\xi)} e^{j2\alpha} + \ldots + e^{j(N-1)kd\cos\xi} e^{j(N-1)\alpha}$$

$$= 1 + e^{j\psi} + e^{j2\psi} + \ldots + e^{j(N-1)\psi} = \sum_{n=0}^{N-1} e^{jn\psi} \tag{12-69}$$

where

$$\psi = kd\cos\xi + \alpha \tag{12-63}$$

Eq.(12-69) is a geometric series and can be summed in closed form:

$$AF = \frac{1 - e^{jN\psi}}{1 - e^{j\psi}} = \frac{e^{j\frac{N\psi}{2}}}{e^{j\frac{\psi}{2}}} \frac{e^{-j\frac{N\psi}{2}} - e^{j\frac{N\psi}{2}}}{e^{-j\frac{\psi}{2}} - e^{j\frac{\psi}{2}}} = e^{j\frac{(N-1)\psi}{2}} \frac{\sin\left(\frac{N}{2}\psi\right)}{\sin\left(\frac{\psi}{2}\right)} \tag{12-70}$$

The array factor in Eq. (12-70) is maximum when $\psi = 0$ and its value is given by

$$\lim_{\psi \to 0} \frac{\sin(N\psi/2)}{\sin(\psi/2)} = \lim_{\psi \to 0} \frac{\frac{N}{2}\cos(N\psi/2)}{\frac{1}{2}\cos(\psi/2)} = N$$

where L'Hopital's rule is used. Then the normalized array factor is given by

$$|AF|_n = \frac{1}{N}\left|\frac{\sin(N\psi/2)}{\sin(\psi/2)}\right| \tag{12-71}$$

As we plot $|AF|_n$ as a function of ψ for different values of N, we observe the following:

1. The function is periodic in ψ, with period 2π.
2. As N increases, the main lobe becomes narrower (more directivity).
3. As N increases, the number of side lobes increases. The total number of lobes in one period is N−1 (one main lobe and N-2 side lobes).
4. The level of side lobes decreases as N increases.
5. The main beam occurs when

$$\psi = kd\cos\xi + \alpha = 0 \quad \text{or} \quad \cos\xi = -\frac{\alpha}{kd} \tag{12-72}$$

When $\alpha = 0$ (all elements are in phase), the main beam occurs at $\xi = 90°$; it is a *broadside array*. When $\alpha = -kd$, the main beam occurs at $\xi = 0°$, which gives an *end fire array*.

6. The side lobes (minor maxima) occur approximately when $\sin(N\psi/2)$ is at maximum, i.e., when $|\sin(N\psi/2)| = 1$.

$$\frac{N\psi}{2} = (2n+1)\frac{\pi}{2}, \quad n = 1, 2, 3, \ldots \tag{12-73}$$

   The first side lobe occurs when $N\psi/2 = 3\pi/2$ (n=1).

7. The nulls occur when

$$\frac{N\psi}{2} = n\pi, \quad n = 1, 2, 3, \ldots \tag{12-74}$$

EXAMPLE 12-5. FOUR-ELEMENT LINEAR ARRAY

Plot the normalized array factor, as a function ψ ($0 \le \psi \le 2\pi$), of a four-element array, uniformly excited and equally spaced. Sketch the radiation pattern of this array (assuming the element pattern is omnidirectional) when $d = \lambda/2$ and $\alpha = \pi$.

SOLUTION:

$$|AF|_n = \left|\frac{\sin(2\psi)}{4\sin(\psi/2)}\right| \quad (N = 4)$$

The side lobes occur approximately at $2\psi = 3\pi/2$, $5\pi/2$, or $\psi = 3\pi/4$, $5\pi/4$. The nulls occur at $2\psi = \pi$, 2π, 3π, or $\psi = \pi/2$, π, $3\pi/2$. $|AF|_n$ is plotted in Figure 12-10(a).

When $d = \lambda/2$ and $\alpha = \pi$, $kd = (2\pi/\lambda)d = \pi$.

$$\psi = kd\cos\xi + \pi = \pi(\cos\xi + 1)$$

The main beam occurs when $\cos\xi + 1 = 0, 2 \rightarrow \xi = 180°, 0°$.
The nulls occur when $\cos\xi + 1 = 1/2, 1, 3/2 \rightarrow \xi = 60°, 90°, 120°$.
The side lobes occur when $\cos\xi + 1 \approx 3/4, 5/4 \rightarrow \xi \approx 75.5°, 104.5°$.

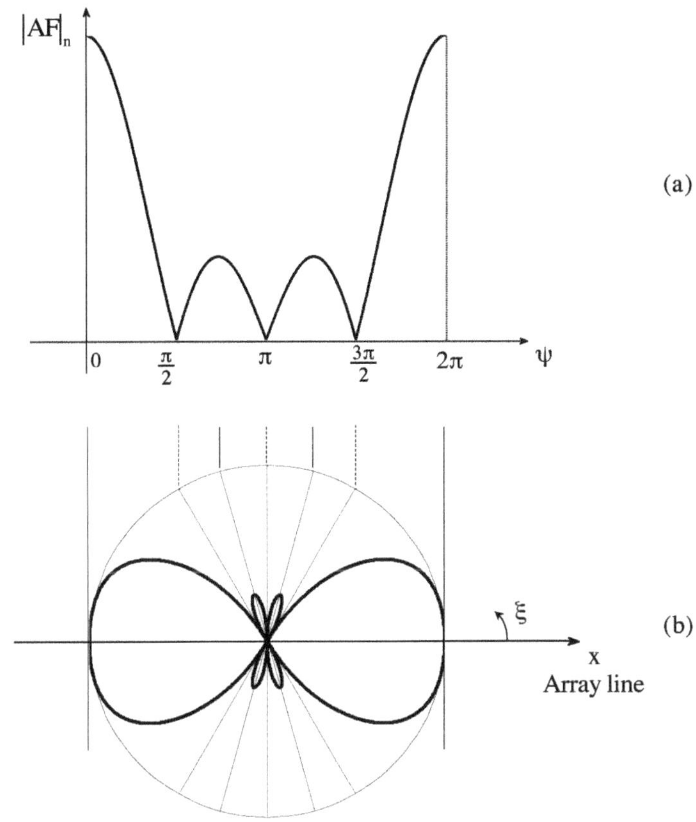

Figure 12-10. Normalized array factor (a) and radiation pattern (b) of a four element linear array with $d = \lambda/2$, $\alpha = \pi$.

The radiation pattern is sketched in Figure 12-10(b). For a more systematic way of sketching the array pattern, see Stutzman and Thiele (1998), pp. 101-106.

We previously mentioned some basic limitations of electrically small antennas. The trend continues. All of the electrical characteristics of an antenna or antenna system such as an array depend on electrical size, not on its physical length in meters but on its electrical size in wavelengths or in square wavelengths.

Furthermore, if we list all of the desirable characteristics of an antenna, such as (1) high efficiency, (2) large bandwidth, and (3) high directivity, then the more characteristics (1, 2, 3) that we ask for and the higher the levels that we demand, the larger the required electrical size of our antenna or array.

PROBLEMS

12-1. Suppose the retarded potentials are given by

$$V = 0, \quad \mathbf{A} = \begin{cases} \mathbf{a}_x \dfrac{K}{v}\left(|z| - vt\right)^2, & \text{when } |z| < vt \\ 0, & \text{when } |z| > vt \end{cases}$$

where K is a constant and $v = 1/\sqrt{\mu\varepsilon}$.

(a) Find the electric field ($\mathbf{E}$) and magnetic field ($\mathbf{H}$).
(b) Confirm that $\mathbf{E}$ and $\mathbf{H}$ satisfy all of Maxwell's equations.
(c) Find the charge and current distributions (ρ_v, ρ_s, $\mathbf{J}$ and $\mathbf{J}_s$) that would produce these fields.

12-2. Suppose the retarded potentials are given by

$$V = 0, \mathbf{A} = \mathbf{a}_z A_o \cos(\omega t - ky)$$

where A_o, ω and k are constants.

(a) Find the $\mathbf{E}$ and $\mathbf{H}$ fields.
(b) Confirm that they satisfy all of Maxwell's equations in *free space* under certain conditions on ω and k. Find the conditions.

12-3. (Oscillating electric dipole)
The Hertzian dipole has a constant-current distribution (I) over the dipole length $\Delta\ell$. This can be achieved by two small metal spheres, separated by a distance $\Delta\ell$ and connected by a thin wire, carrying the charge $\pm\, q(t)$ oscillating in time t:

$$q(t) = q_o \cos \omega t = \mathrm{Re}\{q_o\, e^{j\omega t}\}$$

Show that the current moment ($I\,\Delta\ell$ with I being the complex phasor current) is related to the dipole moment by

$$I\,\Delta\ell = j\omega p$$

where $p = q_o\,\Delta\ell$ is the (phasor) dipole moment.

12-4. **(Short dipole antenna)**
The ideal Hertzian dipole has a uniform current distribution as given by Eq. (12-23). In reality, this would not be a good representation of the current distribution on a *short* straight wire antenna ($\Delta\ell << \lambda$). A better, yet simple, approximation of the current distribution is a triangular function of position along the wire that is maximum at the center and goes to zero at the ends:

$$I(z) = I_o\left(1 - \frac{|z|}{\Delta\ell/2}\right), \quad |z| \le \frac{\Delta\ell}{2}$$

(a) Recalculate the retarded vector potential $\mathbf{A}$.
(b) Find the radiation fields, $\mathbf{E}$ and $\mathbf{H}$.
(c) Calculate the radiation resistance and compare with that of a Hertzian dipole.

12-5. Calculate the total power (in watts) radiated by a short dipole antenna that transmits the AM radio frequency of 2 MHz. Assume that the antenna has the maximum current $I_o = 2$ A and the length $\Delta\ell = 1.5$ m.

12-6. **(Why the sky is blue)**
As shown in Eq. (12-35), the radiated power of a Hertzian dipole is proportional to $|I|^2/\lambda_o^2$ or $|I|^2 f^2$, i.e., strongly dependent on the frequency. Sunlight passing through the atmosphere stimulates small particles to oscillate as tiny dipoles with frequency f of the incident light. These small particles re-radiate as dipole antennas. When you look up in the sky, what you see is this re-radiated light.

(a) The current $I(t)$ is related to the charge q(t) of an oscillating dipole by $I(t) = \frac{dq}{dt}$. Show that the radiated power is proportional to the fourth power of the frequency.
(b) The sunlight consists of a broad range of visible spectrum. Explain why the sky looks blue.

The scattering of electromagnetic waves by small particles is known as *Rayleigh scattering*.

12-7. Consider a Hertzian dipole antenna or a short wire antenna oriented in the z direction and located at the origin. The far (radiation) electric field measured at a point 10 km away in the xy plane is 0.1 V/m.

(a) What is the electric field at a point 15 km away (from the antenna) in the xy plane?
(b) What is the electric field at a point 10 km away in the direction $\theta = 45°$?

12-8. Consider a straight wire VHF TV antenna in your home. In which directions does the antenna receive the signal best (most efficiently) and in which direction does the antenna receive no signal? Assume the antenna is placed vertically.

Note that (i) by reciprocity the receiving property of the antenna is the same as the transmitting (or radiating) pattern of the antenna. (ii) The radiation pattern of the straight wire antenna is similar to that of the Hertzian dipole antenna.

12-9. Figure 12-11 shows two crossed Hertzian dipoles (of length $\Delta\ell$). The currents in the dipoles are, respectively,

$$\mathbf{I}_1 = \mathbf{a}_z I_o \cos \omega t$$
$$\mathbf{I}_2 = \mathbf{a}_x I_o \sin \omega t$$

Change the currents to complex phasors and calculate the radiation electric field of this combination. Show that the electric field is circularly polarized along the y axis, linearly polarized in the xz plane, and elliptically polarized elsewhere.

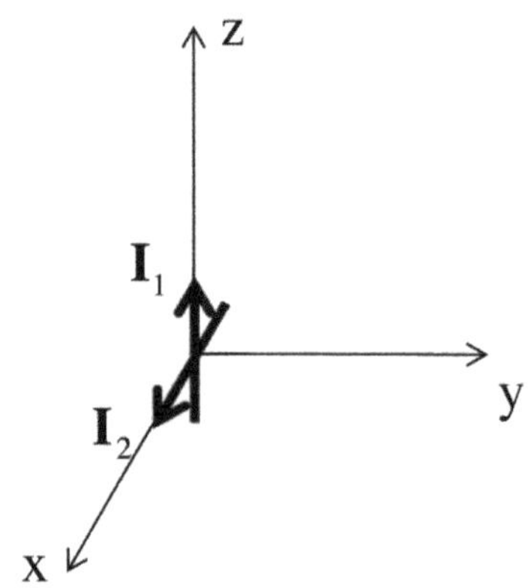

Figure 12-11. Crossed Hertzian dipoles.

12-10. Calculate the total power radiated by the crossed Hertzian dipoles (Figure 12-11) of Problem 12-9.

12-11. In the near fields (Eq. (12-33)) of a Hertzian dipole, the stored electric energy is much greater than the stored magnetic energy. Show that the ratio of the electric energy density ($w_e = ½\, \varepsilon\, |\mathbf{E}|^2$) and the magnetic energy density ($w_m = ½\, \mu\, |H|^2$) is equal to $\dfrac{(4\cot^2\theta + 1)}{(kr)^2}$.

12-12*. **(A small loop antenna)**
Consider a small circular loop of a radius *a* in the xy plane, carrying a uniform sinusoidal current $I(t) = I_o$ cos ωt (see Figure 6-10). Assume that $a << \lambda$, r.

(a) Following the same procedure that was used in Section 6.6, calculate the vector potential from Eq. (12-22). Make use of the approximation:

$$e^{-jk|\mathbf{r}-\mathbf{r}'|} = e^{-jkr}\, e^{-jk(R-r)} \approx e^{-jkr}\, \{1 - jk\,(R-r)\}$$

Show that the vector potential is given by

$$\mathbf{A} = \mathbf{a}_\phi \frac{\mu_o m}{4\pi r^2}(1+jkr)\, e^{-jkr} \sin\theta$$

where $m = I_o \pi a^2$.

(b) Find the electric and magnetic fields.

(c) Show that the electric field $\mathbf{E} = \mathbf{a}_\phi E_\phi$ is identical to the magnetic field $\mathbf{H} = \mathbf{a}_\phi H_\phi$ (Eq. (12-30)) of the Hertzian dipole and the magnetic field $\mathbf{H} = \mathbf{a}_r H_r + \mathbf{a}_\theta H_\theta$ is identical to the negative of the electric field $\mathbf{E} = \mathbf{a}_r E_r + \mathbf{a}_\theta E_\theta$ (Eq. (12-31)) of the Hertzian dipole if you replace $I\,\Delta\ell$ by $-j\omega\mu_o m$. This result reveals the *dual* nature of the electromagnetic fields of a Hertzian dipole and a small loop antenna, which are equivalent to an *oscillating electric dipole* and an *oscillating magnetic dipole*, respectively.

12-13. Using the result of Problem 12-12, calculate the total radiated power and the radiation resistance of a small loop antenna. Show that the ratio of the total radiated power of the magnetic dipole vs. the electric dipole is given by

$$\frac{P_r(\text{loop})}{P_r(\text{Hertzian})} = \left(k\frac{\pi a^2}{\Delta\ell}\right)^2$$

12-14. Suppose you have a combination of a z-directed Hertzian dipole $I_1\,\Delta\ell$ and a small loop antenna of current I_2 and area A in the xy plane. Show that the radiation field due to this combination is circularly polarized everywhere if

$$I_1\,\Delta\ell = k\,I_2 A$$

12-15. Derive Eq. (12-45) (radiation field expression for a dipole of length ℓ) using the following integral formula:

$$\int \sin(a+bx)\, e^{cx}\, dx = \frac{e^{cx}}{b^2+c^2}\left[c\sin(a+bx) - b\cos(a+bx)\right]$$

12-16. Generate the radiation patterns of long dipole antennas shown in Figure 12-6.

12-17. **(A monopole antenna)**
A monopole antenna is a dipole that has been divided in half at its center feed point and fed against a perfectly conducting ground plane (see Figure 12-12). Applying the method of images developed in Chapter 4, the monopole antenna and its image form a dipole antenna. The currents and charges on a monopole are the same as those on the upper half of its dipole counterpart, but the terminal voltage is half that of the dipole and no radiation occurs in the lower half space ($z < 0$). Find the radiation pattern,

radiation resistance and directivity of a quarter-wavelength long monopole antenna. Compare the results with those of a half-wave dipole.

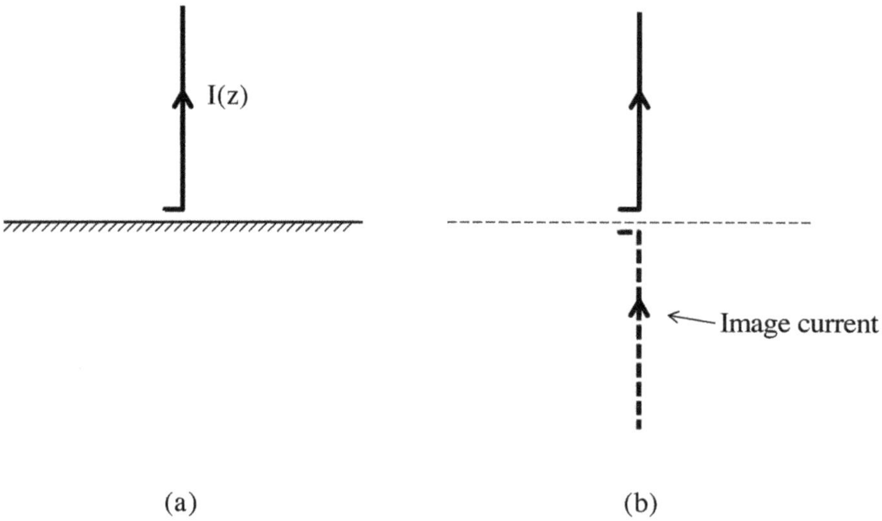

Figure 12-12. Monopole antenna above a ground plane and its equivalent dipole.

12-18. Calculate the total power in watts radiated by a center-fed dipole antenna that transmits the VHF TV signal of 63 MHz (Channel 3). Assume that the antenna has the maximum current 20 A and the length 2.38 m.

12-19. A center-fed dipole antenna of length 1 m operating at 150 MHz transmits a total average power of 200 W. Find the amplitudes of the electric and magnetic fields at a distance 100 m in the horizontal plane, assuming the dipole is vertically oriented.

12-20. Suppose a half-wave dipole antenna has the following volume current density:

$$\mathbf{J} = \mathbf{a}_z J_o \sin\left[k\left(\frac{\lambda}{4} - |z|\right)\right], \quad |z| < \frac{\lambda}{4}$$

Find the volume charge density ρ_v, associated with this volume current, using the equation of continuity. Sketch the current and charge distributions in the antenna as a function of z.

12-21. Calculate the radiation resistance and directivity of a full-wave dipole antenna ($\ell = \lambda$).

12-22. **(Road map—Finding radiation fields)**

If we are interested only in the radiation (far) fields of an antenna, there is a simpler way of finding the radiation fields instead of going through the steps in Table 12-1. First, the radiation field behaves like $1/r$ and the second term of $\mathbf{E} = -j\omega\mathbf{A} - \nabla V$ in Eq. (12-15) or Eq. (12-18) will behave like $1/r^2$ or $1/r^3$; thus it does not contribute to the far field. Secondly, the radiation fields will have only components (E_θ and E_ϕ) transverse to the direction of radiation ($\mathbf{k} = \mathbf{a}_r k$). Therefore, the radiation fields can be obtained by the following steps:

Step 1. Calculate $\mathbf{A}$.

Step 2. Find $\mathbf{E} = -j\omega$ (transverse components of $\mathbf{A}$) $= -\mathbf{a}_\theta j\omega A_\theta - \mathbf{a}_\phi j\omega A_\phi$.

Step 3. Find $\mathbf{H} = \frac{1}{\eta}\mathbf{a}_r \times \mathbf{E} = -\mathbf{a}_\phi j\frac{\omega}{\eta}A_\theta + \mathbf{a}_\theta j\frac{\omega}{\eta}A_\phi$ (plane wave property)

Derive the radiation fields of a long dipole, Eqs. (12-45) and (12-46), using this approach.

12-23. **(Radiation efficiency of an antenna)**
In practice, the radiated power of an antenna may not be equal to the total input power because of Ohmic power loss due to the finite conductivity of the antenna, as well as loss due to other factors. The *radiation efficiency* (e_r) is defined by

$$e_r = \frac{P_r}{P_{in}} = \frac{P_r}{P_r + P_\ell}$$

where P_r is the total radiated power and P_ℓ is the power loss. Consider a Hertzian dipole made of a copper wire of radius a, length $\Delta\ell$, and conductivity σ. Calculate the power dissipated in the dipole ($P_\ell = P_{diss}$) and the radiation efficiency e_r when $a = 1$ mm and $\Delta\ell = 10$ cm at the operating frequency $f = 10$ MHz.

12-24. Most fender-mount car radio antennas are 31 in. long. The fender plays the role of a grounded conducting plane, allowing the car antenna to be modeled as a monopole antenna (see Problem 12-17). Assume that the diameter of the antenna is 0.2 cm and the conductivity of Silicon steel is 2×10^6 [S/m]. Calculate the radiated power, the dissipated power loss, and the radiation efficiency (see Problem 12-23) at the AM frequency $f = 1$ MHz.

12-25. The angle between two half-power points of the main beam of an antenna is called the *half-power beam-width* (HPBW). The half-power points are the directions along which the field strength is $1/\sqrt{2}$ of that in the maximum direction of the radiation pattern. Determine the HPBWs of (i) a Hertzian dipole, (ii) a half-wave dipole, and (iii) a full-wave dipole.

12-26. Sketch the electric field pattern in the xy plane $\left(\theta = \frac{\pi}{2}\right)$ of an array of two vertical (z-directed) half-wave dipoles on the x axis for the following cases: (i) $d = \lambda/2$, $\alpha = \pi/2$, (ii) $d = \lambda/4$, $\alpha = \pi/2$, (iii) $d = \lambda$, $\alpha = \pi$.

12-27. Consider an array of two vertical (z-directed) short dipoles on the x axis, spaced $\lambda/2$ apart and equally excited ($\alpha = 0$). Sketch the radiation pattern of this array (i) in the xy plane, (ii) in the xz plane, and (iii) in the yz plane.

12-28. Consider an array of two horizontal (x-directed) short dipoles on the x axis, spaced $\lambda/2$ apart and equally excited ($\alpha = 0$). Write down the far field expression of this co-linear array. Sketch the radiation pattern of this array (i) in the xy plane, (ii) in the xz plane, and (iii) in the yz plane.

12-29. Consider the four-element linear array of Example 12-5. Sketch the radiation pattern of this array (a) when $d = \lambda/2$, $\alpha = \pi/2$, and (b) when $d = 3\lambda/4$, $\alpha = 0$.

12-30. Plot the normalized array factor as a function ψ $(0 \le \psi \le 2\pi)$, of a five-element array, uniformly excited and equally spaced. Sketch the radiation pattern of this array (a) when $d = \lambda/2$, $\alpha = 0$, and (b) when $d = \lambda/2$, $\alpha = -\pi$. Determine whether each array is broadside or endfire.

12-31. (Side lobe level)
For the uniform linear array, the peaks of side lobes decrease as N increases. A measure of the side lobe peaks is the *side lobe level*, which is defined by

$$\mathrm{SLL} = 20\log\left|\frac{\text{Max value of highest side lobe}}{\text{Max value of main lobe}}\right|$$

Calculate the SLL of the array factor for $N = 5, 10, 20$.

12-32. (A binomial array)
Consider an array of three isotropic radiating elements equally spaced ($d = \lambda/2$). The excitation currents are in phase but have the amplitude ratios 1:2:1. This nonuniform array is equivalent to having a pair of two-element uniform arrays (1:1), spaced $\lambda/2$ apart, i.e., it is a two-element array with each element being a two-element array. Find and sketch the radiation pattern of this three-element array, using the principle of pattern multiplication. This particular array is known as one of *binomial arrays* for which the excitation amplitudes follow the coefficients of a binomial expansion. It has a unique feature of having no side lobes.

12-33* (Nonuniform array)
As shown in Figure 12-10, a uniform linear array tends to have many sidelobes. One way to avoid or reduce sidelobes is to make the elements excited by non-uniform amplitudes. As the current amplitude is tapered (or reduced) more toward the edge of the array, the sidelobes tend to decrease and the beamwidth increases, thus decreasing directivity. Consider an array of five isotropic radiating elements equally spaced ($d = \lambda/2$). Assume the excitation currents are in phase ($\alpha = 0$). Find and sketch the radiation pattern of this array (a) when the excitations have the same amplitude 1:1:1:1:1 (uniform array), and (b) when the excitations have the amplitude distribution 1:4:6:4:1 (binormal array). Compare their patterns and calculate the sidelobe levels (see Problem 12-31).

12-34* (Receiving property of an antenna)

We have mainly considered the transmitting or radiating characteristics of an antenna. The antenna can also be used as a receiving antenna, which extracts energy from the incident electromagnetic wave and delivers it to a load. The receiving pattern of an antenna is defined as the voltage at the antenna terminals due to a plane wave incident upon the antenna. According to the **reciprocity theorem***, the receiving pattern of an antenna constructed of linear matter is identical to its transmitting pattern. A useful measure of receiving property of an antenna is the **effective aperture** or receiving area, which is defined by

$$A_e = \frac{\text{(Power received)}}{\text{(Incident power density)}}$$

It is known that the effective aperture (receiving property) is related to the directive gain (transmitting property) by

$$A_e(\theta, \phi) = \frac{\lambda^2}{4\pi} D(\theta, \phi)$$

Calculate the effective aperture of a Hertzian dipole antenna and confirm the above relationship. Assume that the dipole axis makes an angle θ with the direction of the incident wave.

* R.F. Harrington, *Time-Harmonic Electromagnetic Fields*, McGraw-Hill, 1961; Reissued by IEEE Press, 2001.

BIBLIOGRAPHY

1. A. T. Adams, *Electromagnetics for Engineers*, Ronald Press Co., 1971.
2. D. K. Cheng, *Field and Wave Electromagnetics*, Addison-Wesley, 1989, 2nd Ed.
3. D. K. Cheng, *Fundamentals of Engineering Electromagnetics*, Addison-Wesley, 1993.
4. R. E. Collin, *Foundations for Microwave Engineering*, McGraw-Hill, 1992, 2nd Ed.
5. R. P. Feynman, R. B. Leighton and M. Sands, *The Feynman Lectures on Physics*, Vol. II, Addison-Wesley, 1964.
6. D. J. Griffiths, *Introduction to Electrodynamics*, Prentice-Hall, 1999, 3rd Ed.
7. R. F. Harrington, *Time-Harmonic Electromagnetic Fields*, McGraw-Hill, 1961, reissued by IEEE Press, 2001.
8. H. A. Haus and J. R. Melcher, *Electromagnetic Fields and Energy*, Prentice Hall, 1989.
9. U. S. Inan and A. S. Inan, *Electromagnetic Waves*, Prentice-Hall, 2000.
10. J. D. Jackson, *Classical Electrodynamics*, John Wiley & Sons, 1999, 3rd Ed.
11. J. A. Kong, *Electromagnetic Wave Theory*, EMW Publishing, 2000, 2005, 2008.
12. J. D. Kraus, *Electromagnetics*, McGraw-Hill, 1992, 4th Ed.
13. P. Lorrain, D. R. Corson, and F. Lorrain, *Electromagnetic Fields and Waves*, W. H. Freeman and Company, 1988, 3rd Ed.
14. C. R. Paul, K. W. Whites and S. A. Nasar, *Introduction to Electromagnetic Fields*, McGraw-Hill, 1998, 3rd Ed.
15. D. M. Pozar, *Microwave Engineering*, John Wiley & Sons, 1998, 2nd Ed.
16. J. R. Reitz, F. J. Milford and R. W. Christy, *Foundations of Electromagnetic Theory,* Addison-Wesley, 1993, 4th Ed.
17. M. N. O. Sadiku, *Elements of Electromagnetics*, Oxford University Press, 2010, 5th Ed.
18. L. C. Shen and J. A. Kong, *Applied Electromagnetism*, PWS Publishing Co., 1995, 3rd Ed.
19. D. H. Staelin, A. Morgenthaler and J. A. Kong, *Electromagnetic Waves*, Prentice-Hall, 1994.
20. W. L. Stutzman and G. A. Thiele, *Antenna Theory and Design*, John Wiley & Sons, 1998, 2nd Ed.

APPENDIX A

A-1 Physical Constants

$$\varepsilon_0 = 8.854\times10^{-12} \quad [\text{F/m}]$$

$$\approx \frac{1}{36\pi}\times10^{-9} \quad [\text{F/m}] \qquad \text{(permittivity of free space)}$$

$$\mu_0 = 4\pi\times10^{-7} \quad [\text{H/m}] \qquad \text{(permeability of free space)}$$

$$\frac{1}{\sqrt{\mu_0\varepsilon_0}} = c \approx 3\times10^{8} \quad [\text{m/s}] \qquad \text{(speed of light in vacuum)}$$

$$\sqrt{\frac{\mu_0}{\varepsilon_0}} = \eta_0 \approx 120\pi \text{ or } 377\,[\Omega] \qquad \text{(intrinsic impedance of air)}$$

$$-e = -1.602\times10^{-19}\,[\text{C}] \qquad \text{(charge of an electron)}$$

$$m_e = 9.107\times10^{-31}\,[\text{kg}] \qquad \text{(mass of an electron at rest)}$$

A-2 Prefixes

Prefix	Abbreviation	Meaning
exa	E	10^{18}
peta	P	10^{15}
tera	T	10^{12}
giga	G	10^{9}
mega	M	10^{6}
kilo	k	10^{3}
hecto	h	10^{2}
deka	da	10
deci	d	10^{-1}
centi	c	10^{-2}
milli	m	10^{-3}
micro	μ	10^{-6}
nano	n	10^{-9}
pico	p	10^{-12}
femto	f	10^{-15}
atto	a	10^{-18}

A-3 Frequently Used Symbols and Their Units

Quantity	Symbol	Unit	Abbreviated Unit
angular frequency	ω	radian/second	rad/s
capacitance	C	farad	F
charge	Q, q	coulomb	C
charge density			
volume charge density	ρ_v	coulomb/meter^3	C/m^3
surface charge density	ρ_s	coulomb/meter^2	C/m^2
line charge density	ρ_ℓ	coulomb/meter	C/m
conductance	G	siemens or mho	S
conductivity	σ	siemens/meter	S/m
current	I, i	ampere	A
current density			
volume current density	$\mathbf{J}$	ampere/meter^2	A/m^2
surface current density	$\mathbf{J}_s$	ampere/meter	A/m
dielectric constant	ε_r	(dimensionless)	–
electric dipole moment	$\mathbf{p}$	coulomb-meter	C· m
electric displacement	$\mathbf{D}$	coulomb/meter^2	C/m^2
electric field intensity	$\mathbf{E}$	volt/meter	V/m
electric potential	V	volt	V
electromotive force (emf)	V	volt	V
energy (work)	W	joule	J
energy density	w	joule/meter^3	J/m^3
force	$\mathbf{F}$	newton	N
frequency	f	hertz	Hz
impedance	Z, η	ohm	Ω
inductance	L	henry	H
magnetic dipole moment	$\mathbf{m}$	ampere-meter^2	A· m^2
magnetic field intensity	$\mathbf{H}$	ampere/meter	A/m
magnetic flux	Φ	weber	Wb
magnetic flux density	$\mathbf{B}$	tesla or weber/meter^2	T or Wb/m^2
magnetic vector potential	$\mathbf{A}$	weber/meter	Wb/m
magnetization vector	$\mathbf{M}$	ampere/meter	A/m
permeability	μ, μ_o	henry/meter	H/m
permittivity	ε, ε_o	farad/meter	F/m
polarization vector	$\mathbf{P}$	coulomb/meter^2	C/m^2
power	P	watt	W
Poynting vector (power density)	$\mathbf{S}$	watt/meter^2	W/m^2
propagation constant	k	meter^{-1}	m^{-1}
resistance	R	ohm	Ω
torque	$\mathbf{T}$	newton-meter	N· m
velocity	$\mathbf{v}$	meter/second	m/s
voltage	V	volt	V
wavelength	λ	meter	m

APPENDIX B

TABLES FOR VECTOR ANALYSIS

Table 1-1 Equations of Transformation between Coordinate Systems

$x = \rho \cos\phi = r \sin\theta \cos\phi$ $y = \rho \sin\phi = r \sin\theta \sin\phi$ $z = r\cos\theta$
$\rho = \sqrt{x^2 + y^2} = r\sin\theta$ $\phi = \tan^{-1}\dfrac{y}{x}$ * $z = r\cos\theta$
$r = \sqrt{x^2 + y^2 + z^2} = \sqrt{\rho^2 + z^2}$ $\theta = \tan^{-1}\dfrac{\sqrt{x^2 + y^2}}{z} = \tan^{-1}\dfrac{\rho}{z}$ $\phi = \tan^{-1}\dfrac{y}{x}$ *
* Note that it is important to consider the signs of x and y to determine the appropriate quadrant of angle ϕ.

TABLES FOR VECTOR ANALYSIS

Table 1-2 Differential Elements of Length, Surface, and Volume

(a) Vector Differential Length Elements

$$d\ell = \mathbf{a}_x dx + \mathbf{a}_y dy + \mathbf{a}_z dz$$
$$= \mathbf{a}_\rho d\rho + \mathbf{a}_\phi \rho\, d\phi + \mathbf{a}_z dz$$
$$= \mathbf{a}_r dr + \mathbf{a}_\theta r\, d\theta + \mathbf{a}_\phi r\, \sin\theta\, d\phi$$

(b) Vector Differential Surface Elements

$$d\mathbf{s} = \pm\mathbf{a}_x\, dy\, dz,\ \mathbf{a}_y\, dz\, dx,\ \mathbf{a}_z\, dx\, dy^*$$
$$= \pm\mathbf{a}_\rho\, \rho\, d\phi\, dz,\ \mathbf{a}_\phi\, d\rho\, dz,\ \mathbf{a}_z\, \rho d\rho\, d\phi$$
$$= \pm\mathbf{a}_r r^2\, \sin\theta\, d\theta\, d\phi,\ \mathbf{a}_\theta r \sin\theta\, dr\, d\phi,\ \mathbf{a}_\phi r\, dr\, d\theta$$

(c) Differential Volume Elements

$$dv = dx\, dy\, dz$$
$$= \rho\, d\rho\, d\phi\, dz$$
$$= r^2\, \sin\theta\, dr\, d\theta\, d\phi$$

*The choice of sign $\pm$ depends on the problem.

Table 1-3 Transformation of Unit Vectors

	RECTANGULAR			CYLINDRICAL			SPHERICAL		
	$\mathbf{a}_x$	$\mathbf{a}_y$	$\mathbf{a}_z$	$\mathbf{a}_\rho$	$\mathbf{a}_\phi$	$\mathbf{a}_z$	$\mathbf{a}_r$	$\mathbf{a}_\theta$	$\mathbf{a}_\phi$
$\mathbf{a}_x$	1	0	0	$\cos\phi$	$-\sin\phi$	0	$\sin\theta\cos\phi$	$\cos\theta\cos\phi$	$-\sin\phi$
$\mathbf{a}_y$	0	1	0	$\sin\phi$	$\cos\phi$	0	$\sin\theta\sin\phi$	$\cos\theta\sin\phi$	$\cos\phi$
$\mathbf{a}_z$	0	0	1	0	0	1	$\cos\theta$	$-\sin\theta$	0
$\mathbf{a}_\rho$	$\cos\phi$	$\sin\phi$	0	1	0	0	$\sin\theta$	$\cos\theta$	0
$\mathbf{a}_\phi$	$-\sin\phi$	$\cos\phi$	0	0	1	0	0	0	1
$\mathbf{a}_z$	0	0	1	0	0	1	$\cos\theta$	$-\sin\theta$	0
$\mathbf{a}_r$	$\sin\theta\cos\phi$	$\sin\theta\sin\phi$	$\cos\theta$	$\sin\theta$	0	$\cos\theta$	1	0	0
$\mathbf{a}_\theta$	$\cos\theta\cos\phi$	$\cos\theta\sin\phi$	$-\sin\theta$	$\cos\theta$	0	$-\sin\theta$	0	1	0
$\mathbf{a}_\phi$	$-\sin\phi$	$\cos\phi$	0	0	1	0	0	0	1

[Note] The unit vectors ($\mathbf{a}_x$, $\mathbf{a}_y$, $\mathbf{a}_z$, $\mathbf{a}_\rho$, $\mathbf{a}_\phi$, $\mathbf{a}_r$, $\mathbf{a}_\theta$) can be replaced by the vector components (A_x, A_y, A_z, A_ρ, A_ϕ, A_r, A_θ). The transformation is still valid.

Table 1-4 Vector Derivatives

RECTANGULAR COORDINATES (x, y, z)

$$\nabla V = \mathbf{a}_x \frac{\partial V}{\partial x} + \mathbf{a}_y \frac{\partial V}{\partial y} + \mathbf{a}_z \frac{\partial V}{\partial z}$$

$$\nabla \cdot \mathbf{A} = \frac{\partial A_x}{\partial x} + \frac{\partial A_y}{\partial y} + \frac{\partial A_z}{\partial z}$$

$$\nabla \times \mathbf{A} = \mathbf{a}_x \left(\frac{\partial A_z}{\partial y} - \frac{\partial A_y}{\partial z} \right) + \mathbf{a}_y \left(\frac{\partial A_x}{\partial z} - \frac{\partial A_z}{\partial x} \right) + \mathbf{a}_z \left(\frac{\partial A_y}{\partial x} - \frac{\partial A_x}{\partial y} \right)$$

$$\nabla^2 V = \frac{\partial^2 V}{\partial x^2} + \frac{\partial^2 V}{\partial y^2} + \frac{\partial^2 V}{\partial z^2}$$

CYLINDRICAL COORDINATES (ρ, ϕ, z)

$$\nabla V = \mathbf{a}_\rho \frac{\partial V}{\partial \rho} + \mathbf{a}_\phi \frac{1}{\rho} \frac{\partial V}{\partial \phi} + \mathbf{a}_z \frac{\partial V}{\partial z}$$

$$\nabla \cdot \mathbf{A} = \frac{1}{\rho} \frac{\partial}{\partial \rho} (\rho A_\rho) + \frac{1}{\rho} \frac{\partial A_\phi}{\partial \phi} + \frac{\partial A_z}{\partial z}$$

$$\nabla \times \mathbf{A} = \mathbf{a}_\rho \left(\frac{1}{\rho} \frac{\partial A_z}{\partial \phi} - \frac{\partial A_\phi}{\partial z} \right) + \mathbf{a}_\phi \left(\frac{\partial A_\rho}{\partial z} - \frac{\partial A_z}{\partial \rho} \right) + \mathbf{a}_z \frac{1}{\rho} \left[\frac{\partial}{\partial \rho} (\rho A_\phi) - \frac{\partial A_\rho}{\partial \phi} \right]$$

$$\nabla^2 V = \frac{1}{\rho} \frac{\partial}{\partial \rho} \left(\rho \frac{\partial V}{\partial \rho} \right) + \frac{1}{\rho^2} \frac{\partial^2 V}{\partial \phi^2} + \frac{\partial^2 V}{\partial z^2}$$

SPHERICAL COORDINATES (r, θ, ϕ)

$$\nabla V = \mathbf{a}_r \frac{\partial V}{\partial r} + \mathbf{a}_\theta \frac{1}{r} \frac{\partial V}{\partial \theta} + \mathbf{a}_\phi \frac{1}{r \sin\theta} \frac{\partial V}{\partial \phi}$$

$$\nabla \cdot \mathbf{A} = \frac{1}{r^2} \frac{\partial}{\partial r} (r^2 A_r) + \frac{1}{r \sin\theta} \frac{\partial}{\partial \theta} (A_\theta \sin\theta) + \frac{1}{r \sin\theta} \frac{\partial A_\phi}{\partial \phi}$$

$$\nabla \times \mathbf{A} = \mathbf{a}_r \frac{1}{r \sin\theta} \left[\frac{\partial}{\partial \theta} (A_\phi \sin\theta) - \frac{\partial A_\theta}{\partial \phi} \right]$$
$$+ \mathbf{a}_\theta \frac{1}{r} \left[\frac{1}{\sin\theta} \frac{\partial A_r}{\partial \phi} - \frac{\partial}{\partial r} (r A_\phi) \right] + \mathbf{a}_\phi \frac{1}{r} \left[\frac{\partial}{\partial r} (r A_\theta) - \frac{\partial A_r}{\partial \theta} \right]$$

$$\nabla^2 V = \frac{1}{r^2} \frac{\partial}{\partial r} \left(r^2 \frac{\partial V}{\partial r} \right) + \frac{1}{r^2 \sin\theta} \frac{\partial}{\partial \theta} \left(\sin\theta \frac{\partial V}{\partial \theta} \right) + \frac{1}{r^2 \sin^2\theta} \frac{\partial^2 V}{\partial \phi^2}$$

Table 1-5 Vector Identities

TRIPLE PRODUCTS

(1) $\mathbf{A} \cdot (\mathbf{B} \times \mathbf{C}) = \mathbf{B} \cdot (\mathbf{C} \times \mathbf{A}) = \mathbf{C} \cdot (\mathbf{A} \times \mathbf{B})$

(2) $\mathbf{A} \times (\mathbf{B} \times \mathbf{C}) = \mathbf{B}(\mathbf{A} \cdot \mathbf{C}) - \mathbf{C}(\mathbf{A} \cdot \mathbf{B})$

PRODUCT RULES

(3) $\nabla(V\,W) = V(\nabla W) + W(\nabla V)$

(4) $\nabla(\mathbf{A} \cdot \mathbf{B}) = \mathbf{A} \times (\nabla \times \mathbf{B}) + \mathbf{B} \times (\nabla \times \mathbf{A}) + (\mathbf{A} \cdot \nabla)\,\mathbf{B} + (\mathbf{B} \cdot \nabla)\,\mathbf{A}$

(5) $\nabla \cdot (V\,\mathbf{A}) = V\,(\nabla \cdot \mathbf{A}) + \mathbf{A} \cdot (\nabla V)$

(6) $\nabla \cdot (\mathbf{A} \times \mathbf{B}) = \mathbf{B} \cdot (\nabla \times \mathbf{A}) - \mathbf{A} \cdot (\nabla \times \mathbf{B})$

(7) $\nabla \times (V\,\mathbf{A}) = V(\nabla \times \mathbf{A}) - \mathbf{A} \times (\nabla V)$

(8) $\nabla \times (\mathbf{A} \times \mathbf{B}) = (\mathbf{B} \cdot \nabla)\,\mathbf{A} - (\mathbf{A} \cdot \nabla)\,\mathbf{B} + \mathbf{A}\,(\nabla \cdot \mathbf{B}) - \mathbf{B}\,(\nabla \cdot \mathbf{A})$

SECOND DERIVATIVES

(9) $\nabla \cdot (\nabla \times \mathbf{A}) = 0$

(10) $\nabla \times (\nabla V) = 0$

(11) $\nabla \times (\nabla \times \mathbf{A}) = \nabla\,(\nabla \cdot \mathbf{A}) - \nabla^2\mathbf{A}$

INTEGRAL THEOREMS

(1) Gradient Theorem: $\int_a^b (\nabla V) \cdot d\boldsymbol{\ell} = V(b) - V(a)$

(2) Divergence Theorem: $\iiint_V (\nabla \cdot \mathbf{A})\, dv = \oiint_S \mathbf{A} \cdot d\mathbf{s}$

(3) Stokes' Theorem: $\iint_S (\nabla \times \mathbf{A}) \cdot d\mathbf{s} = \oint_C \mathbf{A} \cdot d\boldsymbol{\ell}$

ANSWERS TO SELECTED PROBLEMS

Chapter 1

1-1 (a) $\sqrt{6}$ (b) $\mathbf{a}_x \frac{1}{\sqrt{6}} - \mathbf{a}_y \frac{1}{\sqrt{6}} + \mathbf{a}_z \frac{2}{\sqrt{6}}$ (c) 6 (d) $\sqrt{6}$ (e) 54.7°

(f) $-\mathbf{a}_x 6 - \mathbf{a}_y 6$ (g) −18

1-3 (a) $5\sqrt{2}$ (b) $\mathbf{a}_x \frac{3}{5\sqrt{2}} - \mathbf{a}_y \frac{4}{5\sqrt{2}} + \mathbf{a}_z \frac{1}{\sqrt{2}}$ (c) 0 (d) 0 (e) 90°

(f) $\mathbf{a}_x 15 + \mathbf{a}_y 20 + \mathbf{a}_z 25$ (g) 50

1-6 For Problem 1-1, (a) $\mathbf{a}_x 7 - \mathbf{a}_y 25 - \mathbf{a}_z 16$ (b) $\mathbf{A}\cdot\mathbf{B} = 6$, $\mathbf{A}\cdot\mathbf{C} = -7$

For Problem 1-3, (a) $\mathbf{a}_x 80 - \mathbf{a}_y 60$ (b) $\mathbf{A}\cdot\mathbf{B} = 0$, $\mathbf{A}\cdot\mathbf{C} = -20$

1-7 $\sqrt{2}$ **1-12** (b) $\mathbf{a}_x \frac{2}{\sqrt{5}} - \mathbf{a}_y \frac{1}{\sqrt{5}}$

1-17 (a) $\phi = 0°$, $\mathbf{a}_\phi = \mathbf{a}_y$ (b) $\phi = 90°$, $\mathbf{a}_\phi = -\mathbf{a}_x$ (c) $\theta = 90°$, $\mathbf{a}_\theta = -\mathbf{a}_z$

1-18 (a) $(\sqrt{2}, 45°, 1)$, $(\sqrt{3}, 55°, 45°)$ (c) $(1, 0°, 1)$, $(\sqrt{2}, 45°, 0°)$

1-19 (a) $\left(\frac{1}{\sqrt{2}}, \frac{1}{\sqrt{2}}, 2\right)$, $(\sqrt{5}, 27°, 45°)$ (c) $(0, -2, 3)$, $(\sqrt{13}, 33.7°, 270°)$

1-20 (a) $\left(\frac{1}{2}, \frac{1}{2}, \frac{1}{\sqrt{2}}\right)$, $\left(\frac{1}{\sqrt{2}}, 45°, \frac{1}{\sqrt{2}}\right)$ (c) $(-1.84, 1.84, -1.5)$, $\left(\frac{3\sqrt{3}}{\sqrt{2}}, 135°, -\frac{3}{2}\right)$

1-24 (a) 0 (b) No **1-27** (a) 7 (b) 7. Yes **1-29** (a) πr_0^4 (b) $\frac{\pi}{2} r_0^4$. No

1-31 (a) $-\mathbf{a}_x e^{-x} \sin 2y \cos 3z + \mathbf{a}_y 2e^{-x} \cos 2y \cos 3z - \mathbf{a}_z 3e^{-x} \sin 2y \sin 3z$

(b) $\mathbf{a}_\rho 2\rho \sin 2\phi + \mathbf{a}_\phi 2\rho \cos 2\phi$ (c) $-\mathbf{a}_r 3r^{-4} \cos^2\theta - \mathbf{a}_\theta r^{-4} \sin 2\theta$

1-33 (a) $\nabla\cdot\mathbf{A} = z^2$, $\nabla\times\mathbf{A} = -\mathbf{a}_x 2yz + \mathbf{a}_y 2y - \mathbf{a}_z(2z + 6y)$

(b) $\nabla\cdot\mathbf{A} = 2\sin\phi + \cos\phi$, $\nabla\times\mathbf{A} = -\mathbf{a}_\rho \frac{z}{\rho} \sin\phi - \mathbf{a}_z \cos\phi$

(c) $\nabla\cdot\mathbf{A} = 4r\sin\theta - r\sin\phi \frac{\cos 2\theta}{\sin\theta}$,

$\nabla\times\mathbf{A} = \mathbf{a}_r \frac{r}{\sin\theta}(\cos 2\theta + \cos\theta\cos\phi) - \mathbf{a}_\theta 3r\cos\theta - \mathbf{a}_\phi r\cos\theta(3\sin\phi + 1)$

1-35 (a) 1 km west, 4 km north (b) 576 m (c) 101.2 m/km

1-39 (a) $3/2$ (b) $3/2$ **1-43** (a) $\frac{\pi}{3}a^3$ (b) $\frac{\pi}{3}a^3$ (c) No

1-45 (a) $\pi/6$ (b) $\pi/6$ **1-47** (a) 2 (b) 2 (c) No

1-49 (a) $6\pi a^2$ (b) $6\pi a^2$ **1-51** (a) $-\frac{2}{3}a^3$ (b) $-\frac{2}{3}a^3$

Chapter 2

2-1 (a) $\frac{4\pi}{3}\left(b^3-a^3\right)\rho_{vo}$ **2-2** (b) $\frac{1}{9}\left(b^3-a^3\right)\left(d^3-c^3\right)\rho_{so}$

2-3 (c) $2\pi a\sin\theta_0\,\rho_{\ell 0}$ **2-4** (a) $\pi a^2\,\ell\,\rho_{vo}$

2-6 (a) $\frac{q}{4\pi\varepsilon_0 a^2}\left(1+\frac{1}{2\sqrt{2}}\right)\left(\mathbf{a}_x+\mathbf{a}_y\right)$ **2-7** $\frac{\sqrt{3}\,q^2}{4\pi\,\varepsilon_0\,a^2}$

2-9 (a) $\mathbf{E}=\mathbf{a}_r\frac{\rho_{vo}\,a^3}{12\varepsilon_0 r^2}\,(r\ge a);\ \mathbf{a}_r\frac{\rho_{vo}\,r}{\varepsilon_0}\left(\frac{1}{3}-\frac{r}{4a}\right)(r\le a)$

2-10 (a) $\mathbf{E}=0\,(r\le a);\ \mathbf{a}_r\frac{\rho_{vo}}{4\varepsilon_0 a}\frac{r^4-a^4}{r^2}\,(a\le r\le b);\ \mathbf{a}_r\frac{\rho_{vo}}{4\varepsilon_0 a}\frac{b^4-a^4}{r^2}\,(r\ge b)$

2-12 (a) $\mathbf{E}=0\,(r\le a);\ \mathbf{a}_r\frac{\rho_{s1}\,a^2}{\varepsilon_0 r^2}\,(a<r<b);\ \mathbf{a}_r\frac{\rho_{s1}\,a^2+\rho_{s2}b^2}{\varepsilon_0 r^2}\,(r>b)$

(b) $\rho_{s2}=-\rho_{s1}\frac{a^2}{b^2}$

2-16 $\mathbf{E}=\mathbf{a}_\rho\frac{-\rho_{vo}\left[e^{-k\rho^2}-1\right]}{2k\,\varepsilon_0\,\rho}\,(\rho<a);\ \mathbf{a}_\rho\frac{-\rho_{vo}\left[e^{-ka^2}-1\right]}{2k\varepsilon_0\,\rho}\,(\rho>a)$

2-18 $\mathbf{E}=0\ (\rho<a);\ \mathbf{a}_\rho\frac{1}{\varepsilon_0\rho}\left[\rho_{s1}a+\rho_{vo}\frac{\rho^2-a^2}{2}\right](a<\rho<b);$

$=\mathbf{a}_\rho\frac{1}{\varepsilon_0\rho}\left[\left(\rho_{s1}a-\rho_{s2}b\right)+\rho_{vo}\frac{b^2-a^2}{2}\right](\rho>b)$

2-20 (a) $\mathbf{E}=\mathbf{a}_x\frac{\rho_{vo}x}{\varepsilon_0}\left(|x|<\frac{d}{2}\right);\ \mathbf{a}_x\frac{\rho_{vo}d}{2\varepsilon_0}\left(x>\frac{d}{2}\right);\ -\mathbf{a}_x\frac{\rho_{vo}d}{2\varepsilon_0}\left(x<-\frac{d}{2}\right)$

2-25 (a) $\mathbf{a}_z\frac{\rho_{\ell 0}\,za}{2\varepsilon_0\left(z^2+a^2\right)^{3/2}}$ (b) $\frac{\rho_{\ell 0}\,a}{2\varepsilon_0\sqrt{z^2+a^2}}$ (c) $\mathbf{a}_z\frac{Q}{4\pi\varepsilon_0\,z^2}$ (d) $\frac{Q}{4\pi\varepsilon_0 z}$

2-31 (a) $\mathbf{a}_z\frac{\rho_{so}}{2\varepsilon_0}\left[\frac{z}{\sqrt{z^2+a^2}}-\frac{z}{\sqrt{z^2+b^2}}\right]$ (b) $\mathbf{a}_z\frac{Q}{4\pi\varepsilon_0 z^2}$

2-32 (a) $\frac{\rho_{so}}{2\varepsilon_0}\left[\sqrt{z^2+b^2}-\sqrt{z^2+a^2}\right]$ (c) $\frac{Q}{4\pi\varepsilon_0 z}$

2-34 $\mathbf{a}_z\frac{\rho_{so}z}{2\varepsilon_0 a}\left[\ln\left(\frac{a+\sqrt{z^2+a^2}}{|z|}\right)-\frac{a}{\sqrt{z^2+a^2}}\right]$

2-36 (a) $\mathbf{a}_z \dfrac{\rho_{\ell 0} a}{\pi\varepsilon_0} \dfrac{z}{\left(z^2+\dfrac{a^2}{4}\right)\sqrt{z^2+\dfrac{a^2}{2}}}$ **2-37** (a) $\dfrac{\rho_{\ell 0}}{\pi\varepsilon_0}\ln\left[\dfrac{\dfrac{a}{2}+\sqrt{z^2+\dfrac{a^2}{2}}}{-\dfrac{a}{2}+\sqrt{z^2+\dfrac{a^2}{2}}}\right]$

2-41 (a) $\mathbf{a}_z \dfrac{\rho_{\ell 0}\, a}{2\varepsilon_0}\left[\dfrac{1}{\sqrt{a^2+(z-\ell/2)^2}} - \dfrac{1}{\sqrt{a^2+(z+\ell/2)^2}}\right]$ (b) $\mathbf{a}_z \dfrac{Q}{4\pi\varepsilon_0 z^2}$

2-50 $\mathbf{E} = \mathbf{a}_\rho \dfrac{\rho_{\ell 0}}{2\pi\varepsilon_0\rho}\,(\rho < a,\ \rho > b);\ 0(a<\rho<b),\ \rho_s = -\dfrac{\rho_{\ell 0}}{2\pi a}\,(\rho = a);\ \dfrac{\rho_{\ell 0}}{2\pi b}\,(\rho = b)$

2-52 (a) $\rho_s = -\dfrac{a^3}{3b^2}\rho_{vo}\,(r = b);\quad \dfrac{a^3}{3c^2}\rho_{vo}\,(r = c)$

(b) $\mathbf{E} = \mathbf{a}_r \dfrac{\rho_{vo}}{3\varepsilon_0} r\ (r<a);\ \mathbf{a}_r \dfrac{\rho_{vo}\, a^3}{3\varepsilon_0 r^2}\ (a < r < b,\ r > c);\ 0\,(b < r < c)$

Chapter 3

3-1 (a) $\rho_{pv} = 0,\ \rho_{ps} = \pm P_0\ (\text{at } z = \pm\dfrac{\ell}{2})$ (b) $\rho_{pv} = 0,\ \rho_{ps} = P_0\cos\theta$

(c) $\rho_{pv} = -3P_0,\ \rho_{ps} = \dfrac{\ell}{2}P_0$ (on all six surfaces)

3-6 $\mathbf{E} = \mathbf{a}_z \dfrac{P_0}{2\varepsilon_0}\left[\dfrac{z+\ell/2}{\sqrt{(z+\ell/2)^2+a^2}} - \dfrac{z-\ell/2}{\sqrt{(z-\ell/2)^2+a^2}}\right]$

3-8 $\mathbf{E} = -\mathbf{a}_z \dfrac{P_0}{3\varepsilon_0}$ **3-9** $\mathbf{E} = -\mathbf{a}_r \dfrac{P_0 r}{\varepsilon_0 a}\,(r < a);\ 0\,(r > a),\ \mathbf{D} = 0$ everywhere

3-19 (a) $\mathbf{D} = \mathbf{a}_\rho \dfrac{\rho_{\ell 0}}{2\pi\rho}$ everywhere, $\mathbf{E} = \mathbf{a}_\rho \dfrac{\rho_{\ell 0}}{2\pi\varepsilon_0\varepsilon_r\rho}\,(\rho < a);\ \mathbf{a}_\rho \dfrac{\rho_{\ell 0}}{2\pi\varepsilon_0\rho}\,(\rho > a)$,

$\mathbf{P} = \mathbf{a}_\rho \dfrac{\rho_{\ell 0}}{2\pi\rho}\left(1-\dfrac{1}{\varepsilon_r}\right)(\rho < a);\ 0(\rho > a)$ (b) $\rho_{pv} = 0,\ \rho_{ps} = \dfrac{\rho_{\ell 0}}{2\pi a}\left(1-\dfrac{1}{\varepsilon_r}\right)$

3-21 (a) $2\pi\left[\dfrac{1}{\varepsilon_1}\ln\dfrac{b}{a} + \dfrac{1}{\varepsilon_2}\ln\dfrac{c}{b}\right]^{-1}$ **3-23** (a) $\dfrac{\pi\varepsilon_1}{\ln\dfrac{c}{a}} + \dfrac{\pi\varepsilon_2\varepsilon_3}{\varepsilon_2\ln\dfrac{c}{b} + \varepsilon_3\ln\dfrac{b}{a}}$

3-24 Fig (a): $w\ell\left[\dfrac{d_1}{\varepsilon_1}+\dfrac{d_2}{\varepsilon_2}\right]^{-1}$ Fig (b): $\dfrac{w}{d}(\ell_1\varepsilon_1 + \ell_2\varepsilon_2)$ **3-31** $W_e = \dfrac{4\pi\rho_0^2 a^5}{15\varepsilon_0}$

Chapter 4

4-5 $V = V_0 \dfrac{\phi}{\phi_0}$, $\mathbf{E} = -\mathbf{a}_\phi \dfrac{V_0}{\phi_0}\dfrac{1}{\rho}$

4-9 $V(r,\theta) = \dfrac{q(2d)\cos\theta}{4\pi\varepsilon_0 r^2}$

4-10 Seven image charges are needed.

4-11 An infinite number of image charges are needed.

4-13 $\rho_s = -\dfrac{\rho_\ell}{2\pi}\left\{\dfrac{a - s\cos\phi}{s^2+a^2-2sa\cos\phi} - \dfrac{a - d\cos\phi}{d^2+a^2-2ad\cos\phi}\right\}$

4-18 (b) Solve $\dfrac{q}{(d-a)} = -\dfrac{q^1}{(a-s)}$ and $\dfrac{q}{(a+d)} = -\dfrac{q^1}{(a+s)}$

4-23 $V(x,y) = \sum\limits_{n=odd} \dfrac{4V_0}{n\pi} \dfrac{\cosh\left(\dfrac{n\pi}{b}x\right)}{\cosh\left(\dfrac{n\pi}{b}a\right)} \sin\left(\dfrac{n\pi}{b}y\right)$

4-25 $V(x,y) = V_0\left\{\sin\left(\dfrac{\pi}{a}x\right)\dfrac{\sinh\left(\dfrac{\pi}{a}y\right)}{\sinh\pi} + \sin\left(\dfrac{2\pi}{a}y\right)\dfrac{\sinh\left(\dfrac{2\pi}{a}x\right)}{\sinh(2\pi)}\right\}$

4-29 $V(\rho,\phi) = \sum\limits_{n=odd} \dfrac{\left(-4V_0\sin\dfrac{n\pi}{2}\right)}{n\pi}\left(\dfrac{\rho}{a}\right)^n \cos n\phi \;(\rho \le a);$

$= \sum\limits_{n=odd} \dfrac{\left(-4V_0\sin\dfrac{n\pi}{2}\right)}{n\pi}\left(\dfrac{a}{\rho}\right)^n \cos n\phi \;(\rho \ge a)$

4-32 $V(\rho,\phi) = \dfrac{\rho_{so}}{4\varepsilon_0 a}\rho^2\sin 2\phi \;(\rho<a);\; \dfrac{\rho_{so}a^3}{4\varepsilon_0}\dfrac{1}{\rho^2}\sin 2\phi \;(\rho>a)$

4-33 $V(r,\theta) = -E_0 r\cos\theta + E_0\dfrac{a^3}{r^2}\cos\theta$

4-37 (a) $\rho_{pv} = 0$, $\rho_{ps} = P_0\cos\theta$ (b) $\mathbf{E} = -\mathbf{a}_z\dfrac{P_0}{3\varepsilon_0}$

(c) $V(r,\theta) = \dfrac{P_0}{3\varepsilon_0} r\cos\theta \;(r<a);\; \dfrac{P_0}{3\varepsilon_0}\dfrac{a^3}{r^2}\cos\theta \;(r>a)$

Chapter 5

5-1 (b) $\mathbf{J} = \mathbf{a}_z\dfrac{3I_0}{2\pi a^3}\rho$

5-3 (a) 9 mΩ (b) 3.54×10^7 S/m (c) 9×10^{-4} V/m (d) 9.0×10^{-5} W

5-4 (a) 1.27×10^6 A/m^2 (b) 5.79×10^7 S/m (c) 8.8×10^{-2} W (d) 9.43×10^{-5} m/s (e) 4.29×10^{-3} m^2/sV

5-5 (a) 8.15 ps **5-6** $\dfrac{a}{\sigma_1 A}+\dfrac{b}{\sigma_2 A}$ **5-7** $\dfrac{d}{(\sigma_1\ell_1+\sigma_2\ell_2)w}$ **5-8** (b) $\dfrac{1}{4\pi\sigma}\left(\dfrac{1}{a}-\dfrac{1}{b}\right)$

5-10 $\dfrac{1}{2\pi}\dfrac{1}{\sigma_1+\sigma_2}\left(\dfrac{1}{a}-\dfrac{1}{b}\right)$ **5-12** $\dfrac{1}{2\pi\ell}\left\{\dfrac{1}{\sigma_1}\ln\dfrac{b}{a}+\dfrac{1}{\sigma_2}\ln\dfrac{c}{b}\right\}$

Chapter 6

6-2 (a) positive (b) $\dfrac{mv_0}{qB_0}$

6-3 $\mathbf{B}=\mathbf{a}_\phi\dfrac{\mu_0 I\rho}{2\pi a^2}\ (\rho<a);\ \mathbf{a}_\phi\dfrac{\mu_0 I}{2\pi\rho}\ (a<\rho<b);\ \mathbf{a}_\phi\dfrac{\mu_0 I}{2\pi\rho}\dfrac{c^2-\rho^2}{c^2-b^2}\ (b<\rho<c);\ 0\ (\rho>c)$

6-5 $\mathbf{B}=\mathbf{a}_\phi\mu_0 J_0\dfrac{e^{-\alpha a}}{\alpha^2}\dfrac{1}{\rho}\left(\alpha\rho e^{\alpha\rho}-e^{\alpha\rho}+1\right)\ (\rho<a);\mathbf{a}_\phi\mu_0 J_0\dfrac{1}{\alpha^2}\dfrac{1}{\rho}\left(\alpha a-1+e^{-\alpha a}\right)\ (\rho>a)$

6-6 $\mathbf{B}=0\ (\rho<a);\ \mu_0 J_0\dfrac{a}{\rho}(\rho-a)\ (a<\rho<b);\ \mu_0 J_0\dfrac{a}{\rho}(b-a)\ (\rho>b)$

6-10 $\mathbf{B}=\mathbf{a}_x\mu_0 J_0 z\left(|z|<\dfrac{d}{2}\right);\ \mathbf{a}_x\mu_0 J_0\dfrac{d}{2}\left(z>\dfrac{d}{2}\right);\ -\mathbf{a}_x\mu_0 J_0\dfrac{d}{2}\left(z<-\dfrac{d}{2}\right)$

6-18 $\mathbf{B}=\mathbf{a}_z\mu_0 I_0\dfrac{\sqrt{3}a^2}{8\pi}\dfrac{1}{\left(z^2+\dfrac{a^2}{12}\right)\sqrt{z^2+\dfrac{a^2}{3}}}$

6-22 Assuming the strip is on the xz plane, $\mathbf{B}=\mathbf{a}_y\dfrac{\mu_0}{2\pi}\dfrac{I}{w}\ln\dfrac{x+w/2}{x-w/2}$ at points on the xz plane $\left(x>\dfrac{w}{2}\right)$

6-24 $\mathbf{B}=\dfrac{\mu_0 I}{4\pi}\dfrac{d}{z^2+(d/2)^2}\left\{\mathbf{a}_z\left(1+\dfrac{\pi d/4}{\sqrt{z^2+(d/2)^2}}\right)-\mathbf{a}_x\dfrac{z}{\sqrt{z^2+(d/2)^2}}\right\}$

6-26 $\mathbf{B}=\mathbf{a}_z\dfrac{\mu_0 J_{so}}{2}\left\{\dfrac{-a}{\sqrt{z^2+a^2}}+\ln\dfrac{\left|\sqrt{z^2+a^2}+a\right|}{|z|}\right\}$

6-28 $\mathbf{B}=\mathbf{a}_z\dfrac{\mu_0 J_0}{2}\left\{\left(z+\dfrac{\ell}{2}\right)\ln\dfrac{a+\sqrt{a^2+\left(z+\ell/2\right)^2}}{\left|z+\ell/2\right|}-\left(z-\dfrac{\ell}{2}\right)\ln\dfrac{a+\sqrt{a^2+\left(z-\ell/2\right)^2}}{\left|z-\ell/2\right|}\right\}$

6-32 (a) $\mathbf{a}_z\mu_0 J_{so}\dfrac{\pi}{4}$ (b) $\dfrac{\mu_0 m}{4\pi r^3}(\mathbf{a}_r 2\cos\theta+\mathbf{a}_\theta\sin\theta)$; $m = J_{so}\dfrac{\pi^2 a^3}{2}$

6-33 (a) $\mathbf{a}_z\mu_0 J_{so}\dfrac{2}{3}$ (b) $m = \dfrac{4}{3}\pi J_{so}a^3$

6-39 $\dfrac{d\mathbf{F}_m}{dz}$ on the left top wire $= \dfrac{\mu_0 I^2}{4\pi a}(-\mathbf{a}_x+\mathbf{a}_y)$ **6-42** $\dfrac{dF_m}{dz} = \dfrac{\mu_0 I^2}{2\pi w}\ln\left|1+\dfrac{w}{d}\right|$

6-43 $\dfrac{dF_m}{dz} = \dfrac{\mu_0 I^2}{2\pi w^2}\{-2(w+d)[\ln(w+d)-1]+d(\ln d-1)+(2w+d)[\ln(2w+d)-1]\}$

Chapter 7

7-1 (a) $\mathbf{J}_m = 0$, $\mathbf{J}_{ms} = \mathbf{a}_z M_0\sin\phi$ (b) $\mathbf{J}_m = 0$, $\mathbf{J}_{ms} = \mp\mathbf{a}_y M_0$ at $z=\pm\dfrac{d}{2}$.

7-3 (a) $\mathbf{J}_m = -\mathbf{a}_\phi\dfrac{M_0}{a}$, $\mathbf{J}_{ms} = \mathbf{a}_\phi M_0$ (b) $\mathbf{J}_m = -\mathbf{a}_\phi M_0\sin\theta\dfrac{2r}{a^2}$, $\mathbf{J}_{ms} = \mathbf{a}_\phi M_0\sin\theta$

7-5 (a) $\mathbf{B}=\mathbf{a}_z\left(\dfrac{2}{3}\mu_0 M_0-\dfrac{2}{3}\mu_0 M_0\right)=0$ (b) $\mathbf{B}=\mathbf{B}_m+\mathbf{B}_{ms}=-\mathbf{a}_z\dfrac{2}{3}\mu_0 M_0+\mathbf{a}_z\dfrac{2}{3}\mu_0 M_0=0$

7-7 (a) $\mathbf{B}=\mathbf{a}_\phi\dfrac{\mu_1 I}{2\pi\rho}$ $(\rho<a)$; $\mathbf{a}_\phi\dfrac{\mu_2 I}{2\pi\rho}$ $(\rho>a)$ (b) $\mathbf{J}_m=0$, $\mathbf{J}_{ms}=\mathbf{a}_z\dfrac{I}{2\pi a}\dfrac{\mu_2-\mu_1}{\mu_0}$

7-9 (a) $\mathbf{B}=\mathbf{a}_\phi\dfrac{\mu I}{2\pi\rho}$ $(a<\rho<b)$; 0 $(\rho<a,\ \rho>b)$ $\mathbf{H}=\mathbf{a}_\phi\dfrac{I}{2\pi\rho}$ $(a<\rho<b)$; $0(\rho<a,\ \rho>b)$

$\mathbf{M}=\mathbf{a}_\phi\dfrac{I}{2\pi\rho}\left(\dfrac{\mu}{\mu_0}-1\right)$ $(a<\rho<b)$; 0 $(\rho<a,\ \rho>b)$

7-10 $\mathbf{B}=\mathbf{a}_\phi\mu\dfrac{NI}{2\pi\rho}$ (in the core); 0 (outside the core)

7-14 (a) $\mathbf{J}_m = 0$, $\mathbf{J}_{ms} = \mp\mathbf{a}_y M_0$ (at $z=\pm\dfrac{d}{2}$) (b) $\mathbf{B} = \mathbf{a}_x\mu_0 M_0\left(|z|<\dfrac{d}{2}\right)$; $0\left(|z|>\dfrac{d}{2}\right)$

(c) $\mathbf{H} = 0$ everywhere

7-26 $\dfrac{\mu N^2}{2\pi}c\ln\left[\dfrac{(a+b)(a+2b)}{a^2}\right]$ **7-27** $\mu_0 N_{\ell 1}N_{\ell 2}\pi a^2$

7-30 $L_{21}\approx\mu_0\dfrac{\pi(a_1a_2)^2}{4d^3}$ **7-31** $L_{21} = \dfrac{\mu_0}{2\pi}N_1N_2\,a_2\ln\left[1+\dfrac{a_2}{b+d}\right]$

7-32 $L_{21}=\dfrac{\mu_0}{\sqrt{3}\pi}\left[(d+b)\ln\left|1+\dfrac{b}{d}\right|-b\right]$, $b=\dfrac{\sqrt{3}}{2}a$

Chapter 8

8-1 (a) $\mathbf{E} = \mathbf{a}_\phi \dfrac{B_0\omega}{3a}\rho^2 \sin(\omega t)\ (\rho < a);\ \mathbf{a}_\phi \dfrac{B_0\omega a^2}{3\rho} \sin(\omega t)\ (\rho > a)$

(b) $V(t) = \dfrac{2\pi B_0\omega}{3a} b^3 \sin(\omega t),\ i(t) = \dfrac{V(t)}{R}$ (c) CCW

8-3 (b) $\mathbf{E}=\mathbf{a}_z \dfrac{\mu_0 I_0}{2\pi}\omega \sin(\omega t) \ln\dfrac{b}{a}\ (\rho<a);\ \mathbf{a}_z \dfrac{\mu_0 I_0}{2\pi}\omega \sin(\omega t) \ln\dfrac{b}{\rho}\ (a<\rho<b);\ 0\ (\rho>b)$

8-6 (a) $V(t) = -\mu_0 H_z\, \omega \cos(\omega t)\, ab$ (b) $i(t)$ flows CW at $\omega t = \pi/4$

8-8 (a) $V_2(t) = -\dfrac{\mu_0 I_1}{2\pi}\omega \sin(\omega t)\ell \ln(1+\dfrac{\ell}{d})$

8-11 (a) $V(t) = vB_0\ell\left[\omega(t+\dfrac{d}{v}) \sin\omega t - \cos\omega t\right]$ (b) $\mathbf{F} = \mathbf{a}_x \dfrac{V}{R} B_0 \ell$

8-12 $V(t) = -\omega B_0\, \ell\, d \sin\omega t$, $i(t)$ CCW at $t = 0^+$

8-14 $V(t) = -N\pi a^2 B$ **8-16** (a) $\mathbf{H}=-\mathbf{a}_\phi \dfrac{\varepsilon}{2d}\dfrac{a^2}{\rho}\dfrac{\partial V}{\partial t}\ (\rho > a)$

8-17 (b) $f > 676.3$ kHz (c) No

8-19 (a) $\mathbf{E} = -\mathbf{a}_y \dfrac{kH_m}{\omega\varepsilon} \sin(\omega t - kz)$ (b) $k^2 = \omega^2\mu\varepsilon$

8-21 (a) $\mathbf{H} = \mathbf{a}_z \dfrac{0.06}{\omega\mu_0} \sin(\omega t - 0.02x)$ (b) $f \approx 955$ kHz

8-23 $E_{1x}=E_{1z}=0,\ E_{1y}=\dfrac{\rho_s}{\varepsilon_0},\ H_{1x}=H_{1y}=0,\ H_{1z} = J_{so}$

8-24 (a) $\mathbf{H} = -\mathbf{a}_\phi \dfrac{\rho}{2}\dfrac{V_0}{d}[\omega\varepsilon \cos\omega t + \sigma \sin\omega t]$ (b) $\mathbf{S} \mathbin{//} -\mathbf{a}_\rho$ (c) $-P_{diss} = \dfrac{dW_e}{dt} + P_{f,out}$

8-27 (a) $\mathbf{B} = \mathbf{a}_z \mu_0 nKt,\ \mathbf{E} = -\mathbf{a}_\phi \dfrac{1}{2}\mu_0 nK\rho\ (\rho \le a)$ (b) $\mathbf{S} \mathbin{//} -\mathbf{a}_\rho$ (c) $\dfrac{dW_m}{dt} + P_{f,out} = 0$

Chapter 9

9-3 $v = 1.88 \times 10^8$ m/s **9-5** (d) $\underline{\mathbf{H}} = H_0\left\{\mathbf{a}_y e^{-j(5x+\frac{\pi}{2})} + \mathbf{a}_z e^{j5x}\right\}$

9-6 (a) $\mathbf{E}(z,t) = \mathbf{a}_x \cos(\omega t-2z) - \mathbf{a}_y 2 \sin(\omega t-2z)$

9-13 (a) $\underline{\mathbf{E}} = \mathbf{a}_z 5 e^{-j(0.5x+\frac{\pi}{2})}$

(b) $\underline{\mathbf{H}} = -\mathbf{a}_y \dfrac{5}{2\omega\mu_0} e^{-j(0.5x+\frac{\pi}{2})},\ \mathbf{H}(x,t) = -\mathbf{a}_y \dfrac{2.5}{\omega\mu_0}\sin(\omega t - 0.5x)$

(c) $\omega = 1.06 \times 10^8$ rad/s (d) $\mathbf{S}_{av} = \mathbf{a}_x \dfrac{25}{\sqrt{2}\,\mu_0 c} = \mathbf{a}_x\, 0.047\ \mathrm{W/m^2}$

9-15 (b) $\lambda = 1$ km **9-16** (b) $f = 5.77$ GHz

9-19 (b) $k_0 = 2\pi/3$ (c) $<\mathbf{S}> = \mathbf{a}_x\, 0.033\ \text{W/m}^2$

9-20 (a) $f = 2.5$ GHz, $\varepsilon = 1.44\,\varepsilon_0$ (b) $n = 1.2$

(c) $\mathbf{E} = \mathbf{a}_z\, 2\sqrt{2}\, e^{jky}$, $\mathbf{H} = -\mathbf{a}_x \dfrac{2\sqrt{2}}{314} e^{jky}$, $H_{rms} = 6.37$ mA/m (d) $314\ \Omega$

9-22 (a) $\mathbf{k} = 5.6\,(\mathbf{a}_x + \mathbf{a}_y 2 + \mathbf{a}_z 3)$ (b) $\mathbf{E} = \sqrt{5}\,(\mathbf{a}_x 2 - \mathbf{a}_y)\, e^{-j(5.6x+11.2y+16.8z)}$

(c) $\mathbf{H} = \dfrac{1}{\eta_0}\sqrt{\dfrac{5}{14}}\,(\mathbf{a}_x 3 + \mathbf{a}_y 6 - \mathbf{a}_z 5)\, e^{-j(5.6x+11.2y+16.8z)}$

9-24 (a) $\alpha = 37$ Np/m, $d_p = 2.7$ cm (b) 4.77 V/m

9-25 (a) $\alpha = 14.87$ Np/m, $d_p = 6.72$ cm (b) L.T.= 0.09 (c) $\alpha = 1.86\times10^{-3}$, $d_p = 538.7$ m

9-27 $d_c > 3.3$ mm **9-31** $v_p = 1.58\times10^6$ m/s, $v_g = 3.16\times10^6$ m/s

9-33 (a) LP (b) LHCP (c) RHEP **9-34** (b) LHCP (c) RHEP

Chapter 10

10-3 $R = -0.8 + j0.002$, $T = 0.2 + j0.002$

10-5 $R = -0.514$, $T = 0.486$, $\mathbf{E} = \mathbf{a}_y\, TE_0\, e^{-j65.2x}\, e^{-0.067x}$

10-8 (a) $3x + 4z = \text{const.}$ (b) $\mathbf{k} = \mathbf{a}_x 3 + \mathbf{a}_z 4$ (c) $f = 238.7$ MHz

(d) $\mathbf{H} = \dfrac{1}{120\pi}(-\mathbf{a}_x 4 + \mathbf{a}_z 3)\, e^{-j(3x+4z)}$, $<\mathbf{S}> = \dfrac{25}{2\omega\mu_0}\mathbf{k}$

10-9 (a) $\theta_i = 36.9°$ (b) $\theta_r = 36.9°$, $\theta_t = 23.6°$

(c) $\mathbf{E}_r = \mathbf{a}_y R_\perp 5 e^{-j(3x-4z)}$, $\mathbf{H}_r = \dfrac{5}{\omega\mu_o} R_\perp (\mathbf{a}_x 4 + \mathbf{a}_z 3)\, e^{-j(3x-4z)}$, $R_\perp = -0.264$

(d) $\mathbf{E}_t = \mathbf{a}_y T_\perp 5 e^{-j(3x+6.87z)}$, $\mathbf{H}_t = \dfrac{5T_\perp}{\omega\mu_0}(-\mathbf{a}_x 6.87 + \mathbf{a}_z 3)\, e^{-j(3x+6.87z)}$, $T_\perp = 0.736$

(e) 6.99% (f) 93%

10-15 (a) $\theta_i = 67.4°$ (b) $\theta_r = 67.4°$, $\theta_t = 8.85°$ (c) $R_\parallel = -0.59$
(d) $T_\parallel = 0.41$ (e) 35% (f) 65%

10-19 Decompose the incident wave into a sum of a perpendicularly polarized wave and a parallel polarized wave.

10-21 (b) $\theta_B = 61°$ when $\alpha = 0.05$ (d) $\theta_B = 29°$, $\theta_c = 33.7°$ when $\alpha = 0.05$

10-24 $65.37\ \text{m}^2$ **10-31** (a) TE (b) Parallel **10-32** (a) Parallel (b) 33°

10-35 (a) $\mathbf{E}_r = -\mathbf{a}_y 5 e^{j(3x-4z)}$, $\mathbf{H}_r = \dfrac{1}{\eta_0}(\mathbf{a}_x 4 + \mathbf{a}_z 3)\, e^{j(3x-4z)}$ (b) 1

10-38 $f = 75$ MHz, $|\mathbf{E}| = \dfrac{\sqrt{3}}{2} E_{max}$ at 1m in front of the plate.

Chapter 11

11-5 (a) 0, 6, 12,18, 24 GHz (b) TM_0, TM_1, TM_2, TM_3

11-8 $k_z = 471.2$ rad/m, $\lambda_g = 13.3$ mm, $v_p = 2 \times 10^8$ m/s, $Z_{TM} = 90.5\ \Omega$,

$\mathbf{S}_{av} = \mathbf{a}_z\, 3\,|E_0|^2 \cos^2(200\pi x)$ mW/m^2

11-11 (a) $0 < f < 1.88$ kHz (b) $1.88 < f < 3.75$ kHz

11-18 (a) 2.39×10^6 W/cm (b) 7.89×10^5 W/cm (c) 7.89×10^5 W/cm

11-19 (a) TE_{10}, TE_{01}, TE_{20}, TE_{11}, TM_{11} (b) $k_z = 124.5$ rad/m, $f_c = 6.95$ GHz,

$v_p = 4.04 \times 10^8$ m/s, $\lambda_g = 5.05$ cm, $Z_{TM} = 124.3\ \Omega$ (c) $2.87 < f < 5.74$ GHz

(d) $5.23 < \lambda < 10.45$ cm

11-21 (a) TE_{10}, TE_{20}, TE_{01}, TE_{11}, TM_{11} (c) $11.7 < f < 23.4$ GHz

11-27 (a) No (b) Yes **11-28** a = 3.2 cm, b = 1.8 cm

11-30 a = 1.5 cm, b = 0.75 cm **11-32** $k_z = -j24.9$, -4.32 dB

11-34 $f_r = 10.6$ GHz, $Q \approx 6.2\times10^3$ **11-35** (a) a = 21.2 cm (b) a = 13.3 cm

11-37 The first three are 10.6 GHz, 16.8 GHz, 18.4 GHz

11-39 $\mathbf{J}_s = \mathbf{a}_z \dfrac{K_0}{\eta a} e^{-jkz}$ (at $\rho = a$), $I_0 = \dfrac{2\pi K_0}{\eta}$

11-46 $v = 1.875 \times 10^8$ m/s, $L = 0.569\ \mu$H/m, $Z_o = 106.7\ \Omega$, $G = 2.21 \times 10^{-14}$ S/m

11-47 $\Gamma_L = \dfrac{1}{3}$, $Z_{in} = 49 - j35\Omega$ **11-48** $10 + j30\ \Omega$

11-52 $\Gamma_L = 0.34\, e^{-j126°}$, $Z_L = 59 - j36\ \Omega$, $z_{max} = -10$ cm

Chapter 12

12-1 (a) $\mathbf{E} = \begin{cases} \mathbf{a}_x\, 2K(|z| - vt), & |z| < vt \\ 0, & |z| > vt \end{cases}$ $\quad \mathbf{H} = \begin{cases} \mathbf{a}_y \dfrac{2K}{\eta}(z - vt), & 0<z<vt \\ \mathbf{a}_y \dfrac{2K}{\eta}(z + vt), & -vt<z<0 \\ 0, & |z| > vt \end{cases}$

(c) $\rho_v = \rho_s = 0$, $\mathbf{J} = 0$, $\mathbf{J}_s = \mathbf{a}_x \dfrac{4K}{\mu} t$ at $z = 0$

12-4 (a) $\mathbf{A} = \mathbf{a}_z \dfrac{\mu I_0 \Delta\ell}{8\pi r} e^{-jkr}$ (b) $\mathbf{E} = \mathbf{a}_\theta \dfrac{I_0 \Delta\ell}{8\pi} j\omega\mu \dfrac{e^{-jkr}}{r} \sin\theta$

12-5 0.039 W **12-7** (a) 0.067 V/m (b) 0.071 V/m

12-9 (a) $\frac{I_0 \Delta\ell\, j\omega\mu}{4\pi r} e^{-jkr} \{\mathbf{a}_\theta (\sin\theta + j\cos\theta\cos\phi) - \mathbf{a}_\phi\, j\sin\phi\}$

12-13 $P_r = \frac{4}{3}\eta\pi^5 \left(\frac{a^2}{\lambda^2}\right)^2 I_0^2$ **12-17** $R_r = 36.5\ \Omega$, $D_{max} = 3.3$

12-18 14.64 kW **12-19** $|E_\theta| = 1.404$ V/m, $|H_\phi| = 3.72$ mA/m

12-21 $R_r = 199\ \Omega$, $D_{max} = 2.41$ **12-23** $e_r = 0.382$

12-25 (i) 90° (ii) 78° (iii) 47°

12-26 (i) $AF = \cos\left(\frac{\pi}{2}\cos\phi + \frac{\pi}{4}\right)$ (iii) $AF = \cos\left(\pi\cos\phi + \frac{\pi}{2}\right)$

12-28 $F(\theta, \phi) = \sqrt{1 - \sin^2\theta\cos^2\phi}\ \cos\left(\frac{\pi}{2}\sin\theta\cos\phi\right)$

12-31 SLL = −12 dB for N=5

INDEX

D

LAWS OF ELECTROMAGNETICS

Maxwell's Equations in Differential Form

(Action of charges on fields)

$$\nabla \times \mathbf{E} = -\frac{\partial \mathbf{B}}{\partial t}$$ (Faraday's law)

$$\nabla \times \mathbf{H} = \mathbf{J} + \frac{\partial \mathbf{D}}{\partial t}$$ (Generalized Ampère's law)

$$\nabla \cdot \mathbf{D} = \rho_v$$ (Gauss' law for electric field)

$$\nabla \cdot \mathbf{B} = 0$$ (Gauss' law for magnetic field)

Maxwell's Equations in Integral Form

$$\oint_C \mathbf{E} \cdot d\boldsymbol{\ell} = -\iint_S \frac{\partial \mathbf{B}}{\partial t} \cdot d\mathbf{s}$$

$$\oint_C \mathbf{H} \cdot d\boldsymbol{\ell} = \iint_S \left(\mathbf{J} + \frac{\partial \mathbf{D}}{\partial t} \right) \cdot d\mathbf{s}$$

$$= I + \iint_S \frac{\partial \mathbf{D}}{\partial t} \cdot d\mathbf{s}$$

$$\oiint_S \mathbf{D} \cdot d\mathbf{s} = \iiint_V \rho_v \, dv = Q$$

$$\oiint_S \mathbf{B} \cdot d\mathbf{s} = 0$$

Lorentz's Force Equation

(Action of fields on charges)

$$\mathbf{F} = q(\mathbf{E} + \mathbf{v} \times \mathbf{B})$$

VECTOR DERIVATIVES

RECTANGULAR COORDINATES (x, y, z)

$$\nabla V = \mathbf{a}_x \frac{\partial V}{\partial x} + \mathbf{a}_y \frac{\partial V}{\partial y} + \mathbf{a}_z \frac{\partial V}{\partial z}$$

$$\nabla \cdot \mathbf{A} = \frac{\partial A_x}{\partial x} + \frac{\partial A_y}{\partial y} + \frac{\partial A_z}{\partial z}$$

$$\nabla \times \mathbf{A} = \mathbf{a}_x \left(\frac{\partial A_z}{\partial y} - \frac{\partial A_y}{\partial z} \right) + \mathbf{a}_y \left(\frac{\partial A_x}{\partial z} - \frac{\partial A_z}{\partial x} \right) + \mathbf{a}_z \left(\frac{\partial A_y}{\partial x} - \frac{\partial A_x}{\partial y} \right)$$

$$\nabla^2 V = \frac{\partial^2 V}{\partial x^2} + \frac{\partial^2 V}{\partial y^2} + \frac{\partial^2 V}{\partial z^2}$$

CYLINDRICAL COORDINATES (ρ, ϕ, z)

$$\nabla V = \mathbf{a}_\rho \frac{\partial V}{\partial \rho} + \mathbf{a}_\phi \frac{1}{\rho} \frac{\partial V}{\partial \phi} + \mathbf{a}_z \frac{\partial V}{\partial z}$$

$$\nabla \cdot \mathbf{A} = \frac{1}{\rho} \frac{\partial}{\partial \rho} (\rho A_\rho) + \frac{1}{\rho} \frac{\partial A_\phi}{\partial \phi} + \frac{\partial A_z}{\partial z}$$

$$\nabla \times \mathbf{A} = \mathbf{a}_\rho \left(\frac{1}{\rho} \frac{\partial A_z}{\partial \phi} - \frac{\partial A_\phi}{\partial z} \right) + \mathbf{a}_\phi \left(\frac{\partial A_\rho}{\partial z} - \frac{\partial A_z}{\partial \rho} \right) + \mathbf{a}_z \frac{1}{\rho} \left[\frac{\partial}{\partial \rho} (\rho A_\phi) - \frac{\partial A_\rho}{\partial \phi} \right]$$

$$\nabla^2 V = \frac{1}{\rho} \frac{\partial}{\partial \rho} \left(\rho \frac{\partial V}{\partial \rho} \right) + \frac{1}{\rho^2} \frac{\partial^2 V}{\partial \phi^2} + \frac{\partial^2 V}{\partial z^2}$$

SPHERICAL COORDINATES (r, θ, ϕ)

$$\nabla V = \mathbf{a}_r \frac{\partial V}{\partial r} + \mathbf{a}_\theta \frac{1}{r} \frac{\partial V}{\partial \theta} + \mathbf{a}_\phi \frac{1}{r \sin\theta} \frac{\partial V}{\partial \phi}$$

$$\nabla \cdot \mathbf{A} = \frac{1}{r^2} \frac{\partial}{\partial r} (r^2 A_r) + \frac{1}{r \sin\theta} \frac{\partial}{\partial \theta} (A_\theta \sin\theta) + \frac{1}{r \sin\theta} \frac{\partial A_\phi}{\partial \phi}$$

$$\nabla \times \mathbf{A} = \mathbf{a}_r \frac{1}{r \sin\theta} \left[\frac{\partial}{\partial \theta} (A_\phi \sin\theta) - \frac{\partial A_\theta}{\partial \phi} \right]$$
$$+ \mathbf{a}_\theta \frac{1}{r} \left[\frac{1}{\sin\theta} \frac{\partial A_r}{\partial \phi} - \frac{\partial}{\partial r} (r A_\phi) \right] + \mathbf{a}_\phi \frac{1}{r} \left[\frac{\partial}{\partial r} (r A_\theta) - \frac{\partial A_r}{\partial \theta} \right]$$

$$\nabla^2 V = \frac{1}{r^2} \frac{\partial}{\partial r} \left(r^2 \frac{\partial V}{\partial r} \right) + \frac{1}{r^2 \sin\theta} \frac{\partial}{\partial \theta} \left(\sin\theta \frac{\partial V}{\partial \theta} \right) + \frac{1}{r^2 \sin^2\theta} \frac{\partial^2 V}{\partial \phi^2}$$

www.ingramcontent.com/pod-product-compliance
Ingram Content Group UK Ltd.
Pitfield, Milton Keynes, MK11 3LW, UK
UKHW052009041026
15590UKWH00008B/2236

9 781516 520657